Heinz-Helmut Perkampus

Encyclopedia of Spectroscopy

Related Titles from VCH:

H. Friebolin
Basic One- and Two-Dimensional NMR Spectroscopy
Second Edition 1993. XXI, 368 pp., 161 figures, 48 tables.
ISBN 3-527-29059-1

J. Manz, L. Wöste (Ed.)
Femtosecond Chemistry
1995. Two volumes, XXVI, 916 pp., 365 figures, 15 tables.
ISBN 3-527-29062-1

D. L. Andrews (Ed.)
Applied Laser Spectroscopy Techniques, Instrumentation, and Applications
1993. IX, 471 pp., 213 figures, 14 tables.
ISBN 3-527-28072-3

J. W. Niemantsverdriet
Spectroscopy in Catalysis – An Introduction
1993. VIII, 289 pp., 180 figures, 26 tables.
ISBN 3-527-28593-8

Distribution:
VCH, P.O. Box 10 11 61, D-69451 Weinheim (Federal Republic of Germany)
Switzerland: VCH, P.O. Box, CH-4020 Basel (Switzerland)
United Kingdom and Ireland: VCH (UK) Ltd., 8 Wellington Court, Cambridge CB1 1HZ (England)
USA and Canada: VCH, 220 East 23rd Street, New York, NY 10010-4606 (USA)
Japan: VCH, Eikow Building, 10-9 Hongo 1-chome, Bunkyo-ku, Tokyo 113 (Japan)

ISBN 3-527-29281-0

Heinz-Helmut Perkampus

Encyclopedia of Spectroscopy

Translated by
Heide-Charlotte Grinter
and Roger Grinter

Weinheim · New York · Basel · Cambridge · Tokyo

Author:
Prof. Dr. Heinz-Helmut Perkampus
Wickrather Str. 43
D-40547 Düsseldorf
Germany

Translators:
Heide-Charlotte Grinter
Dr. Roger Grinter
13 Mark Lemmon Close
Cringleford
Norwich NR4 6UY, GB

First published in German as "parat Lexikon Spektroskopie", H.-H. Perkampus
© VCH Verlagsgesellschaft mbH, Weinheim, 1993.

Editorial Director: Dr. Thomas Kellersohn
Production Manager: Dipl.-Wirt.-Ing. (FH) Bernd Riedel

Library of Congress Card No. applied for.

A catalogue record for this book is available from the British Library.

Deutsche Bibliothek Cataloguing-in-Publication Data

Perkampus, Heinz-Helmut:
Encyclopedia of spectroscopy / Heinz-Helmut Perkampus.
Transl. by Heide-Charlotte Grinter and Roger Grinter. – :
Weinheim ; New York ; Basel ; Cambridge ; Tokyo : VCH, 1995
Dt Ausg. u.d.T.: Perkampus, Heinz-Helmut: Lexikon Spektroskopie
ISBN 3-527-29281-0
NE: HST

Printed on acid-free and chlorine-free paper.

Cover design: Graphik und Text-Studio W. Zettlmeier,
D-93164 Laaber-Waldetzenberg
Composition: Mitterweger Werksatz GmbH, D-68723 Plankstadt
Printing: Strauss-Offsetdruck, D-69509 Mörlenbach
Bookbinding: IVB Heppenheim GmbH, D-64646 Heppenheim
Printed in the Federal Republic of Germany

Preface to the English Edition

As Professor Perkampus notes in his preface to the German edition, spectroscopy spans a wavelength range of 12 powers of ten. But there is another sense in which the scope of spectroscopy is extremely wide; its range of applications. The reader of this Encyclopedia will not fail to note the extraordinary versatility of spectroscopic methods. Indeed, scientific advance in the pure and applied physical sciences has long been critically dependent upon the power of the spectroscopic techniques, in all their forms. And, with notable exceptions, for the first half of this century the impact of spectroscopy was largely confined to the more physical branches of the natural sciences. This is no longer the case. The speed of the advance of spectroscopic methods into the biological sciences and medicine and their significance in these disciplines increases day by day. Many of the great challenges posed by environmental problems would not even have been identified were it not for the sensitivity of spectroscopic analytical methods. Spectroscopy is no longer the esoteric domain of the spectroscopist, it is an indispensible tool for the whole of pure and applied science and technology.

The result of this development is that increasing numbers of scientists, not to mention students and even laymen, are seeking a guide to particular aspects of the subject of spectroscopy; a source for their first steps in a new field which hitherto they had neither the need nor the opportunity to explore. It was this thought which made us feel that Professor Perkampus' masterly survey of this important subject should be more readily available to English speakers and we were pleased when VCH Publishers and Professor Perkampus agreed to our suggestion that we might translate the Encyclopedia. Naturally, the majority of the references to the literature in the original version are to material published in German. We have therefore tried to replace these citations with references to similar sources in English. This proved to be quite a difficult task and, in a few cases, it has not been possible to find suitable alternatives. But we do not believe that this will reduce the value of the Encyclopedia to the wide range of potential readers which we have in mind and we hope that they will find this English-language version useful.

Finally, it is a pleasure to express our thanks to Professor Perkampus and Dr. Kellersohn of VCH for their support during the long haul of ca. 250,000 words.

Cringleford, December 1994

Heide Charlotte Grinter
Roger Grinter

Preface to the German Edition

This book has its origin in discussions between Dr. H.-D. Junge of VCH and the author during which the latter received many valuable and useful ideas with regard to the form of the Encyclopedia.

By means of approximately 1000 key words and ca. 300 figures, the attempt has been made to give as complete an overview as possible of the subject of spectroscopy. However, the author is aware of the fact that there are still gaps here and there which can only be filled by the addition of further entries. Though the thoughtful perusal of the manuscript by members of the publishers' editorial staff revealed missing key words which, at the early stages of typesetting, could still be included.

The subject of spectroscopy encompasses ca. 12 powers of ten in wavelength (wavenumber, frequency, energy) and stretches from Mössbauer to NMR spectroscopy. The Lexicon should reflect this range and all spectroscopic methods which lie within it should be represented with appropriate entries.

The attentive reader will notice that optical spectroscopy, especially molecular spectroscopy, is described in considerable detail. The reason is that the author is by background a molecular spectroscopist and therefore paid particular attention to the associated key words.

But other branches of spectroscopy should not receive too brief a treatment and physical and physical-chemical fundamentals, in so far as they are important to spectroscopy, have also been included among the key words in the Encyclopedia. In this, the author relied upon Bergmann-Schäfer, *Lehrbuch der Experimentalphysik,* vol. 4.1 and also G. Wedler, *Lehrbuch der Physikalischen Chemie.*

The author's colleagues made a wide range of material available to him in the form of reprints and figures and he was also assisted in composing the entries by many booklets from firms. Both colleagues and companies are warmly thanked. Thanks are especially due to the author's co-workers in the Institute, to his secretary who untertook the time-consuming typing, to Herr Werner who devoted much time to drawing the figures and to Herr Bettermann who prepared a number of the entries and made a critical appraisal of others.

Finally, particular thanks are due to the publishers, VCH Weinheim, for the clean and readable printing. Also, I would particularly like to thank Dr. Junge of VCH for his constant assurances that the manuscript would indeed eventually be published.

Düsseldorf, October 1994

Heinz-Helmut Perkampus

A, <*Extinktion*> → absorbance.

Å, abbreviation for Ångström, a unit of length, 1 Å = 10^{-8} cm = 10^{-10} m = 0.1 nm. → Ångström unit.

AAS, <*AAS*> → atomic absorption spectroscopy.

Abbe number, <*Abbesche Zahl*> → dispersion, unit of.

Abbe prism, <*Abbe-Prisma*>, a prism system which gives a constant angle of deviation for each wavelength at the minimum deviation. An Abbe prism with a 90° deviation is shown in Figure 1. With the help of the subsidiary lines, we can imagine this prism to be composed of two 30° prisms, AEB and ACD, and the isosceles 90° prism, BEC. A beam incident at an angle, α, on the prism face, AB, is refracted in such a way that it passes through the prism with the minimum deviation i.e. parallel to AC. It is then totally reflected at surface, BC, and emerges at angle, α, from the surface, AD. Each wavelength requires a different angle of incidence for such a beam path. This angle must be set by rotating the prism. However, the beam emerging from the prism is always perpendicular to the incident beam. It follows that, when the prism is used as dispersing element in a → monochromator, the entrance and exit slit should be arranged perpendicularly to each other. Figure 2 shows another design with a constant deviation of 60°.

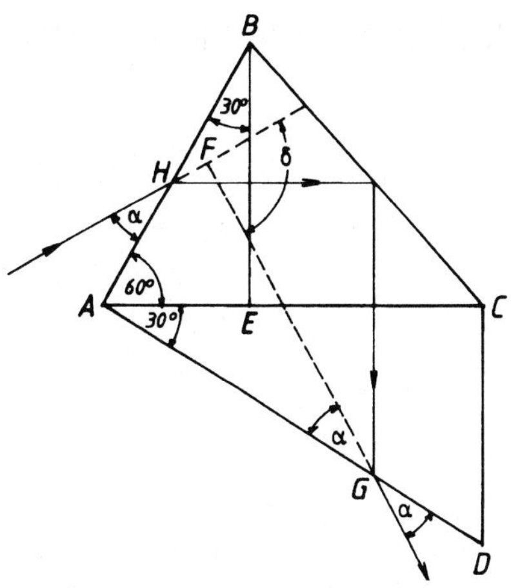

Abbe prism. Fig. 1. Section through an Abbe prism with a 90° deviation

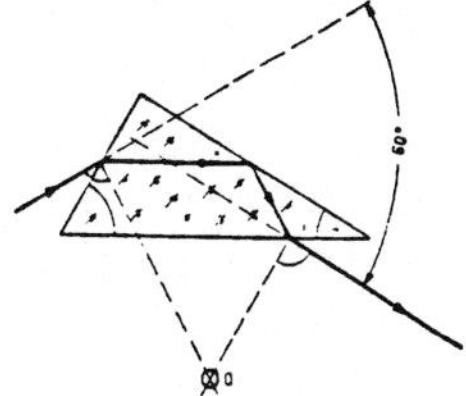

Abbe prism. Fig. 2. A prism with a 60° deviation

The Försterling three-prism arrangement is a combination of an Abbe prism and two 60° prisms; Figure 3. It gives a very high → resolution and → dispersion. Since the deviations produced by the two prisms cancel, only the constant deviation of the Abbe prism remains and the total deviation of the three-prism device is 90°. To further increase the resolution, the three-prism arrangement can be used in → autocollimation.

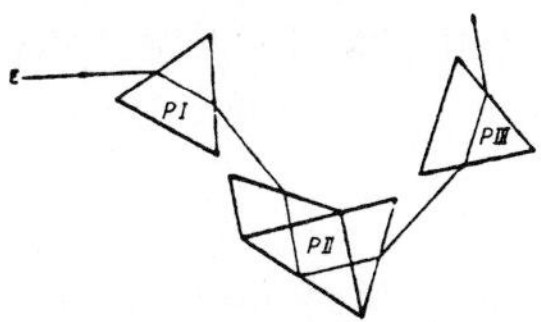

Abbe prism. Fig. 3. Försterling three-prism arrangement

Abney's phenomenon, <*Abneysches Phänomen*>, the change of the color shade of spectral colors (→ spectrum, colors of) when they are mixed with white light, i.e. if their white content is increased. The red and yellow-green shades change towards yellow, but the blues move towards violet. Finally, we can distinguish only yellow, blue-green and violet in the whole → spectral region.

Absorbance, A, <*Extinktion*>, the logarithm of the inverse of the → transmittance (T) or degree of transmission (τ) in → absorption spectroscopy. It is proportional to the path length of the light through the absorbing medium and, for solutions, to the concentration of the absorbing substance (→ Bouguer-Lambert-Beer law). It may be defined in terms of natural:

$$A_n(\lambda) = \ln \frac{1}{\tau(\lambda)} = \ln \left(\frac{\Phi_o}{\Phi_T}\right)_\lambda$$

or decadic logarithms:

$$A_n(\lambda) = \log \frac{1}{\tau(\lambda)} = \log \left(\frac{\Phi_o}{\Phi_T}\right)_\lambda$$

where Φ_o is the intensity of the light entering the absorbing medium, Φ_T the intensity of the transmitted light and λ, the wavelength. $A_n(\lambda) = 2.303 \cdot A(\lambda)$. The absorbance is dimensionless and is the actual quantity of interest in all photometric experiments. In the determination of A it is important that its value depends upon wavelength (or wavenumber or frequency).

Absorbance, apparent, <*konservative Absorption, scheinbare Absorption*>, this is an absorption caused by light scattering in a medium and which is superimposed upon any true absorption. Its presence is usually indicated by a flat tail on the long-wavelength side of the long-wavelength absorption band. The apparent absorption of a turbid sample, when measuring in transmission, is given analogously to the → Bouguer-Lambert-Beer law as:

$$I' = I_o e^{-S'd}$$

or

$$\ln(I_o/I') = S' \cdot d$$

where I_o is the incident radiation intensity, I' the intensity of the transmitted, unscattered radiation, S' the scattering coefficient, also called turbidity, and d the path length. The above relationship represents the basis for photometric light-scattering measurements (photodensitometry).

Absorbance diagram, <*Extinktionsdiagramm*>, a diagram obtained when the dependence of a chemical reaction or equilibrium upon time or pH value is recorded in terms of measurements of the absorbance at a number of different wavelengths, λ_i. In the case of a simple, uniform reaction or a simple equilibrium, straight lines are obtained by this procedure. Curved plots indicate that complicated reactions or equilibria are involved.

Absorbance-difference diagram, <*Extinktionsdifferenzen-Diagramm*>, for a simple reaction, $a + b \rightarrow c$, the absorbance at a wavelength, λ_1, is given by:

$$A_1 = a\varepsilon_{1,a} \cdot d + b\varepsilon_{1,b} \cdot d + c\varepsilon_{1,c} d$$

where a, b and c are the concentrations of reactants and product and d is the path length.

If, because of its stoichiometry, the reaction can be described by the concentration of one variable, x, then the concentrations of a and b can be written as $a_o - x$ and $b_o - x$. x is the transformation variable which depends upon the time in the case of kinetics and upon the concentration for equilibria. Thus, the absorbance, A_1, is given by:

$$\begin{aligned} A_1 &= (a_o - x)\varepsilon_{1,a}d + (b_o - x)\varepsilon_{1,b}d \\ &\quad + x\varepsilon_{1,c}d \\ &= a_o\varepsilon_{1,a}d + b_o\varepsilon_{1,b}d + (\varepsilon_{1,c} \\ &\quad - \varepsilon_{1,a} - \varepsilon_{1,b})xd. \end{aligned}$$

from which the absorbance difference is:

$$A_1 - [a_o\varepsilon_{1,a}d - b_o\varepsilon_{1,b}d] = \Delta A_1 = [(\varepsilon_{1,c} - \varepsilon_{1,a} - \varepsilon_{1,b})d]x = q_1 x.$$

$a_o\varepsilon_{1,a}d + b_o\varepsilon_{1,b}d = A_{1,o}$ is the sum of the absorbance values which would be obtained if there was no reaction between a and b. q_1 is composed of the molar decadic extinction coefficients of all the components involved in the reaction. Analogously, for a wavelength λ_i, $\Delta A_i = q_i x$. Eliminating x we obtain:

$$\Delta A_i = \frac{q_i}{q_1}\Delta A_1 = Z_1 \Delta A_1$$

$(i = 2, 3, 4 \ldots n)$.

This is the equation of the absorbance difference diagram. A graph of the values of ΔA_i for the corresponding wavelengths, λ_i, against ΔA_1 (λ_1 = reference wavelength) gives straight lines passing through zero with gradients of q_i/q_1. This provides a simple way of checking for the uniformity of a reaction or an equilibrium.

Ref.: H.-H. Perkampus, *UV-VIS Spectroscopy and Its Applications*, Springer Verlag, Berlin, Heidelberg, New York, **1992**.

Absorbance-difference-ratio diagram, *<Extinktionsdifferenzen-Quotienten-Diagramm>*. Consider a system in which exactly two linearly independent subreactions with the transformation variables, x_1 and x_2, take place, e.g. $a \rightarrow b$ and $a \rightarrow c$. The absorbance at wavelength λ_1, is:

$$\Delta A_1 = [(a_o - x_1 - x_2)\varepsilon_{a,1} + x_1\varepsilon_{b,1} + x_2\varepsilon_{c,1}]d.$$

In terms of the variables, x_1 and x_2, we have:

$$\begin{aligned} \Delta A_1 &= A_1 - A_{1,o} = d(\varepsilon_{b,1} - \varepsilon_{a,1})x_1 \\ &\quad + d(\varepsilon_{c,1} - \varepsilon_{a,1})x_2 \\ &= q_{11}x_1 + q_{12}x_2 \end{aligned}$$

with

$$q_{11} = (\varepsilon_{b,1} - \varepsilon_{a,1})d$$

and

$$q_{12} = (\varepsilon_{c,1} - \varepsilon_{a,1})d.$$

Analogously, for the wavelengths λ_2 and λ_3:

$$\Delta A_2 = q_{21}x_1 + q_{22}x_2$$

and

$$\Delta A_3 = q_{31}x_1 + q_{32}x_2.$$

After eliminating x_1, x_2 and x_3, the expressions for the absorbance differences can be related as follows:

$$\Delta A_3 = \frac{D_{2,3}}{D_{1,2}}\Delta A_1 + \frac{D_{1,3}}{D_{1,2}}\Delta A_2;$$

Division by ΔA_1 gives:

$$\frac{\Delta A_3}{\Delta A_1} = n + m\frac{\Delta A_2}{\Delta A_1}$$

with

$$n = D_{2,3}/D_{1,2};\ m = D_{1,3}/D_{1,2}$$

and

$$D_{1,2} = q_{11}q_{22} - q_{12}q_{21};$$

$$D_{2,3} = q_{31}q_{22} - q_{32}q_{21};$$

$$D_{1,3} = q_{11}q_{32} - q_{31}q_{11}.$$

These are the equations of the absorbance-difference-ratio diagram. In the case of two linearly independent subreactions, if $\Delta A_3/\Delta A_1$ is plotted against $\Delta A_2/\Delta A_1$, one obtains a straight line which can be used as evidence of the number of subreactions. Ref.: H.-H. Perkampus, *UV-VIS Spectroscopy and Its Applications*, Springer Verlag, Berlin, Heidelberg, New York, **1992**.

Absorbance index, <*Extinktionsmodul*>, a specific proportionality constant in the → Bouguer-Lambert-Beer law which characterizes the material. The natural, $m_n(\lambda)$, and decadic, $m(\lambda)$, values of the index may be distinguished:

$$\frac{A_n(\lambda)}{d} = m_n(\lambda); \quad \frac{A(\lambda)}{d} = m(\lambda).$$

Since absorbance is dimensionless, the dimension of the absorbance index is that of reciprocal length [cm^{-1} or m^{-1}]. For solutions and gas mixtures, the absorbance index is proportional to the concentration:

$$m_n(\lambda) = \varepsilon_n(\lambda) \cdot c; \quad m(\lambda) = \varepsilon(\lambda) \cdot c.$$

ε_n and ε respectively are then the natural and decadic → extinction coefficients which do not depend upon the concentration. In connection with the → Bouguer-Lambert-Beer law (in → photoacoustic spectroscopy for example), the absorbance index is also called the absorption coefficient, β:

$$\beta = \frac{A_n(\lambda)}{d} = 2.303 \cdot \varepsilon(\lambda) \cdot c \ |cm^{-1}|.$$

Absorbance indicator, <*Extinktionsindikator*> → indicator.

Absorbance-time diagram, <*Extinktionszeitdiagramm*>, a diagram in which the measured absorbance, A_λ at wavelength, λ, is plotted as a function of time to follow a chemical reaction. These diagrams are useful in combination with the equations of chemical reaction kinetics (→ kinetics of chemical reactions) in reaction-kinetics studies.

Absorbance, true, <*konsumptive Absorption, wahre Absorption*>. The absorbance, A, which we obtain when measuring light absorption in turbid, i.e. scattering, samples is the sum of the true absorbance, A_t, and the apparent A_a: $A = A_t + A_a$ (cf. → absorbance, apparent). Outside the true absorption range, $A_t = 0$, and therefore, $A = A_a$. If we form $\log A(\lambda) = \log A_a(\lambda)$ and plot $\log A$ as $f(\lambda)$ in the functional form appropriate for → Rayleigh or → Mie scattering, we should obtain a straight line the slope of which is given by the exponent of λ. The straight line obtained in this way can be extended (in approximation) into the absorption region. Thus, the absorbance, $A = A_t + A_a$ measured there can be corrected using $A_t = A - A_a$.

Absorption <*Absorption*> → integral absorption; → absorbance, apparent; → absorbance, true.

Absorption coefficient, absorbance index, <*Absorptionskoeffizient*>, β defined using the → Bouguer-Lambert-Beer relationship as:

$$\beta_\lambda = \frac{A_n(\lambda)}{d} = \varepsilon_n(\lambda) \cdot c$$

or

$$\beta_\lambda = \frac{2.303 \cdot A(\lambda)}{d} = 2.303 \cdot \varepsilon(\lambda) \cdot c$$

The units of β are cm^{-1} if d is given in cm. If the concentration of a solution is given in mol/l, ε is the molar → extinction coefficient with units 1000 cm^2/mol.
If β_λ is introduced into the Lambert-Beer law we obtain: $I = I_0 \exp[-\beta_\lambda d]$. The inverse of β_λ represents a length. If we set $d = \beta_\lambda^{-1}$ then this is the → optical penetration depth at which the intensity of the light is decreased by a factor of e. Such considerations are important for → photoacoustic spectroscopy.

Absorption edge, band edge, <*Absorptionskante*>, the transition between the strong short-wavelength and the weak long-wavelength absorption in the → absorption spectrum of a solid, generally a semiconductor. The spectral position of this edge is determined by the energy separation between the valence and conduction bands (→ band model); hence the term *band edge*.
In the case of transparent solids, the absorption edge can be measured on thin sections by means of transmittance measurements. Diffuse → reflection spectroscopy and particularly → photoacoustic spectroscopy are good methods for opaque, powder-like solids. Figure 1 shows a comparison of such measurements on titanium dioxide powder.

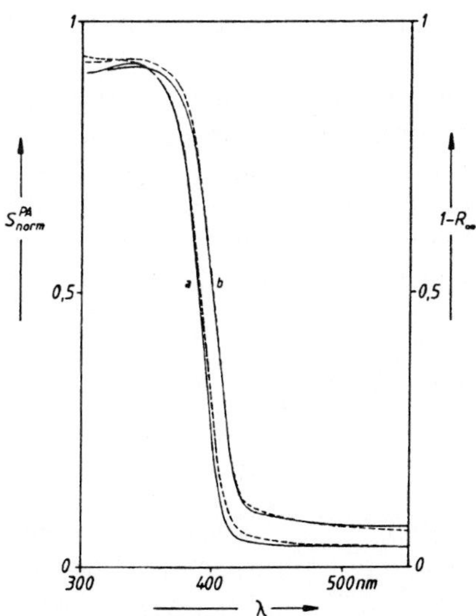

Absorption edge. Fig. 1. TiO_2 a) photoacoustic measurement; b) diffuse reflectance spectrum

Absorption-polarization-fluorescence spectrum, APF spectrum, <*Absorptionspolarisationsfluoreszenzspektrum, APF-Spektrum*> → photoselection.

Absorption-polarization-phosphorescence spectrum, APPh spectrum, <*Absorptionspolarisationsphosphoreszenzspektrum, APPh-Spektrum*> → photoselection.

Absorption spectroscopy, <*Absorptionsspektroskopie*>, generally the quantitative measurement of the absorption of light or electromagnetic radiation by gases, liquids or solids (→ Bouguer-Lambert-Beer law). The quantity measured is known as the → absorbance. On an → energy-level diagram of atomic, molecular or nuclear energy levels, an absorption is always associated with the excitation from an energetically lower lying level to an energetically higher lying level. Depending on the position of the → spectral region, we differentiate in absorption spectroscopy between the → vacuum UV, → UV-VIS, → IR (near, mid, far) and microwave spectrum. (→ Magnetic resonance spectroscopy, → NMR spectroscopy, → microwave spectroscopy).

Absorption spectrum, <*Absorptionsspektrum*>, the graph of the molar decadic → extinction coefficient, ε, (→ Bouguer-Lambert-Beer law) as a function of the wavelength, λ, wavenumber, $\tilde{\nu}$, or frequency, ν, of the electromagnetic radiation. Since ε, a substance-specific quantity, depends only upon the wavelength (wavenumber) the plot gives the characteristic absorption behavior of that substance in the appropriate → spectral region.

Since the molar decadic → extinction coefficient can change by several orders of magnitude ($1 \leq \varepsilon(\lambda) \leq 10^5$) within a spectrum, particularly in the UV-VIS spectral region (→ UV-VIS spectroscopy), it is usual to plot the logarithm of ε against the wavelength instead of $\varepsilon = f(\lambda)$. Figure 1 shows a comparison of the two methods using

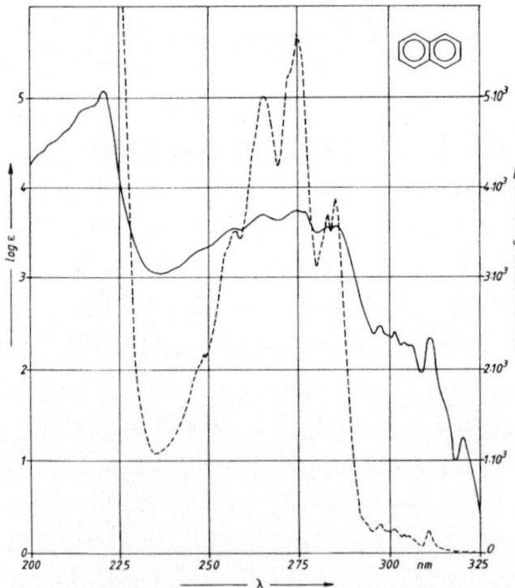

Absorption spectrum. Fig. 1. The UV spectrum of naphthalene; comparison of plotting ε (---) and log ε (—) as $f(\lambda)$

the spectrum of naphthalene in n-heptane as an example. The → wavenumber, $\tilde{\nu}$, is preferred to wavelength, λ, for the abscissa since the positions of the individual bands then correspond directly to the → excitation energy, ΔE. The measured absorbance $A(\lambda)$ is frequently plotted as a function of the wavelength, particularly when the concentration is unknown. In this case, $A = f(\lambda)$ represents the → color curve of the substance (medium) concerned. In → IR spectroscopy, the measured transmittance is recorded as a function of wavenumber, $\tilde{\nu}$. This simple presentation is due to the fact that the concentration and path length are not known exactly in many cases so that the calculation of $\varepsilon(\lambda)$ contains errors. Nevertheless, the IR spectrum as $T(\%) = f(\tilde{\nu})$ provides important information in structure-analytical problems.

In ESR (EPR) spectroscopy (→ magnetic resonance spectroscopy), the first derivative of the absorption is plotted as a function of the varying magnetic field, B. In → NMR spectroscopy, the signal intensity is plotted as a function of the irradiating radiowave frequency, ν, or as a → chemical shift from a chosen standard resonance.

In the UV-VIS and IR spectral regions, the plotting of $A^{1\%}_{1\text{cm}}$ against the wavelength has proved to be practical. The unit is the absorbance of 1 g of the material under investigation dissolved in 100 ml solvent at a path length of 1 cm. It is independent of the relative molecular mass. Absorbance values thus defined are widely used in polymer chemistry and biochemistry.

Achromat, <*Achromat*> → chromatic aberration.

Achromatic color, <*achromatische Farbe*>, non-colored color, describes the colors black or white and the gray shades lying in between. The true colors are known as → hues in color science.

Achromaticity, <*Achromasie*>, indicates the lack of → color error in an optical system. In physiology, achromaticity is also understood to mean the visual refractive color error.

Achromatic prism, <*achromatisches Prisma*>, a deviating prism (→ prism, → Abbe prism) with a compensated → color error usually produced by a combination of two prisms with opposed dispersing actions. The refractive indexes (→ refraction of light) and the Abbe numbers, ν, (→ dispersion) of the two crown or flint glass prisms are chosen so that the color error for two selected colors is zero. It remains insignificantly small for the other colors if suitable types of glass are selected. The prism angles, γ_1 and γ_2, are given by the equations:

$$\gamma_1 = \frac{\delta}{(\nu_1 - \nu_2)\Delta n_1}; \gamma_2 = \frac{\delta}{(\nu_1 - \nu_2)\Delta n_2};$$

where δ is the total deviation, ν_1 and ν_2 are the Abbe numbers of the glasses and Δn_1 and Δn_2 the differences in the refractive indexes of the glasses for the two selected colors.
When achromatizing, i.e. removing the color error, those colors, for which the radiation detector has the highest sensitivity, are paired. For a visual observation, for example, we select the → Fraunhofer lines C (λ = 656 nm) and F (λ = 485 nm).

Acquisition time, <*Akquisitionszeit*>, the time set aside for the gathering and storing of experimental data by an electronic device, e.g. a computer.

Actinochemistry, <*Aktinität*>, derived from the Greek „aktinos“ = ray; the study of the chemical action of light; → actinometer.

Actinometer, <*Aktinometer*>, used in spectroscopy for determining the light flux, I_0, which enters through the front surface into a cuvette. It is more exact to speak of a chemical actinometer since it is, in principle, an integrating detector in which the quantity of radiation, I_0, is determined by the amount of substance transformed in a photochemical reaction. The quantum yield of the reaction concerned must be known for each spectral region from independent measurements. Actinometry is essential for the quantitative study of → photochemical reactions, i.e. for determining the quantum yields of these reactions. The best-known chemical actinometer is the Parker and Hatchard ferrioxalate method (C.A. Parker, *Proc. Roy. Soc. London*, **1953**, *A220*, 104; C.G. Hatchard, C.A. Parker, *ibid*, **1956**, *A235*, 518). The practical execution of measurements with this actinometer has been extensively described in literature (S.L. Murov, *Handbook of Photochemistry*, Marcel Dekker Inc., New York, **1973**, pp. 119–123). In the P & H actinometer, Fe^{++} is produced from the ferrioxalate ion, $[Fe(C_2O_4)]^+$, in an acid solution. The quantum yield of $\Psi_{II}^{III} \cong 1.25$ is almost wavelength-independent in the range of 254–436 nm. After the photoreaction, the concentration of the Fe^{++} is determined photometrically at 510 nm via the 1,10 phenanthroline complex. Operating this actinometer is rather complicated and time-consuming. Furthermore, only one intensity (I_0) value can be determined in each run and the photochemical reaction itself cannot be monitored during actinometry.

For these reasons, actinometers have been designed which make it possible to a) follow the transformation directly photochemically, b) measure the actinometer solution in the apparatus used for the photochemical reaction, c) calculate the result directly after actinometry and d) obtain several intensity values from each actinometric measurement. For this a photoreaction, which is uniform (i.e. there are no complicating factors such as back reactions or branching) and as simple as possible, is required. The following systems are suitable:

Azobenzene in the range of 220–280 nm;
2,2',4,4' tetraisopropylazobenzene (350–390 nm);
heterocoordianthrone (HCD) (300–370 nm);
aberchrome 540 (450–550 nm);
epoxide of HCD (400–440 nm and 470–600 nm);
mesodiphenylhelianthrene (470–620 nm).

Amrein, Gloor and Schaffner have developed an electronically integrating actinometer for determining the quantum yields of photochemical reactions (*Chimia*, **1974**, *28*, 185).

In physics, actinometers are used to measure solar radiation directly. The Michelson-Marten and the Linke-Feussner actinometers were developed for this purpose. The former has a blackened bimetallic strip as the radiation receiver whilst the latter utilizes a thermopile (→ thermocouple).

Action, <*Wirkung*>, is the product of energy and the time during which this energy acts or is effective. The units of action are energy · time, i.e. erg · s or Joule · s. In spectroscopy, Planck's quantum of action is of great importance (→ Planck's constant).

Active Medium, <*aktives Medium*>, the medium in a laser which can be excited to produce the emission of laser light. It consists of the active atoms or molecules which increase the light intensity and a medium which contains these atoms (molecules). Depending on the type of active medium, we distinguish between → gas, dye, semiconductor (→ diode laser) and → solid-state lasers.

Adaptation, <*Adapt[at]ion*>, is the matching of the senses to changes of e.g. light intensity.

ADI filter, <*ADI-Filter*> → all dielectric interference filter.

ADP crystal, <*ADP-Kristall*>, ammoniumdihydrogen phosphate ($NH_4H_2PO_4$), an → optically uniaxial crystal used in the same way as the → KDP crystal.

Aerosol deposition module, <*Aerosol-Deposition-Modul*>, a device for automatic sampling in graphite furnace AAS (→ graphite furnace technique). The sample is transferred as an aerosol and injected into the preheated (150°C) graphite furnace. In comparison with direct injection, this saves the drying stage and increases the sensitivity because smaller crystals are formed.

AES, <*AES, Atomemissionsspektroskopie*> → atomic emission spectroscopy.

AFS, <*AFS, Atomfluoreszenzspektroskopie*> → atomic fluorescence spectroscopy.

Afterglow, <*Nachleuchten*> → phosphorescence.

Albedo, reflecting power, <*Albedo, Rückstrahlvermögen*>, from the Latin „albus“ = white, „albedo“ = white color.

1. Lambert's geometrical albedo is the amount, a, of the incident light diffusely reflected from a perpendicularly irradiated surface element. Values of a are: chalk 1.00, snow 0.79, pumice 0.56, lime 0.42, topsoil 0.08, lava 0.05;
2. Bond's spherical albedo is the ratio, A, of the light diffusely scattered in all directions by a spherical body to the parallel incident radiation.
   For individual materials, the spherical albedo is slightly larger than the geometrical.

Alkali-like ions, <*alkaliähnliche Ionen*>, are those ions which have, after single or multiple ionization of the element, the electron configuration of the alkali metal at the beginning of the appropriate row of the periodic table, i.e. ns^1. For example, the following are the 2nd and 3rd rows.

$2s^1$: Li, Be^{1+}, B^{2+}, C^{3+}, N^{4+}, O^{5+}, F^{6+}, Na^{8+}

$3s^1$: Na, Mg^{1+}, Al^{2+}, Si^{3+}, P^{4+}, S^{5+}, Cl^{6+}.

Therefore, we would expect that the associated → spark spectra (e.g. BeII, BIII, CIV, NV, OVI, FVII) would correspond completely to the → arc spectrum of the lithium atom (LiI) if we take into consideration the effective atomic charge, $Z - p$. p is the number of remaining inner (core) electrons; in this case, $p = 2$ and in the above sequence, $Z - p$ has the values 1, 2, 3, 4 ... Thus, the wavenumbers of the lines or the term energies should differ only by the constant factor $(Z - p)^2$. If the energies are divided by $(Z - p)^2$ we should expect the same values for all the spectra in such a sequence. However, in reality, though the spectra are analogous to each other in all details, the individual energy-level diagrams do not coincide exactly after division by $(Z - p)^2$. The reason is that the field in which the external s electron moves, is not a pure Coulombic central field. Thus, the energies are not simply proportional to $(Z - p)^2$ as is the case for the H atom. By way of illustration, Figure 1 shows the term schemes for the Li-like ions to OVI.

Alkali-metal vapor lamp, <*Alkalidampflampe*>, comprehensive term for sodium, potassium and other alkali-metal vapor lamps (→ metal-vapor discharge lamp). They are preferred for use in → refractometry and polarimetry (→ polarimeter).

Alkali spectra, <*Alkalispektren*>, the absorption and emission spectra of alkali-metal vapors can be classified into four series: the principal series (*PS*) which is observed in absorption and emission plus 2 subseries (*SS*) and the Bergmann series (*BS*) which are only observed in emission. However, a representation of these series using formulas exactly analogous to the → Balmer formula is not possible. Several authors proposed empirical series

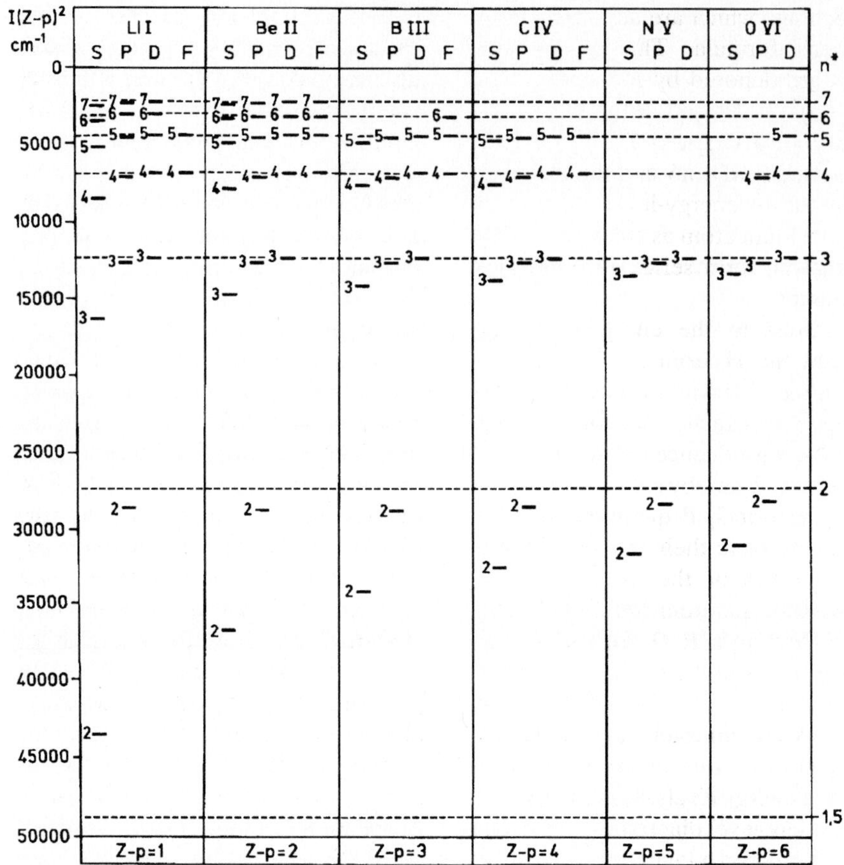

Alkali-like ions. Fig. 1. Energy levels of ions LiI → OVI

formulas before 1900, but Rydberg's general formula for the interpretation of alkali spectra has proved to be the most suitable one. According to Rydberg, the four series can be represented as follows:

$$\text{PS}: \tilde{\nu} = R\left(\frac{1}{(1+s)^2} - \frac{1}{(n+p)^2}\right);$$
$$n = 2, 3, 4, ..., 1S - nP$$

$$1.\ \text{SS}: \tilde{\nu} = R\left(\frac{1}{(2+p)^2} - \frac{1}{(n+d)^2}\right);$$
$$n = 3, 4, 5, ..., 2P - nD; \text{(diffuse)}$$

$$2.\ \text{SS}: \tilde{\nu} = R\left(\frac{1}{(2+p)^2} - \frac{1}{(n+s)^2}\right);$$
$$n = 3, 4, 5, ..., 2P - nS; \text{(sharp)}$$

$$\text{BS}: \tilde{\nu} = R\left(\frac{1}{(3+d)^2} - \frac{1}{(n+f)^2}\right);$$
$$n = 4, 5, 6, ..., 3D - nF.$$

R is the → Rydberg constant (109677.578 cm^{-1}). The *s, p, d* and *f* are decimal fractions, called → Rydberg corrections, which are additions to → Balmer's formula. The three fixed terms are denoted by $R/(1+s)^2 \equiv 1S$, $R/(2+p)^2 \equiv 2P$ and $R/(3+d)^2 \equiv 3D$ whilst the varying terms are given the symbols, *nP, nD, nS* and *nF*. Figure 1 shows the → energy-level diagram of the potassium atom as an example. We see that all term series approach the same limit.

In contrast to the energy-level diagram of the H atom (→ Balmer formula), the different adjacent levels no longer have exactly the same energy. This is a consequence of the loss of → degeneracy whereby states assigned to the same principal quantum number, n, now differ in their energy with different values of the orbital angular momentum quantum number, ℓ. The energy levels, *S, P, D, F* have the following values of the quantum number, ℓ (S: $\ell = 0$; P: $\ell = 1$; D: $\ell = 2$; F: $\ell = 3$). A. Sommerfeld described the connection of this quantum number with the energy levels as early as 1916. The energy-level diagram of the potassium atom shown takes into consideration the → selection rule, $\Delta\ell = \pm 1$. Thus, transitions are only allowed between those terms (energy levels) in which the orbital angular momentum quantum number changes by one unit, i.e. transitions between members of the same term series are forbidden. However, transitions between members of adjacent term series are allowed. The Rydberg correction describes the variations from the hydrogenic energy levels. This correction is greatest for the S energy levels and decreases in the order *P, D* and *F*. The F energy levels are effectively the same as the Balmer energy levels. The expressions, $1 + s$, $2 + p$ and $3 + d$ in the fixed terms and $n + p$, $n + d$, $n + s$, $n + f$ in the running terms, are not true quantum numbers as in the Balmer formula. Therefore, they are called effective quantum numbers. Since the lowest state of all alkali atoms is called the $1S$ energy level, attention must be paid to the assignment of the true principal quantum number for each atom.

Confusingly, S is also used to denote the total spin angular momentum of the atomic electrons, from which the → multiplicity of the atomic state is calculated using the formula, multiplicity = $2S + 1$. The alkali-metal atoms have one unpaired electron so $S = ½$ and the multiplicity = 2. Thus, the alkali-metal atoms are doublet systems (→ multiplet system), i.e. we have the simplest form of the → multiplet structure of atomic spectra. The transitions, $1^2S_{1/2} \to 2^2P_{1/2}$ and $1^2S_{1/2} \to 1^2P_{3/2}$ are assigned to the D lines, the first line of the principal series. The subscripts, 1/2 and 3/2, are quantum numbers associated with the total electronic angular momentum of the atom, the sum of the orbital and spin

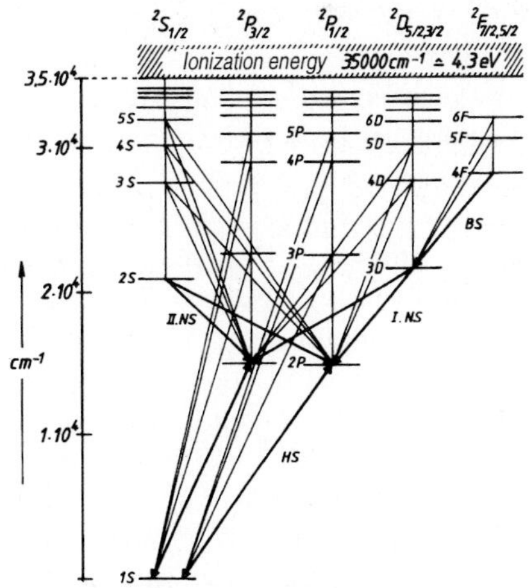

Alkali spectra. Fig. 1. Energy-level diagram for the potassium atom

angular momenta; → Russell-Saunders coupling.

All-dielectric-interference filter, ADI filter, <*All-Dielektrik-Interferenzfilter, ADI-Filter*>, consists of an alternating series of substances with different refractive indexes (→ refraction of light) which do not absorb in the → spectral region of interest. By varying the layer structure, the following devices: mirrors, → edge filters (→ short-wave or → long-wave pass filters as well as color separators), → band-pass filters (line, band and broad-band filters) can be obtained. The half-widths of the last are adjustable in steps to give values between 0.5 % and 40 % of λ_m (→ interference filters).

Alpha radiation, α radiation, <*Alphastrahlung, α-Strahlung*>, consists of bare, i.e. doubly positively charged helium nuclei, also called alpha particles, the velocity of which can lie between $1.5 \cdot 10^9$ and $2.25 \cdot 10^9$ $cm \cdot s^{-1}$. At a mass of approximately $6.6 \cdot 10^{-24}$ g, this corresponds to a kinetic energy of 4.62 to 10.4 MeV (1 MeV = 10^6 eV). Alpha radiation is a product of the decay of radioactive substances and the gradual formation of helium gas from such sources can be established spectroscopically.

Amici prism, Browning prism, <*Amici-Prisma, Browning-Prisma*>, is a → minimum deviation prism. In the simplest case, it consists of a flint-glass prism, i.e. a prism with a large refractive index (n') and two, usually smaller, crown-glass prisms, i.e. prisms with a small refractive index (n) which are glued to the former with Canada balsam (Figure 1). The prism angles for the crown glass, ε, and the flint glass, ε', are selected in such a way that for a given wavelength, e.g. the Fraunhofer D line (588/589 nm), no deviation, ΔD, results. Assuming small prism angles, we have:

$$\frac{\varepsilon}{\varepsilon'} = \frac{n'_D - 1}{n_D - 1}$$

which is the relationship between the prism angles and refractive indexes of the two prisms forming a minimum deviation prism. Furthermore:

$$\bar{\theta}_{F,C} = (n_D - 1)\varepsilon \cdot \left[\frac{n_F - n_C}{n_D - 1} - \frac{n'_F - n'_C}{n'_D - 1}\right]$$

and with the definition of the Abbe number (→ dispersion, unit of), ν, we obtain:

$$\bar{\theta}_{F,C} = (n_D - 1)\varepsilon \cdot \left[\frac{1}{\nu} - \frac{1}{\nu'}\right]$$

$\bar{\theta}_{F,C}$ is the angular separation of the wavelengths of the Fraunhofer lines, F and C, (486.1 nm and 656.2 nm). Thus, a direct vision prism is only possible if the Abbe numbers of the two types of glass used are different.

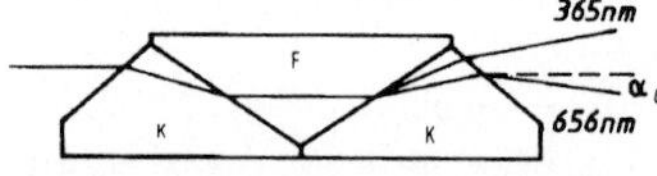

Amici prism. Fig. 1. Construction of an Amici prism, F = flint glass, K = crown glass

Amplitude, <*Amplitude*>, generally the maximum displacement of a vibration.

An amplitude is the displacement from the position of rest, especially in the case of a harmonic vibration. For example, a harmonic vibration can be represented as a sine function: $a = A \cdot \sin(2\pi \nu t)$. Where ν is the frequency of the vibration, t the time and A the maximum amplitude. (→ phase of oscillation, → harmonic oscillator).

Amplitude spectrum, <*Amplitudenspektrum*>, a representation of the amplitudes of the individual oscillations, of which any process, $f(t)$, is composed, as a function of the frequency of each particular vibration. A periodic process, $f(t)$, is frequently represented by a Fourier series:

$$f(t) = \sum_{n=0}^{\infty}(a_n \cdot \sin\omega nt + b_n \cdot \cos\omega nt)$$

The amplitude of the nth component vibration is then given as: $c_n = (a_n^2 + b_n^2)^{1/2}$. The discrete values, c_n, plotted against n give information about the proportion of overtones. This spectrum is called the discrete amplitude spectrum. Such a representation is of particular importance in → Fourier spectroscopy.

Analytical spectral lines <*Analysenlinien*>, are the total of all spectroscopic lines suitable for qualitative and quantitative spectral analysis and particularly for → atomic emission spectroscopy (AES). Most spectral lines used in qualitative analysis are also → resonance lines (→ spectroscopic analysis). If possible, spectral lines used for analysis should be sufficiently sensitive, both absolutely and relatively, and easy to find due to their favorable position in the spectrum. They must also be free from interference by the spectral lines of other elements. There are various tables in which spectral lines for analysis have been ordered by wavelength together with estimated intensities, excitation requirements and degree of ionization as well as a description of their peculiarities (sharp, broad, diffuse, self-absorbed etc.). Many manufacturers of AES or AES-ICP instruments supply tables of the optimum spectral lines and operating conditions for the analysis of each element.
Ref.: G.R. Harrison, *MIT Wavelength Tables*, MIT Press, Cambridge, MA, USA, **1969**; W.F. Meggers, C.H. Corliss and B.F. Scribner, *Table of Spectral Line Intensities*, MBS Monograph 145, US Government Office, Washington, DC, USA, **1975**.

Analytic lamp, <*Analysenlampe*> → black-light lamp.

Analyzer, <*Analysator*>, the optical component with which we establish the polarization of light. An analyzer is generally used in combination with a → polarizer with which it is practically identical, i.e. a polarizer can also be used as an analyzer, and vice versa. → Nicol prism, → Glan-Thompson prism.

Ångström unit, Å, <*Ångström-Einheit*>, a spectroscopic unit of length introduced by A.J. Ångström in 1868 and defined by a number of wavelengths from the solar spectrum: 1Å = 10^{-10} m = 10^{-1} nm = 10^{-8} cm. In 1907, the International Association for Solar Research created a new standard wavelength system based on the red cadmium line (→ wavelength standard). The Å, called the International Ångström, IÅ, was redefined by the equation 1 IÅ = $0.000155316413\lambda_{Cd}$ (λ_{Cd} = 6438.48698Å measured in spectroscopically standard air). Although the Ångström unit is still used in spectroscopy, it is not an SI unit.

Angular dispersion, <*Winkeldispersion*>, is the measure which characterizes the suitability of a → prism as a light → dispersing element. It is defined as follows:

$$\frac{\mathrm{d}\vartheta}{\mathrm{d}\lambda} = \frac{\mathrm{d}\vartheta}{\mathrm{d}n}\frac{\mathrm{d}n}{\mathrm{d}\lambda}.$$

ϑ is the angle of deviation (see Figure 1), $\mathrm{d}\vartheta/\mathrm{d}n$ the dependence of the angle of deviation upon the → refractive index and $\mathrm{d}n/\mathrm{d}\lambda$ the → dispersion of the prism material. $\mathrm{d}\vartheta/\mathrm{d}n$ can be replaced by an expression involving ϑ and the prism angle, φ (see Figure 1):

$$\frac{\mathrm{d}\vartheta}{\mathrm{d}n} = \frac{2\sin\frac{\varphi}{2}}{\cos\frac{\vartheta+\varphi}{2}}.$$

Thus, the angular dispersion depends not only upon the optical properties of the prism material but also upon the shape of the prism. Other factors determining angular dispersion are the base length, b, of the prism and the → aperture, a, of the beam of light to be dispersed which falls onto the face of the prism. Thus, we obtain:

$$\frac{\mathrm{d}\vartheta}{\mathrm{d}\lambda} = \frac{b}{a}\frac{\mathrm{d}n}{\mathrm{d}\lambda}.$$

The greater the angular dispersion of a prism the greater is the distance between two spectral lines which are to be separated (see also → prism monochromator).

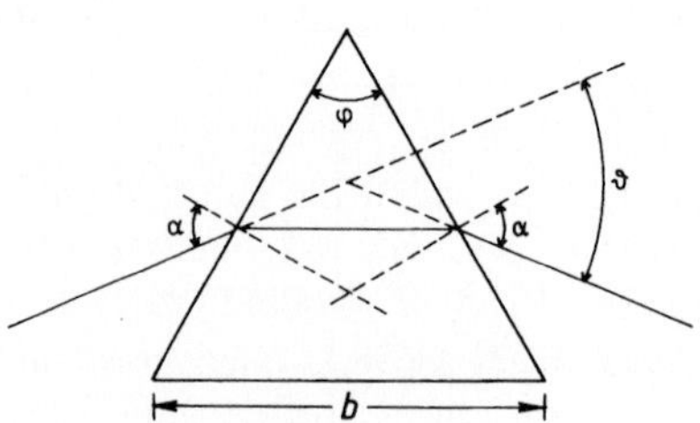

Angular dispersion. Fig. 1. Illustration of angular dispersion

Angular momentum, <*Drehimpuls*>, is given in classical mechanics by the vector product, $\vec{L} = \vec{r} \times \vec{p}$, where $\vec{p} = m \cdot \vec{v}$ is the linear momentum of the particle of mass, m, moving with velocity, $\vec{v}$, about an axis. $\vec{r}$ is the radial vector from the axis to the particle. The angular momentum, $\vec{L}$, is directed perpendicularly to the surface defined by $\vec{r}$ and $\vec{p}$.
According to Bohr's → atomic model, in an atomic system, e.g. an H atom, an electron moves in an orbit around the atomic nucleus. The quantum-mechanical analysis of this problem shows that the orbital angular momentum (→ electronic angular momentum) is quantized.

Anharmonic oscillator, <*anharmonischer Oszillator*>, an oscillator which is not a harmonic oscillator. The → harmonic oscillator is the simplest model for the approximate interpretation of the vibrational excitation of a diatomic molecule. However, this model cannot explain important experimental findings:

1. The occurrence of → overtone vibrations;
2. the dissociation of higher excited vibrational states.

Also, it does not take into consideration the repulsive forces which come into play when the interatomic distance is less than the equilibrium value, r_o, of the ground state.
The true behavior of a diatomic molecule during vibration is better described by a potential curve (Figure 1) which takes account of the above findings. There is no closed mathematical expression for this new potential

curve (but → Morse curve). By way of a trial, we can extend the quadratic potential function of the harmonic oscillator by an additional cubic term and further terms of even higher powers:

$$E_{pot} = fx^2 - gx^3 + \dots$$

x represents the displacement from the equilibrium value and g is considerably smaller than f. Although this formula allows for the anharmonicity of the vibration it is by no means in complete agreement with the true potential curve. An oscillator described by the potential curve shown in Figure 1 is called an anharmonic oscillator. Quantum-mechanical analysis gives the following eigenvalues:

$$E(\upsilon) = hc\omega_e(\upsilon + \frac{1}{2}) - hc\omega_e x_e(\upsilon + \frac{1}{2})^2 + hc\omega_e y_e(\upsilon + \frac{1}{2})^3 + \dots$$

where $\omega_e \gg \omega_e x_e \gg \omega_e y_e$ and ω_e = wavenumber corresponding to the classical vibrational frequency, ν_e.
From this we obtain the energy levels (sometimes also called terms):

$$G(\upsilon) = \frac{E(\upsilon)}{hc} = \omega_e(\upsilon + \frac{1}{2}) - \omega_e x_e(\upsilon + \frac{1}{2})^2 + \omega_e y_e(\upsilon + \frac{1}{2})^3 + \dots .$$

For the → zero-point energy ($\upsilon = o$), we obtain the energy level $G(\upsilon = o) = G(o) = \frac{1}{2}\omega_e - \frac{1}{4}\omega_e x_e + \frac{1}{8}\omega_e y_e$. If we take this as our zero we obtain the following energy levels (terms):

$$G_o\upsilon = \omega_o\upsilon - \omega_o y_o \upsilon^2 + \omega_o y_o \upsilon^3 + \dots$$

and with the internationally used abbreviations:

$$\omega_o = \omega_e - \omega_e x_e + \frac{3}{4}\omega_e y_e + \dots$$

$$\omega_o x_o = \omega_e x_e - \frac{3}{2}\omega_e y_e + \dots$$

$$\omega_o y_o = \omega_e y_e$$

The subscript e refers to the equilibrium state, i.e. to infinitely small vibrational amplitudes about this. Since $\omega_e x_e$ is always positive, but $\omega_e y_e$ can take both positive and negative values, the energy levels are no longer equally spaced, as they are in the harmonic case. They move closer together, with increasing υ, as we approach the dissociation limit, E_D, Figure 1. However, υ_D has a finite value. The quantum-mechanical treatment reveals changes vis-a-vis the harmonic oscillator; in particular in the selection rule for change of the vibrational quantum number. Now $\Delta\upsilon = \pm 1, \pm 2, \pm 3$ ..., i.e. transitions are allowed for which the quantum number changes by 2, 3 or more giving access to the → overtone vibrations. Because the energy levels are no longer equally separated, the overtones are not exact whole-number multiples of the fundamental frequency, $\upsilon'' = 0$

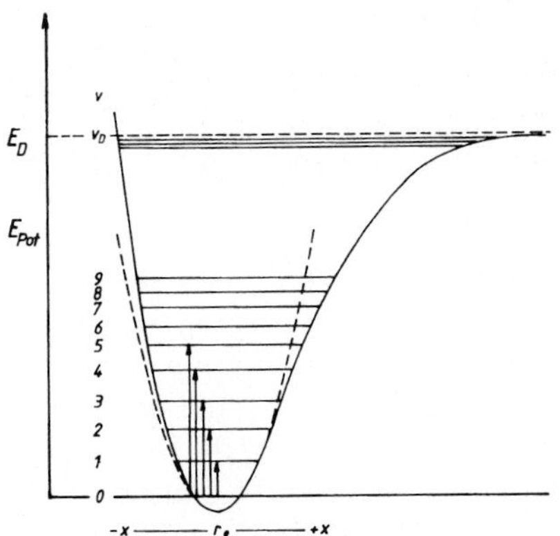

Anharmonic oscillator. Fig. 1. Potential curve of an anharmonic oscillator

$\rightarrow v' = 1$. If we assume that the excitation starts from the state, $v'' = 0$, we find absorption bands at the wavenumbers:

$$\tilde{\nu}_{abs} = G(v') - G(o)$$
$$= \omega_o v' - \omega_o x_o v'^2 + \omega_o y_o v'^3 + \ldots$$

with $v' \geq 1$.
The wavenumbers of the absorption bands give the positions of the energy states above the lowest level ($v'' = o$) directly, see Figure 1. If we ignore the third order terms ($\omega_o y_o = 0$) the separation of adjacent bands (levels) is:

$$\Delta G(v + \frac{1}{2}) = G(v + 1) - G(v)$$
$$= G_o(v + 1) - G_o(v)$$
$$= \omega_e - 2\omega_e x_e - 2\omega_e x_e v$$
$$= \omega_o - \omega_o x_o - 2\omega_o x_o v$$

Forming the second differences, $\Delta^2 G(v + 1)$ we obtain:

$$\Delta^2 G(v + 1)$$
$$= \Delta G(v + \frac{3}{2}) - \Delta G(v + \frac{1}{2})$$
$$= -2\omega_e x_e$$
$$= -2\omega_o x_o.$$

The product of the vibrational constants, $\omega_e x_e = \omega_o x_o$, which is to be regarded as the measure of the anharmonicity of the oscillator, results directly.
For the first transition, $v'' = o \rightarrow v' = 1$, we obtain for the energy-level difference, ΔG, $\tilde{\nu}^{1\text{-}o}_{abs} = \omega_e - 2\omega_e x_e = \omega_o - \omega_o x_o = \Delta G(1/2)$. $\tilde{\nu}^{1\text{-}o}_{abs}$ is the wavenumber of the maximum of the absorption band measured in the IR spectrum; it is not equal to the wavenumber, ω_e, of the classical vibrational frequency, $\nu_e = c\omega_e$. However, if the anharmonicity term, $\omega_e x_e$ or $\omega_o x_o$ is known, ω_e can be calculated using $\tilde{\nu}^{1\text{-}o}_{abs}$. With this value for $\nu_e = c\omega_e$, we can calculate the force constant of the anharmonic oscillator for infinitely small displacements using the equation, $k_e = 4\pi^2 \mu c^2 \omega_e^{\ 2}$.
Ref.: G. Herzberg, *Molecular Spectra and Molecular Structure, I. Spectra of Diatomic Molecules*, D. Van Nostrand Co. Inc., Princeton, New Jersey, **1967**.

Anisotropy of light absorption, <*Anisotropie der Lichtabsorption*>, the transition moment of electronic excitation in molecules is a vector quantity and can, therefore, be written in terms of its three Cartesian components:

$$|\vec{M}_{1,k}| = (\vec{M}_x^2 + \vec{M}_y^2 + \vec{M}_z^2)^{1/2}$$

Consider a planar molecule lying in the x–y plane, then the x and y components, M_x and M_y, lie in this plane and M_z perpendicular to it. In the case of planar molecules, the component $M_z \perp$ to the x–y plane can usually be neglected for singlet transitions ($M_z = o$). In this case, the transition moment corresponds to the excitation of the electrons in the molecular plane. The direction, α, of this transition moment is then given by $\tan\alpha = M_y/M_x$. If one of the components, M_x or M_y, equals zero, then there is only one electronic transition polarized in one direction and the absorption of light is anisotropic. This is frequently the case for condensed aromatic hydrocarbons, for example. An experimental determination of the anisotropy of electronic transitions, as it is related to the molecular geometry, can be made by means of absorption measurements using linearly polarized light. However, it is a

precondition that the spatial orientation of the molecule relative to the plane of polarization of the light is known, or does not change during the measurement. Molecular single crystals are very suitable for such investigations, as are also solutions in solid polymers which can be stretched in one direction. → Photoselection is another technique for investigating the anisotropy of light absorption.
Ref.: H.-H. Perkampus, *UV-VIS Spectroscopy and Its Applications*, Springer Verlag, Berlin, Heidelberg, New York, **1992**, p. 217 ff.

Antiauxochrome, <*Antiauxochrom*> → auxochrome.

Anti-Stokes line, <*Anti-Stokessche Linie*>, spectral line or band (especially in Raman spectroscopy) where → Stokes rule is contravened. In this case the light quantum, $h\nu_e$, emitted following the excitation of an atom or a molecule with light quantum, $h\nu_a$, is of higher energy than that absorbed, i.e. $\nu_e > \nu_a$.

Aperture, <*Apertur, Blende*>.

1. The sine of half the aperture angle of a cone-like beam of light rays, i.e. the angle, α, between the axis of the cone and a generatrix. If the refractive index of the medium, in which the light beam is propagating, differs from 1, the *numerical aperture* [= refractive index · sin(α/2)] is used. For monochromators, the relative aperture is defined as the ratio of aperture, a, to the focal length, f, of the monochromator; $a_{rel} = a/f$. The focal length of a monochromator system is that of the → collimator (lens or concave mirror at the exit slit) which collimates the emerging light.
2. Any cross-sectional limitation (stop, diaphragm) of a light beam forming an optical image. An aperture limits the cross section of the light cone emitted from an object point; it determines the resolving power arising from refraction. We describe the smallest aperture seen from an object point on the axis as the entrance pupil and the image of the entrance pupil as the exit pupil. The ratio *exit pupil to focal length* is the *relative aperture* and determines the image brightness.

APF spectrum, absorption-polarization-fluorescence spectrum, <*APF-Spektrum, Absorptionspolarisationsfluoreszenzspektrum*> → photoselection.

Apochromatic lens, <*Apochromat*> → chromatic aberration.

Appearance potential, <*Auftrittspotential, Auftrittsenergie, Erscheinungspotential*>, in → mass spectroscopy, the minimum energy required to ionize a molecule. In most cases, it corresponds to the ionization energy of the atom or molecule. We can determine the appearance potential by plotting the ion flux against the energy of the ionizing electrons and extrapolating the linear part of the steeply increasing curve to zero ion flux. However, only approximate values of the ionization energy of the molecule concerned are obtained. More reliable results are achieved with ion sources specially designed for such measurements.
Ref.: J.H. Beynon, A.G. Brenton, *An Introduction to Mass Spectroscopy,*

University of Wales Press, Cardiff, **1982**.

APPh spectrum, absorption-polarization-phosphorescence spectrum, <*APPh-Spektrum, Absorptionspolarisationsphosphoreszenzspektrum*> → photoselection.

Arc, <*Lichtbogen*> → arc discharge.

Arc discharge, <*Bogenentladung*>, the electric discharge which arises when two electrodes are bridged by an electric arc. The discharge forms a brightly emitting channel between the two electrodes (continuous arc). The arc reaches temperatures of ca. 5000 K and essentially only → arc spectra are produced. Too rapid a consumption of material, scaling of the sample and melting of the electrodes make the continuous arc useless for quantitative analysis, for which purpose an interrupted arc, formed by a periodic interruption of the discharge, is used. Better detection limits are achieved than with a → spark analysis, though better precision is obtained with the spark. Together with the → spark discharge and the flame, the arc is the oldest source of excitation in → atomic emission spectroscopy.

Arc spectrum, <*Bogenspektrum*>, the spectrum of a neutral atom which requires relatively low energy for its excitation; that provided by a flame or an → arc discharge, for example. The arc spectrum of an element is denoted by a Roman I after the element's symbol, e.g. LiI, CuI, NeI ...

Argon-ion laser, <*Argonionenlaser*> → noble-gas ion laser.

Asterism, <*Asterismus*>, the phenomenon that a light source appears to be distorted when observed through a crystal which contains small acicular regions of another crystal type. Regular and symmetrical figures, in most cases star-like, are frequently formed. The refraction of the light at the deposits is the cause of this phenomenon. Mica with deposits of rutile needles (TiO_2) may be used to demonstrate the 6-rayed star. Light figures can also be formed by fine, parallel, hollow channels or twin laminations. Asterism can also be observed in light reflected from polished crystals of appropriate structure.

Astigmatism, <*Astigmatismus*>, the phenomenon that a point off the axis of a lens is not imaged as a point but, depending on the angle of inclination of the beam, in the form of two image points or (for a large angle of aperture) in the form of two perpendicular lines at different distances from the lens. Astigmatism and image-field curvature are caused by the dissymmetry which a beam acquires during a strongly angled passage through a simple lens with the result that the image points of an object plane lie between two curved limiting surfaces. By combining different lenses, e.g. a convex and a concave lens of suitably selected materials (anastigmats), such image errors can be reduced. However, astigmatism still causes problems and optics manufacturers adapt their lenses, as far as possible, to their different applications.

Astrophotometry, photometry of the stars, <*Astrophotometrie, Photometrie der Gestirne*>, the methods of → photometry, especially modified for astronomical purposes, which are employed to measure brightness differences as exactly as possible. Visual, photographic, photoelectric, thermal and radiotechnical methods are used.

Astrospectrometry, astrospectrography, <*Astrospektrometrie, Astrospektrographie*>, observational aspects of → spectroscopic analysis especially modified for studying the spectra of stars. Star spectra are among the most important results of astrophysical observation since they reveal the star's temperature, speed and direction, its double-star character and many other clues about the development phases of the stars.

Asymmetric top (molecule), <*asymmetrisches Kreiselmolekül*>, a → top molecule where the moments of inertia about the three inertial axes are all different, i.e. $I_A \neq I_B \neq I_C$. In contrast to the → symmetric top molecule, there is no direction in the molecule along which the angular momentum vector, $\vec{P}_J$, has a constant component, i.e. there is no axis within the molecule which executes a simple rotation around $\vec{P}_J$. Therefore, the rotational movement is very complicated and can only be illustrated clearly in the simplest cases. Even rotational terms (energy levels) cannot be given in closed algebraic form as is the case for linear or symmetric top molecules (→ pure rotation spectra of asymmetric top molecules).

Atomic absorption spectrometer, <*Atomabsorptionsspektrometer, Atomabsorptionsspektralphotometer*>, an instrument of relatively simple design for the analytical determination of individual elements. Figure 1a shows the design of a modulated light (DC) instrument. The essential components are: a radiation source (→ hollow cathode lamp, HCL), 1, emitting the spectrum of the element under investigation; a device for → atomization, 2, in which atoms are formed from the sample under investigation – e.g. a

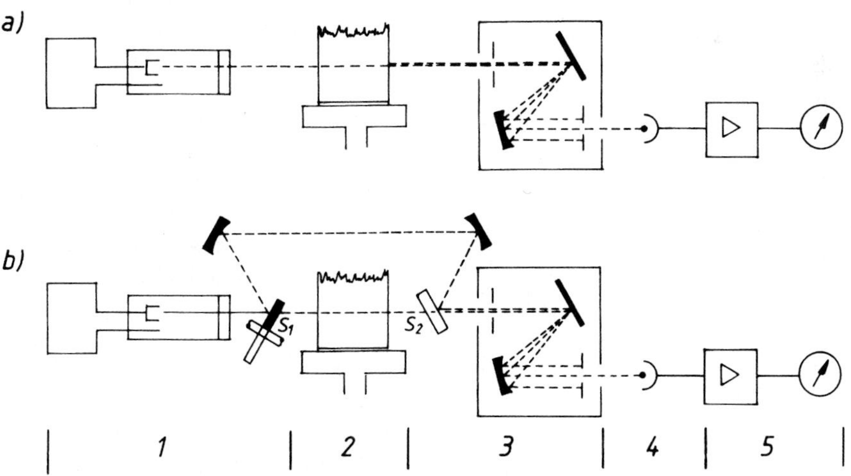

Atomic absorption spectrometers. Fig. 1.
a. Layout of a single-beam, modulated-light spectrometer
b. A double-beam, modulated-light spectrometer

flame of suitable temperature; a → monochromator, 3, for the dispersion of the radiation with an exit slit which separates the resonance line; a → radiation detector, 4, for the measurement of the radiation intensity and connected to it, an amplifier and a display unit, 5, for output of the measured value. Such a modulated-light instrument eliminates practically all interferences which are caused by emission from the flame itself (→ spectral interferences). The radiation from the light source is modulated electronically or mechanically (→ chopper) at a specific frequency. Of the radiation arriving at the detector, the frequency-selective lock-in amplifier amplifies only that which has the modulation frequency whilst the emission of the flame, which is not modulated, is ignored.

The double-beam, modulated-light instrument (Figure 1b) is a refinement. The radiation from the primary source is directed alternately through the flame (the test beam) and around it (the reference beam) by a rotating sector mirror, S_1. A semitransparent mirror, S_2, after the flame is used to recombine the radiation. The emission of the flame passes through the semitransparent mirror, S_2, as unvarying (in time) light and onto the detector. It is ignored in the same way as in the single-beam modulated-light instrument. The electronics of the instrument are designed in such a way that the ratio of the intensity of the two beams is formed. All changes or variations in the lamp intensity, detector sensitivity and the amplifier are eliminated by this double-beam arrangement. In conjunction with the → Zeeman effect, we can construct from the single-beam instrument an ideal double-beam instrument, which also eliminates background absorption. The Zeeman effect is used in this way for background compensation (→ Zeeman background correction). For atomization, the → flame, → graphite furnace, → hydride and → cold-vapor techniques are available. Most modern atomic absorption spectrometers are designed in such a manner that a change from one method to another is relatively quick. Many of the functions are controlled by microprocessors.

Ref.: B. Welz, *Atomic Absorption Spectrometry*, 2nd ed., Verlag Chemie, Weinheim; Deerfield Beach, Florida, Basel, **1986**.

Atomic absorption spectroscopy, AAS, <*Atomabsorptionsspektroskopie, Atomabsorptionsspektrometrie, AAS*>, is a method for the analytical determination of individual elements the basic principles of which were already clear in the classical experiments of R.W. Bunsen and G. Kirchhoff in 1859/1860. Both men are to be regarded as the founders of → spectroscopic analysis. Despite the early success, it was not until 1955 that the pioneering work of A. Walsh created the foundations for the present technique of quantitative atomic absorption spectroscopy and spectrometry. AAS is based on the fact that atoms in the vapor state are capable of absorbing light of a specific wavelength or wavenumber, which therefore reduces the intensity of the incident light. The selection of a suitable line for a particular element is not always simple. A study of → atomic spectra shows that only the lines of the principal series (→ alkali spectra) are observed in absorption. The first line of the principal series, the → resonance line, should be the most intense and most easily excited; but this applies only to alkali and alkaline-earth metals. For a number of other elements, lines which correspond to transitions of the atom from the ground state to higher excited states can be more intense. For atoms of the transition metals, we often have a larger number of low-energy excited states or better, of ground states with slightly increased energy, the occupation of which is temperature-dependent. Only an experimental investigation of the relative intensities can decide in such cases. We must also remember that the profile of an atomic absorption line is very narrow ($\Delta\lambda \leq 0.005$ nm). This means that, for such measurements, the excitation-light source must also transmit very narrow lines. For that reason, Walsh proposed a lamp, which itself emits the spectrum of the element under investigation as radiation source for atomic absorption measurements. In the case of such a light source, it is only necessary to separate the resonance line from the other spectral lines of the same element by means of a → monochromator. → Hollow cathode lamps and electrodeless → discharge lamps have proved to be excellent light sources. However, the disadvantage is that a different lamp is required for every element under investigation.

The absorption is evaluated using the → Bouguer-Lambert-Beer law in the form: $A = 2.303 \cdot \kappa_\nu \cdot N \cdot d$; where the → absorbance, A, is directly proportional to the total number of free atoms, N, present in the light-absorbing path of length, d. κ_ν is the atomic absorption coefficient. The absorbing path is in many cases a flame of suitable temperature in which atoms are formed from the sample under investigation → atomization. In addition to the → flame technique, the → graphite furnace method has proved itself in recent years. d is then the length of the graphite furnace in the interior of which the absorbing atoms are present in the vapor state. In the application of AAS, the elements to be determined are initially present in a sample solution. However, the concentration of the element concerned is measured in the plasma of a flame or in the graphite furnace. This assumes that there is an exact proportionality between the concentration in the sample solution and the measured absorbance. For quantative measurements, the proportionality constant must be determined by measurements on standard solutions of each individual element. In an ideal case, this → standard calibration method is used to obtain a linear correlation over a broad concentration range. The procedure assumes that standard solutions and sample solutions have identical properties. If this is not true, → matrix effects can occur which result in deviations from linearity. In such cases, a calibration using the → standard addition method can be made. In addition to the flame and graphite furnace, a heated quartz cell for the → hydride technique or a cylindrical quartz cuvette for the → cold-vapor method can be used for special analyses. Problems which occur in atomic absorption spectroscopy and which lead to errors in the analyses are usually called → interferences. They are divided, quite generally, into → spectral and → nonspectral interferences. Background radiation can cause further problems and modern → atomic absorption spectrometers are equipped with background compensa-

tion (→ Zeeman background correction).
In recent years, atomic absorption spectroscopy has been shown to be an excellent spectrochemical analytical method and has found extensive application. It is especially valuable for trace-element analysis and extraordinary low limits of detection are achieved with the → graphite furnace technique.

Ref.: B. Welz, *Atomic Absorption Spectrometry*, 2nd ed., Verlag Chemie, Weinheim; Deerfield Beach, Florida, Basel, **1986**; G.F. Kirkbright and M. Sargent, *Atomic Absorption and Fluorescence Spectroscopy*, Academic Press, London, New York, San Francisco, **1974**.

Atomic emission spectrometer, <*Atomemissionsspektrometer, AES-Gerät*>, an AES instrument has the following basic components: an excitation-light source (flame or plasma), a → grating monochromator, → a photomultiplier with a high-voltage supply and amplifier plus a display unit. This setup, with a double monochromator, is shown schematically in Figure 1. In AES, the → monochromator is used to separate the analytical line from all other radiation. In order to reduce → spectral interferences, monochromators with as high a → resolution as possible, i.e. a small spectral → slit width are required. AES instruments for routine measurements have a resolution of ca. 0.02–0.03 nm. The wavelength is set by rotating the plane grating with a computer-controlled stepper motor. This makes automatic, sequential, multielement determination possible and, simultaneously, the elimination of the background. AES instruments can be fitted with a → polychromator for a simultaneous, multielement determination. The dispersing element is a fixed → concave grating by means of which images of the entrance slit for different wavelengths are formed adjacent to each other. The exit slits (secondary slits) of

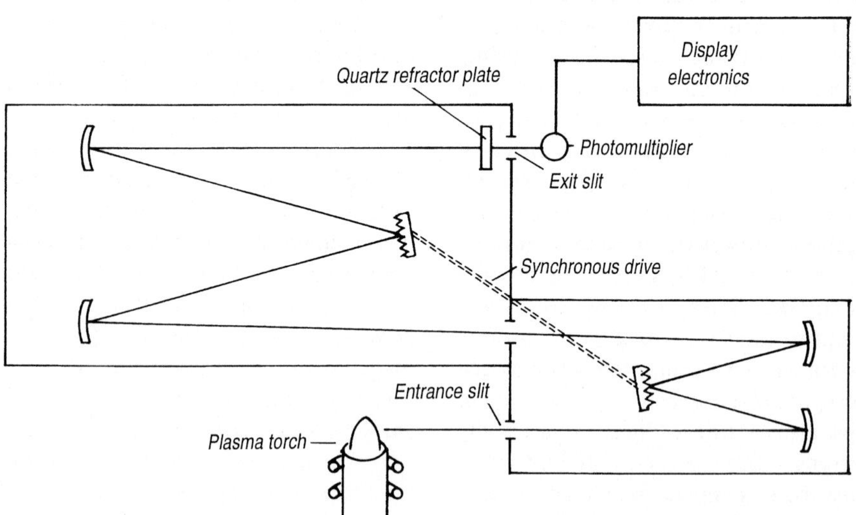

Atomic emission spectrometer. Fig. 1. Schematic view of an atomic emission spectrometer

the system are fixed to transmit specific emission lines. A → photomultiplier with its own high-voltage supply, amplifier and data processing facility is mounted behind each secondary slit. Such an instrument offers the possibility of the simultaneous determination of 40 and more elements in a relatively short time (1–3 minutes). However, an initial decision must be made as to which elements should be measured and at which wavelengths. A subsequent change is costly in time and very expensive. Another disadvantage is the fact that, compared with a stepper-motor-controlled AES instrument, only a limited degree of background correction is possible.

Atomic emission spectroscopy, AES, *<Atomemissionsspektroskopie, Atomemissionsspektrometrie, AES>*, a method of atomic spectroscopy which is based on the phenomenon that, following the excitation or ionization of atoms by high thermal energy or electron impact, a loss of energy from the excited state occurs by emission of light, i.e. with the formation of an → atomic spectrum. In their pioneering work, Bunsen and Kirchhoff developed → spectroscopic analysis, the qualitative assignment of spectral lines to specific elements, which was soon extended to a quantitative method. However, modern technical advances in AES were required to make possible the quantitative determination of very small amounts. In contrast to → atomic absorption spectroscopy, where after → atomization a high concentration of atoms must be present in the ground state, in AES very high concentrations of highly excited or ionized atoms must be available in order to achieve a sufficiently large emission intensity. Since the formation of excited atoms requires a higher energy supply, the temperatures of the excitation sources lie between 5000 K and 10,000 K. Temperatures of ~ 2700 K, which are only suitable for the easily excitable alkali and alkaline-earth metals, can be reached with propane/air or acetylene/air flames (→ flame photometry). Higher temperatures are possible in an → arc discharge or a → spark discharge. These are the oldest excitation sources in AES. Plasmas and particularly the → inductively-coupled plasma (ICP) have recently proved themselves as excitation sources. All excitation sources have a dual function in AES, i.e. → atomization and → excitation. Since the samples to be analyzed have the form of solutions, as in AAS, the mechanism of atomization is the same. However, there is one more step in AES, that which causes the excitation of atoms to higher electronic states, including ionization. It is important for the practical application of AES that the light intensity emitted at a specific wavelength is directly proportional to the number of atoms of an element, i.e. their concentration in a medium not emitting at this wavelength. By measuring the intensity emitted from a known concentration of an element, a calibration factor can be calculated with which unknown concentrations of the same element can be determined, provided that conditions are held constant. The wavelength selected must be such that no other element in the sample also emits there → spectral interference. In AES there is usually a linear correlation between the emission intensity and concentration over a range of 3 to 5 powers of ten. However, with regard to detection limits, AAS and particularly the → graphite furnace technique, is superior to AES. But AES provides good detection limits for elements difficult to atomize in AAS such as boron, niobium, phosphorus, sulfur, tantalum, tungsten and zirconium. There are also → spectral and → nonspectral interferences in AES; the recognition and elimination of these is extremely important. Since an analytical line is measured as light emerging from a glowing flame or plasma, interferences can be caused by the background and several suggestions have been made for their elimination:

Measurement of a blank solution, change to other lines, inclusion of a vibrating quartz refractor plate in the optics, use of a computer program for the stepwise measurement and subtraction of the background. For the effective elimination of → spectral interference an → atomic emission spectrometer must have a monochromator with a high → resolution.

Ref.: Two reviews, *Atomic Absorption, Atomic Fluorescence and Flame Emission Spectrometry* and *Emission Spectrometry* are published every two years in the Fundamental Reviews Issue of *Analytical Chemistry.*

M. Slavin, *Emission Spectrochemical Analysis, Chemical Analysis*, vol. 36, J. Wiley and Sons, Chichester, New York, **1981**.

N. Schrader, Z. Grobensky, H. Schulze, *Introduction to AES with Inductively-Coupled Plasma (ICP), Applied Atomic Spectroscopy,* **1981**, issue 28, 1–37, Bodenseewerk Perkin-Elmer & Co. GmbH, Überlingen.

Atomic fluorescence spectrometer, <*Atomfluoreszenzspektrometer*>, consists basically of the following units: atomization device, excitation-light source, filter or monochromator, fluorescence detector, amplifier and read-out. There are both dispersive and nondispersive instruments. The block diagram of a nondispersive instrument is shown in Figure 1. A modern spectrometer of this type with an → inductively-coupled plasma torch (ICP) is fitted with a maximum of 12 element modules (turrets) which are arranged in a circle around the ICP. These modules are interchangeable so that in total ca. 65 elements can be measured. Each turret contains lenses, interference filters, a photomultiplier and a pulsed → hollow cathode lamp. The details are shown in Figure 2. The point of observation lies 55–75 mm above the torch; in AES it is 15 mm. This means that the temperature for AFS is considerably lower than for AES. Consequently, the atomization is very effective which is very important for detection sensitivity. If the interference filter in Figure 1 is replaced by a monochromator, a dispersive instrument is obtained. Several suggestions for the construction of such a spectrometer, including different lamps, have been made, see: J.C. Van Loorn, *Anal. Chem.*, **1981**, *53*, 332A.

Atomic fluorescence spectroscopy, AFS, <*Atomfluoreszenzspektroskopie, Atomfluoreszenzspektrometrie, AFS*> is a form of atomic spectroscopy which utilizes the fluorescence of free atoms; the reverse of atomic absorption. Although this idea can be traced back to the classical work of Bunsen and Kirchhoff, Wineforder and Vickers recognized its potential for spectrochemical analysis only in

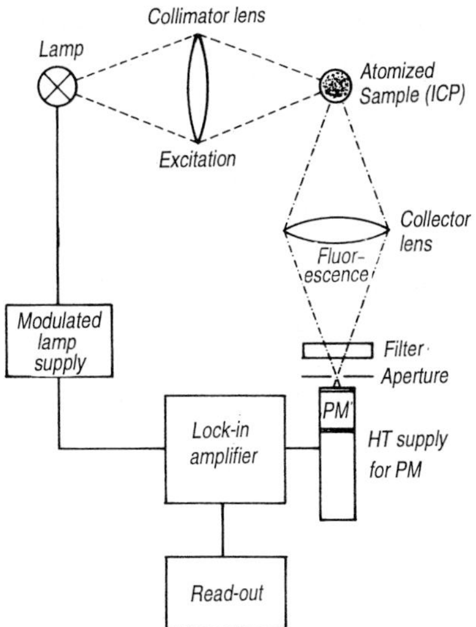

Atomic fluorescence spectrometer. Fig. 1. Block diagram of a nondispersive instrument

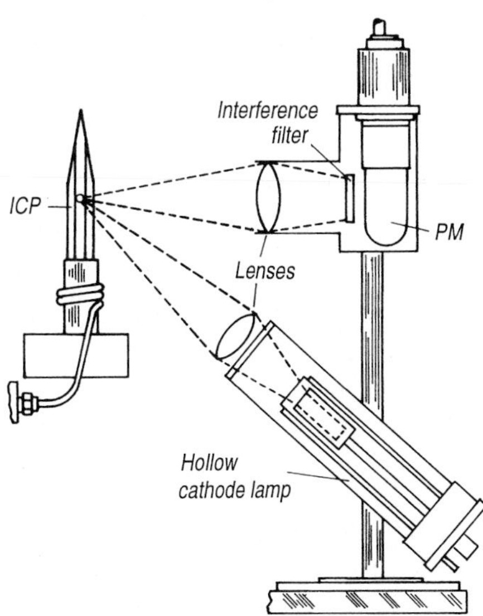

Atomic fluorescence spectrometer. Fig. 2. Element-analysis module for a nondispersive atomic fluorescence spectrometer

1964. Fluorescence is normally the radiative loss of excitation energy gained by absorption of radiation. With some exceptions, atoms in the ground state can only be excited by absorption into the lines of the principal series (→ atomic absorption spectroscopy, → atomic spectrum). The commonest case of fluorescence is → resonance fluorescence where the excitation energy, $h\nu_e$, is immediately emitted unchanged as a fluorescence, $h\nu_f$, i.e. $h\nu_e = h\nu_f$ and the atom returns to the ground state. In addition to the resonance fluorescence, a fluorescence can occur which shows a frequency shift (→ Stokes rule). As in → atomic absorption spectroscopy, a large number of free atoms is required for atomic fluorescence spectroscopy. They are produced by the same mechanism, → atomization, as in AAS. Whilst in atomic absorption, the absorption by atoms in the ground state is made quantitative by means of the → Bouguer-Lambert-Beer law, the absorption in atomic fluorescence is only the primary step of excitation. For that reason, the element-specific radiation of a → hollow cathode lamp (HCL) is always used. The atomic fluorescence is observed at an angle, frequently 90°, to the exciting light. Since the atomization and excitation are made in a flame or ICP, light emission from the flame often interferes with the fluorescence. If the excitation is produced by a pulsed HCL, the fluorescence is modulated by this pulse frequency and the frequency-selective amplifier (lock-in amplifier) amplifies only signals of this frequency coming from the → photomultiplier. Since the emission of the flame itself is not modulated, the amplifier does not register the associated continuous light signal which is therefore eliminated. A pulsed HCL also has the advantage of giving a higher intensity which is extremely important for the excitation of atoms. The → inductively-coupled plasma technique is used in modern equipment as the atomization method. AFS has several advantages when compared with AAS and AES. In theory, the detection limit can be easily reduced to the region of very low concentrations by increasing the intensity of the exciting light. Instabilities in the radiation source are less noticeable in AFS than in AAS. Since the measurement is made at right angles to the direction of irradiation in AFS, no nonabsorbable light falls onto the detector. Thus, calibration curves result which are linear over three to five powers of ten; in AAS, they are frequently linear only over two powers of ten. The geometry of the optical system in AFS makes it possible, in principle, to work with continuum-radiation sources and without monochromators which is not possible in AAS. When using ICP atomization (→ inductively coupled plasma), interfering matrix effects are often extremely small. Despite these advantages, AFS has not won favor vis-a-vis the competing AAS and AES methods.

Ref.: N. Omenetto, J.D. Wineforder, *Prog. Analyt. Spectrosc.*, **1979**, *2*, 1; V. Syehra, V. Svoboda, I. Rubeska, *Atomic Fluorescence Spectroscopy*, Van Nostrand Reinhold Co., London, **1975**.

Atomic model, *<Atommodell>*. Rutherford was the first to develop a dynamic atomic model in which Z negatively charged electrons, e^-, orbit, planet-like, around a nucleus with Z positive elementary charges. The difficulties arising in this model, as seen from the standpoint of classical physics, were eliminated with three decisive additions by Niels Bohr in 1913:

1. There are nonradiating electron orbits.
2. Only a discrete number of such nonradiating orbits is actually possible.
3. Radiation is emitted only upon the transition (jump) of an electron from an outer to an inner orbit and the light quantum emitted, $h\nu$, corresponds to the difference in the electronic energies of these two orbits.

The hydrogen atom is the simplest example of this model. A nucleus with a charge of $+Z \cdot e$ ($Z = 1$) is orbited by a single electron, e^-. The attractive Coulombic force is equal to the centrifugal force:

$$\frac{mv^2}{a} = \frac{Z \cdot e^2}{a^2},$$

whence:

$$mv^2 a = Ze^2,$$

a is the radius of the orbit (→ Bohr radius), v the velocity of the electron and m its mass. This orbital motion gives rise to an → angular momentum, $p_\phi = mva$. Since the angular momentum has the dimensions of action (Js), Bohr proposed that it must be an integer multiple of Planck's quantum of action, h (→ Planck's constant). Bohr introduced a quantum rule which he formulated for the whole orbit, $\phi = 0$ to 2π, as follows:

$$\int_0^{2\pi} p_\phi d_\phi = 2\pi p_\phi = 2\pi mva = nh,$$

with $n = 1, 2, 3 \ldots$

The velocity, v_n, of the electron in the orbit characterized by quantum number, n, and the associated orbital radius a_n can be calculated with this formula. By means of the equation, $E = E_{kin} + E_{pot}$ and the → Bohr- Einstein frequency relation, Bohr was able to derive → Balmer's formula and interpret the spectrum of the hydrogen atom in full agreement with the experimental data.

Further details about Bohr's atomic model may be found in G. Herzberg, *Atomic Spectra and Atomic Structure*, Dover, New York, **1944**, chapt. 1 and A. Sommerfeld, *Atomic Structure and Spectral Lines*, Translator H.L. Brose, E.P. Dutton, New York, **1934**.

Atomic spectrum, <*Atomspektrum*>, if we add thermal energy to free atoms by simply putting them into a hot flame we observe light emission within the flame, the color of which is characteristic of the atom concerned. The → spectral dispersion of the emitted light by Bunsen and Kirchhoff in 1859/1860 showed that quite specific lines can be assigned to each element. In fact, the emission spectrum of an atom consists of a large number of lines and is called a → line spectrum. In a further fundamental experiment, Bunsen and Kirchhoff also showed that light emitted from a sodium flame, for example, is itself absorbed by sodium vapor. This result provided simultaneously the explanation of the absorption lines of the stellar atmospheres discovered by Fraunhofer in 1814 (→ Fraunhofer lines and → line reversal). The detailed investigation of the line spectra of numerous elements following this discovery showed that the lines can be grouped into quite specific series. In 1885, Balmer showed that the wavelengths of the lines of atomic hydrogen observed in the visible region followed a series formula now called → Balmer's formula. By 1900, Rydberg had given a further general series formula which proved useful for the interpretation of the → alkali-metal spectra. Basically, we distinguish between a → principal series and various → secondary series or subseries. The principal series can be observed in absorption and emission whilst subseries can only be observed in emission. The series formulas give the energy differences between a fixed term (fixed energy level) and a running term (variable energy level). The physical explanation of the empirically established series formulas was given by Bohr in 1913 using the → Bohr-Einstein frequency relation. It was followed in the twenties by a quantum-mechanical treatment using → Schrödinger's equation, according to which the energy differences in the series formulas must be assigned to the energy differences between the discrete energy states of the outer electrons of the atom. These states can be defined more exactly by the two → quantum numbers (n = principal quantum number and ℓ = orbital angular momentum quantum number). It is of importance for the theoretical interpretion of atomic spectra that Sommerfeld had formulated the connection of the energy levels in the series formulas with the orbital angular momentum quantum number, ℓ (→ electronic angular momentum) in 1916. It forms the basis of the → term symbol notation. In the case of several outer electrons, the resulting total orbital angular momentum, L, and the term symbols, S, P, D, F ... are used instead of the individual orbital angular momenta, ℓ (s,p,d,f): $L = 0$: S term; $L = 1$: P term; $L = 2$: D term; $L = 3$: F term; $L = 4$: G term. The symbols, S, P, D and F are derived from the characteristics of the observed spectra. The *S* terms are the running energy levels of the sharp subseries, the *P* terms of the principal series, the *D* terms of the diffuse subseries and the *F* terms of the fundamental series (→ alkali spectra). For an interpretation of line spectra, it proved practical to arrange the terms of the series formulas in the form of an → energy-level diagram, in which the individual spectral lines are represented as transitions between the terms according to the → Rydberg-Ritz combination principle. The quantum-mechanical treatment showed that a → selection rule, according to which only those transitions, where $\Delta\ell$ or $\Delta L = \pm 1$ are allowed, must be observed. The systematic investigation of atomic spectra began with the discovery of the new elements, Rb and Cs, by Bunsen and Kirchhoff in 1860. Atomic spectroscopy immediately proved to be an excellent analytical tool which led to → spectroscopic analysis which today, following modern advances in → atomic absorption spectroscopy, → atomic emission spectroscopy and → atomic fluorescence spectroscopy, is very widely applied.

Atomization, <*Atomisierung*>, in → atomic absorption spectroscopy (AAS) the reduction of the ions or molecules present in the sample to metal atoms in their ground states. It occurs in the *absorption cell*, i.e. the flame, graphite furnace or heated quartz cell. The nature of this absorption cell defines the → flame, → graphite furnace and → hydride techniques. Figure 1 shows the individual steps of atomization for a flame or graphite furnace. The nebulized solution is initially present as an aerosol

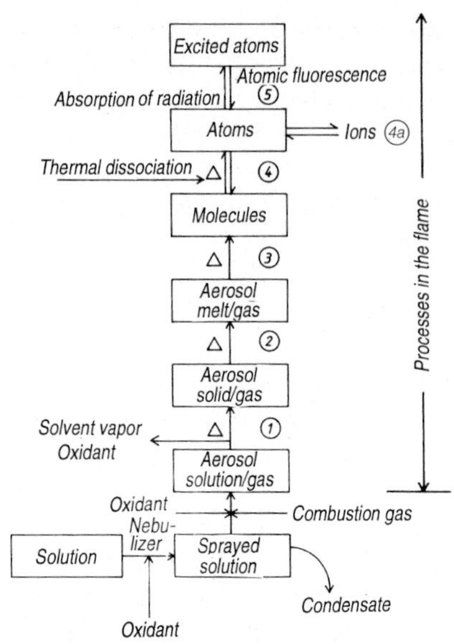

Atomization. Fig. 1. Flow diagram of the atomization steps

solution/gas. Steps 1 to 4 and 4a are thermal processes. The decisive step 4 is the thermal dissociation of molecules in which atoms in their ground states are produced. Step 5 is the actual atomic absorption process which leads to electronically excited atoms. These atoms can release their excitation energy in the form of fluorescence; thus, the reversal of process 5 forms the basis of → atomic fluorescence spectroscopy. If the atoms formed in step 4 are further excited thermally they can be lifted to higher excited states or even ionized and an emission occurs which constitutes the basis of → atomic emission spectroscopy. Subreactions frequently take place between steps 1 to 4 which, in practice, can lead to interferences. In general, compounds difficult to melt or to dissociate can be formed by reactions with atomic oxygen and OH radicals and this reduces the number

of free atoms of the elements to be determined in the flame. These interferences can be partially eliminated by using a higher flame temperature or the → graphite furnace technique.
Ref.: B. Welz, *Atomic Absorption Spectrometry*, 2nd ed., Verlag Chemie, Weinheim, Deerfield Beach, Florida, Basel, **1986**.

ATR, *<ATR-Technik, abgeschwächte Totalreflexionstechnik>* → attenuated total reflection.

Attenuated total reflection, ATR, *<ATR-Technik, abgeschwächte Totalreflexionstechnik>*, total reflection occurs if the → reflection of a light ray takes place at the interface between an optically more dense (refractive index n_1) and an optically less dense (rare) medium (refractive index n_2) with an angle of incidence which is larger than the critical angle, α_c. This critical angle is given by $\sin\alpha_c = n_2/n_1$ ($n_2 < n_1$). Study of this phenomenon shows that radiation is present in the optically rare medium, despite the total reflection. This is due to the fact that diffraction phenomena (→ diffraction of light) occur at the edges of the incident beam of light (cf. Figure 1), so that energy penetrates the rarer medium in the region of *A* and returns to the denser medium at *B*. A transversely damped surface wave (evanescent wave), the amplitude of which is dissipated within a few wavelengths, propagates in the rarer medium between *A* and *B*. Though no energy penetrates the rarer medium between *A* and *B* – i.e. we have a true total reflection – nevertheless, energy is still transported through the interface from *A* to *B* at the edges of the light beam on account of the diffraction. That this explanation is correct was shown experimentally by F. Goos and H. Hänchen in 1949. They demonstrated that the light beam is advanced laterally by a distance of the order of magnitude of one wavelength during a total reflection. (A. von Hippel, *Dielectrics and Waves*, Wiley, New York, **1954**, p. 54.).

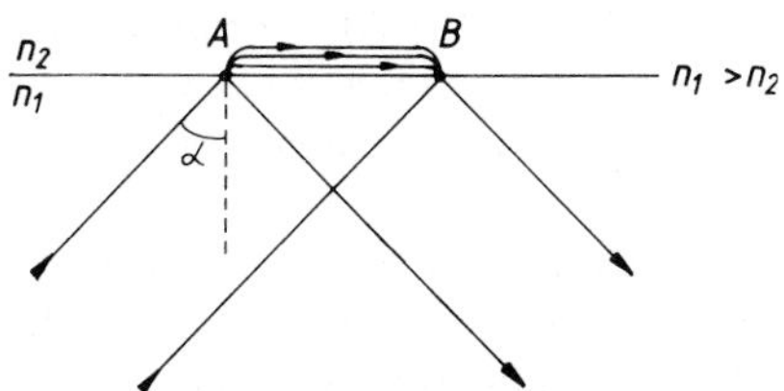

Attenuated total reflection. Fig. 1. Reflection at the interface between an optically dense and an optically rare medium

If the radiation, penetrating the optically rarer medium according to Figure 1, is not absorbed, a total reflection without loss occurs. However, if the optically rarer medium absorbs part of the penetrating radiation the reflected light is attenuated. The measurement of this attentuation and its variation with wavenumber is known as attenuated total reflectance (or reflection) spectroscopy.
A detailed theoretical treatment of ATR shows that, by using such measurements at two different angles of incidence, it is possible to determine the optical constants [refractive index (n_2) and absorption index (κ_2)] of the optically rarer absorbing medium. Numerous reflection elements with variable angles of incidence and ancillary equipment for installation in available spectrometers have been developed for this technique. However, application of the ATR method is frequently limited to the measurement of the apparent absorbance or trans-

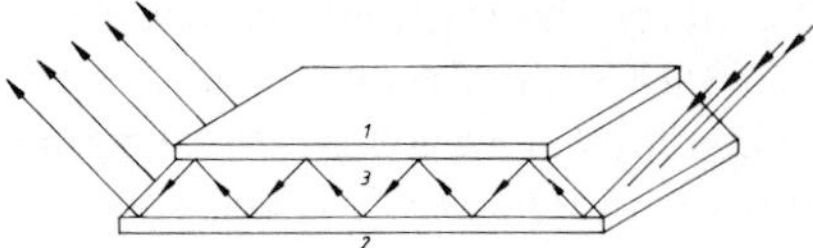

Attenuated total reflection. Fig. 2. Multiple reflection between two plates. In the ATR technique, the sample can be placed on face 1 or face 2

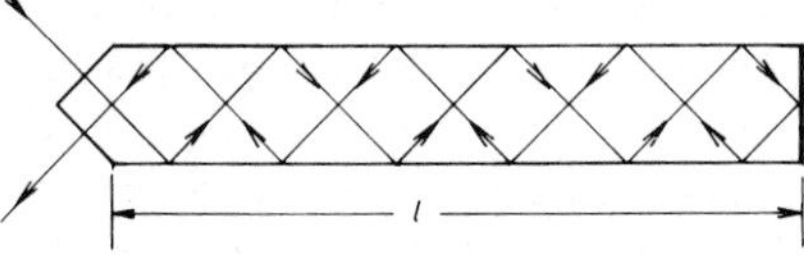

Attenuated total reflection. Fig. 3. Reflection element for immersion

mittance of the optically rare medium as a function of the wavelength (wavenumber) i.e. for obtaining the ATR spectrum of a sample. This special type of ATR is also called internal reflection spectroscopy. In recent years it has been employed with great success in IR spectroscopy and also in the UV-VIS region. Attachments for measuring ATR spectra with commercial spectrometers have been developed by numerous companies. As shown in Figure 2, trapezoidal or parellelopiped-like plates, with a fixed angle of incidence for multiple reflection, are generally used as reflection elements. For a double-beam → IR spectrometer, two reflection elements as uniform as possible are required. For an → FTIR instrument only one is needed. The sample can be placed on either the upper or lower surface of the ATR device.

The reflection element in Figure 3 can be immersed directly in a liquid/solution where both sides become active because of the multiple reflections.

Selectivity plays a considerable role in ATR spectroscopy. For small absorption, the reflecting power, R, is approximately $R \cong 1 - \beta \cdot d_e$ where β is the absorption coefficient in cm^{-1}, defined according to the → Bouguer-Lambert-Beer law. d_e is the effective path length, i.e. the penetration depth of the surface waves into the absorbing medium 2. This is of the order of $5 \cdot 10^{-5}$ cm in ATR. For small β values the product, $\beta \cdot d_e$, is very small and R correspondingly large. For a multiple (N times) reflection, $R^N = (1 - \beta \cdot d_e)^N$, i.e. the effect to be measured will be increased. The question as to how far the spectra obtained with this technique agree with the true spectra measured in transmission was discussed in detail by N.J. Harrick in the sixties.

In the event that the optically rarer medium with the refractive index, n_2, does not absorb, the penetration depth, d_p of the surface wave is proportional to the wavelength, λ_1, in the optically denser medium. Since a proportionality exists between d_p and the effective path length, d_e, the latter also increases with λ. This, for example, explains why long-wavelength bands are relatively more intense in ATR spectra than in transmission spectra. Furthermore, d_e depends inversely on the angle of incidence, α, and for angles of incidence approaching the critical angle, α_c, d_e becomes indeterminately large. These statements about d_e apply only for weak absorption. Since d_e decreases with an increasing angle of incidence, strong absorptions may be reduced by using large angles of incidence. Thus, we can obtain well resolved multiple reflection ATR spectra, even in the case of strongly absorbing media.

The ratio of the refractive indexes, n_2/n_1, of the adjacent media is of particular importance and $n_1 > n_2$ applies

here. Therefore, the reflection element, also called the ATR crystal, must be made of a material which has a large refractive index, e.g.: KRS5 $n = 2.37$; germanium $n = 4$; silicon $n = 3.4$; silver chloride $n = 2$ in the IR region; and quartz $n = 1.15$; sodium chloride $n = 1.5$; flint glass $n = 1.7$; aluminium oxide $n = 1.8$; magnesium oxide $n = 1.8$ in the UV-VIS region.
It is important for the practical application of this method that the penetration depth of the radiation into the optically rarer material is extremely small. Thus, very thin surface layers can be investigated such as metals, plastics, paints, paper and textile laminates, concentrated solutions etc. It is sometimes necessary to adapt the ATR crystal to the particular sample state.
Ref.: N.J. Harrick, *Internal Reflection Spectroscopy*, Harrick Scientific Corporation, John Wiley and Son Inc., Ossining, New York, **1979**; G. Kortüm, *Reflectance Spectroscopy*, Translator, J.E. Lohr, Springer Verlag, Berlin, Heidelberg, New York, **1969**, chapt. VIII.

Auer burner, <*Auer-Brenner*>, is occasionally used for investigations in the long-wavelength IR region (above 50 μm or below 200 cm^{-1}). It consists of thorium oxide heated by a gas flame to ca. 1800 K. It is characterized by a very low emission at wavelengths below 10 μm (> 1000 cm^{-1}), whilst above this figure, an emission closely approaching black-body radiation is observed.

Auger process, <*Auger-Prozeß*>, a process competing with the generation of → X-radiation and → X-ray spectra named after its discoverer, P.V. Auger. Characteristic X-ray lines are produced when an electron jumps from a higher, fully occupied shell into a hole in a lower lying shell which was created by some primary process, e.g. by bombardment with high-energy electrons $\sim > 10^4$ V. The corresponding energy difference, ΔE, is emitted as electromagnetic radiation in accordance with the → Bohr-Einstein frequency relation. The process can also occur radiationlessly if the energy, ΔE, is transferred to a second electron which then leaves the atom with a kinetic energy equal to the difference between its ionization energy and ΔE. This phenomenon forms the basis of Auger spectroscopy and the ejected electron is called the Auger electron. It carries the extra energy in the form of kinetic energy. However, this extra energy is not identical with the energy which is determined in the X-ray spectrum. A singly ionized atom is produced in the X-ray process whereas a doubly ionized one results from the Auger sequence. This is because not only is the electron ejected in the primary process missing, but also the Auger electron. The X-ray and Auger processes are competing processes. The latter is dominant for lighter atoms and the former for heavier ones. Figure 1a shows the Auger process in oxygen. In each of three cases, an electron is knocked out of the K shell by a primary electron. This hole can be filled by an electron from the L_I or L_{II} shell. The energy released is transferred to an electron from the L_I or $L_{II,III}$ shell which can then leave the atom. The Auger process is denoted with letters which are assigned to the associated shells: the first letter denotes the hole created by the primary electron (here K), the second the ori-

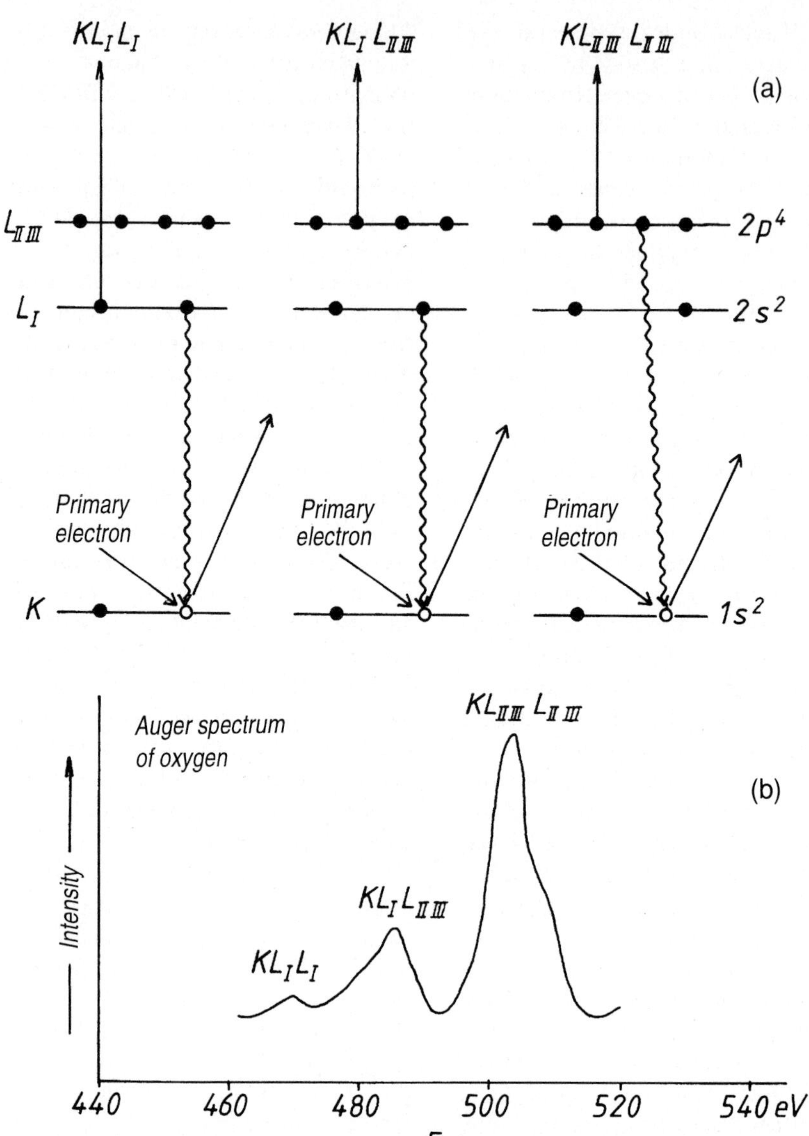

Auger process. Fig. 1. a) Energy-level diagram for the Auger process; b) Auger spectrum of oxygen

gin of the electron which fills the hole (here L_I, $L_{II,III}$) and the third where the Auger electron comes from (here L_I, $L_{II,III}$). The processes shown in Figure 1a are therefore a KL_IL_I, a $KL_IL_{I,III}$ and a $KL_{II,III}L_{II,III}$ process. Taking account of energy conservation, the energy of the Auger electron, E_A, is given by:

$$E_A = (E_H - E_J) - E_B$$

E_H is the energy of the level in which the hole was created, E_J the energy of the electron which jumps into the hole and E_B the binding energy of the Auger electron before taking up energy and corrected for a doubly positively charged atom. The energy of the primary electron is not added to this sum; it must only satisfy the condition that its energy is greater than the binding energy of the electron which is to be knocked out. Since all the energies on the right-hand side of the above equation are characteristic for a specific atom and a specific Auger process, the Auger electron from such a process has a specific kinetic energy which can be measured using a retarding field. By continuous variation of the field, we obtain an Auger electron spectrum in which the intensity of the Auger electrons is shown as a function of their energy. Thus, an Auger spectrum reflects the energy relationships in the inner electron shells. Since these are characteristic of a specific atom, Auger spectroscopy is a valuable method for identifying atoms. Its most important application is in surface analysis.

Ref.: D. Briggs, M.P. Seah, *Practical Surface Analysis*, vol. I, *Auger and X-Ray Photoelectron Spectroscopy*, J. Wiley, New York, **1991**.

Autocollimation, *<Autokollimation>*, an optical method in which an image or mark is imaged visibly upon itself (→ autocollimation prism).

Autocollimation prism, *<Autokollimationsprisma>*, a → prism which belongs to the class of prisms having constant deviation, i.e. it produces the same angle of deviation for each wavelength at the minimum deviation. This results from the fact that, in addition to the refraction, a reflection also occurs at a rotated surface. The autocollimation prism is the simplest prism of this type. It is generally a 30° prism with a mirrored back surface, see Figure 1. A light beam which falls perpendicularly onto this mirror is reflected into itself and thus undergoes a deviation of 180°. The wavelength assigned to this light path can be varied at will by rotating the prism vis-a-vis the fixed direction of the incident light. A quartz prism of this form is frequently installed in prism monochromators in a → Littrow mounting.

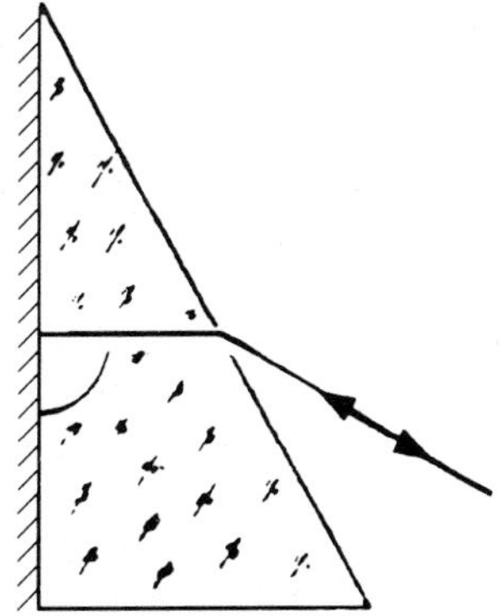

Autocollimation prism. Fig. 1. Autocollimation prism

Autocollimation spectrograph, *<Autokollimationsspektrograph>* generally a → prism spectrograph.

Autocollimation spectroscope, Pulfrich's autocollimation spectroscope, <*Autokollimationsspektroskop, Pulfrichsches Autokollimationsspektroskop*>, a → spectroscope the operation of which is basically that of an autocollimation spectrograph.

Autocorrelation technique, <*Autokorrelationstechnik*>, a technique for the indirect measurement of the duration of laser pulses in the region of $\leq 10^{-10}$ s. The chronological course of the pulse, $p(t)$, is not measured directly. The autocorrelation function, $G(\Delta)$, of the pulse is measured from which we can infer the actual pulse time. $G(\Delta)$ is given by:

$$G(\Delta) = (1/T) \int_{o}^{T} p(t) \cdot p(t + \Delta t) dt$$

For an experimental implementation we must:

1. Divide the pulse, $p(t)$, into two equal parts;
2. delay one of these parts by any time, Δt, vis-a-vis the other;
3. determine the product of $p(t)$ and $p(t + \Delta t)$;
4. integrate this product from $t = 0$ to $t = T$ (T = laser-pulse interval) and
5. vary the delay from $-\infty$ to $+\infty$.

This is achieved by an optical unit, an autocorrelator, the basic construction of which is shown in Figure 1. It consists of:

a) A beam splitter;
b) a rotating glass cube, *W*, in one of the divided beams (by rotating the cube we vary the optical path of the pulse in the glass and thus the distance through the glass);
c) a → KDP crystal which, when suitably adjusted, forms a signal with the frequency, $\omega_1 + \omega_2$, from two incident frequencies, ω_1 and ω_2. In this case, we use the fact, that the intensity at the sum frequency is proportional to the product of the intensities of the two entering signals. If suitable filters, F_1 and F_2, are used, only this summed frequency is transmitted and its intensity is measured using a photomultiplier.

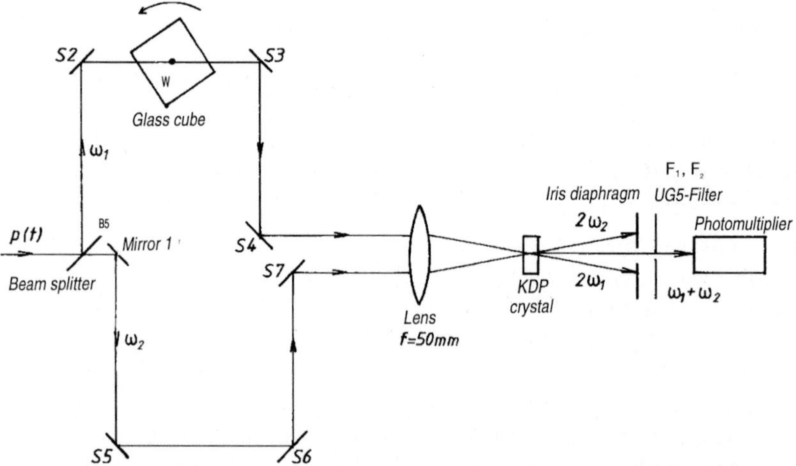

Autocorrelation technique. Fig. 1. Schematic construction of an autocorrelator

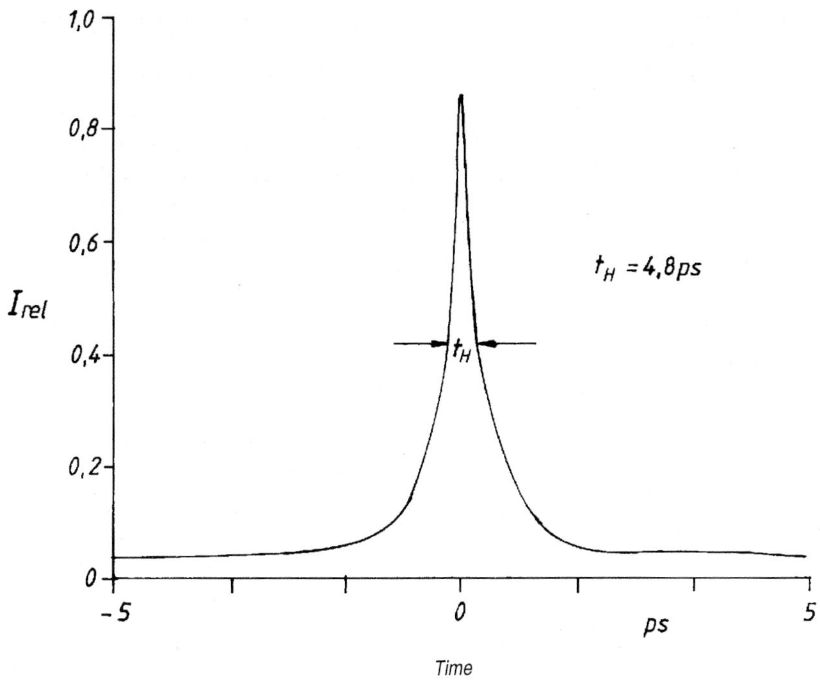

Autocorrelation technique. Fig. 2. Oscilloscope signal for a pulse of 4.8 ps

d) a → photomultiplier which, due to its inertia, cannot follow the fast signals. An integration occurs in practice. The rotation of the glass cube produces the variable pulse, $[p(t + \Delta t)]$. If we trigger an oscilloscope in phase with this rotation a typical signal results. Such a signal is shown in Figure 2 for the pulse time, $t_p = 4.8$ ps.

Ref.: S.L. Shapiro, Ed.: *Ultrashort Light Pulses*, Topics in Appl. Phys., vol. 18, Springer Verlag, Berlin, Heidelberg, New York, Toronto, 2nd edtn., **1984**; C.V. Shank, F.P. Ippen, S.L. Shapiro, *Picosecond Phenomena*, Springer Series in Chemical Physics, vol. 4, Springer Verlag, Berlin, Heidelberg, New York, **1978**.

Auxochrome, *<Auxochrom>*, a substituent which, though not a chromophore itself, acts upon a → chromophore or a chromophoric system by donating (electron donor) or withdrawing (electron acceptor) electrons. Substituents of the first type are called auxochromes. Typical examples in the order of their effectiveness are:

$-F < -CH_3 < -Cl < -Br < -H < -OR < -NH_2 < -O^-$.

Substituents of the second type are called antiauxochromes. Typical examples are:

$-^+NH_3 < -SO_2NH_2 < -COO^- < -CN < -COOH < -COCH_3 < -CHO < -NO_2$.

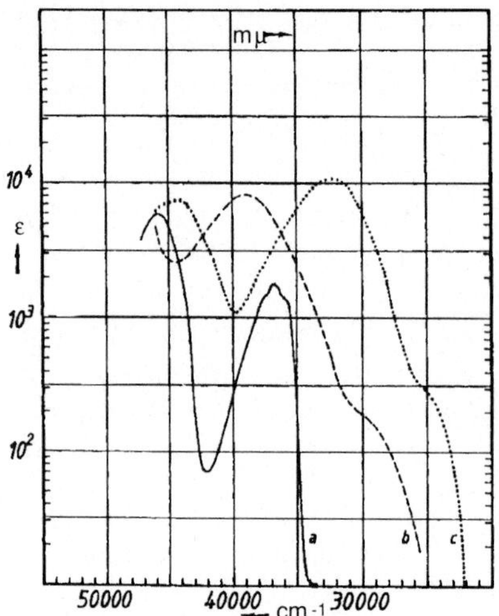

Auxochrome. Fig. 1. UV spectra of a) phenol, b) nitrobenzene and c) p-nitrophenol

The influence of auxochromes or antiauxochromes as substituents has been investigated in great detail for benzene as chromophore.

Combination of an auxochrome and an antiauxochrome produces a particularly strong → bathochromic shift. To illustrate this effect, the UV spectra of phenol, nitrobenzene and p-nitrophenol are compared in Figure 1.

B

Babinet's rule, <*Babinetsche Regel*>, a rule, discovered by J. Babinet in 1837, according to which, the direction of the larger principal absorption index of optically uniaxial crystals is the same as that of the larger principal refractive index. In the case of tourmaline, this rule applies to the whole visible spectral region. However, in general it applies only to those wavelengths which are not too distant from the electronic absorption maxima.

Background correction, <*Untergrundkorrektur*>, → Zeeman background correction. Background correction in → AAS according to Smith and Hieftje. The self-reversal and line broadening of a pulsed hollow cathode lamp are used to provide its own background correction. A structured background and interferences are accurately corrected because the measurement of the background is made in the immediate proximity of the resonance line. The method is recommended by the EPA.

Back-scattering power, <*Rückstrahlvermögen*> → albedo.

Bäckström filter, <*Bäckström-Filter*>, a filter once commonly used for isolating mercury emission lines.

Balmer's formula, <*Balmer-Formel*>, a formula for line spectra. The spectrum of the H atom extends from the IR spectral region into the far UV. In 1885, the Swiss physicist J.J. Balmer proposed the following series formula for the lines observed in the visible region:

$$\lambda = 3645.6\frac{n^2}{n^2 - m^2}$$

The wavelength, λ, is given in → Ångström units [Å] (1 Å = 10^{-8} cm); and $n = 3, 4, 5 \ldots$ with $m = 2$. In 1888, C. Runge replaced wavelength with frequency and J.R. Rydberg gave the Balmer formula its final form in 1890:

$$\tilde{\nu} = R\left(\frac{1}{m^2} - \frac{1}{n^2}\right)$$

R is the Rydberg constant (= 109677.578 cm^{-1}), $m = 2$ remains constant whilst n can take all integer values between $m + 1$ and ∞.

Later, other hydrogen atom series were observed, in addition to the Balmer series, and there are now five series which can be interpreted by means of a Balmer formula, viz:

$$\tilde{\nu} = R\left(\frac{1}{1^2} - \frac{1}{n^2}\right);\ n = 2, 3, 4 \ldots$$

Lyman series (1906)

$$\tilde{\nu} = R\left(\frac{1}{2^2} - \frac{1}{n^2}\right);\ n = 3, 4, 5 \ldots$$

Balmer series (1885)

$$\tilde{\nu} = R\left(\frac{1}{3^2} - \frac{1}{n^2}\right);\ n = 4, 5, 6 \ldots$$

Paschen series (1908)

$$\tilde{\nu} = R\left(\frac{1}{4^2} - \frac{1}{n^2}\right);\ n = 5, 6, 7 \ldots$$

Brackett series (1922)

$$\tilde{\nu} = R\left(\frac{1}{5^2} - \frac{1}{n^2}\right);\ n = 6, 7, 8 \ldots$$

Pfund series (1924)

The first term (energy level) in these series formulas is the fixed term and the second the running term. The table shows the fixed terms and the first line of each series in order to make the spectral regions of the series clear.

The five series are summarized in an → energy-level diagram in Figure 1.

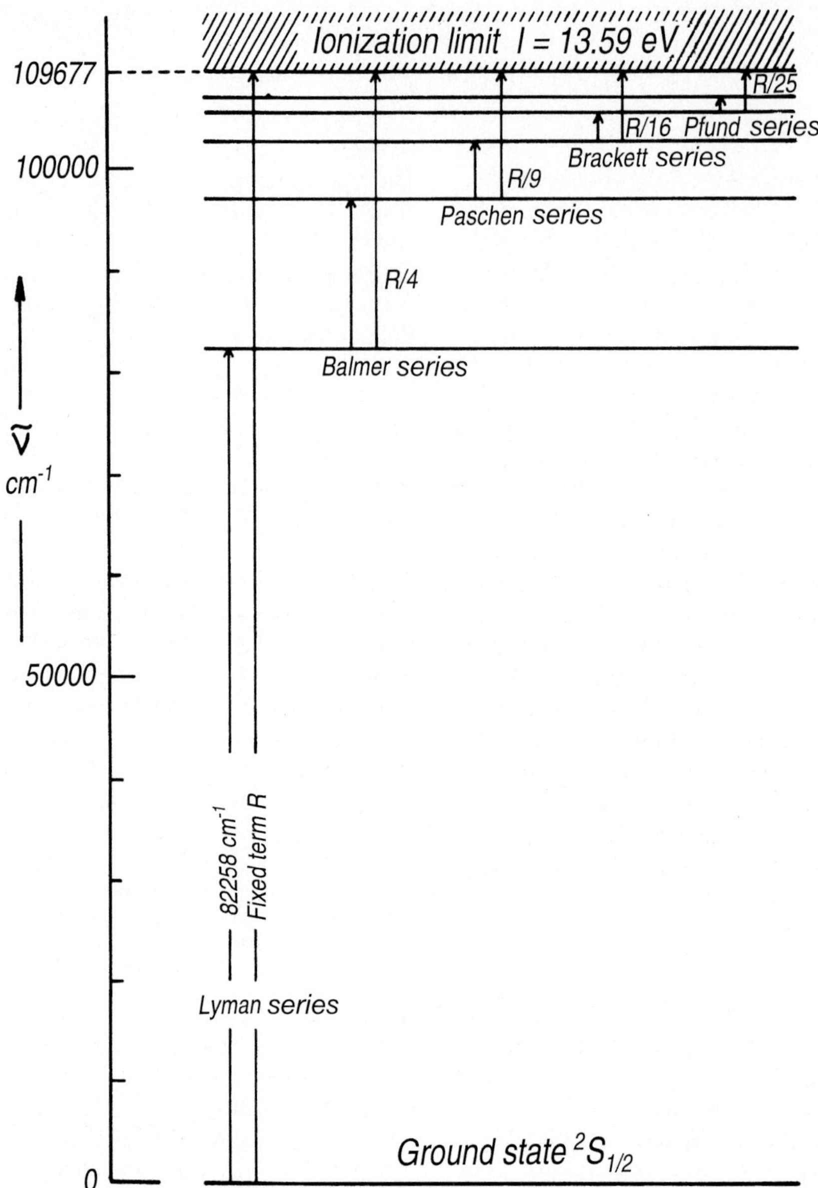

Balmer's formula. Fig. 1. Energy-level diagram for the hydrogen atom

Series	Fixed term			1st. Line of series		Spectral region
	R/m^2	$\tilde{\nu}$; cm^{-1}	λ; nm	$\tilde{\nu}_1$; cm^{-1}	λ_1; nm	
Lyman	R	109677.58	91.2	82258.2	121.6	Vacuum UV
Balmer	$R/4$	27419.39	364.7	15232.9	656.5	VIS
Paschen	$R/9$	12186.40	820.1	5331.6	1875.6	Near IR
Brackett	$R/16$	6854.85	1458.8	2467.8	4052.2	IR
Pfund	$R/25$	4387.10	2279.4	1340.5	7459.9	IR

Only the first line and the series limit for each series are plotted. The series limit corresponds to the promotion of the electron into the continuum, i.e. the ionization of the H atom. Since R is the energy of the series limit of the Lyman series, we obtain the ionization energy of the H atom in its ground electronic state as $I = 13.59$ eV (1 cm^{-1} $\approx 1.24 \cdot 10^{-4}$ eV). The theoretical interpretation of the line spectrum of the H atom was made initially by means of the Bohr theory (Bohr 1913) and later using the Schrödinger equation. According to both theories, the electronic energy of the H atom is given by:

$$E(n) = -\frac{m \cdot e^4}{8\pi^2 h^2} \cdot \frac{Z^2}{n^2} \quad \text{in eV}$$

where the principal quantum number, $n = 1, 2, 3 \ldots$ (→ quantum number). The nuclear charge, Z, is 1 for the H atom. The difference between the two energy states, E_1 and E_2 with $n_2 > n_1$, is given by the → Bohr-Einstein frequency relation as:

$$\begin{aligned}\Delta E &= E_2 - E_1 \\ &= \frac{me^4}{8\pi^2 h^2} \cdot \frac{1}{n_2^2} - \frac{me^4}{8\pi^2 h^2} \cdot \frac{1}{n_1^2} \\ &= \frac{me^4}{8\pi^2 h^2} \left(\frac{1}{n_1^2} - \frac{1}{n_2^2} \right)\end{aligned}$$

Dividing by hc, we obtain the energy difference as a → term difference given in wavenumbers by:

$$\tilde{\nu} = \frac{me^4}{8\pi^2 h^3 c} \left(\frac{1}{n_1^2} - \frac{1}{n_2^2} \right) = R \left(\frac{1}{n_1^2} - \frac{1}{n_2^2} \right)$$

The factor outside the brackets corresponds exactly to the → Rydberg constant, R, the numerical value of which was determined empirically by Balmer's formula. If we insert $n_1 = 1$ and $n_2 = 2, 3, 4 \ldots$ we obtain the Lyman series; with $n_1 = 2$ and $n_2 = 3, 4, 5 \ldots$ the Balmer series etc. Thus, the fixed number, m, and the running numbers, n, in Balmer's formula are identical with the principal quantum numbers which determine the energy states of the H atom.

Baly tube, *<Baly-Rohr>*, a → cuvette with a variable path length for absorption measurements on solutions. The Baly tube consists of two quartz tubes with fused windows and ground so as to fit tightly together. A reservoir is attached as shown schematically in Figure 1. The outer tube has a mm scale and the inner movable tube a scribed ring mark. The path length can be set exactly to within ca. 0.1 mm without parallax errors. When the path length has been adjusted, the position is fixed by means of a screw in

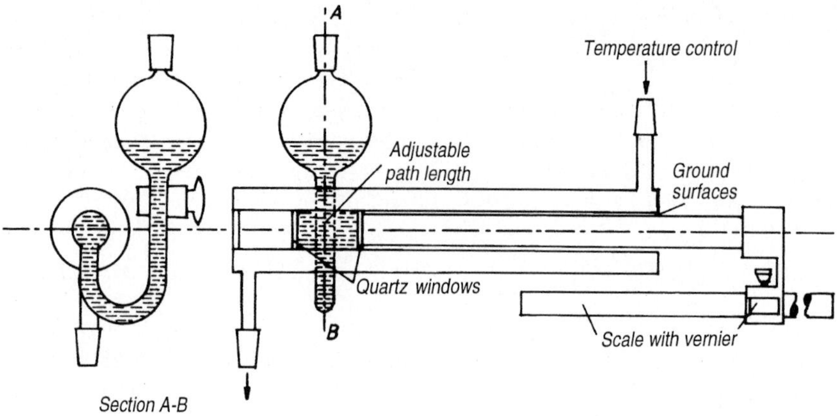

Baly tube. Fig. 1. Construction of a Baly tube

a guide rail. Commercial Baly tubes normally have a usable length of 100 mm, but some are produced which have longer path lengths. They are advantageous for absorption measurements where a wide range of log ε values is to be studied without changing the concentration.

Band, <*Band*>, the rotational energy-level structure of an energy level, which is required to explain the → rotation-vibration spectrum of nonlinear molecules, classified as → asymmetric top molecules, is complicated. Since the three moments of inertia about the three principal axes of inertia are all different, all three → rotational constants, $A \neq B \neq C$, enter the energy expressions. By convention, we always choose the axes so that $I_A < I_B < I_C$, which implies that $A > B > C$. The → selection rule is again $\Delta J = 0, \pm 1$. Depending on whether the direction of the dipole-moment change during a vibration lies parallel to the axis of the smallest, intermediate or largest moment of inertia, a different band type results. For molecules of the → symmetry point groups, C_2, V or V_h, we differentiate between type A, type B and type C bands. In the case of even lower symmetry, mixed band types (→ hybrid bands) occur. It is advantageous, when describing the individual band types, to introduce the ratio of the moments of inertia as $\varrho = I_A/I_B = B/A$.

Type A bands: The changing dipole moment lies along the axis of the smallest moment of inertia. The band resembles the → parallel band of a symmetric top molecule, if the resolution is not too high. Therefore, it has an unresolved central maximum with a branch of equidistant lines on both sides, provided that ϱ is small. For large ϱ values, the lines of the Q branch separate, and with decreasing ϱ, they move toward the band center with a simultaneous intensity decrease.

Type B bands: The changing dipole moment lies along the axis of the intermediate moment of inertia. These bands have no strong central branch, and the lines of the Q branch are often superimposed on those of the P and R branches. For $\varrho = 1$, type B is like type A. In the limiting case, $\varrho = 0$, the structure corresponds to the → perpendicular bands of a symmetric top molecule.

Type C bands: The changing dipole moment lies along the axis of the largest moment of inertia. The structure of this band is almost identical to that of type B for small ϱ values. As $\varrho \rightarrow 1$, it tends to the → parallel band of a symmetric top molecule with a strong Q branch. If the available dispersion of an IR spectrometer is insufficient for a resolution of the rotational structure of the bands the envelope is sufficient for deciding the band type.

Band analysis, <*Bandenanalyse*>, the deconvolution of the superimposed bands which occur in UV-VIS and IR absorption spectra measured in solution. On account of the superposition, the frequency (wavelength) of maximum absorption, $\tilde{\nu}_{max}$ (λ_{max}) and ε_{max}, cannot be read from the spectrum. In particular, the integral → absorbance is no longer a true measure of the intensity since the degree of overlapping cannot be determined. Thus, a band analysis is required. In many cases, the component bands can be described by a Gaussian profile (→ band shape):

$$A = A_{max} \exp\{-B(\tilde{\nu}_{max} - \tilde{\nu})^2\};$$

with

$$B = \frac{4 \ln 2}{\Delta\tilde{\nu}_{1/2}}$$

$$\sqrt{\ln \frac{A_{max}}{A}} = \tilde{\nu}_{max}\sqrt{B} - \tilde{\nu}\sqrt{B}.$$

The last equation can be used in a linear regression to fit the experimental curve. A wavenumber or wavelength interval is required in the spectrum which can only be assigned to a single band and is not affected by other bands. Figure 1a shows the absorption spectrum of the dye morin in 0.1M HCl. The illustration shows that we can assume that the long-wavelength side of the first absorption band is not affected by the shorter-wavelength bands. The steep decrease of the long-wavelength absorption edge also indicates that the band can be represented to a good approximation by a Gaussian profile. This suggests the following procedure. We start with a maximum at the lowest wavenumber in the spectrum and look on the low-energy edge for the region which is not affected by other bands. We then vary the following three parameters:

1. The band maximum A_{max}.
2. The start- and
3. the finish wavenumber of the interval selected.

The calculated absorbance values, A, are then fitted to the experimental curve using a linear regression. We obtain the center of the band, $\tilde{\nu}_{max}$ (λ_{max}), from the intercept of the regression line and the → half-width of the Gaussian curve from the slope. The sum of the squared errors is calculated as a measure of the fit. A_{max} and the wavenumber interval are varied in the next step (see above). Each variation of the fitting parameters produces a new value for the sum of the squared errors. A good fit is obtained when the sum is an absolute minimum. When a wavenumber interval has been dealt with in this way and the first Gaussian curve has been produced it is subtracted from the experimental spectrum and the procedure is repeated for the (calculated) rising edge of the second band. Four bands can be found in the absorption spectrum of morin, Figure 1a. The Lorentzian function (→ band shape) is also frequently used for

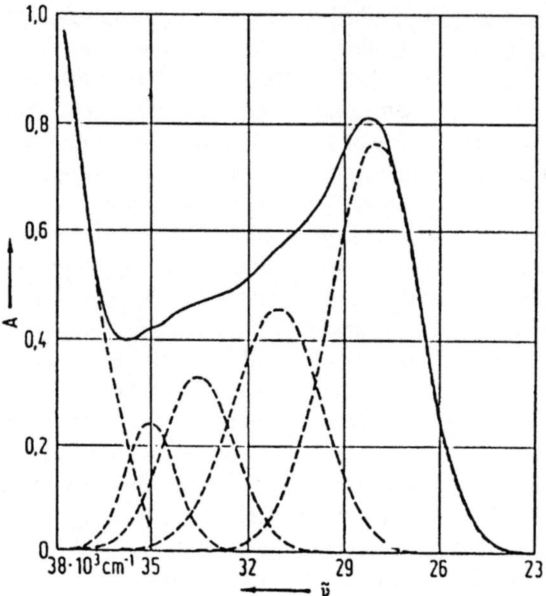

Band analysis. Fig. 1a. Absorption spectrum of morin in 0.1M HCl with a band analysis

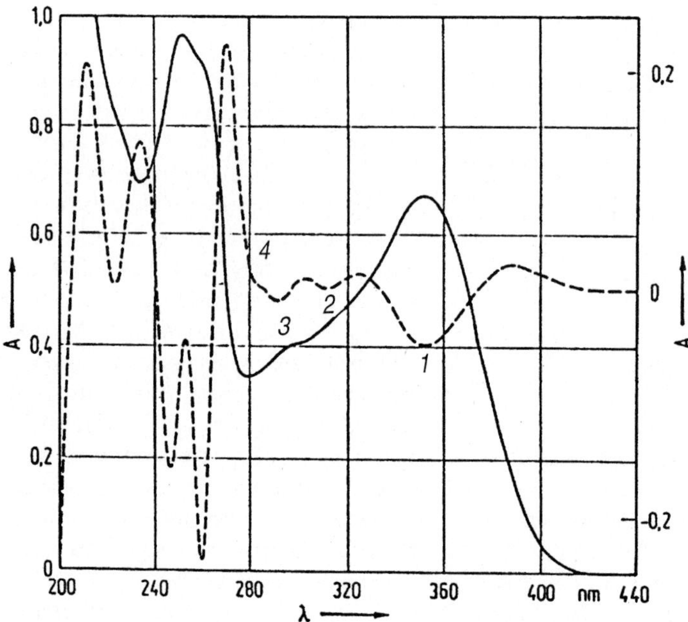

Band analysis. Fig. 1b. The morin spectrum on a wavelength scale and its derivative

band analysis. → Derivative spectroscopy is a good complement to band analysis since it offers a simple experimental method of checking on the number of superimposed bands. Figure 1b shows the second order derivative spectrum of morin. The minima 1–3 and the shoulder 4 indicate at least 3 or 4 contributing bands. This example also shows that it is advisable to combine mathematical band analysis with another independent method.
Ref.: H.-H. Perkampus, *UV-VIS Spectroscopy and Its Applications*, Springer Verlag, Berlin, Heidelberg, New York, **1992**, chapt. 8.2, p. 220 ff.

Band edge, <*Bandkante*> → electronic band spectrum, → absorption edge.

Band-head table, band-head scheme <*Kantenschema*>, a tabular help in the analysis of the vibrational fine structure of the → electronic spectral bands of diatomic molecules or molecular fragments. It is a special form of the → Deslandres table. Generally, the observations are made in emission. If the wavenumber of the pure electronic transition is $\tilde{\nu}_{oo}$, then taking account of the first anharmonic correction, the wavenumbers of the vibronic transitions, are given by:

$$\tilde{\nu} = \tilde{\nu}_{oo} + (\omega_o' v' - \omega_o' x' v'^2) - (\omega_o'' v'' - \omega_o'' x_o'' v''^2)$$

The single prime denotes the upper and the double prime the lower electronic state. ω_o' and $\omega_o' x_o'$ as well as ω_o'' and $\omega_o'' x_o''$ are defined as for the → anharmonic oscillator, i.e. they refer to the lowest vibrational state, v' or $v'' = 0$, and are corrected for the zero-point energy. This also applies to $\tilde{\nu}_{oo}$.

This equation represents all possible transitions between the different vibrational levels of the two electronic states involved. Every vibrational state, v', of the upper electronic state can be combined with every state, v'', of the lower state. For a constant v', v'' can assume all values 0, 1, 2, 3 ...; this is a v'' progression. Analogously, for a constant v'', a series of bands with a variable v' forms a v' progression. In a band-head table the wavenumbers of these progressions are arranged in rows for wavenumbers corresponding to changing v'' at constant v' and in columns for wavenumbers corresponding to changing v' at constant v''. Figure 1 shows a band-head table schematically.
Thus, the rows (v'' progressions) extend from $\tilde{\nu}\,(v' = v'' = 0)$ to smaller and the columns (v' progressions) to larger wavenumbers. Furthermore, the differences between adjacent entries in the rows or the columns are nearly constant. It is obvious from the above formulas that the $\Delta\tilde{\nu}_{v''}$ or $\Delta\tilde{\nu}_{v'}$ values still depend linearly upon v'' or v' according to the equation, $\Delta G = \pm\,[\omega_o - 2\omega_o x_o(v + 1/2)]$ (→ anharmonic oscillator, → vibrational spectrum). However, the second difference, $\Delta^2 G$, is constant and gives $2\omega_o' x_o' = 2\omega_e' x_e'$ or $2\omega_o'' x_o'' = 2\omega_e'' x_e''$. Figure 2 shows the transitions of the v' and v'' progressions in an energy-level diagram. In conjunction with the band-head table, we see immediately that the differences between two adjacent entries, one above the other along two rows, must be constant; e.g. for $v' = 0$ and $v' = 1$ with a variable v'' the constant difference corresponds to the fundamental vibrational quantum of the excited electronic state, $\omega_o' - \omega_o' x_o'$. This applies analogously to a

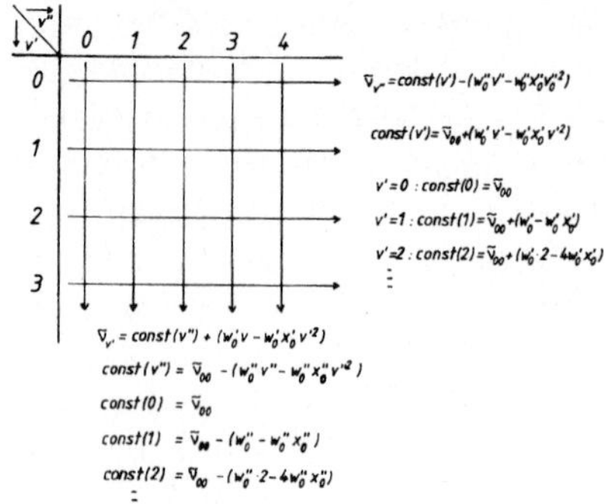

Band-head table. Fig. 1. The layout of a band-head table for electronic band spectra, see text

pair of neighboring horizontal figures, $v'' = 1$ and 2 say, $v' = 2, 3$ ... This constancy is a criterion for the correct assignment of the measured wavenumbers to their place in the band-head table. It is obvious that the exact analysis of the spectrum provides the quantities, ω_0 and $\omega_0 x_0$ for the electronic ground state.

Band model, <*Bändermodell*>, in isolated metal atoms, the energy states of electrons (s, p, d ...) can be regarded as discrete levels separated from each other, as shown in their energy-level diagrams. When atoms come together to form a solid, the interaction of the electrons, which increases with decreasing distance between the atoms, causes the discrete energy states of the electrons to spread out into energy bands. The width of the bands increases with decreasing interatomic distance until, finally, the individual bands overlap. Each band consists of as many very closely spaced levels as the number of atoms, N. In the lowest energy state, the available electrons occupy the lowest energy levels in accordance with the → Pauli principle. In the case of lithium ($1s^2 2s^1$), for example, all N energy states arising from the interaction of the $1s$ orbitals will be occupied with 2 electrons each and the narrow

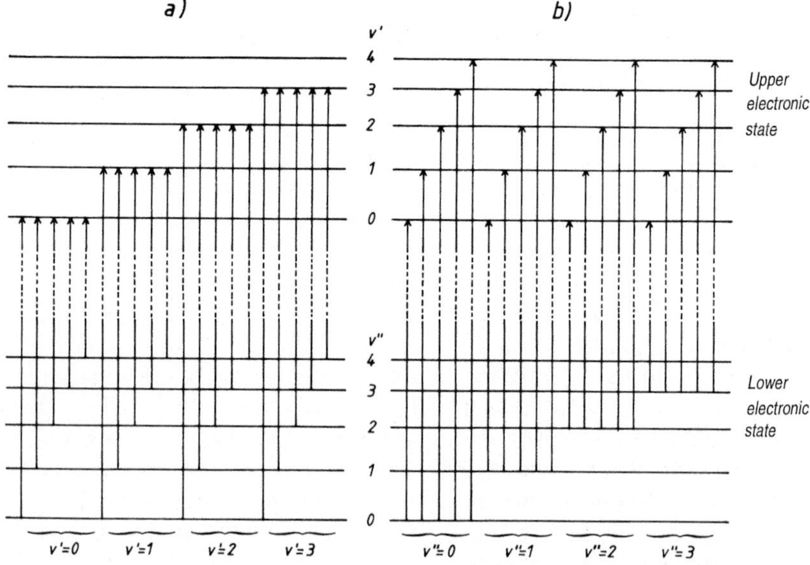

Band-head table. Fig. 2. Transitions and progressions between two electronic states, see text

1*s* band will be fully occupied with 2 *N* electrons. The 2*s* band would be half occupied because only one electron is available from each lithium atom, i.e. the lower *N*/2 energy states in the band should be occupied. In reality, the situation in lithium metal is rather more complicated because of overlap of the 2*s* and 2*p* bands.

It is very significant that energy gaps (band gaps), which have no energy levels and therefore contain no electrons, occur between the individual bands. This property of the band model leads to an easily visualizable classification of solids as insulators, conductors and semiconductors. If a crystal has just so many electrons that the allowed energy bands are either fully occupied or completely unoccupied, no electron can move under the influence of an electric field and the material is an insulator. If one or more bands are partially (i.e. between 10% and 90%) occupied, the substance is an electrical conductor. If all bands are fully occupied, apart from one or two bands which are either minimally occupied or almost completely full, the material is a semiconductor or a semimetal.

A semiconductor becomes an insulator at the absolute zero. The difference between an insulator and a semiconductor is expressed by the magnitude of the band gap, E_g. This is 5.33 eV for a typical insulator, diamond. In contrast, band gaps of only 1.14 eV and 0.67 eV respectively separate the fully occupied valence band and the unoccupied conduction band of Si and Ge. If the temperature of a semiconductor is raised, starting from the absolute zero, the intrinsic electrical conductivity increases. This may be distinguished from the impurity conductivity by its temperature dependence and it can be used for determining the band gap, E_g. A second method for determining E_g is based upon optical absorption. Since semiconductor crystals frequently possess impurities, they show impurity conductivity, in addition to their intrinsic conductivity. The former can be influenced in a planned manner by doping with foreign atoms to give, for example, *n* or *p* conductivity. (→ photoelement, → photodiode, → phototransistor).

Ref.: C. Kittel, *Introduction to Solid-State Physics*, 6th ed., John Wiley & Sons, Inc., New York, **1986**, chapt. 9, 10 and 11.

Band-pass filter, <*Bandpaßfilter*>. A band-pass filter has the property that a region of high transmission (transmission region) is bordered on both sides by regions of low transmission (blocked region), i.e. the filter transmits only a selected part of the spectrum. Some optical glass filters (→ colored glass filters) and particularly → interference filters are of this type. Band-pass filters are characterized more exactly by the following details (cf. Figure 1):

τ_{max} = → transmittance at the maximum of the transmitting region;

λ_m = spectral position (central wavelength of the transmitting region).

If $\lambda_{1/2}'$ and $\lambda_{1/2}''$ are the wavelengths at which the transmittance has fallen to half its maximum value = $\tau_{max}/2$, then the central wavelength is defined by:

$$\lambda_m = \frac{\lambda'_{1/2} + \lambda''_{1/2}}{2}.$$

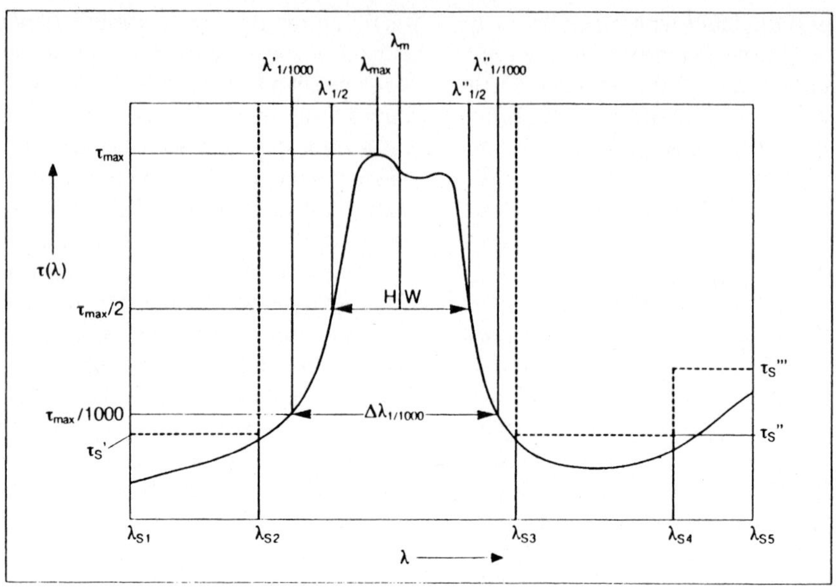

Band-pass filter. Fig. 1. Definition of the characterizing quantities

where

$\Delta\lambda_{1/2}$ $= HW$ (half-width)

$=$ band width at $\tau_{max}/2$

HW $= \Delta\lambda_{1/2} = \lambda''_{1/2} - \lambda'_{1/2}$

$\Delta\lambda_{1/10}$ $= ZW$ (one tenth-width)

$=$ band width at $\tau_{max}/10$

$\Delta\lambda_{1/000}$ $=$ one thousandth-width

$=$ band width at $\tau_{max}/1000$

$$Q \text{ value: } Q = \frac{\text{one tenth-width } ZW}{\text{half-width } HW}$$

$$q \text{ value: } q = \frac{\text{one thousandth-width}}{\text{half-width}} = \frac{\Delta\lambda_{1/1000}}{HW}$$

τ_s' and τ_s'' characterize the upper limits of the transmittance in the blocked regions. The Q and q values are used specially for more exact characterization of interference filters.

Band shape, <*Bandenform*>, in the → band analysis (deconvolution) of spectral bands it is necessary to approximate the band shape with a meaningful mathematical function. The Lorentzian band shape and the Gaussian error-distribution function are particularly suitable for this purpose and we frequently speak of the Lorentzian or Gaussian shape of an absorption band. The equation of a Lorentzian band is:

$$A' = A'_{max} \frac{\Delta\tilde{\nu}_{1/2}}{4(\tilde{\nu}_{max} - \tilde{\nu})^2 + \Delta\tilde{\nu}_{1/2}};$$

$$A' = \ln \frac{\Phi_o}{\Phi} = 2.303\,A$$

where A' = the natural (log to base e) absorbance, A'_{max} = the natural absorbance at the absorption maximum,

$\Delta\tilde{\nu}_{1/2}$ = the half-width of the band, $\tilde{\nu}_{max}$ = the wavenumber of maximum absorption and $\tilde{\nu}$ the variable wavenumber over the absorption band. The equation of a Gaussian band is:

$$A = A_{max} \exp\{-B(\tilde{\nu}_{max} - \tilde{\nu})^2\};$$

$$B = \frac{4\ln 2}{\Delta\tilde{\nu}_{1/2}}$$

Clearly, B depends on the half-width. Using the → Bouguer-Lambert-Beer law, we can express the absorbances, A', A'_{max}, A and A_{max}, in terms of the molar decadic → extinction coefficient.

Lorentzian band:

$$\varepsilon = \varepsilon_{max} \frac{\Delta\tilde{\nu}_{1/2}}{4(\tilde{\nu}_{max} - \tilde{\nu})^2 + \Delta\tilde{\nu}_{1/2}}$$

Gaussian band:

$$\varepsilon = \varepsilon_{max} \exp\{-B(\tilde{\nu}_{max} - \tilde{\nu})^2\}$$

The Gaussian and Lorentzian band shapes are compared in Figure 1. In both cases, A_{max} was set = 1 and the same value for $\Delta\tilde{\nu}_{1/2}$ was used. The difference between the Gaussian curve (b) and the Lorentzian curve (a) is the considerably steeper slope of the sides in the former. The long-wavelength decrease of the Gaussian band represents a particularly good approximation to the decrease on the long-wavelength side of an absorption band in a UV-VIS spectrum. Both band profiles are symmetrical which is not the case for real absorption bands. Therefore, additional terms are required in the above equations. The addition of a cubic term has proved useful in the case of the Gaussian function. The equation is used in the form:

$$\ln\frac{A_{max}}{A} = b(\tilde{\nu}_{max} - \tilde{\nu})^2[1 + a(\tilde{\nu}_{max} - \tilde{\nu})];$$

$$b = B\frac{2}{1 + a\Delta\tilde{\nu}_{1/2}};$$

a must be determined from the experimental curve.

Band superposition, *<Bandenüberlagerung>*, the superposition of two adjacent absorption bands (also vibrational bands) is a phenomenon fre-

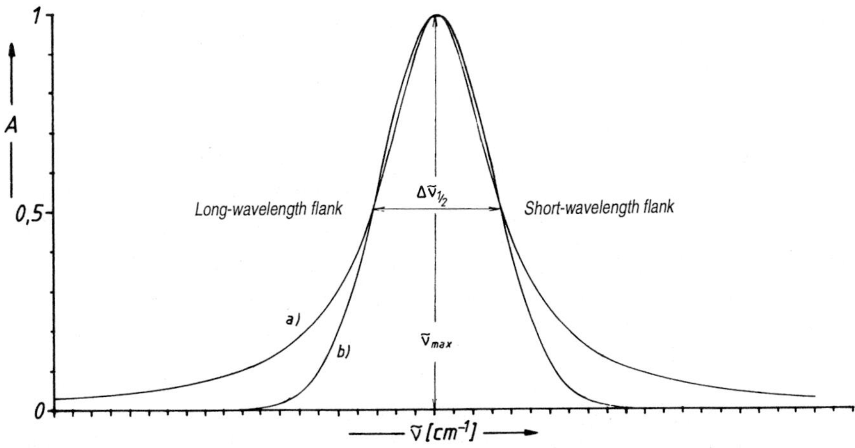

Band shape. Fig. 1. a) Lorentzian profile, b) Gaussian profile

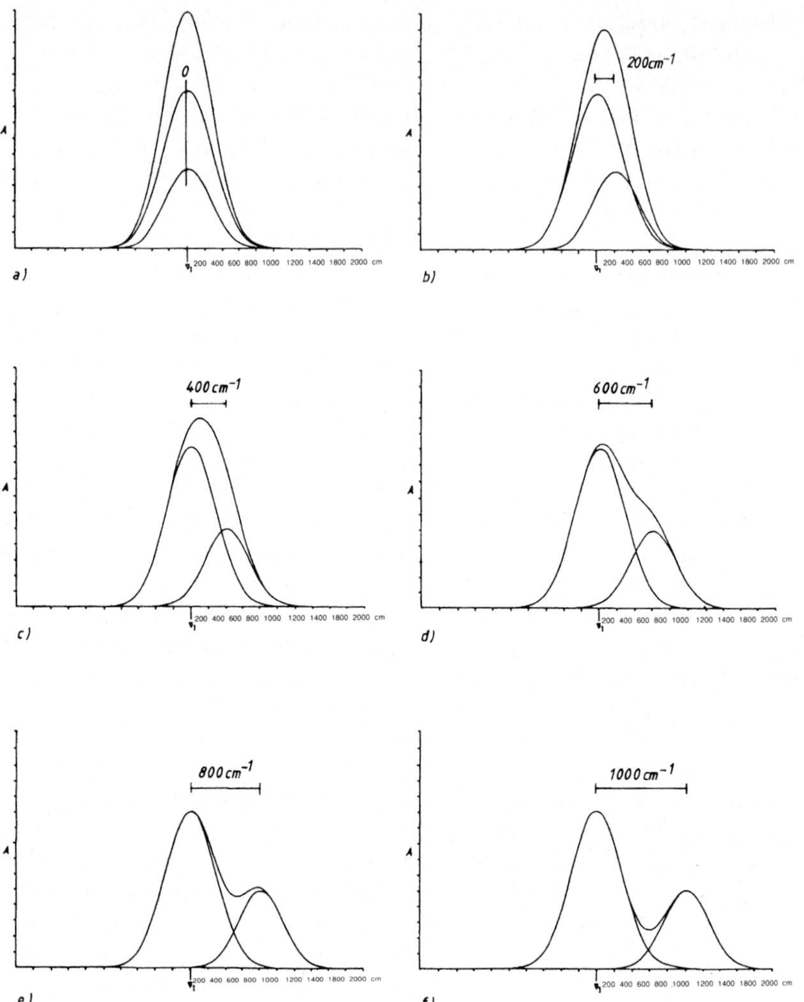

Band superposition. Fig. 1. Superposition of two Gaussian bands, see text

quently observed in UV-VIS spectroscopy. In an extreme case, we observe only a single band the position of which is no longer characteristic of either constituent band ($\tilde{\nu}_{1max}$, $\tilde{\nu}_{2max}$). At greater separation of the overlapping bands, shoulders occur for which ε_{max} and $\tilde{\nu}_{max}$ cannot be given exactly. Figure 1 a – f shows examples of the superposition of two bands using Gaussian curves for the → band shape. The absorbances have a ratio of 2:1, and the half-width was assumed to be the same for both bands. The separation of the maxima, $\tilde{\nu}_1$ and $\tilde{\nu}_2$, was varied in 200 cm^{-1} steps from 200 to 1000 cm^{-1}. Figures 1 b and c ($\Delta\tilde{\nu}$ = 200 and 400 cm^{-1}) show both bands superimposed giving a single band maximum which is shifted from $\tilde{\nu}_1$. In case *d* ($\Delta\tilde{\nu}$ = 600 cm^{-1}) there is a shoulder and in case *e* ($\Delta\tilde{\nu}$ = 800 cm^{-1}), both bands are clearly separated although band 2 is still influenced by the underlying band 1. Only when $\Delta\tilde{\nu} = 1000$ cm^{-1}, are the two bands effectively independent. Figure 1 shows clearly that → band analysis is required to extract exact data from such overlapping bands.

The decisive factor for the superposition of adjacent bands is the → half-width, $\Delta\tilde{\nu}_{1/2}$. The smaller this is the better separated (resolved) are the bands when measured. The half-width of absorption bands can be reduced by measuring at liquid nitrogen (77 K) or liquid helium (4.2 K) temperature. The broadening of absorption bands at room temperature in solution is essentially caused by hindered rotation, but these motions are frozen at low temperature so that the contributing bands become sharper (pure vibrational structure). In such measurements the → resolution in $\Delta\tilde{\nu}$ must be smaller than the separation between two adjacent bands. Therefore, in addition to the band parameters described here, the experimentally selectable spectral → slit width has an influence upon band resolution. Band overlap is frequently caused by too great a spectral slit width.

Barium sulfate, <*Bariumsulfat*>, a → white standard. Barium sulfate is frequently used as reference standard in the → integrating sphere or photometer sphere in → reflectance spectroscopy.

Barrier-layer photovoltaic effect, <*Sperrschichtphotoeffekt*>, if a *p* type semiconductor is brought into contact with an *n* type semiconductor charge migration occurs at the interface, the *p-n* junction. Until a state of equilibrium is reached, *p* conducting holes diffuse into the conducting *n* region and *n* conducting electrons into the conducting *p* region and extensively neutralize each other at the interface. During this process, fixed charge carriers remain on both sides of the junction in the form of ionized donor atoms in the *n* conducting layer and ionized acceptor atoms in the *p* conducting layer. An electric double layer (space-charge zone) is formed from a positive charge distribution on the *n* side and a negative charge distribution on the *p* side of the junction. The electric field, thus established at the *p-n* junction, extracts all movable charge carriers from the immediate region and creates a transitional layer of strongly reduced conductivity which acts as a barrier to the flow of current in one direction.

If the junction is irradiated, electrons and positive holes are formed by the internal photoeffect. Depending on their charge, these particles migrate across the junction and into both sides of the double layer. A photovoltage (photo EMF) results at the junction which generates a photocurrent when the external circuit is closed. The *p* semiconductor is the positive and the *n* semiconductor the negative electrode of the → photoelement.

This barrier-layer photovoltaic effect can only occur if minority carriers are produced by the radiation. For that reason, it is limited in practice to the spectral region in which the host lattice absorbs (internal → photoelectric effect).

→ Photodiodes and → phototransistors are other radiation detectors in which the barrier-layer photovoltaic effect is utilized.

Base, <*Basis*> in optics, the base of a → prism.

Bathochromic shift, <*Bathochromie, Rotverschiebung*>, a shift of the UV-VIS absorption spectrum to longer wavelengths or lower wavenumbers when substituents are introduced into a → chromophore. Substituents which induce such a shift are called → auxochromes and antiauxochromes. If two or more chromophores are combined in a chromophoric system, as in the polyenes for example, an extraordinarily strong bathochromic shift occurs. A bathochromic shift can also be caused by the solvent, particularly if the pH value is changed. The opposite effect is called a → hypsochromic shift.

Beam splitter, <*Strahlenteiler*>, an optical component which splits a beam of radiation into two beams. In the case of quartz plates, glass plates and partially transparent mirrors (on glass or quartz), a specific proportion of the incident light is reflected whilst the remainder is transmitted. For partially transparent mirrors, the ratio of reflected to transmitted light can be varied over a wide range by controlling the thickness of the vacuum-deposited metal-mirror layer. A beam splitter is usually set at an angle of 45° to the path of the radiation so that the two resulting beams lie at an angle of 90° to each other. When using beam splitter plates, attention must be paid to the facts that the divided beams traverse different optical paths and that the optical axis undergoes a parallel shift which depends upon the angle of incidence, ε, and the plate thickness, d. This parallel shift, v, (Figure 1a) is given by the equation:

$$v = d\frac{\sin(\varepsilon - \varepsilon')}{\cos\varepsilon'}$$

with

$$\sin\varepsilon' = \frac{\sin\varepsilon}{n'}.$$

n' is the refractive index of the plate material. Wedge-shaped beam splitter plates are also obtainable. As shown in Figure 1b, this causes rear-side reflected rays to be deflected from the light path. With these plates, beam deflection occurs in addition to the axis shift. If the wedge angle is γ (Figure 1b), the deflection angle, α, for an angle of incidence of 45° is given by the equation, $\alpha = \gamma\ (n' - 1)$. Beam splitters of the above types are called physical beam splitters. They preserve the undivided → aperture in both the resulting beams, i.e. the incident light is divided according to the degree of transmission, τ, and the degree of reflection, R, in the same way across the whole cross-section of the incident

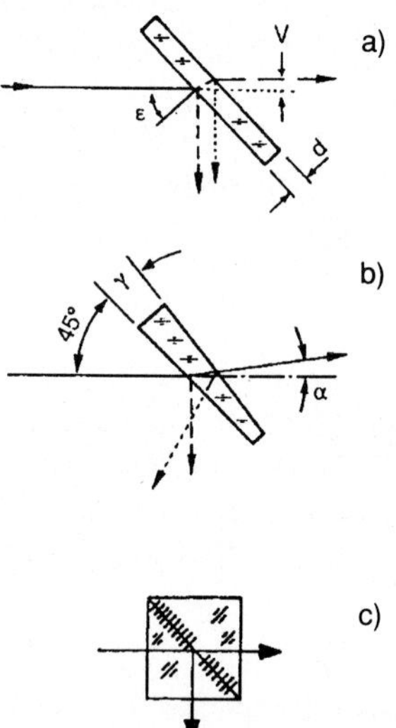

Beam splitter. Fig. 1. Physical beam splitters: a) beam splitter plate, b) wedge-shaped beam splitter c) beam splitter cube.

beam. Therefore, resolving power is not diminished due to reduction of aperture. In order to obtain a separation without losses the degree of absorption of a beam splitter must be as small as possible. For that reason, thin metal layers are now very rarely used. Multiple dielectric layers (→ all dielectric filter) are preferable.

By suitable construction of the dielectric layers, the intensity ratio of the separated beams can be set to any value with almost no absorption while, at the same time, demanding spectral requirements can also be met. Dichroic beam splitters, which give a separation without loss of the spectral region (e.g. short-wavelength reflection, long-wavelength transmission) are possible. The beam splitter cube (Figure 1c) is important as a physical beam splitter for spectroscopic purposes. It consists of two rectangular prisms of crown glass BK 7 (→ glass, optical) or of quartz glass (Suprasil) and a metallic splitting layer which are cemented together. The outer surfaces are covered with broad-band antireflection coatings. For special applications, they are produced with dielectric, nonabsorbing splitter plates. In addition to physical beam splitters, geometrical beam splitters should also be mentioned. These divide the incident beam cross section into two or more larger areas or also into more, smaller areas (strips, mirror spots) with aperture separation, cf. Figures 2a, b, c, d. Separation into very small elements must be avoided or resolution may be reduced by diffraction. Beam splitters of the *a*, *b* and *d* types can be used for the production of two beams in double-beam spectrometers. Periodic beam splitters constitute a third group. They produce two interrupted beams, with chronologically shifted half-cycles (180° phase shift), from a continuous beam. The incident light beam can be alternately transmitted or deflected by means of a rotating, mirrored sector placed at 45° to the beam path. The intensity ratio can be adjusted by choice of sector width.

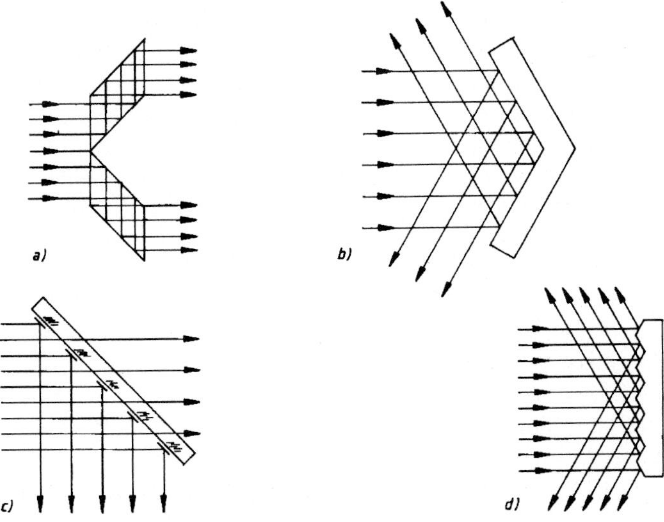

Beam splitter. Fig. 2. Geometrical beam splitters, see text.

A vibrating mirror acts in the same way (→ UV-VIS and → IR spectrometer). Very thin, taught plastic films, → pellicle, have been used as beam splitters in recent years.
Ref.: H. Naumann, G. Schröder, *Bauelemente der Optik*, Hanser Verlag, Munich, Vienna, 5th edtn., **1987**, p. 185ff.

Beer's formula, <*Beersche Formel*>, named after A. Beer, gives the reflection coefficient, R, for a perpendicularly incident light beam upon a strongly absorbing body. It is an extension of Fresnel's formula for → reflection at a plane mirrored surface in which the refractive index, n, is replaced by the complex refractive index, $n' = n \cdot (1 - i\kappa)$, which takes → absorption into consideration. κ is the absorption index. Beer's formula is:

$$R = (n_1)^2 + (n\kappa)^2/(n+1) + (n\kappa)^2$$

(→ Reflection).

Beer's law, <*Beersches Gesetz*>, found by A. Beer, states that the absorbance of a solution is proportional to the product of the concentration and the path length of the radiation through the solution. Therefore, the absorbance depends only upon the total number of the absorbing centers and not upon their concentration. The proportionality constant is the molar decadic extinction coefficient, ε_λ. (→ Bouguer-Lambert-Beer law).

Bending vibration, deformation vibration, <*Deformationsschwingung*>, in contrast to a → valence vibration, this is a form of molecular vibration in which the atoms move at right angles to the bond(s) joining them. The simple description *bending vibration* is only used if the structure of the group under discussion is so simple that further differentiation is not required. However, if necessary we can classify bending vibrations further in terms of → vibrational modes.

a) The simple bending or deformation vibration (symbol δ) is a vibration which is associated with only one changing bond angle (Figure 1a).
b) A bending vibration, in which a group of atoms does not undergo a change of its own bond angles, but vibrates to and fro as an entity, perpendicular to the plane of the group is called a wagging vibration (symbol w or ω, Figure 1a).
c) A rocking vibration (symbol r or ϱ) is similar to bending vibration b, but here the group of atoms vibrates to and fro, as a whole, in the plane of the group (Figure 1a).
d) A twisting or torsion vibration (symbol t or τ) is a bending vibration in which the group concerned vibrates as a whole, in a twisting motion, about the bond or atom which connects the group with the rest of the molecule (Figure 1a).

In specific atomic groupings, e.g. $-YX_3$, we must further distinguish between symmetric, δ_s, and asymmetric, δ_{as}, combinations of simple angle-bending vibrations. This is shown in Figure 1c using the $-CH_3$ group as example. In addition, bending vibrations can occur in plane (*i.p.*) or out of plane (*o.p.*) so that these descriptions allow further classification. The out of plane, i.e. the nonplanar bending vibrations of the C-H group are denoted with the symbol γ(C-H). The symbols most frequently used in the literature for bending and stretching vibrational modes are compiled in the following table:

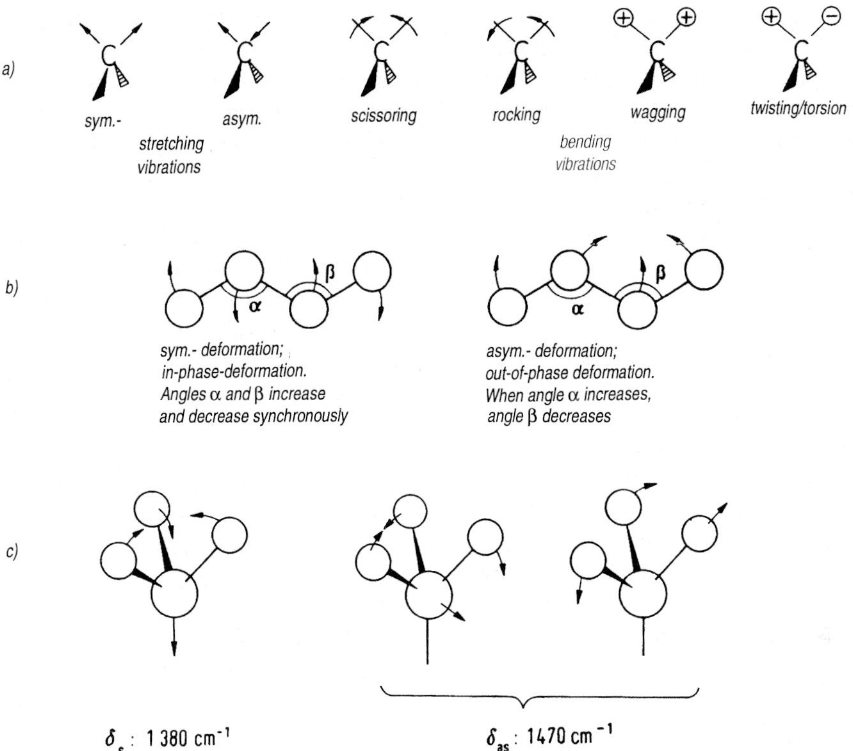

Bending vibrations. Fig. 1. a) Bending vibrations of the >CH_2 group, see text for valence vibrations; b) symmetric and asymmetric skeletal bending vibrations; c) symmetric and asymmetric bending vibrations of a -CH_3 group.

α i.p. bending vibration
β i.p. bending vibration
Γ o.p. bending vibration of skeletal atoms
γ o.p. bending vibration of a C-H bond
Δ i.p. bending vibration of skeletal atoms
δ i.p. bending vibration of an X-H bond
δ_s symmetric bending vibration
δ_{as} asymmetric bending vibration
δ' twisting, rocking vibration
κ o.p. wagging vibration of an XH_2 group (X = C)
r rocking vibration
r_β i.p. rocking vibration
r_γ o.p. rocking vibration
ϱ i.p. rocking vibration of an XH_2 group (X = C)
ν stretching vibration of an X-H bond
ν_s symmetric stretching vibration
ν_{as} asymmetric stretching vibration
ν_β i.p. stretching vibration
ν_γ o.p. stretching vibration
t twisting vibration
τ torsional vibration as in H-C-C-H
τ twisting vibration of an XH_2 group (X = C)
Φ o.p. ring deformation
ω wagging vibration
ω stretching vibration of skeletal atoms without H participation

Ref.: N.B. Colthup, L.H. Daly and S.E. Wiberley, *Introduction to Infrared and Raman Spectroscopy,* 3rd ed., Academic Press Inc., Boston, New York, London, Sydney **1990**.

Bergmann series, <*Bergmann-Serie*> → alkali spectra.

Beta radiation, <*Betastrahlung, β-Strahlung*>, a beam of fast moving electrons formerly known as cathode rays. The speed of the electrons is extremely varied and can be as high as 99% of the velocity of light. The kinetic energy of β radiation lies between 0.01 and 12 MeV. A distinction is made between nuclear and secondary beta radiation.

Binary combination, <*Binäre Kombination*> → combination vibrations.

Binormal, <*Binormale*>, the direction in optically, doubly refracting crystals in which only a single wave-normal velocity exists in contrast to the adjacent directions. It is the axis of optical isotropy of the electric field vector, $\vec{E}$, and is called a primary optic axis or an optic binormal. The $\vec{E}$ vector can vibrate in any direction perpendicular to the binormal. Crystals of the orthorhombic, monoclinic and triclinic systems have two optic binormals and are optically biaxial crystals. → Biradial.

Bioluminescence, <*Biolumineszenz*>, a special case of → chemiluminescence, which has been known ever since man first observed and described natural phenomena such as the emission of light by glow worms, rotting wood and sea organisms. Bioluminescent systems which have been investigated in great detail are, for example, the American firefly (*Photinus* species), the Japanese ostracod (*Cypridina hilgendorfii*) and the sea feather (*Renilla reniforms*). Bioluminescence is considerably more complex than chemiluminescence since enzymes are always involved in chemiluminescent reactions *in vivo* and very small amounts of the reactants are present. However, bioluminescent reactions are of particular interest since they show higher quantum yields than have yet been observed in chemiluminescence.

Ref.: E. N. Harvey, *Bioluminescence*, Academic Press, New York, **1952**; *A History of Luminescence*, American Philosophical Soc., Philadelphia **1957**; W.R. Seitz, *Chemiluminescence and Bioluminescence Analysis: Fundamentals and Biochemical Applications*, CRC Crit. Rev. Anal. Chem., **1981**, *13*, 1.

Biot's law, <*Biotsches Gesetz*> → optical rotation.

Biradial, <*Biradiale*>, the direction in optically doubly refracting crystals of the orthorhombic, monoclinic and triclinic systems in which there is only one velocity of light in contrast to the adjacent directions. This direction is called a secondary optic axis, a line of single ray velocity or an optic biradial. It is the axis of optical isotropy of the electric field vector, $\vec{E}$, which can vibrate in any direction perpendicular to the biradial. Crystals of the above systems are characterized by having two biradials and are therefore called optically biaxial crystals; → binormal.

Birefringence, electrical, <*Doppelbrechung, elektrische*> → Kerr effect.

Birefringence, induced, <*Doppelbrechung, künstliche*>, optically isotropic materials such as glass can be made anisotropic, and thus birefringent, in many ways. Induced birefringence can result from externally imposed pressure, bending, temperature differences, electrical fields, etc.

Birefringence, magnetic, <*Doppelbrechung, magnetische*> → magnetooptical rotation.

Birefringence, natural, <*Doppelbrechung, natürliche*>, the phenomenon observed by E. Bartholinus in 1669 whereby a ray of light falling onto a crystal of Iceland spar (→ calcite) is simultaneously refracted in two different directions so that two separated rays emerge from the crystal. Birefringence occurs, to a greater or lesser degree, in all crystals with the exception of those belonging to the cubic system. It is a result of their anisotropy and it arises from the fact that different directions through a point in the crystal are not equivalent physically.

A further study of this phenomenon shows that one of the rays behaves according to Snell's law (→ refraction of light) but the other does not, i.e. the two rays have different refractive indexes. The ray obeying Snell's law is called the ordinary ray (o ray) and the other is the extraordinary ray (e ray). It is important to note that the two rays are linearly polarized, perpendicular to each other in doubly refracting, optically uniaxial crystals. This is the basis of the application of natural birefringence in the construction of → polarizers. Doubly refracting, optically uniaxial crystals are distinguished by their optical character which is defined as the difference of the refractive indexes, $n_e - n_o$. If $n_e < n_o$, the character is optically negative, if $n_e > n_o$ it is optically positive. The table lists the refractive indexes, n_e and n_o, of a few doubly refracting crystals.

Birefringence, Table.

Crystal	n_o	n_e	Optical Character
Calcite	1.6584	1.4864	$n_e - n_o < 0$
Corundum	1.7682	1.6598	
$NaCO_3$	1.5874	1.5361	negative
Tourmaline	1.625	1.6220	
Beryl	1.5740	1.5674	
Quartz	1.5442	1.5533	$n_e - n_o > 0$
Rutile	2.6158	2.9029	
K_2SO_4	1.4550	1.5153	positive
Cinnabar	2.854	3.201	
Ice	1.309	1.313	

Bisectrix, <*Bisektrix*>, optically biaxial crystals, i.e. crystals of the orthorhombic, monoclinic and triclinic systems, have two directions, known as binormals, which show special optical properties. In particular, in these directions the crystals are not doubly refracting. The line which bisects the acute angle between the two binormals is the acute bisectrix or first median line; the line dividing the obtuse angle is the obtuse bisectrix or second median line. The angle between the binormals is a property of the material and can be used to characterize the crystal. If the first bisectrix and the semimajor axis (largest refractive index) of the refractive index ellipsoid (indicatrix) are coincident then the crystal is said to have a positive optical character. If the second bisectrix lies along this ellipsoid axis, the crystal has a negative character.

Bjerrum double band, <*Bjerrumsche Doppelbande*>, is observed when the lines of the P and R branches of a → rotation-vibration spectrum are not resolved. The positions of the maxima of the P and R branches depend upon the occupation of the rotational states which in turn depends on the rotational constant, B, and the absolute temperature, T, according to Boltzmann's energy-distribution law → Boltzmann statistics. Using the Boltzmann distribution, we can obtain a simple expression for the separation of the two maxima, $\Delta\tilde{\nu}_{P,R}$:

$$\Delta\tilde{\nu}_{P,R} = \sqrt{\frac{8BkT}{hc}} = 2.3583\sqrt{BT}\ [\mathrm{cm}^{-1}].$$

At a known temperature, a measurement of the splitting of the two bands provides an approximate value of the → rotational constant, B, and also of the moment of inertia. The first, spectroscopic estimates of → moments of inertia were obtained in this way.

Black-body radiator, <*schwarzer Körper*>, a → thermal radiator with a black surface which absorbs all incident electromagnetic radiation, whatever its wavelength. The radiation of a thermal radiator depends upon its temperature, size and the nature of its surface. A body with a completely black surface emits more radiation than any other body having the same temperature and size. Since there is no body with a completely black surface, the black body used in practice is a cavity with a small radiation exit, as shown in Figure 1. Radiation entering through this orifice is reflected several times within the cavity and undergoes absorption by the walls at each reflection. Thus, the probability that the radiation will emerge again through the orifice is extremely small. The radiation exiting through the orifice of such a cavity is identical with the radiation of a black body. It is independent of the chemical nature of the radiator and is solely a function of the temperature. The spectral energy distribution of a black body is described by → Planck's radiation law.

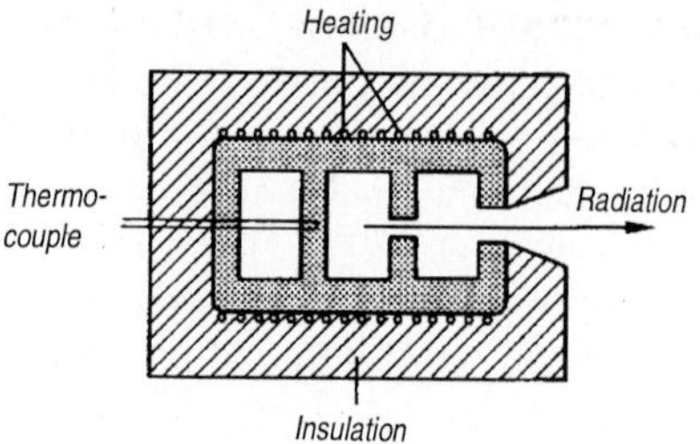

Black-body radiator. Fig. 1. The practical implementation of a black-body radiator

Black-body temperature, brightness temperature, radiation temperature, <*schwarze Temperatur*>. The black-body temperature of a → thermal radiator is the temperature at which a → black-body radiator would emit the same radiation density as the thermal radiator in a particular waveband.

Black box, <*Black Box*>, a modern measuring instrument the function of which is unknown to the user who only knows how to operate the knobs on the front. The connection between the construction of the instrument and the physical or chemical measurements being made is hidden from him.

Black-light lamp, analytic lamp, <*Analysenlampe, Schwarzlichtlampe*>, a UV source with filters which transmit only UV radiation.

Thus, observation of the characteristic luminescence of a substance is not complicated by light coming directly from the source. The UV source is housed in an airtight casing with a black filter (Wood's filter) fitted to the exit. This is a glass filter which is made almost opaque to visible light by adding nickel oxide. It transmits predominantly radiation of the wavelengths between 400 nm and 320 nm with a transparency maximum at 380 nm. This region is called the black-light (bl) region. Hg high- and low-pressure lamps are used as the UV sources. The analytic lamp is used for the → luminescence analysis of chromatograms. In physics, it is used for exciting fluorescent and phosphorescent materials.

Black standard, <*Schwarzstandard*>, a → black body, which has a very large → absorption coefficient, β, in the UV-VIS region can be used as a black standard. In → photoacoustic spectroscopy (PAS), such a body can be approximated with a carbon standard which is most easily produced by blackening a glass or quartz plate with soot. It is also possible to embed high-quality carbon black in a plastic foil. Such black standards are relatively insensitive and very stable over long periods of time. Black drawing ink can be used as a liquid black standard.
A comparison of the different black standards used in PAS shows that it is just as difficult to produce an ideal black standard as an ideal → white standard.

Blaze angle, <*Blaze-Winkel*> → echelette grating, → grating.

Blaze wavelength, <*Blaze-Wellenlänge*> → echelette grating, → grating.

Blocking, <*Blockung*>, in optical filtering the use of additional filters to attenuate radiation, especially that outside the filter's own blocking range. It is usually achieved by absorption and/or reflection of the undesired radiation.

Blocking filter, <*Sperrfilter*>, a filter for → blocking undesired radiation in the → UV-VIS region when using continuum and line sources. It is usually an → edge filter which absorbs UV radiation strongly and transmits only long-wavelength light unattenuated (→ long-wave pass filter). Filters which block long-wavelength light, especially IR radiation, and transmit short-wavelength light are called → short-wave pass filters. Blocking filters are frequently used in UV-VIS spectrometers operating in the visible spectral region in order to remove scattered or stray UV light. In grating spectrometers, they have the additional purpose of blocking out higher orders.

Blooming, <*Oberflächenvergütung*>, the coating of a surface or a → surface layer with reflection-reducing layers (→ dielectric layers).

Blue shift, <*Blauverschiebung*> → hypsochromic shift

Bohr-Einstein frequency relation, <*Bohr-Einstein-Beziehung*>, the basic equation of spectroscopy. The energy difference between two discrete energy states ($E_2 > E_1$) is related to the frequency of the emitted spectral line by the equation:

$$E_2 - E_1 = \Delta E = h\nu.$$

Following the indroduction of → Planck's constant, h, in 1905, Einstein formulated the equivalence of energy, E, and electromagnetic radiation of frequency, ν, with the equation, $E = h\nu$, which Bohr applied to the line spectra of atoms in 1913.
If we replace the frequency by the → wavelength, λ, using the equation, $\nu = c/\lambda$, we obtain:

$$\Delta E = E_2 - E_1 = h\frac{c}{\lambda},$$

Use of the definition of the → wavenumber now leads to the expression for the term difference and a → term

$$\tilde{\nu} = \frac{E_2}{hc} - \frac{E_1}{hc} = T_2 - T_1$$

is defined as E/hc which has the units cm^{-1} or m^{-1}.

Bohr magneton, <*Bohrsches Magneton*>, the magnetic moment, μ, of an electron with a nonzero orbital angular momentum ($\ell \neq 0$) is given by:

$$\mu_\ell = \frac{e}{2m_e} \cdot \frac{h}{2\pi} \sqrt{\ell\,(\ell + 1)}$$

e is the elementary charge, m_e the mass of the electron and h → Planck's constant. The equation shows that the magnetic moment is measured in multiples of $(e/2m_e) \cdot (h/2\pi)$. The Bohr magneton, μ_B, was introduced to represent this quantity:
$\mu_B = eh/4\pi m_e = 9.274078 \cdot 10^{-24}$ A m^2

Bohr radius, <*Bohrscher Radius*>, the radius of the Bohr orbit of lowest energy ($n = 1$) for the H atom. $a_0 = \varepsilon_o h^2/\pi m_e e^2 = 52.917 \times 10^{-12}$ m = 52.917 pm. The notation is the same as in the entry above with the addition of ε_o, the vacuum permittivity. a_0 is the atomic unit of length (→ atomic model).

Bolometer, <*Bolometer*>, a → thermal detector which utilizes the temperature dependence of electrical resistance for measuring temperature. When used as a simple IR detector, it consists of two thin blackened metal strips which form two arms of a Wheatstone bridge. One of these strips is exposed to the radiation to be measured. The change in resistance due to the resulting temperature difference can be measured by balancing the bridge or by measuring the voltage drop if a constant current flows. Due to the required stability of the zero reading, extreme demands are made upon the current supply to the bridge. The sensitivity of the bridge depends upon the relative temperature coefficient of the resistance. This is particularly high in nickel which is frequently used in bolometers. Other metals used are antimony, bismuth, gold and platinum. To produce bolometer strips these metals are evaporated in a high vacuum or deposited electrolytically onto carrier materials such as Al_2O_3 or plastic foils. Sensitivities up to 11800 V/W (antimony) are possible with such bolometers. The detection limit is about $10^{-9} - 10^{-10}$ W and the time constant is in the region of milliseconds.
Bolometers have been largely replaced by thermopiles (→ thermocouple) as radiation detectors in IR spectroscopy. In addition to metal-strip bolometers, semiconducting (thermistors) and superconducting bolometers should be mentioned.
Ref.: R.A. Smith, F.E. Jones, R.P. Chasmar, *The Detection and Measurement of Infrared Radiation*, Oxford University Press, Oxford, **1957**.

Boltzmann's constant, <*Boltzmann-Konstante*>, the constant, k, discovered by L.E. Boltzmann which occurs in the radiation laws (→ spectral energy distribution) and in statistics. It is defined in terms of the gas constant, R, and the Loschmidt or Avogrado number, N_L, as:

$$k = R/N_L \text{ where } R = \frac{P_o V_o}{T_o}$$

$k = 1.380662 \cdot 10^{-23}\ \mathrm{JK}^{-1}$

Boltzmann's energy-distribution law, <*Boltzmannscher Energieverteilungssatz*> → Boltzmann statistics.

Boltzmann statistics, <*Boltzmann-Statistik*>, classical statistics established by L.E. Boltzmann which is based upon the following model for determining a distribution function, i.e. the function which describes the distribution of individual particles in individual states (also called cells in statistics):

a) The system under observation consists of N particles;
b) the particles are distinguishable, i.e. we might imagine that they are numbered;
c) the particles are independent of each other, i.e. no interaction occurs between them;
d) any particle can occupy any one of the energy states ε_0, ε_1, ε_2 ... ε_i ... ε_{n-1}. The number of particles in energy state ε_0 is denoted with N_0 and in general those in energy state ε_i with N_i;
e) n different energy states are possible in total. Thus we have:

$$\sum_0^{n-1} N_i = N_0 + N_1 + \ldots N_{n-1} = N$$

(total number of particles)

$$\sum_0^{n-1} \varepsilon_i N_i = \varepsilon_0 N_0 + \varepsilon_1 N_1 + \ldots \varepsilon_{n-1} \cdot N_{n-1} = E$$

(total energy)

According to the rules of the theory of combinations, for this model the statistical weight of a distribution is given by:

$$G = \frac{N!}{N_0! N_1! \ldots N_{n-1}!}$$

The most probable distribution will be that which has the maximum value for G. A detailed calculation gives the most probable distribution as:

$$N_i = N \frac{e^{-\varepsilon_i/kT}}{\sum e^{-\varepsilon_i/kT}} = N \frac{e^{-\varepsilon_i/kT}}{Z}$$

which is known as the Boltzmann distribution. The denominator, Z, of this distribution function is the → partition function.

Boltzmann statistics are important in spectroscopy because the ε_i values of the distribution function correspond to the molecular translational, rotational, vibrational and electronic energies, which are connected through the partition function with the thermodynamic quantities. An understanding of the distribution of atoms and molecules in the different quantized energy states as a function of the temperature is very important for many problems in spectroscopy. Classical Maxwell-Boltzmann statistics make it possible to obtain information about this distribution, which is called Boltzmann's energy distribution, and describes a thermal energy distribution. The number of particles (atoms or molecules), N_i, which are present in an excited state, $\Delta\varepsilon_i$, at an absolute temperature, T, is given by:

$$N_i = N\frac{g_i e^{-\Delta\varepsilon_i/kT}}{\sum_i g_i e^{-\Delta\varepsilon_i/kT}} = N\frac{g_i e^{-\Delta\varepsilon_i/kT}}{Z}.$$

N is the total number of particles in the volume (mol volume) considered, Z the partition function and g_i the degeneracy of the excited state of energy $\Delta\varepsilon_i$. The sum in the denominator runs over all energy states. The partition function can be written explicitly as:

$$Z = g_0 1 + g_1 e^{-\Delta\varepsilon_1/kT} + g_2 e^{-\Delta\varepsilon_2/kT} + g_3 e^{-\Delta\varepsilon_3/kT} + \ldots$$

If the second term in this sum is small vis-a-vis g_0 we can write to a good approximation:

$$N_i = N \cdot \frac{g_i}{g_o} \cdot e^{-\Delta\varepsilon_1/kT}.$$

This is always true if the excitation energy of the first excited state, $\Delta\varepsilon_1$, is very large. This requirement is almost always satisfied for electronic excitation of atoms and molecules and for vibrational excitation at room temperature.

If we consider the first excited state, neglecting the degree of degeneracy, at temperature T the Boltzmann distribution function gives the population ratio, N_1/N_0, i.e. the ratio of the number of particles in the first excited state to the number of particles in the ground state directly. At room temperature (300 K), $N_L \cdot kT = RT = 200$ cm^{-1} (N_L is Loschmidt's number). Thus we obtain the values of N_1/N_0 in the table for the different orders of magnitude of the various spectroscopic excitation energies.

At decreasing excitation energies, the Boltzmann factor, a = exp[-$\Delta\varepsilon_i/kT$], tends to a value of 1, i.e.

lim a as $\Delta\varepsilon_1 \rightarrow 0 = e^0 = 1$; and the difference in population continuously decreases. Thus, higher rotational states are already populated to a considerable extent at room temperature, which is the reason for the intensity progression of rotational lines. In the case of vibrational excitation, the number of molecules in the first excited vibrational state is very small and we can assume that vibrational and electron excitation always starts from the vibrational ground state (v = 0), though we do see some → hot bands. The ground and excited states are almost equally populated in → NMR spectroscopy. When irradiating at an appropriate resonance frequency, therefore, (→ resonance) a saturation of the signal can easily occur. Relaxation phenomena are the processes which try to re-establish the original state of population. A flame

Boltzmann statistics. Table.

$\Delta\varepsilon_1$ [cm^{-1}]	Excitation of	$e^{-\Delta\varepsilon_1/200}$	$a = N_1/N_0$
20000	Electrons	e^{-100}	$3.7 \cdot 10^{-45}$
200	Vibration	e^{-10}	$4.5 \cdot 10^{-5}$
10	Rotation	$e^{-0.05}$	$9.5 \cdot 10^{-1}$
1	ESR	$e^{-0.005}$	0.995
10^{-2}	NMR	$e^{-0.00005}$	0.99995

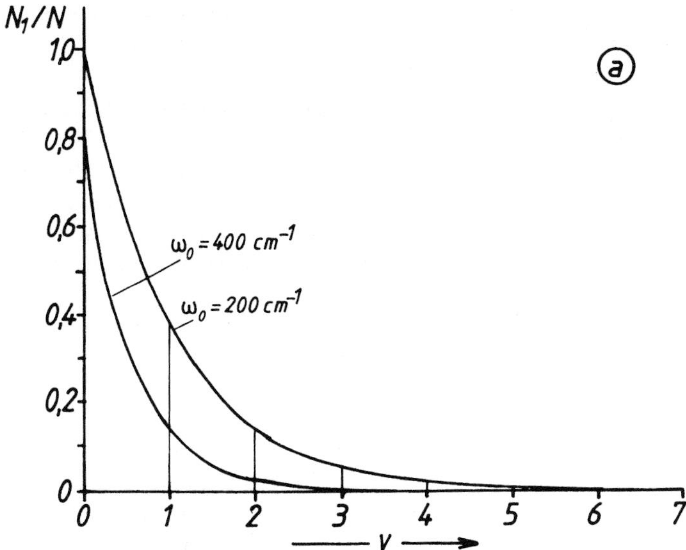

Boltzmann statistics. Fig. 1a. Dependence of the population of vibrational states upon the vibrational quantum number, v

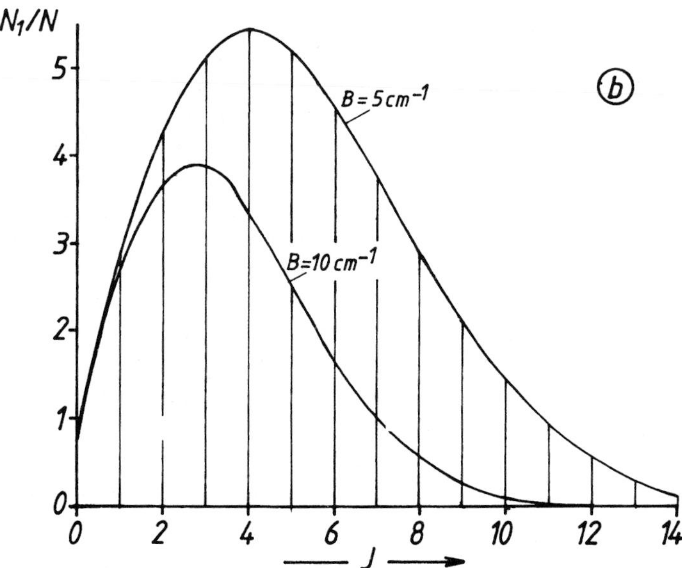

Boltzmann statistics. Fig. 1b. Dependence of the population of rotational states upon the rotational quantum number, J

temperature of 3000 K, available in → atomic absorption spectroscopy, is an appropriate reference temperature for the electronic excitation of free atoms by absorption of radiation. A Boltzmann factor of $a = e^{-10} = 4.5 \cdot 10^{-5}$ results, i.e. even at this temperature, the atoms are predominantly in the ground state which is of considerable importance in the practical application of AAS. At infinitely high temperatures, at thermal equilibrium, only an approximately equal population density in both states can be achieved (however, see → population inversion).

For electronic and vibrational excitation, we can assume a value of 1 for the partition functions, Z_{el} and Z_{vib}, making allowance also for the degree of degeneracy of the lower state concerned. However, the partition function for rotation must be calculated because, as mentioned above, higher rotational states contribute to it markedly. For the rigid rotator (→ dumbbell model), we obtain:

$$Z_{rot} = \frac{kT}{Bhc}.$$

and for the energy distribution:

$$N_J = \frac{NhcB}{kT}(2J+1)e^{-BhcJ(3+1)/kt}.$$

For a → harmonic oscillator with a partition function, $Z_{vib} = 1$, we have:

$$N_v = Ne^{-hc\omega_o v/kT}$$

If the excitation energies and thermal energy per mol ($N_L kT = RT$) are expressed in wavenumbers (cm^{-1}), with $RT \cong 200\ cm^{-1}$ at room temperature (300 K), and $\omega_0 = 400$ or $200\ cm^{-1}$ and $B = 10$ or $5\ cm^{-1}$, the distribution functions and dependence upon quantum numbers, v and J shown in Figures 1a and 1b are obtained. A rapid exponential decrease is found for vibrational excitation whilst an initial increase to a maximum value results for rotational excitation, due to the degeneracy factor, $2J + 1$. This is followed by an exponential decrease for higher values of J. The position of the maximum depends upon the rotational constant, B, and the temperature, T. This characteristic distribution of the population of rotational states is responsible for the intensities of the individual rotational lines in the rotation-vibration spectrum of simple molecules. If the spectra are insufficiently resolved, the envelopes of the R and P branches can be observed, in the absence of a Q branch, as the → Bjerrum double band. The maxima are given by the distribution shown in Figure 1b.

Born-Oppenheimer approximation, <*Born-Oppenheimer-Näherung*>, an approximation which separates the electronic and nuclear motion of a molecule and describes the total wave function as a product of an electronic and a nuclear part:

$$\Psi_{ne}(x, X) = \Psi_e(x, X)\chi_{ne}(X). \quad (1)$$

Ψ_{ne} is the total wave function and Ψ_e the electronic wave function; both are dependent upon the electronic coordinates, x, and nuclear coordinates, X. χ_{ne} is the nuclear wave function which depends only upon the nuclear coordinates, X. A separation of the electronic and nuclear motion is possible because the mass of the electron is much smaller than the mass of the nucleus, e.g. $M_H/m_e = 1836$. We can interpret this classically by saying that the velocity of the electrons is so much greater than that of the nuclei that the electrons instantaneously adjust their positions to changes in the positions of the nuclei.

The Hamilton operator, H, (→ Schrödinger's equation) then separates into two parts: $H_e(x,X)$ and $H_n(X)$. $H_n(X)$ is the operator for the kinetic energy of the nuclei which is solely a function of nuclear coordinates, X. $H_e(x, X)$ is the Hamilton operator for fixed nuclei. It is a function of the nuclear coordinates, X, and the electronic coordinates, x. Thus, the basic equation of quantum mechanics (→ Schrödinger's equation), $H\Psi = E\Psi$, can be separated into two when the Born-Oppenheimer approximation applies:

$$H_e(x,X)\Psi_e(x,X) = E_e(X)\Psi_e(x,X) \quad (2)$$

$$(H_{ne}(X) + E_e(X))\chi_{ne}(X) = E_{ne}\chi_{ne}(X) \quad (3)$$

The electronic wave function, $\Psi_e(x, X)$ is obtained by solving wave equation (2) for fixed nuclear positions. It can be shown, using quantum mechanics, that the Born-Oppenheimer approximation applies only if the electronic wave function is not a rapidly changing function of the nuclear coordinates. This is generally the case for electronic excitation, so that it is of great importance for the theoretical study of electronic spectra. With the help of the Born-Oppenheimer approximation, the → Franck-Condon principle can be easily derived; see also → oscillator strength. The nuclear wave function, $\chi_{ne}(X)$, can be further separated into a product of the rotational and vibrational wave functions which is fundamental to the study of rotational and vibrational spectra.

Ref.: J. Murrell, *The Theory of The Electronic Spectra of Organic Molecules*, Methuen, London, **1963**.

Bose-Einstein statistics, *<Bose-Einstein-Statistik>* → quantum statistics.

Bouguer-Lambert-Beer law, *<Bouguer-Lambert-Beersches Gesetz>* In 1729, P. Bouguer discovered that the attenuation of the intensity, $-d\Phi$, of monochromatic light passing through a light-absorbing medium is directly proportional to the light intensity (radiation flux), Φ, and the path length (sample thickness), d. Lambert formulated the mathematical relationship in 1760:

$$d\Phi = -m_n(\lambda)\Phi dx.$$

Integration between the limits ($x = 0 \rightarrow x = d$ and $\Phi_0 \rightarrow \Phi_T$) gives:

$$\ln(\Phi_0/\Phi_T)_\lambda = m_n(\lambda)\cdot d$$

or

$$\log(\Phi_0/\Phi_T)_\lambda = m(\lambda)d.$$

The expressions, $\ln(\Phi_0/\Phi_T)_\lambda$ or $\log(\Phi_0/\Phi_T)_\lambda$, define the → absorbance: $A_n(\lambda) = m_n(\lambda)d$ or $A(\lambda) = m(\lambda)\cdot d$. The quantities $m_n(\lambda)$ and $m(\lambda)$ are defined as the natural and decadic → absorbance indexes. For gas mixtures, Lambert formulated the relationship, $m_n(\lambda) = \varepsilon_n(\lambda)\cdot c$ or $m(\lambda) = \varepsilon(\lambda)\cdot c$. ε_n and ε are the natural and decadic → extinction coefficients, respectively. In 1852, Beer showed by means of absorption measurements on solutions that the absorbance, A, is constant if the product, $c\cdot d$, is constant. Together these results comprise the Bouguer-Lambert-Beer law:

$$A_n(\lambda) = \ln(\Phi_0/\Phi_T)_\lambda = \varepsilon_n(\lambda) \cdot c \cdot d;$$

$$A(\lambda) = \log(\Phi_0/\Phi_T)_\lambda = \varepsilon(\lambda) \cdot c \cdot d.$$

The absorbance and the → transmittance (degree of transmission), $\tau = \Phi_T/\Phi_0$ are related by:

$$A(\lambda) = -\log(1/\tau)_\lambda = \varepsilon(\lambda) \cdot c \cdot d.$$

In the form, $A(\lambda) = \varepsilon(\lambda)c \cdot d$, this law is the basic law of quantitative → absorption spectroscopy. $\varepsilon(\lambda)$ is the molar decadic → extinction coefficient which is a substance-specific constant at wavelength, λ.
The Bouguer-Lambert-Beer law applies only in the limit of dilute solutions, $c < 10^{-2}$ mol l^{-1}. In terms of the concentration, c, we have the basic equation of the analytical applications of spectrometry (→ photometry):

$$c = \frac{A(\lambda)}{\varepsilon(\lambda)d}.$$

For many purposes (e.g. → photokinetics, → derivative spectroscopy, → photoacoustic spectroscopy), the law is applied in its exponential form. Then, from $A_n(\lambda) = 2.303 \cdot A(\lambda)$ we obtain:

$$\Phi_T = \Phi_0 \cdot e^{-2,303 \cdot \varepsilon(\lambda) \cdot c \cdot d}$$

or

$$\Phi_T = \Phi_0 \cdot 10^{-\varepsilon(\lambda) \cdot c \cdot d}.$$

Ref.: J.D. Ingle, Jr., S.R. Crouch, *Spectrochemical Analysis*, Prentice-Hall International, Inc., Englewood Cliffs, N.Y., **1988**.

Brackett series, <*Brackett-Serie*> → Balmer's formula.

Bragg condition, <*Braggsche Bedingung*> → X-ray diffraction.

Branley-Lenard effect, <*Branley-Lenard-Effekt*>, the ionization of air or other gases upon irradiation by short-wavelength UV light. The ion pairs produced are formed in the whole air volume and the effect is known as volume ionization. It was discovered by Branley and studied more closely by Lenard. It is directly related to the external → photoelectric effect.

Bremsstrahlung, <*Bremsstrahlung*> → X-ray radiation.

Brewster angle, <*Polarisationswinkel, Brewster-Winkel*> → reflection of light.

Brewster interference, <*Brewstersche Streifen*>, interference fringes produced by multireflection between two mutually inclined plane-parallel glass plates.

Brewster's law, <*Brewstersches Gesetz*>, the law discovered by D. Brewster. Following reflection at the polarizing angle, α_p (the Brewster angle → reflection of light, → polarized light), the reflected beam and the refracted beam are perpendicular to each other. A simple relationship then exists between α_p and the refractive indexes of the two adjacent media, n_a and n_b ($n_a < n_b$):

$$\tan \alpha_p = \frac{n_b}{n_a}.$$

If $n_a = 1$ (as for air), then $\tan \alpha_p = n_b$. Thus, the Brewster angle of a medium in air is determined solely by the → refractive index of the medium.

C

Cadmium red line, <*rote Cadmiumlinie*>, the spectral line of the chemical element, cadmium (Cd), the wavelength of which, $\lambda_{Cd} = 643.84696$ nm was defined as the standard of length in 1927. Accordingly, 1 m = $1.5531641 \cdot 10^{-3}\ \lambda_{Cd}$ (λ_{Cd} in nm!) at 15°C, 760 Torr and g (the acceleration due to gravity) = 980.665 $cm \cdot s^{-2}$. However, in 1958 the international meter prototype was again adopted and in 1960 the red-orange line of ^{86}Kr replaced Cd as the → wavelength standard.

Calcite, <*Kalkspat*>, calcium carbonate ($CaCO_3$) crystallizing in the trigonal system. It is found in a particularly pure form in Iceland (Iceland spar); though it does contain $FeCO_3$, $ZnCO_3$ and $SrCO_3$ in measurable amounts. On account of its symmetry properties, calcite exhibits the phenomenon of optical double refraction (→ birefringence). The optically uniaxial crystal is used for the production of → polarizers (→ Nichol prism) which are used to produce and analyze linearly polarized light. Attempts to synthesize calcite have been unsuccessful, but single crystals of the isomorphic sodium nitrate ($NaNO_3$), also used in polarizing prisms, can be produced artificially (→ crystals, optical).

Cameron bands, <*Cameron-Banden*>, the band system of carbon monoxide between 226 and 258 nm.

Canada balsm <*Kanadabalsam*> → optical cement.

Candela, cd, <*Candela*>, the internationally defined unit of light intensity (→ photometric units). It is equal to the radiation emitted by $(1/6) \cdot 10^{-5}$ m^2 of the surface of a black-body radiator perpendicular to its surface at the temperature at which platinum solidifies at a pressure of 101,325 Nm^{-2}.

Candle-light turbidity meter, <*Kerzenlicht-Trübungsmesser*> → turbidity meter.

Carbon dioxide laser, <*Kohlendioxidlaser*> → CO_2 laser.

CARS, coherent anti-Stokes Raman spectroscopy <*CARS, kohärente Anti-Stokes-Ramanspektroskopie*>, this spectroscopy is one of those methods in which the signals are due to nonlinear optical effects. The nonlinearity of the polarization can be developed in a series of increasing powers of the electric field strength:

$$\overline{P} = \chi^{(1)}\vec{E} + \chi^{(2)}\vec{E}^2 + \chi^{(3)}\vec{E}^3 + \ldots$$

In this equation, the expressions $\chi^{(i)}$ ($i > 1$) are the higher order electric polarizability tensors which link the dyadic products of field strength, $\vec{E}$, with the polarization. The second order polarizability term describes the optical effects caused by two electric field vectors and is responsible for → frequency doubling. The third order polarizability term couples three electric field vectors. When two electromagnetic waves (two lasers) with frequencies ω_L and ω_S are combined, this term produces polarizations in the medium which oscillate with frequencies $3\omega_L$, $3\omega_S$, $(2\omega_L + \omega_S)$, $(2\omega_S + \omega_L)$, $(2\omega_L - \omega_S)$, $(2\omega_S - \omega_L)$, ω_L and

ω_S. While the first four frequencies can provide coherent radiation in the far UV, the remaining frequencies form the basis of CARS, coherent Stokes Raman spectroscopy (CSRS), the inverse Raman effect and the Raman gain effect.

The contribution of the third order polarizability

$$\chi^{(3)}_{\text{CARS}} \sim N \cdot \left(\frac{\partial \alpha}{\partial q}\right)^2 \frac{1}{\omega_V^2 - (\omega_L - \omega_S)^2 - i\Gamma},$$

to the CARS effect is a maximum when $(\omega_L - \omega_S)$ coincides with a vibrational frequency ω_V of the molecule. (Γ is a measure of the mean lifetime of the state, $\omega_L - \omega_S$, $i = \sqrt{(-1)}$). The intensity, I, of the resulting radiation of frequency $2\omega_L - \omega_S$ is proportional to the square of the polarizability contribution and, thus, to the square of the number of particles, N. Furthermore, there is a proportionality to $I^2(\omega_L)$ and $I(\omega_S)$. Because of these relations, a photon of frequency ω_L and a photon of frequency ω_S are required to produce a vibrational excitation which, together with a second photon of frequency ω_L, gives the CARS scattering signal of frequency, $2\omega_L - \omega_S$. Figure 1 shows the states involved. Although the CARS frequency is $\omega_A = 2\omega_L - \omega_S$, the associated → wave vector, $\vec{K}$, is not necessarily equal to $2\vec{K}_L - \vec{K}_S$ because of the dispersion of the medium. The correct angle between the crossing laser beams for matching their wave vectors can be calculated using the cosine equation:

$$\cos\theta = \frac{4n_L^2\omega_L^2 + n_S^2\omega_S^2 - n_A\omega_A^2}{4n_L n_S \omega_L \omega_S},$$

n_L, n_S and n_A are the refractive indexes at the frequencies in question. When recording coherent anti-Stokes spectra, the angle, θ, at which the laser beams cross must not remain constant. On account of the dispersion

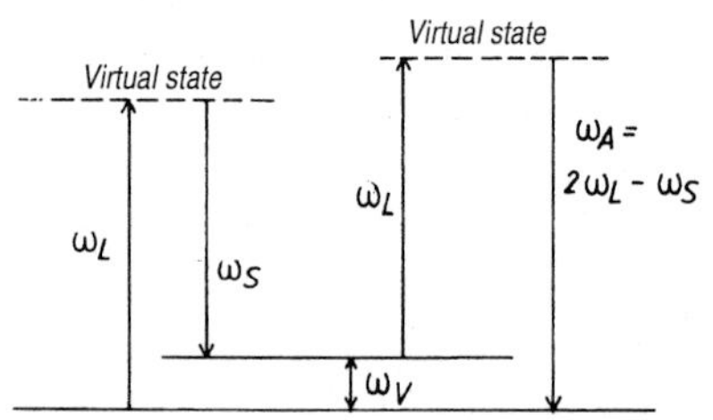

CARS. Fig. 1. The states associated with CARS

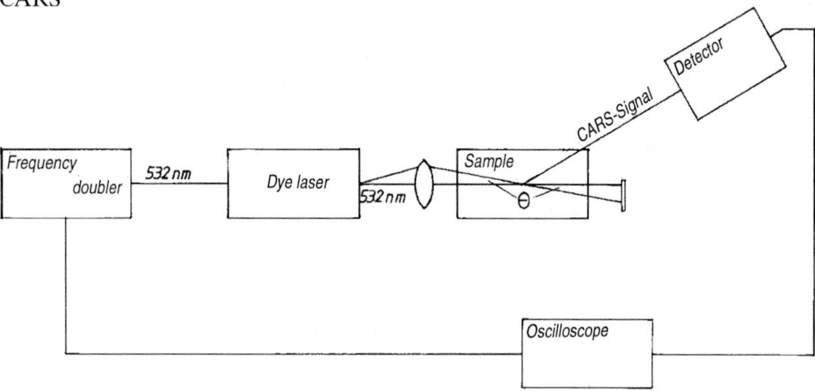

CARS. Fig. 2. Simplified schematic of a CARS apparatus

of the substance under investigation, it must be adjusted during the measurement of the spectrum.

Since the refractive indexes of gases change more slowly with wavelength than those of liquids and solids, CARS is predominantly used in the investigation of gases. In recent years, CARS has been especially used for studying the combustion processes in engines.

Figure 2 shows a CARS apparatus schematically.

Ref.: R.J.H. Clark, R.E. Hester, Eds.: *Advances in Non-Linear Spectroscopy,* Advances in Spectroscopy, vol. 15, J. Wiley and Sons, Chichester, New York, Brisbane, Toronto, Singapore, **1988**.

Cassegrainian objective, <*Cassegrain-Objektiv*>, a mirror objective especially used in microspectroscopy (→ microscope spectrophotometer). The numerical aperture (NA) for this objective is defined as the product of the refractive index of the material between the objective and sample (usually air, $n \cong 1$) and the sine of the maximum aperture angle, α, of the objective:

$$\mathrm{NA} = n \sin\alpha.$$

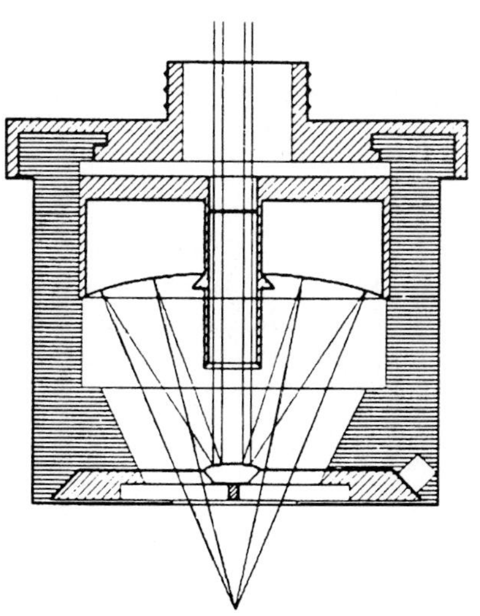

Cassegrainian objective. Fig. 1. Cassegrainian objective

Figure 1 shows a cross section through a Cassegrainian objective. Examples of applications of IR microspectroscopy can be found in: *The Design, Sample Handling and Applications of Infrared Microscopes* (Ed.: P.B. Roush), (*ASTM STP 949*), American Society for Testing and Materials, Philadelphia, **1987**.

Cavity, <*Cavity*>, a word with different but related meanings in optics and spectroscopy.

In → ESR spectroscopy, it forms part of a cavity resonator into which the sample tube is inserted.

In laser technology, the resonator is termed a laser cavity in a similar way.

In → interference filters, the three elements, mirror – dielectric layer – mirror, are together called a cavity.

CD instrument, <*CD-Gerät*>, an instrument for measuring the differential absorption of → circularly polarized light (CPL) as a function of wavelength. Since the optical device for producing CPL, a → quarter-wave plate, produces pure CPL at one wavelength only and elliptically polarized light elsewhere, no really satisfactory CD instrument was constructed until the early '60s when CD spectrometers using a Pockels cell (→ Pockels effect), the optical properties of which can be varied with wavelength, became commercially available. In the late sixties, following the work of Billardon and Badoz, the Pockels cell was replaced by the →

photoelastic modulator (PEM), a much more robust optical device, which enabled the technique of CD spectroscopy to be extended into the near and later the mid infrared regions. When used in a CD instrument, the PEM is driven in such a way that it produces left-circularly polarized light (LCP) at one extreme of the voltage cycle and RCP at the other. Thus, when the light emerging from a circularly dichroic sample is detected by a photomultiplier (PM), a signal with an AC component is produced and this can be selectively amplified by a lock-in amplifier taking its reference signal from the PEM. This is a very sensitive detection mechanism which accounts for the quality of the CD spectra which are obtained in spite of the inherent weakness of the signal in most cases. The CD measurement is a single-beam measurement, so to correct for the changing optical properties of the instrument over the spectral range, e.g. lamp intensity, transmission of the optics, detector sensitivity etc., the DC signal emerging from the PM is monitored and held constant by means of the a feedback loop which controls the dynode voltage supplied to the PM. For the UV-VIS region, the source is usually a → xenon lamp.

Originally, CD spectroscopy was applied in the UV-VIS region only, but during the last 20 years notable advances have been made in extending the measurements into the near and mid infrared. For these regions PEMs of more complex construction, with zinc selenide as the optical element, are used. The signal in the IR is inherently much weaker than in the UV-VIS and the measurement of IR CD spectra is much more difficult.

By making use of synchrotron radiation, CD spectroscopy has been extended into the vacuum UV, using calcium fluoride as the PEM material, mostly in connection with the measurement of magnetically induced circular dichroism, → magnetic circular dichroism.

Centrifugal term, *<Zentrifugalglied, Dehnungsglied>*. The → dumbbell model is used to describe the rotation of a simple linear molecule, especially in the simplest case of a diatomic molecule. With a fixed bond length, this model represents the rigid rotator and its rotational energy is given quantum-mechanically as:

$$E_{\text{rot}} = \frac{h^2}{8\pi^2 I_B} J(J+1) = BhcJ(J+1)$$

where the → rotational constant, $B = h/8\pi^2 I_B c$ in cm^{-1}, I_B is the moment of inertia with respect to the axis of rotation, $I_B = \mu r^2$ (μ is the → reduced mass), and J is the rotational quantum number, $J = 0, 1, 2, 3$ ... By means of the rotational energy levels, $F(J) = BJ(J+1)$, and the selection rule, $\Delta J = \pm 1$, the → rotational spectra can be explained; only excitation need be considered. A series of equidistant rotational lines with a separation of $2B$ results. Experiment shows that the lines close together with increasing J value, i.e. this simple representation of a rotational energy level requires correction which is achieved by adding a second term: $F(J) = BJ(J+1) - DJ^2(J+1)^2$. The constant, D, allows for the fact that the rotator is not rigid. The centrifugal forces due to the rotation lead to an elongation of the bond. $DJ^2(J+1)^2$ is therefore called the centrifugal term and D the → elongation factor.

The centrifugal term must always be considered in a complete interpretation of → pure rotation or → rotation-vibration spectra, although its influence is often neglected because $D \ll B$. For linear molecules, $D = 4B^3/\omega_o^2$ (→ pure rotation spectra of linear molecules).

Chappius band, <*Chappius-Bande*>, an ozone band of low intensity in the VIS spectral region.

Characteristic curve, <*Schwärzungskurve, charakteristische Kurve*>. When using a photographic emulsion as a radiation detector, the determination of the blackening or density of the developed and fixed emulsion is an important task. This blackening measures the whole, photometrically effective radiant energy, $\Phi \cdot t$, to which the emulsion was exposed during the time t. Here, Φ denotes the intensity of the radiation per unit area element. The blackening, D, is given by $D = \gamma \log \Phi t$. If we plot D against $\log \Phi t$ we obtain the characteristic curve which should be a straight line in an ideal case. The central part of this curve is approximately linear and should be regarded as the photometrically suitable interval, since a direct proportionality between D and Φt is found only there. The factor γ, i.e. the contrast of the plate, can be determined from the slope of the characteristic curve.
In the days when photographic plates were used as detectors, in both emission and absorption spectroscopy, their characteristic curves were of considerable importance in the quantitative evaluation of plates.
Ref.: G.R. Harrison, R.C. Lord, J.R. Loofbourow, *Practical Spectroscopy*, Prentice-Hall, Inc., New York, **1948**.

Characteristic frequency, <*charakteristische Frequenz*> → group frequency.

Character table, <*Charakterentafel*>, a table of the characters of all → symmetry species or irreducible representations of all → symmetry operations. Character tables, derived from group-theoretical considerations of the symmetry properties of bodies, are an important tool for chemists and physicists. Particularly in the vibrational spectroscopy of polyatomic molecules (→ vibrational spectrum) they facilitate the enumeration of the IR- and Raman-active vibrations and their symmetry species. A character table has the form shown in Figure 1. The → symmetry point group chosen as the example is D_{2h}, the group to which the naphthalene molecule belongs. In the heading, the name of the → symmetry group and the associated → symmetry operations are listed. Their sum gives the order of the group, N_g. The left-most column lists the irreducible representations of the symmetry group (also called → symmetry species) and classified, in general, according to a proposal by Placzek. The entries in the column under symmetry operation, s, are the characters, $\chi_s(\gamma)$ of the associated symmetry operations for each symmetry species, γ. Most character tables show, in a further column (see Figure 1), the symmetry species to which the translations, T_z,T_y,T_x, and rotations, R_z,R_y,R_x, belong. The components of the polarizability tensor, α_{ik} $(i,k = x, y, z)$, are also frequently given. This information indicates which symmetry species contain (span) the IR-active and Raman-active vibrations. Species containing the translations span the

Point group ← Symmetry operations $\Sigma = N_g$ →

Irreducible representation = Symmetry species

D_{2h}	E	C_{2z}	C_{2y}	C_{2x}	i	σ_{xy}	σ_{zx}	σ_{yz}		
A_g	1	1	1	1	1	1	1	1		α_{xx} α_{yy} α_{zz}
B_{1g}	1	1	-1	-1	1	1	-1	-1	R_z	α_{xy}
B_{2g}	1	-1	1	-1	1	-1	1	-1	R_y	α_{zx}
B_{3g}	1	-1	-1	1	1	-1	-1	1	R_x	α_{yz}
A_u	1	1	1	1	-1	-1	-1	-1		
B_{1u}	1	1	-1	-1	-1	-1	1	1	T_z	
B_{2u}	1	-1	1	-1	-1	1	-1	1	T_y	
B_{3u}	1	-1	-1	1	-1	1	1	-1	T_x	

Characters of the irreducible representations γ for each symmetry operation $i = \chi_i(\gamma)$

Character table. Fig. 1. Example of D_{2h}, the point group of naphthalene

IR-active vibrations. The components of the polarizability indicate the species which contain the Raman-active vibrations. The relations are particularly simple in the nondegenerate → symmetry point groups where the characters can take only the values +1 and –1. The symmetry groups, C_2, C_{2v}, D_2, C_{2h} and D_{2h}, are of this type.

From the total of $3N$–6 normal vibrations, the number, $\bar{n}(\gamma)$, belonging to symmetry species, γ, can be easily determined using the character table and the following equation:

$$\bar{n}(\gamma) = g\frac{1}{N_g}\sum_i N_{g,i}\chi_i(\gamma)\overline{\chi}(S).$$

g is the degree of degeneracy of the species, N_g, the order of the symmetry group, $N_{g,i}$ the number of symmetry operations in the ith class of the symmetry operations and $\chi_i(\gamma)$ is the character of the symmetry operation. $\chi(S)$ is given for the proper symmetry operations which are rotation, $C_p^{\,k}$ (= rotation by $2\pi k/p$ about a p-fold axis of rotational symmetry) and the identity ($E = C_p^{\,p}$) by:

$$\overline{\chi}(S) = \left[1 + 2\cos\frac{2\pi k}{p}\right](\alpha - 2)$$
$$= \chi(C_p^k)(\alpha - 2)$$

and for improper symmetry operations which are reflection in a mirror plane, σ, inversion, i, and the special case of rotation reflection, $S_p^{\,k}$ by:

$$\overline{\chi}(S) = \left[-1 + 2\cos\frac{2\pi k}{p}\right]\alpha = \chi(S_p^k)\cdot\alpha$$

Here, α is the number of atoms which do not change their positions under the corresponding symmetry operation. The values for naphthalene have been compiled stepwise in the following table. The first line gives the character for a basis of three Cartesian vectors placed on an atom which does not change its position under the operation at the head of the column. The total number of such atoms is given for each operation in row two. The last line gives $\chi(S)$, the characters of the reducible representation which has as its basis the 48 true vibrations of the naphthalene molecule; the rotations and translations having been removed in the formulas used to calculate $\chi(S)$. The number, $n(\gamma)$, of vibrations in each symmetry species, γ, can now be determined using the equation for $\bar{n}(\gamma)$, $\chi(S)$ and the characters, $\chi_i(\gamma)$, from the character table. The answer is; $a_g = 72/8 = 9$; $b_{1g} = 64/8 = 8$; $b_{2g} = 24/8 = 3$; $b_{3g} = 4$; $a_u = 4$; $b_{1u} = 4$, $b_{2u} = 8$; $b_{3u} = 8$; giving a total of $48 = 3N–6$. The question as to whether the vibrations of the symmetry species, γ, are Raman- and/or IR-active can also be answered using the character table. First we need to know how the components of the polarizability (Raman) and dipole moment (IR) transform.

For the Raman spectrum (polarizability) we use:

$$n_\alpha(\gamma) = g\frac{1}{N_g}\sum_i N_{g,i}\chi_i(\gamma)\chi_\alpha(S)$$

in which $\chi_\alpha(S)$ is the character for the change of the polarizability under the symmetry operation, S. This is given by:

$$\chi_\alpha(S) = 2\cos\frac{2\pi k}{p}\left[\pm 1 + 2\cos\frac{2\pi k}{p}\right].$$

The plus sign in the square brackets applies for the proper symmetry operations and the minus sign for the improper operations. We find, in agreement with Figure 1, that the polarizability tensor has three a_g components and one each from b_{1g}, b_{2g} and b_{3g}.

The corresponding equations for the infrared spectrum (dipole moment or translation) are:

$$n_M(\gamma) = g\frac{1}{N_g}\sum_i N_{g,i}\chi_i(\gamma)\chi_M(S)$$

where

$$\chi_M(S) = \pm 1 + 2\cos\frac{2\pi k}{p}$$

The translations are found to span b_{1u}, b_{2u} and b_{3u}.

In the case of infrared spectra, the excitation of a vibration from the low-

Character table. Vibrations of the naphthalene molecule

	E	C_{2z}	C_{2y}	C_{2x}		i	σ_z	σ_y	σ_x
$\chi(C_p^k)$	3	–1	–1	–1	$\chi(S_p^k)$	–3	1	1	1
α	18	0	2	0		0	18	0	2
α–2	16	–2	0	–2					
$\bar{\chi}(S) = (\alpha–2)\chi(C_p^k)$	48	2	0	2	$\bar{\chi}(S) = \alpha\chi(S_p^k)$	0	18	0	2

est level ($v = 0$) to the first excited state ($v = 1$) is allowed if the vibration belongs to the same symmetry species as a translation, T_x, T_y or T_z. Further, we can say that the infrared radiation absorbed in exciting this vibration must be polarized along the corresponding x, y or z direction in the molecule.
In the case of Raman spectroscopy, a vibrational transition from $v = 0$ to $v = 1$ is allowed if the symmetry species of the vibration is the same as that of one of the components of the polarizability tensor α_{ij} ($i, j = x, y, z$). The incident light must then be polarized along either the i or the j direction in the molecule and the scattered light will be polarized along the other.
In the most general terms, we can say that if a spectroscopic transition between the states Ψ_g and Ψ_e is to be allowed then the direct product, i.e. the product of the characters of the species to which Ψ_g and Ψ_e belong, must contain the symmetry species of the operator, P, responsible for the interaction with radiation in that particular branch of spectroscopy. In IR and UV-VIS spectroscopy, P is the dipole moment, in Raman spectroscopy it is the polarizability. The group-theoretical requirement ensures that the integral, $<\Psi_g|P|\Psi_e>$, contains a part belonging to the totally symmetric symmetry species which *may* be nonzero. An integral over a function of any other symmetry species must be zero. In addition to the application of character tables in vibrational spectroscopy described here, these tables are important for the classification of the symmetry of atomic and molecular orbitals where they can be used to establish selection rules for the electronic spectra of atoms and molecules.

Ref.: D. Schonland, *Molecular Symmetry*, D. Van Nostrand Company, London, Toronto, New York, **1965**. J.M. Hollas, *Symmetry in Molecules*, Chapman and Hall, London, **1972**. G. Davidson, *Group Theory for Chemists*, McMillan, London, **1991**.

Charge-injection device, (CID) optics, *<CID-Optik>*, a detector in spectroscopy which, in conjunction with echelle optics, permits the simultaneous measurement of 30,000 lines.

Charge-transfer complex, *<Charge-Transfer-Komplex>*, a molecular compound in which an electron is transferred from a donor to an acceptor molecule due to intermolecular interactions between the two.
The appearance of an intense absorption band, usually lying in the long-wavelength region, accompanies this complex formation. This band is called a charge-transfer band or CT band.
In general terms, this is an example of an electron-donor-acceptor interaction, and an EDA complex. Aromatic molecules are electron donors. Tetrachlorobenzoquinone is an electron acceptor.

Chemical ionization, CI, *<chemische Ionisation, CI>*, ionization in → mass spectroscopy brought about by chemical reactions. In chemical ionization, an excess ($\varrho \sim 1$ Torr) of additional gas, the reactant gas, is added to the substance to be analyzed ($\varrho \sim 10^{-6}$ Torr) in the electron-impact ionizing source of the mass spectrometer. Methane and ammonia are preferred as reactant gases. The high pressure in the ion source leads to the formation of positive and negative molecular ions of the reactant gas due to electron impact and ion-molecule reactions. In methane it is usually CH_5^+ and, in small amounts, $C_2H_5^+$ and in *iso*-butane t-$C_4H_9^+$. Generally, we speak of a CI plasma. The ionization of the substance to be analyzed is then caused by a transfer of protons, combination with molecular ions of the reactant gas or by charge exchange: e.g. $M + CH_5^+ \rightarrow (M\text{-}H)^+ + CH_4$; $M + C_2H_5^+ \rightarrow (M\text{-}C_2H_5)^+$; $M + Ar^+ \rightarrow M^+ + Ar$. The ions of the analyte produced by a proton transfer and complexation with molecular ions are called quasi-molecular ions since their mass deviates from the molecular mass. On account of the great versatility of the experimental conditions (pressure, reactant gas), it is possible to obtain a very wide range of information with CI.

Chemical kinetics, <*Reaktionskinetik*> → kinetics.

Chemical shift, <*chemische Verschiebung*>. This phenomenon, initially observed in 1950 for the nuclei ^{31}P, ^{19}F and ^{14}N and a little later in proton resonances, is of decisive importance for → NMR spectroscopy in chemistry. The resonance condition, $\nu_0 = (\gamma_I/2\pi)B_0$, implies that the resonance frequency of a specific type of nucleus is determined only by the → magnetogyric ratio, γ_I, at a constant magnetic flux density, B_0; i.e. only one resonance line is expected for each nuclear species. However, experiment shows that nuclei of the same type within one molecule give as many resonance absorptions as there are differently bound nuclei of the same species present in the molecule. In ethyl alcohol (CH_3-CH_2-OH) for example, there are three types of proton and therefore three proton-resonance signals. The difference, in Hz, between two resonance lines, $\Delta\nu$, is called the chemical shift. The reason for the dependence of the resonance frequency on molecular structure does not lie in a change of the magnetogyric ratio, γ_I, which is a nucleus-specific constant. The shift is due to the electric currents, induced in the circulating electrons when the sample is placed in the magnetic field, which results in a magnetic field opposing the applied field. The magnitude of this diamagnetic field is proportional to the applied field, B_0, which causes it and depends upon the electron density around the nucleus involved. Therefore, at the site of the nucleus a weaker local field, B_{loc}, acts, i.e. the nuclei are *shielded* from the applied field and:

$$B_{loc} = B_0 - \sigma B_0 = B_0(1 - \sigma).$$

σ is the → shielding constant, a dimensionless quantity, which can, in principle, be calculated for free atoms and very simple molecules (→ Lamb's formula). For a better theoretical interpretation, σ is usually split into three components:

$$\sigma = \sigma_{dia} + \sigma_{para} + \sigma_A.$$

σ_{dia} represents the contribution of local, diamagnetic electron currents at the site of the nucleus observed. σ_{para} allows for the nonspherically symmetrical electron distribution. σ_A is that part of the shielding constant, which accounts for the → magnetic anisotropy of adjacent groups. Thus, the resonance condition must be written:

$$\nu = \frac{\gamma}{2\pi} B_0(1 - \sigma).$$

The magnetic shielding for chemically nonequivalent nuclei of the same type, such as protons at various positions within the molecule, is different and results in separate resonance signals in the nuclear magnetic resonance spectrum. The absolute value of σ for protons is small, of the order of 10^{-5}. It is larger for heavier nuclei because the shielding is larger due to the greater number of electrons around the nucleus.

The above, modified resonance condition shows that the chemical shift, $\Delta\nu$, defined as the difference in the positions of two resonance lines, also depends upon the applied magnetic field, B_0. → NMR spectra measured with instruments having different B_0 and ν_0 cannot be compared directly. For that reason, in NMR spectroscopy we always refer to a reference signal and define a dimensionless unit, δ, as the chemical shift:

$$\delta = \frac{\nu_{\text{Sample}} - \nu_{\text{Refer.}}}{\nu_{\text{Refer.}}} \cdot 10^6$$

$$= \frac{\Delta\nu}{\text{Measurement frequency}} \cdot 10^6$$

No significant error is caused by setting ν_{Refer} equal to the average measurement frequency. The factor of 10^6 is introduced to obtain more practical numerical values and the δ values are therefore given in parts per million (ppm). However, ppm is not a dimension and is frequently not mentioned when citing δ values. By definition, the reference substance has the value, $\delta = 0$. TMS, tetramethylsilane $(CH_3)_4Si$, has proved to be particularly suitable as the proton, ^{13}C and ^{29}Si resonance standard. The following standards are used for other important nuclei: ^{11}B : $^{11}BF_3 \cdot OEt_2$; ^{15}N : $^{15}NH_4{}^{15}NO_3$ (aqueous solution); ^{17}O : $H_2{}^{17}O$; ^{19}F : $C^{19}FCl_3$; ^{31}P : $H_3{}^{31}PO_4$ (85 % as external standard). The reference substance is usually added directly to the solution containing the sample as an internal standard. However, this is not always possible, e.g. TMS is not soluble in water, or the standard may interact/react chemically with the sample. An external standard must then be used. In such cases, the reference substance is sealed into a capillary which is placed in the NMR sample tube and measured with the sample. When evaluating the spectrum it must be recognized that the magnetic flux densities at the position of the standard and sample solution are different. The chemical shift, δ, is plotted as abscissa and, by general agreement, increasing from right to left. Figure 1 shows the relationship between the magnetic flux density, B, the shielding, σ and the frequency of the resonance, ν. The data output facilities of an NMR spectrometer are calibrated in δ values so that the results determined with different instruments or at different frequencies can be compared directly. The chemical shift, δ, is important because the resonances of the differently bound nuclei can be assigned to specific and characteristic → expectation regions, as with IR spectroscopy. This makes it possible to assign the δ values to specific groups of nuclei or structural elements within a molecule. The resonances of almost all the differently bound protons lie in a narrow range of ca. 20 ppm; and more than 95 % in region of $\delta = 0$–10.

Figure 2 shows some of the characteristic regions of proton chemical shifts and Figure 3 the experimentally determined resonance ranges for ^{13}C. In comparison to the ^{1}H shift in Figure 2, we notice that ^{13}C resonances extend

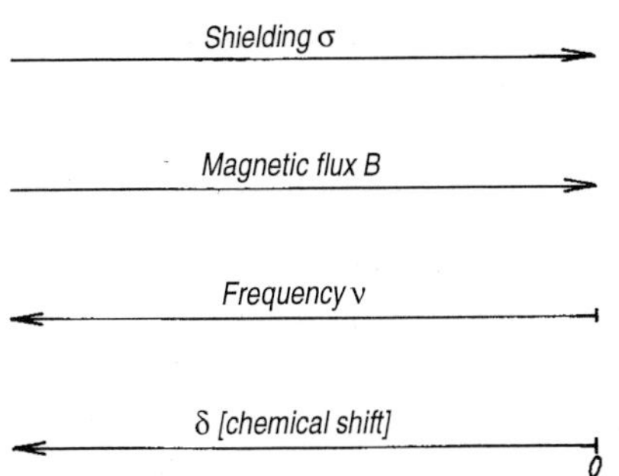

Chemical shift. Fig. 1. The relationships between B, ν, δ and σ

over a range of ca. 200 ppm, 10–20 times larger than the ^{1}H resonances. However, expressed in frequency or energy units, the factor is approximately 5. Thus, we can expect an even greater sensitivity of the chemical shift to changes in molecular structure in ^{13}C NMR spectroscopy than in proton NMR.

In general, the following chemical shift ranges (δ_{TMS}) can be given for ^{13}C nuclei:

δ = 0–100: aliphatic compounds, substituted alkynes,

δ = 100–150: olefins, aromatic and hetero-aromatic molecules, nitriles,

δ = 150–230: carbonyl and thiocarbonyl compounds.

Analogous, characteristic regions can be identified for the chemically interesting nuclei, ^{11}B, ^{15}N, ^{17}O, ^{19}F and ^{31}P. In addition to shielding by bonding electrons, other inductive influences such as → magnetic anisotropy and → ring-current effects must be taken into

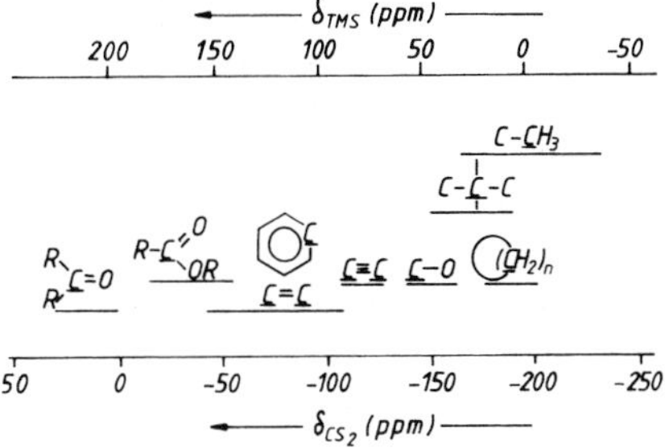

Chemical shift. Fig. 3. δ for ^{13}C

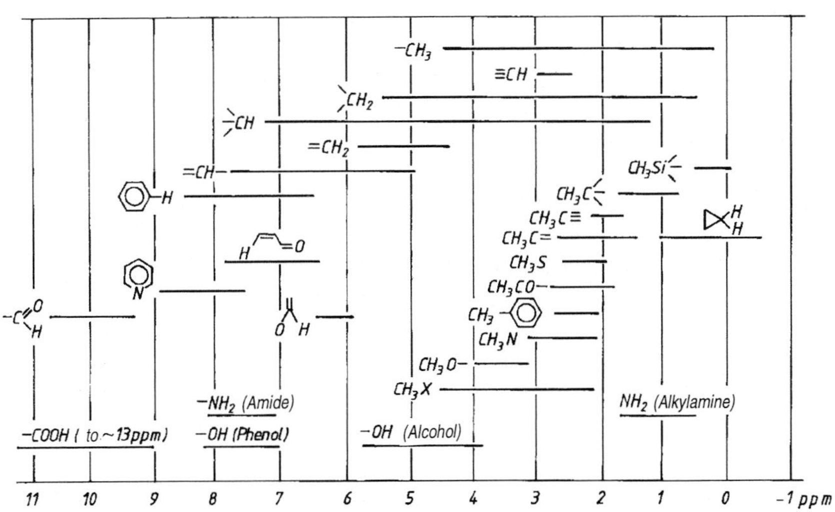

Chemical shift. Fig. 2. δ for protons

consideration when discussing chemical shifts.
The occurrence of peaks at slightly different energies, depending upon the chemical structural environment, is an important phenomenon in → Mössbauer spectroscopy and in → X-ray photoelectron spectroscopy. In both cases, the observed peak displacements are sometimes called chemical shifts, though the term is much more established in the context of NMR spectroscopy.
Ref.: R.K. Harris, *Nuclear Magnetic Resonance Spectroscopy*, Pitman, London, Marshfield **1983**. D.H. Williams, I. Fleming, *Spectroscopic Methods in Organic Chemistry*, 4th edtn. McGraw-Hill, London, New York, **1989**.

Chemiluminescence, <*Chemilumineszenz*>, the radiation of cold light due to chemical reactions. This means that the radiation emitted from a chemiluminescent substance in a specific spectral region has a higher intensity than it would if it came from a black body at the same temperature. Chemiluminescence can be observed from the UV to the IR region. The first observations date back to Radziszewski, who in 1877 noticed that light is emitted during the oxidation of simple organic compounds with oxygen in an alkaline solution. In the visible region (400–700 nm), the emission of luminescence requires an excitation energy of 14,000–25,000 cm^{-1} which corresponds to ~ 160–300 K J/mol^{-1}. Therefore, in chemiluminescence only those reactions need be considered in which correspondingly high reaction enthalpies are released; and in one single step. In most cases, chemiluminescence arises from an excited singlet state so that the emission of the excited molecule is identical with its fluorescence. A knowledge of the structure of the emitting species and the → quantum yield are required for the clarification of the mechanism and the kinetics of chemiluminescent reactions. By this is meant the number of photons formed per molecule of substrate converted in the chemiluminescent reaction. The quantum yield is very small in chemiluminescence, e.g. 0.05 for luminol. In → bioluminescence, it can be nearly 1. Organic compounds which show chemiluminescence under suitable reaction conditions include, carbonic-acid chlorides, anhydrides, esters and nitrides, tetrakis(dimethylamino)ethylene (TDE), tetracyanoethylene, luminol (3-aminophthalhydrazide), lucigenin (N,N'-bismethyl acridiniumdinitrate), lophine (2,4,5-triphenylimidazole), pyrrole, indole and carbazole derivatives. An alkaline peroxide solution provides suitable reaction conditions, i.e. these are oxidation reactions. For TDE, an autooxidation reaction with atmospheric oxygen occurs in water. The reaction enthalpy released is frequently transferred to another molecule (e.g. 9,10-diphenylanthracene) the fluorescence of which is then observed. This other molecule does not participate in the actual reaction. The autooxidation of hydrocarbons is accompanied by a chemiluminescence reaction.
Ref.: R.P. Wayne, *Principles and Applications of Photochemistry*, Oxford University Press, Oxford, New York, **1988**.

Chemiluminescence indicator, <*Chemilumineszenz-Indikatoren*> → indicator.

Chiroptical methods, *<chiroptische Methoden>*, spectroscopic methods which give opposite signs for the results of measurements on two enantiomers. These are the methods based upon the natural, optical activity of chiral molecules; → optical rotary dispersion (ORD), → circular dichroism (CD) and, primarily, → optical rotation (OR).

Chopper, *<Chopper, Zerhacker>*, an instrument for chopping a light beam. A chopper is constructed like a → rotating sector but, unlike the latter, its purpose is not to attenuate the light. It is used instead to modulate a continuous beam of light by generating alternating light/dark phases. A mechanical chopper can produce modulated frequencies from several Hz up to 3000 Hz.
Figure 1 shows a commercially available chopper. Using three chopper blades, the overlapping ranges 4 Hz – 200 Hz, 20 Hz – 1 KHz and 60 Hz – 3 KHz can be selected.

Chromatic aberration, *<chromatische Aberration>*, results because a lens material shows a normal → dispersion, i.e. the refractive index (→ refraction of light) increases with decreasing → wavelength. Since the focal length of a lens depends upon the refractive index (it is inversely proportional to $[n - 1]$), the convergence of the emerging light rays is slightly different for each wavelength and neither image position nor image size are coincident for different wavelengths. A lens may be achromatized by combining a convex and a concave lens of different materials with different dispersions ($dn/d\lambda$). However, such an achromat can have the same focal length for only two wavelengths, e.g. for the Fraunhofer *C* and *F* lines (656.3 nm and 486.1 nm respectively). The focal length depends upon the wavelength for all other wavelengths (secondary chromatic aberration) although the variation is generally considerably less than for simple lenses. By combining three lenses of different materials with different dispersions (an apochromat), three wavelengths can be brought to the same focal length and the chromatic aberration curve (focal length position as a

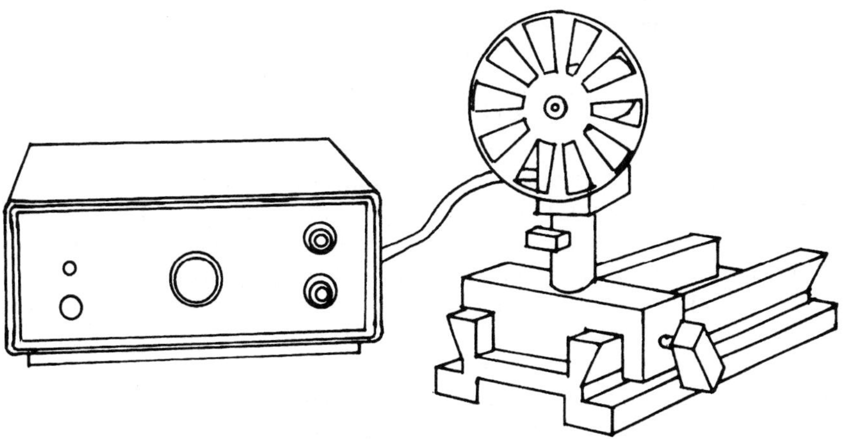

Chopper. Fig. 1. Chopper model 230 from H.M. Strasser,

function of λ) flattened even further. Such apochromats are rarely used for photometric measurements in practice. For the visible region, crown glass and flint glass are used as materials with different dispersions; for the UV, quartz and fluorite. Instead of fluorite, which is expensive and difficult to obtain in large pieces, artificially grown lithium fluoride crystals have been used recently. Achromats made of CaF_2 and LiF have proved valuable for the far UV ($\lambda < 180$ nm). Quartz-water achromats have been used occasionally. Their disadvantage is the large temperature dependence of the refractive index of liquids. In the IR region, lenses are generally replaced by mirrors which are always achromatic.

Chromophore, <*Chromophor, chromophores System*>, an atomic grouping capable of absorbing light above 180 nm, i.e. below 55,500 cm^{-1}. All unsaturated groups which contain π electrons belong to the classical chromophoric systems. These are groups with double or triple bonds which have low-lying π^* states so that excitation is possible at wavelengths above 180 nm. Typical chromophoric systems are:

>C=C<; –C≡C–; >C=N–; >C=O; >C=S

Groups containing lone electron pairs are also included in the broader meaning of the word *chromophore*, because they generally absorb light above 180–200 nm. These are, for example: -C-Br; -C-I, -C-OH; -C-SH; -C-NH_2. Because of their particular bonding, both benzene and pyridine are regarded as chromophores. The important fact for the UV-VIS spectroscopy of organic unsaturated compounds is that several chromophores can be joined together in a chromophoric system. The light absorption is then moved to longer wavelengths vis-a-vis an individual chromophoric system (→ bathochromic shift) and, in general, there is also a simultaneous intensity increase (→ hyperchromism). Examples of such systems are polyenes, poly-ynes, polyenes/ynes, acenes, phenes and pericondensed aromatic substances, diphenyl polyenes, azobenzenes, Schiff bases and quinones. Organic dyestuffs are particularly characteristic chromophoric systems.

Ref.: J.N. Murrell, *The Theory of The Electronic Spectra of Organic Molecules*, Methuen, London, **1963.**

Circular dichroism, CD, <*Circulardichroismus, CD*>, the difference between the molar decadic → extinction coefficient for left- (LCP) and right- (RCP) circularly polarized light, $\Delta\varepsilon = \varepsilon_l - \varepsilon_r$. In the region of an absorption band, an abnormal ORD curve (→ optical rotatory dispersion) occurs the form of which corresponds to the refractive index, n, of natural, not → polarized light. Since the refractive indexes of RCP and LCP light (n_r and n_l) are slightly different, the rotatory dispersion is proportional to the difference ($n_l - n_r$). Furthermore, the absorption of RCP and LCP light is different, i.e. $\varepsilon_l \neq \varepsilon_r$. The two effects, the abnormal ORD and CD which are always found together, are called the Cotton effect. ORD curves have peaks, troughs and turning points or shoulders in their abnormal regions whilst positive and negative maxima occur in CD curves.

The maximum of the CD curve coincides with the absorption maximum and approximately with the turning point of the ORD curve for a simple ORD curve of abnormal rotatory dispersion. The direction of rotation may or may not be reversed in the region of abnormal dispersion. Every optically active substance must have absorption bands with a positive Cotton effect and bands with a negative Cotton effect. Figure 1 shows the ORD and CD spectra, together with the associated UV absorption bands, for a positive and negative Cotton effect. Because of the differential absorption of LCP and RCP light, the incident linearly polarized light is elliptically polarized on emerging from the sample. The ellipticity, Ψ, can be measured directly with an ellipsometer. In analogy with the specific rotation (→ optical rotation), we define a specific ellipticity $\Psi = \Psi(c \cdot d)$ or $\Psi = 100\Psi c'd$ with $c' = g/100\ cm^3$. We also use a molar ellipticity $\Theta = \Psi \cdot M/100$ where M is the molecular weight. In the case of small ellipticities, the connection with CD is given by $\Psi = 33 \cdot \Delta A$ or $\Theta = 3300\Delta\varepsilon$ ($\Delta A = A_l - A_r$ and $\Delta\varepsilon = \varepsilon_l - \varepsilon_r$). The → chiroptical methods of ORD (normal and abnormal) and CD spectroscopy are

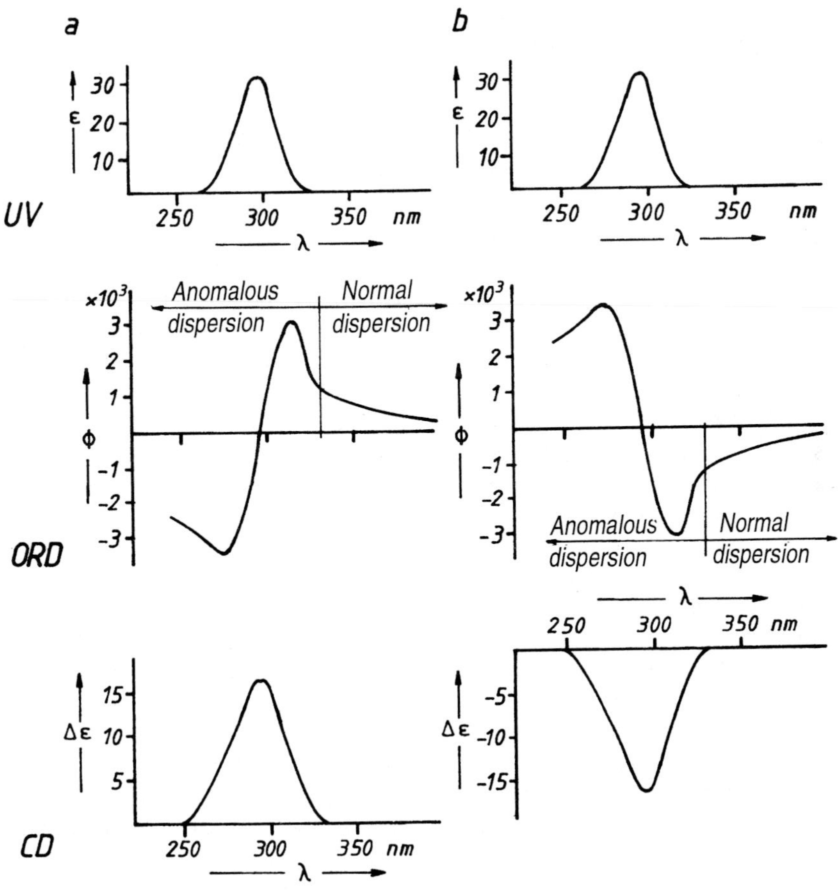

Circular dichroism. Fig. 1. Comparison of UV, ORD and CD spectra

of particular importance in the investigation of the structure of optically active compounds.
Ref.: K. Nakanishi, N. Berova and R.W. Woody (Eds.), *Circular Dichroism: Principles and Applications,* VCH Publishers, New York, Weinheim, Cambridge, **1994**.

Clar classification, Clar's nomenclature, <*Clar-Klassifizierung, Clarsche Nomenklatur*>. The first classification of the absorption bands of the aromatic hydrocarbons published by E. Clar in 1952 (see literature). The classification is based primarily upon the band intensity, but the vibrational structure and frequency shifts on changing solvent also play a role. Clar distinguished:
α-bands: weak bands having $\varepsilon = 10^2 - 10^3$ and frequently possessing a complicated vibrational structure. The first band in the spectrum is usually of this type and if it is absent it is assumed that it is obscured by more intense bands which overlie it.
p-bands: bands of moderate intensity, $\varepsilon \sim 10^4$, which usually show a very regular vibrational structure. This type of band is either the first or second in the spectrum, depending upon whether an α-band is present.
β-bands: strong bands with $\varepsilon \sim 10^5$, and with rather less vibrational structure than either of the other two types. A second, very intense β'-band is found on the short-wavelength side of the β-band in the spectra of the larger hydrocarbons.
The usual order of energy of the three band types is $\beta > p > \alpha$, but the α-band disappears under the *p*-band and then emerges between the *p*- and β-bands in the polyacene series. The ratio of the absorption frequencies, ν_β/ν_α, lies between 1.25 and 1.45. α-bands move to shorter wavelengths on lowering the temperature whereas *p*- and β-bands show a red shift. Also, α-bands show a smaller red shift on going from the gas phase to alcohol or hexane solution than do *p*- and β-bands. Clar's nomenclature is essentially an empirical one; the → Platt classification has a more theoretical basis.
Ref.: E. Clar, *Aromatische Kohlenwasserstoffe*, Springer Verlag, Berlin, Heidelberg, New York, **1952**. H.H. Jaffe and M. Orchin, *Theory and Applications of Ultraviolet Spectroscopy,* J. Wiley and Sons, Inc., New York, London, **1962**.

Coherence length, <*Kohärenzlänge, Kohärenzzeit*> → coherent light.

Coherent light, <*kohärentes Licht*>, light in which a fixed phase relationship exists between the wave trains emitted from a light source. Two wave trains are coherent if the difference of their phase factors, $\delta_2 - \delta_1$, is constant, whilst in incoherent wave trains the two phases, δ_2 and δ_1, change in a random manner. Conventional → light sources such as the → thermal radiator, → metal-vapor and → gas-discharge lamps are → incoherent light sources. This also applies to spontaneous fluorescence and generally to the → luminescence of atoms and molecules. The only coherent light sources are → lasers.
Physical optics teaches that incoherent light sources do not show the phenomenon of → interference. However, we can produce coherent wave trains if a wave train from a single origin is split by mirrors, plates, prisms, apertures, slits, gratings etc. Interference is then caused by the recombination of wave

trains resulting from the splitting. The classical mirror experiment by Fresnel in 1821 is an example of this. The concept of coherence includes the coherence length, L, and the coherence time, τ. The first concerns the maximum path difference between coherent wave trains for which interference can still be observed. It is given by $L \approx \lambda_o^2/\Delta\lambda$ or $L = c/\Delta\nu$ (c is the velocity of light and $\Delta\lambda$ or $\Delta\nu$ the spectral line width). The associated time, which the light requires to travel the coherence length, L, is the coherence time, τ: τ G $L/c \approx 1/\Delta\nu$. Both quantities depend upon the spectral line width, $\Delta\nu$ or $\Delta\lambda$. In the case of very sharp spectral lines ($\Delta\lambda \sim 10^{-3} - 10^{-4}$ nm), we obtain for λ_o, in the visible spectral region, a coherence length, L, of ca. 1 m from which a coherence time, τ, of the order of 10^{-8} results.

Ref.: B. Curnutte, J. Spangler, L. Weaver, *Theory of Radiation and Radiative Transitions*, Methods of Experimental Physics, Ed.: D. Williams, Academic Press, New York, San Francisco, London, **1976**.

Coherent light source, <*kohärente Lichtquelle*> → laser.

CO_2 laser, carbon dioxide laser, <*CO_2-Laser, Kohlendioxidlaser*>, a highly efficient molecular laser of great practical importance the emission of which lies in the IR spectral region. Figure 1 shows a simplified energy-level diagram of the CO_2 laser. Nitrogen is used as an auxiliary gas which is primarily excited to the vibrational state $\upsilon' = 1$ by electron impact in a low-pressure discharge. The required excitation energy is 2335.1 cm^{-1}. This metastable state of the N_2 molecule has a long life ($\tau > 0.1$ s) so that the upper laser level of the CO_2, which corresponds to the state $\upsilon' = 1$ of the asymmetric stretching vibration, $\nu_3 = \nu_a$ can be populated by collision with a vibrationally excited nitrogen molecule. The required excitation energy is 2349.3 cm^{-1}, i.e. slightly greater than the energy transferred during the collision. The difference ($\Delta E \sim -14$ cm^{-1}) is made up by thermal energy. The lower laser levels, unoccupied due to the excitation mechanism, are the vibrational state $\nu_1 = \nu_S$ (1388.3 cm^{-1}), i.e. the symmetric stretching vibration, and the overtone $2\nu_2$ ($\upsilon' = 2$; 1285.5 cm^{-1}) which is split due to → Fermi resonance with ν_1. ν_2 (667 cm^{-1}) is the doubly degenerate deformation vibration of the CO_2 molecule. Thus, the laser lines correspond to the differences, $\tilde{\nu}_3 - \tilde{\nu}_1$ and $\tilde{\nu}_3 - 2\tilde{\nu}_2$. The influence of the rotational states has not been taken into consideration in the energy-level diagram shown here. However, this influence cannot be neglected in practice. At T = 400 K, the rotational levels with $J \cong 20$ correspond to the population maximum. The experimentally measured laser lines are therefore found at 943.4 cm^{-1} ($\tilde{\nu}_3 - \tilde{\nu}_1$) and 1041.7 cm^{-1} ($\tilde{\nu}_3 - 2\tilde{\nu}_2$). Helium and water are added as further auxiliary gases. Their function is to increase the formation of metastable nitrogen levels and to speed the emptying of the lower laser levels. A typical laser mixture consists of 1 Torr CO_2, 1 Torr N_2, 6 Torr He and 0.1 Torr H_2O. The theoretical efficiency of the CO_2 laser is ca. 40 %. This is due to the fact that, to produce a laser photon with an energy of ~ 1000 cm^{-1}, corresponding to ~ 0.12 eV, only ~ 2400 cm^{-1}, corresponding to 0.3 eV, are required ($0.12/0.3 \cong 40\%$). In con-

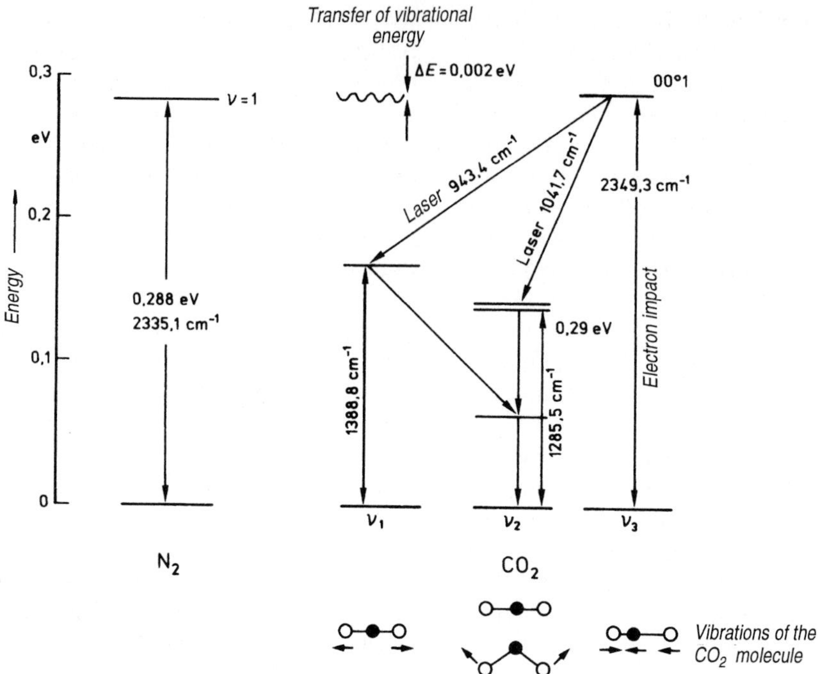

CO_2 laser. Fig. 1. Vibrational energy levels of CO_2 and laser transitions

trast, in a → noble-gas ion laser more than 36 eV must be expended to generate a photon with an energy of 2.5 eV (~ 20,000 cm^{-1} = 500 nm) which corresponds to a theoretical efficiency of ~ 7%. However, the practical efficiency of the CO_2 laser is lower than 40% and during routine operation only ca. ≥ 10%.

Resonator lengths of $L \cong 3$ m are required because of the necessity of tuning the resonator mode to the

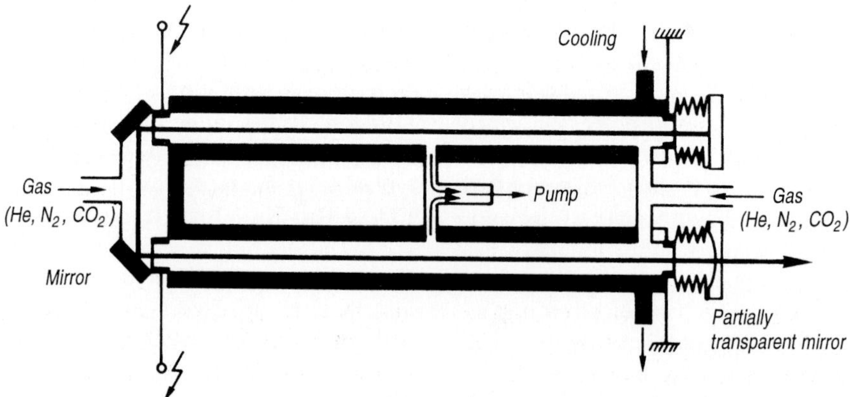

CO_2 laser. Fig. 2. Cross section of a CO_2 laser

resonator length. In order to avoid inpractically long constructions one or more resonators are used. Figure 2 shows a cross section of a CO_2 laser.
Carbon dioxide lasers are preferred in the field of material processing. Another excitation mechanism for the CO_2 laser utilizes the differing lifetimes of the laser levels (→ gas-dynamic laser).
Ref.: D.L. Andrews, *Lasers in Chemistry*, 2nd edtn., Springer Verlag, Berlin, Heidelberg, New York, **1990**.

Cold-vapor technique, <*Kaltdampftechnik*>, by far the most sensitive and reliable method for determining mercury in → atomic absorption spectroscopy. The fact that this is the only metal which has a marked vapor pressure at room temperature is utilized. A mercury salt is initially reduced to the metal with a reducing agent such as a tin(II) chloride or sodium borhydride. It is then blown out of the solution with a gas stream and measured in an absorption cell mounted in the optical path of an → atomic absorption spectrometer. If required, the sensitivity of this method can be increased by first collecting the evaporated mercury as an amalgam on a gold/platinum gauze. The collected mercury is then driven off, all at once by rapid heating, and measured. By this method it is possible to determine 0.001 μg Hg/l; a mercury concentration which is far less than that in most ultrapure chemicals.

Collector, <*Kollektor*>, in → optical spectroscopy, an optical element which has an imaging or focusing (collecting) function. In a → spectrograph or → monochromator, the collector images the spectrum produced by the → dispersing element in the focal plane (slit plane). Depending on the spectral region, glass, quartz or CaF_2 lenses were used for this purpose in spectrographs. However, they have the disadvantage that the focal length depends upon the wavelength and this can cause → image errors. Concave mirrors, which do not suffer from this disadvantage, are used in modern monochromators as collectors and → collimators, particularly in → grating monochromators. However, these mirrors must be of a high optical quality. A → concave grating in a monochromator has itself the function of a collector.

Collimator, <*Kollimator*>, in → optical spectroscopy an optical element which produces a parallel beam of light. The collimator in a → monochromator is used to irradiate the → dispersing element. Depending upon the wavelength, glass or quartz and occasionally also CaF_2 or KBr lenses were used in the classical → spectrographs. However, lenses have the disadvantage that their focal lengths depend upon the wavelength and this can cause → image errors. Following the increased use of grating instruments (→ grating monochromators), concave mirrors, which do not have this disadvantage, have been used more widely as collimators and → collectors in recent years. However, high demands are made upon the quality of the mirrors.

Collision broadening, <*Stoßverbreiterung von Spektrallinien*>, an important effect which, in addition to → Doppler broadening, contributes to the broadening of spectral lines. We differentiate between elastic and inelastic collisions when two atoms or molecules colide. No energy, or more exactly, no energy from internal degrees of freedom is transferred to the collision partner during an elastic collision; although the kinetic energy changes (translational energy is an external degree of freedom). In an inelastic collision, however, the excitation energy of an atom or a molecule, A^*, is partially or completely transferred to a collision partner B. Figure 1 shows the situation in an elastic collision. The potential curves, $V(R_{A,B})$, for the approach of collision partner B to atom A which has the energy states E_i and E_k are represented.

At distance $R_{A,B}(\infty)$, atom A is not perturbed, i.e. the vertical difference, (1) $E_k - E_i = h\nu_{i,k}$, corresponds to the frequency of the resonance line between the two states of atom A. As B approaches A more closely a collision occurs at $R_{A,B} \leq R_{co}$ where $2R_{co}$ is the collision cross section. The potential energy falls as a result of the approach of both partners and it would tend to a potential minimum if a stable collision complex were formed in the collision. The partners generally separate again after a collision time, τ_{co}. τ_{co} is given by $\tau_{co} = R_{co}/v$, where v is the relative velocity. For $v = 5 \cdot 10^2$ m s^{-1} and $R_{co} = 1$ nm we obtain $\tau_{co} = 2 \cdot 10^{-12}$. If only one radiating transition occurs between states, E_k and E_i, we have:

$$E_k(R) - E_i(R) = h\nu_{i,k},$$

where distance, $R(t)$, depends upon the time of the transition, t. If the radiating transition occurs so rapidly that R does not change during this time we have the vertical transition (2) drawn in Figure 1 which is marked only by a shift vis-a-vis $\nu_o = \nu_{i,k}(\infty)$.

In a gaseous mixture of atoms, A and B, the interatomic distance $R(A,B)$ has a statistical variation about the average value, R, which depends upon the temperature and pressure. This leads to a corresponding distribution of the frequency (wavenumber) about that of the most probable value, $\tilde{\nu}_{i,k}(R_m)$ which accounts for collision broadening, $\delta\tilde{\nu}_{co}$. The value, $\tilde{\nu}_{i,k}(R_m)$ is shifted vis-a-vis $\tilde{\nu}_o$ of the unperturbed atom, A. This shift, $\Delta\tilde{\nu}_{co} = \tilde{\nu}_o - \tilde{\nu}_{i,k}$ depends upon the relative shifts of the energy states, E_i and E_k, at interatomic distance, $R_m(A,B)$, where the emission probability has its maximum value. The intensity profile, $I(\nu)$ of the collision broadened and shifted emission line is then obtained from the equation:

$$I(\nu) = \int A_{i,k}(R)P_{co}(R) [E_i(R) - E_k(R)]\mathrm{d}R.$$

$A_{i,k}$ is the Einstein coefficient for spontaneous emission, A. It depends upon R because the electronic wave functions of the collision pair (AB) depend upon R. P_{co} is the probability per unit time that the distance, A-B, lies between R and $R + \mathrm{d}R$.

This relationship also shows that the intensity profile, $I(\nu)$, reflects the potential curves; we have:

$$E_i(R) - E_k(R) = V[A(E_i), B] - V[A(E_k), B].$$

In addition to elastic collisions, inelastic collisions, in which energy such as the electronic excitation energy of an excited atom or molecule, A^*, is transferred to collision partner, B, must also be considered. These inelastic or quenching collisions effect a decrease in the number of excited atoms or molecules and hence to the quenching of fluorescence. The total transition probability, A_i, of the depopulation of the excited state, E_k, must then be regarded as the sum of spontaneous and collision-induced transition probabilities. In this case also, the number of collisions in unit time is dependent upon the pressure of any foreign gas and the total transition probability is given by the following equation:

$$A_i^{\text{eff}} = \sum_k A_{i,k} + aP_B$$

where

$$a = \sigma_{AB} \cdot \frac{(2kT)^{3/2}}{(\pi\mu)^{1/2}}.$$

σ_{AB} is the collision cross section and μ the → reduced mass, $\mu = M_A \cdot M_B/(M_A + M_B)$.

Since A_i^{eff} is always greater than A_i we obtain from the ratio $\delta\tilde{v}_{\text{co}} = A_i^{\text{eff}}/2\pi c$ a correspondingly greater contribution to the half-width. Since this additional, collision-induced line broadening depends upon the pressure we frequently speak of the pressure broadening of spectral lines. Both the elastic and inelastic collisions cause a line broadening. In addition, in the former a shift of the line occurs which depends on the forms of the potential curves of the associated states, $E_i(R)$ and $E_k(R)$, see Figure 1.

For practical applications and particularly for IR gas spectra, it is possible to start with the number of collisions which is given by elementary gas kinetics as:

$$Z_{AB} = 3.37 \cdot 10^{10} \cdot \sigma_A \sigma_B \sqrt{\frac{1}{M_A} + \frac{1}{M_B}} \cdot \frac{P_i}{\sqrt{T}} \left[s^{-1}\right];$$

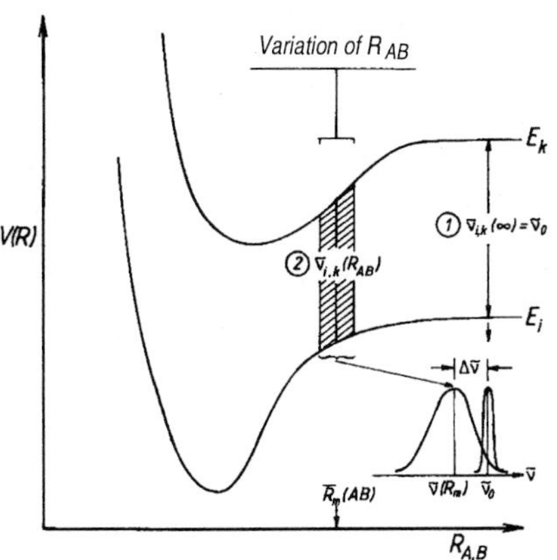

Collision broadening. Fig. 1. Potential curves for the explanation of collision broadening, see text

σ_A and σ_B are the collision cross sections of molecules, A and B, M_A and M_B their masses, T is the temperature and P_i the pressure of the component considered.

The collision broadening, $\delta\tilde{v}_{\text{co}}$ is given by:

$$\delta\tilde{v}_{\text{co}} = \frac{Z_{AB}}{c} \ [\text{cm}^{-1}]$$

(c = the speed of light).

If no foreign gas is present, i.e. $M_A = M_B$, we obtain the self-broadening, $\delta\tilde{v}_E = Z_C/c$ from the above formula. If a foreign gas is added an additional foreign-gas pressure width, $\delta\tilde{v}_F$, results.

The total line broadening of a spectral line is therefore the sum of the → natural line width, $\delta\tilde{v}_n$, the → Doppler broadening, $\delta\tilde{v}_D$ and the collision broadening, $\delta\tilde{v}_{\text{co}}$: $\delta\tilde{v} = \delta\tilde{v}_n + \delta\tilde{v}_D + \delta\tilde{v}_{\text{co}}$. The collision broadening must be divided into self-broadening, $\delta\tilde{v}_E$ and foreign gas broadening, $\delta\tilde{v}_F$. Since $\delta\tilde{v}_n \ll \delta\tilde{v}_D + \delta\tilde{v}_{\text{co}}$, $\delta\tilde{v}_D + \delta\tilde{v}_E + \delta\tilde{v}_F$ determine the line width. $\delta\tilde{v}_D + \delta\tilde{v}_E$ are taken together as $\delta\tilde{v}_o$, the line width without foreign gas. The foreign gas-broadening is experimentally accessible: $\delta\tilde{\nu}_F = [A_F/A_o)^2 - 1]\delta\tilde{\nu}_o$. A_F and A_o are respectively the measured absorbances with and without the foreign gas. The → Doppler broadening, $\delta\tilde{\nu}_D$, can be calculated using the equation, $\delta\tilde{\nu}_D = 7.16 \cdot 10^{-7} \cdot \tilde{\nu}_o \cdot \sqrt{(T/M)}$ [cm^{-1}] and the self-broadening, $\delta\tilde{\nu}_E$, using the collision number (see above). Measurements on ethylene bands near $\tilde{\nu}_o = 950\ cm^{-1}$ at 300 K and $P_A = 10$ Torr gave the following data: $\delta\tilde{\nu}_D = 0.0035\ cm^{-1}$; $\delta\tilde{\nu}_E = 0.0020\ cm^{-1}$, $\delta\tilde{\nu}_o = \delta\tilde{\nu}_D + \delta\tilde{\nu}_E = 0.0055\ cm^{-1}$. Foreign gas broadening, $\delta\tilde{\nu}_F = 0.0113\ cm^{-1}$, is found for $P_F = 50$ Torr of N_2. Detailed investigations of the effect of pressure broadening on gases in the IR region have given a result of importance for quantitative analysis. In the region, where pressure broadening begins the → Bouguer-Lambert-Beer law breaks down and must be replaced by a power law, in particular a square-root dependence. After the disappearance of the rotational structure, the Bouguer-Lambert-Beer law becomes valid again, i.e. quantitative measurements are possible under these conditions. If the values of $\delta\tilde{\nu}_F$ are known, we can calculate the collision cross sections, σ_F, of the foreign gases vis-a-vis the gas being measured using the equation for the collision number. Thus, investigations of the collision broadening of spectral lines are also important for gas kinetics.

Ref.: W. Demtröder, *Laser Spectroscopy*, Springer Series in Chemical Physics, Springer Verlag, Berlin, Heidelberg, New York, **1981**, vol. 5, p. 89 ff; C.H. Townes, A.L. Schawlow, *Microwave Spectroscopy*, McGraw-Hill Book Co. Inc., New York, Toronto, London **1955**, chapt. 13. H. Luther, R. Germershausen, *Ber. Bunsenges. physikal. Chemie*, **1963**, *67*, 571, where further literature references can be found.

Colorants, <*Farbmittel*>, in → color theory the broad general concept which brings together soluble → dyes and insoluble → pigments as substances producing a specific color or impression of color.

Color curve, <*Farbkurve*>. The → Bouguer-Lambert-Beer law states:

$$\tilde{A}_{\tilde{\nu}} = \log\left(\frac{I_o}{I_{\tilde{\nu}}}\right) = \varepsilon_{\tilde{\nu}} c \cdot d.$$

$A_{\tilde{\nu}}$ is the absorbance at wavenumber, $\tilde{\nu}$, I_o and $I_{\tilde{\nu}}$ the light intensity before and after absorption, $\varepsilon_{\tilde{\nu}}$ the molar decadic extinction coefficient in $l \cdot mol^{-1} \cdot cm^{-1}$, c the concentration in $mol \cdot l^{-1}$ and d the sample path length in cm. The graph of log ε as a function of wavenumber, $\tilde{\nu}$, or of wavelength, λ, is the UV-VIS absorption spectrum. However, concentration, c, and path length, d, must be known to calculate ε and c is frequently unknown. In these circumstances, a graph of log A, against the wavenumber or wavelength is useful; we have:

$$\log A = \log \varepsilon + \log\ (c \cdot d).$$

Such a plot then corresponds to the absorption spectrum with a constant shift of the ordinate by log $(c \cdot d)$. It is called a typical color curve. → Reflectance spectroscopy is used to obtain the color curves of insoluble solids.

Color disk, <*Farbkreisel*>, an arrangement of the spectral colors (→

spectrum, colors of) on a disk introduced by Newton in 1666 and improved by Maxwell. If the disk is rotated rapidly additive → color mixing produces a white shade. A color disk produces the same effect if only the complementary colors yellow and blue or red and green are applied to it.

Colored glass filter, *<Farbglasfilter, optisches Glasfilter>*, a colored glass characterized by its selective absorption of light in the UV-VIS and NIR region. These filters are available from a variety of sources. As an illustration of their range and properties, the products of the manufacturer Schott are described. Schott classify their glass filters into the following groups:

- UG black and blue filters with UV transmission,
- BG blue, blue-green and band filters,
- VG green filters,
- GG yellow filters,
- OG orange filters,
- RG red and black filters with IR transmission,
- NG neutral filters,
- WG almost colorless filters with varying positions of the UV absorption edge,
- KG almost colorless filters with absorption in the IR region (cold-light or heat-protection filters),
- FG blue and brown colored filters.

Numbers are added to the two-letter group reference, e.g. BC 12, RG 610. One- and two-digit numbers generally refer to the chronological order of development, three- and four-digit numbers give the spectral position of the absorption edge in nm. Glass filters can be classified into three principal groups: simple filters, ion-colored filters and tempered filters. Simple filters are colorless, technical and optical glasses with different positions of their absorption edge in the UV. The position of this edge depends upon the composition of the glass.

Ion-colored filters owe their color to the absorption of genuinely dissolved metal ions with relatively broad absorption bands in the UV-VIS region. Copper, chromium, manganese, iron, cobalt, vanadium, titanium, neodymium and praseodymium are the preferred metals. Usually, the metal oxide is added to the molten glass base, not unlike the method used to make colored glass in ancient times. The ions of the above metals give the following colors:

Copper	Cu^{2+},	weak blue
Chromium	Cr^{3+},	green
	Cr^{6+},	yellow
Manganese	Mn^{3+},	violet
Iron	Fe^{3+},	yellow-brown
	Fe^{2+},	blue-green
Cobalt	Co^{2+},	intense blue; in borate glass pink; in phosphate glass violet
	Co^{3+},	green
Nickel	Ni^{2+},	depending on the glass base gray-brown, yellow, green, blue to violet
Vanadium	V^{3+},	green in silicate glass, brown in borate glass
Titanium	Ti^{3+},	violet (reducing melt)
Neodymium	Nd^{3+},	red-violet
Praseodymium	Pr^{3+},	weak green.

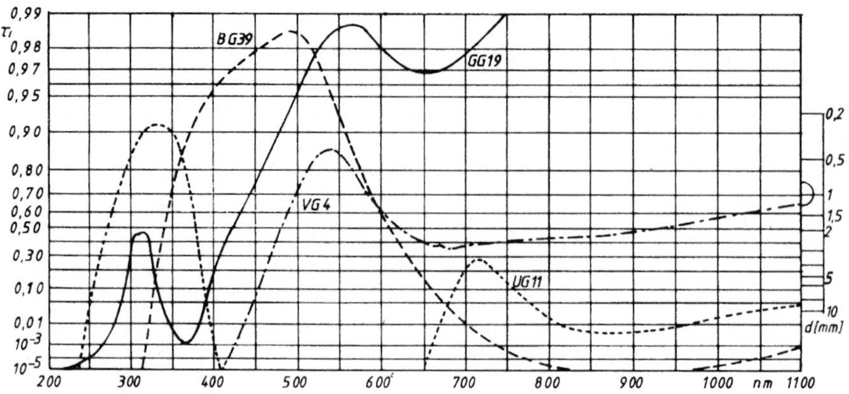

Colored glass filters. Fig. 1. Transmission curves for the colored glass filters UG 11, BG 39, VG 4, GG 19.

The table shows that the color depends upon the glass base and the oxidation state of the metal ions. The color also depends upon the metal-ion concentration and, in the case of several metals, on the proportion of the colored metal ions. Figure 1 shows typical transmission curves for the glass filters UG 11, BG 39, VG 4 and GG 19. Colored filters containing the rare-earth ions, Nd^{3+} and Pr^{3+} have much narrower absorption bands than the examples shown (→ didymium-glass filters). The transmission versus wavelength, λ, curves have been plotted in double-logarithmic form (→ diabatic scale), and the regions of very small and large transmission values are correspondingly spread out.

→ Schott glass filters usually obtain their color following heat treatment of the, initially almost colorless, glass. The yellow, orange and red sulfide and selenide filters, with very steep absorption edges (→ edge filters), are particularly important filters in this group.

Ref.: H.G. Pfaender, *Schott Guide to Glass*, Van Nostrand Reinhold Co., New York, **1983**; Schott Glaswerke catalogue: *Optical Glass Filters*.

Color error, <*Farbfehler*>, the undesired dispersion of white light into a spectrum on traversing an optical element, e.g. a lens or a prism. Optical components can be designed so as to correct or reduce this problem (→ achromatic prism, → achromaticity).

Colorimeter, <*Kolorimeter*>, a colorimeter using the principles of visual → colorimetry, which consists of two cuvettes with variable path length which can be set within 0.1 mm by means of a micrometer screw and a vernier scale. Figure 1 shows the simplest, single-stage colorimeter.

Colorimetry, visual, <*visuelle Kolorimetrie*>, the comparison of the light density of two adjacent radiating surfaces with the same spectral composition of the test light. The eye is used as the evaluating detector. If the eye recognizes the same light density in two light beams of the same intensity

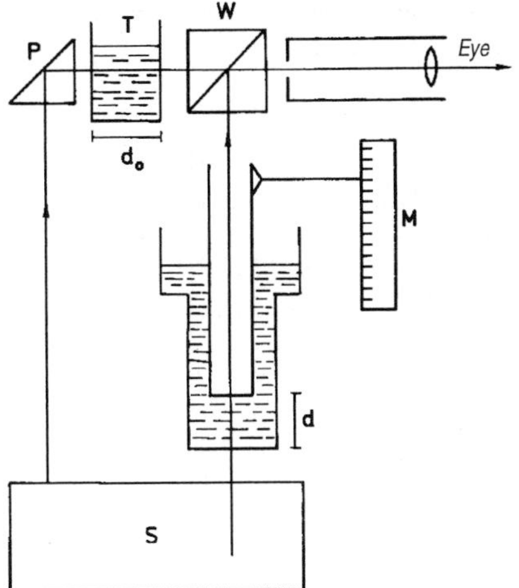

Colorimeter. Fig. 1. Duboseq's colorimeter

and spectral composition, which have traversed two solutions of the same colored substance having different concentrations, c, and different path lengths, d, then both solutions have the same average → absorbance and consequently:

$$A_1 = \bar{\varepsilon} \cdot c_1 \cdot d_1 = A_2 = \bar{\varepsilon} \cdot c_2 \cdot d_2.$$

The basic equation of colorimetry results:

$$c_2 = c_1 \frac{d_1}{d_2}.$$

If the concentration of one solution is known, the other unknown concentration can be calculated from the path length ratio at which equal light density is observed. Prior to the introduction of photoelectric detectors, colorimetry was a widely used method. It assumes that the → Bouguer-Lambert-Beer law holds.

Ref.: A.B. Calder, *Photometric Methods of Analysis*, Adam Hilger Ltd., London, **1969**, chapt. 5.

Color measurement, <*Farbmessung*>, the measurement of the three normalized tristimulus values, X, Y and Z, which characterize a color (→ color theory).

We are usually concerned with the assessment of → pigments which are perceived as reflected light after irradiation of the pigment with white light. For that reason, we use a reflectance spectrophotometer, which records the reflectance, R_∞ of the sample as a function of wavelength, λ, or wavenumber, $\tilde{\nu}$, for color measurement (→ reflectance spectroscopy, → integrating sphere). The normalized tristimulus values, X, Y and Z, are given by:

$$X = k \int_0^\infty S_\lambda \bar{x}_\lambda \cdot R_{\infty,\lambda} \mathrm{d}\lambda$$

or

$$X = k \cdot \sum_{\lambda=380}^{770} S_\lambda \bar{x}_\lambda \cdot R_{\infty,\lambda} \Delta\lambda;$$

$$Y = k \int_0^\infty S_\lambda \bar{y}_\lambda \cdot R_{\infty,\lambda} \mathrm{d}\lambda$$

or

$$Y = k \cdot \sum_{\lambda=380}^{770} S_\lambda \bar{y}_\lambda \cdot R_{\infty,\lambda} \Delta\lambda;$$

$$Z = k \int_0^\infty S_\lambda \bar{z}_\lambda \cdot R_{\infty,\lambda} \mathrm{d}\lambda$$

or

$$Z = k \cdot \sum_{\lambda=380}^{770} S_\lambda \bar{z}_\lambda \cdot R_{\infty,\lambda} \Delta\lambda.$$

S_λ is the spectral intensity distribution of the light source. $\bar{x}_\lambda$, $\bar{y}_\lambda$ and $\bar{z}_\lambda$ are the normalized spectral value curves or color-matching functions, (see Figure

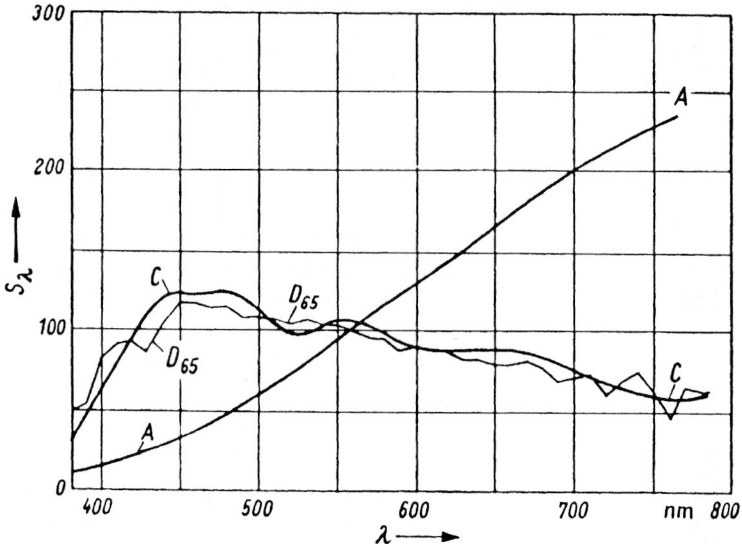

Color measurement. Fig. 1. The spectral intensity distribution of standard CIE illuminants

3 under → color theory). The product, $S_\lambda \cdot R_{\infty\lambda} = \Psi_\lambda$ is the color-stimulus function of a pigment and depends upon the light source selected, i.e. upon S_λ. For technical applications, the standard illuminants, *A, C* and *D*65 have been agreed internationally and their spectral intensity distributions are given in Figure 1. *k* is a normalizing factor which is defined by:

$$k = \frac{100}{\sum_{380}^{770} S_\lambda y_\lambda \Delta\lambda}.$$

Thus, the normalized color value, *Y*, gives the percentage reflectance relative to an ideally reflecting standard under the same illumination directly.

For practical applications, we use the sum in the expression above instead of the integral. Since tables of the values of $S_\lambda \bar{x}_\lambda$, $S_\lambda \bar{y}_\lambda$ and $S_\lambda \bar{z}_\lambda$ in increments of 5 or 10 nm are available for the common standard light sources, the sum can be calculated from the reflectance measurements and the normalized color values, *X, Y* and *Z*, obtained (Judd and Wyszecki, see literature, give examples.)

Such measurements are of particular importance to the technical problem of color matching and the establishment of color recipes. The measurement of reflectance spectra using a sequentially recording reflectance spectrometer is generally very time-consuming. Furthermore, the surface available for fixing the sample in the → integrating sphere is limited and this curtails the applications. The introduction of fiber optics improved matters considerably because the integrating sphere can now be placed outside the instrument and the sample surfaces can be scanned. Attachments for commercial double-beam spectrometers are available. However, this did not eliminate the disadvantage of the sequential measurement of pigment and white standard.

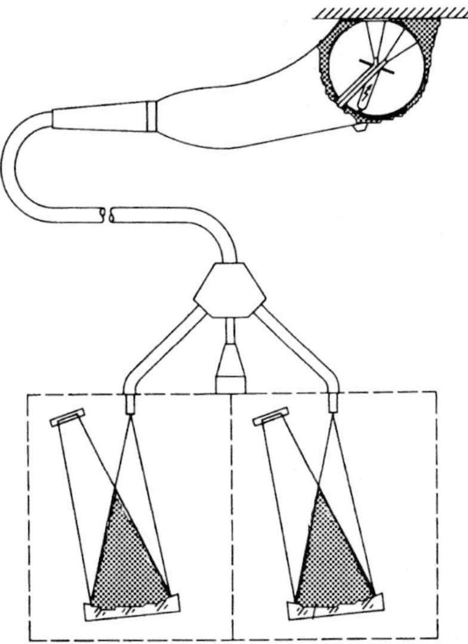

Color measurement. Fig. 2. Color meter; the Zeiss instrument, MCS2 × 512

The MCS 2X512 Zeiss color meter is a considerable advance which allows the simultaneous measurement of complete reflectance spectra with subsequent mathematical evaluation according DIN 5033 or DIN 6174 in 0.3 s. Figure 2 shows the setup of this double-beam instrument with an integrating sphere. Two identical spectrometer units (→ multichannel spectrometer, MCS 512) are connected via two flexible light pipes (Φ = 10 mm, 2–5 m long) to the measuring unit which has roughly the size and shape of a hand-held shower head. It contains the integrating sphere (Φ = 60 mm, covered with a white standard) which contains a flashbulb. One light pipe carries the light reflected by the sample during each flash whilst the other conducts the light from the white internal wall of the sphere and directs it to the corresponding spectrometer unit. This is a double-beam method in which differing radiative powers and spectral distributions are eliminated. If the instrument has been calibrated using a white standard we can calculate the spectrum for each sample, relate it to the white standard and determine the spectral reflectance, $R_{\infty\lambda}$. The whole instrument is computer-controlled. The advantages of this instrument are principally its flexible, convenient measuring head which permits the selection of test site and the measurement of the color of large sample surfaces. The time required for a spectrum (a few milliseconds) eliminates practically any movement of the sample.

Ref.: D.B. Judd, G. Wyszecki, *Color in Business, Science and Industry*, J. Wiley & Sons, New York, London, Sydney, Toronto, **1975**; G. Kortüm, *Reflectance Spectroscopy*, Springer Verlag, Berlin, Heidelberg, New York, **1969**, chapt. VIIg; H. Schlemmer, M. Mächler, *Color Measurements with Diode Array Spectrometers*, die Farbe, **1985/86**, *32/33*, 69. K. McLaren, *The Color Science of Dyes and Pigments*, Adam Hilger, Bristol and Boston, **1986**.

Color mixing, additive, <*Farbmischung, additive Farbmischung*>, the mixing of the primary colors. Additive color mixing can be demonstrated with a simple experiment. We overlap three circular spots on a white screen using three projectors, one having been fitted with a blue, one with a green and one with a red filter. We see where the green and red spots overlap the color yellow, where red and blue overlap the color purple, where blue and green overlap the color blue-green and where all colors overlap the color white. If we change the relative intensities of the red and green lights from zero to a given maximum we can produce not only yellow but also all shades lying between red and green, i.e. from pure green through yellowy-green, a rather pure yellow and orange to pure red. The same applies if the intensities of the green and blue or blue and red lights are varied. Figure 1 shows the relationships in a schematic form. Two color stimuli are added in this experiment which is therefore described as additive color mixing. It is based on the fact that the radiation from two (or three) projectors is transmitted simultaneously to the same site on the retina after reflection by the screen. The same effect can be produced if the light is transmitted not simultaneously but periodically to the same site on the retina by using, for example, a → color disk. At a frequency > 25 Hz, the eye can no longer resolve the individual color stimuli and registers a constant impression of uniform color. In a third variant of additive color mixing, the eye is presented with adjacent spots of different colors. These must be so small and so close together that the eye can no longer resolve them. This gives the impression of a uniformly colored surface,

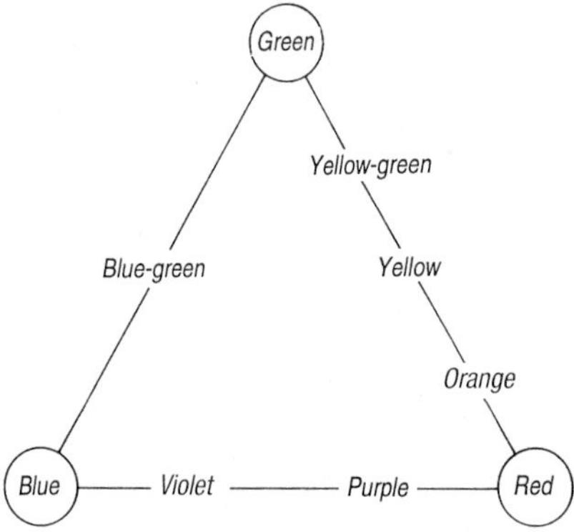

Color mixing, additive. Fig. 1. The mixing of three colored lights, see text

the color of which is the additive mixed color of the unresolved surface elements. Here, closely adjacent sites on the retina, rather than the same site, are struck simultaneously by different radiation. This implementation of additive color mixing is used in color printing and color television.
Ref.: W.D. Wright, *The Measurement of Colour*, 4th ed., Adam Hilger Ltd., London **1969**.

Color mixing, subtractive, <*Farbmischung, subtraktive Farbmischung.*>. Radiation (light) passing through a medium is generally attenuated due to absorption processes. The intensity, I, or the spectral intensity function, $S(\lambda)$, has the smaller value, $S(\lambda) \cdot \tau(\lambda)$, determined by the → transmittance. $\tau(\lambda)$ is to be regarded as the spectral → degree of transmission of the medium (filters or solutions in cuvettes). The color-stimulus function, $\Psi_\lambda = S(\lambda)$ (→ color measurement), is changed to $\Psi_\lambda = S(\lambda) \cdot \tau(\lambda)$, i.e. the eye perceives an attenuated or changed color stimulus after transmission through the medium. An observer might, for example, see a solution as yellow because the blue component was absorbed from the white light. Clearly, if several different colored glass filters or solutions are placed in series, as shown in Figure 1, the stimulus function, $\Psi_\lambda = S(\lambda) \cdot \tau_1(\lambda) \cdot \tau_2(\lambda) \cdot \tau_3(\lambda)$ results, i.e. different radiation with another color. The effect of a series of n color filters or solutions upon the spectral radiation distribution is determined by the product of the n transmittance values, $\tau_\lambda = \tau_1(\lambda) \cdot \tau_2(\lambda) \ldots \cdot \tau_n(\lambda)$. A color change caused in this way is termed a subtractive color mixing. However, this is not a mixing of colors as in additive → color mixing.

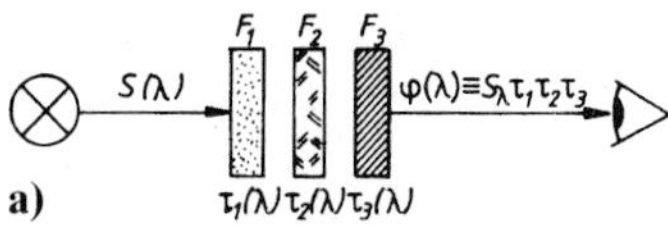

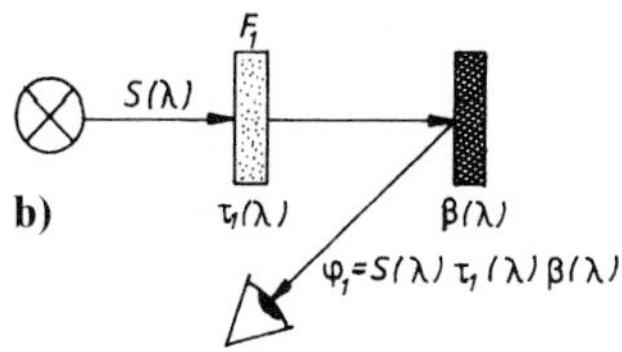

Color mixing, subtractive. Fig. 1. The principle of subtractive color mixing in transmission (a) and transmission plus reflection (b)

Primarily, there is mutual spectral influence with the result that different colors appear; the influence is not subtractive but multiplicative. A subtractive color mixing also results if a reflecting surface is placed after the first filter, as shown in Figure 1. Taking the reflection into consideration, we have $\Psi_\lambda = S_\lambda \cdot \tau_1(\lambda) \cdot \beta(\lambda)$. The principle of subtractive color mixing can be demonstrated nicely using two solutions. We take a → tandem cuvette and fill the first chamber with a yellow and the second chamber with a blue solution. This cuvette acts like a filter combination in series. With a spectrometer, we can measure either the product of the transmittances, $\tau(\lambda) = \tau_1(\lambda) \cdot \tau_2(\lambda)$ or the absorbance $A = A_1 + A_2$. If the contents of the two chambers are mixed the same result is obtained, i.e. $A = A_1 + A_2$ If the second chamber is filled with a pure solvent, $\tau_1(\lambda)$, $\tau_2(\lambda)$, A_1 and A_2 can be measured separately and $A = A_1 + A_2$ can be compared with the above result. We find that there is complete agreement. However, this applies only if no change of the dissolved dyes occurs following the concentration change when the solutions are mixed. Figure 2 shows the results of such an

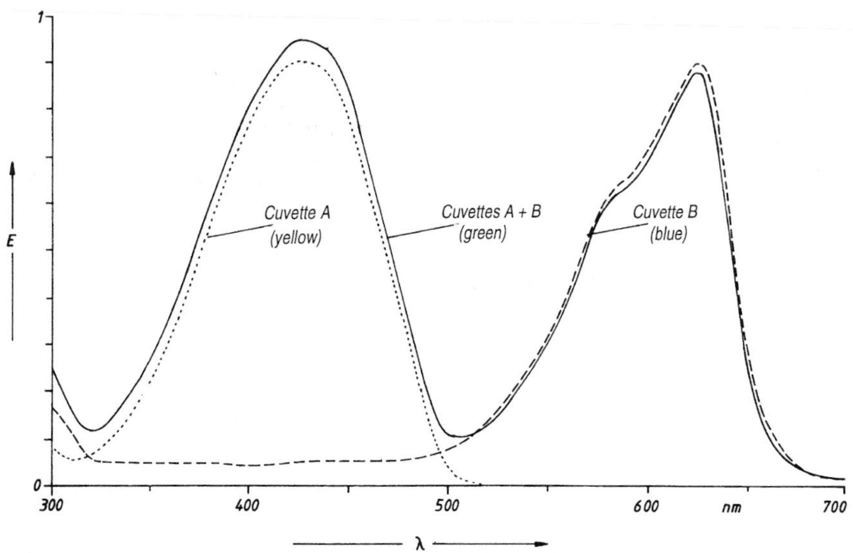

Color mixing, subtractive. Fig. 2. An example of subtractive color mixing: absorbance measurements on two solutions A (tartrazine ···); B (brilliant cresol blue ----); $A + B$ (the two solutions in series ——)

experiment. Using a tandem cuvette has the advantage that reflection losses at the cuvette surfaces remain constant during all measurements.
Subtractive color mixing is particularly important in color photography.

Colors due to thin films, <*Farben dünner Blättchen*> → interference at thin films.

Color theory, <*Farbenlehre*> a science concerned generally with the perception of colors in their total diversity. Color is a sensory perception which is usually released by electromagnetic radiation in the visible spectral region. The radiation from self-luminous or illuminated bodies reaches the eye and is converted by the sensory cells into a neural impulse which is transmitted to the brain and there perceived as color. The perception of a color is sensory-physiological processing of electromagnetic radiation. But white light (radiation) can be split into its spectral colors by, for example, a prism and can be characterized by stating a wavelength or a wavelength region, so the primary production of a color is a physical process, although the eye must be used as a subjective detector. Newton showed that the recombination of all spectral colors results in pure white and that a number of intermediate color shades can be produced using additive → color mixing. This lead to Newton's color theory which was a very controversial subject. The great German author and poet J.W. von Goethe, in particular, opposed Newton's theory all his life, often expressing his opposition in poems. He also wrote a large book on the nature of light.
The reason for this controversy and the associated misunderstandings was surely the word *color* which was used with different meanings. In physical color science the word *color* should only be understood in the sense of an optical phenomenon, i.e. a sensory perception. All material media used to give color should be described in such a way that the *materiality* is clearly recognizable. Generally, we speak of color agents of which the soluble → dyes and insoluble → pigments, suspended in their bonding agents, are the most important. Color science was developed further in the last century; in particular the triple color theory of vision, i.e. the physiological interpretation by Young and Helmholtz, advanced strongly. At the beginning of this century and based upon the findings of the previous century, color quantification was developed as the most important element of color science. This should be regarded as the science of the mensural relationships between the colors themselves which provides quantitative expressions suitable for practical applications.
E. Schrödinger developed the theoretical fundamentals of color quantification in 1920 (*Ann. Physik (IV)*, **1920**, *63*, pp. 397–456; 481–520). Schrödinger introduced a vectorial representation since it had been shown that additive → color mixing obeys the rules of vector addition. A local vector beginning at the black point, S, and having its own direction is assigned to each color type. The length of the vector represents the color value. The vector obtained in this way describes a color valence and, accordingly a mixture, M, of red and blue would be described by the vectorial color equation:

$$\vec{M} = R_M \cdot \vec{R} + B_M \cdot \vec{B}$$

R_M and B_M are the color values with which the red, $\vec{R}$, and the blue, $\vec{B}$, participate in the additive mixing. The diagram in Figure 1 shows M as an internal additive color mixing. External additive color mixing gives another mixed color, C, expressed in vector notation by:

$$\vec{C} = B_c \cdot \vec{B} - \vec{R}_c \cdot R.$$

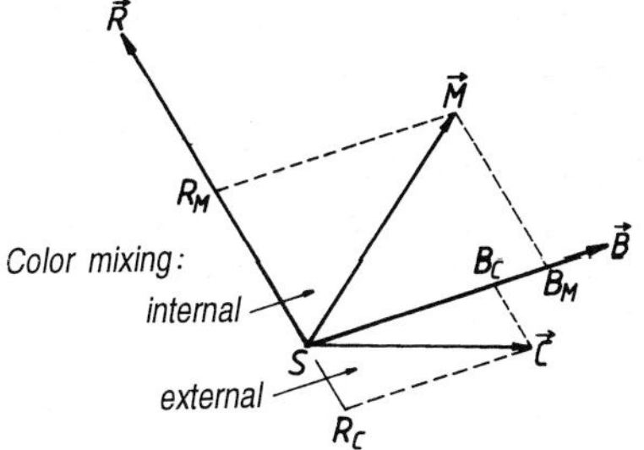

Color theory. Fig. 1. Internal and external color mixing, vector diagram

However, only a limited number of colors can be produced from color valences, $\vec{R}$ and $\vec{B}$ (including $\vec{C}$); from red-violet to blue-violet in fact. Their brightness can be changed continuously by varying the color values, R_M and B_M and also B_c and R_c. However, yellow and green cannot be produced using additive color mixing in a two-dimensional representation with color valences, $\vec{R}$ and $\vec{B}$. If we introduce a third component such as green ($\vec{G}$) we obtain yellow shades from $\vec{R}$ and $\vec{G}$, and blue-green from $\vec{B}$ and $\vec{G}$. And, as shown above, violet shades from $\vec{R}$ and $\vec{B}$. Added to this there are the mixed colors formed from all three components. According to the rules of internal or external color mixing, a definite numerical relationship exists between a color valence, $\vec{F}$, and any three arbitrarily selected, but then held constant, primary valences, e.g. for $\vec{R}$, $\vec{G}$, $\vec{B}$:

$$\vec{F} = R_F\vec{R} + G_F\vec{G} + B_F\vec{B}.$$

It has been found in fact, that three primary valences are always sufficient to reproduce all imaginable color valences. This is, in effect, the first Grassmann law of 1853 which states that there is always a definite linear relationship between four color valences. A color valence, therefore, can only be described using three independent numbers, i.e. color is a three-dimensional quantity. The vector diagrams therefore require a three-dimensional construction which is inconvenient for practical use.

On the basis of numerous experimental facts, M. Richter formulated the fundamental law of quantitative color theory which states, "The light-adapted, trichromatic eye evaluates incident radiation linearly and continuously according to three independent spectral functions of reaction and the individual reactions are combined to form an unseparable total reaction". The spectral sensitivity curves of the eye must be understood to be the functions of reaction. This law combines the three-color theory of color vision with the three-component theory of color quantification.

The CIE system (named after the Commission Internationale de l'Eclairage) was established by the International Luminescence Commission as the color-quantification system for practical application. The primary valences, $\vec{R}$, $\vec{G}$ and $\vec{B}$, chosen for this system, are denoted as normalized valences (imaginary primary colors), $\vec{X}$, $\vec{Y}$ and $\vec{Z}$ with the corresponding tristimulus values, X, Y and Z. The tristimulus values are selected in such a way that the Y standard color value agrees with the spectral sensitivity curve of the eye of a standard observer (ca. 95% of all humans). The following expression describes a color valence, $\vec{F}$, in the CIE system:

$$\vec{F} = X_F\vec{X} + Y_F\vec{Y} + Z_F\vec{Z}.$$

However, usually only the tristimulus values, X, Y and Z, are given. The graphical representation of the tristimulus values would require a three-dimensional coordinate system. To avoid this problem, normalized color value fractions, or chromaticity coordinates, are defined as:

$$x = \frac{X}{X+Y+Z}; \quad y = \frac{Y}{X+Y+Z}; \quad z = \frac{Z}{X+Y+Z}.$$

Since $x + y + z = 1$ the three chromaticity coordinates represent only two independent quantities. If we select the normalized color contributions, x and y, and a right-angled isosceles triangle for the graphical representation, we obtain a chromaticity diagram as shown in Figure 2. Any color position can be plotted on it using Cartesian coordinates.

For determining the chromaticity coordinates, x and y, the tristimulus values, X, Y, Z are required. They can be obtained using the equations:

$$X = k\int_S^L \phi_\lambda x(\lambda)\,d\lambda;$$

$$Y = k\int_S^L \phi_\lambda y(\lambda)\,d\lambda;$$

$$Z = k\int_S^L \phi_\lambda z(\lambda)\,d\lambda.$$

$\phi_\lambda = S_\lambda$ for self-luminous light sources,

$\phi_\lambda = S_\lambda \cdot \tau(\lambda)$ for absorbing solutions (transparent colors) where $\tau(\lambda)$ = transmittance and

$\phi_\lambda = S_\lambda \cdot \beta(\lambda)$ for pigments where $\beta(\lambda)$ = reflectance.

S_λ is the spectral radiation density of the light source. $x(\lambda)$, $y(\lambda)$ and $z(\lambda)$ are the tristimulus spectral value curves shown in Figure 3. k is a normalizing factor. S and L are the lower (short wavelength) or the upper (long wavelength) integration limits. The color values for pigments are normal-

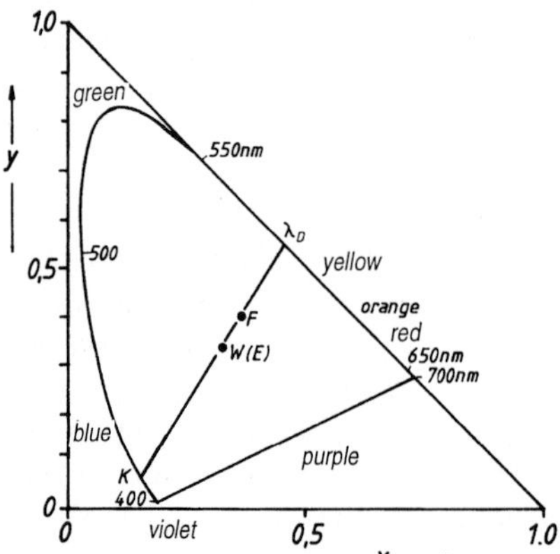

Color theory. Fig. 2. $x - y$ chromaticity diagram with spectrum locus and purple line

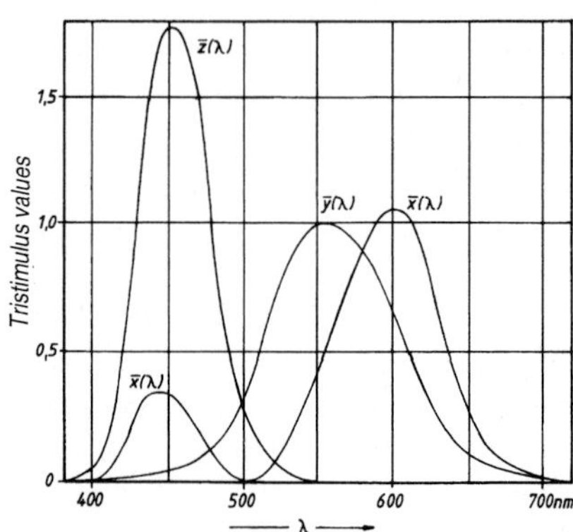

Color theory. Fig. 3. Spectral tristimulus value curves

ized by selecting k such that normalized color value, $Y = 100.00$ (ideal white surface or perfectly reflecting diffuser; $\beta(\lambda) = 1$):

$$k = \frac{100}{\int S_\lambda \cdot y(\lambda)d\lambda}.$$

Since $x(\lambda)$, $y(\lambda)$, $z(\lambda)$ are known functions and the determination of S_λ for the light source is easy, absorbance or reflectance measurements may be used for the determination of X, Y and Z. The values of $S_{\lambda,N} \cdot x(\lambda)$ for different wavelengths and a standard light source are conveniently tabled. In the case of pigments, these values must be multiplied by β_λ and for transparent colors by τ_λ. By summation with normalized k, we obtain the tristimulus value, X, or analogously, Y and Z. Figure 2 shows a rectangular coordinate system with abscissa, x and ordinate, y used for the graphical representation of colors in the CIE system. Pure spectral colors lie on a parabolic curve the ends of which are connected by the straight purple line. The noncolored colors (white, gray, black → achromatic colors) lie above each other roughly in the middle of the figure at the no-color point, E (color-type center point). Here, $x = y = z = 0.333$. This is the white point, W, in the x–y plane. The straight line drawn from the white point, W, through a color point, F, (given by x and y) gives the dominant wavelength, λ_D of the color shade. The closer the color point lies to the periphery the greater is the spectral color contribution, P_e. The extrapolation of the straight line in the other direction gives the complementary wavelength of F, K.

The CIE system is a stimulant-measurement system and takes account of the fact that the sensitivity of the eye varies with spectral region. Although the CIE system is capable of describing each color clearly, it is not adequate to define tolerances in different patterns of color differences. A sensitivity measurement system is used in this case.

Ref.: K. McLaren, *The Color Science of Dyes and Pigments*, Adam Hilger, Bristol and Boston, **1986**. D.B. Judd, G. Wyszecki, *Color in Business, Science and Industry*, J. Wiley & Sons, New York, London, Sydney, Toronto, **1975**.

Coma, <*Koma*>, an image error in light beams which have traversed a lens at an angle. Coma is caused by the fact that simple lenses cannot develop a clear image of objects which lie off the optic axis. The resulting image becomes unsymmetrical and a point is distorted to an elongated blur similar to the tail of a comet; hence the name. Since the coma is reversed if the curvature of the lens is reversed, this phenomenon can be reduced by the selection of suitable curvatures in the components of a compound lens.

Combination lamp, <*Verbundlampe*>, an electric lamp in which a continuum and a line source are combined to form one lamp. Lamps of this type are also called mixed-light lamps and sun lamps. The most important example is the combination of a → tungsten and a → mercury-vapor discharge lamp. The tungsten coil and mercury discharge are connected in series so that the tungsten coil has the function of a preresistance or balast resistance for the mercury lamp. The load on the tungsten coil (filament) is greatest immediately after switching on the lamp. The mercury is then not

completely vaporized in the lamp and the discharge voltage is very low. The filament must be designed in such a way that it survives this run-up period without adverse consequences. When the final mercury-vapor pressure has been reached (ca. 3 min after switching on) the voltage on the filament is low. To the degree with which the mercury lamp takes up the required voltage, the voltage drop is divided between the two lamps. Therefore, the operation of this lamp requires no pre-switching equipment. The color of the light changes from filament yellow to daylight white as the lamp comes into operation. The → spectral energy distribution corresponds to the radiation of the pure mercury lines plus the continuous tungsten radiation. The proportion of red lacking in the mercury radiation is supplied by the tungsten filament. When fitted with a glass envelope, these lamps are preferred for street and shop-floor lighting. When fitted with a UV-transparent envelope (usually for the UV-A, → UV radiation) combination lamps are used for medical and cosmetic applications. They are sold commercially under the names Ultra-Vialux (Osram), Ultraphil (Philips) or Sunlamp (General Electric).
Ref.: J.E. Kaufman, Ed., *IES Lighting Handbook*, Illuminating Engineering Society, New York, **1981**.

Combination principle, <*Kombinationsprinzip*> → Rydberg-Ritz combination principle.

Combination vibrations, <*Kombinationsschwingungen*>. In addition to those due to the → fundamental vibrations, ω_i, weaker bands occur in an → IR spectrum the wavenumbers of which are of the form of $\omega_1(v_1) \pm \omega_2(v_2) \pm \omega_3(v_3) \pm \ldots$, i.e. the sum of two or more → normal vibrations. v_1, v_2, and v_3 are the quantum numbers of the 3*N*-6 or 3*N*-5 possible vibrations.

$$G_o(v_1, v_2, v_3 \ldots) = \sum \omega_i v_i.$$

The associated vibrations are called combination vibrations (→ vibrational spectrum). They are described as binary combinations if the total change of the vibrational quantum number is 2. This occurs if one quantum number jumps by two units or if two v_i jump one unit each, e.g. (v_2 and v_3; v_1 and v_2; v_1 and v_3). The description *ternary combinations* follows analogously.

Combined glass filter, glass filter combinations, <*Farbglaskombination*>, a combination of several → colored glass filters which restricts the transparent region.
Numerous combinations, which generally refer to the use of Schott colored glass filters, have been described, especially for isolating individual lines in line sources (→ metal-vapor discharge lamps), (see H.-H., Perkampus, *UV-VIS Spectroscopy and its Applications*, Springer Verlag, Berlin, Heidelberg, New York, London, **1992**, pp. 10–11). Permanently cemented combined glass filters formed by combinations of suitable → long-wave pass, → short-wave pass and → band-pass filters drawn from a selection of ca. 90 Schott colored glass filter types have been available commercially for several years. They have been produced as standard filter combinations (abbreviated *SFK*) for various closely restricted wavelength regions or for use with mercury high-pressure lamps. The table gives a summary. Filters

Combined glass filters. Table 1. Standard Filter combinations for radiation sources with a continuous spectrum

SFK	λ_m [nm] ca.	τ_{max}	HW [nm]	ZW/HW ca.	Composition		Thickness [mm] ca.
1	320	0.20–0.30	25–30	1.9	UG GG WG	11 19 320	9
2	340	0.30–0.40	45–50	1.6	UG UG	1 11	9
3	360	0.25–0.35	25–30	1.6	UG BG WG	11 38 360	9
4	380	0.30–0.40	27–32	1.7	UG BG WG KG	1 38 1 3	8
5	400	0.25–0.35	40–45	1.6	UG BG GG KG	3 38 385 3	15
6	420	0.30–0.40	55–65	1.6	BG BG GG KG	3 12 400 3	10
7	450	0.35–0.45	50–60	1.6	BG BG GG KG	37 38 435 3	9
8a	470	0.29–0.39	30–40	1.7	BG BG GG KG	25 38 455 3	13
9	496	0.25–0.35	55–65	1.9	BG GG KG	23 10 3	15
10	523	0.20–0.30	55–65	1.9	VG KG	9 3	8

Combined glass filters. Table 1. Standard Filter combinations for radiation sources with a continuous spectrum

SFK	λ_m [nm] ca.	τ_{max}	HW [nm]	ZW/HW ca.	Composition		Thickness [mm] ca.
					BG	18	
11	552	0.25–0.35	25–30	1.4	BG	36	11
					OG	530	
					KG	3	
					BG	18	
12a	576	0.20–0.28	30–35	2.0	OG	570	9
					KG	3	
					BG	18	
13	601	0.20–0.30	35–45	2.1	OG	590	7
					KG	3	
					BG	38	
14	625	0.25–0.35	50–60	2.2	RG	610	7
					KG	3	
15	650	0.30–0.40	55–65	2.0	RG	630	11
					KG	3	
					RG	645	
16	670	0.30–0.40	55–65	2.1	KG	3	10
					RG	665	
17	695	0.30–0.40	60–70	2.0	KG	3	8
17a	720	0.22–0.32	50–60	1.8	UG	11	4
					RG	630	

*SFK*1 to *SFK*17a are designed for radiation sources with a continuous spectrum and Hg Filters, *SFK*18 to *SFK*22a for Hg lines, 313, 365, 436, 546 and 578 nm
Ref.: See text and Schott Glaswerke catalogue: *Interference Filters and Special Filters*.

Compensator, <*Kompensator*>, an optical device for compensating light-path differences or rotations of the plane of polarization. Compensators, using a phase shift, produce a measurable path difference which has the same magnitude as that caused by the double refraction but the opposite sign. Thus, the path difference to be measured is compensated. Mica compensators, → quarter-wave plates, → half-wave plates, → lambda plates (gypsum red, first order) are compensators with a fixed path difference. Quartz wedges, the Michel-Lévy and Babinet quartz-wedge compensators, Soleil double plates and the quartz-

Combined glass filters. Table 2. Hg filters for use with mercury high-pressure lamps

SFK	λ_m [nm]	τ_{max}	HW [nm]	ZW/HW ca.	Composition		Thickness [mm] ca.
20	436	0.30–0.40	30–40	1.7	BG BG BG GG	3 12 38 435	9
21	546	0.40–0.50	20–30	1.5	BG BG OG	18 36 530	6
22a	578	0.22–0.30	30–40	1.8	BG OG	18 570	8
18	313	0.20–0.30	25–30	1.8	UG GG	11 19	6
19	365	0.30–0.40	25–30	1.7	UG WG	1 360	6

Combined glass filters. Table 3. Transmission of desired and undesired spectral lines by Hg filters.

SFK	Average value of τ for the following Hg lines								
	Hg [nm]	313	333	365	405	436	546	578	1014 [nm]
18	313	0.30	0.09	0.002	–	–	–	–	2×10^{-5}
19	365	–	–	0.25	10^{-5}	–	–	–	5×10^{-3}
20	436	–	–	–	–	0.25	–	–	5×10^{-4}
21	546	–	–	–	–	–	0.50	–	–
22a	578	–	–	–	–	–	–	0.25	5×10^{-4}

– = $<10^{-7}$

combination wedge of Wright are compensators with a variable path difference. Rotational compensators are those which compensate for the path difference or the rotation of the plane of polarization by rotating a suitable wedge plate.

Complementary colors, <*Komplementärfarben*>, a color pair consisting of a pure spectral color (except in the region of 492 nm to 570 nm; blue-green to green-yellow) and its associated second pure spectral color which, when mixed in a specific intensity ratio, result in a pure white (→ spectrum, colors of). Examples of complementary color pairs are blue-yellow and green-red. They are of special importance for the sensory perception *color.* A solution appears yellow because the blue part of the white light has been absorbed so that the complementary color yellow is no longer *made up* to white. Similarly, a blue solution absorbs the yellow light, a red solution the green light and a green solution the red light. The same is true for → pigments when illuminated by white light. The complementary color of the corresponding color pair is no longer present in the reflected light due to its absorption by the pigment. The complementary color pair blue-yellow is important in → optical (fluorescent) brighteners.

Compton effect, <*Compton-Effekt*>. In 1923 the physicist A.H. Compton discovered fundamental evidence for the dualism of corpuscle and wave, i.e. the particle and wave models of electromagnetic radiation. He studied the interaction between photons with energy, hv, and the loosely bound electrons of a scattering material of low atomic weight such as graphite. Because of their low work functions, these electrons can be regarded as quasi-free. If short-wavelength X-ray radiation of specific wavelength, λ_o, or frequency, v_o, i.e. a photon of specific energy, falls onto such an electron there is a collision. The photon transfers momentum and energy to the electron, which is ejected from the material by the collision, and is itself diverted (scattered) from its original direction according to the laws of elastic collision; Figure 1. The photon loses energy so the frequency, v', of the scattered photon is smaller than the irradiating frequency, v_o. ($\lambda' > \lambda_o$) The difference, Δv or $\Delta\lambda$, increases with increasing angle of scattering, measured relative to the direction of the transmitted primary beam. Δv or $\Delta\lambda$ is called the Compton shift. If the

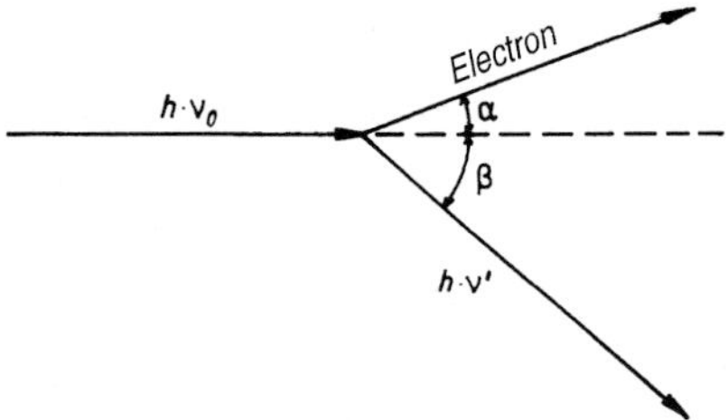

Compton effect. Fig. 1. Collision diagram for the Compton effect

conservation of energy and momentum is applied in Figure 1, once the direction of the incident photon and once perpendicular to it, the following expressions are obtained for the Compton shift.

$$\Delta\nu = \frac{2h\nu^2}{m_e c^2} \cdot \sin^2\frac{\beta}{2}$$

or

$$\Delta\lambda = \frac{2h}{m_e c} \cdot \sin^2\frac{\beta}{2}$$

In agreement with the experiment, we find that the wavelength shift, $\Delta\lambda = \lambda' - \lambda_o$, is solely a function of the scattering angle, β (or α), and independent of the wavelength of the irradiating light. The quantity, $h/m_e c = \lambda_e$ is called the Compton wavelength of the electron. More exact evaluation of the experiment gives a value of λ_e = 0.0024 nm. If h and c are known the mass, m_e, of the electron can be determined.

Ref.: R.M. Eisberg, *Fundamentals of Modern Physics*, J. Wiley and Sons Ltd., New York, London, Sydney, **1961**.

Compton scattering, <*Compton-Streuung*>, the scattering of X-rays by free or loosely bound electrons discovered by A.H. Compton (→ Compton effect).

Concave grating, <*Konkavgitter, Rowland-Gitter*>, a → grating proposed by Rowland the rulings of which are drawn on a concave spherical mirror. The grating grooves are ruled parallel at a constant interval along an imagined chord. Concave gratings have imaging properties and can perform, wholly or in part, the functions of an entrance and exit collimator. They are used in spectrometers in the → Paschen-Runge, → Eagle or → Seya-Namioka mountings which mostly utilize the → Rowland circle. The Paschen-Runge mounting is preferred for → polychromators.

Concentration quenching of fluorescence → self-quenching of fluorescence.

Continuous wave (CW) method, <*Continuous-Wave-Verfahren*>, a spectroscopic method in which a spectrum is measured chronologically, point by point, i.e. a sequential method. This description is particularly frequently used in → NMR spectroscopy.

Continuum source, <*Kontinuumsstrahler*>, a radiation source which emits a continuum of electromagnetic radiation. → Tungsten lamps, → Nernst glowers, → globars, → deuterium lamps in the UV region, → xenon lamps → noble-gas discharge lamps etc. are continuum sources.

Contrast, optical, <*Kontrast, optischer Kontrast*>, a collective name for brightness and color contrast. We must distinguish between photometric contrast (the brightness difference between two radiating surfaces relative to the brightness of one of them) and physiological contrast (a contrast function of the eye).
The definition of photometric brightness contrast is based on the radiation densities, B_1 and B_2, of the two surfaces according the equation, $K = (B_1 - B_2)/B_2$ or less commonly, $K = (B_1 - B_2)/B_1$. If $B_1 = B_2$, $K = 0$. Most experimental methods in visual photometry are based on the fact that B_1 can be varied during the measurement in such a way that the brightness contrast is reduced to zero, the unkown B_2 can then be assumed to be equal to the known B_1. The brightness contrast of a (small) internal field with radiation density, B_1, is usually observed against the (larger) surrounding field with light density, B_2. If B_1 is larger than B_2, there is a positive brightness contrast; if B_1 is smaller than B_2 a negative contrast. The first can take all values between 0 and ∞ ($B_2 = 0$) and the latter all values between 0 and 1 ($B_1 = 0$).

Conversion, <*Umwandlung*> → internal conversion.

Conversion filter, color-correcting filter, light-balancing filter, <*Konversionsfilter*>, a filter which is used to convert the spectral distribution of a light source having a color temperature, T_{f1} (→ thermal radiator) into another distribution with color temperature, T_{f2}. Conversion filters show a gradually increasing or decreasing spectral transmission from the blue to the red. According to → Wien's radiation law, an approximation to → Planck's radiation law, the spectral energy density, $S_{\lambda 1}$, is given by:

$$S_\lambda = \frac{c_1}{\pi} \cdot \lambda^{-5} \cdot e^{-\frac{c_2}{\lambda T_f}} \cdot \frac{1}{\Omega_o}$$

$[c_1 = 3.741832 \cdot 10^{-16}$ $(\mathrm{W \cdot m^2})$; $c_2 = 1.438786 \cdot 10^{-2} \cdot \mathrm{m \cdot K}$; λ = wavelength; T_f = color temperature; Ω_o = unit solid angle]. If the spectral distribution, $S_{\lambda T1}$, is to be transformed into distribution, $S_{\lambda T2}$, the conversion filter must have the transmission function:

$$\tau(\lambda)_{rel} = (S_{\lambda T_2}/S_{\lambda T_1})/(S_{\lambda T_2}/S_{\lambda T_1})_{max}$$

If λ_o is the wavelength at which the ratio $S_{\lambda T2}/S_{\lambda T1}$ assumes a maximum value, the optical density $D = \ln(1/\tau(\lambda))$ is given by:

$$D(\lambda) = \ln S_{\lambda_o T_2} - \ln S_{\lambda_o T_1} - \ln S_{\lambda T_2} + \ln S_{\lambda T_1},$$

and substituting for S_λ we have:

$$D(\lambda) = c_2 \left(\frac{1}{\lambda} - \frac{1}{\lambda_o}\right) \cdot \left(\frac{1}{T_2} - \frac{1}{T_1}\right) + \text{const.}$$

The expression $(1/T_2 - 1/T_1)$ is defined as the conversion value, ΔM, in K^{-1}. Using the smaller unit of micro-reciprocal degree μrd (mired); 1 μrd = 10^{-6} K^{-1}, we obtain $\Delta M = (10^6/T_2) - (10^6/T_1)\mu\mathrm{rd}$ with T_1 and T_2 in degrees Kelvin. Yellowish filters have a positive value of ΔM and reduce the color temperature; λ_o lies at ca. 700–800 nm. An increase of the color temperature can be achieved with blue filters which have a negative value of ΔM and λ_o lies at ca. 350–400 nm. The table shows the ΔM values in mired for Schott conversion filters.

Conversion filter. Table.

Filter type	ΔM [mired] for 2 mm filter thickness = ΔM_2; 20 °C
BG 34	−150
FG 6	− 30
FG 13	+200
FG 15	+120
FG 16	+ 60

The approximation, $d_x = 2\Delta M_x/\Delta M_2$, can be used to estimate the required filter thickness, d_x, for a desired conversion value, ΔM_x.
Figure 1 shows the transmission curves, in double logarithmic plotting of τ_i (→ diabatic scale), of the glass filters BG 34 and FG 6 used for increasing, and FG 15 and FG 16 used for reducing the color temperature.
Conversion filters are used for the production of different standard light

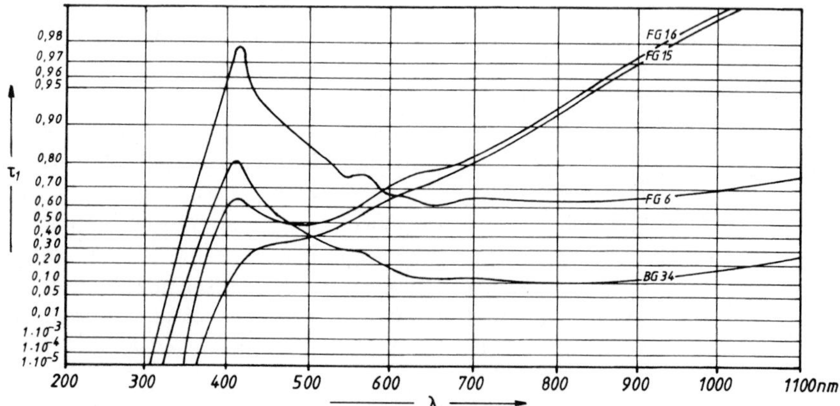

Conversion filter. Fig. 1. Transmission curves for various filters from the UV to NIR

types from filament lamps or for adapting daylight to artificial-light color films.
Ref.: F. Goldstein in, *Geometrical and Instrumental Optics*, (Ed.: D. Malacara), Methods of Experimental Physics, Academic Press Inc., New York, London, Toronto, **1988**, vol. 25, p. 273; Schott Glaswerke catalogue: *Optical Glass Filters*.

Coolidge tube, <*Coolidge-Röhre*> → X-ray fluorescence.

Coriolis coupling, <*Coriolis-Kupplung*>, the interaction between the rotation and vibration of a molecule. If we stand in a rotating frame of reference, which has a constant angular velocity, ω and observe a particle of mass, m, which moves at right angles to the rotational axis of the reference system at velocity, V_m, then this particle undergoes an extra acceleration in addition to the centrifugal acceleration. The extra acceleration is called the Coriolis acceleration after its discoverer (in 1835). According to the laws of classical mechanics the centrifugal and Coriolis forces which occur in such a rotating system are given by:

$$F_{\text{Centrif.}} = mr^2\omega \quad \text{and}$$
$$F_{\text{Coriolis}} = 2mV_m\omega \sin\alpha$$

α is the angle between the rotational axis and the direction of the velocity vector, V_m, of the particle of mass, m. $\alpha = 90°$ in the above example and the Coriolis force assumes its maximum value. A comparison of the two formulas shows that the Coriolis force applies only to a moving particle ($V_m \neq 0$). Its direction is perpendicular to the direction of the movement and perpendicular to the rotation in the sense of a screw with a right-handed thread. If we transfer these observations to a molecule, we always have moving masses in a rotating reference system when there is simultaneous excitation of vibration and rotation. Consequently, the influence of the Coriolis forces must be taken into consideration. Figure 1 shows an example of this using the symmetric X-Y → valence (stretching) vibration, $\nu_{x\text{-}y}$, of a linear, triatomic molecule, Y-X-Y. The direction of rotation is given by the direction of the arrow from the atomic position, 1 to 2 and the rotational axis is perpendicular to the plane of the figure ($\sphericalangle\ \alpha = 90°$). If the molecule was

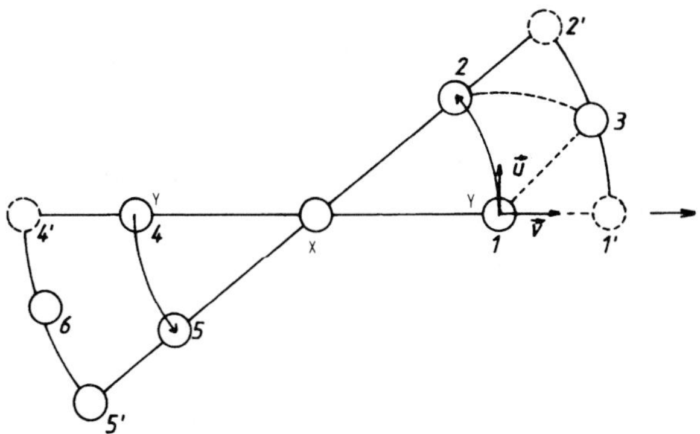

Coriolis coupling. Fig. 1. Coriolis effect on ν_s (CO_2), see text

not rotating the Y atom would move to atomic position 1', because of the vibration. If the molecule rotates simultaneously we would expect to find atom Y in atomic position 2'. The actual position of the vibrating and rotating atom, Y, is position 3 or on the opposite side, position 6. An observer outside the rotating molecular system perceives atom Y to have a tangential velocity component, u in addition to the radial velocity, V_m. For an observer in the rotating system, atom Y moves along the dashed parabolic curve (2 → 3). This is only possible if an additional force is available perpendicular to the direction of movement of the Y atom which corresponds to the Coriolis force defined above.

The consequence is an additional interaction between rotation and vibration which is called Coriolis coupling. In general, it is greater than the centrifugal interaction because the velocity during a vibration, V_m, is usually considerably larger than with rotation. Although the centrifugal interaction occurs with a pure rotation, Coriolis coupling occurs only in a vibrating molecule. Figure 2 shows a linear triatomic molecule Y-X-Y as an aid to understanding Coriolis coupling. The heavy arrows represent shifts from the equilibrium position (movement of the point masses). The dashed arrows represent the Coriolis forces perpendicular to the direction of movement. Figure 2 shows immediately that vibration ν_3 contains components of vibration ν_2 and vice versa, on account of the Coriolis forces. In other words, vibration ν_3 can excite vibration ν_2 but at frequency ν_3 and correspondingly vibration ν_2 can excite vibration ν_3 with frequency ν_2. This becomes very apparent when we consider the situation after a rotation of 180°. If frequencies ν_2 and ν_3 lie close together a strong excitation of one occurs when the other is excited. However, the mutual excitation is weak if the two vibrations have considerably different frequencies. In the case of vibration ν_1 used in Figure 1 for the explanation of the Coriolis forces, no coupling occurs with the other vibrations, ν_2 and ν_3. A coupling with the rotational motion results in this case. This is the same effect as that which is responsible for the deviation of the average value of $1/I$ for diatomic molecules from the equilibrium value, $1/I_\varepsilon$, even for strictly harmonic vibrations. This is not usually thought of as a Coriolis coupling. Coriolis coupling is particularly important for degenerate vibrations. In that case, the rotational energy for a → symmetric top molecule is given by:

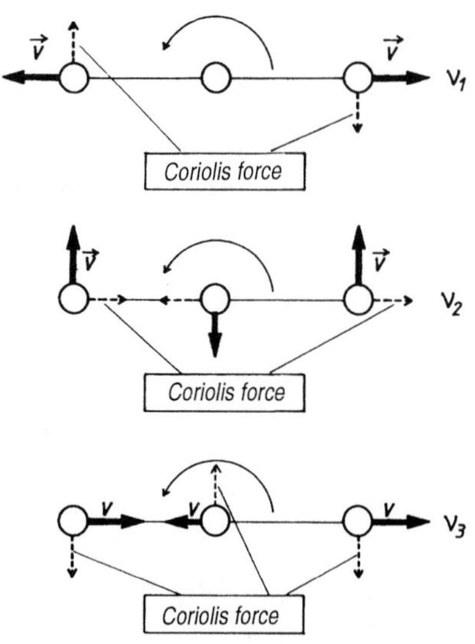

Coriolis coupling. Fig. 2. The normal vibrations of CO_2 (ν_2 is twofold degenerate) and their interaction with Coriolis forces

$$F[v](J,K) = B[v]J(J+1) + (A[v]- -B[v])K^2 \pm 2A[v]\zeta_i K$$

ζ_i, (zeta) is the Coriolis constant or Coriolis coefficient, $0 \leq \zeta \leq 1$. The value of ζ depends upon the form of the degenerate normal vibration, ν_i; the sign is determined by the relative directions of vibration and rotation: – for the same and + for the opposite direction. This results in a splitting of the degenerate vibrational states. The subscript, i, on the Coriolis constant refers to the ith degenerate normal vibration. The individual ζ_i values are coupled by summation expressions which are correlated directly with the moments of inertia, I_A and I_B, or the rotational constants, A and B, of a symmetric top. For the molecule $HCCl_3$ with degenerate vibrations, ν_4, ν_5, ν_6 we have:

$$\zeta_4 + \zeta_5 + \zeta_6 = \frac{I_A}{2I_B} = \frac{B}{2A}.$$

In a spherical top (→ top molecules), the triply degenerate vibrations of the → symmetry species, T_1 and T_2, are subject to a strong Coriolis coupling which leads to a triple splitting. The summation formula for the two degenerate normal vibrations of symmetry species, T_2, ν_3 and ν_4, of a molecule, XY_4, is:

$$\zeta_3 + \zeta_4 = \frac{1}{2}, \text{ since } I_A = I_B\ (A = B),$$
vide supra.

The question as to whether a Coriolis interaction occurs between two vibrations is answered by a general rule due to Jahn. Two vibrations in a rotating molecule can only interact as a consequence of Coriolis forces if the direct product of their symmetry species contains the symmetry species of a rotation.

Ref.: G. Herzberg, *Molecular Spectra and Molecular Structure II Infrared and Raman Spectra of Polyatomic Molecules*, D. Van Nostrand Company, Inc. Princeton, New Jersey, **1966**, 12th ed., chapt IV, p. 370 ff.

Cornu prism, <*Cornu-Prisma*>, a quartz prism, proposed by A. Cornu in 1881 which eliminates the interfering circular double refraction of crystalline → quartz. In order to avoid the usual double refraction, the light must travel parallel to the optic axis of the quartz. Even then, with quartz, all spectral lines are split into two closely spaced, oppositely polarized lines due to the circular polarization. This double refraction of quartz in the direction of the optic axis can be eliminated if, following Cornu, we construct the prism in two halves, one of right and the other of left rotating quartz. Cornu prisms are no longer of importance because modern quartz prisms are produced from synthetic (fused) quartz (→ quartz glass) which shows no double refraction.

Corpuscular radiation, <*Korpuskularstrahlung*>, radiation consisting of corpuscules, individual particles of a specific type which propagate in a straight line and at uniform velocity, as long as they are not subjected to an external force or force field. → Alpha or → beta radiation is corpuscular radiation but → gamma radiation, also emitted by radioactive substances is not corpuscular radiation, in the narrow meaning of the term, but rather electromagnetic radiation with a very high frequency. The electron beams used in electron optics, and also ion beams, are corpuscular radiation. Because of the duality of waves and particles we can treat corpuscular radiation as waves and wave radiation as corpuscular. The analogy between the optics of light and electrons is based on this observation.

Cotton effect, <*Cotton-Effekt*> → circular dichroism.

Cotton-Mouton effect, <*Cotton-Mouton-Effekt*>, the magnetic double refraction (birefringence), discovered by Cotton and Mouton in 1907, which occurs when a magnetic field is applied to an isotropic body. It is much smaller than the analogous → Kerr effect. With refractive indexes, n_1, parallel and, n_2, perpendicular to the magnetic field the experimental result is:

$$\Delta n = n_1 - n_2 = K'\lambda \cdot H^2.$$

K' is the Cotton-Mouton constant, λ, the wavelength of the light used (in m) and H the magnetic field strength. The optic axis is parallel to H.

Coupling constant, <*Kopplungskonstante*> → spin-spin coupling.

Critical angle, <*Grenzwinkel*> → total reflection.

Crown flint glass, <*Kronflint*> → glass, optical.

Crown glass, <*Kronglas*> → glass, optical.

Crown glass prism, <*Kronglas-prisma*> → dispersion element in → prism monochromators for the visible region.

Cryophosphorescence, <*Tieftemperaturphosphoreszenz*>, the → phosphorescence observed at low temperatures which is characterized by a → bathochromic shift in its spectrum vis-a-vis the → fluorescence spectrum. The low temperature (liquid nitrogen) is not alone sufficient. A glassy solidified matrix, which can also be made by embedding the material in a polymer, is also necessary. The phenomenon can be explained using a → Jablonski diagram (→ phosphoresence).

Crystal, optical, <*Kristalle, optische Kristalle*>, crystals with specific optical properties used, especially in spectroscopy, as optical materials for prisms and windows in the UV and IR spectral regions. Today, most optical crystals are grown in large, optically pure blocks. The alkali halides especially can be grown very well as pure crystals. Fused quartz of good optical quality has been produced for some time. In contrast to natural quartz crystals, it shows no → birefringence or optical rotatory dispersion. Fused quartz glass is sold commercially under names such as Homosil, Ultrasil and Suprasil and has a greater transparency than natural quartz (→ prism materials).

Cut-off filter <*Kantenfilter*> → edge filter.

Cuvette, cell, <*Küvette*>, a container which is filled with the solution or gas under spectroscopic investigation. Plane-parallel windows, which must be optically transparent to the measuring light, are fitted perpendicularly to the direction of the traversing beam. The windows are fused, cemented or held by pressure to the cuvette body. Rectangular cuvettes, externally 45 mm high × 12.5 mm wide with a path length of 10 mm are used for most

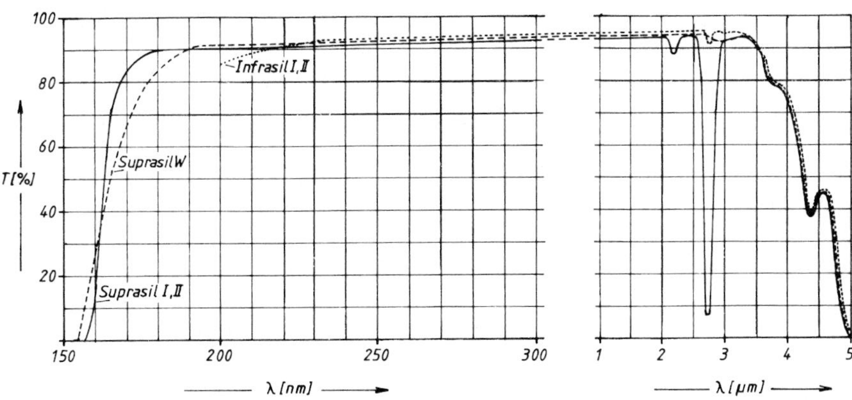

Cuvettes. Fig. 1. Transmission curves for various cuvette materials from the UV to the NIR

measurements in the UV-VIS to the NIR region. They have a capacity of 4–5 ml. Such cuvettes can be produced with path lengths from 1 to 100 mm. The following materials are used as windows in the region from 180 nm to 4 μm (55,000 cm^{-1} to 2500 cm^{-1}): Suprasil, Infrasil, Herasil, optical special glass, optical glass and pyrex (Duran 50). The transmission curves of some of these materials are shown in Figure 1. Cuvettes of this type are effectively resistant to all media, with the exception of hydrogen fluoride, and therefore, meet the further condition that the cuvette material must not interact with the solution in any way. Special cuvettes have been developed for numerous applications in UV-VIS spectroscopy. The product brochure, 1985 D, 2nd edtn. of Hellma GmbH & Co. Müllheim, Baden, lists ca. 150 different cuvettes, some of which are equipped with a thermostatted jacket. Many spectrophotometer manufacturers also supply cuvettes, e.g. Perkin-Elmer.

CaF_2 is a suitable window material in the wavelength region below 180 nm (55,000 cm^{-1}) and it can also be used in the IR region down to 1600 cm^{-1}. In the central IR region from 4000 cm^{-1}, sodium chloride down to 600 cm^{-1}, potassium bromide down to 400 cm^{-1} and cesium iodide down to ~ 200 cm^{-1} are used as window materials. In IR spectroscopy, the path lengths are usually set with spacer rings in liquid cuvettes and for gas measurements by glass cylinders of specific length, with the windows pressed onto their plane-ground ends. In many cases, particularly with IR liquid cuvettes (cells), it is necessary to determine the path length. This, as with any checking required, can be done by various methods. Where the windows are not fused on, the thickness of the whole cuvette and both windows can be measured with a micrometer. In cuvettes with fused windows, the → Bouguer-Lambert-Beer law can be used with the measured → absorbance of a substance with an exactly known extinction coefficient at a given wavelength. This assumes the validity of the Bouguer-Lambert-Beer law and a spectrally very pure radiation (→ UV-VIS standards). Very small path lengths, *d*, can be determined from the interference fringes which are formed by multireflections between the parallel windows of the cuvette, analogously to the → Fabry-Perot interferometer, and are superimposed on the spectrum of the solution under investigation. Interference maxima occur if the path difference is a whole multiple, n, of wavelength, λ and $2d = n \cdot \lambda$. It follows that:

$$d = \frac{n}{2} \cdot \frac{\lambda_1 \lambda_2}{\lambda_1 - \lambda_2}$$

or

$$d = \frac{n}{2} \cdot \frac{1}{\tilde{\nu}_2 - \tilde{\nu}_1},$$

where n equals the number of interference maxima between wavelengths λ_1 and λ_2 (in μm) or wavenumbers $\tilde{\nu}_1$ and $\tilde{\nu}_2$ and d is the path length to be determined. In addition to possible application in the UV-VIS, this method is used very frequently in the IR spectral region (→ interference spectroscopy). In the → vacuum UV and → Schumann UV region, cuvettes are rarely used. Instead, the gas to be measured is released at a reduced pressure into the evacuated spectrometer. Fluorescence cuvettes are usually rectangular with four optically transparent sides and sometimes a transparent base. Users develop their own cuvettes, e.g. low-temperature cuvettes (cells), for many applications in the UV-VIS, NIR and IR spectral regions.

Fine tubes replace the cuvette in classical Raman spectroscopy. Larger sample tubes are used in → ESR and → NMR spectroscopy.

Czerny-Turner monochromator, <*Czerny-Turner-Monochromator*>, a grating monochromator, the plane grating of which is mounted between the entrance and exit slits. This gives a relatively large distance between the two slits which is important in practical applications. The collimator and collector mirrors are separate as Figure 1 shows. Image errors can be kept small by an unsymmetrical construction (angle, focal length) and we distinguish between a symmetrical and unsymmetrical positioning of the plane grating. The Czerny-Turner monochromator mounting differs from the Ebert mounting (→ Ebert monochromator) in that two separate mirrors are used. It is therefore also called a modified Ebert construction. Modern monochromators have this design almost exclusively.

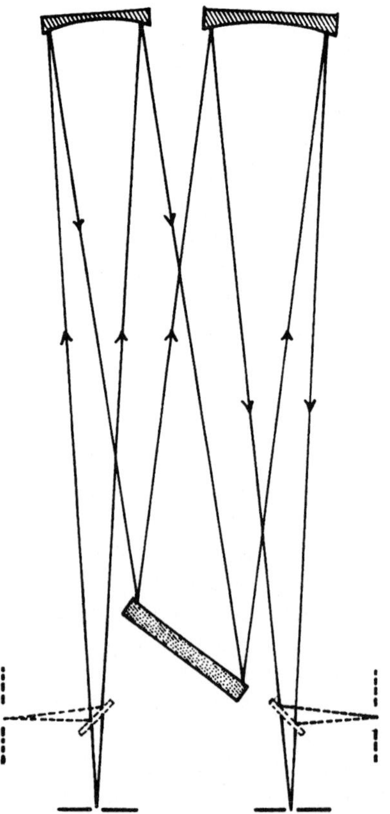

Czerny-Turner monochromator. Fig. 1. The light path in a Czerny-Turner monochromator

D

Davydov splitting, <*Davydov-Aufspaltung*>, a phenomenon which occurs in the electronic excitation of molecular crystals. The interpretation is based on the theory of molecular → excitons, developed by A.S. Davydov at the end of the '40s and beginning of the '50s, which has been especially applied to the spectroscopic properties of crystals of aromatic hydrocarbons. The starting point is the fact that an excited molecule in a real crystal interacts with nonexcited molecules by Coulombic and electron-exchange interactions. As a consequence of these interactions, exciton states occur in a crystal and their number is given by the number of non-translationally equivalent molecules in the unit cell. In a molecular crystal with two molecules in the unit cell, naphthalene and anthracene are representative examples, each excited state in the molecule produces two exciton states in the crystal. The energy difference between these two states is called the Davydov splitting which corresponds to the interaction energy of the molecule with the non-translationally equivalent molecules. The excitation energy of an excitonic state is given by: $E = E_0 + D \pm B$ whence $\Delta E = 2B$. E_0 is the excitation energy of the free molecule. D can be positive or negative and is a parameter which takes into consideration the shift of the excited state with respect to the position in the gas phase or solution and also the interaction with the translationally equivalent molecules. $2B$ is the Davydov splitting, i.e. the excited level of the molecule is shifted and split into two states. Two bands, which also show opposed polarizations, are observed in absorption or emission. The magnitude of the Davydov splitting depends upon the transition dipole moment, M, of the associated transition. The quantity, B, can be determined more precisely by evaluating the appropriate matrix elements. For allowed electric dipole transitions, e.g. (→ Clar classification, → Platt classification) the β bands in aromatic hydrocarbons ($^1A \rightarrow {}^1B_{a,b}$), $2B \approx$ 20,000 cm^{-1}; for the transitions, $^1A \rightarrow {}^1L_a$, (p band of anthracene) and $^1A \rightarrow {}^1L_b$, (α band of naphthalene) $2B \approx$ 200 cm^{-1}. For the singlet-triplet transition, $^1A \rightarrow {}^3L_a$, in anthracene and naphthalene $2B \approx 10$ cm^{-1}. The Davydov splitting of vibrational transitions (0–0, 0–1, 0–2 ...) in an electronic band depends upon the vibrational transition moment which, in turn, is determined by the → Franck-Condon principle. Thus, $2B$ falls in the order 0–0 > 0–1 > 0–2 > 0–3 ... in a normal case.

Ref.: A.S. Davydov, *Theory of Molecular Excitons*, Translators M. Kasha and M. Oppenheimer jr., McGraw-Hill Book Company Inc, New York, San Francisco, Toronto, London, **1962**. D.S. McClure, *Electronic Spectra of Molecules and Ions in Crystals*, in Solid State Physics, vol. 8 and 9., Eds.: F. Seitz, D. Turnbull, Academic Press Inc. **1959**; J.B. Birks, *Photophysics of Aromatic Molecules*, Wiley Interscience, London, New York, Sydney, Toronto, **1970**.

de Broglie's relationship, <*de Broglie-Beziehung*>, the fundamental discovery, made by de Broglie in 1924, which forms the foundation of wave or quan-

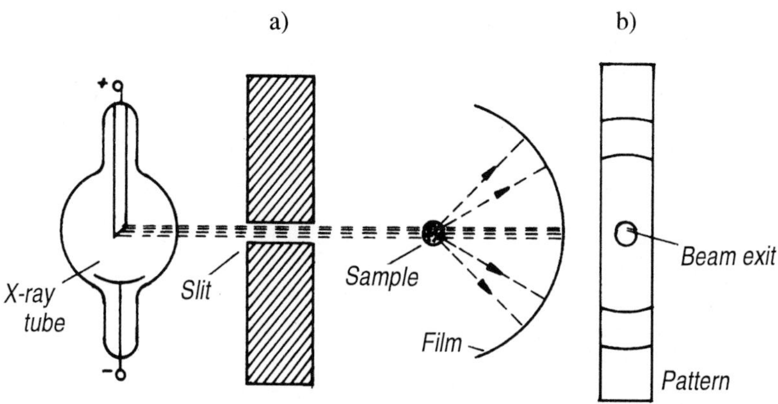

Fig. Debye-Scherrer technique. Fig. 1. a) Schematic experimental arrangement; b) simplified Debye-Scherrer pattern

tum mechanics. According to de Broglie, the translational motion of a material corpuscle is associated with a wave motion the wavelength of which, λ, is given by $\lambda = h/mv$. m is the mass and v the velocity of the particle (mv = momentum), h is → Planck's constant. Shortly after its announcement, Davisson and Germer and also Stern impressively confirmed de Broglie's postulate by diffracting beams of electrons, atoms and molecules. The postulate states that every atomic and molecular movement is associated with a wave process which can be represented by a corresponding wave function, Ψ. Ψ is a function which depends periodically upon the time at every position in space. In 1926, Schrödinger developed these ideas further in the equation, which bears his name (→ Schrödinger's equation) and forms the basis of the theoretical study of atomic and molecular spectra.

Ref.: L. de Broglie, *Ann. de Phys.*, **1925**, *3*, 22,; E. Schrödinger, *Ann. d. Physik*, **1926**, *79*, 361, 489; *80*, 437; *81*, 109.

Debye-Scherrer technique, <*Debye-Scherrer-Verfahren*>, an important method for the qualitative determination of crystal structures by means of X-rays. It was described independently by P. Debye and P. Scherrer in 1916 and by A.W. Hull in 1917. It has the great advantage that a powder of the material under investigation can be used (crystal powder method), rather than a good single crystal which is difficult to produce from many substances. The powder consisting of many crystallites is pressed into a small bar and irradiated by a monochromatic X-ray beam (Figure 1a). Since the crystallites have all possible orientations in the bar, the X-rays are reflected, or more precisely refracted, according to Bragg's condition (→ X-ray diffraction) in such a way that the refracted rays lie on different cone surfaces but the axes of all the different cones coincide with the direction of the incident ray. If the powder bar is surrounded by a circle of photographic film we obtain a Debye-Scherrer pattern on the film (Figure 1b).

The Straumanis technique is a precision variant of this technique. The film is inserted asymmetrically in the exposure chamber so that the X-rays enter through the first quarter of the film and exit through the third quarter. Thus, the position of entrance and exit are fixed exactly on the film and the refraction angles can be determined independently of humidity variations in the gelatine and length variation of the film. In principle, this is a rotating crystal method where the crystal need not be rotated since all orientations of the crystallites are statistically available.

Debye-Sears effect, *<Debye-Sears-Effekt>*, the diffraction of light by ultrasonic waves discovered by Debye and Sears and, approximately simultaneously, by Lucas and Biquard. Debye was the first to explain the phenomemon theoretically.

Decay luminenscence, *<Zerfallsleuchten>*, an emission continuum resulting from the transition from a stable, discrete upper state to an unstable, continuous lower state. Figure 1 shows the associated potential curves. Transitions from the lowest vibrational state, $v' = 0$, of the excited electronic state to positions on the lower potential curve occur according to the → Franck-Condon principle, and roughly between A and B which lie above the asymptote, i.e. within the continuous region. Thus, as the system decays to the lower state, a continuous spectrum is emitted and a simultaneous dissociation occurs. The atoms fly apart with a kinetic energy which is determined by the magnitude of the radiated quantum, i.e. the height above the asymptote reached

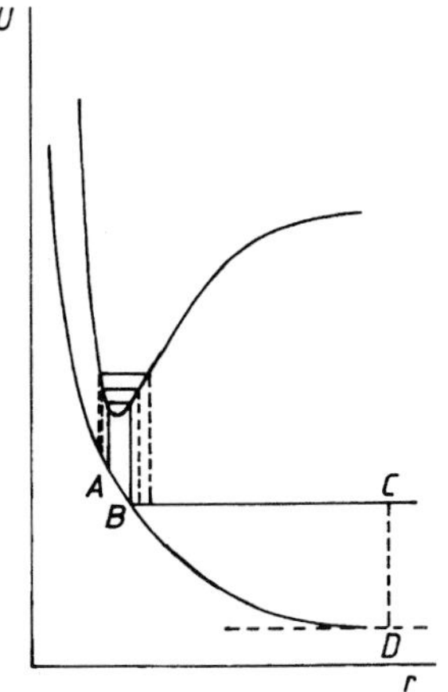

Decay luminenscence. Fig. 1. Potential energy curves for the explanation of decay luminescence

after the light-emitting quantum jump. It can have rather large values under certain circumstances. If various vibrational states, in the upper electronic state, are excited the extent of the continuum is considerably enlarged as shown in Figure 1. The well-known continuous emission spectrum of the hydrogen molecule, H_2, which occurs in almost every electric discharge, is the most important example of such a continuum. It extends from roughly 160 nm to ca. 400 nm and is used as a continuous light source in the UV region in form of the → hydrogen lamp. The continua of → noble-gas discharge lamps can also be attributed to decay luminescence. Stable, excited molecules can be formed by interaction between an excited and a nonexcited atom. These species then decay, with the emission of a continuum, to the unstable ground state. The continuum observed in every Hg lamp is an example of this. In the case of larger molecules, the interaction between an excited molecule, M^* and a molecule in the ground state, M^o, can also lead to a stable, excited dimer $[MM]^*$, or → excimer, the broad and structureless fluorescence spectrum of which can be assigned to the same process. Excited complexes called exciplexes can form between two different molecules. Since the complexes are also not stable in the ground state, broad and structureless fluorescence spectra result. The phenomenon is utilized in exciplex lasers; wrongly called → excimer lasers.

Déchêne effect, <*Déchêne-Effekt*> → electrophotoluminiscence.

Deformation birefringence, <*Deformationsdoppelbrechung*>, the optical birefringence in deformed solids which is primarily due to the shift of the atoms from their equilibrium positions; in particular, to changes in valence angles and the internuclear distances between atoms which are not directly bonded. D. Brewster observed the correlation between birefringence and deformation in glasses in 1815. Deformation birefringence, which is the same as the stress birefringence in materials with a low molecular weight, is proportional to the extension for uniaxial stress and, within the limits of Hooke's law, to the stress. This correlation is the basis of stress optics in which stresses are determined by measuring the → stress birefringence. In addition to deformation birefringence there is → orientation birefringence in high polymers (see explanation in → stress birefringence). Deformation birefringence itself depends only slightly upon the temperature, in fact, only in so far as the material elastic constants and polarizabilities change with temperature in response to the variation of the interatomic distances. In contrast to orientation birefringence, deformation birefringence follows the applied load immediately and no relaxation phenomena can be observed.

Deformation vibration, <*Deformationsschwingung, Knickschwingung*> → bending vibration.

Degeneracy, <*Entartung*>, characterized in quantum mechanics by the occurrence of several eigenfunctions belonging to a single energy eigenvalue of → Schrödinger's equation. Examples are the functions assigned to a single value of ℓ, the orbital angular momentum quantum number, which have different values of the magnetic quantum number, m_ℓ. This is a $(2\ell + 1)$-fold degeneracy. The same also applies to the total orbital angular momentum, L, and the total angular momentum (spin and orbital) J (→ electronic angular momentum). Two vibrations, which have the same energy, the most common case, are known as degenerate vibrations. Every state characterized by the rotational quantum number, J, is $(2J + 1)$-fold degenerate.

Degree of absorption, (absorptivity), <*Absorptionsvermögen, Absorptionsgrad*>, a numerical factor, α, stating what fraction of the radiation incident upon a body is absorbed. $\Phi_a = \alpha \cdot \Phi_0$ is the fraction of radiation absorbed from a quantity of incident radiation, Φ_0. The remainder, $\Phi_r = (1 - \alpha)\Phi_0$ is reflected and α cannot, therefore, exceed 1. Its value depends upon the material and surface properties of the body and varies with the wavelength of the radiation. The spectral variation of the degree of absorption, which is important for the color (→ pigment) of the material concerned, can only be determined by measurement (→ reflection spectroscopy). Bodies are called gray if the degree of absorption has the same value for radiation of all wavelengths but is smaller than 1. Those which absorb radiation of all wavelengths completely ($\alpha = 1$) are called black bodies. Since the → emissivity is always the same as the degree of absorption (→ Kirchhoff's law), a black body also has the highest possible radiation emission at a given temperature; → black-body radiator.

Degree of depolarization, <*Depolarisationsgrad*>, the fraction, ϱ, of the total intensity of a mixture of polarized and unpolarized light due to the unpolarized light. $\varrho = I_{unpol}/(I_{unpol} + I_{pol})$. If we investigate partially linearly polarized light using an analyzer (→ Nichol prism) we determine two positions, differing by 90°, in which maximum and minimum intensities are observed. We can now calculate the degree of depolarization from $\varrho = 2I_{min}/(I_{min} + I_{max})$. If the intensity ratio, $i = I_{min}/I_{max}$, is introduced the following relationships exist between i and ϱ:

$$\rho = \frac{2i}{1+i}, \quad i = \frac{\rho}{2-\rho}.$$

The determination of the degree of depolarization is of particular importance in investigations of the scattering of light by very small particles, e.g. in turbid media (the → Tyndall effect) or in → Raman spectroscopy.

Degree of freedom, <*Freiheitsgrade*> → kinetic degrees of freedom.

Degree of transmission, <*Transmissionsgrad*> → transmittance.

Dejardin window, <*Dejardin-Fenster*>, a photocell window consisting of a thin spherical glass membrane with large transmittance in the UV spectral region.

Delayed fluorescence, <*verzögerte Fluoreszenz*> → photoluminescence the spectral distribution of which corresponds to that of → fluorescence but is chronologically delayed vis-a-vis standard fluorescence. There are two possible causes of delayed fluorescence.

a) If the triplet state, T_1, of a molecule is energetically close to the lowest excited singlet state, S_1, the T_1 state populated by → intersystem crossing, $S_1 \rightarrow T_1$, can, during its lifetime, undergo the reverse transition, $T_1 \rightarrow S_1$, by thermal excitation of vibrational states. The fluorescence observed then has a decay

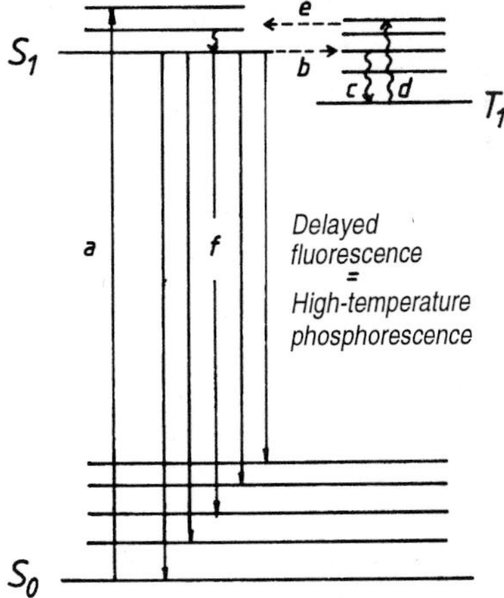

Delayed fluorescence. Fig. 1. Energy-level diagram for high-temperature phosphorescence

time (→ fluorescence lifetime) which is determined by the lifetime of T_1; → high-temperature phosphorescence. This type of delayed fluorescence was initially observed in eosine, whence the denotation *E type* (Figure 1).

b) Another mechanism can occur if the T_1 states are considerably lower than the S_1 states. Two excited triplet molecules can meet during the lifetime of the T_1 state and redistribute their energy so as to produce one molecule in the first excited singlet state, S_1, and one in the ground state, S_o. This mechanism, called triplet-triplet annihilation, can be described as follows:

$$^3M^* + ^3M^* \rightleftharpoons ^1M^* + ^1M_o$$
$$^1M^* \longrightarrow ^1M_o + h\nu_f.$$

The radiative deactivation appears as fluorescence with a delay which depends upon the lifetime of the triplet state. The mechanism assumes that the sum of the triplet energies of two $^3M^*$ molecules is equal to or greater than the singlet excitation energy of one $^1M^*$ molecule. This relationship applies to aromatic hydrocarbons where the energy difference, $S_o \rightarrow T_1$ frequently corresponds to just half the excitation energy, $S_o \rightarrow S_1$. This mechanism for delayed fluorescence was initially observed in pyrene, whence the denotation *P type*.

Densitometry, <*Densitometrie*>, the measurement of the → absorbance of a sample which absorbs light; also called the measurement of the optical density of the sample.

Derivative spectroscopy, <*Derivativspektroskopie, Ableitungsspektroskopie*>, the representation of the first, second or higher derivatives of a spectroscopic observable, *y,* with respect to an independent variable, *x*, as a function of this variable. In light absorption, the observable is the → absorbance $A_{\tilde{\nu}}$, for emission (fluorescence) it is the corresponding intensity (more exactly the measured increased photocurrent) and for electron-spin resonance (ESR → magnetic resonance spectroscopy) the intensity of the microwave signal. The independent variable, *x*, is wavenumber, $\tilde{\nu}$, or wavelength, λ, in the first two cases and magnetic field strength, *H*, in the last. The direct recording of the first derivative is the norm in ESR spectroscopy. Derivative spectroscopy has only become common in → UV-VIS and → fluorescence spectroscopy since ca. 1970. This was primarily a consequence of the introduction of computer control in spectrometers which made differentiation of the signal simple. Using the → Bouguer-Lambert-Beer law, in UV-VIS spectroscopy the expressions to be differentiated are:

$$A_{\tilde{\nu}} = \varepsilon_{\tilde{\nu}} \cdot cd$$

or

$$A_{\lambda} = \varepsilon_{\lambda} \cdot cd.$$

ε is the molar extinction coefficient, *d* the path length and *c* the molar concentration of the sample. The first and second derivatives are:

$$\frac{dA_{\tilde{\nu}}}{d\tilde{\nu}} = cd\frac{d\varepsilon_{\tilde{\nu}}}{d\tilde{\nu}};$$
$$\frac{dA_{\lambda}}{d\lambda} = cd\frac{d\varepsilon_{\lambda}}{d\lambda}$$

and

$$\frac{d^2 A_{\tilde{\nu}}}{d\tilde{\nu}^2} = cd\frac{d^2 \varepsilon_{\tilde{\nu}}}{d\tilde{\nu}^2};$$

$$\frac{d^2 A_{\lambda}}{d\lambda^2} = cd\frac{d^2 \varepsilon_{\lambda}}{d\lambda^2}.$$

Starting with the exponential form, $\Phi = \Phi_o \exp[-\varepsilon' cd]$ where ε' = is the natural molar extinction coefficient we obtain:

$$\frac{1}{\Phi}\frac{d\Phi}{d\tilde{\nu}} = -cd\frac{d\varepsilon'}{d\tilde{\nu}}$$

$$\frac{1}{\Phi}\frac{d^2\Phi}{d\tilde{\nu}^2} = -cd\frac{d^2\varepsilon'}{d\tilde{\nu}^2} + c^2 d^2\left(\frac{d\varepsilon}{d\tilde{\nu}}\right)^2.$$

For the analysis of absorption spectra, the table of important correlations between the absorption spectrum (zero order derivative) and the derivative spectra results:

Figure 1 shows the typical behavior of first and second order derivative spectra, which can assume both positive and negative values. For this reason, the zero value of the ordinate (absorbance) is set to half the value of the maximum ordinate. Since $dA_{\tilde{\nu}}/d\tilde{\nu}$ $(d\varepsilon_{\tilde{\nu}}/d\tilde{\nu})$ or $d^2A_{\tilde{\nu}}/d\tilde{\nu}^2$ $(d^2\varepsilon_{\tilde{\nu}}/d\tilde{\nu}^2)$ are extremely sensitive to any change in the gradient of the absorption spectrum, this method is very well suited to the analysis of shoulders or overlapping absorption bands (→ band analysis).

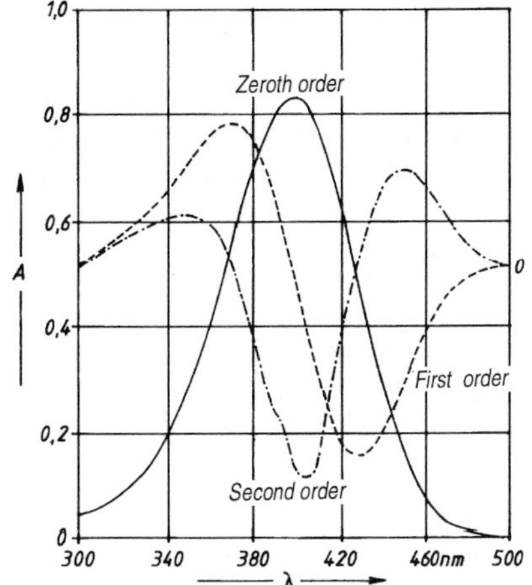

Derivative spectroscopy. Fig. 1. First and second order derivative spectra compared with the zero order absorption spectrum

Derivative spectroscopy has important applications in analysis both in trace-element detection and in quantitative work. The tangent method, peak-peak method and peak-zero method are used in the evaluation of the data.

Derivative spectroscopy. Table.

Absorption spectrum 0th Order	1st Derivative 1st Order	2nd Derivative 2nd Order
$A_{\tilde{\nu},max} \triangleq \varepsilon_{\tilde{\nu},max}$	$\frac{dA_{\tilde{\nu},max}}{d\tilde{\nu}} = 0$	$\frac{d^2A_{\tilde{\nu},min}}{d\tilde{\nu}^2_{min}} =$ Minimum
$A_{\tilde{\nu},min} \triangleq \varepsilon_{\tilde{\nu},min}$	$\frac{dA_{\tilde{\nu},min}}{d\tilde{\nu}} = 0$	$\frac{d^2A_{\tilde{\nu},min}}{d\tilde{\nu}^2} =$ Maximum
$A_{\tilde{\nu},w} \triangleq \varepsilon^{1)}_{\tilde{\nu},w}$	$\frac{dA_{\tilde{\nu},w}}{d\tilde{\nu}} =$ Minimum or Maximum	$\frac{d^2A_{\tilde{\nu},w}}{d\tilde{\nu}^2_w} = 0$

[1)] The index, w, refers to a point of inflexion on an absorption band

Ref.: G. Talsky, L. Mayring, H. Kreuzer, *Angew. Chem. Int. Ed.*, **1978**, *17*, 532; S.M. Kimbrell, K. Booksh, R.J. Stolzberg, *Applied Spectroscopy*, **1992**, *46*, 704; H.-H. Perkampus, *UV-VIS Spectroscopy and its Applications*, Springer Verlag, Berlin, Heidelberg, New York, **1992**.

Deslandres table, <*Deslandres-Tafeln*>. The vibrational structure of the → electronic band spectra of polyatomic molecules ($N \geq 3$) is caused by the superimposition of the excitation of several vibrations upon the electronic excitation. The series formula for nondegenerate vibrations is:

$$\tilde{\nu} = \tilde{\nu}_{00} + [\sum_i \omega_{o,i} v_i' - \sum_i \sum_{k \geq i} x'_{0,i,k} v_i v_k'] - [\sum_i \omega_{o,i} v_i'' - \sum_i \sum_{k \geq i} x''_{0,i,k} v_i v_k'].$$

$\tilde{\nu}_{oo}$ is the wavenumber of the band origin for which $v_i' = v_i'' = 0$, and $\omega_{o,i}$ and $x_{o,i,k}$ are defined as for the → vibrational spectra of polyatomic molecules. The superscript with one or two primes refers to the upper and lower electronic states respectively. The subscripts, i and k, refer to the ith and kth normal vibrations.

As with diatomic molecules, each vibrational state, v_i', can combine

v_1' \ v_1''		0				1				2			
	v_2' \ v_2''	0	1	2	3	0	1	2	3	0	1	2	3
0	0	●	●	●	●	○							
	1	●	●	●	●		○						
	2	●	●	●	●			○					
	3	●	●	●	●				○				
1	0					●							
	1						●						
	2							●					
	3								●				
2	0									●			
	1										●		
	2											●	
	3												●

Deslandres table. Fig. 1. Example of a Deslandres table, see text

with states, v_i'' or vice versa. In addition, every v_i' or v_i'' can combine with v_k'', v_k' or vice versa. These relationships can be summarized in a Deslandres table. Figure 1 shows a case where only two normal vibrations participate in an electronic transition. Each individual square corresponds in principle to the → band-head table for an individual vibration. Figure 1 shows that for polyatomic molecules not only the v' or v'' progressions occur. The scheme can be appropriately extended for more than two vibrations. However, we see that the relationships then become very complicated.
Ref.: G. Herzberg, *Molecular Spectra and Molecular Structure, III. Electronic Spectra and Electronic Structure of Polyatomic Molecules*, D. Van Nostrand, Comp. Inc., Princeton, New Jersey, **1966**, chapt. II, 2, p. 142 ff.

Destriau effect, <*Destriau-Effekt*> → electroluminescence.

Detector, <*Detektor, Empfänger*>, in spectroscopy, an instrumental component which converts incident electromagnetic radiation directly or indirectly into an electrically measurable signal. In the → optical spectroscopy region, radiation detectors are collectively known as → optoelectronics components. They generate an intensity-proportional electrical signal. It is frequently possible, in other spectral regions and when using other spectroscopic methods, to convert the primary process into electromagnetic radiation so that the common radiation detectors can be utilized. Typical examples are the → scintillation counters used in X-ray spectroscopy (→ ESCA) and → secondary electron multipliers in → mass spectrometry. The → Golay detector used in → IR spectroscopy is also a direct radiation detector. The table gives a summary of the different detector types and their applications.

Deuterium lamp, <*Deuteriumlampe*> → hydrogen lamp.

Diabatic scale, log(log) scale, <*Diabatie, spektrale Diabatie*>, a transmittance unit, $\theta(\lambda)$, defined by the equation:

$$\theta(\lambda) = 1 - \log(\log 1/\tau_i)$$

where τ_i is the → transmittance (internal). Since, by definition, the → absorbance, $A_\lambda = \log(1/\tau_i)$:

$$\theta(\lambda) = 1 - \log A_\lambda.$$

τ_i is to be inserted as pure fraction, $0 \leq \tau_i \leq 1$. $A = \infty$ or $A = 0$ correspond to the limiting values $\tau_i = 0$ and $\tau_i = 1$ and hence the limiting values for $\theta(\lambda)$ are:

$$A = \infty \rightarrow \theta = -\infty$$
$$A = 0 \rightarrow \theta = +\infty$$

In accordance with the definition above, $\theta(\lambda)$ provides a double logarithmic correlation with the transmittance, τ_i. The result is that in the typical transmission curves for → colored glass, → edge and → interference filters the regions of low transmission ($\tau < 0.1$) and also those of high transmission ($\tau > 0.9$) are expanded. Figure 1 shows the relationship between θ and τ_i. The value, $\theta = 0$, corresponds to transmittance, $\tau_i = 10^{-10}$ or $A = 10$.

Dichroism, <*Dichroismus*>, the phenomenon that two different absorption spectra are found in doubly refracting substances because an opt-

Detector. Table. Summary of detectors used in different regions

Region	Detector	Principle
X-ray spectrum	pin-diode	internal photoelectric effect
	gas-flow counter	ionization electron current
	→ scintillation counter	conversion of high-energy radiation into visible light, measurement with SEM
Vacuum UV	photomultiplier, secondary electron multiplier (without window)	external photoelectric effect
	photographic plate	blackening
UV	photomultiplier, photocell	external photoelectric effect
	photoelement (Se, Si), photodiode array	internal photoelectric effect
VIS	photomultiplier, photocell	external photoelectric effect
	photoelement, photodiode, photodiode array	internal photoelectric effect
NIR	photocell	external photoelectric effect
	photoelement (Ge), photodiode (Ge), photoresistive cell	internal photoelectric effect
IR	thermocouple, thermopile	thermal voltage
	bolometer	temperature dependence of the resistance
	Golay cell (pneumatic detector)	pressure change with optical detection system
ESR	crystal diode	barrier-layer (*p-n*) rectifier
NMR	high-frequency coil	induction process
Mass spectroscopy	Faraday cup	charge, voltage decrease
	secondary electron multiplier (without window)	photoelectric effect
	photographic plate	blackening

ically uniaxial crystal gives a characteristic absorption curve for both the ordinary and the extraordinary ray. The absorption maxima of these curves can be very different, whilst in an optically isotropic medium the

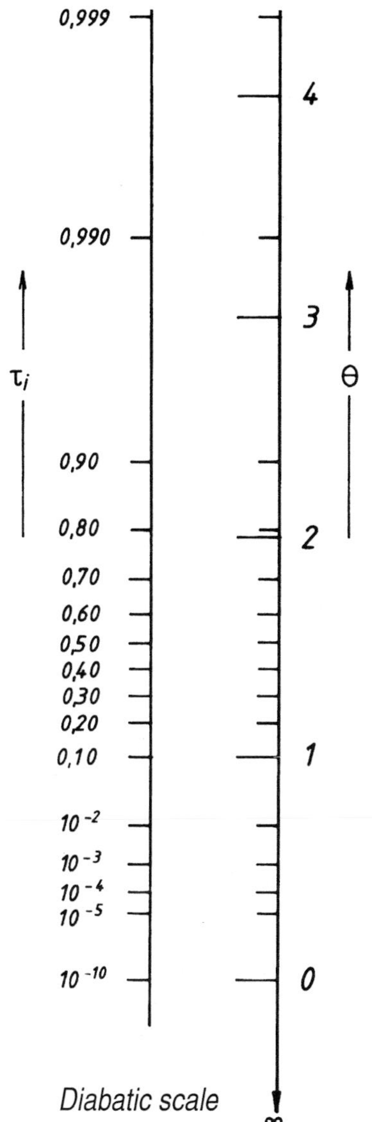

Diabatic scale. Fig. 1. Relationship between diabatic scale, θ, and transmission, τ_i

absorption of electromagnetic radiation is independent of the direction of the incident ray and its polarization. In most doubly refracting materials used for the construction of → polarizers, such as → calcite or quartz, the absorption lies in the UV and/or IR regions of the spectrum. It is for this reason that they are preferred for use in the central, near UV and VIS regions. Suppose that the absorption maxima lie at different wavelengths in the VIS region, e.g. at 650 nm for the ordinary ray and at 500 nm for the extraordinary ray. Thus, if the path length is sufficient, wavelengths around 650 nm or 500 nm will be missing in the transmitted light depending on whether the vibrational plane lies perpendicular or parallel to the optic axis. This means that the substance is colored, even in natural light, and that the color depends upon the direction of vibration of the (polarized) incident light. This phenomenon is called dichroism and in optically biaxial crystals trichroism. Crystals, in which the ordinary and extraordinary rays show extremely different degrees of absorption, are of particular interest. In this case for example, the ordinary ray polarized perpendicular to the optic axis may be completely absorbed by a small path length so that only the extraordinary ray remains when transmitting daylight. Such a crystal is a polarizer for light within a specific spectral region and crystals of this type are called dichroic polarizers. The best-known example is the → tourmaline plate.

Didymium-glass filter, <*Didymglasfilter*>, a filter made of a glass to which rare-earth oxides (neodymiun and praseodymium oxide) have been added. In contrast to other → colored glass filters, didymium filters have narrow absorption bands in the VIS and NIR region. They are therefore frequently used for calibrating and checking → UV-VIS-NIR spectrophotometers.

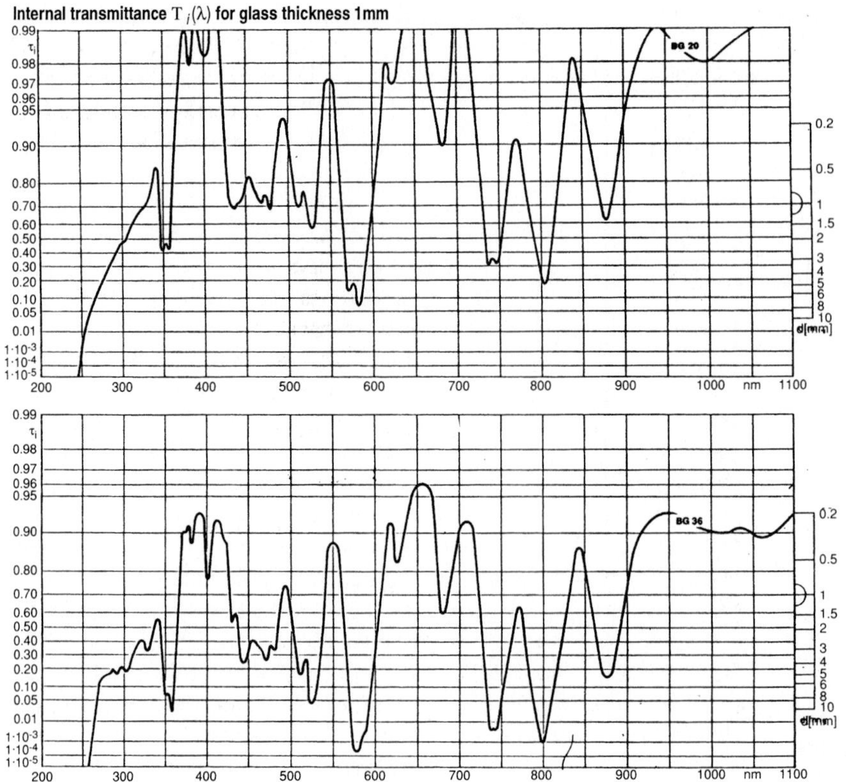

Didymium-glass filter. Fig. 1. Transmission curves of filters BG 20 and BG 36

Figure 1 shows the transmittance curves of the Schott special filters BG 20 and BG 36. The name *didymium* is based on the fact that its discoverer at first thought it to be a single rare-earth element. Subsequently, it was found to be a mixture of praseodymium and neodymium.

Dielectric layer, <*dielektrische Schichten*>, surface layers for reducing reflection at glass surfaces and therefore very important in the production, for example, of → interference filters. Thin metallic layers show a high degree of reflection if the thickness is sufficient but are partially transparent and absorb strongly in very thin layers. In contrast, dielectric layers are practically absorption free. However, the absorption-free spectral region depends upon the material used. If dielectric layers are inserted between two media such as air and glass they are exceptionally well suited for the redistribution of energy between reflection and transmission. For an intermediate layer with refractive index, n_s, we obtain a minimum total reflection at the interfaces air/dielectric (n/n_s) and dielectric/glass (n_s/n') when $n_s = \sqrt{n \cdot n'}$, where n and n' are the refractive indexes of air and glass. The relation, $n_s = \sqrt{n \cdot n'}$, is also called the amplitude condition.

The following dielectric materials are used most frequently:

Cryolite (Na_3AlF_6);	n_{550}	= 1.35; transparent 200–14,000 nm
Magnesium fluoride (MgF_2);	n_{550}	= 1.38; transparent 120– 8,000 nm
Cerium fluoride (CeF_3);	n_{550}	= 1.63; transparent 300– 5,000 nm
Zinc sulfide (ZnS);	n_{550}	= 2.32; transparent 390–14,000 nm
Titanium dioxide (TiO_2);	n_{550}	= 2.40; transparent 400–12,000 nm

together with silicon, germanium and lead telluride which have high refractive indexes. MgF_2 is the most important material for dielectric layers and produces, when evaporated onto the substrate, a hard and quite resistant layer which is transparent in the short-wavelength UV region. Zinc sulfide provides an adequately hard and resistant layer with a high refractive index in the VIS and NIR regions. Titanium dioxide is also transparent in the VIS and IR region and provides hard and resistant layers with a high refractive index.

By carefully tuning the thickness of the layers using → interference, a considerable improvement in the energy distribution between reflection and transmission can be achieved. Figure 1 illustrates this using a single layer. However, two conditions must be adhered to, i.e.:

the phase condition

$$n_s d_s = (2m + 1)\lambda/4$$

and the amplitude condition

$$n_s = \sqrt{n \cdot n'}$$

n_s and d_s are the refractive index and thickness of the dielectric layer, m = 0,1,2,3 ... the order number of the interference and λ = the wavelength in air. The phase condition states that the optical path, $n_s \cdot d_s$, in the layer must be an odd multiple of $\lambda/4$. Since a ray reflected on the rear surface passes through the layer twice (for perpendicular incidence) the two reflected beams are shifted with respect to each other by an odd multiple of $\lambda/2$, i.e. an

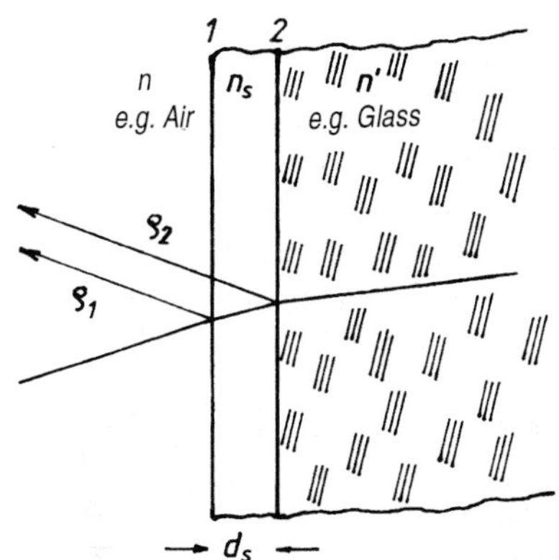

Dielectric layer. Fig. 1. Construction of a dielectric layer

interference minimum results. This minimum becomes zero, i.e. the reflection is annihilated if the amplitudes of the two reflected waves are equal. The interference of many reflections is utilized in thin-layer → interference filters. In an → MDI filter, a transparent dielectric layer (e.g. MgF_2) having $d = \lambda/2$ lies between two partially transparent metal-mirror layers. Alternating dielectric layers with different, high and low refractive indexes are combined in → ADI filters.

Ref.: H. Naumann, G. Schröder, *Bauelement der Optik* (*Optical Components*), Carl Hanser Verlag, Munich, Vienna, **1987**, 5th ed., p. 71 ff. Z. Knittl, *Optics of Thin Films*, J. Wiley and Sons, London, New York, Sydney, Toronto, **1976**.

Diffraction dispersion, <*Beugungsdispersion*>, the → dispersion of light caused by diffraction by a diffraction grating (→ grating).

Diffraction edge, *<brechende Kante>*, the line where the two transparent faces of a → dispersing prism intersect.

Diffraction grating, *<Beugungsgitter>* → grating, → diffraction of light.

Diffraction of light, *<Beugung von Lichtwellen>*. Departure from the laws of geometrical optics in the propagation of light. Diffraction always occurs if the free propagation of waves is hindered by obstacles in the light beam such as apertures, slits, gratings etc. This departure from the normally observed rectilinear propagation, which is the basis of geometrical optics, was explained by A. Fresnel in 1818 using a combination of Huygens principle of elementary wavelets and the interference principle of Young. According to Huygens principle, an illuminated aperture (hole or slit) acts as a secondary light source which sends out light in all directions. However, the aperture must not be large in comparison with the wavelength of light emitted by the primary light source (coherence condition). The secondary light source differs from a luminous surface in that the light emitted from all points is coherent and thus capable of → interference on account of its origin in the same light source. Diffraction phenomena which are caused by a regular array of slits are of particular importance, especially in spectroscopy. Fraunhofer first carried out investigations of diffraction at a → grating in 1821. Consider an arrangement of p parallel slits of width b, lying in a plane. The distance between two slits, measured from center to center, is denoted by a and called the period of the grating or grating constant, see Figure 1. For diffraction at angle φ, the path difference of the beams from the edges of slit 1 is $d_1 = b \cdot \sin\varphi$, which corresponds to a phase difference of $\delta_1 = 2\pi d_1/\lambda = 2\pi b(\sin\varphi)/\lambda$. This applies to every slit. The phase difference, δ, with respect to the edge of the beam at slit 1, increases for the other slits in a whole-number multiple of the grating constant, a, which can be easily seen at d_2 and d_4 in Figure 1. All beams are therefore in a fixed phase relationship and coherent. If the p beams diffracted at angle φ are brought together by a lens then interference occurs, a multiple interference in fact. Alternating interference maxima and minima, which occur at quite specific intervals, are observed. The salient fact can be

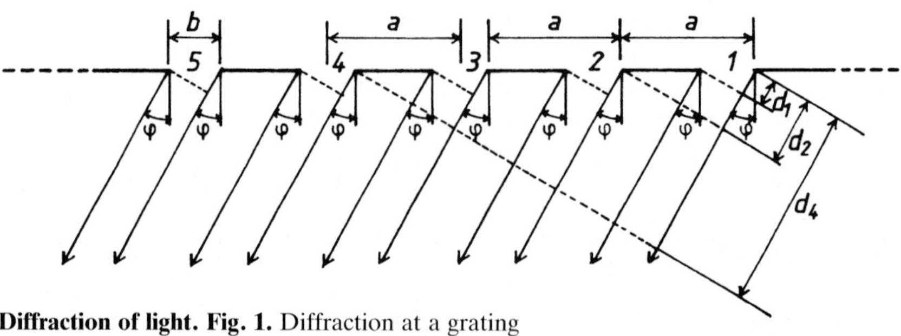

Diffraction of light. Fig. 1. Diffraction at a grating

expressed as follows. When light is diffracted at many equally spaced, adjacent apertures (slits) the principal diffraction maxima are found at the points:

$$\sin \varphi_k = \frac{k \cdot \lambda}{a} \quad \text{with } k = 0, 1, 2, 3 \ldots$$

The grating constant, a, gives the separation of the centers or homologous points of two adjacent apertures (slits); the greater the number of diffracting apertures the more intense and narrow are the diffraction maxima. Figure 2 shows this for $p = 2$, 4 and 8. We find that $(p - 2)$ submaxima, which become more numerous but smaller with increasing p, lie between the principal maxima. For large p, which is the case for gratings used in optics, only the principal maxima remain and the intervals are dark. The k values indicate the order of the diffraction maxima. Zero order, $k = 0$, is the undiffracted beam the intensity of which is proportional to p^2. For the

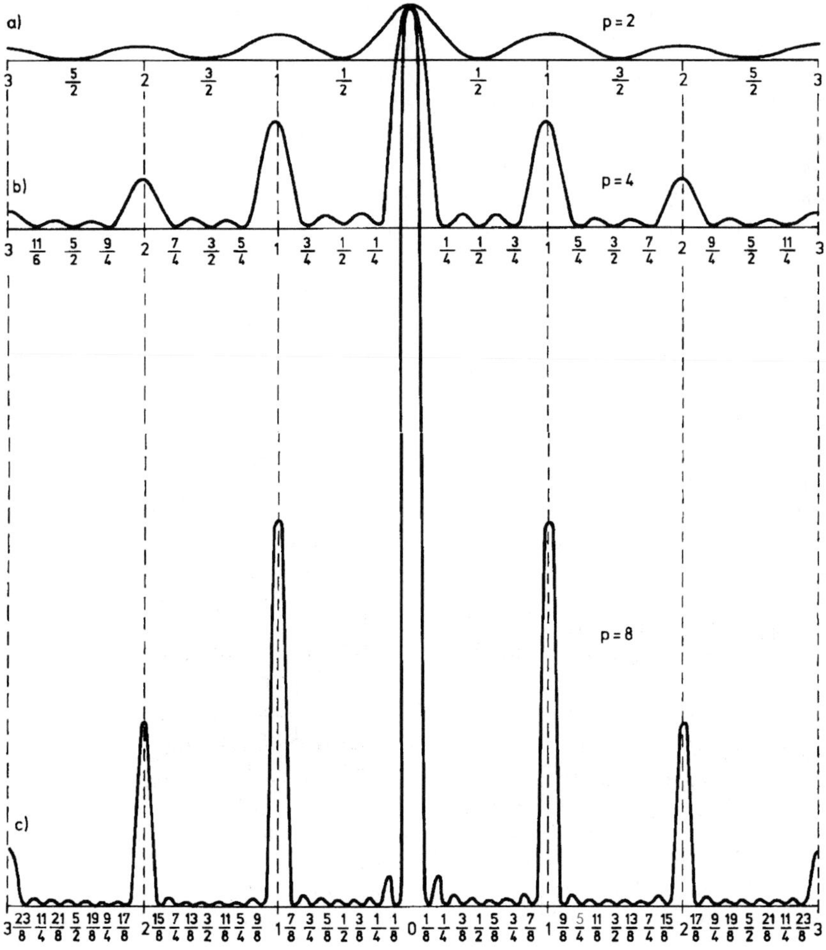

Diffraction of light. Fig. 2. Interference after diffraction at 2, 4 and 8 slits

higher orders, $(k > 0)$, pairs of diffraction maxima are symmetrically placed on either side of the intense maximum of the zero order. From the first order onwards, therefore, the intensity of the principal maxima is distributed equally to two spatially separated sites. The equation above also shows that the intervals between the principal maxima are proportional to the wavelength, i.e. a grating acts as a → dispersing element. The importance of gratings in → optical spectroscopy is based on this fact.

The grating or transmission grating discussed here has the disadvantage that half of the radiant energy is lost when it is used in optical instruments. The relationship formulated above applies also to the case where we observe interference phenomena from a grating using reflected rather than transmitted light. Rutherford and Rowland developed reflection gratings at the end of the last century and Woods the → echelette grating in 1910. These are the types of grating used in most modern grating monochromators.

Diffuse reflection, <*diffuse Reflexion*> → reflection of light → reflection spectroscopy.

Diffuse series, first subseries, <*diffuse Nebenserie, 1. Nebenserie*> → alkali spectra.

Dilution method, <*Verdünnungsmethode*>, an important experimental method in → reflection spectroscopy. It is always used when regular (specular) reflection must be eliminated. The powder (pigment) under investigation is diluted with an inert, nonabsorbing standard such as MgO, $NaCl$, $BaSO_4$, SiO_2 or TiO_2 in such a large excess that, within the accuracy of the method, the regular reflection is eliminated by a relative measurement against the same pure standard. This extremely well-proven method, developed in the mid fifties, made possible the experimental testing of the Kubelka-Munk theory under conditions which are consistent with the assumptions made in that theory (→ reflection spectroscopy). In the case of solid, insoluble substances, dilution is achieved by grinding with an excess of the standard in a mortar or a small ball mill until the mixture is homogeneous. Either a simple homogeneous mixture of the crystallites is obtained or the sample is absorbed as a molecular dispersion on the surface of the standard. The latter is usually the case when organic solids are ground with inorganic standards and the reflection spectrum then obtained is that of the absorbed substance, which can differ considerably from that of a nonabsorbed sample. Soluble substances can also dissolve into the surface of the standard.

The dilution method has a number of advantages for investigations in reflection spectroscopy:

1. If the substance under investigation has a large absorption coefficient the measured R_∞ value can be moved into the range, $0.1 \leq R_\infty \leq 0.7$. Figure 1 shows this for successive dilutions of the pigment Hostapermrot E3B with MgO. The plot for the undiluted pigment (f) shows only a weak absorption maximum in the region of ~ 460 nm to 370 nm. However, with increasing dilution the peak is seen to be strong. As in → photoacoustic spectroscopy, we have here a → saturation

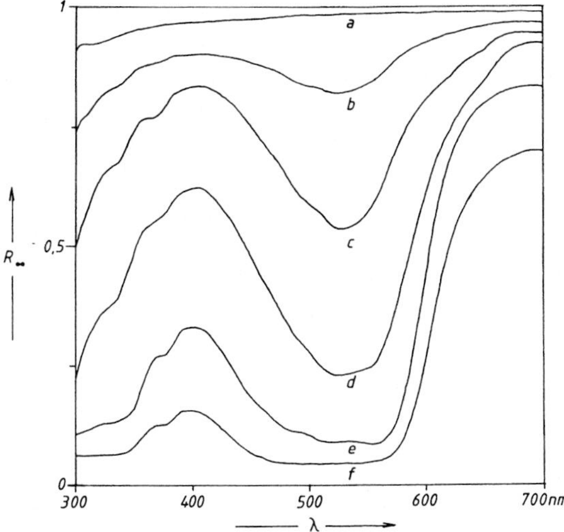

Dilution method. Fig. 1. Reflection spectra of the pigment Hostapermrot E3B diluted with MgO; a) pure MgO; b, c, d, e) pigment: MgO = 1:1000, 1:100, 1:10, 1:1; f) pure pigment

effect with the pure pigment, i.e. R_∞ is independent of the → absorption coefficient.

2. The scattering coefficient of the mixture is determined almost exclusively by that of the dilution agent, which can always be measured in these nonabsorbing or weakly absorbing substances.
3. Possible deviations from an isotropic scattering distribution are eliminated in a relative measurement against the pure dilution agent so that the method is independent of the experimental configuration.
4. The contribution of regular reflection is of no importance in the measurement of the diluted sample against the pure dilution agent.
5. If the substance under investigation is molecularly dispersed on the standard, the particle-size dependence of the absorption coefficient is also eliminated.

The dilution method is also used in → photoacoustic spectroscopy. The two methods are complementary since it can be assumed as an initial approximation that the photoacoustic signal is proportional to $(1 - R_\infty)$.

Ref.: G. Kortüm, *Reflection Spectroscopy*, Springer Verlag, Berlin, Heidelberg, New York, **1969**.

Diode array detector, <*Diodenarraydetektor*>, a number, usually a power of two such as 256 or 512, of → photodiodes and capacitors connected in parallel. The condenser is charged to a specific operating voltage, usually 5 V, prior to application in measurement. When light falls onto the diode, a photocurrent is generated which discharges the capacitor. The degree of discharge depends upon the exposure time and intensity. This process is the functional principle of a diode array detector. An external electronic unit measures the degree of discharge whilst the capacitor is recharged to the

operating voltage. The electronic signal is transmitted via an A-to-D converter to a computer which calculates the absorption.
The whole measurement procedure is continuous and every photodiode is controlled via a shift register and switching transistor. The switching transistors connect the diodes in sequence to a common data line for a fixed cycle time which is usually characteristic for each diode array. Commercial diode arrays require 4 μs per diode to collect information. A total time of 0.25 ms to ca. 2 ms is therefore required for recording the complete information from 512 diodes.

Diode array spectrometer, *<Diodenabsorptionsspektrometer>* → UV-VIS diode array spectrometer.

Diode laser, *<Diodenlaser>*, a semiconductor laser the construction of which corresponds to that of a → light-emitting diode. Light-emitting diodes with a suitable geometrical construction (resonator) become diode lasers at a high current density. They are light-emitting diodes with a *p-n* barrier layer and parallel end faces arranged so as to form an optical resonator. By applying a voltage in the conducting direction, charge carriers are injected on both sides of the *p-n* junction and radiation is emitted when they recombine. However, laser activity (self-excitation) occurs only above a certain current, the threshold current, which is strongly dependent upon the temperature. The higher the temperature the higher the threshold current; typical values are 100–500 mA, maximum 2A. Diode lasers must therefore be operated at low temperatures; $T < 100$ K. Figure 1 shows the

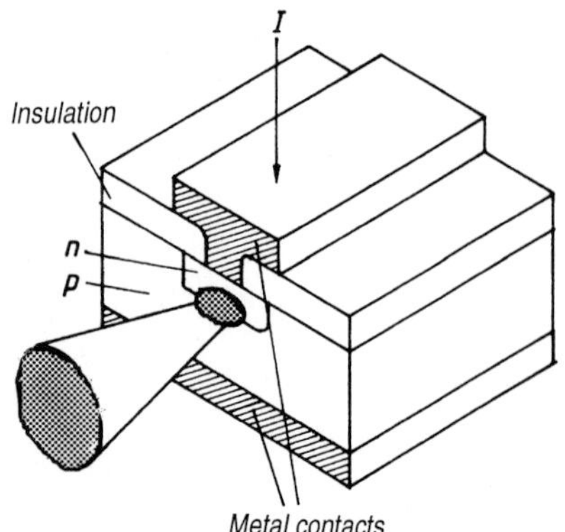

Diode laser. Fig. 1. Schematic drawing of a diode laser

schematic construction of a diode laser. The wavelength emitted by a diode laser is determined by its band gap which, in turn, depends upon the composition of the semiconductor. Figure 2 shows the emission ranges of various semiconductors. IR diode lasers, which emit in the wavelength region of 3–30 μm, are of particular importance in spectroscopy.
The basic materials of these lasers are semiconducting compounds of lead, tin, sulfur, selenium and tellurium; the lead-(tin) chalcogenides: $Pb_{1-x}Sn_xSe$ (Figure 2). Depending on the stoichiometry, a *p* or *n* type material is obtained.
Figure 3a shows that, depending upon the type, the emitted wavelength of an IR diode laser can be varied over a range of ca. 200 cm^{-1} by means of the temperature. However, more exact observation shows that a continuous tuning of an individual mode can only be observed over a range of 1 cm^{-1}. Figure 3b shows this for the rectangle drawn in Figure 3a. The reason is the temperature dependence of the refractive index, dn/dT, and also the value of $\Delta E = E_2 - E_1$. A change in

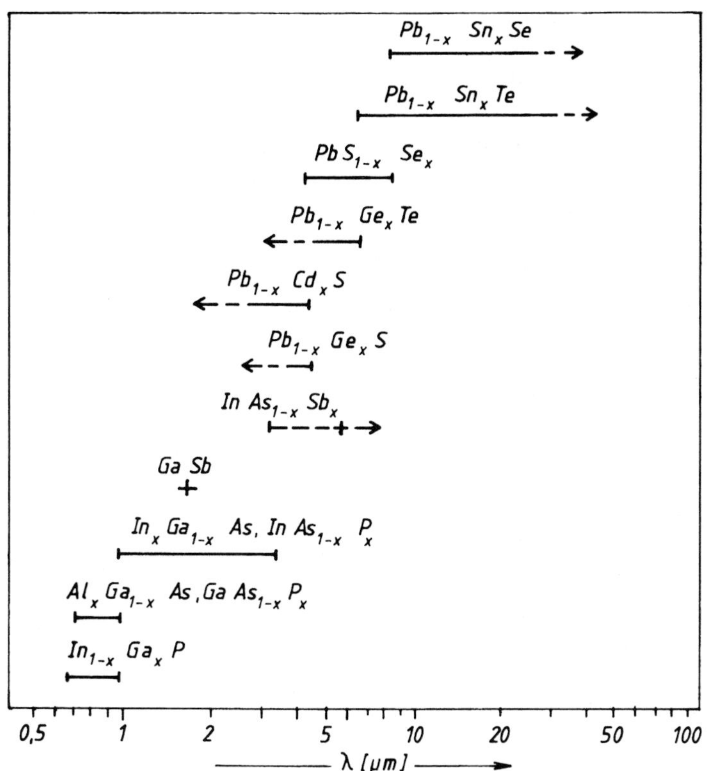

Diode laser. Fig. 2. Emission regions of various semiconductors

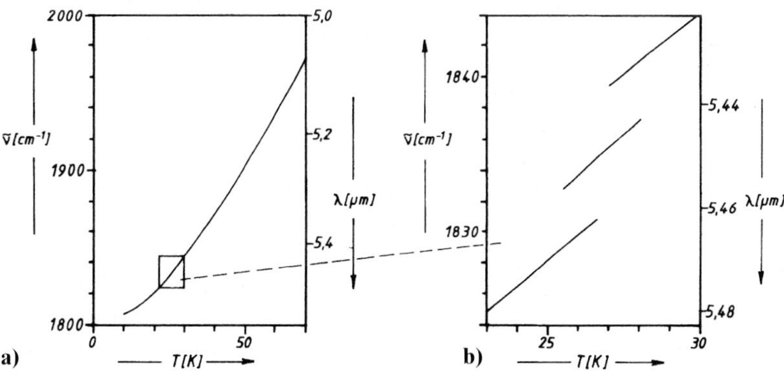

Diode laser. Fig. 3. a) Dependence of the emitted wavenumber upon temperature; b) enlarged section of Figure 3a

temperature of ΔT changes the energy difference, ΔE, and the refractive index by $\Delta n = (\mathrm{d}n/\mathrm{d}T)\Delta T$. However, the resulting shift, $\Delta\tilde{\nu} = \tilde{\nu}\Delta n/n$ of the resonator's natural wavenumber, $\tilde{\nu}$, is only ca. 10–20% of the shift, $\Delta\tilde{\nu}_g =$

$\Delta E/hc$. As soon as the maximum of the gain profile reaches the next resonator mode, the gain for this mode becomes larger than that of the original and the laser wavenumber jumps to a new mode (mode hopping). However, the actual laser temperature is not determined solely by the temperature of the heat sink but also by the heating due to the diode current flow. Thus, variation of the current flow is the simplest and quickest means of fine tuning. Here we find a picture analogous to Figure 3b. Gaps of ca. 2 cm^{-1} occur between the individual straight tuning lines. The positions of the spectral gaps can be easily shifted by changing the diode current and/or temperature. However, a complete closing of the gaps generally requires another diode laser. When operated with current pulses, the emitted frequency changes during the pulse due to the heating. A rapid, very fine tuning in the μs region is achieved during this operation. However, it is vital that the line widths of the individual modes do not exceed 10^{-4} cm^{-1}. With a tuning range in the individual modes of ca. 1 cm^{-1}, IR diode lasers can be used for high-resolution measurements of the → rotation-vibration spectra of gases. In practical applications, attention must be paid to the particular properties of IR diode lasers. They must be cooled to low temperatures, the laser current supply must be extremely stable (< 10 μA noise) and a → monochromator should be available for mode selection. This concept underlies the development of IR diode laser spectrometers, → tunable diode laser absorption spectrometer.

Diode laser absorption spectrometer, <*Diodenlaser-Absorptionsspektrometer*> → tunable diode laser absorption spectrometer.

Direct vision prism, <*Geradsichtsprisma*>, generally a combination of two or more prisms of materials having different refractive indexes. A beam of light incident upon a direct vision prism is dispersed but, for a specific wavelength, does not incur a deflection. Such prisms are preferred for the construction of manual spectroscopes or where compactness of the instrument is desirable. The → Amici, Wernicke and → Zenker prism are direct vision prisms.

Discharge lamp, <*EDL, Entladungslampe*> → electrodeless discharge lamp (EDL).

Dispersing element, <*Dispersionselement*>, a prism or grating used for the → spectral dispersion of light. Dispersion elements are essential components of → monochromators (→ prism monochromator and → grating monochromator).

Dispersing prism, <*Dispersionsprisma*>, a prism in prism spectrographs or prism spectrometers for the spectral dispersion of UV, VIS, NIR or IR radiation. Specific → prism materials must be used depending on the spectral region.
The most important types of dispersing prisms are → direct vision prisms such as the → Amici, → Rutherford and → Zenker prims and also prism systems with constant deviation such as → autocollimation and → Abbe prisms which are generally used in prism spectrometers.

Dispersion, <*Dispersion*>, the phenomenon that, in the visible region, the → speed of light continuously decreases on going from long wavelengths (red) to short wavelengths (violet), while the refractive index increases in the same direction. The dependence of the refractive index of a substance upon → wavelength, λ, is based upon the fact that the → phase velocity of an electromagnetic wave depends upon the wavelength in all substances (media) except a vacuum. White light, which can be regarded as a mixture of all spectral colors, is split into those colors, which have different wavelengths, upon refraction by a substance (→ spectral dispersion). In → optical spectroscopy, dispersion is generally understood to be the spectral dispersion of electromagnetic radiation (→ light).

Dispersion curve, *<Dispersionskurve>*, the graphical representation of the → refractive index as a function of wavelength. The refractive index, n, of a substance generally increases continuously from longer to shorter wavelengths in the visible spectral region and this is called normal dispersion. There are also regions of anomalous dispersion which are characterized by an absorption of electromagnetic radiation, i.e. a → resonance occurs between a natural frequency of the substance and the frequency of the electromagnetic radiation. Figure 1 shows a dispersion curve in the region of an absorption band. The region of anomalous dispersion contains not only an absorption maximum but also a reflection maximum. If the damping is small, the reflection in an otherwise transparent substance can become as large as in metals whilst it is much smaller in directly adjacent regions. In the IR, the → residual ray method of Rubens is based on this. Absorption, and hence anomalous dispersion by the → prism materials used, is also reached with increasing wavelength in the IR spectral region (see Figure 2).

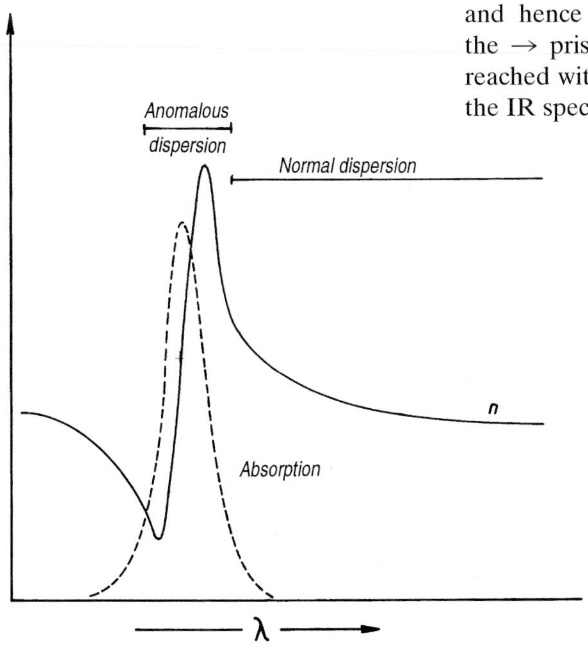

Dispersion curve. Fig. 1. An anomalous dispersion curve

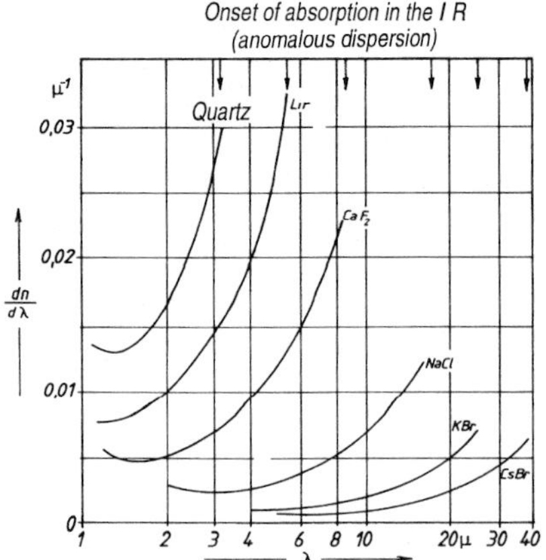

Dispersion curve. Fig. 2. Dispersion curves for different prism materials in the IR

Dispersion filter, *<Dispersionsfilter>*, a filter for isolating several spectral regions. If we use a lens with a strong dispersion of refraction for imaging a continuous light source a → chromatic aberration occurs, i.e. the position and size of the image are not coincident for different wavelengths. We can make use of single or multiple intermediate imaging of such radiation by means of uncorrected lenses. By means of a series of lenses which are immersed in a liquid with a similar refractive index but different dispersion, it is possible to produce simple monochromators. The Christiansen dispersion filter of 1884 is obtained if glass or quartz, powder instead of lenses, is immersed in a liquid with a similar refractive index. Only those wavelengths for which powder and liquid have the same refractive index pass through without deviation, the remainder is scattered as in a turbid medium. If the temperature is raised the maximum of transmission, λ_{max}, can be shifted to shorter wavelengths because of the large change of the refractive index of the liquid. Therefore, the filter requires highly accurate temperature control.
Ref.: G. Kortüm, *Kolorimetrie, Photometrie and Spektrometrie, (Colorimetry, Photometry and Spectrometry)*, Springer Verlag, Berlin, Göttingen, Heidelberg, **1962**, 4th ed. p. 86 ff.

Dispersion spectroscopy, *<Dispersionsspektroskopie>*, a branch of spectroscopy which utilizes the dependence of → refractive index, n, upon wavelength, λ (→ dispersion and → dispersion curve). The dispersion curve as provider of spectroscopic information has found little use in optical spectroscopy to date, but the method has become increasingly important with the development of new techniques.
The method is based upon the interference spectrum (→ interference spectroscopy) which can be measured on very thin plates or foils (< 50 μm) or in cuvettes with extremely small path lengths (< 10 μm). With polychromatic light, an interference pattern is obtained which is determined both by the optical path length, $n \cdot d$, and the wavelength dependence of the refractive index, n. Taking account of the phase change of $\lambda/2$, the interference condition for order m is:

$$\frac{2\pi}{\lambda}(2nd + \lambda/2) = 2\pi m$$

or

$$2nd + \frac{\lambda}{2} = m\lambda.$$

For two adjacent interference maxima with wavelengths, λ_1 and λ_2, we have:

$$2nd + \lambda_1/2 = m\lambda_1 \text{ and}$$
$$2nd + \lambda_2/2 = (m + \delta_m)\lambda_2$$

or

$$\frac{2nd}{\lambda_1} + \frac{1}{2} = m \text{ and } \frac{2nd}{\lambda_2} + \frac{1}{2} = m + \delta_m.$$

The difference gives the well-known equation:

$$2nd = \frac{\lambda_1 \cdot \lambda_2}{\lambda_1 - \lambda_2} \delta_m \ (\delta_m = 1, 2, 3 \ldots),$$

with which the path length of a cuvette or a thin layer can be determined. If path length, d, and order, m, are known, the refractive index as a function of the wavelength is given as $n(\lambda) = (m \cdot \lambda)/(2d)$. The order, m, can also be obtained from the interference spectrum because for two adjacent interference maxima, λ_m and λ_{m+1}, i.e. $\delta_m = 1$ and a known path length, d, we have:

$$m = \frac{\lambda_{m+1}}{\lambda_m - \lambda_{m+1}} \cdot \delta_m \ (\delta_m = 1!).$$

However, this equation is strictly true only if the interference spectrum is measured with a cuvette filled with air ($n \sim 1$).

Figure 1 shows the interference spectrum for water at 25 °C (a), and the dispersion spectrum obtained from it (b). The experimental arrangement is shown in Figure 2. A specially developed flow cell with a path length of $\sim 6\ \mu$m and a volume of ca. 30 nl is the central feature. The cell is connected to the light source and a photodiode array spectrometer (→ multichannel spectrometer) via Y-shaped fiber optics.

The dispersion spectrum shown in Figure 1b is normal dispersion (→ dispersion curve). However, it is important to note that dispersion spectra in the region of anomalous dispersion can be measured with this instrumental setup. Since these spectra correspond to normal absorption spectra they are important analytically. Because of the small sample volume of the flow cell and the rapid recording of spectra using multichannel spectrometers, the future principal applications would appear to lie in HPLC (high-performance liquid chromatography) as shown recently by G. Gauglitz.

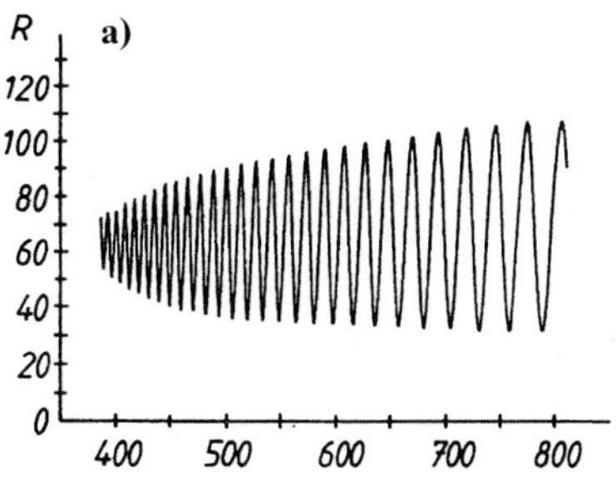

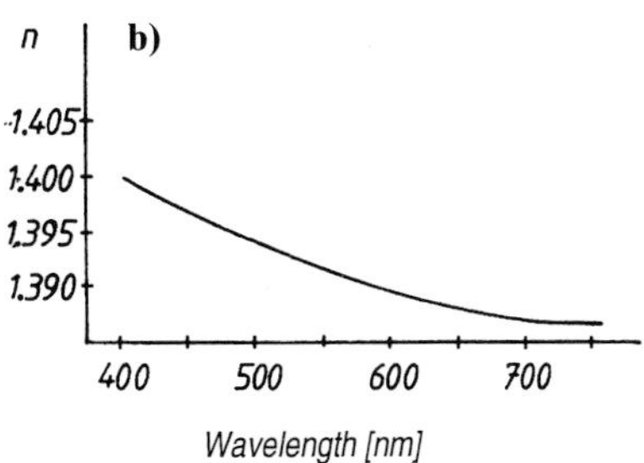

Dispersion spectroscopy. Fig. 1. a) Interference spectrum of water at 25 °C; b) dispersion curve of water

Ref.: H.H. Schlemmer, M. Mächler, *J. Phys. E., Sci. Instrum.*, **1985**, *18*, 904; K.-P. Koch, H. Schlemmer, M. Mächler, *Fresenius Z., Anal. Chem.*, **1986**, *325*, 544; G. Gauglitz, J. Krause-Bonte, H. Schlemmer, A. Matthes, *Anal. Chem.*, **1988**, *60*, 2609; M. Mächler, H. Schlemmer, *Zeiss Inform.*, **1989**, *30*, 16.

Dispersion, unit of, <*Dispersionsmaß*>, the unit, dn/dλ. If the → dispersion curve with n as a function of wavelength λ is determined experi-

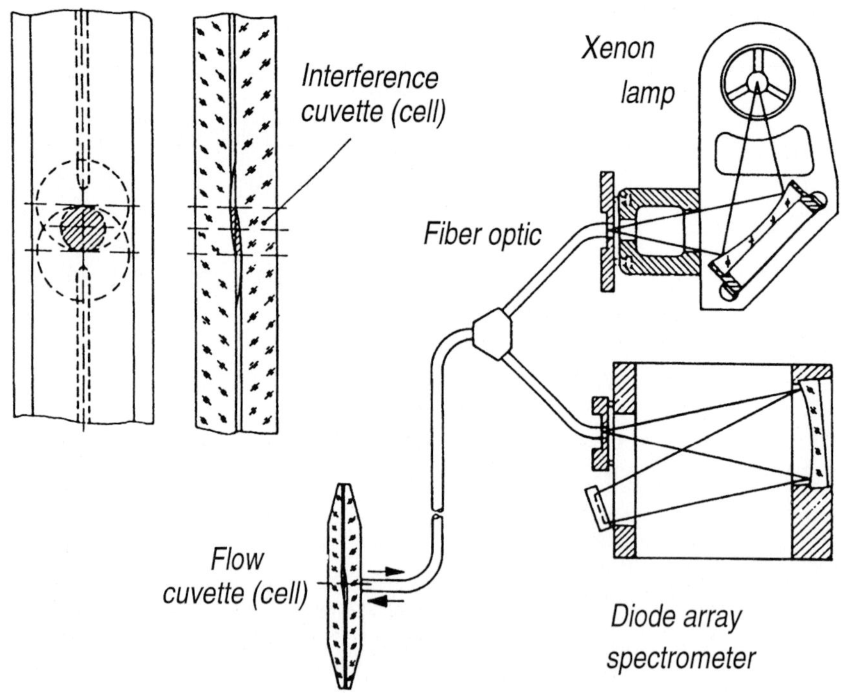

Dispersion spectroscopy. Fig. 2. Apparatus for measuring interference spectra

mentally, dn/dλ is the magnitude of the dispersion, i.e. the inclination of the tangent to the curve at every point n, λ. In addition to this exact unit, practical units are used which relate to specific wavelengths and are generally determined at the → Fraunhofer lines. The table gives the refractive indexes of various substances for the most important Fraunhofer lines.

The deviation, which a light beam of a specific wavelength undergoes in a prism with prism angle, ε, is given by: $\delta_\lambda = (n_\lambda - 1)\varepsilon$. The "total dispersion,

Dispersion, unit of. Table.

Material	The dispersion of selected materials				
	$\lambda_A = 760.8$	$\lambda_B = 686.7$	$\lambda_C = 656.3$	$\lambda_D = 589.3$	$\lambda_E = 527.0$
	n_A	n_B	n_C	n_D	n_E
Water	1.3289	1.3304	1.3312	1.3330	1.3352
Fluorite	1.4310	1.4320	1.4325	1.4338	1.4355
Boron crown glass BK1	1.5049	1.5067	1.5076	1.5100	1.5130
Heavy crown glass SK1	1.6035	1.6058	1.6070	1.6102	1.6142
Flint glass F3	1.6029	1.6064	1.6081	1.6128	1.6190
Calcite (o. ray)	1.6500	1.6529	1.6544	1.6584	1.6634

Dispersion, unit of. Table, continued.

Material	The dispersion of selected materials				
	$\lambda_F = 486.1$	$\lambda_G = 430.8$	$\lambda_H = 396.9$		
	n_F	n_G	n_H	$n_F - n_C$	$\frac{n_D - 1}{n_F - n_C}$
Water	1.3371	1.3406	1.3435	0.0059	56.4
Fluorite	1.4370	1.4398	1.4421	0.0045	96.4
Boron crown glass BK1	1.5157	1.5205	1.5246	0.0081	62.9
Heavy crown glass SK1	1.6178	1.6244	1.6300	0.0108	56.5
Flintglass F3	1.6246	1.6355	1.6542	0.0165	37.0
Calcite (o. ray)	1.6679	1.6761	1.6832	0.0135	48.8

θ", is defined by forming the difference $\delta_H - \delta_C = (n_H - n_C)\varepsilon$ related to the Fraunhofer lines H (396.85 nm) and C (656.27 nm). The difference, $n_H - n_C$, is called the specific dispersion, σ_{spec} of the material concerned; it follows that $\theta = \sigma_{spec} \cdot \varepsilon$. A partial dispersion is defined as the difference for other Fraunhofer lines. Dispersion, $n_F - n_C$, determined for the most intense part of the visible spectrum between lines C (656.27 nm) and F (486.1 nm) is the average dispersion listed in the penultimate column of the table. The dispersive power or relative dispersion, σ_{rel}, is defined by the equation:

$$\sigma_{rel} = \frac{n_F - n_C}{n_D - 1}.$$

where n_D is the refractive index at the D line (589.9 nm). The Abbe number, v, is the inverse of σ_{rel}:

$$\nu = \frac{n_D - 1}{n_F - n_C};$$

(see last column of the table).

The Abbe number measures the relationship between refraction and dispersion. If the average refractive index, n_D, and Abbe number, v, are known, an approximate characterization of the optical properties of a glass can be given. Glass types with high dispersion have a large average dispersion ($n_F - n_C$) and thus a small Abbe number whilst glasses with low dispersion have a small average dispersion and a high Abbe number.

Ref.: Bergmann-Schäfer, *Lehrbuch der Experimentalphysik, (Textbook of Experimental Physics)*, vol. III, Optik, 7th edtn., Ed.: H. Gobrecht, Verlag Walter de Gruyter, Berlin, New York, **1978**, p. 204 ff.

Dispersive spectrometer, *<dispersives Spektrometer>*, operates exclusively with spectrally dispersed radiation. Consequently, a → monochromator is a particularly important central part of the whole spectrometer. Similarly, its → dispersing elements such as → prisms or → gratings are essential components. A dispersing instrument allows measurements over a continuous wavelength range within the corresponding → UV-VIS, NIR and IR spectral region.

Displacement coordinate, *<Displacement-Koordinaten>*, coordinates which describe the movements (displacements) of atoms from their equilibrium positions during a harmonic vibration. These movements may be expressed in terms of Cartesian coordinates, x, y and z where the origins of the coordinate systems are placed at the equilibrium positions of the atoms. $3N$ coordinates are obtained for an N atomic system, i.e. in addition to the internal movements of the molecule (vibrations), this representation also includes the rotations and translations.

Dissociation energy, spectroscopic, <*Dissoziationsenergie, spektroskopische Dissoziationsenergie*>. The vibrational spectrum of a diatomic molecule may be described by the → anharmonic oscillator model (also → Morse curve). The corresponding potential energy curve in Figure 1 shows an asymptotic behavior for large departures from the equilibrium bond length which corresponds to the dissociation of the molecule. The height of the asymptote above the lowest vibrational level, $v = 0$, indicates the energy which is required to dissociate the molecule, i.e. the dissociation energy, which is denoted by D_o relative to $v = 0$ or by D_e relative to the potential minimum. D_e is slightly larger by the amount of the → zero-point energy. The potential curve is filled with vibrational quantum states $G(v + ½)$ up to the asymptote. The sum of these vibrational quanta corresponds to the dissociation energy:

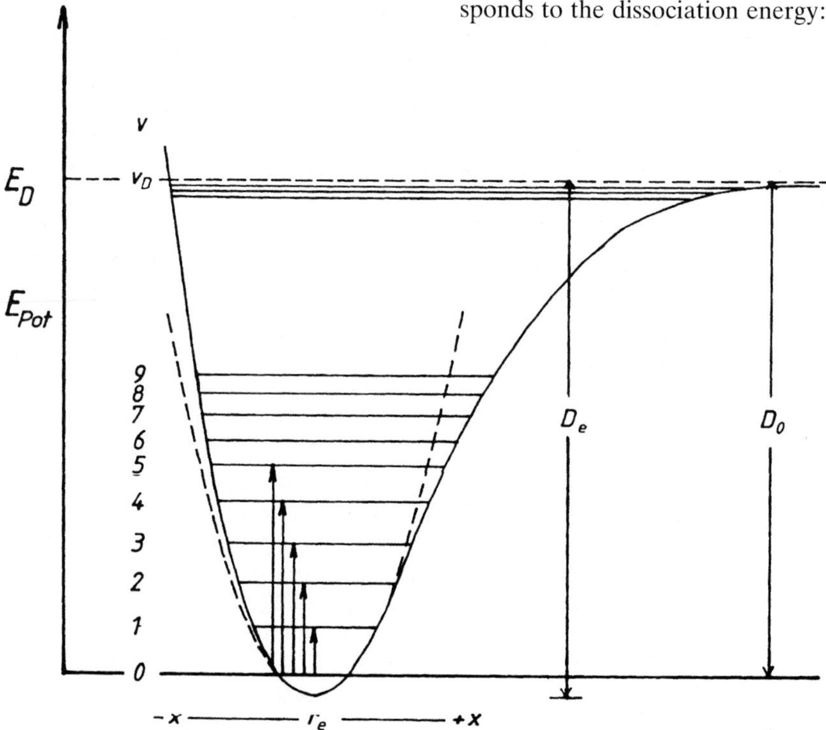

Dissociation energy, spectroscopic. Fig. 1. Potential curve of an anharmonic oscillator, E_{pot} as $f(r)$

$$D_e = \sum_v \Delta G(v + \frac{1}{2}).$$

D_e and D_o are related by the equation:

$$D_e = D_o + G_v(0)$$
$$= D_o + \frac{1}{2}\omega_e - \frac{1}{4}\omega_e x_e \pm \ldots$$

where the expression:

$$G_v(0) = \frac{1}{2}\omega_e - \frac{1}{4}\omega_e x_e \pm \ldots$$

takes the zero-point energy into consideration. Neglecting the cubic terms, the energy levels of a diatomic molecule are given by: $G_o(v) = \omega_o v - \omega_o x_o v^2$. The difference between adjacent energy levels, $\Delta v = 1$, is:

$$\Delta G_o(v) = \omega_o x_o - 2\omega_o x_o - 2\omega_o x_o v$$
$$= \omega_o - 2\omega_o x_o(v + \frac{1}{2}).$$

However, the second difference, $\Delta^2 G_o(v) = 2\omega_o x_o$, is a constant and gives the anharmonicity term directly. Since there are no discrete vibrational levels above the asymptote, D_e must be the maximum value of $G(v)$ and D_o the maximum value of $G_o(v)$. Also, $\Delta G_o(v)$ equals zero at this limit. If the quadratic expression, $\omega_o v - \omega_o x_o v^2$, represents all vibrational levels accurately, i.e. if $\Delta G_o(v)$ is a linear function of v, we can obtain the vibrational quantum number, v_D, at the dissociation limit from $\Delta G_o(v) = 0$:

$$v_D = \frac{\omega_o}{2\omega_o x_o} - \frac{1}{2}.$$

The dissociation energy, D_o, is then found to be:

$$D_o = G_{o,\,max}(v) = \omega_o v_D - \omega_o x_o v_D^2$$
$$= \frac{\omega_o^2}{4\omega_o x_o} - \frac{1}{4}\omega_o x_o.$$

In the case of HCl, with $\omega_o = 2937.6$ cm^{-1} and $2\omega_o x_o = 103.12$ a value of 28 is found for v_D. This also shows that only a limited number of vibrational levels exist below the dissociation limit. The relation between D_e and D_o can be written:

$$D_e = D_o + \frac{1}{2}\omega_o = \frac{\omega_e^2}{4\omega_e x_e}$$

For $H^{35}Cl$ we have: $\omega_e = 2989.2$ cm^{-1}, $\omega_o = 2937.6$ cm^{-1} and $2\omega_e x_e = 2\omega_o x_o = 103.12$ cm^{-1}, $\omega_o x_o/4 = 12.9$ cm^{-1}, giving: $D_o = 41{,}829$ cm^{-1}; $D_e = 43{,}325$ cm^{-1}. The difference, $D_e - D_o$, is 1496 cm^{-1} and corresponds to $\omega_e/2$. This method for determining the dissociation energy of diatomic molecules dates back to Birge and Sponer in 1926 and assumes that we can extrapolate to v_D from measurements for smaller v values. However, extreme care must be taken when using the extrapolation method since generally only a few vibrational levels are observed in the case of the IR spectrum. A D_e value of 37,212 cm^{-1} is obtained by thermodynamic measurements on $H^{35}Cl$. The evaluation of the first five lines of the $H^{35}Cl$ spectrum in the IR and NIR region provides an upper limit for D_o or D_e. The dependence of ΔG upon v is illustrated schematically in Figure 2. The continuous curve corresponds to the experimental data whilst the linear extrapolation considers only the first data points and consequently gives too high a value for v_D. If the experimental data can be obtained or well approximated by a nonlinear extrapolation, D_o can be calculated from the area under the curve, i.e. as a sum of all the energy levels up to the limit, v_D. These observations are not particularly important in IR spectroscopy. However, they play an important role

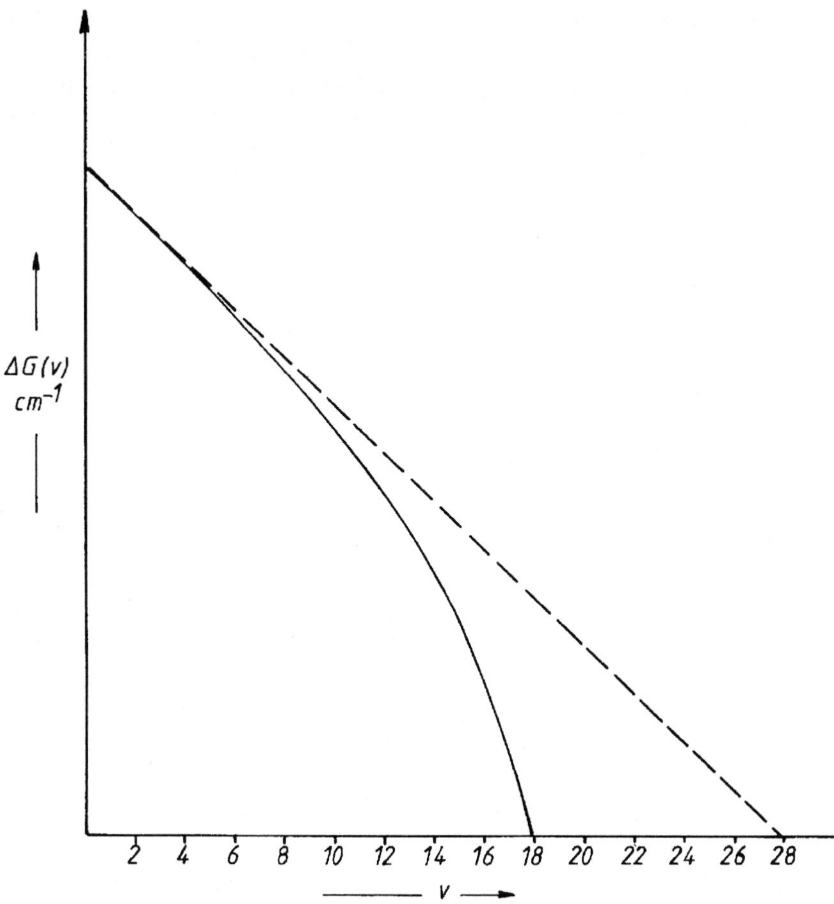

Dissociation energy, spectroscopic. Fig.2. $\Delta G(v)$ as $f(v)$

in → electronic band spectra where higher v values can be observed for the different electronic states.

Doppler broadening, *<Dopplerverbreiterung>*. The emission (absorption) of radiation by atoms or molecules in the gas phase always occurs from (to) a moving system where the average velocity of the particles is given, according to gas kinetics, by $V = (2RT/M)^{1/2}$. Therefore, if there is interaction with electromagnetic radiation of frequency, ν_o, propagating with the speed of light, c, the → Doppler effect must always be taken into consideration. For a single velocity, $\vec{v}$, this leads initially to a shift of the spectral lines given by $\nu_e = \nu_o + \vec{k} \cdot \vec{v}$ where $\vec{k} = 2\pi/\lambda$, is the → wave vector and $\vec{v}$ is the velocity vector, $\vec{v} = \{v_x, v_y, v_z\}$. ν_e increases if the molecule moves towards the observer ($\vec{k} \cdot \vec{v} > 0$) and decreases if it moves away ($\vec{k} \cdot \vec{v} < 0$). However, since at thermal equilibrium a Maxwellian velocity distribution must be considered, we must integrate over all velocities. This leads to a

Doppler broadening of all lines under normal conditions and low pressures. Detailed study leads to the following expression for the → half-width, $\delta\nu_D$, of the Doppler broadened lines:

$$\delta\nu_D = 7,16 \cdot 10^{-7} \cdot \nu_o\sqrt{T/M}; [s^{-1}]$$

or

$$\delta\tilde{\nu}_D = 7,16 \cdot 10^{-7} \cdot \tilde{\nu}_o\sqrt{T/M}; [cm^{-1}].$$

The factor $\sqrt{T/M}$ originates from the expression for average speed, $\overline{V}$. All numerical values, including R and c, are collected in the numerical factor. The Doppler broadening, $\delta\nu_D$ or $\delta\tilde{\nu}_D$, is therefore proportional to $T^{1/2}$ and also depends upon the value of the average frequency, ν_o or wavenumber $\tilde{\nu}_o$.

For benzene with molar mass $M = 78$ g and a wavenumber for the $S_1 \rightarrow S_o$ transition of $\tilde{\nu}_o = 40{,}000$ cm^{-1} at $T = 300$ K, $\delta\tilde{\nu}_D = 0.056$ cm^{-1}. For a lifetime of $\tau = 10^{-9}$ s, the natural → line width for this transition is $\sim 3.3 \cdot 10^{-4}$ cm^{-1}. Therefore, the Doppler width is ca. 170 times larger than the natural line width.

Doppler broadening of vibrational transitions in the electronic ground state is of the order of 10^{-3} cm^{-1}. When we consider that the natural → line width is $10^{-6} - 10^{-8}$ cm^{-1}, it is clear that the Doppler broadening is several orders of magnitude greater and, together with the → collision broadening, it determines the line profile.

Ref.: W. Demtröder, *Laser Spectroscopy*, Springer Series in Chemical Physics. vol. 5, Springer Verlag, Berlin, Heidelberg, New York, **1983**, chapt. 3, p. 78 ff.

Doppler effect, <*Doppler-Effekt*>, an effect which occurs with moving optical or acoustic sources. For example, if a light source, which emits light of specific frequency, ν_o, or wavelength, λ_o, moves towards or away from a stationary observer with velocity, $\pm v$, a frequency or wavelength shift, $\Delta\nu$ or $\Delta\lambda$, occurs vis-a-vis the original frequency or wavelength. According to the wave model, the frequency, ν', measured by the observer is given by:

$$\begin{aligned}\nu' &= \nu_o/(1 \pm [v/c]) \\ &= \nu_o(1 \pm [v/c] + [v/c]^2 \ldots)\end{aligned}$$

Restriction to small values of, v/c, and termination of the series expansion at the second term provides the simple equation:

$$\nu' = \nu_o(1 \pm \frac{v}{c})$$

or

$$\nu' - \nu_o = \Delta\nu = \pm\nu_o \cdot \frac{v}{c}$$

or in wavelength:

$$\Delta\lambda = \pm\lambda_o\frac{v}{c}.$$

The Doppler effect has become very important in astrophysics. If a star approaches the earth $(+v)$ a decrease in the wavelength of its emitted light is observed, i.e. a shift of the spectral lines towards the violet. However, if the star moves away from the earth $(-v)$ a shift to longer wavelengths occurs, i.e. towards red. Using the Doppler effect it is possible to determine the velocity at which stars move away from the earth. In general, the Doppler shift is also dependent upon the angle of observation, α, and:

$$\Delta\lambda = \nu_o \cdot \frac{v}{c} \cdot \cos\alpha;$$

The case, $\alpha = 0$ was discussed above. For applications in spectroscopy it is important that $\Delta\lambda$ depends upon the frequency, ν_o. For the visible spectral region, if we take $\nu_o = 6 \cdot 10^{14}\ s^{-1}$, corresponding to a wavenumber $\tilde{\nu}$ = 20,000 cm^{-1}, then with $\upsilon = 3$ cm s^{-1} and $c = 3 \cdot 10^{10}$ cm s^1, we obtain $\Delta\nu = 6 \cdot 10^4$ s^{-1} or in wavenumbers, $\Delta\tilde{\nu} = 2 \cdot 10^{-6}$ cm^{-1}.

In the region of γ spectroscopy, ν_o is of the order of 10^{18} to 10^{19} s^{-1}. For a speed, $\upsilon = 3$ cm s^{-1} and $\nu_o = 10^{18}$ s^{-1} we obtain $\Delta\tilde{\nu} = 6 \cdot 10^9$ s^{-1} or in wavenumbers, $\Delta\tilde{\nu}$ = $2 \cdot 10^{-2}$ cm^{-1}. The Doppler effect is used experimentally in → Mössbauer spectroscopy. In emission spectroscopy it must be remembered that emitting atoms in flames or plasmas have a considerable velocity due to their thermal motion. Because of the Doppler effect this leads to a → Doppler broadening of the emitted spectral lines. → Collision broadening also contributes to the line broadening.

Doppler-free spectroscopy, *<dopplerfreie Spektroskopie>*, a modern spectroscopic technique by means of which → Doppler broadening is eliminated. Doppler-free → multiphoton spectroscopy is one of the most important methods and the case of two-photon spectroscopy will be described qualitatively. A molecule simultaneously absorbs two photons with energies, $(h/2\pi)\omega_1$ and $(h/2\pi)\omega_2$ ($\omega = 2\pi\nu$), if the resonance condition, $E_k - E_i = (h/2\pi)(\omega_1 + \omega_2)$ is fulfilled for the transition between the two states, i and k. We assume that the molecule moves with its gas-kinetic velocity, $\overrightarrow{\upsilon}$, in the experimental coordinate system. Due to the → Doppler effect, the frequency, ω, of an electromagnetic wave is shifted to ω' in the coordinate system of a moving molecule:

$$\omega' = \omega(1 \pm \vec{\upsilon}/c).$$

If $\omega = 2\pi/\lambda$ and $\lambda\nu = c$ we can also write $\omega' = \omega - \overrightarrow{k} \cdot \overrightarrow{\upsilon}$ with $\overrightarrow{k} = 2\pi/\lambda$, the wave vector. If we use this in the above resonance condition for the two simultaneously absorbed photons we have:

$$(E_k - E_i)/(h/2\pi) = \omega_1' + \omega_2'$$
$$= \omega_1 + \omega_2 - \vec{\upsilon}(\vec{k}_1 + \vec{k}_2).$$

For two light waves with frequencies, $\omega_1 = \omega_2 = \omega$, which propagate in opposite directions through the absorbing system $\overrightarrow{k}_1 = -\overrightarrow{k}_2$, and the above equation shows immediately that the Doppler shift becomes zero. This means that all molecules, independently of their velocity, absorb at the same sum frequency, $\omega_1 + \omega_2 = 2\omega$. If the moving molecule interacts simultaneously with several wave fronts which have wave vectors $\overrightarrow{k}_i$, the total Doppler shift is $\upsilon \cdot \sum_i \overrightarrow{k}_i$. This expression becomes zero if the condition, $\sum_i \overrightarrow{k}_i = 0$, is fulfilled.

A possible experimental arrangement for Doppler-free, three-photon spectroscopy is one in which all three beams intersect at an angle of 120°.

Doppler-free two-photon absorption in atoms was first achieved in 1974. During 1980/82, H.J. Neusser *et al.* were able to show that Doppler-free, two-photon spectroscopy is a practical method of obtaining rotationally resolved, Doppler-free spectra of larger molecules such as benzene. Figure 1 shows their experimental arrangement. A continuous wave, single-mode ring dye laser (→ ring laser) provides light of high-frequency precision ($\Delta\nu$ = 1 MHz corresponding to

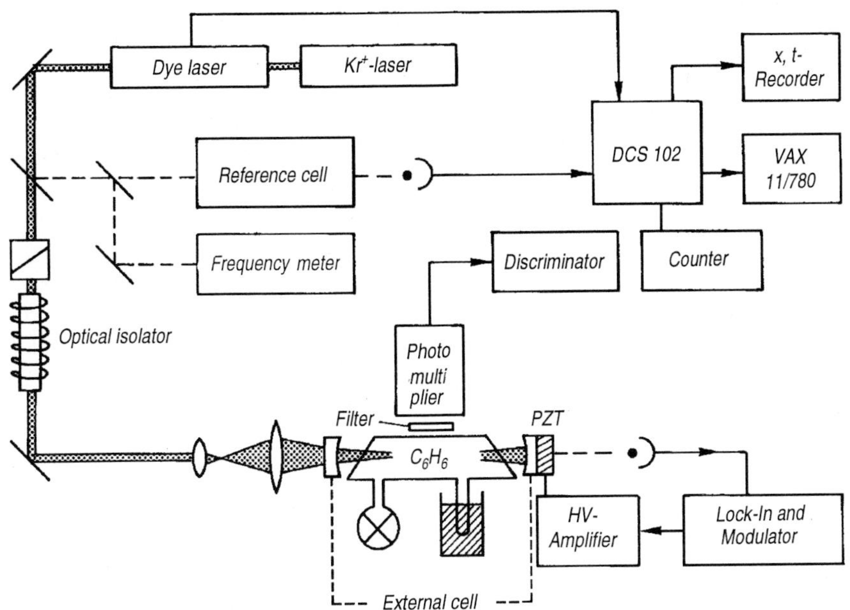

Doppler-free spectroscopy. Fig. 1. A spectrometer for Doppler-free spectroscopy

$\Delta\tilde{\nu} = 3 \cdot 10^{-5}$ cm^{-1}). The absolute wavelength of the laser light is measured using a wavemeter and relative frequency changes are determined very accurately with an interferometer (reference cavity). In order to eliminate intermolecular interaction and → collision broadening, the molecular gas under investigation is contained in a fluorescence cell at a pressure of less than 1 Torr. This cell is placed in an external resonator so that the Doppler-free, two-photon absorption can take place in a standing light wave which is composed of two beams moving in opposite directions ($\vec{k}_1 = -\vec{k}_2$). If a molecule absorbs a photon from each of the opposed beams the Doppler shift of the transition frequency is compensated for each molecule, independently of its velocity component, and a Doppler-free spectrum is observed at the sum of the two-photon energies. By using an external resonator, the light intensity can be increased considerably. And this leads to an increase of several orders of magnitude in the two-photon signal, which depends upon the square of the light intensity.

The resulting two-photon absorption is monitored by detecting the UV fluorescence subsequently emitted with a → photomultiplier (56 DUVP, Valvo). As the laser light frequency is changed, the length of the external resonator is adjusted synchronously via an electronic control using a piezoelectric device (PZT). This ensures that the resonance condition is always met for the wavelength concerned. A data-processing system stores the signal and prints out the Doppler-free, two-photon spectra at the required scale. Figure 2 shows an example of the high-resolution rotational struc-

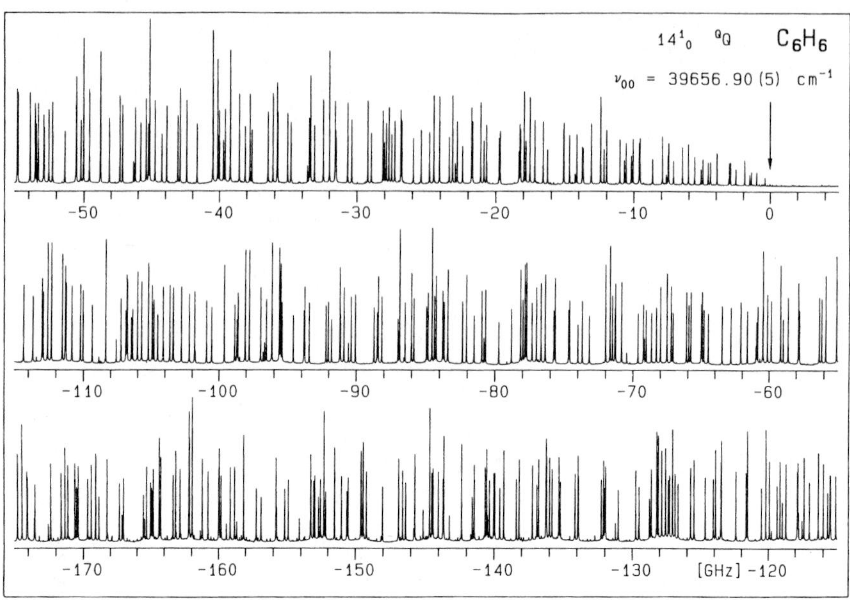

Doppler-free spectroscopy. Fig. 2. Fluorescence spectrum of benzene; rotational structure of the transition, $\tilde{\nu}_{00} = 39656.9\ \text{cm}^{-1}$

ture of benzene measured by Neusser. Ref.: W. Demtröder, *Laser Spectroscopy,* Springer Series in Chemical Physics, vol. 5, Springer Verlag, Berlin, **1981**, chapt. 10, 6, p. 525 ff; H.J. Neusser, *Chimia*, **1984**, *38*, 379, and further references given there.

Double monochromator, *<Doppelmonochromator>*, an instrument for the dispersion of continuous radiation in the UV-VIS region. It operates as follows; (see → UV-VIS spectrometer). The light dispersed in the first → monochromator is imaged at an intermediate slit which is also the entrance slit of the second monochromator. This arrangement achieves a higher monochromaticity of the light leaving the exit slit of the second monochromator and the proportion of stray light is reduced by several powers of ten. Because of the greater number of optically reflecting surfaces, a double monochromator always has a lower transmittance than a single monochromator consisting of the same optical parts (mirrors, lenses, prisms or gratings). Thus, the higher spectral purity of the dispersed light must be paid for by a loss of intensity. But this loss is compensated by improved sources and detectors and by modern electronics.

Double pendulum interferometer, *<Doppelpendelinterferometer, DPI>*, a modified → Michelson interferometer (Figure 1). Two mirrors in the form of cube corners (K) are rigidly connected in a pendulum structure. The path difference of the beams is produced by oscillating this arrangement about the axis, A, perpendicular to the plane of the drawing. At the pendulum-reversal point, one of the

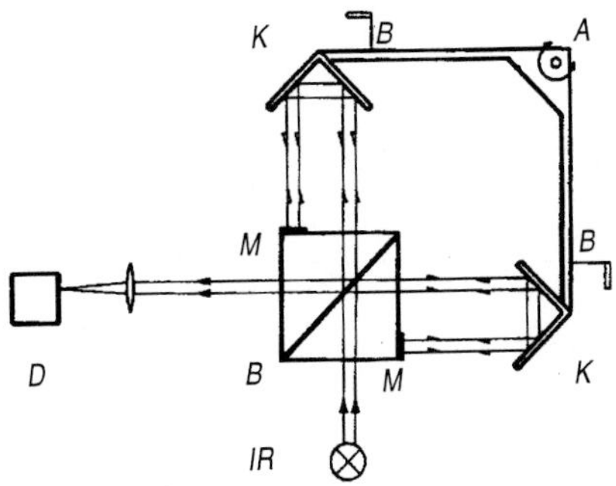

Double-pendulum interferometer. Fig. 1. Cross section showing the optical paths

interfering beams has its shortest and the other its longest path to the beam splitter, B. The mirrors, M, for reversing the beams are mounted on the beam splitter. FT-IR spectroscopy is the field of application for double-pendulum interferometers. A → FT-IR spectrometer constructed in such a way is very robust and can be used both in the laboratory and in the field. Manufacturer: Kayser-Threde GmbH, Wolfratshauser Str. 48, 8000 Munich 70, Germany.

Double refraction, *<Doppelbrechung, natürliche Doppelbrechung>* → birefringence, natural.

Double resonance (technique), *<Doppelresonanztechnik>*, a method in → NMR spectroscopy in which the → spin-spin coupling between individual nuclei is removed thereby simplifying complex spectra. In order to understand this method, we consider a spin system consisting of two nuclei, A and X, such as two protons with nuclear spin, $I_H = 1/2$, the resonance frequencies of which, for an applied magnetic field B_o, lie at ν_A and ν_X. Since the spins of the two nuclei are coupled together via the bonding electron system, the signals at ν_A and ν_X are both split into doublets. If the resonance of nucleus X is excited by a second intense radio-frequency field, B_2, whilst the signal of nucleus A is measured, the spin coupling of A and X is removed and the resonance signal, ν_A, is observed as a singlet. A simple explanation of this phenomenon may be based on the fact that the spin-spin interaction is proportional to the scalar product of the nuclear moments concerned, μ_A and μ_X according to the equation, $E_{AX} = J_{AX} \cdot \vec{I}(A) \cdot \vec{I}(X)$. The strong radio-frequency field of frequency, ν_x, keeps nucleus X permanently at resonance, i.e. in continuous precession around the x axis. But nucleus A precesses around the z axis, so that μ_A and μ_X are orientated at right angles or orthogonally to one another. The nuclear spin vectors, I_A and I_X, are then quantized along the z and x axes respectively, i.e. they are also orthogonal and their scalar product or the scalar spin-spin coupling disappears according to the above equation. Expressed somewhat more descriptively; in a double resonance experiment, an intense radio-frequency field, B_2, is applied to a specific multiplet so that in all the nuclei giving the multiplet the orientation of the nuclear spins is very rapidly interchanged between $M_I = +½$ and $M_I = -½$. Consequently, no two additional fields corresponding the parallel and antiparallel orientations of the nuclei in the applied magnetic field can be experienced by the adjacent nuclei which only sense the average value of the different orientations. This average is exactly zero so that the coupling is zero, too. This interpretation also applies to rapid exchange processes; the proton of an alcoholic OH-group

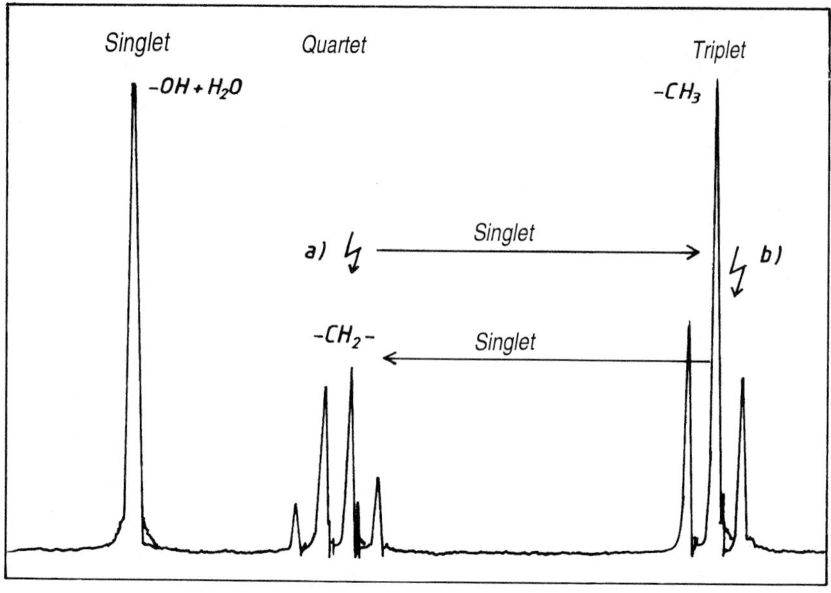

Double resonance (technique). Fig. 1. The NMR spectrum of ethanol in acid solution with decoupling possibilities a) and b), see text

in acid solution, for example. Figure 1 shows the NMR spectrum of ethanol in acid solution and the remaining decoupling possibilities.

Case a: The methyl-proton resonance is reduced to a singlet by irradiating into the quartet of the methylene protons.

Case b: By irradiating into the methyl-proton triplet the methylene-proton resonance is reduced to a singlet. The technique of double resonance can be implemented in the → frequency-sweep, the → field-sweep and the Fourier-transform methods. The example of ethanol shows that double resonance is not limited to two-spin systems. If nucleus *X*, in any complex system, is excited by an intense radio-frequency field, B_2, all possible couplings to nucleus *X* are reduced to zero. Spin decoupling is possible not only between nuclei of the same element (homonuclear decoupling) but is frequently used for heteronuclear decoupling. The spin decoupling of protons and ^{13}C nuclei is of particular importance.

Ref.: R.K. Harris, *Nuclear Magnetic Resonance Spectroscopy*, Pitman, London, Marshfield, **1983**.

Drude's formula, <*Drudesche Formel*>, an equation proposed by P. Drude at the beginning of this century for representing the optical rotation, α, of a substance or solution as a function of wavelength, λ, outside the absorption range of that substance:

$$|\alpha|_{20\leftrightarrow}^{\lambda} = \sum_i \frac{A_i}{\lambda^2 - \lambda_i^2}.$$

A_i and λ_i are constants. λ_i corresponds to the wavelength of an absorption maximum of the optically active substance. Wavelengths are entered in μm (10^{-6} m). In applications to solutions of optically active substances (→ optical rotatory dispersion) it is frequently sufficient to use only the first term of the above formula:

$$|\alpha|^{\lambda}_{20\leftrightarrow} = \frac{A_o}{\lambda^2 - \lambda_o^2}.$$

λ_o can be found by measuring the UV absorption spectrum or from the CD spectrum in the region of anomalous dispersion. The inverse of the last equation provides the expression:

$$\frac{1}{|\alpha|} = \frac{\lambda^2}{A_o} - \frac{\lambda_o^2}{A_o}.$$

A graph of $1/|\alpha|$ against λ^2 (Biot-Lowry plot) should be a straight line of gradient A_o^{-1} and intercept, $\lambda_o^2 \cdot A_o^{-1}$. This graphical method is used to check the validity of Drude's formula in its simplified form and gives an empirical value for the constant A_o. Optically active substances frequently do not absorb above 200 nm (0.2 μm) and a determination of λ_o by UV measurements is not possible. In such cases, the above method can be used to determine, λ_o. For a cane-sugar solution (dextrorotatory) in the region from $\lambda = 0.302$ μm to $\lambda = 0.547$ μm and a concentration range of 10–26 g/100 g solution:

$$|\alpha|^{\lambda}_{20\leftrightarrow} = \frac{21.648}{\lambda^2 - 0.0213}$$

in which

$$\lambda_o = 0.146\ \mu\text{m} \mathrel{\hat{=}} 146\ \text{nm}.$$

In more complicated cases, a relationship with two terms must be used:

$$|\alpha|^{a}_{20\leftrightarrow} = \frac{A_o}{\lambda^2 - \lambda_o^2} + \frac{A_1}{\lambda^2 - \lambda_1^2}.$$

λ_o and λ_1 are the two absorption maxima of the optically active substance with $\lambda_o > \lambda_1$. Since $\lambda_1 < \lambda_o$ and *a fortiori* $< \lambda$, λ_1^2 can frequently be neglected; and we obtain $|\alpha|^{\lambda}_{20\leftrightarrow} = A_o/(\lambda^2 - \lambda_o^2) + A_1/\lambda^2$. Many examples of the use of this formula have been reported (A.E. Lippman, E.W. Foltz, C. Djerassi, *J. Am. Chem. Soc.* **1955**, *77*, 4364). The application of Drude's formula without complications is only possible in the region of normal rotatory dispersion. Its use in the anomalous region is strongly disputed, see C. Djerassi, *Optical Rotatory Dispersion, Applications to Organic Chemistry*, McGraw-Hill Book Company Inc., New York, Toronto, London, **1960**, p. 5 ff.

Dual-wavelength spectrometer, <*Doppelwellenlängenspektrometer, Doppelwellenspektralphotometer*>, an instrument typically fitted with two → monochromators. Figure 1 is a simplified diagram of the optics of the Hitachi-Perkin-Elmer model 557. This instrument can also be operated as a conventional double-beam spectrometer with sample and reference cuvettes. In that case, the shutter of monochromator λ_1 is closed and that following the mirror S_8 is opened. The sector mirror switches the monochromatic light beam from monochromator λ_2 alternately through the sample and reference cuvettes. In the case of a dual-wavelength instrument, both beams coming from the source (→ hydrogen, [UV], or tungsten halogen, [VIS]) pass through both monochromators; the shutter of monochromator λ_1 being open. If the shutter following S_8 is closed, the light from monochromator λ_2 can only traverse the sample cuvette and the beam from

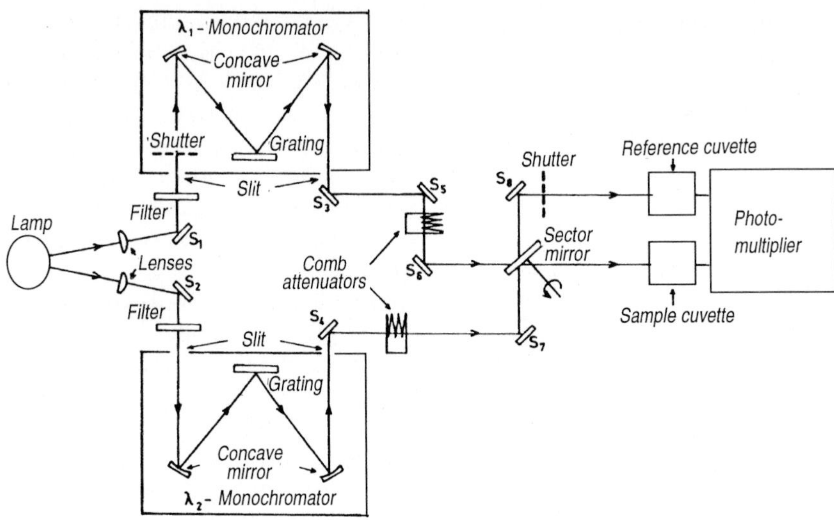

Dual-wavelength spectrometer. Fig. 1. The optical path in the Hitachi-Perkin-Elmer dual-wavelength spectrometer, model 557

monochromator λ_1 also traverses the sample cuvette, after a 180° rotation of the sector mirror. The two beams, chronologically shifted by the rotational frequency of the sector mirror, converge on the → photomultiplier where they generate the responses to be processed in the sequence λ_1, zero signal, λ_2, zero signal, λ_1 ... which are transmitted to the channels provided for the individual signals. Microprocessors control the process. Modern electronics have reduced the noise by a factor of approximately $2 \cdot 10^4$.

Dual-wavelength spectroscopy, <*Doppelwellenlängenspektroskopie, Zweiwellenlängenspektroskopie*>, uses two wavelengths, λ_1 and λ_2, and thus makes it possible to determine the absorbance difference:

$$\Delta A = A_{\lambda_2} - A_{\lambda_1}$$

It is important in analysis that the absorbance difference, ΔA, in a binary mixture of components a and b is proportional to the concentration of only one component if the correct wavelength, λ_1, has been selected. If the concentrations are c_a and c_b, the absorbances at wavelengths λ_1 and λ_2, for $d = 1$ cm, are given by:

$$A_1 = \varepsilon_{1a}c_a + \varepsilon_{1b}c_b = A_{1,a} + A_{1,b},$$

$$A_2 = \varepsilon_{2a}c_a + \varepsilon_{2b}c_b = A_{2,a} + A_{2,b}.$$

ε_{1a} and ε_{1b} are the molar decadic → extinction coefficients of components a and b at wavelength λ_1; similarly ε_{2a} and ε_{2b} at wavelength λ_2. The absorbance difference obtained with a dual-wavelength measurement is:

$$A_2 - A_1 = \Delta A = (\varepsilon_{2a} - \varepsilon_{1a})c_a + (\varepsilon_{2b} - \varepsilon_{1b})c_b.$$

If component b has the same extinction coefficient, ε, at wavelengths λ_1

and λ_2, the expression in brackets equals zero:

$$\varepsilon_{2b} \cdot c_b - \varepsilon_{1b} \cdot c_b = A_{2,b} - A_{1,b} = 0.$$

and:

$$\Delta A = (\varepsilon_{2a} - \varepsilon_{1a})c_a. \qquad (a)$$

Thus, the absorbance difference, ΔA, depends only upon component *a*, and the influence of component *b* can be eliminated. If component *a* does not absorb at wavelength λ_1, $\varepsilon_{1a} = 0$ and equation (a) simplifies to:

$$\Delta A = \varepsilon_{2a} c_a = A_{2,a}. \qquad (b)$$

Equations (a) and (b) are the basis of the → equiextinction method. Wavelengths λ_1 and λ_2 can be selected on the basis of the known absorption spectra of components *a* and *b*. Figure 1 illustrates the method with the absorption spectra of components *a* and *b*, and their superimposition in a mixture. It is a precondition, when applying the equiextinction method, that the absorption maximum of component *a* overlaps the short-wavelength flank of the absorption of the accompanying component *b* whilst the spectrum of component *a* does not underly that of component *b* completely (Figure 1). In this case, component *b* can be determined directly by spectrophotometric methods, but the measurement of component *a* would lead to a large positive error since a partial absorption of *b* is also measured. Use of dual-wavelength spectroscopy and the equiextinction method allows the elimination of the interfering influence of component *b*. We use only one cuvette in this method, i.e. a reference solution is not

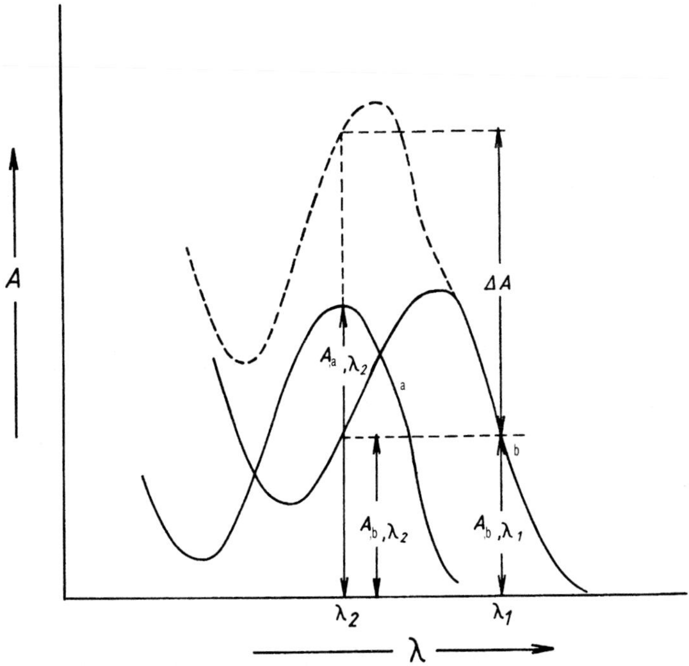

Dual-wavelength spectroscopy. Fig. 1. The equiextinction method, see text

required. Thus, interferences in the baseline, especially those due to turbidity and chronological changes, can be eliminated. This is difficult in conventional absorption spectroscopy. Dual-wavelength spectroscopy was initially developed for the quantitative analysis of turbid solutions of biological and physiological samples, but a general application to chemical-analytical problems developed from it. Since the sample and reference cuvettes are identical, it is possible to determine reliably the smallest absorbance values in a full-scale deflection range of $0.0001 \leq A \leq 0.010$. The quantitative determination or detection of substances can be extended into the ppb region and this is of particular value in environmental analysis. The detection limit for mercury with dithizone in aqueous solution is 1.8 ppb. If the above precondition for the use of the equiextinction method is not met, the method of → signal amplification can be used.

Ref.: H.-H. Perkampus, *UV-VIS Spectroscopy and Its Applications*, Springer Verlag, Berlin, Heidelberg, New York, **1992**.

Dumbbell model, <*Hantelmodell*>, a model of a diatomic molecule with unequal masses, m_1 and m_2, separated by internuclear distance, r (cf. Figure 1). Because $m_1 \neq m_2$, their molecular

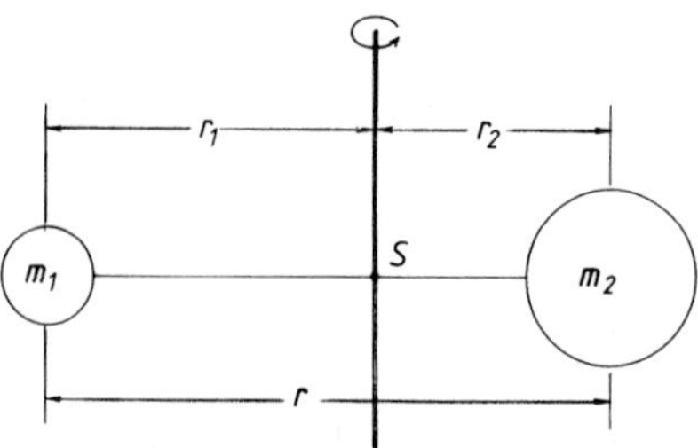

Dumbbell model. Fig. 1. See text

center of gravity, S, does not lie in the center of r but is shifted in the direction of the greater mass. The distances, r_1 and r_2, from the center of gravity, S, may be calculated from the classical mechanics of levers and $r = r_1 + r_2$ as:

$$r_1 = \frac{m_2}{m_1 + m_2} r$$

and

$$r_2 = \frac{m_1}{m_1 + m_2} r.$$

For a rotation of the dumbbell with a fixed distance, r, between the two masses, the classical rotational energy is given by $E_{rot} = (1/2) \cdot I \cdot \omega^2$. I is the moment of inertia and ω the angular velocity of the rotation. In this simple case, the moment of inertia can be expressed:

$$I = \sum m_i r_i^2 = m_1 r_1^2 + m_2 r_2^2.$$

Using the above expressions for r_1 and r_2 we obtain:

$$I = \frac{m_1 m_2}{m_1 + m_2} r^2 = \mu r^2.$$

The expression

$$\mu = \frac{m_1 m_2}{m_1 + m_2}$$

gives μ, the → reduced mass. Thus, instead of the rotation of the dumbbell model, we can equally well consider the rotation of a replacement mass, μ, about an axis at a fixed distance r. Such a system is described as a simple rigid rotator. If we assume that m_1 and m_2 are coupled by elastic springs to the center of gravity and perform harmonic vibrations about their equilibrium positions, r_1 and r_2, then the conservation of momentum applies, i.e.:

$$m_1 v_1 = m_2 v_2.$$

since

$$v_1 = \frac{dr_1}{dt}$$

and

$$v_2 = \frac{dr_2}{dt}$$

we have:

$$m_1 \frac{m_2}{m_1 + m_2} \frac{dr}{dt} = m_2 \frac{m_1}{m_1 + m_2} \frac{dr}{dt} = \mu \frac{dr}{dt}.$$

Here, the harmonic vibration of the dumbbell model has also been replaced by the motion of the reduced mass, μ, about an equilibrium position. This model is the → harmonic oscillator.

Duplex interferometer, <*Duplex-Interferometer*>, two → Fabry-Perot interferometers arranged in series with different interplate intervals, d_1 and d_2 which must have an integral relationship to one another → interference spectroscopy.

Duschinsky effect, <*Duschinsky-Effekt*>, the different spatial alignment of the vibrational amplitudes in the excited electronic states of molecules. In contrast to a ground state molecule, molecules in excited electronic states have not only a changed equilibrium geometry but also a modified intramolecular force field. This results not only from the specific binding energies of the excited state but also from the fact that normal vibrations in the excited state transform according to different → symmetry species and take up other vibrational directions when compared with the ground state.

If Q^s is the vector of the normal coordinates in the excited state and Q^l that of the normal coordinates in the ground state, the two vectors are related by the equation:

$$Q^s = W \cdot Q^l + K^s$$

The vector K^s describes the shift of the equilibrium positions of the normal vibrations in a system of normal coordinates in the excited state. W is the Duschinsky matrix. Since the normal coordinates of the excited state are linear combinations of the normal coordinates of the ground state, the elements of the Duschinsky matrix represent the proportions of the ground state coordinates in the normal coordinates in the electronically excited state. In the electronic states under consideration, the normal coordinates are orthogonal to each other and W has the action of a rotation matrix (unitary transformation).

In a representation using internal coordinates, the above relationship assumes the following form:

$$R^s = x \cdot R^l + \Delta R^s$$

R^l and R^s are the internal coordinates of states, l and s and x is the transformation matrix. The vector, R, describes the changes in the equilibrium bond angles and lengths between the electronic ground and excited states. But the normal coordinates can be transformed into internal coordinates using the L matrix; i.e. $R^{s,l} = L^{s,l} Q^{s,l}$. We, therefore, obtain the Duschinsky matrix as a product of the inverse L matrix of the excited state and the L matrix of the ground state.

The Duschinsky effect is used for the calculation of multidimensional Franck-Condon integrals which cannot be split into a series of one-

dimensional Franck-Condon integrals due to the fact that *W* is usually not equal to the unit matrix. For that reason, multidimensional Franck-Condon integrals are determined with the help of generating functions.
Ref.: F. Duschinsky, *Acta Physicochem.*, U.S.S.R., **1937**, *7*, 551.

Dye, Dyestuff, <*Farbstoffe*>, an organic compound capable of absorbing energy from that part of the spectrum of electromagnetic radiation which is visible to the human eye (→ spectral regions, → UV-VIS spectrum). Therefore, the color perceived by the human eye corresponds essentially to the → complementary color of the absorbed light, e.g. blue/yellow, green/red. Dyestuffs are soluble → colorants and are thus characterized by being soluble in solvents and/or binders, in contrast to → pigments. In some cases their color, or more exactly their absorption spectrum, depends strongly upon the solvent (→ solvatochromism). But it may also depend upon the pH value and because of this dyes are used as → indicators (indicator dyes). Recently, → laser dyes have gained considerably in significance. In photography, the → polymethine dyes are very important as → sensitizers, and the vast majority of dyestuffs and pigments are characterized by their technical applications. Those dyes used in technology are described in the color index (C.I.) which classifies the dyestuffs according to the characteristic properties of their colors, their trade names and their chemical properties and constitution.
For more details see: *The Colour Science of Dyes and Pigments*, K. Mclaren, 2nd edtn. Adam Hilger Ltd., Bristol, **1986**.

Dye laser, <*Farbstofflaser*>, a liquid laser the active medium of which consists of dyestuff molecules dissolved in ethanol, water or some other solvent. Their greatest advantage, in comparison with other lasers, lies in the fact that they are easy to tune. The fluorescence of organic dye molecules, which lies from the visible through to the NIR and can be readily excited by visible or ultraviolet light, is exploited. At the present time, about 100 dyes are known whose → fluorescence spectra are distributed over the region between 400 nm and just above 1000 nm. The best-known dye laser is the rhodamine-6G-laser (ethanol solution). In Figure 1 the absorption and fluorescence spectra of rhodamine-6G are shown. For dye lasers, as with all lasers, a → population inversion must be present. This is easy to explain using the → Jablonski diagram in Figure 2. On excitation, the molecule goes from the ground state, S_0, to the excited state, S_1, both of which have their associated vibrational and rotational structure. Here only the vibrational states, v, are shown. When the molecule reaches a particular vibrational state, v', in the excited state, S_1, it is first deactivated by → internal conversion (10^{-12} s) and falls to $v' = 0$. During its lifetime of 10^{-8} s, S_1 stabilizes to a slightly lower energy state, S_1' (shown with a dotted line). The fluorescence takes place from this state and therefore appears red-shifted by $\Delta\tilde{\nu}$ with respect to the absorption spectrum. Now, at room temperature the vibrational levels $v'' \geq 1$ of the ground state, S_0, are all unoccupied,

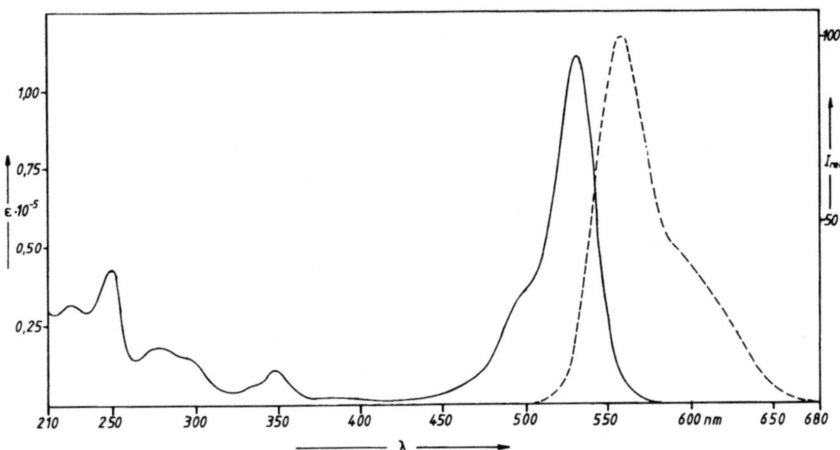

Dye laser. Fig. 1. Absorption and fluorescence spectra of rhodamine-6G

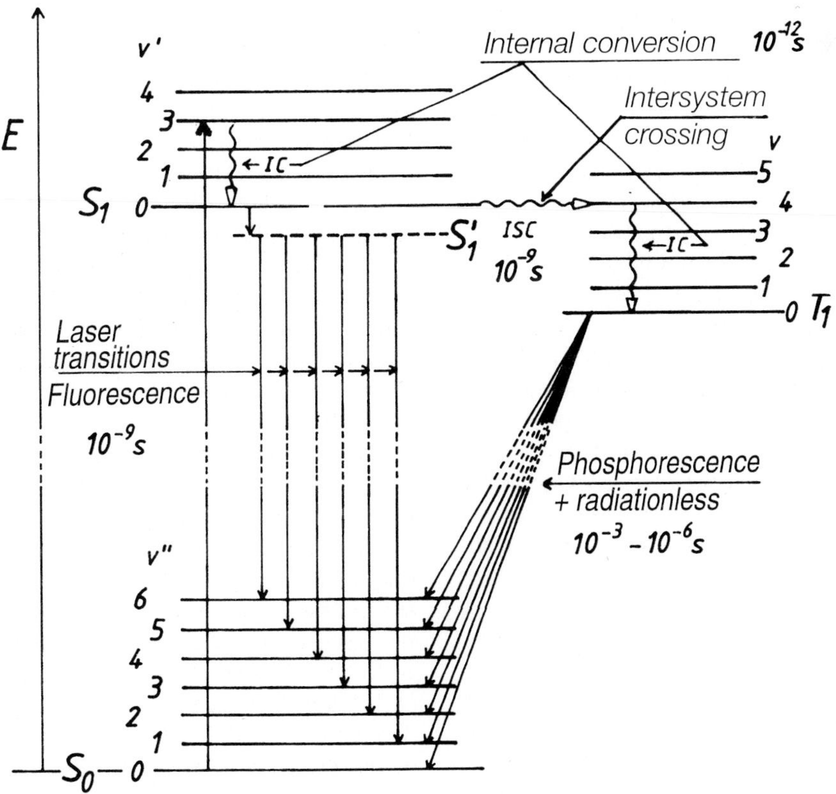

Dye laser. Fig. 2. Jablonski diagram and laser transitions

(or almost unoccupied), on account of the Boltzmann distribution (→ Boltzmann statistics) so that, *a priori*, there is a population inversion. Thus, if the exciting light is sufficiently intense, that is we use → optical pumping and build the system into a resonator, it can act as a → laser. On account of the surrounding solvent molecules, the rotational states suffer extreme → collision broadening and the vibrational states are also broadened by this effect. Thus, in general, only a broad fluorescence band or laser emission with a half-width of between 50 and 100 nm results. This represents the approximate tuning range of the dye laser. From Figure 2 we can also see that the triplet state, T_1, is undesirable for the laser process because it has a lifetime of 10^{-3} – 10^{-6} s and molecules which enter it, as a result of intersystem crossing (~ 10^{-9} s), sit there as if in a trap and therefore contribute to a diminution of the population in the upper laser level, S_1'. The result is that the continuous operation of the laser stops after a short time because of the lack of active molecules. Therefore, it is an objective in the synthesis of laser dyes to remove this competing process, at the start, by placing suitable substituents in the laser dye molecule. Dye lasers are used either continuously (CW laser) or pulsed. In most cases, other CW lasers are used as pumps, for example a → ruby, → neodymium or → argon-ion laser (→ noble-gas ion laser). For pulse work, a pulsed → xenon-pulse discharge lamp or very frequently a pulsed → nitrogen laser is used. The pulse width of the latter lies between 10^{-8} and 10^{-9} s.

The principles of the construction of a flashlamp pumped dye laser are shown

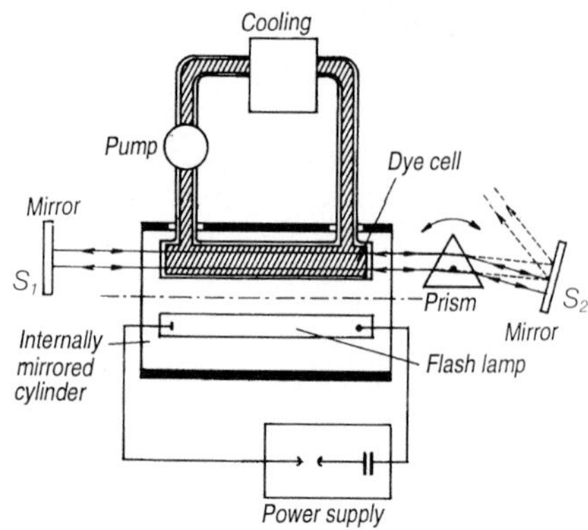

Dye laser. Fig. 3. Construction of a flashlamp pumped dye laser

in Figure 3. The dye solution flows through a capillary which is closed at its ends with plane parallel glass plates. Parallel to the capillary there is the exciting flashlamp. Both are located within an elliptical reflector in such a way that they lie exactly at the two foci. Outside this arrangement, there are two resonator mirrors, S_1 and S_2. To tune the dye laser there is a prism inside the → cavity in front of the mirror, S_2. By rotating the prism, a sequence of different frequencies is brought to resonance and the color of the laser light can be changed from red to green, i.e. tuned. Apart from this conventional construction, other ways of building a dye laser have been described in the literature. One of the most important is the jet-stream laser in which the stream of dye solution is projected from a carefully prepared and polished jet through the focus of the pump laser. At a flow velocity of $10 \text{ m} \cdot \text{s}^{-1}$, the time-of-flight of the dye molecules through the focus of the pump laser, which has a spatial extent of 10^{-6} m, is about 10^{-6} s. During this short time, no great concentration of triplets can be achieved by intercombination and the triplet losses are therefore small. The jet stream must be continuously supplied by circulation of the dye solution with a pump. Figure 4 shows the construction of this dye laser schematically.

Figure 5 shows the spectral distribution of the power of a jet-stream laser; model 375B from the company Spectra Physics. These spectra are called tuning curves and for every dye they are determined by the particular

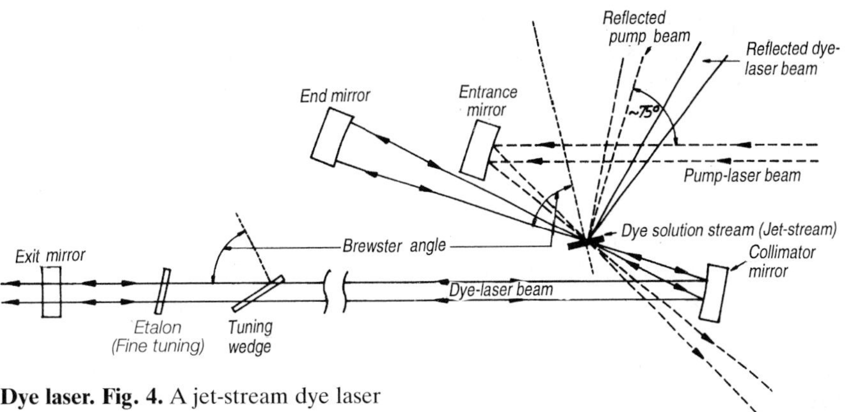

Dye laser. Fig. 4. A jet-stream dye laser

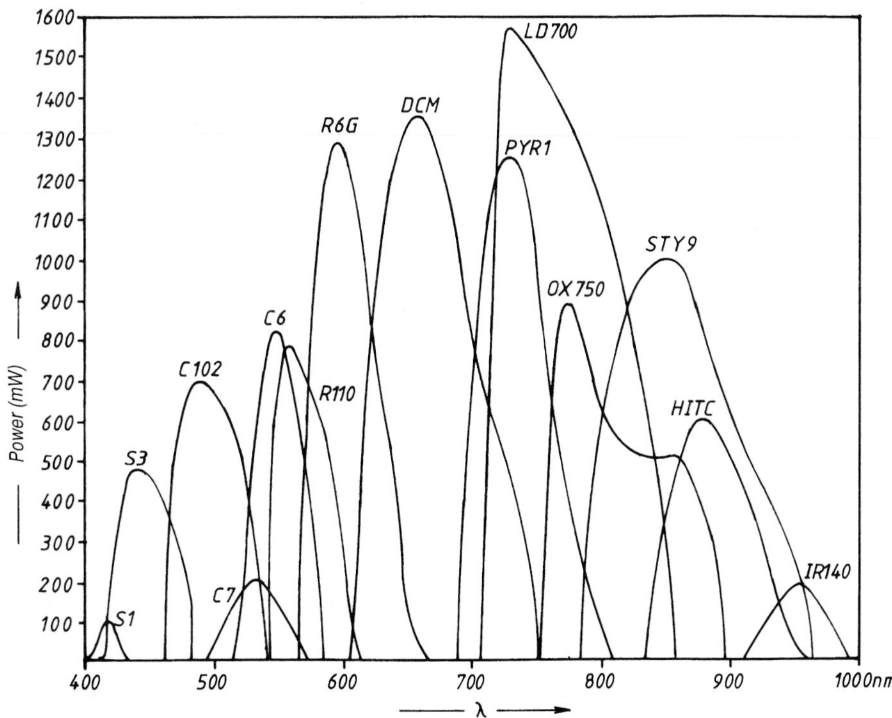

Dye laser. Fig. 5. Spectral distribution of the output power of various laser dyes

performance of the pumping-light source. They are not transferable to other dye laser systems and they therefore represent specification characteristics of the various commercial lasers. In the table, the appropriate data are assigned to the individual tuning curves of the above illustration.
Narrow line widths of the tunable laser lines are required for many applications of dye lasers. This can be achieved by the incorporation of → Fabry-Perot interferometers or → Lyot filters. With a single-mode ring dye laser, line widths of $\delta\tilde{\nu} = 3 \cdot 10^{-5}$ cm^{-1} can be achieved which, with reference to electronic excitation, lie in the range of the natural line width (→ Doppler-free spectroscopy).
Ref.: W. Demtröder, *Laser Spectroscopy,* Springer Series in Chemical Physics, vol. 5, Springer Verlag, Berlin, Heidelberg, New York, **1981**.

Dynamic nuclear polarizition, nuclear polarization, *<dynamische Kernpolarisation, Kernpolarisation>* → Overhauser effect.

Dynode, *<Dynode>*, an electrode which is used in a series of 10 to 14 units in → secondary electron multipliers (SEM). The name is derived from the dual function of the electrode which acts primarily as an anode, for the incident electrons but also as a cathode for the emitted electrons. If several dynodes are fitted in an SEM every successive unit must have a higher potential than the previous one. This is achieved by means of a voltage-splitting circuit.

Dye laser. Table.

Dye		Pump power	Wavelength	Laser
S1	Stilbene 1	1.25 W	351–364 nm	Argon, all UV-lines
S3	Stilbene 3	2.0 W	351–364 nm	Argon, all UV-lines
C102	Coumarin 102	2.25 W	407–423 nm	Krypton, all violet lines
C7	Coumarin 7	2 W	476 nm	Argon –
C6	Coumarin 6	4 W	488 nm	Argon –
RMO	Rhodamine 110	4 W	458–514 nm	Argon, all lines
R6G	Rhodamine 6G	4 W	458–514 nm	Argon, all lines
	DCM	4 W	458–514 nm	Argon, all lines
Pyrl	Pyridine 1	7,5 W	458–514 nm	Argon, all lines
	LD 700	4 W	647 und 676 nm	Krypton, all red lines
OX750	Oxazine 50	4 W	647 und 676 nm	Krypton, all red lines
Sty 9	Styryl 9	7.5 W	458 und 514 nm	Argon, all lines
	HITC-P	4 W	647 und 676 nm	Krypton, all red lines
	IR 140	1.6 W	752 und 799 nm	Krypton, all NIR-lines

E

Eagle mounting, *<Eagle-Aufstellung>*, in contrast to the → Paschen-Runge mounting of a → concave grating this is a → Rowland mounting with a moving concave grating. The grating, the center point of the → Rowland circle and the spectrum change their position for different wavelength settings and only the entrance slit, *b-c*, remains fixed (Figure 1). Using a simple mechanism, the grating, *a*, can be moved along a straight line between the entrance slit and the grating and simultaneously rotated, in such a way that the Rowland condition is always fulfilled. In this application, the entrance and exit slit are either arranged close together or narrowly separated above and below the Rowland circle. By moving the center of the Rowland circle, *d-d'*, by moving the grating, different wavelengths can be selected at the exit slit (Figure 1). The arrangement can be mounted in a small, evacuable housing, and it is therefore frequently used in vacuum spectrographs.

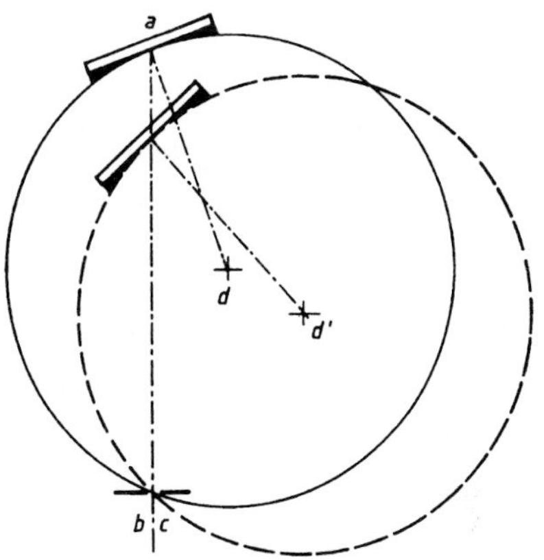

Eagle mounting. Fig. 1. Arrangement of a concave grating in the Eagle mounting

Ebert monochromator, *<Ebert-Monochromator>*, the mounting, invented by Ebert in 1889, of a → plane grating in a → grating monochromator. The incident light beam lies on one side of the grating and the exiting light beam on the other (side by side). The grating is mounted between the entrance and exit slits of the monochromator which therefore have a relatively large distance between them. This is important in practical applications. Figure 1 shows

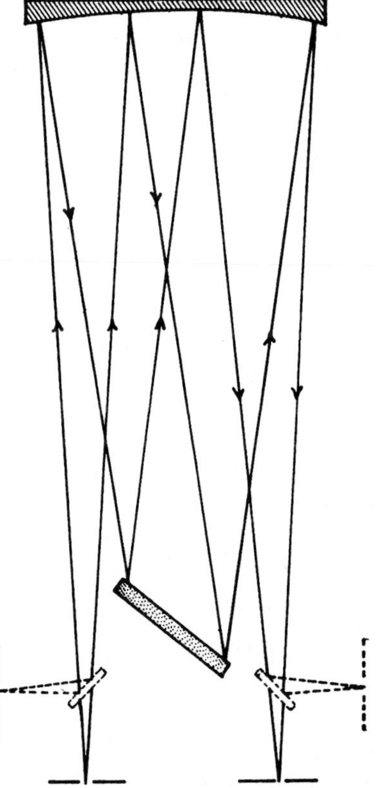

Ebert monochromator. Fig. 1. Cross section of a monochromator with the Ebert mounting of the grating

the Ebert mounting. The use of a single spherical mirror, which is not only the collimator but also the collector mirror, is the characteristic feature. The slits can be relocated laterally by fitting plane mirrors, as indicated by the dashed lines.

Echelette grating, <*Échelette-Gitter*>, a grating, introduced by Wood in 1910 and based upon his systematic work concerning the influence of the form of the grooves on the efficiency of reflection gratings (from the French échelle = ladder, sequence of steps). The grating has unsymmetrical V-shaped grooves, one wall of which is as long and flat as possible and the other one as short and steep as possible. Figure 1 shows that the angle, φ, of the flat side to the macroscopic surface of the grating is selected in such a way that the radiation specularly reflected from this side has the same direction, for a specific wavelength, as the desired diffracted radiation. In this way it is possible to concentrate almost the whole intensity into a spectrum of any order, i.e. the spectrum of this order has preferential intensity vis-a-vis all other orders. The blaze angle, φ, can be varied, so that the angle of reflection and refraction change, to select another optimum wavelength. This setting of the optimum wavelength is called blazing the grating and hence the name blaze angle for φ and blaze wavelength for the optimum wavelength. Today, blazed gratings are used almost exclusively. They are optimally matched to specific regions from the UV to IR by setting the blaze angle (→ gratings).

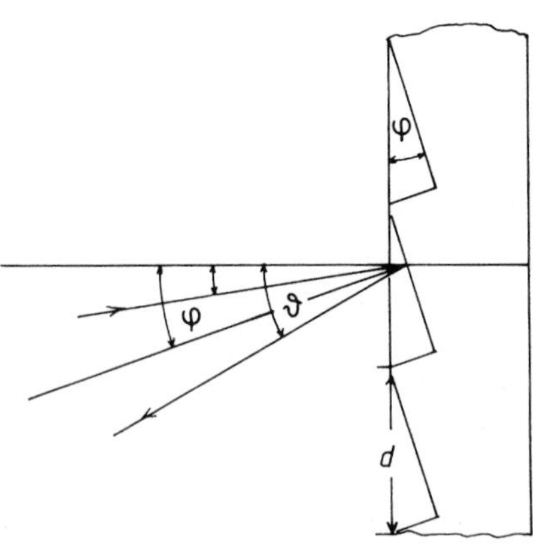

Echelette grating. Fig. 1. Schematic view of an echelette grating

Edge filter, cut-off filter, <*Kantenfilter*>, a filter characterized by the fact that a region of low transmission (blocking region) lies adjacent to a region of high transmission (transmitting region) or vice versa. We distinguish between two edge filter types, i.e. → short-wave pass filters in which the transmitting region is of shorter wavelength than the blocking region and → long-wave pass filters where the transmitting region is of longer wavelength than the blocking region.

Figure 1 shows the characterizing features of edge filters:

$\lambda_c(\tau)$ the edge wavelength at which transmittance, $\tau(\lambda)$, reaches a specific established value, e.g. 0.46 of [τ_D', τ_D''] etc. minimum degree of transmission in the transmitting regions, λ_{D1} to λ_{D2}, λ_{D2} to λ_{D3} etc.

τ_S', τ_S'' etc. the upper limits of the degree of transmittance in the blocking regions, λ_{S1} to λ_{S2}, λ_{S2} to λ_{S3} etc.

Edge filters based on → Schott glass filters or → glass-plastic laminated filters are always → long-wave pass filters. With → interference filters, or more exactly → ADI filters, → short-wave pass or long-wave pass filters can be produced, depending upon the con-

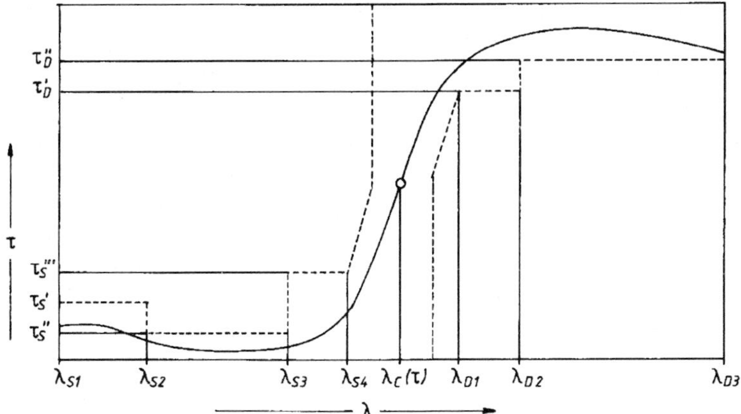

Edge filter. Fig. 1. The characterizing features of an edge filter

struction. Optical glass filters used as → heat filters are short-wave pass filters.

Einstein, <*Einstein*>, the name given to 1 mol of photons = 1 mol light quanta; 1 Einstein = $N_L \cdot h\nu$ (photon energy).

Einstein coefficients, <*Einstein-Koeffizienten*>, describe the transition probability of the absorption and emission processes in atoms and molecules.

The Einstein *A* coefficient defines the probability of spontaneous emission from an excited state, *kn*, to the ground state, *lm*: $A_{kn \to lm}$.

The Einstein *B* coefficient defines the probability of an absorption process from ground state, *lm*, to a higher state, *kn*, induced by a radiation field, $B_{lm \to kn}$. For the Einstein *B* coefficient:

$$B_{lm \to kn} = B_{kn \to lm}.$$

The following relationship exists between *A* and *B*:

$$A_{kn \to lm} = 8\pi h^3 \nu^3_{kn \to lm} n^3 c^{-3} B_{kn \to lm},$$

where *h* is Planck's constant, *n* the refractive index of the medium, *c* the velocity of light and $\nu_{kn \to lm}$ the frequency of the transition, $kn \to lm$. When denoting the state with *lm* and *km*, the letters *l* and *k* refer to the electronic states and the letters *m* and *n* to the vibrational states superimposed on those states. If the vibrational ground state in state *l* is designated with $m = 0$, the Einstein *B* coefficient can be represented as the sum over all vibrational states, *n*, of the excited electronic state, *k*:

$$B_{lk} = \sum_n B_{lo \to kn}.$$

B_{lk} is connected with the experimental data as follows:

$$B_{lk} = \frac{2303 \cdot c}{hnN_L} \int_{\text{band}} \frac{\varepsilon(\tilde{\nu})}{\tilde{\nu}} \mathrm{d}\tilde{\nu}.$$

Furthermore, the Einstein *B* coefficient is directly related to the → transition dipole moment, M_{lk}, of the electronic transition:

$$B_{lk} = \frac{8\pi^3}{3h^2} G |\vec{M}_{lk}|^2.$$

G in this equation is the statistical weight of the states involved which is one for pure singlet electronic states ($S_o \rightarrow S_p$).

The Einstein coefficients introduce a differentiation between spontaneous emission (A coefficient) and induced emission (B coefficient). The Einstein A coefficient is important for all fluorescence processes (→ fluorescence spectrum, → fluorescence lifetime of excited states), i.e. for spontaneous emission processes. The Einstein B coefficient of induced emission plays a decisive role in → laser radiation since this is an induced emission of radiation.

Electrodeless discharge lamp, EDL, <*Entladungslampe, elektrodenlose Entladungslampe, EDL*>, a radiation source used in → atomic absorption spectroscopy and → atomic fluorescence spectroscopy which has the highest radiation intensity and smallest line width. These lamps were investigated as early as 1935 as radiation sources for high-resolution spectroscopy. However, suitable lamps have only been used in AAS applications since 1973. Electrodeless discharge lamps consist of a quartz tube, closed at both ends, which is several centimeters long and has a diameter of ca. 5–10 mm. This tube contains a few milligrams of the element of interest under an argon pressure of a few mbar. The contents can be added as a pure metal, as the corresponding halide or as metal and iodine. The quartz tube is placed in the coil of a high-frequency generator and excited with a power of a few Watt to 200 Watt. Figure 1 shows that, in contrast to the earlier lamps, the modern AAS quartz tube, filled with the element under investigation, is fixed rigidly to the high-frequency coil and mounted in a well-insulated case. This facilitates the handling of the lamps and they show a very good stability after a relatively short running-in time. It is a further advantage that electrodeless discharge lamps can be operated with only 27 MHz (older types with 2500 Hz). In comparison with hollow cathode lamps, electrodeless discharge lamps have a better sensitivity and provide lower limits of detection for elements such as arsenic, selenium and tellurium. Phosphorus, rubidium and cesium are other elements suitable for special applications of electrodeless discharge lamps.

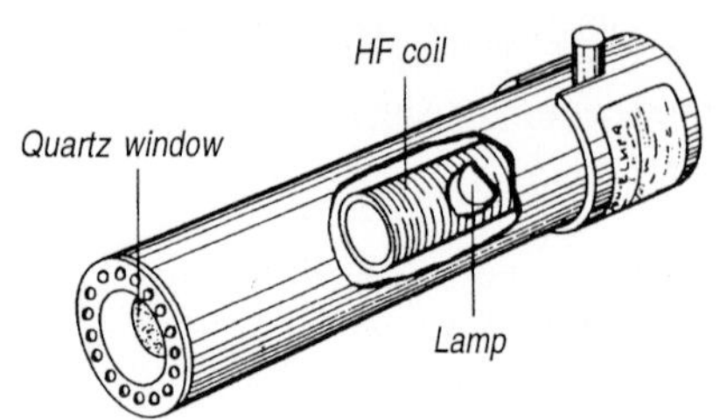

Electrodeless discharge lamps. Fig. 1. Schematic view and section through an electrodeless discharge lamp

Electroluminescence, <*Elektrolumineszenz*>, the generation of luminescence by application of an electric field to a phosphor, discovered by G. Destriau. Characteristically, the effect begins above a certain, not clearly definable, threshold of the field strength and increases rapidly with increasing field strength. However, this threshold finally depends upon the sensitivity of the detection equipment. Studies of electroluminescence have been carried out on phosphors of the ZnS, ZnO and $ZnSiO_4$ types. However, they contain ten times as much copper activator as the usual luminophores, which can be excited by UV light.

Electromagnetic radiation, <*elektromagnetische Strahlung*> → light.

Electron, <*Elektron*>, an electrically, negatively charged elementary particle with the following physical properties: m_e = 9.10953410^{-31} kg, electric charge, e = $-1.6021892 \cdot 10^{-19}$ C, specific charge, $e/m = -1.758805 \cdot 10^{+11}$ C kg^{-1}. It has a diameter of the order of 10^{-13} cm.
There is also an elementary particle with a positive electric charge called a positron (positive electron). Apart from the sign, its charge is equal to that of the electron. (However, the equality with regard to mass and size is quite uncertain and as far as the size is concerned even improbable.) Both electrons and positrons have a spin, s = 1/2 and a magnetic moment of one → Bohr magneton in magnitude.
Electrons are released by glowing filaments and by the → photoelectric effect. They are very important in all optical instruments which image objects using electron beams (cathode-ray tube, electron microscope). Electrons are the carriers of electric currents in metals, high vacuums and gases (gas discharge). They are components of atoms (→ atomic model) and the change of their energy states, i.e., their bond to the atomic nucleus, gives rise to atomic spectra. Electrons are also released by the decay of certain radioactive elements (→ beta radiation). Free electrons are those in metals which are not bound to specific atoms and are capable of moving through the metal in response to an electric field. They are responsible for electrical conductivity and incandescent emission and are the reason for various characteristic properties of metals.

Electron beam, <*Elektronenstrahl*>, electrons which are brought together and accelerated by an electric field after leaving a cathode. Although electron beams are corpuscular or particle beams they have a wave character (wave-particle duality). The wavelength of electron beams is determined by the velocity of the electrons and hence by the accelerating potential.

$$\lambda = \frac{h}{m \cdot v} = \frac{12.3 \cdot 10^{-3}}{\sqrt{u}} \text{ [cm]}.$$

λ = wavelength, h = Planck's constant, m = mass, v = velocity, u = accelerating potential. The wavelength of electrons accelerated with 50 kV is 0.00055 nm (electron microscope).

Electron-beam apparatus, <*Elektronenstrahlgeräte*>, equipment which includes an imaging device. The cathode-ray oscillograph, image converter, instruments for telerecording and playback and the electron microscope are such instruments.

Electron-bombardment ion source, <*Elektronenbeschuß-Ionenquelle*>, an ion source the surface of which is bombarded with electrons. The impinging electrons heat the sample to temperatures of 3000 K at the point of impact and evaporate the sample material which is ionized by further electrons. Electron-bombardment ion sources can be used for a local analysis of microregions of sample surfaces if the electron beam is highly focused by electron optics.

Electron emission, <*Elektronenemission*>, the emission of electrons from the surface of a solid. We distinguish between four types of electron emission, depending upon how the emission is induced:

1. Photoemission utilizing the external photoelectric effect,
2. secondary electron emission,
3. thermal electron emission,
4. field-electron emission.

The first two types are important to → photocells and → photomultipliers and thus also to spectroscopy.

Electronic angular momentum, <*Elektronendrehimpuls*>. According to the quantum mechanics of one-electron problems such as the hydrogen atom, (→ atomic model), the stationary states of an electron are characterized by the principal quantum number, n, and a second quantum number, ℓ. The latter is called the orbital angular momentum quantum number and corresponds to Bohr's postulate that action can only occur in whole-number multiples of $h/2\pi$ (→ Planck's constant). This postulate gives the orbital angular momentum as:

$$p_\psi = \ell \cdot \frac{h}{2\pi} \psi ,$$

where $\ell = 0, 1, 2, 3 \ldots n-1$. The exact quantum-mechanical result for the orbital angular momentum is:

$$|\vec{\ell}| = \frac{h}{2\pi}\sqrt{\ell(\ell+1)}.$$

A correlation exists between the principal quantum number, n (= 1, 2, 3 ...) and the orbital angular quantum number, ℓ; $\ell = 0, 1, 2, 3 \ldots n-1$. Hence, n states, which differ in ℓ, exist for each state characterized by the principal quantum number, n. In the case of the H atom, all states with different ℓ values associated with a particular n value, are degenerate (→ degeneracy). This is no longer the case for heavier atoms such as the alkali-metal atoms (→ alkali spectra). When we move to multielectron systems, the orbital angular momenta of the individual electrons are added vectorially to give a resultant orbital angular momentum, $\vec{L}$, where:

$$|\vec{L}| = \frac{h}{2\pi}\sqrt{L(L+1)}$$

L is the quantum number of the resulting total orbital angular momentum. The individual values of L, like those of ℓ, must differ by 1. For two electrons we have:

$$\vec{L} = \vec{\ell}_1 + \vec{\ell}_2,\ \vec{\ell}_1 + \vec{\ell}_2 - 1, \ldots, \vec{\ell}_1 - \vec{\ell}_2.$$

For $\ell_1 = 1$ and $\ell_2 = 2$, $L = 2, 1, 0$. In addition to the orbital angular moment of the electrons, electron spin angular momentum must be considered. The spin quantum number is s = 1/2 and the contribution of the spin angular momentum is given by:

$$|\vec{s}| = \frac{h}{2\pi}\sqrt{s(s+1)}.$$

In the case of multielectron systems, the spins of the individual electrons are added to give the total spin, S, and the total spin angular momentum is given by:

$$|\vec{S}| = \frac{h}{2\pi}\sqrt{S(S+1)}.$$

The solution of → Schrödinger's equation for the H atom reveals another quantum number, the magnetic quantum number m_ℓ, which can assume $2\ell + 1$ values for a given value of ℓ. The relationship between ℓ and m_ℓ is:

$$m_\ell = +\ell, \ell-1, \ldots, 0, \ldots, -\ell+1, -\ell;$$

i.e. for $\ell = 2$, $m_\ell = 2, 1, 0, -1, -2$. These $2\ell + 1$ values are degenerate in the absence of a magnetic field.
A quantization of the directions of $\vec{\ell}$ and $\vec{s}$ or $\vec{L}$ and $\vec{S}$ occurs in a magnetic field. The possible $2\ell + 1$ or $2L + 1$ values of m_ℓ or M_L, the values $M_s = \pm 1/2$ resulting for s and the $(2S + 1)$ values of M_S represent the components of $\vec{\ell}$, $\vec{L}$, $\vec{s}$ or $\vec{S}$ in the direction of the magnetic field (→ Zeeman effect). The degeneracy is lifted and the $(2\ell + 1)$ or $(2L + 1)$ values of m_ℓ or M_L have different energies. This also applies to m_s or M_s.
The orbital angular momentum, $\vec{\ell}$ ($\vec{L}$) and spin angular momentum, $\vec{s}$ ($\vec{S}$), are coupled and add vectorially resulting in a total angular momentum, $\vec{j}$ ($\vec{J}$) which is given by $\vec{j} = \vec{\ell} \pm \vec{s}$ in the case of an electron (H atom, alkali term) or $\vec{J} = \vec{L} + \vec{S}, \vec{L} + \vec{S} - 1 \ldots |\vec{L} - \vec{S}|$ (→ Russell-Saunders coupling) for many electrons. The coupling to $\vec{j}$ in the simplest case or to $\vec{J}$ for several electrons is the origin of the → multiplet structure of → atomic spectra. ℓs or LS coupling is replaced by jj coupling in heavier atoms.

Electronic band spectrum, <*Elektronenbandenspektrum*>, a spectrum which is observed, particularly in molecular gases, as a result of electric discharges, flames or fluorescence and occurs in addition to the characteristic line spectra. At low → dispersion, an electronic band spectrum does not consist of individual sharp lines but rather of more or less broad bands, hence the name. We observe exclusively bands in the absorption spectra of molecular gases in the UV-VIS region. It was recognized that the bands were due to the simultaneous coexcitation of vibrations and rotations which were superimposed upon the excitation of the electronic state. The bands usually have a sharp edge at one end which is called the band head. There the intensity drops suddenly to zero whilst it decreases more or less slowly at the other end. Depending upon whether this slower intensity drop occurs towards shorter or longer wavelengths, bands are said to be shaded (degraded) towards the violet or the red.
The overall term (energy in wavenumber units), T, for a diatomic molecule can be written as the sum of an electronic term, T_e, a vibrational term, G, and a rotational term, F, as:

$$T = T_e + G + F.$$

According to the Bohr-Einstein frequency relation, the wavenumbers of the emitted or absorbed spectral lines (prime = upper, double prime = lower electronic state) are:

$$\begin{aligned}\tilde{\nu} &= T' - T'' \\ &= (T'_e - T''_e) + (G' - G'') + (F' - F'')\end{aligned}$$

or

$$\tilde{\nu} = \tilde{\nu}_e + \tilde{\nu}_v + \tilde{\nu}_r.$$

$\tilde{\nu}_e = T_e' - T_e''$ is a constant for a specific electronic transition. The variable component, $\tilde{\nu}_v + \tilde{\nu}_r$, has a form similar to that for a → rotation-vibration spectrum, but with the difference that G' and G'' originate from different vibrational term series. Similarly, F' and F'' belong to different rotational term series. G' and F' always refer to the upper and G'' and F'' to the lower electronic state. For the vibrational structure of an electronic transition in a diatomic molecule ($F' = F'' = 0$), we obtain, including the first anharmonicity term:

$$\tilde{\nu} = \tilde{\nu}_{oo} + (\omega_o' v' - \omega_o' x_o' v'^2) - (\omega_o'' v'' - \omega_o'' x_o'' v''^2)$$

$\tilde{\nu}_{oo}$ is the wavenumber of the pure electronic transition, $\tilde{\nu}_e$ corrected by the zero-point energy of the upper and lower electronic states. ω_o and $\omega_o x_o$ are defined under → vibrational spectra and the → anharmonic oscillator. This equation represents all possible transitions between the different vibrational levels of the two participating electronic states. There is no strict selection rule for the vibrational quantum numbers, v, for electronic transitions. In principle, every vibrational state, v', of the upper electronic state can combine with every vibrational state, v'', of the lower electronic state. Thus, we can expect a large number of lines as Herzberg, Birge, Brice *et al.* showed in detailed investigations of the emission and absorption spectra of diatomic molecules between 1927 and 1933. The → band-head table is an effective means of analyzing the vibrational structure. However, a large number of lines is seen only in the emission spectrum. In absorption, in contrast, only one progression is intense. It corresponds to the change of the vibrational quantum number, v', in the upper electronic state. In the ground state, $v'' = 0$, because of the Boltzmann distribution (→ Boltzmann statistics). Higher vibrational levels, $v'' = 1, 2, \ldots$, are occupied only to the smallest extent at room temperature. Since $v'' = 0$, the wavenumbers of the absorption bands are given by the above formula as:

$$\tilde{\nu}_{abs} = \tilde{\nu}_{oo} + \omega_o' v' - \omega_o' x_o' v''^2.$$

Figure 1 shows the absorption spectrum of iodine vapor. Here, the vibrational quantum with $\tilde{\nu}_o = 213.1\ \mathrm{cm}^{-1}$ is relatively small. At room temperature, therefore, 27% of all molecules are already in the excited vibrational state, $v'' = 1$. Because of the anharmonicity term, $\omega_o x_o$, the separation between neighboring vibrational levels decreases with increasing v and finally approaches zero. Thus, a con-

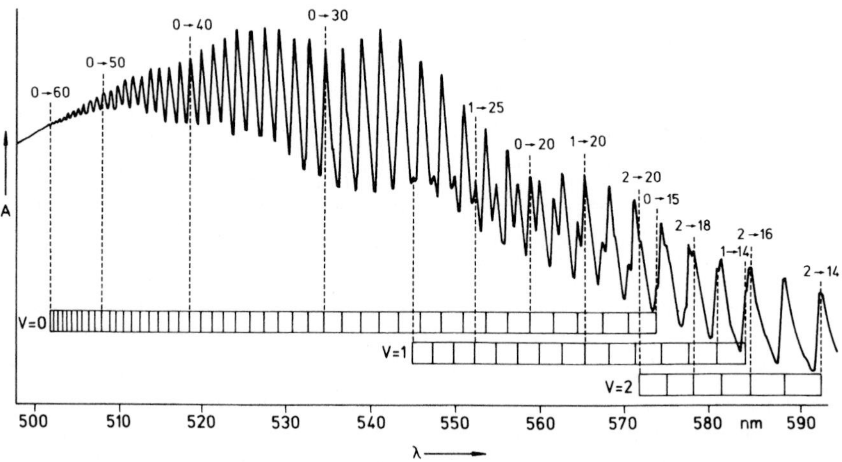

Electronic band spectrum. Fig. 1. Spectrum of I_2 vapor in the visible region

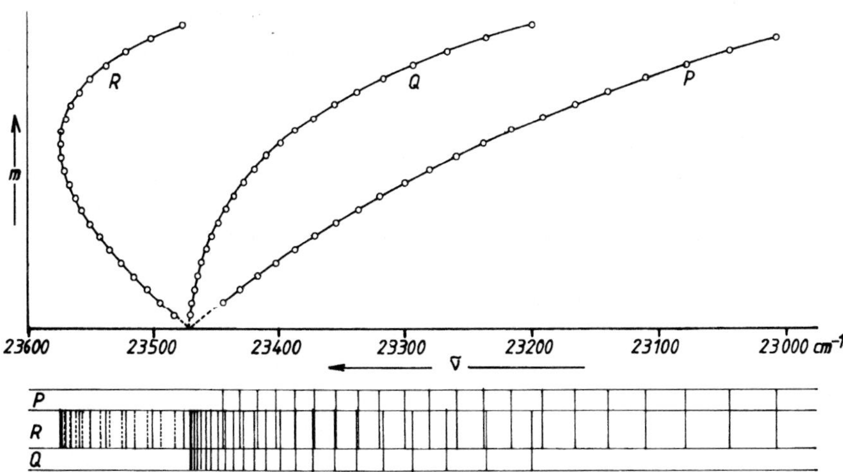

Electronic band spectrum. Fig. 2. Fortrat parabolas, AlH bands, $B' < B''$

tinuous absorption spectrum, which corresponds to the dissociation of the molecule, follows the discrete vibrational levels. In favorable cases, it is possible to determine the dissociation energy of a diatomic molecule from such a spectrum.

The vibrational structure alone is not sufficient to explain an electronic band spectrum. The simultaneous rotational excitation must also be considered. The wavenumbers of the rotational bands are given by the equation:

$$\tilde{\nu} = \tilde{\nu}_o + F'(J') - F''(J''),$$

where $\tilde{\nu}_o = \tilde{\nu}_e + \tilde{\nu}_v$ is constant. In general, the rotational energy levels of a diatomic molecule, which is a symmetric top, are given by;

$$F(J) = B_v J\,(J+1) + (A - B_v)\,\Lambda^2 - D_v J^2\,(J+1)^2$$

The subscript, v, takes account of the interaction between vibration and rotation → rotation-vibration spectrum. The term in Λ^2 is constant for a specific vibrational state of a specific electronic state. For the calculation of the possible rotational transitions, this constant term can be included in $\tilde{\nu}_o$. The selection rules for the symmetric top, $\Delta J = J' - J'' = 0, \pm 1$, with the limitation $J' = 0 \rightarrow J'' = 0$ is forbidden, and neglecting the elongation term D, give analogously to the rotation-vibration spectrum, three rotational branches:

$$\begin{aligned}
\Delta J &= +1:\ \text{R-branch}\ \tilde{\nu} = \tilde{\nu}_0 + 2B' + (3B' - B'')J + (B' - B'')J^2 \\
\Delta J &= 0:\ \text{Q-branch}\ \tilde{\nu} = \tilde{\nu}_0 + (B' - B'')J + (B' - B'')J^2 \\
\Delta J &= -1:\ \text{P-branch}\ \tilde{\nu} = \tilde{\nu}_0 - (B' + B'')J + (B' - B'')J^2.
\end{aligned}$$

The P and R branches can be represented by a single formula:

$$\tilde{\nu} = \tilde{\nu}_o + (B' + B'')m + (B' - B'')m^2,$$

where m is equal to $-J$ for the P branch and $J + 1$ for the R branch. When Δ is equal to zero in both electronic states ($^1\Sigma \rightarrow {}^1\Sigma$ transition), the transition $\Delta J = 0$ is forbidden and there is no Q branch. This corresponds to the analogous relationship in rotation-vibration spectra, but with the difference that, $\tilde{\nu}_o = \tilde{\nu}_e + \tilde{\nu}_v$ and that B' and B'' can now be very different. As a result, the quadratic term can be very much larger, i.e. we obtain a much faster change of the separation of successive lines. The above equations represent parabolas. The evaluation of the rotational transitions is, therefore, frequently carried out by constructing a → Fortrat parabola. Because of the quadratic term, a reversal occurs in either the P or the R branch resulting in a band head. In Figures 2 and 3, Fortrat parabolas for P, Q and R branches are shown. At the vertex there is a bunching of the rotational lines but not a continuum. Whether a band head appears in the P or the R branch depends upon the sign of the difference, $B' - B''$.

$B' - B'' < 0$, i.e. $B' < B''$: The quadratic term has a negative sign and thus, in the R branch, the opposite sign to that of the linear term. Therefore, at larger m values there is a reversal in the R branch. The spectrum shows a band head on the short-wavelength side and is degraded to long wavelengths (to the red), Figure 2.

$B' - B'' > 0$, i.e. $B' > B''$: The quadratic term has a positive sign and therefore the opposite sign to the linear term in the P branch. This produ-

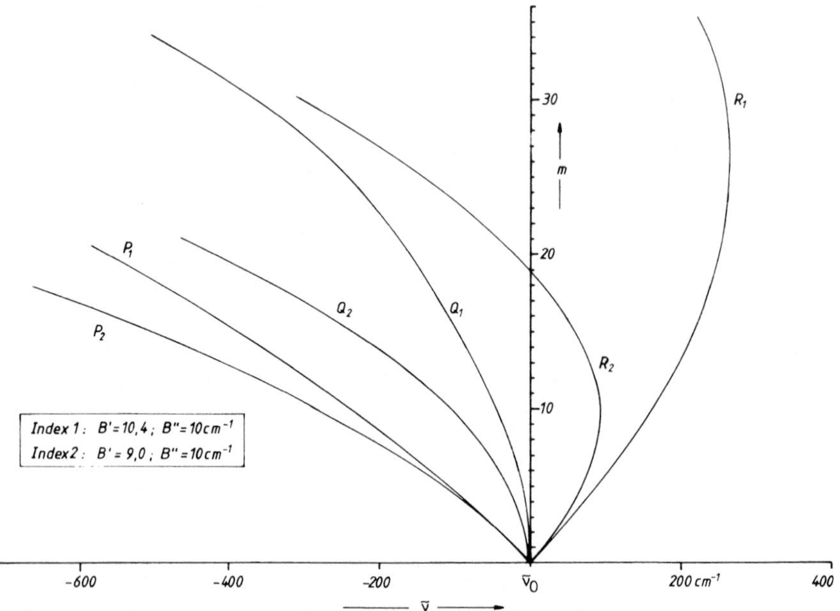

Electronic band spectrum. Fig. 3. Calculated Fortrat parabolas for rotation-vibration spectra, see text

ces a reversal at larger m values in the P branch. The spectrum shows a band head on the long-wavelength side and is degraded to short wavelengths (to the blue).
The m value at the reversal of the parabola can be determined from $d\tilde{\nu}/dm = 0$ to be:

$$m_{\text{vertex}} = -\frac{B' + B''}{2(B' - B'')}.$$

Thus the separation between the zero line and the vertex is:

$$\tilde{\nu}_{\text{vertex}} - \tilde{\nu}_o = -\frac{(B' + B'')^2}{4(B' - B'')}.$$

If B' is only slightly different from B'', i.e. $B' - B''$ is very small, then the band head lies at a very large m value and can frequently not be observed because the intensity of the line has effectively fallen to zero. This is almost always the case in pure rotation-vibration spectra in the IR where B' and B'' are the rotational constants for the upper and lower vibrational levels. Here, they refer to the upper and lower electronic states. In Figure 3 these relationships are illustrated for $B' = 10\ \text{cm}^{-1}$ and $B'' = 10.4\ \text{cm}^{-1}$ and also for $B' = 9\ \text{cm}^{-1}$ and $B'' = 10\ \text{cm}^{-1}$.
The situation in polyatomic molecules is considerably more complicated than in diatomics. The $3N$–6 normal vibrations superimposed upon the electronic excitation must all be considered. Of the $3N$–6 vibrations, those which couple with the electronic excitation can be determined with the help of group theory. The wavenumbers of the transitions between these vibrational levels of the upper (prime) and lower (double prime) electronic states are given by the equation:

$$\tilde{\nu} = \tilde{\nu}_{oo} + \left[\sum_i \omega_{o,i} v'_i + \sum_i \sum_{k \geq i} x'_{o,i,k} \cdot v'_i \cdot v'_k\right] - \left[\sum_i \omega''_{o,i} v''_i + \sum_i \sum_{k \geq i} x''_{o,i,k} \cdot v''_i \cdot v''_k\right]$$

$\tilde{\nu}_{oo}$ is the band origin referred to $v'_i = v''_i = 0$, $x_{o,i,k}$ are the anharmonicity terms (→ anharmonic oscillator). Further, it is assumed that all vibrations are nondegenerate. As with diatomic molecules in emission, progressions are found, which have a more complicated composition due to the possibilities of combination among the normal vibrations. The whole band system can be described in the form of a → Deslandres table. This table is an extension of the → band-head table for diatomic molecules. For larger molecules, the rotational constants B, A and C are very small, because of the relatively large moments of inertia. Therefore, the separation between the rotational levels is relatively small and spectrometers with high resolving powers are required. Grating spectrometers working in high orders, sometimes the tenth or even fourteenth, are usually used for such studies of larger molecules in the gas phase.

Ref.: G. Herzberg, *Molecular Spectra and Molecular Structure I: Diatomic Molecules*, Prentice Hall, New York, **1939**. G. Herzberg, *Molecular Spectra and Molecular Structure III: Electronic Spectra and Electronic Structure of Polyatomic Molecules*, D. Van Nostrand Co. Inc., Princeton, New Jersey, **1966**.

Electronic spectrum, <*Elektronenspektrum*> → electronic band spectrum, → UV-VIS spectrum.

Electron-impact ionization, <*Elektronenstoßionisation*>, the removal of electrons (ionization) from atoms or molecules in the gas phase at reduced pressure by electron bombardment. The electrons are accelerated by a potential difference, U_B, giving them a kinetic energy, e_oU_B, which can be transferred, completely or partially by means of an inelastic impact, to an atom or molecule. In the case of an ionization, the energy of the impinging electrons must be equal to the ionization energy of the corresponding atom or molecule. If the impact energy is smaller the atom will only be excited, and the existence of discrete excitation energies of atoms may be demonstrated (→ Franck and Hertz experiment). The electron flow (current), i, measured in this type of experiment can be plotted as a function of the variable accelerating potential, U_B. Initially, the current increases steeply but it drops suddenly when the energy of the electrons reaches that value which is required for the excitation or ionization of a gas contained in the system. Figure 1 shows a typical plot. However, it is not possible to differentiate between excitation and ionization since we observe the sudden drop of the current for both processes. Following Hertz, the space charge effect of the positive ions formed in electron-impact ionization is used for the detection of the beginning of ionization. At the same applied potential, the space charge effects a strong increase in the current with which the ionization energy can be measured.

In addition to electron-impact ionization, ion-impact ionization is an important method for the ionization of gases. It is used in the ion sources of → mass spectrometers for the generation of positive ions, M^+. If we measure the ion current at the exit of the analyzer as a function of the accelerating potential of the bombarding electrons, the ionization energy can be measured via the → appearance potential.

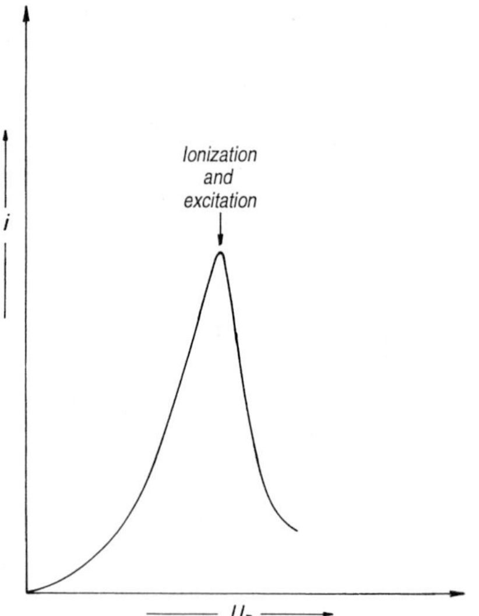

Electron-impact ionization. Fig. 1. i as a function of the accelerating potential, U_B

Electron-impact ion source, <*Elektronenstoßionenquelle*>, an important ion source for the generation of positive ions. Electron-impact ion sources are used in → mass spectrometers. Figure 1 shows the principle of the electron-impact ion source. A glowing filament cathode, GK, emits electrons which are accelerated to the collector (anode A) by an attractive potential, U_A and develop a corresponding kinetic energy, e_oU_A. This energy can be transferred in inelastic collisions to atoms or molecules contained in the ionization chamber, which leads to ionization of these species. A detailed investigation shows that the ionization

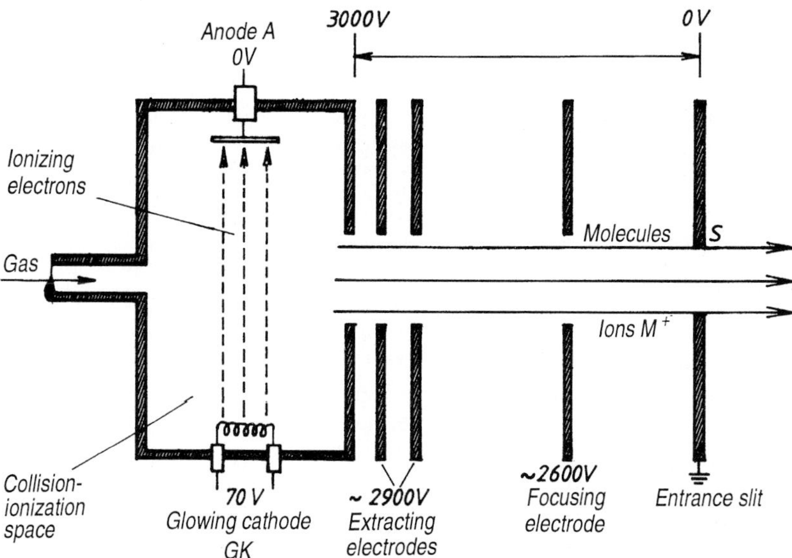

Electron-impact ion source. Fig. 1. Schematic cross section of an electron-impact ion source

reaches a maximum value at an accelerating potential of ~ 70 eV. Under these conditions, the ion current, i^+, of the ions formed is proportional to the partial pressure, p. Like every other ion source, the electron-impact ion source in a mass spectrometer has the task of accelerating the ions, M^+, generated in the ionization chamber with a voltage U (2000–3000 V) and of imaging them at the entrance slit of the analyzer with a focusing electrode.

Electron spin resonance, <*Elektronenspinresonanz*> → ESR

Electron volt, <*Elektronenvolt*>, an energy unit widely used in spectroscopy and also in atomic and nuclear physics. However, it is not an SI unit. It corresponds to the kinetic energy which a free electron obtains by passing through a potential difference of 1 V, i.e.:

$$1\ \mathrm{eV} = 1.602 \cdot 10^{-19}\ \mathrm{Ws}$$
$$1.602 \cdot 10^{-12}\ \mathrm{erg}.$$

(→ energy units)

Electrooptics, <*Elektrooptik*>, a branch of physics dealing with phenomena which can be traced back to the interaction of an electric field with an optical process: → Stark effect, → Kerr effect, → Pockels effect.

Electrophotoluminescence, <*Elektrophotolumineszenz*>, an effect which occurs when an electric field is applied to a phosphor excited to radiate by UV or X-ray radiation (see also → electroluminescence). We distinguish between the Gudden-Pohl and the Déchêne effects which were discovered in 1921 and 1935 respectively. In the former, when an electric field is applied to a phosphor, during the decay process or during continuous excitation, a short-term increase of luminescence intensity, the Gudden-Pohl flash, is observed. In the latter, we observe a luminescence intensity decrease when applying an electric field. Thus, the Déchêne effect is described as a quenching process whilst the Gudden-Pohl effect is an electrical stimulation of the luminescence.

Elongation factor, D, <*Dehnungskonstante*>, a factor which allows for the nonrigid behavior of a rotating molecule. The name *elongation factor* is not in common use in the English-language scientific literature where D is usually called a rotational constant. For the → harmonic oscillator, the two rotational constants, B and D, are related by $D = 4B^3/\omega^2$, where ω is the wavenumber of the fundamental frequency of the oscillator. Therefore, D is always very much smaller than B. (→ centrifugal term).

Emission center, <*Emissionszentrum*>, a general term for a radiating dipole.

Emission continuum, <*Emissionskontinuum*> → recombination continuum, → decay luminescence.

Emissivity, <*Emissionsvermögen*>, a measure, characteristic of the material, of the emission of radiation by a body (→ thermal radiator). It indicates the fraction of the theoretically possible maximum emission which the radiator concerned emits at a given temperature and wavelength. The emissivity, ε, is equal to 1 for the whole spectrum of a → black body, i.e. it is the optimum radiation source with the highest radiative power at any given temperature. For all other thermal radiators, ε is smaller than 1 and usually dependent upon temperature and wavelength. Thermal radiators with an emissivity in the visible spectral region which is constant but smaller than 1, are called gray radiators. The radiation of coal is almost gray in the visible spectral region. Its emissivity is fairly constant and lies between 0.8 and 0.9, depending upon the type and surface of the material. According to → Planck's radiation law, the total emissivity, ε_g, in the spectral region from λ_1 to λ_2 is given by:

$$\varepsilon_g = \frac{\int_{\lambda_1}^{\lambda_2} \varepsilon_\lambda S_\lambda \mathrm{d}\lambda}{\int_{\lambda_1}^{\lambda_2} S_\lambda \mathrm{d}\lambda};$$

ε_λ is the spectral emissivity and S_λ the spectral radiation density of a black body at wavelength, λ.

Emitter, <*Emitter, aktiver Emitter*> → field ionization.

Enantiomeric form, optical isomer, <*Antipode, optischer, enantiomere Form, optisches Isomer*>, an optically active compound which differs from another compound having the same chemical and physical properties by the fact that it rotates the plane of → polarized light by the same angle but in the opposite direction to the isomeric substance. The right-rotating compound is called the *d* form (from *dextro*) and the left-rotating compound the *l* form (from *laevo*). The two molecules are mirror images of each other.

Energy-level diagram, term scheme, <*Termschema*>, the representation of the → terms or energy levels of an atom or a molecule on a vertically increasing scale in cm^{-1}. The term scheme is a one-dimensional representation and presents all the energy states relative to the lowest state, generally the ground state or ground term. All transitions between the different energy states can be expressed as term differences, though the → selection rules may considerably reduce the apparent number of transitions. In particular, transitions between states (terms) of different multiplicity are forbidden. For molecules, where the spins are always paired in the ground state, two sets of terms always appear, singlets and triplets, (→ Jablonski diagram). The triplet states arise when an electron is excited and changes its spin to give a total spin, $S = 1$. The → multiplicity of such a state is 3, i.e. there is a triplet energy-level scheme of which the lowest term is of higher energy than the lowest singlet term. This lowest triplet is sometimes known as a metastable state.

The states corresponding to other forms of energy, e.g. vibrational, rotational, nuclear spins in a magnetic field, etc. may also be arranged in a term scheme. Such schemes are invariably called energy-level schemes or diagrams in the modern English-language scientific literature and use of the word *term* appears to be decreasing even in atomic spectroscopy, the field for which it was originally coined.

Energy spectrum → spectral energy distribution.

Energy units, <*Energieeinheiten*>, the units cm^{-1}, $cal \cdot mol^{-1}$, $J \cdot mol^{-1}$, electron volt (eV) and $erg \cdot mol^{-1}$ are used in spectroscopy as measures of the energy. The table gives the conversion factors.

Although the units, cal and erg, are no longer used in the modern system of units, they have been included in view of the references to the older literature.

The following approximations are easily remembered for *ball-park* calculations: 10,000 cm^{-1} = 1.25 eV = 120 kJ or 28 kcal: 1 eV = 8000 cm^{-1} = 100 kJ or 23 kcal. In addition to the $erg \cdot mol^{-1}$, the unit, erg/molecule has been frequently used. In this case, the conversion factors must be divided by Loschmidt's number, $N_L = 6.0228 \cdot 10^{23}$ $molecule \cdot mol^{-1}$.

Equiextinction method, <*Äquiextinktionsmethode*>, a specific mode of operation in → dual-wavelength spectroscopy. If the absorbance of a two-component system (*a*, *b*) is measured at two suitable wavelengths, λ_1 and λ_2,

Energy units. Table.

Unit	cm^{-1}	$cal \cdot mol^{-1}$	$Joule \cdot mol^{-1}$	eV	$erg \cdot mol^{-1}$
1 cm^{-1}	1	28575	11.9612	$1.2395 \cdot 10^{-4}$	$11.9612 \cdot 10^{7}$
1 cal mol^{-1}	0.34966	1	4.1859	$4.3379 \cdot 10^{-5}$	$4.1859 \cdot 10^{7}$
1 Joule mol^{-1}	0.0836	0.2389	1	$1.0363 \cdot 10^{-5}$	$1 \cdot 10^{7}$
1 eV	8067.5	23053	96497	1	$96497 \cdot 10^{7}$
1 erg mol^{-1}	$0.0836 \cdot 10^{-7}$	$0.2389 \cdot 10^{-7}$	$1 \cdot 10^{-7}$	$1.0363 \cdot 10^{-12}$	1

and the absorbance difference, $\Delta A = A_2 - A_1$, is formed, then it is frequently possible to set the difference $(\varepsilon_{2b} - \varepsilon_{1b})c_b \cdot d$, i.e. the absorbance difference for component b only, equal to zero. The two wavelengths, λ_1 and λ_2, must be selected in such a way that the absorbances, A_{2b} and A_{1b}, are the same; whence the name of the method. It is a precondition for the application of the method that the absorption maximum of component a overlaps the short-wavelength flank of component b, but does not underly the spectrum of component b completely, (→ dual-wavelength spectroscopy).

Error of observation, <*Beobachtungsfehler*>, an error which occurs when measuring a quantity. We distinguish primarily:

a) Gross errors which occur due to carelessness;
b) systematic errors which, by their nature, falsify the measurement in one direction;
c) random (accidental) errors which are unvoidable und result from the unreliability of the human senses, the instrument, the method of observation and the external circumstances when the observation is made.

If an excess of observations has been made, i.e. more observations than are required for a definite solution of the problem, the gross and systematic errors must be eliminated before any compensation is possible. An unavoidable error of observation made when reading a scale is a reading error. It is characterized by the mean error, m_a, of a reading which results from a series of observations. If we take n readings, $a_1 \ldots a_n$, from a reading device, the arithmetic mean, M, can be formed as $M = [a]/n$. (Square brackets, [], are generally used to indicate a sum when calculating errors.)

If the difference between the arithmetic mean, M, and the individual reading, a, is V, i.e. $V_1 = M - a_1$, $V_2 = M - a_2$, etc. we obtain the mean error, m_a, of a reading as $m_a = \{[\mathbf{VV}]/(n-1)\}^{1/2}$. The mean error, m_M, of the arithmetic mean, M, of the series of observations is given by:

$$m_M = \sqrt{\frac{[VV]}{n(n-1)}}.$$

The error, which occurs when estimating the position of a pointer on a scale is the error of estimate. If n equal quantities are each measured twice with the same care independently of each other, the mean error, m', of one observation and the mean error, m, of the arithmetic mean of two observations can be calculated from the differences, d, of the observations (observational differences) using the following formulas:

$$m' = \sqrt{\frac{[dd]}{2n}}, \quad m = \sqrt{\frac{[dd]}{4n}}.$$

ESCA, electron spectroscopy for chemical analysis <*ESCA*> → X-ray photoelectron spectroscopy.

ESR spectrometer, electron spin resonance spectrometer, <*ESR-Spektrometer, ESR-Spektralphotometer, Elektronenspinresonanzspektrometer*>, an ESR spectrometer is shown as a block diagram in Figure 1. It operates in the → continuous wave mode. The sample to be measured is placed into a cavity resonator (c) between the pole caps of an electro-

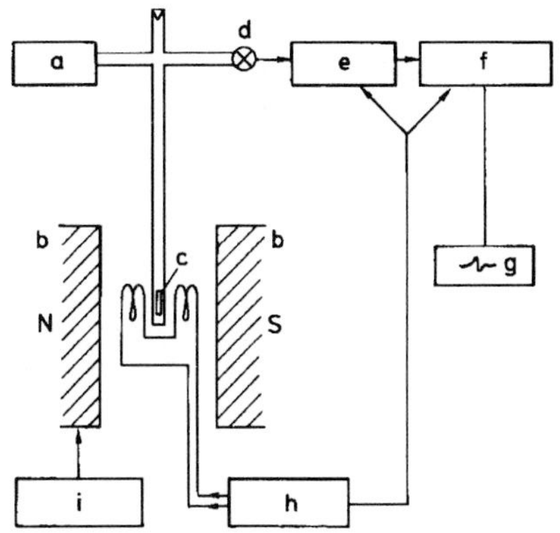

ESR spectrometer. Fig. 1. Block diagram of an ESR spectrometer

magnet (b). A klystron (a) provides the high-frequency, electromagnetic field which is conducted to the sample location through a waveguide. A crystal detector (d) is used to detect the resonances. The ESR spectrum is normally measured at a constant klystron frequency while the magnetic field, B, is varied. Thus, the resonance curve is obtained as a function of B. A better signal-to-noise ratio is achievable if the magnetic field is modulated, e.g. at 100 KHz (h). The current recorded by the crystal detector (d) on passing

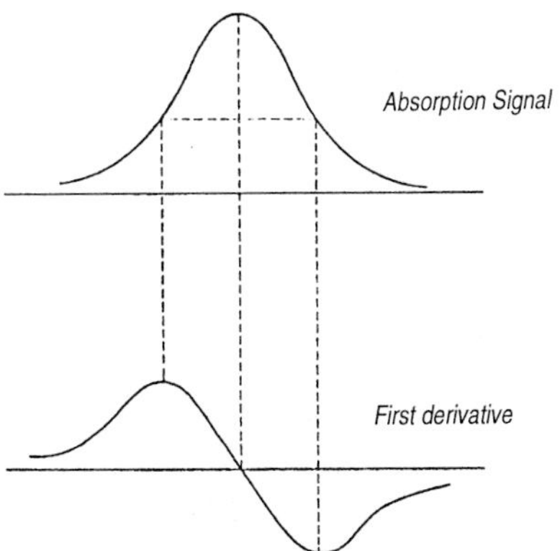

ESR spectrometer. Fig. 2. Absorption spectrum and first derivative

through a resonance signal is also modulated at 100 KHz. An amplifier (e) and a phase-sensitive rectifier (f) convert the incoming signal into a form which corresponds to the first derivative of an absorption signal (g); Figure 2. As in → derivative spectroscopy, the first derivative has the advantage that nuances in an absorption curve can be more easily recognized.

ESR spectroscopy, EPR spectroscopy, *<ESR-Spektroskopie, Elektronenspinresonanzspektroskopie>* → magnetic resonance spectroscopy.

Etalon, *<Etalon>*, generally a very accurate standard of length but in particular a plane-parallel plate made of glass or fused quartz with reflecting surfaces. In → laser spectroscopy, etalons are mounted in the laser resonator primarily as wavelength-selective transmission filters. They are used to narrow the laser band width. Their effectiveness is based on the multiple reflection of light within the plate and on the interference caused by this. Figure 1 illustrates, in a simplified manner, the conditions in a plate of thickness, d, with refractive index, n. A beam of light from a source, LS, is incident with angle of incidence, α. At point, A, it is partially reflected at the surface as beam, a, and partially refracted at an angle of refraction, β, into the plate. It is again reflected at the back of the plate, B, towards C and finally exits into the air as beam b, parallel to the first beam a reflected at A. The other part of the beam leaves the plate at its lower surface and enters the air at angle, α, as beam, a' (transmitted fraction). The same applies to the beams, $B - C$, $C - D$,

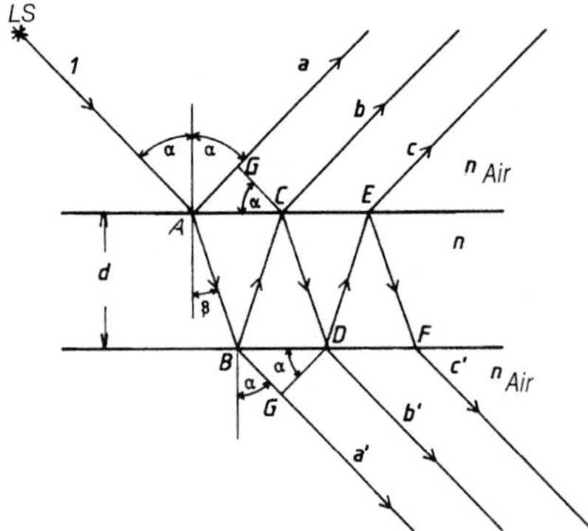

Etalon. Fig. 1. Cross section of an etalon with the interfering beams, see text

$D - E$, $E -$ etc. This means that beam 1 can be split by multiple reflection into an infinite number of parallel beams, a, b, c ... and parallel refracted beams, a', b', c' ... All these beams are coherent, on account of the manner of their generation, and therefore capable of → interference. The path difference between beams, a and b (in reflection) or a' and b' (in transmission), is decisive. The optical path difference, Δs, between a and b is, from Figure 1, $\Delta s = n(AB + BC) - AG$. In terms of the angles, α and β and the → refractive index, $n = \sin\alpha/\sin\beta$, we obtain for Δs:

$$\Delta s = 2nd \cos \beta = 2d\sqrt{n^2 - \sin^2\alpha}.$$

A phase shift of π, which corresponds to a path difference of $\lambda/2$, must be added in the case of the reflected light (beams a and b ...), and the path difference is then:

$$\Delta s = 2d\sqrt{n^2 - \sin^2\alpha} + \frac{\lambda}{2}.$$

However, the first formula applies to the transmitted light since there is no phase shift of π for beams, a', b', c' ...

If the path difference, Δs, corresponds to a whole-number multiple, k, of wavelength, λ, we obtain at the recombination of all beams incident at angle α, a maximum brightness (transmission):

$$\Delta s = 2d\sqrt{n^2 - \sin^2\alpha} = k \cdot \lambda$$

(k = 0, 1, 2, 3 ...) when Δs corresponds to half an odd-number multiple of λ, darkness results since all beams are eliminated by interference:

$$\Delta s = 2d\sqrt{n^2 - \sin^2\alpha} = \frac{2k+1}{2}\lambda$$

The expression for the wavelength of maximum transmittance for the kth order is:

$$\lambda_k = \frac{2d}{k}\sqrt{n^2 - \sin^2\alpha} = \frac{2nd}{k} \cos \beta.$$

This formula shows that by changing the angle, α, the wavelength of maximum transmission can be varied. An etalon is mounted in the beam path of a laser → cavity in such a way that it can be tilted with respect to the cavity thus tuning the laser.

Evanescent wave, *<Grenzschichtwelle>*, a light wave occurring in the second medium, the medium with the smaller refractive index, in → total reflection. However, the light must be incident upon the common (plane) interface between the media in the optically denser medium, i.e. in the medium with the larger refractive index. Also, the angle of incidence must be equal to the critical angle for total reflection, or larger. Evanescent waves hardly penetrate the medium with the lower refractive index. The depth of penetration depends upon the wavelength. The surfaces of the same amplitude lie parallel to the interface, i.e. the surfaces of the same phase are perpendicular to the interface. However, these statements are only approximations. When absorption occurs in the optically less dense medium there is a reduction of the totally reflected light which is utilized in the important → ATR method.

Excimer, <*Excimer*>, the formation of a dimer between a molecule in excited state, ${}^1M^*$, and the same species in the ground state, ${}^1M$:

$$ {}^1M^* + {}^1M \rightleftharpoons ({}^1M^* \cdot {}^1M) = {}^1D^*. $$

In aromatic hydrocarbons in solution, the formation of excimers leads to quenching of the monomer fluorescence with increasing concentration of the dissolved aromatic (→ self-quenching). At the same time, a change in the fluorescence is observed. A broad, structureless fluorescence band, shifted to longer wavelengths vis-a-vis the original fluorescence of the monomers, occurs which is called → excimer fluorescence. The formation of excimers is a diffusion-controlled process. Furthermore, it is important to note that an excimer in the ground state dissociates immediately and, for that reason, no change in the absorption spectrum can be observed. Thus, there is a significant difference between a normal dimer in an excited state and an excimer.
Ref.: J.B. Birks, *Photophysics of Aromatic Molecules*, Wiley Interscience, John Wiley and Sons, London, New York, Sydney, Toronto, **1970**, chapt. 7.

Excimer fluorescence, <*Excimerfluoreszenz*>, a concomitant phenomenon in solutions of aromatic hydrocarbons. Depending upon the concentration of the dissolved substance, → self-quenching of the fluorescence and a new, broad and structureless fluorescence, which is shifted to longer wavelengths vis-a-vis the monomer fluorescence of dilute solutions, is observed. This excimer fluorescence is due to the formation of dimers in the excited state (→ excimer). The formation and radiative deactivation of the excimers can be described as follows:

$$ {}^1M + h\nu \rightarrow {}^1M^* $$

$$ {}^1M^* + {}^1M \rightarrow {}^1D^* \rightarrow 2\,{}^1M + h\nu_{FD}. $$

The various competing deactivation processes for ${}^1M^*$ and ${}^1D^*$ and their associated rate constants, k, are shown in an energy-level diagram (Figure 1) where the radiating processes are denoted by full lines and the radiationless processes by dashed lines. Figure 2 shows the fluorescence spectra of pyrene in n-heptane for different concentrations. The spectra are all normalized to the monomer peak at $\tilde{\nu}$ = 25,100 cm^{-1} and the broad, structureless excimer fluorescence and its concentration dependence can be easily recognized. Excimer fluorescence is observed not only in concentrated solutions of aromatic substances, but

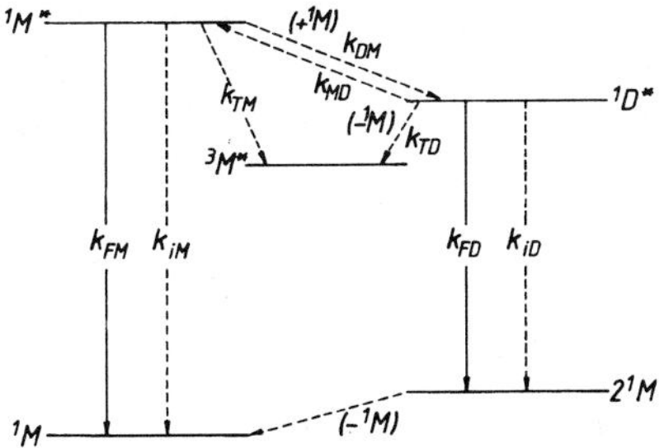

Excimer fluorescence. Fig. 1. Energy-level diagram for excimer fluorescence

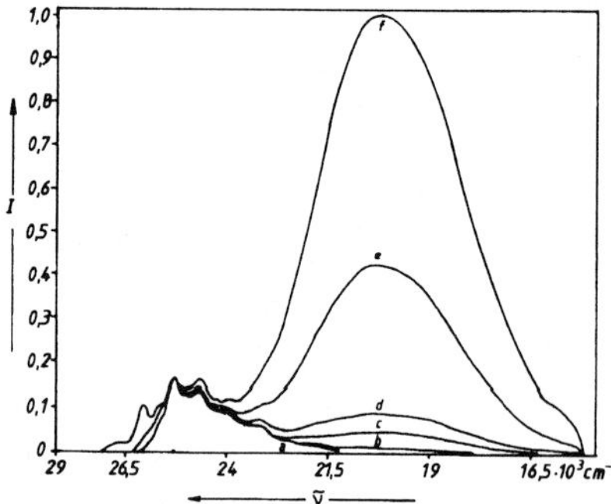

Excimer fluorescence. Fig. 2. Concentration dependence of the fluorescence of pyrene in n-heptane

also in their crystals where it depends upon the arrangement of the molecules in the unit cell.
If an excimer fluorescence is present the → quantum yield of the monomer fluorescence, Φ_{FM}, decreases with increasing concentration of ^1M whilst the quantum yield of the excimer fluorescence, Φ_{FD}, increases correspondingly. According to the → Stern-Volmer relation for the quantum yield, the → self-quenching of the monomer fluorescence is given by:

$$\Phi_{FM} = \frac{q_{FM}}{1 + [^1M]/[^1M]_h} = \frac{q_{FM}}{1 + K[^1M]}.$$

where $[^1M]_h$ = the half-value concentration at which the quantum yield has dropped to half its value: $\Phi_{FM} = q_{FM}/2$. q_{FM} is the molecular quantum yield when no excimer formation occurs, i.e. in very dilute solutions. $[^1M]_h^{-1} = K$ is the Stern-Volmer coefficient of the self-quenching.
Correspondingly, the quantum yield of the excimer fluorescence is:

$$\Phi_{FD} = \frac{q_{FD}}{1 + [^1M]_h/[^1M]} = \frac{q_{FD}}{1 + K^{-1}/[^1M]}.$$

q_{FD} is the quantum yield for the limiting value of Φ_{FD} when $1/[^1M] \to 0$ (high concentration; $\Phi_{FM} \approx 0$).
Using the rate constants, k (in s^{-1}), defined for the deactivation of $^1M^*$ and $^1D^*$ (see Figure 1), the solution of the kinetic equations gives an expression for the Stern-Volmer coefficient, K, in the photostationary case:

$$K = \frac{1}{[M]_h} = \frac{k_D}{k_M} \cdot K_e$$

with

$$k_D = k_{FD} + k_{TD} + k_{ID} = \frac{1}{\tau_D} = \frac{k_{FD}}{q_{FD}};$$

$$k_M = k_{FM} + k_{TM} + k_{IM} = \frac{1}{\tau_M} = \frac{k_{FM}}{q_{FM}};$$

$$K_e = \frac{k_{DM}}{k_{MD} + k_D} = \frac{[^1D^*]}{[^1M^*][^1M]}.$$

K_e is the formation constant for the excimers. The resulting expression for the ratio of the quantum yields is:

$$\frac{\Phi_{FD}}{\Phi_{FM}} = \frac{q_{FD}[^1M]}{q_{FM}[^1M]_h} = \frac{k_{FD}}{k_{FM}} \cdot K_e[^1M] = k_1[^1M].$$

This equation includes all the parameters which are accessible by measuring the fluorescence quantum yields, q_{FM}, q_{FD}, Φ_{FM} and Φ_{FD} and the constants, K_e and k_1, in combination with measurements of the fluorescence-decay times, τ_M and τ_D.
Ref.: J.B. Birks, *Photophysics of Aromatic Compounds*, Wiley Interscience, John Wiley and Sons, London, New York, Sydney, Toronto, **1970**, chapt. 7.

Excimer laser, <*Excimer-Laser*>, or more exactly exciplex laser, a laser with excited dimers (→ excimers) or (usually) excited complexes. Excimer lasers were described by different researchers, independently of each other, for the first time in 1975. In recent years, they have proved themselves as intense and stable radiation sources for the ultraviolet spectral region. The most important exciplexes are the excited noble-gas-halogen complexes such as (ArF)*, (KrF)*, (XeCl)* and (XeF)* which are formed and deactivated according to the scheme: $A + B^* \rightarrow (AB)^* \rightarrow A + B + h\nu_e$ or $A^* + B \rightarrow (AB)^* \rightarrow A + B + h\nu_e$. The deactivation of the exciplexes, $(AB)^*$, occurs with the emission of intense UV radiation showing characteristic lines between 193 nm and 351 nm. Figure 1 shows the positions of the lines emitted by the above exciplexes. The emission lines of the F_2 laser at 157 nm, → N_2 laser at 337 and → He-Cd^+ laser at 441.5 nm are also plotted. With the most commonly used Ar^+ laser (→ noble-gas-ion laser), lines are only obtained in the UV region by means of → frequency doubling (224 nm and 257 nm). If excimers or exciplexes are to be effective as the → active medium it is important that they are unstable in the ground state, i.e. the ground-state potential curve must be repulsive. In comparison with the corresponding potential curve in the excited state $(AB)^*$ and its potential minimum, the ground state is practically unpopulated, i.e. a → population inversion is present after the formation of an exciplex. Since excimers have only a short lifetime of the order of a few nanoseconds, they must be excited rapidly. This is achieved by a rapid electric discharge at a high pressure. The discharge takes place in a gas mixture which contains small amounts of the noble gas and halogen diluted with helium or neon as buffer gases. The short UV laser pulses generated leave the discharge cavity with a low divergence and have a characteristic rectangular profile with a cross section of a few hundred mm^2. Such lasers produce individual pulse energies in the range of 10 mJ to 2 J and have repetition rates from single shot up to 200 Hz. The pulse widths are only 10–20 ns from which peak powers of 1 to 50 MW result. Modern, fifth generation excimer lasers, such as the PX100 and LPX200 models, are fully microprocessor-controlled. They are used as

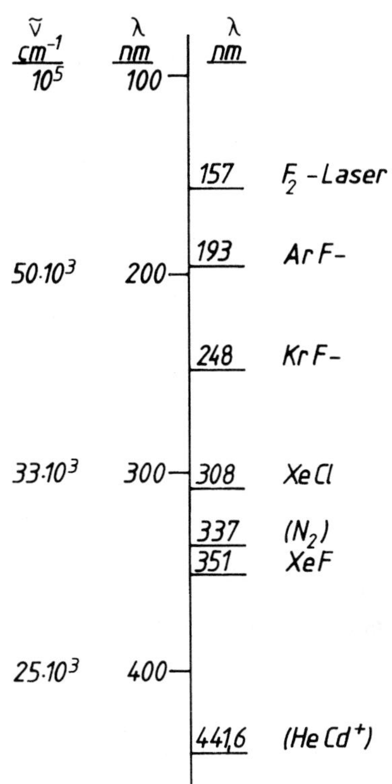

Excimer laser. Fig. 1. Summary of the positions of excimer laser lines

pump lasers for → dye lasers for XUV production, in photochemistry for the amplification of femto-second pulses, in → Raman spectroscopy and in material processing and medicine.
Manufacturer: Lambda Physik GmbH, Hans-Boeckler-Str. 12, D-3400 Göttingen, Germany.

Exciplex laser, <*Exciplex-Laser*> → excimer laser.

Excitation, <*Anregung*>, a process in which an atom or a molecule, in general a system, is excited from a lower to a higher energy level. The required energy, the excitation energy, can be supplied by:

a) Absorption of electromagnetic radiation, $h\nu$;
b) electron and ion bombardment, eV;
c) thermal energy, kT;
d) reaction enthalpy, ΔH.

Excitation according to (a) is the basis of the majority of spectroscopic methods which can be interpreted in terms of the → Bohr-Einstein frequency relation.
The excitation of discrete energy states by electron bombardment was first demonstrated by the → Franck and Hertz experiment. → Ionization of atoms or molecules can take place if the energy of the bombarding electrons is sufficiently high. This is used, for example, in → mass spectrometry.
Thermal excitation generally occurs in flames, electric arcs (→ arc discharge), sparks and plasmas and is the primary process in → atomic emission spectroscopy (AES). The higher rotational states and, to some extent, the higher vibrational states of molecules are excited with thermal energies of the order of room temperature (300–400 K) where the population of the higher states is determined by → Boltzmann statistics. If the reaction enthalpy released in the course of a reaction is transferred to a product molecule, the molecule may be promoted to an electronically excited state which is manifested, for example, by the occurrence of → chemiluminescence.

Excitation spectrum (fluorescence), <*Fluoreszenzanregungsspektrum, FA-Spektrum*>, a spectrum which is obtained if the wavelength for the excitation of the fluorescence, λ_E, is varied over the absorption spectrum whilst the wavelength at which fluorescence is observed, λ_F, is kept constant. Since the → fluorescence intensity is proportional to the → extinction coefficient in dilute solutions, this spectrum reflects the absorption spectrum. However, we must take account of the spectral energy distribution of the exciting light, $I_o(\lambda)$. For such measurements, λ_F is generally selected as the maximum of the fluorescence spectrum or the total fluorescence light may be measured using a suitable, nonfluorescent → edge filter; see also → synchronous fluorescence-excitation spectroscopy.
Ref.: H.-H. Perkampus, *UV-VIS Spectroscopy and Its Applications*, Springer Verlag, Berlin, Heidelberg, New York, **1992**, chapt. 5.5.

Exciton, <*Exciton, Exziton*>, an electron-hole pair consisting of an excited electron and the vacated positive hole. The electron and hole cannot move independently but only together and in the same direction.

Exclusion principle, <*Ausschließungsprinzip*> → Pauli exclusion principle.

Expectation region, <*Erwartungsbereich*>, in spectroscopy, a frequency or wavenumber region in which a quite specific and characteristic spectroscopic absorption or emission can be expected. → IR spectroscopy offers the best example with the expectation regions of the → group frequencies. However, the → chemical shifts in → NMR spectroscopy or the masses in the → fragmentation patterns in → mass spectrometry can also be tabulated in terms of expectation regions.

Extinction coefficient, molar decadic, <*Extinktionskoeffizient, molarer, dekadischer*>, a characteristic and substance-specific unit which was introduced with the → Bouguer-Lambert-Beer law and is defined as follows:

$$\varepsilon(\lambda) = \frac{A(\lambda)}{c \cdot d}.$$

With the concentration, c, in $\mathrm{mol \cdot l^{-1}}$ and the path length, d, in cm, the units of ε are $\mathrm{l \cdot mol^{-1} \cdot cm^{-1}}$. For $1\,\mathrm{l} = 1000\,\mathrm{cm^3}$ we obtain: $1000\,\mathrm{cm^2 \cdot mol^{-1}}$. Although the molar decadic extinction coefficient is not dimensionless, it is almost always given as a pure number in the literature. Its dependence upon the wavelength of the absorbed light is its most important property. If we plot ε as ordinate against the wavelength (or wavenumber) as abscissa we obtain the → absorption spectrum which is characteristic of the system involved. The importance of → absorption spectroscopy in chemical structural analysis is based on this. It is also important that the experimentally accessible molar extinction coefficient, $\varepsilon(\lambda)$, is related to the theoretically established → oscillator strength, which is also easily accessible experimentally. This is significant for the theory of → molecular spectroscopy.
If $\varepsilon(\lambda)$ is known the concentration can be determined directly by measuring the absorbance, $A(\lambda)$, at a given path length, d (→ photometry). The magnitude of $\varepsilon(\lambda)$ determines the sensitivity and limit of detection of the photometric determination.

Extinction coefficient, molar natural, <*Extinktionskoeffizient, molarer, natürlicher*>, the → Bouguer-Lambert-Beer law may be written with the → absorbance, A, in natural logarithms as:

$$\varepsilon_n(\lambda) = \frac{A_n(\lambda)}{c \cdot d}.$$

The relationship with the → molar decadic extinction coefficient is simply:

$$\varepsilon_n = 2.303 \cdot \varepsilon.$$

Extraction photometry, <*Extraktionsphotometrie*>, a special photometric, analytical method for the determination of cations. Extraction photometry has an increased detection sensitivity. It is based on the fact that complexed cations can be extracted from an aqueous solution using an organic solvent. Basic or acidic dyes which form ion associates with complexed cations are used as complexing agents. Generally, the ion associates have very high extinction coefficients since several dye molecules can be bound in one ion associate. For example, the zinc-1,10-phenanthroline complex forms an ion associate with the acid dye eosin in aqueous solution which can be extracted with chloroform. Referred to the metal cation, the extracted complex has an extinction coefficient of $\varepsilon_{max} = 120 \cdot 10^3\ l \cdot mol^{-1}\ cm^{-1}$. In this case, the absorbance of two dye molecules per gram atom of the cation under investigation is measured.

Ref.: H.-H. Perkampus, *UV-VIS Spectroscopy and Its Applications*, Springer Verlag, Berlin, Heidelberg, New York, **1992**.

Extraordinary ray, <*außerordentlicher Strahl*>, in the refraction of light by optically anisotropic crystals, the description of that refracted, linearly polarized ray which does not obey Snell's law of refraction. Optically uniaxial crystals generate one ordinary and one extraordinary ray. However, two extraordinary rays with differing velocities are produced by optically biaxial crystals.

F

Fabry-Perot interferometer, Fabry-Perot etalon, <*Fabry-Perot-Interferometer, Interferenzspektrometer, F.P.I.*>, an instrument, described as an interference spectroscope for the first time by Fabry and Perot in 1889. It uses multireflections and interferences between two plane-parallel glass plates, P_1 and P_2, with a separation, d. Air, with the refractive index, $n \cong 1$, fills the space between the plates. In order to avoid interfering reflections at the external faces of the glass plates, these are often made slightly wedge-shaped. Both plates have transmitting mirrors with a reflection coefficient, R, of 0.90 to 0.95 on the inside faces. Figure 1 shows the construction of a Fabry-Perot interferometer schematically. Plate, P_1, is mounted rigidly and P_2 can be adjusted very accurately parallel to the first plate and also moved along the common normal to P_1 and P_2, thereby changing d. In the case of very small distances, d, the Fabry-Perot interferometer behaves as an → interference filter with air as the dielectric layer. Plate intervals of 1 to 10 cm are used for interferometric purposes. A Fabry-Perot interferometer of this form is only used in transmission.

After penetrating the mirror layer of the first plate a beam, l, coming from one point of the light source is multiply reflected, to and fro. At each reflection on P_2, a fraction of the energy exits downwards. Constructive or destructive interference of the beams occurs, depending upon the path difference of the rays. The optical

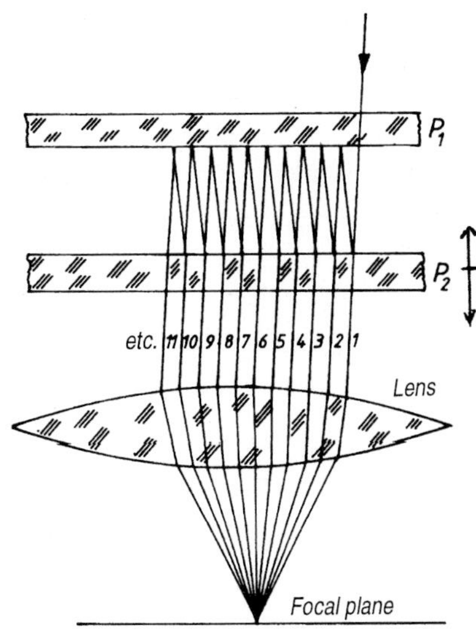

Fabry-Perot interferometer. Fig. 1. Principle of the Fabry-Perot interferometer

path difference Δs of the exiting rays is given, analogously to the interference at a thin plate (→ etalon), by:

$$\Delta s = 2d\sqrt{n^2 - \sin^2\alpha}.$$

With $\Delta s = k \cdot \lambda$ ($k = 0, 1, 2, 3 \ldots$) the wavelengths, λ_k, of the beams of increased intensity are:

$$\lambda_k = \frac{2d}{k}\sqrt{n^2 - \sin^2\alpha}.$$

When divergent light (all values of α) is used, circles corresponding to the same value of α are formed in the focal plane around the normal to the plate. If d is known, the wavelengths of the spectral lines can be determined from the radii of adjacent circles, or from the corresponding angle, α, using the equations $\Delta s = k \cdot \lambda$ and $\Delta s = (k + 1)\lambda$.

If $\alpha = 0$, $\Delta s = k \cdot \lambda = 2nd$. In this case, d can be varied and the intensity change in the center of the circle mea-

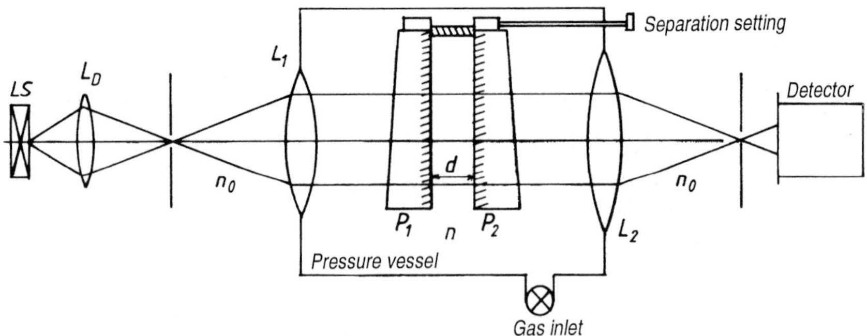

Fabry-Perot interferometer. Fig. 2. Construction for variable pressure operation

sured as a function of the plate separation. Successive intensity maxima for the path differences, $k \cdot \lambda$ and $(k + 1) \cdot \lambda$ then correspond to a difference of the distance between the plates of $\lambda/2n$.

The above relation shows further that the intensity can also be changed by changing n, which can be achieved by varying the air pressure between the two interferometer plates.

If the Fabry-Perot interferometer is operated with varying plate separation or varying air pressure, it is best to work with parallel light. By varying the plate interval or pressure, the device alternates between transparent ($2nd = k$) and nontransparent ($2nd = (2k + 1)/2$). Figure 2 illustrates a Fabry-Perot interferometer for variable pressure.

The resolving power of the Fabry-Perot interferometer is:

$$\frac{\lambda}{\Delta\lambda} = \frac{4\pi nd}{(1-R)\lambda}.$$

When $R = 0.95$, $d = 1$ cm, $n = 1$ and $\lambda = 600$ nm we have a resolving power of $4 \cdot 10^6$ which cannot be achieved with prism or grating spectrometers. However, the actual dispersion range, $\Delta\lambda = \lambda^2/2nd$, is very narrow. With $d = 1$ cm, $n = 1$ and $\lambda = 600$ nm: $\Delta\lambda \sim 1.8 \cdot 10^{-2}$ nm, i.e. in wavenumbers at $\lambda = 600$ nm, ($\tilde{\nu} = 16{,}666.67$ cm^{-1}) a $\Delta\tilde{\nu}$ range of 0.56 cm^{-1}.

Consequently, the Fabry-Perot interferometer is only suitable for → interference spectroscopy if the radiation is restricted to a narrow wavelength range naturally, or artificially by means of interference filters, prism or grating spectrographs.

The → hyperfine structure of many spectral lines has been investigated using a Fabry-Perot interferometer. Furthermore, the FPI is a very important tunable, optical component in laser spectroscopy, especially in dye lasers.

Ref.: W. Demtröder, *Laser Spectroscopy*, Springer Series in Chemical Physics, vol. 5, Springer Verlag, Berlin, Heidelberg, New York, **1981**, chapt. 4.2.9, p. 167 ff.

Faraday cup, <*Faraday-Auffänger*>, one of the most commonly used detectors for recording ions in mass spectroscopy. The ions fall into a box-like trap which is mounted behind the exit slit of the analyzer. Figure 1a shows a diagram of a single cup with the subsequent amplifying devices. The geome-

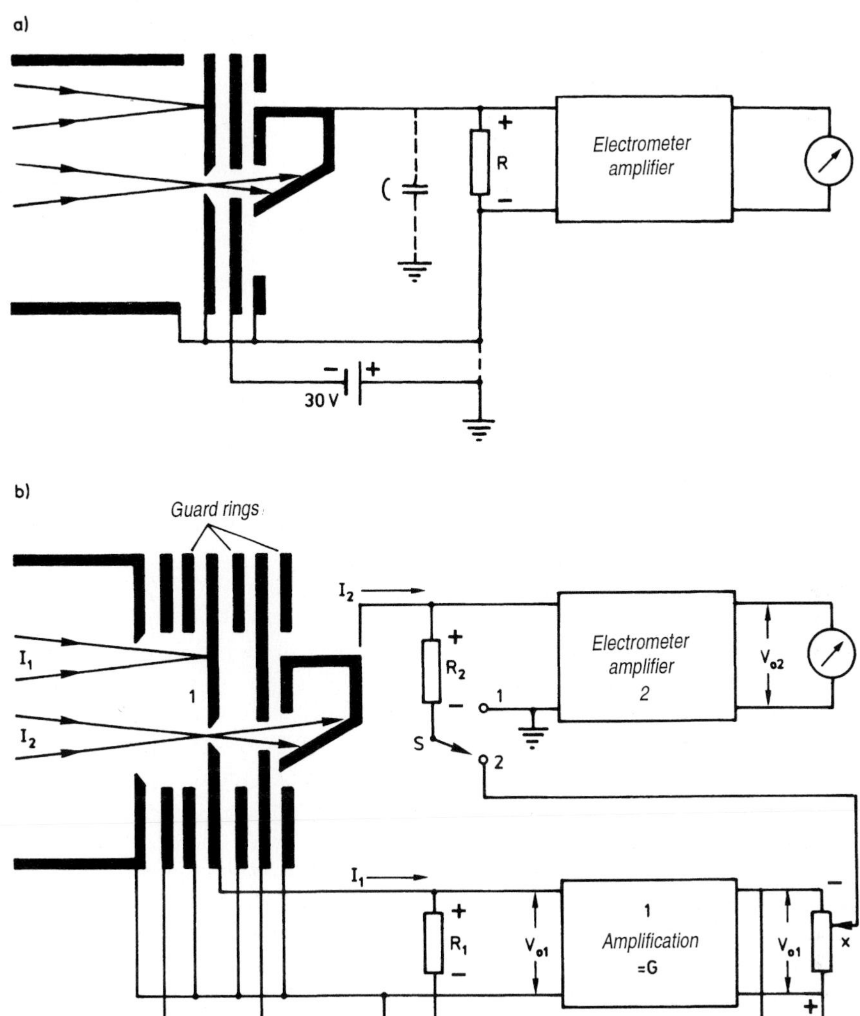

Faraday cup. Fig. 1. Diagram of a Faraday cup;
a) single cup and amplifier;
b) double cup and amplier circuit

try of the Faraday cup is such that reflected ions cannot leave it again. This is achieved by the angled impact site and the semiclosed design. A secondary electron aperture, which is maintained at a negative potential relative to the cup, is mounted in front of the cup in order to suppress the secondary electrons (→ secondary electron emission) released by the impact. The ion current, i, incident in the Faraday cup is conducted away via a

high ohmic resistance, R. The voltage drop, $U = i \cdot R$, across this resistance is then amplified and measured. In addition to the single cup, double cups are also used for special methods; Figure 1b. With this arrangement, the ion current of two ion beams of different mass can be measured simultaneously, i.e. a direct measurement of the relationship of two ion currents.
Ref.: J. Hoefs, *Stable Isotope Geochemistry*, Springer Verlag, Berlin, Heidelberg, New York, **1973**.

Faraday effect, magnetooptical rotation, MOR, <*magnetooptische Drehung, MOR, Faraday-Effekt*>, the rotation of the plane of polarization of linearly polarized light induced by a magnetic field when the light traverses a transparent, isotropic body (e.g. glass) parallel to the magnetic lines of force. Faraday discovered magnetooptical rotation in 1846. The angle of rotation, α, is proportional to the magnetic field strength, H, (in Oersted) and the path length, l, (in cm) of the light through the irradiated body: $\alpha = \omega \cdot l \cdot H$.
The constant, ω, which is usually given in minutes of arc/(cm·Oersted), is called the Verdet constant (after the physicist E. Verdet); it depends upon the substance, the temperature, θ, and the wavelength of the radiation. For water, for example, $d\omega/d\theta = -30.5 \cdot 10^{-6}$. The direction of the magnetooptical rotation is the same as the direction in which the current in the coil generating the magnetic field circles around it.
Magnetooptical rotation has effectively no inertia; the rotation follows the magnetic field in less than $5 \cdot 10^{-10}$ s. Therefore, it is used for the construction of inertia-free optical shutters and for the modulation of light beams.
The wavelength dependence of MOR is called magnetooptical rotatory disperson (MORD). The Verdet constants of a few substances are listed in the Table.
Ref.: C.E. Waring, R.L. Custer, in *Techniques of Organic Chemistry*, 3rd edtn., Ed.: A. Weissberger, Interscience Publishers, New York, London, **1960**, vol. I, part III, chapt. XXXVII.

Faraday effect. Table. Verdet constant for the magnetooptical rotation of various substances

	θ [°C]	λ [nm]	Verdet constant
Water	20 °	589	0.0131
		546	0.0155
Benzene	20 °	589	0.0302
		436	0.0589
		405	0.0765
α-Bromonaphthalene	20 °	589	0.0819
		496	0.1315
		453	0.1722
		405	0.2250
Rock salt	20 °	589	0.0328
		404.8	0.0775
TeO_2-glass (20 % TeO_2, 80 % PbO)		1060	0.048
		700	0.127
SF59, Schott optical glass		1060	0.028
		632.8	0.089
		589.3	0.107
		500	0.160
		435.8	0.24

Faraday modulator, <*Faraday-Modulator*>, a practical application of the → Faraday effect. It is used especially in the design of → spectropolarimeters, → Faraday effect, magnetooptical rotation.

Fast atom bombardment, FAB, <*Fast-Atom-Bombardment, Schnellatombeschuß*> a method of obtaining ions for → mass spectroscopy from samples which are difficult to volatilize, e.g. sugars, peptides and nucleotides. A beam of ions, typically $Xe^{+\cdot}$, is obtained by ionizing xenon atoms and accelerating the ions by passage through an electric field. The resulting (fast) ions are directed through a chamber containing further xenon where charge exchange takes place:

$Xe^{+\cdot}$ (fast) + Xe (thermal) → Xe (fast) + $Xe^{+\cdot}$ (thermal)

The fast atoms formed in this way retain most of the original energy of the fast ions and continue in the same direction. Any remaining fast xenon ions are deflected away with an electrode. Other gases, e.g. helium and argon can be used to obtain a beam of fast atoms.

The sample is usually applied to a copper-tipped probe as a layer of solution in an inert, involatile solvent (*matrix material*) such as glycerol. The presence of the solvent aids ionization and it is important that the sample is dissolved in the matrix; a suspension of the sample gives inferior spectra. When the beam of fast atoms impinges on the sample layer the large amount of kinetic energy in the atomic beam is dissipated in various ways, some of which cause volatilization and ionization of the sample. Both positive and negative ions are formed and either species can be directed into the mass spectrometer by means of electrodes having a suitable potential with respect to the probe tip. The spectra produced usually provide relatively abundant molecular or quasi-molecular ions and also show some structurally important fragment ions. An important advantage of the FAB over the → field desorption ion source is that the ion beams from the former persist longer (10–15 minutes). A further advantage of the source is its simplicity and robustness. A disadvantage is the fact that the spectrum of the matrix is always present in the spectrum and must be allowed for.

FD, <*FD*> → field desorption.

Fermi-Dirac statistics, <*Fermi-Dirac-Statistik*> → quantum statistics.

Fermi resonance, <*Fermi-Resonanz*>, a phenomenon observed in IR and Raman spectra which is due to the accidental → degeneracy of two excited vibrational states. This resonance is present if the fundamental of one vibration and the overtone of another, different vibration have the same energy, i.e. if $\tilde{\nu}_1 \approx 2\tilde{\nu}_2$. If combination vibrations, rather than overtones meet this condition, i.e. if $\tilde{\nu}_1 \approx \tilde{\nu}_2 + \tilde{\nu}_3$, Fermi resonance also occurs. In both cases, the overtone/combination obtains (*borrows*) intensity from the fundamental vibration so that the intensity of the overtone/combination increases whilst that of the fundamental decreases. A further consequence of this accidental degeneracy is the mutual repulsion of the energy levels, i.e. a separation of the frequencies of the two vibrations and thus a splitting of the band. The vibration with the higher energy moves to greater wavenumbers and the one with the lower energy to smaller wavenumbers.

The Raman spectrum of CO_2 provides a typical example. Of the four fundamental vibrations, ν_1, ν_2 (doubly degenerate) and ν_3, only the symmetric stretching vibration, ν_1, is Raman active. However, in the → Raman spectrum we observe two strong lines which are separated by ~ 100 cm^{-1}. E. Fermi explained this phenomenon in 1931. The first overtone of the degenerate deformation vibration, (ν_2 at $\tilde{\nu}$ = 667 cm^{-1}) lies at $2\tilde{\nu}$ = 2×667 = 1334 cm^{-1} and is coincident with the fundamental, ν_1. The two vibrational states then split and we observe two bands at 1286 and 1389 cm^{-1}, shifted by ±50 cm^{-1} with respect to the center of gravity of the bands.

CCl_4 provides another example. $\nu_{as}(\nu_3)$ is in Fermi resonance with ν_s and the deformation vibration, $\delta_{as}(\nu_4)$. $\nu_{as} + \delta_{as}$ = 459 cm^{-1} + 315 cm^{-1} = 774 cm^{-1} which leads to a Fermi resonance between a valence and a combination vibration.

The bands at 3061 cm^{-1} and 3098 cm^{-1} in the IR spectrum of benzene are also a Fermi doublet. They are formed by the interaction of the fundamental vibration, $\nu_{12} = \nu_{CH}$ with the combination of $\nu_{16} + \nu_{13}$; ν_{16} at 1585 and 1606 cm^{-1} is itself a resonance doublet.

A frequently observed case in practice are the aldehydes in which the wavenumber of the first overtone vibration, $2\delta_{OCH}$ (O=C-H angle-deformation vibration), is coincident with the CH valence vibration, ν_{CH} of the aldehyde group and gives rise to a Fermi doublet at 2770 and 2830 cm^{-1}. An important restriction applies to the occurrence of Fermi resonance; the overtone or combination vibration must be of the same symmetry species as the fundamental vibration with which it interacts. The wavenumbers of the unperturbed vibrations can be calculated from the wavenumbers of the Fermi doublet with the following approximate formula:

$$\tilde{\nu}_{corr} = \frac{\tilde{\nu}_1 + \tilde{\nu}_2}{2} + \frac{\tilde{\nu}_1 - \tilde{\nu}_2}{2} \cdot \frac{s-1}{s+1}.$$

$\tilde{\nu}_1$ and $\tilde{\nu}_2$ are the wavenumbers of the observed doublet and $s = I_1/I_2$ is their intensity ratio.

Fiber optics, <*Faseroptik*>, the general term for all products which are manufactured using optical fibers. Optical fibers consist of a highly refracting glass core sheathed (clad) in a glass of low refractive index. Light incident at one end of the fiber is conducted through the core by → total reflection at the interface between core and cladding, following the bends of the fiber, to exit at the other end. The most important parameters of optical fibers are the numerical → aperture, the optical transmittance and the fiber core diameter.

The numerical aperture (NA) of optical fibers depends upon the refractive indexes of the glasses forming the core and cladding. For light which passes through the fiber in a plane containing the fiber axis:

$$NA = n_o \cdot \sin\alpha_o = \sqrt{n_1^2 - n_2^2}$$

n_o = refractive index of the surrounding medium, e.g. air, n_1 = refractive index of the fiber core, n_2 = refractive index of the fiber cladding and α_o = critical angle to the optical axis which is shown in Figure 1. All rays incident with an angle, $\alpha \leq \alpha_o$, upon the polished front face, which is perpendicular to the axis, are transmitted. The

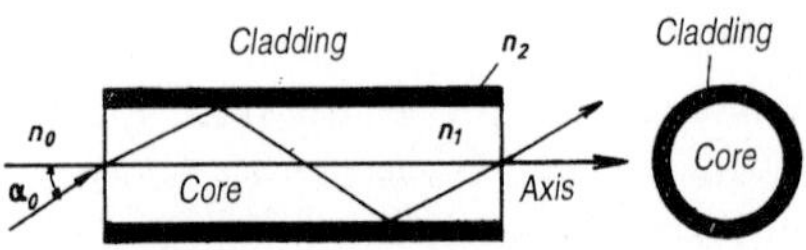

Fiber optics. Fig. 1. Cross section through a light conducting fiber, see text

equation for *NA* also applies approximately to a light conductor, i.e. to optical fibers gathered into a bundle. The optical transmittance is reduced by ca. 34% on account of the fiber-optical construction. The individual cylindrical fibers are jacketed. The jacket, like the outer glass cladding, does not contribute to the conduction of the light. Losses also occur due to

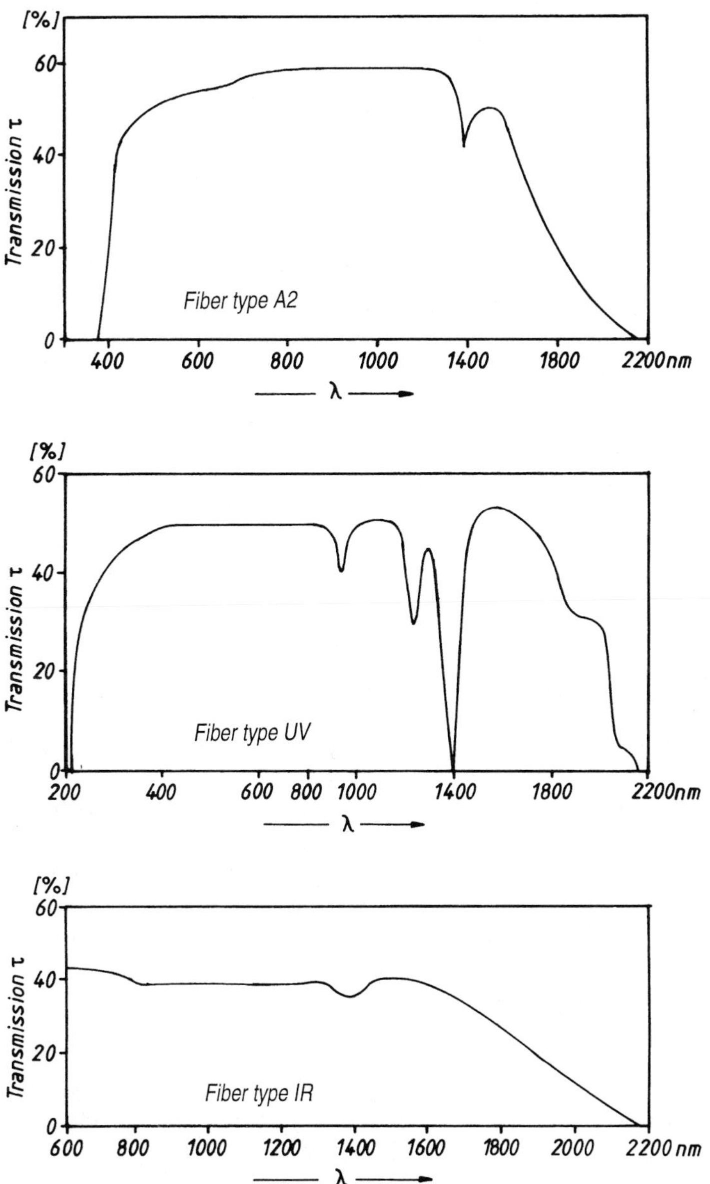

Fiber optics. Fig. 2. Transmittance of different types of fiber in the UV, VIS and NIR

reflection where the coupling and uncoupling of the light beams take place. The spectral transmittance of light conductors is determined primarily by the glass core used and its light-absorbing properties. Figure 2 shows the spectral transmittance for 1 m lengths of the Schott flexible light conductors of fiber types A2, UV and IR. In order to ensure total reflection in the visible region, the fiber cladding must have a thickness of at least 2 μm. For constant cladding thickness, the optically useful cross section of a light conductor becomes more favorable with increasing fiber diameter. However, at the same time the fiber flexibility decreases. For most applications, an optimum relationship between transmittance and flexibility is obtained at a fiber diameter of 70 μm. Flexible light conductors (light pipes) consist of fiber bundles the ends of which are closely packed and glued into a nickel-silver sleeve. The fiber bundles are covered with protective tubes and their front faces are polished. A complete light conductor has a total diameter of between 7 and 16 mm and a maximum fiber bundle diameter between 3 and 10 mm.
Cross section transformers are important for the application of optical fibers in spectroscopy. They facilitate the optimum utilization of a light source since the cross section of the light conductor can be matched to the image of the lamp coil. They distribute the light uniformly over the entrance slit of a spectrometer, → monochromator or → polychromator with a relatively low loss.
Flexible light conductors have found increasing application in experimental spectroscopy in recent years. They make feasible measurements which, formerly, were hardly possible due to the rigid construction of single- and double-beam spectrometers, e.g. remission spectra with the integrating sphere outside the spectrometer, fluorescence and absorption measurements on samples located outside the actual spectrometer.
Ref.: H.-G. Unger, *Planar Optical Waveguides and Fibres*, Oxford University Press, Oxford, **1977**.

Field curvature, *<Bildfeldwölbung>* → astigmatism.

Field desorption, FD, *<Felddesorption, FD>* a variant of → field ionization which has gained great analytical significance as an ion source in mass spectrometry.
A sample can be placed onto the activated emitter by evaporation of a solution (→ field ionization). However, in an electric field, a molecule may be desorbed as an ion without prior evaporation of the sample (field desorption). Field desorption is a method which has the advantage that molecules are accessible to mass-spectroscopic analysis which would decompose on evaporation. This is the reason why the method is important in biochemistry and medicine. Since the samples are always presented as solutions, a combination of mass spectrometry and high-performance liquid chromatography (HPLC) is possible.

Field ionization, FI, *<Feldionisation, FI>*, an effect in which ions are formed without excitation energy. A molecule is placed in a very high electric field ($\sim 10^8\ \mathrm{V \cdot cm^{-1}}$) so that, due to the tunnel effect, an electron can leave the molecule. Field ionization is one method for the construction of ion sources in → mass spectrometers. The field strength required can be generated on spikes, thin wires or sharp edges which are subjected to high voltages (8 to 12 kV). In practice, an arrangement known as an active emitter is used. This is a 20 μm thick tungsten wire with a dense coating of carbon microneedles approximately 30 μm long produced by activation with benzonitrile at 1200°C. The emitter is mounted on a probe which is brought into the ion source through a vacuum port. However, the sensitivity of field ionization is very low in comparison to the usual MS methods and the instrumentation is not easily handled. Thus, its use remains restricted to rather special problems.

Field-sweep method, *<Feld-Sweep-Methode>*. The resonance condition, $\nu_o = (\gamma_I/2\pi)B$ applies in → NMR spectroscopy. It may be achieved either by varying the frequency of the transmitter at constant magnetic flux density, (frequency sweep) or by varying the flux density at constant transmitter frequency (field sweep). At constant magnetic flux density, B, each nucleus has a specific resonance frequency, ν_o. If B = 1.41 T ($1.41 \cdot 10^4$ Gauss) the nuclei ^{1}H, ^{19}F, ^{31}P, ^{13}C and ^{14}N have the following resonance frequencies, $\nu_o(^1H)$ = 60 MHz; $\nu_o(^{19}F)$ = 56.4 MHz; $\nu_o(^{31}P)$ = 24.3 MHz; ν_o (^{13}C) = 15.1 MHz; $\nu_o(^{14}N)$ = 4.3 MHz. However, the resonances do not lie at exactly the same frequency for all nuclei of the same type e.g. for all protons (^{1}H) or for all ^{13}C nuclei. The frequencies differ depending upon the position of the atom concerned in the molecule and the type of bonding (→ chemical shift). For example, not all ^{1}H resonances lie at exactly 60 MHz but within a range of approximately 1000 Hz near that frequency. The frequency of the transmitter at constant magnetic flux density or, alternatively, the magnetic flux density at constant transmitter frequency, must be varied over this narrow range in order to obtain the resonances. The field changes are minimal; only a few (10^{-5} T) which can easily be achieved with sweep coils (→ NMR spectrometer).

Figure axis, *<Figurenachse>*, the unique member of the three, mutually perpendicular, principal inertial axes of a → symmetric top molecule. The moment of inertia with respect to this axis is always designated I_A in the literature so that $I_A \neq I_B = I_C$; or $A \neq B = C$. This particular rotation axis is the figure axis and, at the same time, the symmetry axis of highest order, e.g. C_6 in benzene, C_4 in SF_6 and C_3 in $ClCH_3$, NH_3 and BF_3.

Filter, optical, *<Filter, optisches Filter>*, a filter, which is used in the UV-VIS and NIR spectral regions because of its selective light transmittance, to select out specific spectral regions of a continuous spectrum (→ thermal radiator) or a discontinuous spectrum (→ line source). Optical filters are used instead of → monochromators when high spectral purity of the radiation is not important, or if particularly high intensities are desirable. The selectivity of absorbing optical filters is based on suitably colored substances which are molecularly dispersed or colloidally dissolved in the form of submicroscopic crystals in a matrix stable to external influences. Glasses, solvents, gelatine and other high polymers are used as matrices. For that reason, we distinguish between colored glasses

(→ colored glass filter), liquid filters and plastic filters (→ glass-plastic laminated filter). In the case of line sources, → combined glass filters or → interference filters are used, since it is generally not possible to filter out an individual line with a single colored glass filter. In addition to these filters, polarization and dispersion filters for special applications should be mentioned. → Colored glass and interference filters are preferred in → filter photometers, in fluorescence spectroscopy and in → photokinetics and photochemistry.

Filter photometer, <*Filterphotometer*>, a type of photometer which consists of a continuous light source, a → colored glass or → interference filter, a sample chamber with cuvette changer and a display unit, as shown in Figure 1. This version is called a broad-band photometer. The quality of a filter photometer is determined by the → half-width of the filter. If a → metal-vapor discharge lamp, which emits individual spectral lines is used instead of a continuous light source and the color filter is replaced by a filter combination capable of selecting the individual spectral lines, this special version of a spectral line photometer is called a narrow-band → photometer.

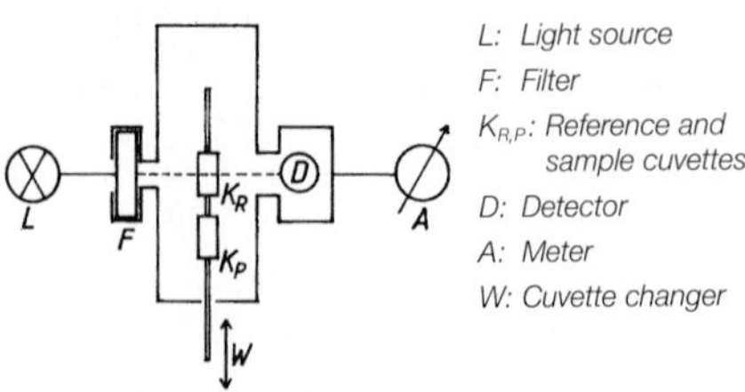

Filter photometer. Fig. 1. Diagram of a filter photometer

Filter photometers are preferred for analytical purposes in the near UV and VIS spectral region. Special versions with suitably selected filters are used in color measurement, lighting technology, photobiology and clinical chemistry.

Fine structure of spectral lines, <*Feinstruktur der Spektrallinien, FS*>, in → atomic spectra the → multiplet structure of the spectral lines; in molecular → electronic spectra the rotational structure of the individual, excited vibrational bands. In general, the fine structure of molecular spectral lines can only be observed in the gas-phase spectra of smaller molecules (→ electronic band spectra). It is often not possible to get larger molecules into the gas phase without decomposition; particularly if they are solids at room temperature. Therefore, they must be measured in solution. Because of the obstruction of molecular rotation by the solvent molecules, solution spectra in the UV-VIS region show a structure which is due to vibrations coexcited with the electronic excitation (→ vibrational structure) but no rotational fine structure.

Flame photometry, <*Flammenphotometrie*>, the old version of modern → atomic emission spectroscopy for which chemical flames such as air and town gas or air and acetylene are used. Thus, the principles of atomic emission spectroscopy (AES) are directly applicable to flame photometry. The disadvantage of the → flame technique lies in the small number of elements which can be excited to emission thermally in flame temperatures between 2200 and 3300 K. The alkali

☒ Determination by absorption											☐ Determination by flame emission							
	H																H	He
He	Li ☐☒	Be ☒											B ☐☒	C	N	O	F	Ne
Ne	Na ☐☒	Mg ☒											Al ☒	Si ☒	P	S	Cl	Ar
Ar	K ☐☒	Ca ☐☒	Sc	Ti	V ☒	Cr ☒	Mn ☒	Fe ☒	Co ☒	Ni ☒	Cu ☒	Zn ☒	Ga ☒	Ge ☒	As ☒	Se ☒	Br	Kr
Kr	Rb ☐	Sr ☐☒	Y	Zr ☒	Nb	Mo ☒	Tc	Ru	Rh ☒	Pd ☒	Ag ☒	Cd ☒	In ☐☒	Sn ☒	Sb ☒	Te ☒	I	Xe
Xe	Cs ☐	Ba ☐☒	La	Hf	Ta	W	Re	Os	Ir	Pt ☒	Au ☒	Hg ☒	Tl ☒	Pb ☒	Bi ☒	Po	At	Rn
Rn	Fr	Ra	Ac															
Lanthanides					Ce	Pr	Nd	Pm	Sm	Eu	Gd	Tb	Dy	Ho ☒	Er	Tm	Yb	Lu

Flame photometry. Fig. 1. Overview of the elements which can be determined using AES and AAS

metals Li, Na, K, Rb and Cs, the alkaline earths Ca, Sr and Ba, and B and In are suitable elements. For these elements, there is also always the possibility, after modification of the equipment, of measuring the absorption spectra quantitatively. Figure 1 gives an overview of the elements which can be measured in emission and/or absorption using flames. The successful development of inductively coupled plasma (ICP) AES in recent years has effectively displaced the flame method from analytical emission spectroscopy.

Flame spectrophotometer, <*Flammenspektralphotometer, Flammenspektrometer*>, an apparatus which corresponds to a single-beam UV-VIS photometer in which the lamp housing has been replaced by a flame attachment. Modular photometers, constructed of components, such as the Zeiss PMQ III are particularly suitable for this purpose. Two flame options are available for this instrument; one for atomic emission and one for atomic absorption. The latter can be switched to emission if desired. The flame attachment for emission measurements contains a direct atomizer burner with a stainless-steel canula through which the sample solution is syphoned into the flame. Figure 1 illustrates the construction schematically. The flame attachment for absorption measurement has a slot burner

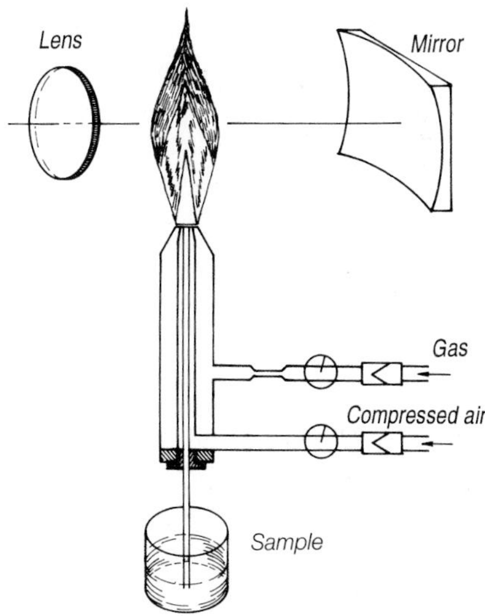

Flame spectrophotometer. Fig. 1. Flame emission option for the PMQ III

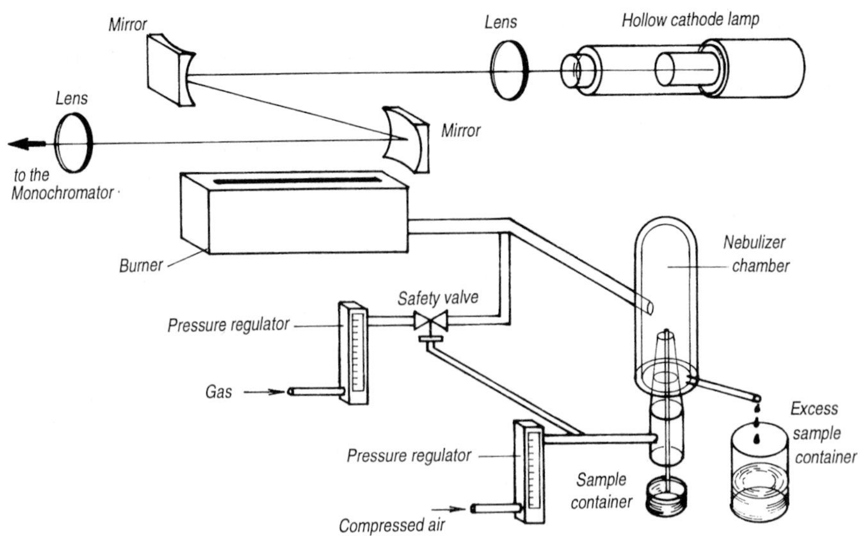

Flame spectrophotometer. Fig. 2. Atomic absorption option for the PMQ III

with a 5 cm long flame which is similar to a → mixing-chamber burner. Mirrors direct the radiation from the → hollow cathode lamp through the flame three times so that its effective length is 15 cm. Figure 2 shows the design of this flame attachment schematically. The flexible component arrangement has the advantage that the → monochromator can be replaced without interfering with the optical layout. For example, a → prism monochromator can be easily replaced by a → grating monochromator and a quartz-prism monochromator by glass-prism monochromator for the visible spectral region.

Flame technique, <*Flammentechnik*>, an older technique in which the sample in the form of a solution is sprayed into the flame, where → atomization takes places, with the help of a pneumatic nebulizer. Today, → mixing-chamber burners are used exclusively in → atomic absorption spectroscopy. Their laminar flow flames are excellently suited to exact analytical measurement. With these burners, the solution of the sample is sprayed, using a pneumatic nebulizer, into a mixing chamber. There, the sample aerosol is thoroughly mixed with the oxidant before it passes through the burner slot above which the flame is burning. The flame, depending on the construction of the burner slot, is from 5 to 10 cm long and only a few millimeters wide; the radiation from the primary light source passes through its complete length. The use of an atomizer to spray the sample into the flame produces a constant, time-independent signal the magnitude of which is proportional to the concentration of the element to be determined. It can be observed as long as the sample solution is drawn in. Thus, the measurement can be repeated several times and a mean value determined.

The most widely used flame in atomic absorption spectroscopy (AAS) is the

air-acetylene flame, the average temperature of which is ca. 2570 K. This flame offers a suitable environment for many elements and a temperature sufficient for → atomization. In only a few cases (alkali metals and occasionally alkaline earths) does ionization occur (→ atomization, step 4a).
Another flame common in AAS is the nitrous oxide-acetylene flame the mean temperature of which is ca. 3000 K. The low combustion speed of this hot flame offers a favorable chemical, thermal and optical environment for all metals which are difficult to analyze in the air-acetylene flame. Two disadvantages are the high ionization rate of many atoms at ~ 3000 K and the relatively strong emission of the flame itself. The interference from ionization can often be eliminated by the addition of an excess of another, readily ionizable, element.
Other flames have been suggested and tested, but they have not become established for routine work in AAS.
Ref.: B. Welz, *Atomic Absorption Spectroscopy*, 2nd edtn., Verlag Chemie, Weinheim, Deerfield Beach, Florida, Basel, **1986**.

Flashlamp, flashtube, <*Blitzlampe*>, a lamp giving a light pulse with a small half-width in the UV-VIS region. It is important for practical applications that the light pulse can be emitted with a constant repetition frequency which, in addition, should be variable over a wide range. The most important and most widely used lamps are → xenon-pulse discharge lamps the pulse half-widths of which lie between 10 and several hundred μs. This type of discharge lamp is used, for example, in → flash photolysis, photochemistry, stroboscopes and in → fluorescence spectrometers for excitation purposes. Shorter pulse half-widths, in the region of nanoseconds, can be obtained with → nitrogen-discharge lamps or pulsed → nitrogen lasers and also with a hydrogen flashlamp. Light pulses with a half-width of picoseconds can be achieved with the help of active mode coupling in lasers. A very intense → spark discharge between magnesium electrodes gives a flash with a duration of 100–200 ms which shows two intense lines at 383 and 280 nm.
Ref.: F.E. Carlson and C.N. Clark in *Applied Optics and Engineering*, (Ed.: R. Kingslake), Academic Press, New York, London, **1965**, vol. I, section 2.F.

Flash photolysis, <*Flash-Photolyse, Blitzlichtphotolyse*>, a technique for the characterization of short-lived intermediates in photochemical processes first developed by G. A. Porter. The principle of this technique is the production of a high concentration of a short-lived intermediate by a short flash of very high intensity light. Immediately after this flash, the system is analyzed with the help, among other methods, of emission and absorption spectroscopy. Depending upon timing requirements, the flash source can be a flashlamp (e.g. a xenon-pulse discharge lamp, duration of flash $\geq$ 1 ns), a spark discharge, duration of flash $\geq 1\ \mu$s or an exploding wire, duration of flash ~ 100 to 200 μs. The detection techniques depend upon the nature of the system to be analyzed and the information required. The emission spectrum of an intermediate can be obtained using a spectrograph and the simultaneous exposure of a photographic plate. The visible absorption spectrum can be taken by means of an analyzing light beam which passes through the reaction cell and is triggered by the primary flash at a specific time after the flash itself. Alternatively, the changes in the reaction cell can be followed kinetically if the emission or absorption signal from a photomultiplier at a specific wavelength is recorded with an oscilloscope. This method gives only limited information but it does allow the lifetime of the intermediate to be determined.

The use of gas-discharge lamps with polychromatic radiation has a disadvantage which can be removed by the use of a pulsed laser.

Ref.: J.A. Barltrop and J.D. Coyle, *Principles of Photochemistry*, John Wiley & Sons, New York, Brisbane, Toronto, **1978**; G.A. Porter and M.A. West in *Investigation of Rates and Mechanism of Reactions*, part II, Ed.: G.G. Hammes, 3rd edtn. Wiley Interscience, New York, **1974**, chapt. 11; R.P. Wayne, *Principles and Applications of Photochemistry*, Oxford University Press, Oxford, New York, Tokyo, **1988**.

Flint glass, <*Flintglas*> → glass, optical.

Flotation photometry, <*Flotationsphotometrie*>, a high-sensitivity photometric analysis technique for elements. The method is based upon the fact that, in → extraction photometry, some low-solubility ion associates of basic dyes cannot be extracted with organic solvents which have low dieletric constants. The associates coagulate into flakes which collect at the phase boundary or adhere to the walls of the containing vessels. The solvent and the aqueous solution are separated by decanting or filtering, after which the precipitate is dissolved, usually in acetone or alcohol. The complex then dissociates. The high extinction coefficients determined with this method depend on the dissociation of the sparingly soluble ion associates upon dissolution. Finally, the absorbance of the dye transferred into solution is measured in that solution. Since several dye molecules can be bound in one complex (up to six) and the measured absorbance is related to the molecular weight of the complex, extremely high extinction coefficients are obtained per gram atom of the element to be determined. Whole-number multiples of the extinction coefficient of the dye in the same solvent are involved. For example, for silicon in the form of molybdatosilicic acid associated with the basic dye rhodamine-B, an extinction coefficient of $\varepsilon = 500 \cdot 10^3\,\mathrm{l \cdot mol^{-1}}$ is found in ethanol.

Ref.: H.-H. Perkampus, *UV-VIS Spectroscopy and Its Applications*, Springer Verlag, Berlin, Heidelberg, New York, **1992**.

Flow method, <*Strömungsmethode*>, in every kinetic experiment there is a dead time between the mixing of the various starting reagents and the first accurately measurable experimental point. In a spectroscopic study of such an experiment, the mixing of the two solutions may take place directly in the cuvette for which the dead time is ca. 5–10 sec. To reduce this time, special mixing procedures based upon flow methods are used. The dead time then lies between 1 and 3 ms. This provides the link to the time scale of slow reactions, since reactions, which lie in the range $10^{-3} < t < 60$ s, can be measured by means of flow methods. Thus, a reaction time range of $10^{-3} - 10^{+5}$ s is accessible. However, the upper value is essentially limited by the long-term stability of the measuring apparatus used. There are basically two different flow techniques. Either the reaction of a mixture is observed at a constant flow rate at a specific position in the flowing stream or the flow is interrupted suddenly. The latter is the → stopped-flow method in which the further progress of the reaction in the volume under observation can be monitored. In both cases, the reactants must be mixed rapidly and effectively in specially designed mixing chambers.

The flow-through cuvette (flow tube) is an open reaction system through which a reaction mixture of known initial composition is passed at constant velocity. The composition of the mixture is determined analytically at a specific position, the reaction or observation zone, in the flowing stream.

For a first order reaction we have the relationship:

$$k_\lambda = \frac{u}{V} \cdot \ln \frac{c_o}{c}$$

where u is the flow rate, V the volume, c_o the initial concentration of the reactant and c its concentration at the observation point.

Since the units of u are $cm^3\ s^{-1}$ and those of V cm^3, u/V has the dimension of reciprocal time and the above equation corresponds to the rate law of a first order reaction:

$$k_\lambda = \frac{1}{t} \cdot \ln \frac{c_o}{c}.$$

In general, it can be shown that for reactions of any order, the integrated rate law for a closed system also applies to reactions in a flow tube, provided that there is no back-mixing. However, instead of the time, t, the quotient, u/V, is the variable. Since $V = Q \cdot t$, where Q is the cylindrical cross section, one can also say that in place of the time, the distance traveled, l, is the variable. This shows clearly that, in order to follow the rate of reaction in a flow tube, the reaction must be measured at several points at specific distances, l, from the mixing chamber (corresponding to the timed interval, t) under conditions of constant flow rate, u. This means that measurements with a single flow tube are very time-consuming. The flow method briefly described here was introduced into spectrophotometry in 1923.

Instead of moving the zone under observation, another version of this method uses a continuous variation of the flow rate in which the course of a reaction can be followed as a concentration-time curve.

Ref.: H.-H. Perkampus, *UV-VIS Spectroscopy and Its Applications*, Springer Verlag, Berlin, Heidelberg, New York, **1992**.

Fluorescence, *<Fluoreszenz>*, an emission phenomenon which was investigated in fluorite by Brewster (1833) and by Herschel (1845). It characterizes the radiative deactivation, in accordance with the → Bohr-Einstein frequency relation, of an excited state of an atom or molecule following primary excitation by the absorption of photons. The phenomenon had already been observed in the 16th and 17th centuries as a result of solar irradiation of solutions of organic substances. However, no one was able to describe it in more detail. Fluorescence, like → phosphorescence, is one of the phenomena known as → photoluminescence. In contrast to phosphorescence, fluorescence involves a spin-allowed transition, i.e. a transition without change of multiplicity. Figure 1 shows a simplified energy-level diagram for a molecule in solution to explain the fluorescence transitions. Here, the → Kasha rule, according to which fluorescence results from state S_1, has been taken into consideration. The stabilization of the S_1 state to S_1', due to the interaction with the solvent, is also indicated. The radiative transitions, $S_1' \rightarrow S_o$ and $T_2 \rightarrow T_1$ are spin-allowed. The radiationless deactivations, $S_1' \rightarrow S_o$ ($T_1 \rightarrow S_o$), must be considered as processes which compete with the fluorescence (phosphorescence). The fluorescence of a molecule or an atom is characterized in more detail by the spectral distribution of the emitted light (→ fluorescence spectrum), by the lifetime of the fluorescent state (→ fluorescence lifetime) and by the relationship of the emitted light quanta to the absorbed quanta (→ fluorescence quantum yield). Fluorescence can be

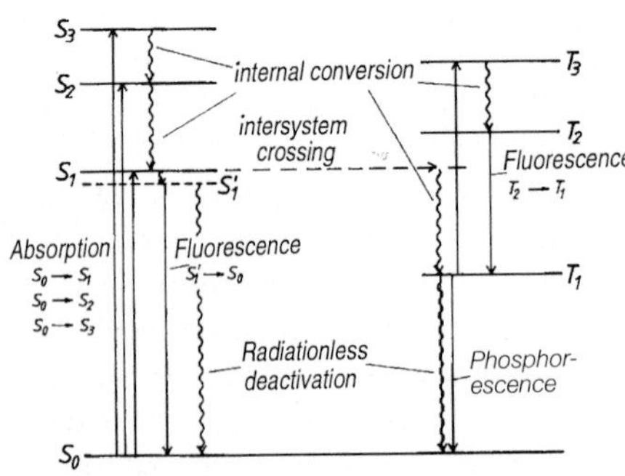

Fluorescence. Fig. 1. An energy-level scheme to illustrate fluorescence

observed and measured in gases, solutions and solids. The observation of a → resonance fluorescence is also possible in gases at reduced pressures. Inorganic and organic compounds, as well as free atoms (→ atomic fluorescence spectroscopy), can be investigated in the same way, although not all molecules which absorb electromagnetic radiation fluoresce. To date, the correlation between fluorescence capability and molecular structure has not been investigated in such detail as that between absorbance and molecular structure. In recent decades, solutions of organic compounds and also inorganic and organic solids have been investigated intensively using fluorescence spectroscopy.

In 1922 G. Cario observed the sensitized fluorescence, predicted by Franck, for the first time when irradiating Hg vapor with the Hg → resonance line, $\lambda = 253.7$ nm, in the presence of thallium metal vapor. Tl vapor is transparent to the Hg line but, in addition to the fluorescence of the Hg resonance line, the green fluorescence of the Tl line at $\lambda = 535$ nm was also observed. The explanation is as follows. The excitation energies of the Tl lines, and the lines of some other metals (e.g. Na, Ag) which are suitable as additives, are smaller than that of the Hg resonance line, i.e. the excited state of the Hg atom lies higher in energy than the excited state corresponding to the Tl line. At a sufficiently high vapor density, excited Hg atoms can collide with Tl atoms and lift them into the excited state from which the green fluorescence of Tl results after a short time. The process can be written in the simplified manner:

$$Hg^* + Tl \rightarrow Hg + Tl^* + E_{kin}.$$

The kinetic energy released in this process is distributed according to the relative masses of the two colliding atoms. Consequently, the green Tl line is broadened due to the → Doppler effect because the Tl atom experiences a considerable increase in its velocity as a result of the energy transfer during the collision.

Ref.: J.R. Lakowicz, *Principles of Fluorescence Spectroscopy*, Plenum, New York, London, **1983**.

Fluorescence indicator, <*Fluoreszenzindikator*> → indicators.

Fluorescence intensity, <*Fluoreszenzintensität*>, due to the competition between radiative and radiationless processes, the fluorescence intensity is not always simply proportional to the light intensity absorbed according to the → Bouguer-Lambert-Beer law. The general expression for the fluorescence intensity of an isotropically distributed quantity of the emitting species, k, in a solution of low viscocity is:

$$F_k(\lambda_i, \lambda_j) = K \cdot q_k(\lambda_i, \lambda_j) \cdot \frac{2.303 \varepsilon_k(\lambda_i) \cdot c_k}{S_k(\lambda_i)} \cdot I_o(\lambda_i) \cdot \{1 - \exp[-S_k(\lambda_i) d]\}$$

with

$F_k(\lambda_i,\lambda_j)$ the fluorescence intensity in Einstein/s after excitation at the wavelength, λ_i, and a band width, $\Delta\lambda_i$;

$q_k(\lambda_i,\lambda_j)$ the spectral quantum yield of the component, k;

$\varepsilon_k(\lambda_i)$ the molar decadic → extinction coefficient of the component, k ($m^2 \cdot mol^{-1}$);

c_k the concentration of the component, k ($mol \cdot m^{-3}$);

$S_k(\lambda_i)$ $= 2.303 \sum_{k=1}^{r} \varepsilon_k(\lambda_i) c_k$ where ε_k is the molar extinction coefficient at wavelength λ_i and c_k is the concentration of component k in $mol \cdot m^{-3}$;

$I_o(\lambda_i)$ the irradiating light intensity in Einstein/s at wavelength, λ_i;

d the path length in cm;

λ_i the wavelength of the exciting radiation;

λ_j the observation wavelength (λ_F).

K is an instrumental constant which, for the above formulation in terms of wavelength, is independent of the wavelength for → grating monochromators and a fixed slit width. When formulated in wavenumbers, K varies as λ^{-2}. If the exponential function is expanded in a series it can be terminated after the second term for dilute solutions since c_k is very small:

$$F_i(\lambda_i, \lambda_j) = I_o(\lambda_i) \cdot 2.303 q_k(\lambda_i, \lambda_j) \cdot \varepsilon_k(\lambda_i) \cdot c_k \cdot d \cdot K.$$

The fluorescence intensity is directly proportional to the extinction coefficient, $\varepsilon_k(\lambda_i)$, in this equation and the correlation with the absorption spectrum can be seen immediately. The fluorescence intensity is also proportional to the concentration, c_k, and this provides the basis for quantitative → fluorimetry. However, the value of the product $2.303 \cdot \varepsilon_k(\lambda_i) \cdot c_k \cdot d$ must be smaller than 0.1. If, in contrast, the exponent is very large (e.g. for large concentrations and large extinction coefficients) the expression, exp $[-S_k(\lambda_i) \cdot d]$ approaches zero and all the light is already completely absorbed at a small penetration depth. Fluorescence then occurs only in a thin layer of the solution. No fluorescence → excitation spectrum can be measured under these conditions and an analysis is not possible. → Self-absorption can be expected in concentrated solutions and is always particularly troublesome if the 0-0 transitions are only slightly separated from each other in absorption and fluorescence. Thus, the above equations apply only to dilute solutions.

Ref.: J.N. Miller, Ed.: *Standards in Fluorescence Spectroscopy*, Chapman and Hall, New York, **1981**; D. Rendell, *Fluorescence and Phosphorescence*, J. Wiley and Sons, Chichester, New York, **1987**.

Fluorescence lifetime, fluorescence decay time, <*Fluoreszenzlebensdauer, Fluoreszenzabklingzeit*>, the time required for the fluorescence intensity to decay to a value of 1/e times the initial intensity. If the fluorescence of a molecule is excited by a light flash of negligible length (a δ pulse), a finite concentration $[^1M^*]_o$ of excited molecules is produced. At time, $t > 0$, the molecules are deactivated by fluorescence with the rate constant, k_{FM} in s^{-1} and nonradiatively with the rate constant, k_{IM} in s^{-1}. The rate law for the decrease of $[^1M^*]$ in this case is:

$$\frac{d[^1M^*]}{dt} = -(k_{FM} + k_{IM})[^1M^*] = -k_M[^1M^*].$$

Integration from $t = 0$ to t gives:

$$\frac{[^1M^*]}{[^1M^*]_o} = e^{-k_M \cdot t}.$$

The equation for the measurable fluorescence intensity, i_M, as a function of time is:

$$i_M(t) = k_{FM}\frac{[^1M^*]}{[^1M^*]_o}.$$

whence it follows that:

$$i_M(t) = k_{FM}e^{-k_M \cdot t} = k_{FM}e^{-t/\tau_M}.$$

The molecular fluorescence lifetime is defined by:

$$\tau_M = \frac{1}{k_M} = \frac{1}{k_{FM} + k_{IM}}$$

which gives the time required for the fluorescence intensity to decay to 1/e times its initial value, $i_M(t = 0)$, hence the name *fluorescence decay time*. The reciprocal value of k_{FM}, $\tau_{FM} = (1/k_{FM})$, is the radiative lifetime which is also called the natural fluorescence lifetime. The quantity k_{FM} corresponds to the transition probability for the radiating transition and is equal to the Einstein A coefficient (→ Einstein coefficient):

$$\frac{1}{\tau_{FM}} = k_{FM} = A_{ko \to 1} = \sum A_{ko \to lm}.$$

Since it is generally not possible to measure the natural lifetime directly, relationships which offer the possibility of an indirect determination of τ_{FM} or k_{FM} are of particular importance. k_{FM} can be calculated from the absorption spectrum, using the mirror-image symmetry of fluorescence and absorption, by means of the → Einstein coefficient (→ fluorescence spectrum):

$$\frac{1}{\tau_{FM}} = k_{FM} = 2.88 \cdot 10^{-9} n^2 \int \frac{(2\tilde{\nu}_o - \tilde{\nu})^3}{\tilde{\nu}} \varepsilon(\tilde{\nu}) d\tilde{\nu}.$$

Here, $\tilde{\nu}_o$ is the wavenumber of the mirror-image point and n the refractive index of the solvent. For → resonance fluorescence this equation can be simplified:

$$\frac{1}{\tau_{FM}} = k_{FM} = 2.88 \cdot 10^{-9} \cdot \tilde{\nu}_{kl}^2 \cdot n^2 \int \varepsilon(\tilde{\nu}) d\tilde{\nu}.$$

This approximate expression is frequently used to estimate τ_{FM} from the → integral absorption of long-wavelength absorption bands. $\tilde{\nu}_{k \to l}$ then corresponds to the 0-0 transition. According to the definition of → fluorescence quantum yield, q_{FM}, the fraction τ_M/τ_{FM} is given by:

$$q_{FM} = \frac{k_{FM}}{k_{FM} + k_{IM}} = \frac{k_{FM}}{k_M} = \frac{\tau_M}{\tau_{FM}} = \int_o^{\infty} F(\tilde{\nu}) d\tilde{\nu}.$$

Since τ_M can be measured and τ_{FM} can be calculated estimation of the molecular quantum yield is possible or, conversely, τ_{FM} can be calculated from q_{FM} and τ_M.
Ref.: J.B. Birks, *Photophysics of Aromatic Molecules*, Wiley Interscience, John Wiley and Sons, London, New York, Sydney, Toronto, **1970**.

Fluorescence measurement, *<Fluoreszenzmessung>*. Since no absolute intensity measure, comparable with → absorbance in absorption spectroscopy, exists for fluorescence, the results of different researchers are only comparable when generally accepted calibration methods are used. A method of comparison must therefore be created. For absorption measurements in the UV-VIS region, a linear correlation between the → absorbance, A, and concentration, c, is generally found over a large concentration range $10^{-2} > c > 10^{-6}$ mol · l^{-1} of the dissolved substance according to the → Bouguer-Lambert-Beer law. In contrast, there is only an approximate relationship between the → fluorescence intensity, $I_f(\lambda)$, and the concentration c: $I_f = I_o \cdot \phi_f \cdot 2.303 \cdot \varepsilon \cdot c \cdot d$. I_o is the intensity of the excitation source at wavelength, λ_e, ϕ_f the → fluorescence quantum yield, ε the decadic molar → extinction coefficient in 1 · mol^{-1} · cm^{-1}, c the concentration in mol · l^{-1} and d the path length through the solution in cm. This equation applies only if the product, $\varepsilon \cdot c \cdot d \ll 0.05$ or generally $\leq$ 0.01. Under these conditions, the fluorescence intensity is proportional to the concentration of the dissolved fluorescent substance for constant $I_o(\lambda_e)$, so that a calibration curve for the determination of concentration can be drawn up. In the case of a solution with constant concentration, $I_f(\lambda)$ is proportional to the excitation intensity, $I_o(\lambda_e)$. $I_f(\lambda)$ plotted against wavelength, λ, at constant I_o, provides an uncorrected fluorescence spectrum. The true → fluorescence spectrum is determined by taking account of the → spectral transmittance of the monochromator and the → spectral sensitivity of the → photomultiplier. In the case of → fluorescence spectrometers, with which the fluorescence → excitation spectrum (FE spectrum) and also the fluorescence spectrum may be measured, the spectral energy distribution of the excitation source, which is usually a → xenon lamp, must also be known. For this a → quantum counter such as a solution of rhodamine-B or -101 in ethyl alcohol, ethylene glycol or trifluoroethanol is used. Another possibility lies in the fact that the FE spectrum should correspond completely to the absorption spectrum in the same solvent. By comparing the two spectra, the wavelength-independent correction factors for constructing the excitation spectrum can be obtained. However, this method applies only to dilute solutions in which re-absorption is negligibly small. Some fluorescence spectrometers are fitted with a rhodamine-B or -101 quantum counter so that the corrected FE spectrum can be recorded directly. The signal-to-noise ratio, *S/N*, is frequently given as an indication of the sensitivity of a fluorescence spectrometer. Following Parker's proposal, a Raman band of the pure solvent, preferably bidistilled water, should be measured. For water, there is a suitable band at about 3380 cm^{-1}. For excitation with $\tilde{\nu}_{ex}$ = 28,570 cm^{-1} (λ_{ex} = 350 nm), the Raman band lies at $\tilde{\nu}_R$ = 25,190 cm^{-1} (λ_R = 397 nm). Taking the experimental conditions into consideration, the different types of spectrometer give an *S/N* value of ≥ 35:1 for the Raman signal. This value should be checked at regular intervals. Three sources of sharp spectral features are generally used for calibrating the wavelengths of the monochromator:

a) Lines of the excitation source, with use of a suitable filter, if required;
b) lines of an additional source (→ UV-VIS standards);
c) fluorescence peaks of organic and inorganic compounds in solution or in transparent polymer matrices or, in the case of inorganic compounds, in genuine glasses (→ fluorescence standards).

The → self-absorption of fluorescent light (→ internal filter effect) and the influence of the solvent must also be considered when measuring fluorescence. This influence is more pronounced in fluorescence spectra than in UV-VIS spectra (→ solvent). The → fluorescence quantum yield is an important quantity. Various methods for its determination have been described in the literature.

Ref.: J.N. Miller, Ed.:, *Standards in Fluorescence Spectrometry*, Chapman and Hall, London, New York, **1981**.

Fluorescence polarization spectrum, *<Fluoreszenzpolarisationsspektrum, FP-Spektrum>* → photoselection.

Fluorescence quantum yield, *<Fluoreszenzquantenausbeute>*, in accordance with the definition of the → quantum yield, the ratio of the number of photons emitted per unit time to the number of the photons absorbed per unit time. The number of photons emitted per unit time is $k_{FM}[^1M^*]$; k_{FM}, in s^{-1}, is the rate constant of the radiative deactivation (fluorescence) and $[^1M^*]$ is the concentration of the excited molecules. The number of the photons absorbed per unit time is given by the intensity of the irradiating light, I_o, in Einstein $\cdot l^{-1} \cdot s^{-1}$. We obtain the quantum yield as:

$$q_{FM} = \frac{k_{FM}[^1M^*]}{I_o}.$$

Under continuous irradiation the rate of formation of $^1M^*$ is:

$$\frac{d|^1M^*|}{dt} = I_o - (k_{FM} + k_{IM})[^1M^*],$$

where k_{IM} is the rate constant for the radiationless deactivation. Under photostationary conditions (→ photostationary state), $d[^1M^*]/dt$ = 0 and therefore $I_o = (k_{FM} + k_{IM})[^1M^*]$, i.e. the absorbed intensity, I_o, is exactly equivalent to the sum of the deactivation processes so that:

$$q_{FM} = \frac{k_{FM}}{k_{FM} + k_{IM}} = \frac{k_{FM}}{k_M};$$
$$k_M = k_{FM} + k_{IM}.$$

The quantum yield defined in this way is called the molecular fluorescence quantum yield. The molecular fluorescence spectrum, $F(\tilde{\nu})$, is defined by the relative fluorescence quantum intensity at wavenumber, $\tilde{\nu}$, normalized by the integral:

$$q_{FM} = \int_o^\infty F(\tilde{\nu})\mathrm{d}\tilde{\nu}$$

It corresponds to the molecular fluorescence quantum yield. The reciprocal of k_M is defined as the molecular → fluorescence lifetime, $\tau_M = k_M^{-1}$ and as radiative lifetime in accordance with $\tau_{FM} = k_{FM}^{-1}$. Thus, we obtain the following relationship:

$$q_{FM} = \frac{k_{FM}}{k_{FM} + k_{IM}} = \frac{\tau_M}{\tau_{FM}} = \int_o^\infty f(\tilde{\nu})\mathrm{d}\tilde{\nu}.$$

If other processes influence the fluorescence intensity, in addition to the two competing processes of radiative and radiationless deactiviation, the fluorescence quantum yield will be reduced. It is then useful to denote the quantum yield with Φ_{FM} (→ fluorescence quenching). Absolute determinations of the fluorescence quantum yield, Φ_{FM}, can be carried out using various methods, e.g. photocalorimetry, → photoacoustic spectroscopy, → thermal blooming etc. Since these methods are in some cases difficult and cumbersome, we frequently use a relative method and compare the fluorescence intensity of the sample solution with a reference solution for which the fluorescence quantum yield, Φ_{FR}, is known. A solution of quinine sulfate in 0.5 M H_2SO_4 has proved to be a suitable reference solution. For a concentration of $5 \cdot 10^{-3}$ mol·l^{-1}, $\Phi_{FR} = 0.51$ and for an infinitely dilute solution, $\Phi_{FR} = 0.55$ at 25 °C. If $F(\tilde{\nu})$ and $F_R(\tilde{\nu})$ are respectively the corrected → fluorescence spectra of the solution and the reference, the ratio of the fluorescence quantum yields is given by:

$$\frac{\Phi_{FM}}{\Phi_{FR}} = \frac{n^2 \int_o^\infty F(\tilde{\nu})\mathrm{d}\tilde{\nu}}{n_R^2 \int_o^\infty F_R(\tilde{\nu})\mathrm{d}\tilde{\nu}}.$$

This method of comparison assumes that the excitation of the fluorescence of both solutions was made under identical conditions and that the fluorescence was observed at right angles to the cuvette entrance window. The integrals represent the total fluorescence intensities of both solutions and are easily determined. n and n_R are the refractive indexes of the sample and reference solution. If the fluorescence quantum fluxes are measured directly, i.e. by interposing a → quantum counter, the relative method can be considerably simplified. The fluorescence intensity in dilute solutions is:

$$F(\tilde{\nu}) = K \cdot I_o \Phi_{FM} \cdot \varepsilon_{\tilde{\nu}} \cdot c \cdot d.$$

For the sample ($F_P(\tilde{\nu})$) and the reference solution ($F_R(\tilde{\nu})$) under identical experimental conditions ($K \cdot I_o$ = constant: the same wavenumber for the excitation):

$$\frac{F_P(\tilde{\nu})}{F_R(\tilde{\nu})} = \frac{\Phi_{FM}\varepsilon_P c_P}{\Phi_{FR}\varepsilon_R c_R}$$

or

$$\Phi_{FM} = \frac{F_P(\tilde{\nu})}{F_R(\tilde{\nu})} \cdot \frac{\varepsilon_R c_R}{\varepsilon_P c_P} \cdot \Phi_{FR}.$$

If both solutions have the same → optical density then $\varepsilon_R c_R = \varepsilon_P c_P$ and the above equation simplifies to:

$$\Phi_{FM} = \frac{F_P(\tilde{\nu})}{F_R(\tilde{\nu})} \cdot \Phi_{FR}.$$

Ref.: J.B. Birks, *Photophysics of Aromatic Molecules*, Wiley Interscience, John Wiley and Sons Ltd., London, New York, Sydney, Toronto, **1970**.

Fluorescence quenching, <*Fluoreszenzlöschung*>, the decrease of the molecular → fluorescence quantum yield, $q_{FM} = k_{FM}/(k_{FM} + k_{IM})$ in the excited state, which occurs as soon as other competing processes are possible, in addition to the radiative and radiationless processes (k_{FM} = rate constant for radiative decay; k_{IM} = rate constant for radiationless decay). The fluorescence decay can be caused by:

a) Foreign molecules (→ fluorescence quenching by impurities);
b) the fluorescent molecule itself in solutions of higher concentration (→ self-quenching of fluorescence);
c) the transfer of excitation energy to another molecule contained in the solution (energy-transfer quenching);
d) → self-absorption or reabsorption.

In fluorescence studies account must be taken of all these phenomena.

Fluorescence quenching by impurities <*Fremdlöschung der Fluoreszenz*>, a type of → fluorescence quenching which is frequently caused by contamination of the solvent. Dissolved molecular oxygen is one of the commonest contaminants. It has a concentration of $\sim 2 \cdot 10^{-3}$ mol·l^{-1} at room temperature in aliphatic solvents and exceeds the concentration of the dissolved, potentially fluorescent molecules by 1 to 2 powers of ten. If a solution contains a molar concentration, $[Q]$, of a foreign molecule Q, (a contaminating → quencher) a diffusion-controlled collision can occur between $^1M^*$ and Q which leads to a radiationless deactivation of the $^1M^*$ state. The rate constant of the process can be denoted with k_{QM}(mol^{-1}·s^{-1}). Then, with k_{FM} and k_{IM} as the rate constants for the radiative and radiationless deactivation, from the rate law in the → photostationary state we obtain:

$$\frac{d[^1M^*]}{dt} = I_o - (k_{FM} + k_{IM} + k_{QM}[Q])[^1M^*] = 0$$

and for the → fluorescence quantum yield, the expression:

$$\Phi_{FM} = \frac{k_{FM}}{k_{FM} + k_{IM} + k_{QM}[Q]} = \frac{q_{FM}}{1 + \tau_M k_{QM}[Q]};$$

where $k_{FM}/(k_{FM}/k_{IM}) = q_{FM}$ (molecular → fluorescence quantum yield) and $(k_{FM} + k_{IM})^{-1} = k_M^{-1} = \tau_M$ (molecular → fluorescence lifetime). In this case, the fluorescence lifetime is defined as:

$$\tau_Q = \frac{1}{k_{FM} + k_{IM} + k_{QM}[Q]} = \frac{\tau_M}{1 + \tau_M k_{QM}[Q]}$$

and depends, like Φ_{FM}, upon the concentration, $[Q]$, of the foreign molecule.

Furthermore, $\Phi_{FM} < q_{FM}$ and $\tau_Q < \tau_M$. In practical applications, we rearrange the equations to:

$$\frac{q_{FM}}{\Phi_{FM}} = 1 + \tau_M k_{QM}[Q];$$

$$\frac{\tau_M}{\tau_Q} = 1 + \tau_M k_{QM}[Q].$$

Under constant excitation conditions, q_{FM} can be replaced by the intensity, I_o, of the fluorescence of a solution without foreign molecules or → quencher and Φ_{FM} by the intensity, I_Q, with the quencher:

$$\frac{I_o}{I_Q} = 1 + K[Q]; \; K = \tau_M \cdot k_{QM}.$$

If this quenching mechanism is present we obtain a straight line the slope of which provides $\tau_M \cdot k_{QM}$. If τ_M is known, e.g. by determining the → fluorescence lifetime in the solution without foreign molecules, k_{QM} can also be obtained. Stern and Volmer (1919) carried out this type of study for the first time in fluorescence measurements on gases and Vavilov (1929) applied it to solutions. In addition to this simple, dynamic quenching mechanism, other mechanisms, including static ones, have been proposed for quenching by impurities.

Ref.: J.B. Birks, *Photophysics of Aromatic Molecules*, Wiley Interscience, John Wiley and Sons, London, New York, Sydney, Toronto, **1970**.

Fluorescence spectrometer, luminescence spectrometer, <*Fluoreszenzspektrometer, Fluoreszenzspektralphotometer, Lumineszenzspektrometer, Lumineszenzspektralphotometer*>, a spectrometer which, as a minimum, consists of the following components; an excitation source, a fluorescence cuvette, a → monochromator to disperse the fluorescence, a detector plus amplifier and a display unit. If the monochromator is equipped with a continuous drive which simultaneously drives a recorder, then an uncorrected → fluorescence spectrum can be recorded directly. (Regarding the correction of the → fluorescence spectrum, see calibration of → fluorescence measurements and → fluorescence standards).

As excitation source mercury-vapor discharge lamps are most often used. Suitable excitation wavelengths can be selected from their line spectra by means of → interference filters or filter combinations.

In some cases, it is useful to work on the excitation side with the dispersed spectrum of a continuous source; for example, when the fluorescence → excitation spectrum is to be measured in addition to the fluorescence spectrum. In this case, it is useful to have a monochromator on the excitation side and also a light source of high intensity, e.g. a → xenon lamp. Figure 1 shows schematically the optics of the Perkin-Elmer LS5 fluorescence spectrometer. The light source is a pulsed xenon flashlamp (8 W, 50 Hz, light pulse maximum ca. 1.2 kW) focused by means of an ellipsoidal mirror, via a toroidal mirror onto slit 2 of the excitation monochromator. Within the monochromator, the light beam is dispersed by a grating having 1200 lines/mm. A spherical mirror forms an image with the dispersed light in the plane of slit 1 so that, depending upon the position of the grating, a narrow band of wavelengths passes through slit 1, and via a plane mirror and a following toroidal mirror, is focused on the sample cuvette. The fluorescence light produced in the cuvette follows a mirror-image path via toroidal mirror, plane mirror and slit 1 into the emis-

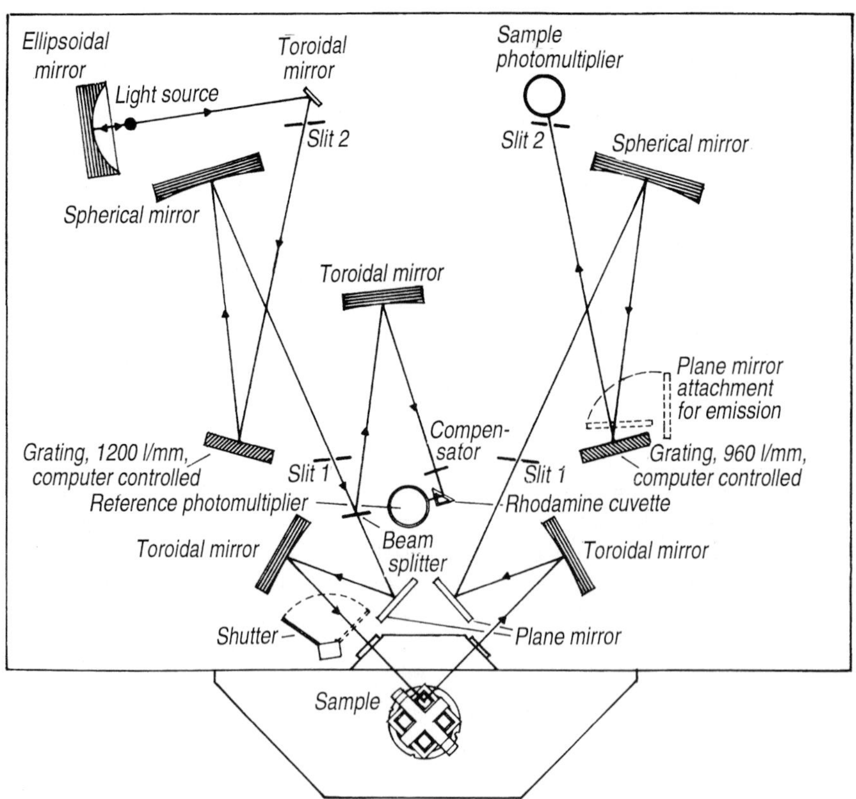

Fluorescence spectrometer. Fig. 1. Optical diagram of the Perkin-Elmer spectrometer LS5

sion monochromator. A → photomultiplier for the detection of the dispersed fluorescence is located behind the exit slit 2. When recording a fluorescence excitation spectrum, a portion of the exciting light from slit 1 is taken with a → beam splitter and focused onto a cuvette containing a rhodamine-101 solution. The fluorescence intensity of this solution is detected by a reference photomultiplier and used to correct the excitation spectrum (→ quantum counter). The correctable spectral region ranges from 230 to 630 nm, or with another dye to 700 nm. The excitation and emission monochromators are driven with microcomputer-controlled stepper motors. The possible combinations of the instrument permit the measurement of → excitation spectra, → fluorescence spectra and → synchronous fluorescence excitation spectra. Since the instrument uses a xenon-pulse lamp as excitation source, the signal of the photomultiplier is gated to give a limited measurement window during the duration of the flash (ca. $< 10\ \mu$s). If this window is moved in time, away from the exciting flash of the xenon lamp, then the → phosphorescence can also be measured. For → bioluminescence and → chemiluminescence measurements,

the excitation monochromator is closed with a shutter. In order to gather the total fluorescence, the grating in the emission monochromator can be covered with a mirror so that the fluorescence passes through the exit slit 2 and onto the detector without spectral dispersion. A similar but more advanced fluorescence spectrometer, the LS-50, is also produced by Perkin-Elmer.
Ref.: Perkin-Elmer Corp., Norwalk, CT, USA.

Fluorescence spectrum, <*Fluoreszenzspektrum*>, the spectrum which one obtains when the fluorescence of a sample, k, is dispersed and the fluorescence intensity, $F_k(\lambda)$, is plotted as a function of wavelength, λ, or wavenumber, $\tilde{\nu}$. This is not the true relative energy spectrum of the fluorescence because it is distorted by the → spectral sensitivity of the detector, the variable → dispersion of the monochromator and, in some cases, by reflection losses in the optical system. For the necessary correction, the fluorescing sample is replaced by a tungsten-filament lamp of known color temperature (→ thermal radiator), the spectral energy distribution of which is therefore known. The signal $F_s(\lambda)$ is recorded under the same experimental conditions (slit width, focusing, detector, etc.) and related to the function, $R(\lambda)\mathrm{d}\lambda$ to which the spectral energy distribution curve of the lamp must correspond. This then gives the relative intensity of the fluorescence through the expression:

$$B(\lambda)\mathrm{d}\lambda = F_k(\lambda)\frac{R(\lambda)\mathrm{d}\lambda}{F_s(\lambda)}.$$

In terms of wavenumber, $\tilde{\nu}$, the relationship is:

$$B(\tilde{\nu})\mathrm{d}\tilde{\nu} = F_k(\tilde{\nu})\frac{R(\lambda)\lambda^2\mathrm{d}\tilde{\nu}}{F_s(\tilde{\nu})}.$$

If the factor, $R(\lambda)\mathrm{d}\lambda/F_s(\lambda)$, is plotted as a function of λ a curve is obtained from which the correction factors, with which the signal, $F_k(\lambda)$, must be multiplied in order to obtain the true relative energy spectrum of the fluorescence, can be determined. This formerly very time-consuming process presents no problems today, thanks to computer-controlled → fluorescence spectrometers. Use of the standard lamps (e.g. Osram Wi15) presupposes adherence to the conditions (voltage, current) specified by the manufacturer since only then is the given color temperature guaranteed. Because the lamps, being thermal radiators, have only a small intensity below 400 nm, the calibration is only suitable for the visible region, $\lambda \geq 360$ nm. Continuation into the UV is possible using → UV standards. Apart from this widely used procedure for the calibration of → fluorescence spectrometers, other methods have been proposed:

a) Determination of the factor by means of a reference solution the fluorescence spectrum of which is known exactly; e.g. a $2 \cdot 10^{-6}$ M solution of quinine sulfate in 0.5 M H_2SO_4 or a $5 \cdot 10^{-6}$ M solution of anthracene in ethanol;
b) measurement with a thermopile which has a constant spectral sensitivity;
c) measurement with a fluorescing solution which behaves as a → quantum counter. In this case, the relative quantum spectrum of the fluorescence is obtained directly.

Ref.: E. Lippert, W. Nägele, I. Seibold-Blankenstein, U. Staiger and W. Voss, *Z. anal. Chem.*, **1959**, *170*, 1.

If the fluorescence spectrum of a molecule is plotted against the wavenumber, $\tilde{\nu}$, a structure is frequently seen which, in general, is the result of the participation of vibrational transitions. Similarly, the absorption spectrum plotted as $\varepsilon_{\tilde{\nu}} = f(\tilde{\nu})$ shows vibrational structure. Figure 1 shows the origin of this structure in a simplified energy-level diagram for the $S_o \rightarrow S_1$ absorption and the $S_1 \rightarrow S_o$ fluorescence transition. It is clear to see that the structure in the absorption spectrum is determined by the vibrational levels of the excited electronic state, S_1, while for the fluorescence spectrum it is the vibrational levels of the ground state, S_o. This assumes that, at room temperature, the great majority of the molecules in the ground electronic state, S_o, are also in the vibrational ground state, $v = 0$, so that transitions from the excited vibrational states, $v \geq 1$, are very improbable (→ Boltzmann statistics). At low temperature, the vibrational structure is generally clearer. Also, it is frequently observed that the fluorescence spectrum is a mirror image of the absorption spectrum. This is easy to understand in the light of the energy-level scheme (Figure 1). This mirror symmetry can be interpreted theoretically in terms of the relationship between the → Einstein coefficients $B_{lm \rightarrow kn}$ and $A_{kn \rightarrow lm}$.

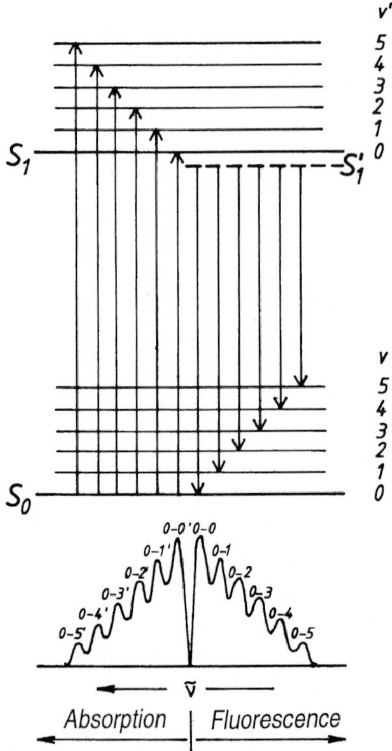

Fluorescence spectrum. Fig. 1. An energy-level scheme illustrating the vibrational structure of the absorption and fluorescence spectra of a molecule

Ref.: J.B. Birks, *Photophysics of Aromatic Molecules*, Wiley Interscience, John Wiley and Sons Ltd., London, New York, Sydney, Toronto, **1970**, p 84 ff.

Fluorescence standard, *<Fluoreszenzstandard>*, a solution of a strongly fluorescing compound in a solvent, polymer matrix or glass.

Depending upon the purpose of the fluorescence measurement, a fluorescence standard must fulfill a variety of conditions. For wavelength calibration, for example, the standard must have a number of relatively sharp fluorescence bands. For this purpose, glasses containing rare-earth oxides are particularly suitable. The color and fluorescence are determined by the rare-earth ions in the glass. Glasses containing thulium and terbium are particularly suitable; the former shows maxima at 355, 456 and 705 nm and the latter at 486 and 541 nm. However, the exact position depends on the composition of the glass (→ colored glass filter). These glasses have the advantage that they can be easily manufactured in the form of a cuvette. Aromatic hydrocarbons are also suitable wavelength standards. They usually show an intense fluorescence with a clearly defined vibrational structure. In addition to their use in solution they can also be dissolved in acrylic polymers from which, again, small blocks in the form of a cuvette can be cut. Several aromatic hydrocarbons and other molecules exhibit phosphorescence as well as fluorescence in such solid matrices and can, therefore, be used over a wider spectral range. A typical example is coronene which shows fluorescence maxima at 426, 434, 446, 453 and 474 nm and phosphorescence maxima at 499, 511, 529, 522 and 566 nm.

In addition to wavelength calibration, fluorescence standards are also used in the correction of fluorescence spectra for the determination of → fluorescence quantum yields. They are called secondary standards because their fluorescence properties can only be determined using a calibrated detector. Such standards must fulfill the following criteria:

1. A broad fluorescence spectrum without fine structure;
2. a large Stokes shift (→ Stokes rule) so that the overlap of the absorption and fluorescence spectra, and hence → self-absorption, is minimal;
3. a relatively high quantum yield;
4. a fluorescence spectrum which is independent of the excitation wavelength in both the form of the band and the quantum yield;
5. a minimal sensitivity to oxygen quenching;
6. minimal concentration quenching (and excimer formation);
7. a completely isotropic fluorescence (depolarized);
8. ready availability in a very pure state and also chemically and photochemically stable.

In addition to quinine sulfate in sulfuric acid solution which is the most widely studied standard in solution, many other organic compounds have been suggested. However, many of these have been found to be unsuitable in that they do not fulfill some of the above criteria. Those with limited application include: 3-aminophthalimide in sulfuric acid, 2-aminopyridine in sulfuric acid, tryptophan in aqueous solution (especially for biochemical applications), 9,10-diphenylanthracene (quantum yield ~ 1.0), p-terphenyl, 2-(4-biphenylyl)-5-phenyl-1,3,4-oxadiazole and 1,1,4,4-tetraphenylbutadiene. Standards in the form of solid solutions in acrylic copolymers have also been described. Their good photochemical stability makes cuvette-form polymer blocks containing p-terphenyl, tetraphenylbutadiene, perylene and rhodamine-B particularly useful. Glasses containing heavy metal ions such as Tl^{+}, Ce^{3+}, Pb^{2+} and Cu^{+} are also suitable fluorescence standards. However, the position of the fluorescence maximum (λ_{max}) depends upon the nature of the glass, as the table shows. The fluorescence intensity of glasses of this type is linear in the concentration of the ions over a range of three powers of ten.

Ref.: J.N. Miller, Ed.:, *Standards in Fluorescence Spectrometry*, Chapman and Hall, London, New York, **1981**.

Fluorescence standards. Table.

Ion	Base glass	λ_{max}[nm]	HW [nm]
Tl^+	Phosphate	302	90
Ce^{3+}	Phosphate	334	55
Ce^{3+}	Borate	365	65
Ce^{3+}	Silicate	380	85
Pb^{2+}	Phosphate	390	70
Pb^{2+}	Borate	425	165
Cu^+	Phosphate	445	120

Fluorimetry, <*Fluorimetrie*>, the quantitative, analytical application of fluorescence spectroscopy which is based on the fact that the → fluorescence intensity of a dilute solution, F_k, is directly proportional to the concentration of dissolved species, k:

$$F_k(\lambda_a, \lambda_s) = K \cdot q_k \cdot I_o(\lambda_a) \cdot 2.303 \cdot \varepsilon_k(\lambda_a) \cdot c_k \cdot d.$$

This relationship shows that the fluorescence intensity is also directly proportional to the intensity, I_o, of the irradiating light at the absorption wavelength, λ_a and the associated molar decadic extinction coefficient, $\varepsilon_k(\lambda_a)$. Furthermore, the → fluorescence quantum yield, q_k, is also involved. λ_F, is the wavelength at which the fluorescence is observed. In contrast to → photometry where the quantity measured is the → transmittance, $T(\tau) = I/I_o$, which does not depend upon the absolute intensity, I_o, the fluorescence intensity increases with the increasing excitation intensity, I_o. High → extinction coefficients, ε_k, and fluorescence quantum yields close to $q_k = 1$ are important for the sensitivity of fluorimetry and the detection limit of the method. Although nearly all inorganic and organic compounds absorbing in the UV-VIS region can be analytically determined using photometry, the number of substances capable of fluorescing is considerably smaller. This means that the selectivity of fluorimetry is much larger than that of photometry. In practice, the simple equation $\varepsilon \cdot c_{\max} \cdot d \leq 0.01$ has proved useful for the estimation of the maximum concentration, $c_{\max}$, up to which a linear correlation between fluorescence intensity and concentration can be expected. It states that the absorbance of the solution to be analyzed should not exceed a value of $A = 0.01$ at the excitation wavelength, λ_a. $c_{\max} \leq 10^{-6}$ $mol \cdot l^{-1}$ is found for an extinction coefficient, $\varepsilon = 10^4 \, l \cdot mol^{-1} \cdot cm^{-1}$ and a path length of $d = 1$. Thus, we can expect a linear dependence of the fluorescence intensity upon the concentration for all concentrations below 10^{-6} $mol \cdot l^{-1}$.

Fluorimetry is used as an analytical method in inorganic and organic chemistry for the determination of both single and multiple components. It is also used in biochemistry, clinical chemistry and foodstuff analysis and as a detector in HPLC.

Ref.: D. Rendell, *Fluorescence and Phosphorescence*, J. Wiley and Sons, Chichester, New York, **1987**; T.G. Dewey, Ed.:, *Biophysical and Biochemical Aspects of Fluorescence Spectroscopy*, Plenum Press, New York, London, **1991**.

Fluorophore, <*Fluorophor*>, a → chromophore which fluoresces, i.e. one which, after the absorption of electromagnetic radiation, gives out its excitation energy in the form of radiation. Not all molecules, which absorb in the UV-VIS region, fluoresce.

Focus, <*Fokus*>, (from the Latin focus = hearth, fire) generally the junction of a converging cone of rays, in optics the focal point.

Focusing, <*Fokussierung*>, the reunion of a diverging pencil of rays, e.g. light, electron, X-rays at one point. The geometric optical conditions are called the focusing conditions.

Form birefringence, <*Formdoppelbrechung*>, the optical birefringence of nonspherical particles built up from isotropic volume elements, parallelly orientated and embedded in an optically isotropic substance (O. Wiener 1904). If the embedded particles are cylindrically symmetrical, the direction of the axes of the cylinders is the optic axis and the whole is optically uniaxial. The form birefringence disappears when the refractive indexes of the particles and the embedding substance are equal. If the optical birefringence does not disappear completely, the particles have an optical → anisotropy from the outset, i.e. an intrisic birefringence which, in this case, is conditioned not only by the form but also by the internal, anisotropic structure of the particles. Therefore, form birefringence and intrinsic birefringence are generally superimposed. Depending upon the special form of the embedded particles, we speak of rod birefringence in the case of rods and of lamellar birefringence in the case of lamellas. The former is always optically positive and the latter always optically negative (→ birefringence, natural), if the particles and the embedding substance do not absorb.

Försterling three-prism set, <*Försterlingscher Dreiprismensatz*> → Abbe prism.

Fortrat parabola, <*Fortrat-Parabel*>, the selection rule, $\Delta J = \pm 1$, governs the appearance of the *R* and *P* branches in the → rotation-vibration spectrum of diatomic molecules. The associated series formulas are:

$$\tilde{\nu}_R = \tilde{\nu}_o + 2B'_v = (3B'_v - B'_v)J + (B'_v - B''_v)J^2$$

$$\tilde{\nu}_P = \tilde{\nu}_o - (B'_v + B''_v)J + (B'_v + B''_v)J^2.$$

B_v' and B_v'' are the → rotational constants of the upper and lower vibrational states respectively; the subscript, v, takes account of the rotation-vibration interaction. The same relationships apply to the → electronic band spectrum. However, $\tilde{\nu}_o = \tilde{\nu}_e + \tilde{\nu}_v$ is the band origin for the superimposition of the electronic and vibrational excitation. Since a diatomic molecule is a symmetric top, another, generally constant, term appears and the selection rule, $\Delta J = 0$, also applies, i.e. a *Q* branch appears with the series formula:

$$\tilde{\nu}_Q = \tilde{\nu}_o + (B'_v - B''_v)J + (B'_v + B''_v)J^2.$$

The equations for the *R* and *P* branch can be given as a single equation:

$$\tilde{\nu} = \tilde{\nu}_0 + (B'_v - B''_v)m + (B'_v + B''_v)m^2.$$

with $m = 1, 2, 3 \ldots$ for the *R* branch ($m = J + 1$) and $-1, -2, -3 \ldots$ for the *P* branch ($m = -J$). The formulas given are the equations of parabolas. The graph of, $\tilde{\nu} = f(\mathrm{m})$ is a parabola which is called a Fortrat parabola. In how far the graph of the experimental $\tilde{\nu}$ values as a function of m (Fortrat diagram) actually shows a parabola depends upon two factors:

1. The difference, $B_v' - B_v''$ must be sufficiently large so that the quadratic term can become effective. The difference between B_v' and B_v'' is usually very small in pure rotation-vibration spectra so that very high m values are required to obtain a complete parabola with an apex. In contrast, B_v' and B_v'' are usually very different in → electronic band spectra so that the required condition is realized (see Figure 1).
2. The intensity of the transitions decreases with increasing m value (J value) so that higher rotational transitions frequently cannot be measured.

The illustration of the P, R and Q branches in the form of Fortrat parabolas is very useful for the analysis of → electronic band spectra since it provides diagrammatic information about the band edges and the shading (degrading) of the bands. In general, the above equations also apply to polyatomic linear molecules. Thus, we can plot Fortrat diagrams for those also.

Fourier spectroscopy, *<Fourier-Spektroskopie>*, a method, increasingly used in recent times, for the spectrophotometric analysis of radiation; particularly in the IR region. The principle of this method was initially used by Rubens to analyze the long-wavelength IR radiation of the quartz-mercury lamp. Michelson used it to analyze the fine structure of spectral lines by means of his dual-beam interferometer.

By Fourier analysis (harmonic analysis) we understand the mathematical fact that any periodic or nonperiodic function, $y = f(x)$, can be written as a sum or an integral of sine and cosine functions of the argument. Figure 1 shows the simple example of the saw-tooth curve, $y = x$ from $-\pi$ to $+\pi$, approximated by three partial sums of the series:

$$y = 2 \sum_{n=1}^{\infty} (-1)^{n-1} \frac{\sin(nx)}{n}$$

The curves a, b and c correspond to the sums of the first two, three and four terms of the series respectively.

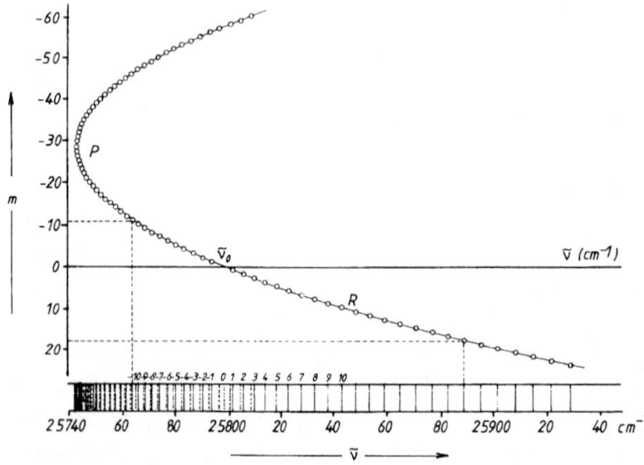

Fortrat parabola. Fig. 1. *P* and *R* branches of a diatomic molecule as a Fortrat parabola; after Herzberg

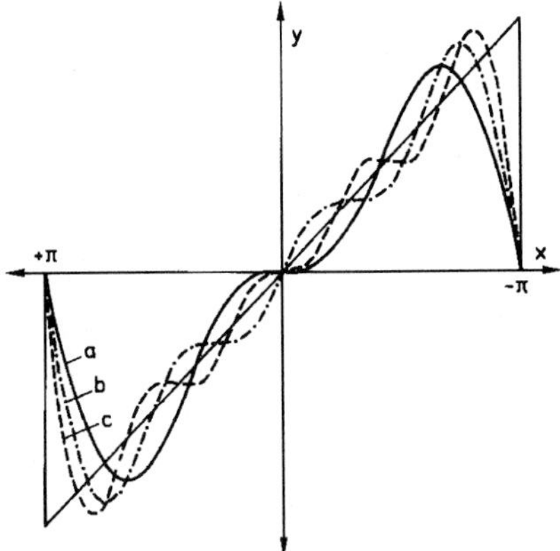

Fourier spectroscopy. Fig. 1. Fourier analysis of a sawtooth curve into sine curves, see text

This procedure can be continued indefinitely so that the sawtooth curve can be approximated with any desired accuracy by the superimposition of sine functions. Fourier analysis and synthesis play an important role in physics.

When applying this principle in spectroscopy, we use a chronological Fourier decomposition. The polychromatic radiation under investigation is passed through a dual-beam interferometer with a path difference which varies in time, e.g. the position of one arm of a → Michelson interferometer is changed smoothly. Each individual frequency of the light mixture gives a light intensity at the exit aperture of the interferometer modulated periodically with time, i.e. a periodic function in the time domain. The modulation frequency is proportional to the speed of the path difference change and to the frequency of the radiation. If we place a radiation receiver at the exit aperture we obtain a sinusoidal signal, changing with time, which can be plotted spatially as a sine wave on a recorder. For polychromatic radiation containing several frequencies, the recorded curve is a Fourier-synthetized representation of the spectrum. The Fourier components, together with their amplitudes, can be determined computationally or electronically from such a curve by harmonic analysis. These are a direct measure of the frequency and intensity of the components of the radiation. This mathematical process corresponds to a transformation from the time into the frequency domain. There are two fundamental differences between this type of spectroscopy and the dispersive. Firstly, we do not require a prism or grating to disperse the radiation. Secondly, we obtain an instantaneous statement about the whole spectrum under investigation, as with photographic spectrophotometry (→ multichannel spectroscopy). The resolving power obtainable with Fourier-transform spectroscopy depends upon the quality of the interferometer and the maximum path difference between the beams. Since we work with undispersed radiation, the detection sensitivity is high. Fourier spectroscopy is applied in the IR region (Fourier-transform IR spectroscopy, → Fourier-transform infrared spectrometer) and also in → NMR spectroscopy (→ Fourier-transform NMR spectroscopy). In recent times, instruments for the UV-VIS spectral region (→ Fourier-transform UV-VIS spectrometer) have been constructed.

Fourier-transform (FT) infrared spectrometer, FT-IR spectrometer, *<Fourier-Transform(FT)-Infrarot-Spektrometer, Fourier-Transform-Infrarot-Spektralphotometer, FT-IR-Spektrometer, FT-IR-Spektralphotometer>*, a nondispersive IR spectrometer (→ IR spectrometer) the essential component of which is an → interferometer, usually a → Michelson interferometer. Such an instrument primarily provides an interferogram in the time domain which, when converted into the frequency domain by a Fourier transformation (→ Fourier spectroscopy), gives the IR spectrum. In addition to the optics, a computer is always an integral component of an FT-IR spectrometer. Figure 1 shows the optics of the Perkin-Elmer model 1600 schematically. The radiation from the IR light source (Q) passes via a toroidal mirror, S_1, into a modified Michelson interferometer and is divided between the two beam paths, 1 and 2, of the interferometer by the beam splitter, T. Beam 1 falls onto a fixed interferometer mirror (*IS* 1) and is reflected there and in the beam splitter, T, at 60°. Beam 2 is reflected by the movable interferometer-mirror system (*IS* 2) so that a chronologically varying optical path difference between the two beams of the interferometer is generated. Figure 2 shows schematically the beam path 2 from the beam splitter to the interferometer-mirror system 2. The optical path difference is set by a motor-controlled inclination of the mirrors, S_1 and S_2 around the axis D. It is given, as shown in Figure 2, by $a' + b' + c' < a + b + c$, the optical path length at the zero position of the system. The two rays, 1 + 2, are focused by means of an adjustable toroidal mirror, S_2, at the position of the sample, and finally arrive at the IR detector via two other toroidal mirrors, S_3 and S_4. The beam of a He-Ne laser follows the IR beam through the interferometer (dashed line) and falls onto a separate detector. This beam path is used to control the position of the movable interferometer-mirror system (*IS* 2) by means of its interference pattern and provides an exact time basis. Each interference maximum of the laser beam provides a trigger signal which marks a defined position of the mirror system (*IS* 2) and determines simultaneously that the reading of the IR detector should be recorded. The sequence of these experimental data points provides the interferogram in the time domain which is stored in the computer and is available for transformation into the frequency domain, i.e. into the IR spectrum. The whole optical system is hermetically sealed in order to eliminate, as far as possible the influences of water vapor and CO_2. To permit entry and exit of the IR radiation, the optical system is closed off from the sample chamber by KBr windows. The entry window for the He-Ne laser beam is glass. A lithium-tantalate detector is commonly used as the IR detector.

Compared with a sequential (dispersive) → IR spectrometer, an FT-IR spectrometer has a number of advantages:

1. The complete IR spectrum is measured simultaneously. This is known as the multiplex or Fellgett advantage. As a result, a complete spectrum can be recorded in a very short time which makes possible IR-GC coupling, for example.
2. The signal-to-noise ratio (S/N) is considerably greater so that even small absorbances can be reliably recorded.
3. The intensity of the IR radiation is not influenced either by apertures or by monochromator slits. This improvement in optical throughput

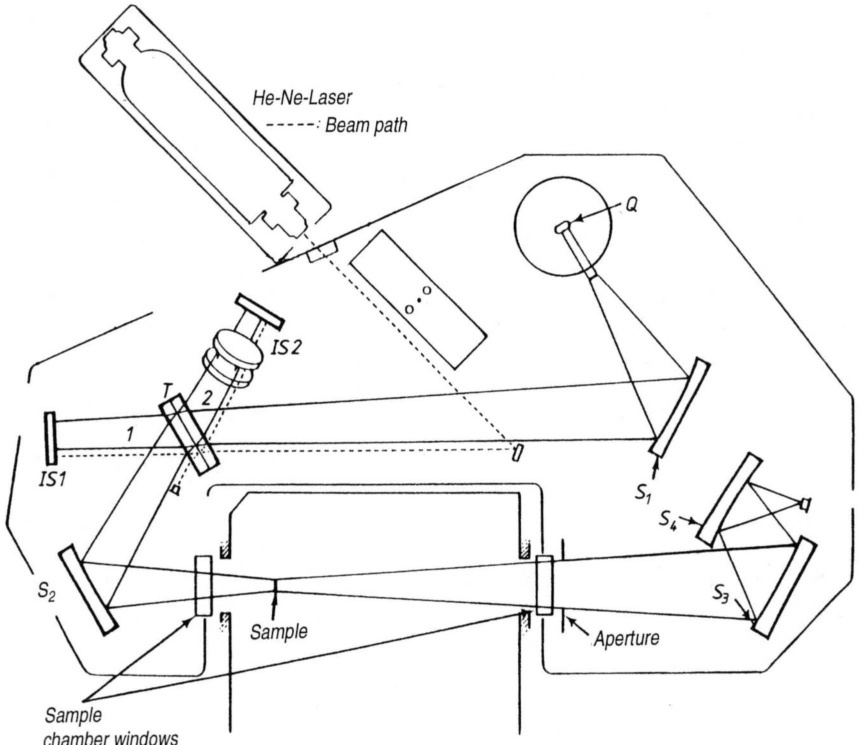

FT-IR spectrometer. Fig. 1. Beam paths in an FT-IR spectrometer, Perkin-Elmer model 1600

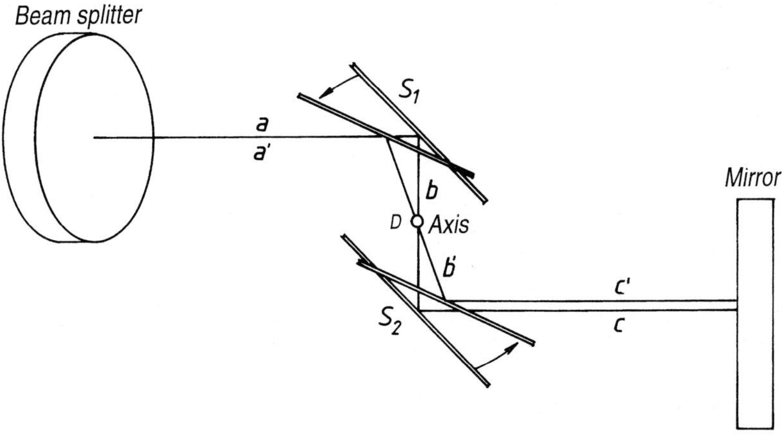

FT-IR spectrometer. Fig. 2. Interferometer system in the model 1600

is known as the Jacquinot advantage.

4. The IR spectrum is calculated from the interferogram using a computer and is available in digital form. This permits a further processing of the spectrum using simple programs which are generally part of the computer software.
5. Where, as is usually the case, a laser is used to monitor the position of the moving interferometer mirror, a large improvement in the accuracy of the measured frequencies is obtained. This is known as the Connes advantage.

Various instrument manufacturers have developed FT-IR spectrometers in recent years. In addition to the single-beam instrument illustrated in Figure 1, dual-beam instruments, which have a reference-radiation path, have been developed. Examples are the Perkin-Elmer model 1800, the Bruker IFS 88 and IFS 113 v and the Digilab FTS 15 and FTS 20. Figure 3 shows the optical layout of the IFS 88 the details of which can be recognized directly from the diagram. It has two light sources and six detectors (D_1 – D_6) which are adjusted to the particular spectral regions. In addition to the two beam paths for the reference and sample, a third path can be utilized which passes through the sample chamber as a parallel beam and is provided for special sample attachments. Furthermore, a gas chromatograph or an IR microscope can be coupled to the instrument via two parallel beam ports in the interferometer component. The range of the instrument can be extended into the UV-VIS spectral region by exchange of the light source, beam splitter and detector. If all its

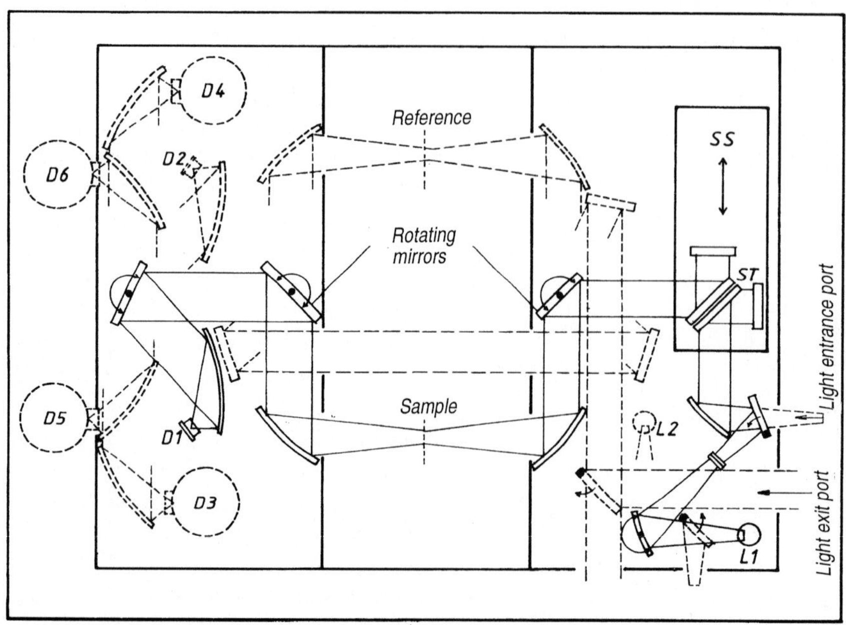

FT-IR spectrometer. Fig. 3. Beam paths in the Bruker IFS 88

possibilities are utilized, the instrument spans a spectral range from 40,000 to 10 cm^{-1}. This also applies to the IFS 113 v.

Fourier-transform (FT) NMR spectroscopy, FT-NMR spectroscopy, pulsed Fourier-transform NMR spectroscopy, *<Fourier-Transform(FT)-NMR-Spektroskopie, FT-NMR-Spektroskopie, PFT-NMR-Spektroskopie>*, a technique which was first developed to reduce the time required to obtain an NMR spectrum. But it also offers many other advantages over the → continuous wave method (→ NMR spectroscopy) to the NMR spectroscopist seeking to interpret complex spectra, determine physical properties or make quantitative analyses. The basic principle is relatively simple. A high-frequency pulse generated by a brief switching-on of a high-frequency generator contains, in addition of the transmitter frequency, ν_1, a whole band of frequencies (the Fourier components) given by $\nu_1 \pm (1/\tau_p)$ where τ_p is the pulse width (cf. Figure 1a). The pulse contains frequencies between ν_1 + 20,000 Hz and ν_1 −20,000 Hz if $\tau_p = 50\ \mu s = 5 \cdot 10^{-5}$ s. If

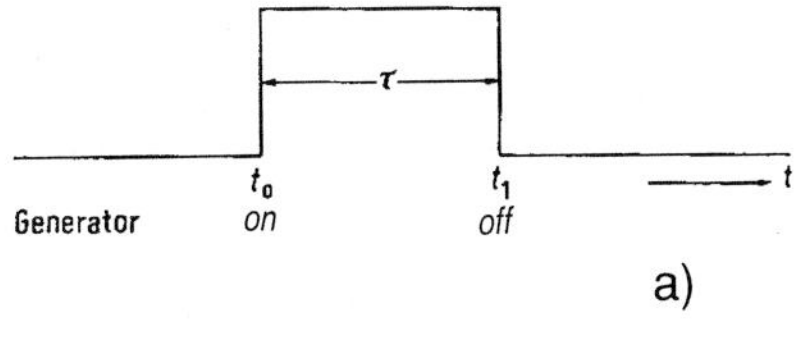

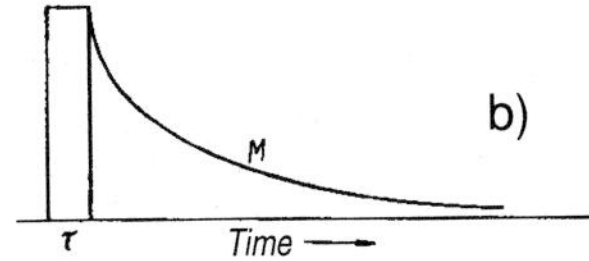

FT-NMR spectroscopy. Fig. 1. a) Pulse width, τ_p, b) free induction decay, FID

the mean or carrier frequency, ν_1, is selected correctly, the frequency band contains all the frequencies of an NMR spectrum. The transmitter of the NMR instrument has become a polychromatic transmitter and all resonance frequencies are excited simultaneously. At a magnetic flux density, B_o of 1.41 T (14,100 Gauss) ν_1 = 60 MHz for protons and ν_1 = 15.1 MHz for ^{13}C. The pulse time, τ_p, is determined by the width of the spectrum to be measured. The measurement of the simultaneous excitation is made using the free induction decay (FID) which is caused by the relaxation processes of the excited nuclei. According to the classical description, at resonance the macroscopic magnetization, M, is tipped through the angle Φ from the z direction (field direction) by the pertubation B_1. This perturbation arises from the magnetic field polarized in the x direction, which is due to the high-frequency field generated by the transmitter coil. Thus, a transverse magnetization, M_t, is induced in the x-y plane where it rotates on account of the precession of M. The same applies for each resonance and the corresponding component of the pulse. The angle, Φ (pulse angle), through which M is tipped depends upon the pulse length, τ_p, and the amplitude of B_1: $\Phi = B_1\tau_p$. If τ_p is selected such that $\Phi = 90°$ we speak of a 90° pulse. The magnetization in the x-y plane is a maximum in this case and a maximum signal is recorded in the receiver coil on the y axis. After switching off the pulse, equilibrium is re-established by the relaxation of the freely precessing nuclei and the magnetization component in the x-y plane, M_t, falls exponentially. When the transmitter and Larmor frequency are

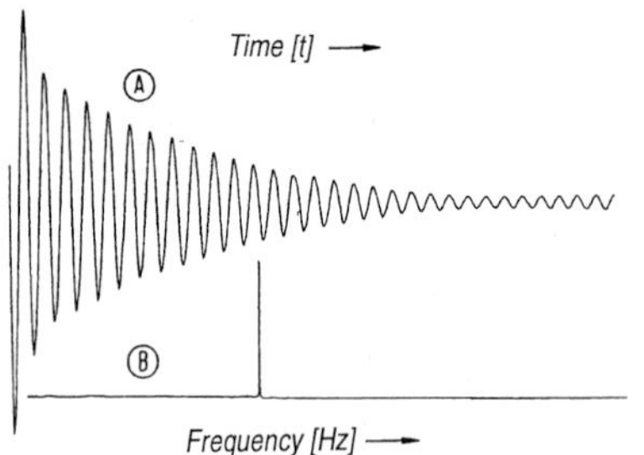

FT-NMR spectroscopy. Fig. 2. FID when the transmitter frequency, ν_1, is not equal to the resonance frequency, ν_A

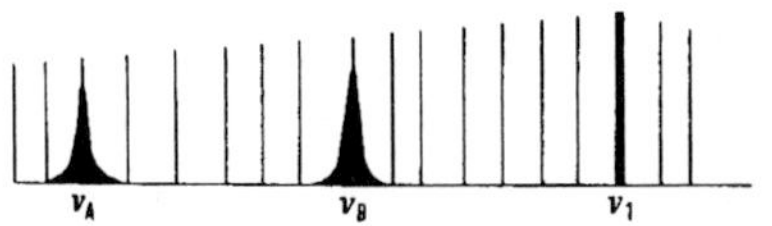

FT-NMR spectroscopy. Fig. 3. Two nuclei with ν_A and $\nu_B \neq \nu_1$

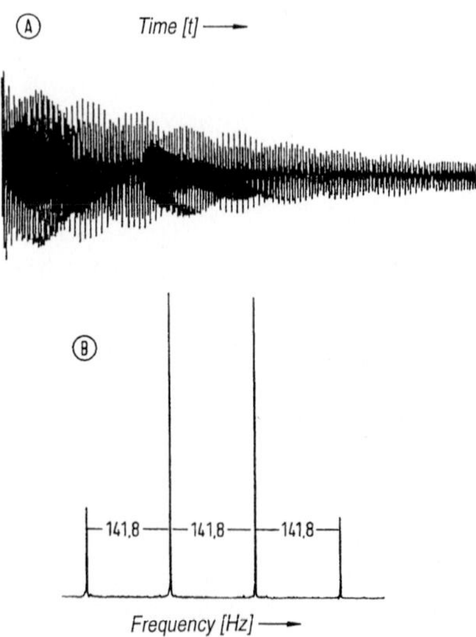

FT-NMR spectroscopy. Fig. 4. Superimposed FIDs of several nuclei, interferogram and spectrum

the same, the decay curve (FID) shown in Figure 1b results. If the two frequencies are not equal, we obtain a damped sinusoidal oscillation (Figure 2) as the free induction decay or FID. In the case of several nuclei with different resonance frequencies (e.g. ν_A and ν_B in Figure 3) the FIDs of the various magnetizations superimpose and generate a very complex FID known as an interferogram, as shown in Figure 4. These interferograms in the time domain contain all the important information such as the resonance frequencies and intensities. However, they cannot be analyzed directly since a spectrum is usually interpreted in the frequency domain and not in the time domain. But the two spectra can be interconverted, one into the other, by Fourier transformation.

The Fourier transformation of the time spectrum into the frequency spectrum requires a computer. Furthermore, the interferogram (FID) must be converted into a computer-compatible form in order to be stored. This is done by an analog-to-digital converter (ADC) which is a basic component of the computer.

The intensity of the FID of a single pulse is usually so weak that the signal after the Fourier transformation is very small in the relation to the noise. This is always the case if nuclei of low natural abundance and sensitivity, such as ^{13}C and ^{15}N, or samples having very low concentrations are measured. For that reason, the FIDs of many pulses, sometimes several hundred-thousand, are accumulated in the computer and only then transformed into the frequency spectrum. The accumulation of so many FIDs over a long period of time, sometimes hours or days, presupposes a very precise

field/frequency lock and an exact storing of the information from each FID at the same memory location in the computer.
The free induction decay requires time. How quickly the FID disappears depends upon the relaxation time T_2, i.e. the transverse or → spin-spin relaxation time, and also upon the field inhomogeneities. This decay time is particularly important with respect to the interval between pulses because the system must be completely relaxed prior to each new pulse for exact measurement of the intensities. For protons and ^{13}C nuclei this is usually the case in a few seconds, but it can sometimes take 50 s or more. In practice, since the inaccuracy does not affect the resonance frequencies but only the signal intensities, a compromise is made between the measurement time and the accuracy. Pulse intervals of approximately 1 s are generally used in ^1H and ^{13}C NMR spectroscopy. The interferogram is stored digitally in the computer during this interval between two pulses.
Ref.: R.K. Harris, *Nuclear Magnetic Resonance Spectroscopy*, Pitman, London, Marshfield, **1983**. C.H. Yoder, C.D. Schaeffer, Jr., *Introduction to Multinuclear NMR*, Benjamin/Cummings Publishing Co. Inc., Menlo Park (CA), Reading (MA), Wokingham (UK), Sydney, Tokyo, **1987**.

Fourier-transform (FT) photoluminescence spectrometer, FT-PL spectrometer, <*Fourier-Transform(FT)-Photolumineszenz-Spektrometer, FT-Photolumineszenz-Spektrometer, FT-Photolumineszenz-Spektralphotometer, FTPL-Spektrometer*>, an instrument developed by Messrs Bruker as a unit, complementary to the → Fourier-transform infrared spectrometer IFS 88, for the investigation of the photoluminescence of semiconductor materials the luminescence of which lies preferentially in this region. Figure 1 shows the construction of an FT-PL spectrometer. A → noble-gas ion laser, which is passed through a filter to eliminate the plasma lines in the NIR, is used to excite the photoluminescence of the sample. The light beam is focused on the surface of the sample which is normally contained in a bath cryostat with liquid helium (4 K) so that it can be kept at a constant temperature even during intense irradiation. For temperature-dependent investigations, the cryostat can be heated to 250 K using an He gas flow. A parabolic collecting lens couples the photoluminescence into the IFS 88 via an intermediate image and a filter. Figure 1 shows this part of the FT-PL spectrometer at the upper left. The lower part shows the optical path of

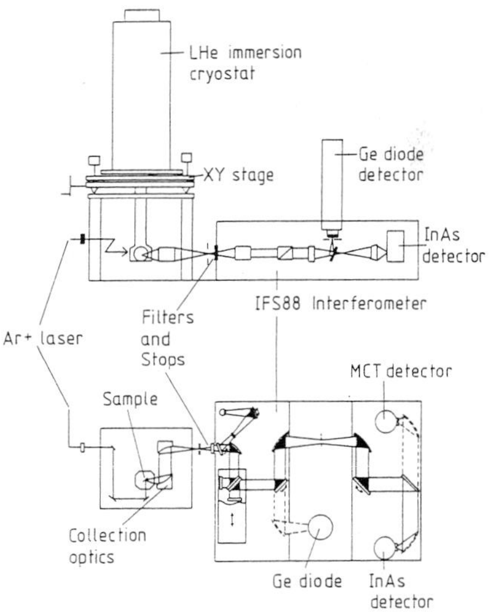

FT-photoluminescence spectrometer. Fig. 1. A block diagram of the Bruker instrument with the IFS 88

the FT-IR spectrometer, after coupling the two units. There are three selectable detectors, InAs, MCT and a high-sensitivity Ge diode, which is operated at 77 K. The cryostat for the operation of this detector is drawn schematically in the upper part of Figure 1. The instrument is designed in such a way that all normal applications of FT-IR spectroscopy are possible in transmission or reflection in the mid and near IR regions. Use of an FT-IR microscope is also possible. The transition to photoluminescence measurements requires only the selection of the relevant mirrors.

Ref.: M.B. Simpson, *Bruker Report*, no. 2, **1988** pp. 38–41.

Fourier-transform (FT) Raman spectrometer, FT-Raman spectrometer, <*Fourier-Transform(FT)-Raman-Spektrometer, FT-Raman-Spektrometer*>, an instrument for → Raman spectroscopy which operates in the → IR region and is usually designed as an optional attachment for a → Fourier-transform infrared spectrometer. A → neodymium-YAG laser with a wavelength λ = 1,064 nm, corresponding to $\tilde{\nu}$ = 9,398 cm^{-1}, is used to excite the Raman scattering. A germanium detector operating at 77 K or an InGaAs detector operating at 250 K is used as the detector for the scattered radiation which lies in the NIR region. Figure 1 shows the beam path of a → Fourier-transform infrared spectrometer optically coupled to a Perkin-Elmer model 1710 TIR spectrometer. An effective collecting lens for the Raman radiation is essential. In Figure 1 it is an elliptical reflector which sends the Raman radiation into the FT-IR instrument. Other possible collecting lenses have been proposed by

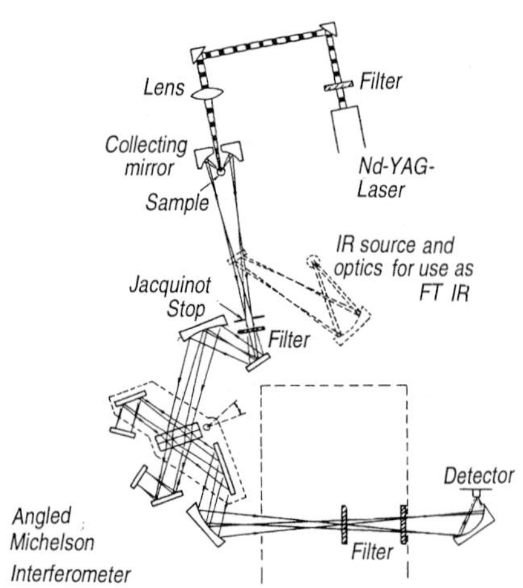

FT-Raman spectrometer. Fig. 1. Block diagram of the Raman module for the Perkin-Elmer FT-IR 1600

B. Schrader (Ref.). Bruker have implemented a similar design in conjunction with their IFS 66 FT-IR instrument. Figure 2 is a block diagram of this spectrometer with the Raman light path shown as a full line. The great advantage of Raman spectroscopy in the NIR region lies in the fact that the interfering fluorescence of samples excited with visible light is practically nonexistent with NIR excitation, i.e. fluorescing or colored samples are much less of an experimental problem. Furthermore, all the advantages of an FT-IR instrument (→ Fourier-transform infrared spectrometer) are also present in this form of FT-Raman spectroscopy. The power of the FT-method is very important in overcoming the fact that scattering is proportional to the inverse fourth power of the wavelength of the incident light (→ Rayleigh scattering). This severely reduces the available Raman signal on going from VIS to NIR excitation.

Ref.: P. Hendra, H. Mould, *International Laboratory*, September, **1988**;

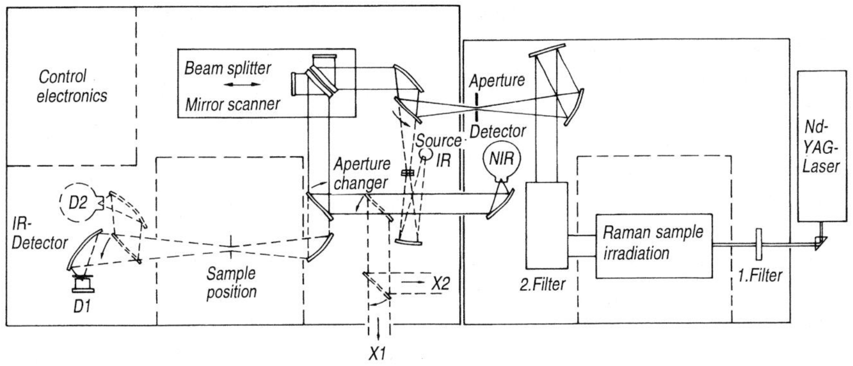

FT-Raman spectrometer. Fig. 2. Block diagram of the Raman module for the Bruker IFS 66

B. Schrader, A. Simon, *Bruker Report*, **1987**, *2*, p. 13 ff. An insight into the many applications of FT-Raman spectroscopy may be obtained from three special issues of Spectrochimica Acta edited by P.J. Hendra: *Spectrochim. Acta*, **1990**, *46A*, pp. 121–338; **1991**, *47A*, pp. 1133–1494; **1993**, *49A*, pp. 611–887.

Fourier-transform (FT) UV-VIS spectrometer, FT-UV-VIS spectrometer, *<Fourier-Transform(FT)-UV-VIS-Spektrometer, FT-UV-VIS-Spektrometer, FT-UV-VIS-Spektralphotometer>*, a development with which the Fourier-transform technique has been transferred to the UV-VIS spectral region. In the case of → Fourier-transform infrared spectroscopy, it is possible to extend the wavelength range into the UV-VIS spectral region by an appropriate selection of the optical components such as the light source, beam splitter or detector. An example of this is the Bruker FT-IR spectrometer IFS 120HR. Chelsea Instruments have developed the FT 500 UV-VIS spectrometer especially for the UV-VIS region. Figure 1 shows the light path and the optical components. The crucial device is a modified → Michelson

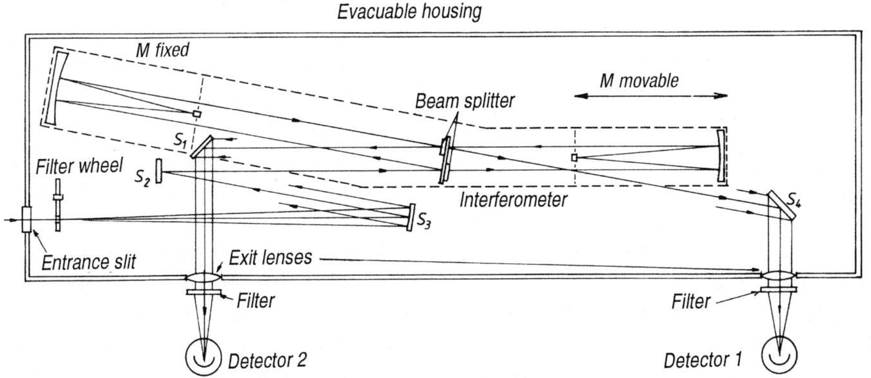

FT-UV-VIS spectrometer. Fig. 1. Light path in the Chelsea Instruments FT 500 UV

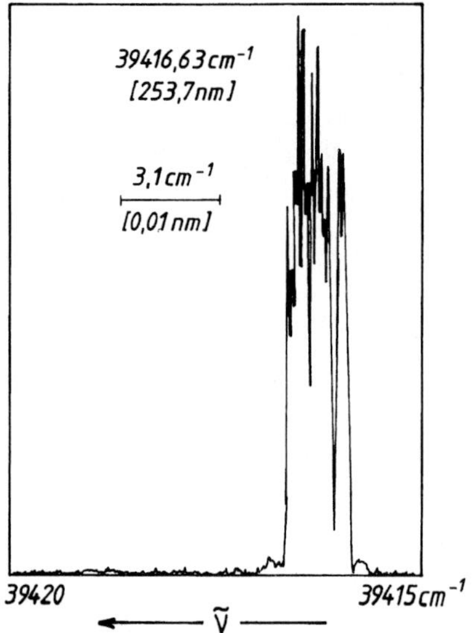

FT-UV-VIS spectrometer. Fig. 2. Emission spectrum of an Hg low-pressure lamp measured with the FT 500 UV

interferometer in which the parallel beam of light from the source is split by a → beam splitter into two equal parts which fall onto a fixed and a movable mirror. The radiation reflected by these mirrors passes back to the beam splitter, is recombined and falls onto the detector. Since one of the mirrors is moved relative to the other, interference results at the detector and an interferogram of the light source is recorded. A conventional spectrum is obtained by Fourier transformation. The instrument can be evacuated and covers the spectral region from ~ 175 nm to ~ 1000 nm (57,150 cm^{-1} to 10,000 cm^{-1}). The range can be extended to 120 nm (~ 83,300 cm^{-1}) by using a MgF_2 beam splitter. The resolving power is ca. $2 \cdot 10^6$ at 200 nm. The emission spectrum of a low-pressure mercury lamp in the region of the intercombination line at 253.7 nm (39,417 cm^{-1}) is shown in Figure 2. The spectrum illustrates the resolved → hyperfine structure of this Hg line. In addition, there is further splitting of the lines due to self-reversal (→ line reversal, → self-absorption). This example shows that the FT 500 spectrometer is preferable for the analysis of highly resolved emission spectra of the elements in ICP AES (→ ICP). AAS measurements (→ AAS) are possible in conjunction with → hollow cathode lamps.

Ref.: D. Snook, A. Grillo, *Amer. Laboratory*, **1986**, *18*, 28, 30, 32; A. Thorne, *J. Anal. Atomic Spectrom.*, **1987**, *2*, 227.

Four-level system, <*Vierniveausystem*> → neodymium laser.

FPI, <*F.P.I.*> → Fabry-Perot interferometer.

Fragmentation, <*Fragmentierung*>, the occurrence of numerous other peaks at smaller *m*/*e* values, in addition to the more or less intense molecular ion peak in the → mass spectrum of a molecule. These peaks can be assigned to the masses of fragments of the original molecule. The fragmentation is connected with reactions which are a consequence of the ionization. Under normal operating conditions, the bombarding electrons have an energy of 70 eV, which corresponds to ca. 6700 $kJ \cdot mol^{-1}$. Since an average bond energy is ca. 400 $kJ \cdot mol^{-1}$, sufficient energy is theoretically available to break more than ten bonds simultaneously. However, this process, which is definitely observed, is of no great importance. In reality, the molecules absorb, on average, far less

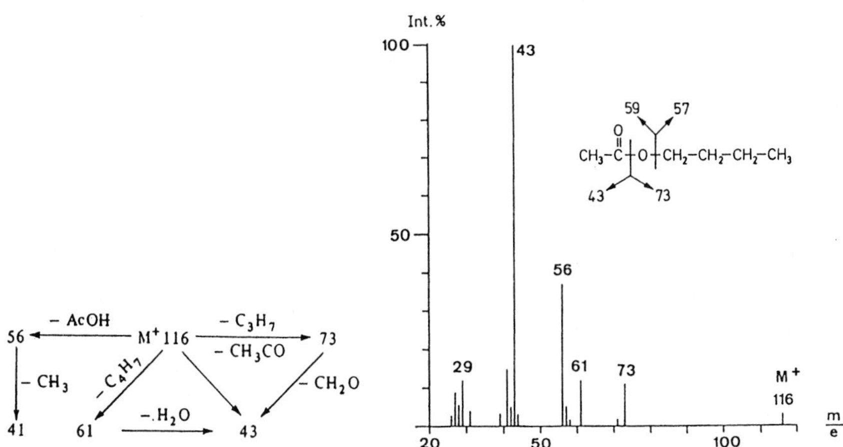

Fragmentation. Fig. 1. Mass spectrum and fragmentation of n-butyl acetate

excitation energy during ionization than is theoretically available. However, they do take up sufficient energy to induce chemical reactions which lead to a fragmentation of the molecular ions and thus to the formation of fragments of smaller masses. The type and course of these chemical reactions, and therefore the mass and abundancy distribution of the fragments formed, are, as is always the case in chemistry, structure-dependent. It is therefore possible to draw conclusions about the structure of a compound from its fragmentation pattern. Figure 1 shows, as an example, the → mass spectrum and fragmentation pattern of n-butyl acetate as it might appear in the literature. For simple fragmentation, empirical regularities have been combined to formulate several rules. They are based upon the influence of branching, double bonds, hetero atoms and the importance of steric factors.

Ref.: F.W. McLafferty, *Interpretation of Mass Spectra*, W.A. Benjamin, Inc., New York, Amsterdam, 4th ed., **1994**.

Franck and Hertz experiment, <*Franck-Hertz-Versuch*>, an experiment carried out by Franck and Hertz in 1914 which resulted in the experimental confirmation of Bohr's frequency relation, $\Delta E = E_2 - E_1 = h\nu = hc\tilde{\nu}$ proposed in 1913. Since the equivalence of energy, E, and electromagnetic radiation, $h\nu$, initially formulated by Einstein, is contained implicitly in this relation, it is frequently also called the → Bohr-Einstein frequency relation. The discrete energy states of an atom were connected to electromagnetic radiation for the first time in this relation which made possible the interpretation of → atomic spectra in absorption and emission.

The experimental confirmation of Bohr's frequency relation was achieved with the experimental arrangement shown in Figure 1. The electrons emitted from a hot cathode, K, in an evacuated tube are accelerated by a variable voltage, U_b, applied between K and the grid, G. The current intensity rises rapidly with increasing acceleration voltage, U_b

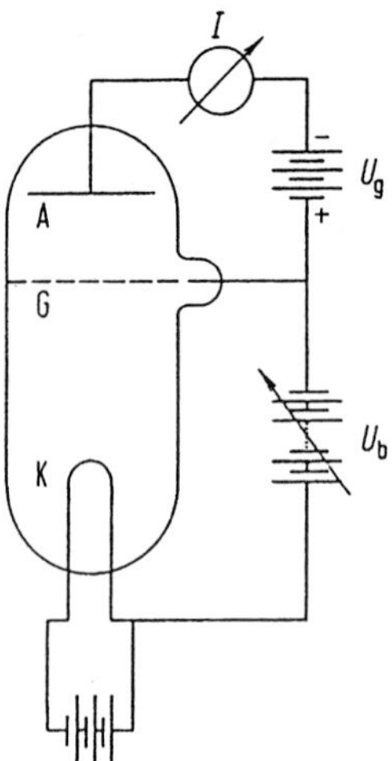

Franck and Hertz experiment. Fig. 1. The experimental arrangement

(dashed line in Figure 2). If mercury atoms are present in the evacuated tube, the electrons collide with the Hg atoms. The electrons arriving at G pass through the grid and approach the collecting electrode, A, which is at a negative potential of ca. 0.5 V with respect to the grid, G. If their energy is greater than 0.5 eV, the electrons can reach the electrode, A, and the resulting current, i, is measured with an ammeter. Those electrons which have lost their kinetic energy completely, or a large part of it, during the inelastic collisions with the Hg atoms cannot overcome the countervoltage, U_g, of 0.5 V so that the current intensity, i, decreases strongly. Figure 2 shows the result of such an experiment as the directly recorded plot of a lecture demonstration experiment. We see that the electron flux, i, decreases strongly at an acceleration voltage of 4.9 V. This is explained by the fact that the electrons have transferred their kinetic energy to the Hg atoms as excitation energy, $\Delta E = E_2 - E_1$, by inelastic collisions and do not possess sufficient energy to overcome the countervoltage of 0.5 V on the collecting electrode, A. If the voltage, U_b, is increased further the current intensity also rises again until a sharp drop occurs at 9.8 V. The phenomenon is repeated at 14.7 V .... At voltages of 9.8 and 14.7 V the electrons can excite Hg atoms two and three times respectively on their way from K to G. Thus, the first excitation energy of the Hg atom is 4.9 eV. Since excited electronic states have only a short lifetime, the excited Hg atoms should return to the ground state by emitting a photon of the energy corresponding to Bohr's frequency relation. If the excitation energy of 4.9 eV is converted into wavenumbers, we obtain $\tilde{\nu}$ = 39,531 cm^{-1} which corresponds to a wavelength, λ = 253 nm. If the experiment is carried out in an evacuated quartz tube, the emission of a line at λ = 253.7 nm can be observed. It corresponds to the transition from the first excited state to the ground state of the

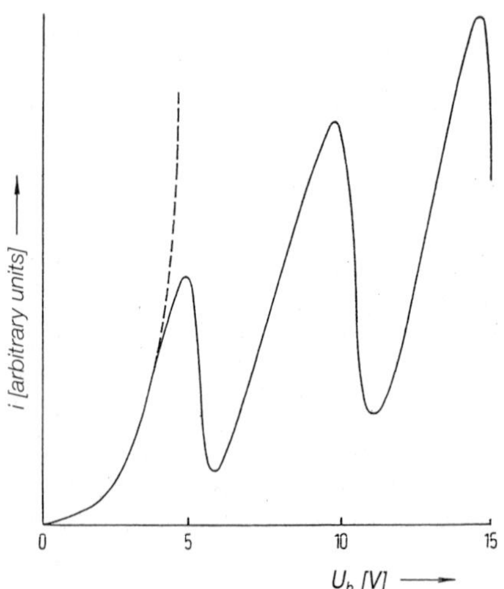

Franck and Hertz experiment. Fig. 2. Current-voltage plot, see text

Hg atom, i.e. the → intercombination transition $6^3P_1 \rightarrow 6^1S_o$. In summary, the excitation energy $\Delta E \approx 4.9$ eV, was transferred to the Hg atom by an inelastic electron collision raising the atom to an excited state of energy E_2. Upon the return of the atom to the ground state energy, E_1, the energy difference, $E_2 - E_1 = \Delta E$ was radiated in the form of a photon of wavelength, $\lambda = 253.7$ nm corresponding to a wavenumber of 39,417 cm^{-1} or 4.89 eV. This confirms Bohr's frequency relation.

Franck-Condon principle, FC principle, *<Franck-Condon-Prinzip, FC-Prinzip>*, the principle which explains the variation of the intensity of the vibrational structure of an electronic absorption band. The Franck-Condon principle states that an electronic transition in a molecule occurs so rapidly, compared to the vibrational motion that immediately afterwards the nuclei have not noticeably changed either their relative positions, i.e. their separations or their velocities, vis-a-vis the state immediately before the transition. Another way of expressing this is to say that on a diagram in which the potential-energy curves of the two electronic states are plotted, the transition between them may be represented by a vertical line. Quantum-mechanically, the Franck-Condon principle results from the Born-Oppenheimer approximation and can be expressed as follows:

$$M_{oo,jk} = \vec{M}_{oj} S_{oo,jk}$$

$M_{oo,jk}$ is the → electronic transition-dipole moment of the combined electronic ($0 \rightarrow j$) and vibrational transition ($v'' = 0 \rightarrow v' = k$; $k = 0, 1, 2 \ldots$). $\vec{M}_{oj}$ is a mean value of this moment for the pure electronic transition which is approximately independent of the nuclear, i.e. vibrational, coordinates, (cf. the above formula). $S_{oo,jk}$ represents the overlap integral of the vibrational eigenfunctions associated with the transition. Since the intensity expressed by the → oscillator strength, $f_{oo,jk}$ is proportional to the square of the transition-dipole moment we obtain:

$$f_{oo,jk} = 4.7 \cdot 10^{29} \cdot v_{jk} \cdot S^2_{oo,jk} \cdot |\overline{M}_{oj}|^2.$$

The mean value, $\vec{M}_{oj}$, for a specific electronic transition can be assumed to be constant, consequently the intensity depends upon the value of the overlap integral, $S_{oo,jk}$. In order to illustrate the Franck-Condon principle we consider the potential curves of both electronic states and distinguish between the three cases shown in Figure 1:

Case a. The minima of the potential curves lie at the same equilibrium distance. The transition $v'' = 0 \rightarrow v' = 0$ satisfies the Franck-Condon principle and $S_{oo,jo}$ is large. The transitions $v'' = 0 \rightarrow v' = k$, with $k \geq 1$, correspond to a change of internuclear distance during the transition and the overlap integral $S_{oo,jk}$ is less than $S_{oo,jo}$, i.e. $f_{oo,jk}$ is less than $f_{oo,jo}$ and thus the transitions to the higher vibrational states, $v' > 1$, show intensities decreasing as k increases from 0.

Case b. The minimum of the potential curve for the excited electronic state lies at a slightly greater equilibrium distance. The transition $v'' = 0 \rightarrow v' = 0$ is now not in accord with FC principle and the overlap integral, $S_{oo,jo}$, is small. However, the transition $v'' = 0 \rightarrow v' = 2$ fulfills the FC requirement,

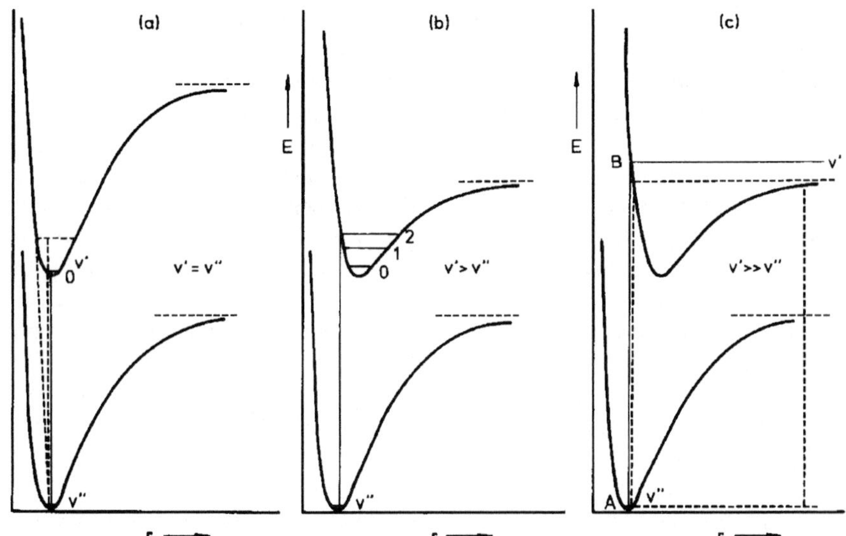

Franck-Condon principle. Fig. 1. Three relative positions of the potential curves

the overlap integral, $S_{oo,j2}$, is greater than $S_{oo,jo}$. The intensity of the individual vibronic bands first increases from $v'' = 0 \rightarrow v' = 0$ to $v'' = 0 \rightarrow v' = 2$ and then decreases for $v'' = 0 \rightarrow v' = k, k \geq 3$.

Case c. The minimum of the upper potential curve lies at even greater internuculear distance than in case b. The FC principle is strictly fulfilled for the transition A-B. However, point B lies above the asymptote of the upper potential curve and therefore corresponds to a dissociation of the molecule following the electronic transition. Transitions to points on the potential curve below B are possible, but the overlap integral gets smaller

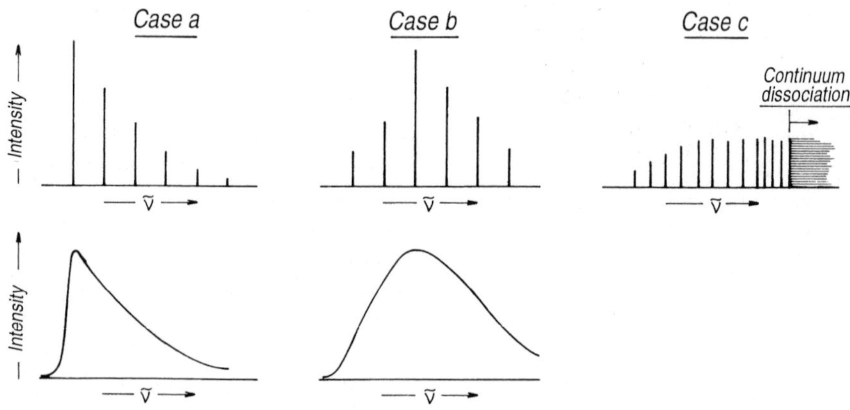

Franck-Condon principle. Fig. 2. Intensity variation in electronic band spectra

with decreasing v' values and, therefore, the intensities of the individual vibronic bands decrease also.
These three cases of intensity variation are illustrated in Figure 2.
Although the Franck-Condon principle is only valid for diatomic molecules, it is frequently applied to polyatomic molecules. From the intensity variation of the individual vibrational bands of an electronic absorption band, that is the envelope, conclusions may be drawn concerning possible changes of geometry in the excited state. In Figure 2, the envelopes corresponding to cases a and b under certain conditions of poor resolution are illustrated.

Fraunhofer lines, *<Fraunhofersche Linien>*, the dark lines always found at the same position in the continuous spectrum of the sun which were first observed in 1802 by Wollaston and investigated more thoroughly by Fraunhofer in 1814/15. The Fraunhofer lines have definite wavelengths; the 8 strongest of over 500 lines were given the initials *A* to *H* by Fraunhofer and were of decisive importance for technical optics (→ dispersion, unit of). The table lists these lines and their wavelengths.

Fraunhofer lines. Table.

Line	[nm]
A	760.8
B	686.7
C	656.3
D	589.3
E	527.0
F	486.1
G	430.8
H	396.9

Although Fraunhofer was unable to explain the origin of these lines, he contributed greatly to the development of spectroscopy by his systematic investigation of them. Thus, he knew that the bright emission line in the spectrum of an oil lamp lay at exactly the position where, in the spectrum of the sun, the dark *D* line was found. Not until 1860 was the line identified as the sodium *D* line by Bunsen and Kirchhoff, who recognized that the dark lines in the sun's spectrum arose as the result of the absorption of emission lines (→ line reversal).

Free spectral range, spectral range of a spectrometer, *<freier Spektralbereich, Durchlässigkeitsbereich eines Spektrometers>*, the wavelength interval, $\Delta\lambda$, of incident radiation for which a definitive relationship exists between the wavelength, λ, and the position $X(\lambda)$ of the image of the entrance slit. For prism spectrometers, the free spectral range coincides with the complete region of normal → dispersion of the → prism material. For grating spectrometers, $\Delta\lambda$ is limited by the order m of the spectrum. For the diffraction angle, β, in which direction a constructive → interference takes place, and a grating constant of d we have: $m\lambda_m = d \cdot \sin\beta$ or $\lambda_m = (d \cdot \sin\beta)/m$. Thus, two wavelengths, $\lambda_m = (d \cdot \sin\beta)/m$ and $\lambda_{m+1} = (d \cdot \sin\beta)/(m + 1)$, of adjacent orders, appear at the same angle, β, at the exit slit of the spectrometer. For example, a wavelength, $\lambda_1 = 500$ nm ($m = 1$) is overlapped by $\lambda_2 = 250$ nm ($m = 2$). $\Delta\lambda$ is therefore given by:

$$\Delta\lambda = d \cdot \sin\beta\left(\frac{1}{m} - \frac{1}{m+1}\right)$$
$$= d \cdot \sin\beta/[m(m+1)].$$

Thus, $\Delta\lambda$ decreases with increasing order, *m*. For this reason, UV-VIS-NIR spectrometers using → grating monochromators are fitted with order-sorting filters in the VIS and NIR regions. These filters operate as → long-wave pass filters and block off the short-wavelength radiation.
In general, → interferometers are used in very high orders ($m = 10^4 - 10^8$, → Fabry-Perot interferometers); they have a very high resolving power but a narrow free spectral range. Therefore, in order to make use of the high resolution and still have a certain knowledge of the wavelength, preliminary dispersion is necessary.

Frequency, <*Frequenz, Schwingungszahl*>, a measure of the number of complete vibrations of a wave per unit time (1 s). A simple relationship exists between wave velocity, *c*, wavelength, λ, and frequency ν; $\nu = c/\lambda$. For electromagnetic radiation, there is the special expression, $h\nu$, which gives the energy of a → photon of that radiation in terms of → Planck's constant.

Frequency band, <*Frequenzband*>, description of a frequency range in high-frequency physics or in → high-frequency spectroscopy. Frequency bands are denoted by one or two letters; the table gives a summary.

The boundaries are important in → microwave and → ESR spectroscopy.

Frequency doubling, <*Frequenzverdopplung*>, the doubling of the frequency of light as a direct consequence of → nonlinear optics. For the explanation of this phenomenon we must remember that every light wave of angular frequency, $\omega = 2\pi\nu$, is associated with an oscillating electric field, $E = E_o \cdot \sin\omega t$. When light passes through a medium, this field generates in the medium a polarization, *P*. This polarization is due to the influence of the electric field upon electrons, atoms or molecules and, at low light intensities, it is proportional to the field strength, *E*, (linear optics); $P = \varepsilon_o\chi \cdot E_o\sin(\omega t - \phi)$. Therefore, the electric wave is accompanied by a polarization wave which is delayed with respect to the electric field because of the phase shift, ϕ. The quantity ε_o is a constant (vacuum permittivity) with a numerical value of $8.8 \cdot 10^{-14}$ $A \cdot s \cdot V^{-1} \cdot cm^{-1}$; χ is the electric susceptibility, a material constant which can be determined experimentally. At higher light intensities, such as are available in laser light sources, we must take account of higher terms in *E* in the relation between *P* and *E*. In the simplest case, including the term in E^2 and neglecting the phase shift, ϕ, we obtain:

Frequency band. Table.

Band symbol	Frequency range ν [GHz(10^9 Hz)]	Wavelength range λ [cm]	Wavenumber range $\tilde{\nu}$ [cm^{-1}]
C	3.95–5.85	7.60–5.15	0.132–0.195
XN	5.85–8.20	5.13–3.66	0.195–0.273
X	8.20–12.4	3.66–2.42	0.273–0.413
KU	12.4–18.0	2.42–1.67	0.413–0.599
K	18.0–26.5	1.67–1.30	0.599–0.855
Ro .V	26.5–40.0	1.13–0.74	0.885–1.33

$$P = \varepsilon_o(\chi \cdot E_o \sin\omega t + \chi' E_o^2 \sin^2\omega t);$$

using $2\sin^2\omega t = 1 - \cos 2\omega t$:

$$P = \frac{\varepsilon_o}{2}\chi' E_o^2 + \varepsilon_o \chi E_o \sin\omega t - \frac{\varepsilon_o}{2}\chi' E_o^2 \cos 2\omega t.$$

or

$$P = P(o) + P_L(\omega) - P_{NL}(2\omega),$$

i.e.

$P(o)$ = $(\varepsilon_o/2)\chi' E_o^2$ is the constant component,

$P_L(\omega)$ = $\varepsilon_o \chi E_o \sin\omega t$ is the linear component at the fundamental frequency, and

$P_{NL}(2\omega)$ = $(\varepsilon_o/2)\chi' E_o^2 \cos 2\omega t$ is the nonlinear part at double the frequency.

The polarization waves at frequencies, ω and 2ω, radiate electric fields of the same frequencies. Thus, because of the nonlinearity, light of frequency, ω, has been converted into light of double the frequency, 2ω, i.e. red light, $\lambda \sim 700$ nm becomes violet light, $\lambda \sim 350$ nm. In an exact mathematical treatment we must remember that the three components of the polarization vector, $\vec{P} = \{P_x, P_y, P_z\}$ are linked to each of the three components of the field strength vector, $\vec{E} = \{E_x, E_y, E_z\}$. For $P = \varepsilon_o \chi E + \varepsilon_o \chi' E^2$ this leads to the following complicated expression:

$$P_x = \varepsilon_o(\chi_{11}E_x + \chi_{12}E_y + \chi_{13}E_z) + \varepsilon_o(\chi_{111}E_xE_x + \chi_{112}E_xE_y + \chi_{113}E_xE_z + \chi_{121}E_yE_x + \chi_{122}E_yE_y + \chi_{123}E_yE_z + \chi_{131}E_zE_x + \chi_{132}E_zE_y + \chi_{133}E_zE_z);$$

$$P_y = \varepsilon_o(\chi_{21}E_x + \chi_{22}E_y + \chi_{23}E_z) + \varepsilon_o(\chi_{211}E_xE_x + \chi_{212}E_xE_y + \chi_{213}E_xE_z + \chi_{221}E_yE_x + \chi_{222}E_yE_y + \chi_{223}E_yE_z + \chi_{231}E_zE_y + \chi_{232}E_zE_y + \chi_{233}E_zE_z);$$

$$P_z = \varepsilon_o(\chi_{31}E_x + \chi_{32}E_y + \chi_{33}E_z) + \varepsilon_o(\chi_{311}E_xE_x + \chi_{312}E_xE_y + \chi_{313}E_xE_z + \chi_{321}E_yE_x + \chi_{322}E_yE_y + \chi_{323}E_yE_z + \chi_{331}E_zE_x + \chi_{332}E_zE_y + \chi_{333}E_zE_z);$$

Linear Fraction $P_L(\omega)$	Nonlinear Fraction $P_{NL}(2\omega)$
$\chi_{ij} = \chi_{ji}$	$\chi_{ijk} = \chi_{ikj}$.

In the case of anisotropic media, χ generally stands for 9 and χ' for 27 constants (see above). However, because of the interrelationships indicated, only 6 remain for χ and 18 for χ'. In optically uniaxial crystals, which have a very high symmetry and are of special importance to frequency doubling, most elements of the nonlinear susceptibility, χ', disappear or they are identical. For example, only the following remain for potassium dihydrogen phosphate (KDP):

$$\chi_{123} = \chi_{132} = \chi_{213} = \chi_{231} = 1.01 \cdot 10^{-1}\ \text{cm/V}$$

$$\chi_{312} = \chi_{321} = 1.00 \cdot 10^{-11}\ \text{cm/V}.$$

The polarization can be split into a linear fraction, $P_L(\omega)$ and a nonlinear fraction, $P_{NL}(2\omega)$ at the double frequency, 2ω: $P = P_L(\omega) + P_{NL}(2\omega)$. Both fractions are caused by the field amplitude $E_1(\omega)$, i.e. the polarization and the field amplitude both spread with velocity $c_1 = E_o/n_1(\omega)$. The linear polarization, P_L, traverses the medium with the velocity c_1, and generates at each location in the medium an electric field amplitude E_1 of the same frequency, ω, which also propagates with c_1. All contributions to the field amplitude produced in all different regions of the medium are superimposed in phase and a field amplitude of frequency, ω, results.

The nonlinear polarization, $P_{NL}(2\omega)$, spreads with velocity c_1, because it is also caused by the field $E_1(\omega)$ spreading with c_1. It generates at each location of the medium an electric field amplitude $E_2(2\omega)$, but with double the frequency, 2ω. However, this harmonic wave, E_2, propagates with velocity $c_2 = c_o/n_2(2\omega)$, corresponding to a refractive index n_2 for the frequency 2ω. Since $n_2(2\omega) > n_1(\omega)$, $c_2 < c_1$, i.e. the harmonic wave spreads more slowly than the polarization, P_{NL}, which generates it. The consequence is that the contributions to the harmonic wave get out-of-phase after a certain distance in the medium, e.g. a nonlinear crystal, and disappear by destructive interference.

In order to obtain intense light of the double frequency we must obviously have: $c_1(\omega) = c_2(2\omega)$ or, since the velocity is determined by the refractive index, the following condition must be fulfilled: $n_1(\omega) = n_2(2\omega)$. Since the refractive index, n, increases with the frequency in all crystals, it appears that this condition cannot be met. However, if we consider a uniaxial, doubly refracting crystal such as KDP we see that the condition can be fulfilled. If the fundamental wave, $E_1(\omega)$, is incident at an angle ϑ upon such a crystal, so that the electric field vector is perpendicular to the optic axis z, of the KDP crystal, this corresponds to an ordinary ray, the refractive index of which, n_o, is independent of the angle, ϑ. This ordinary ray produces a nonlinear polarization P_{NL} which is parallel to the optic axis, z, i.e. it radiates an extraordinary ray of light of doubled frequency, $E_2(2\omega)$. The refractive index, n_e, of this beam of light depends upon the angle of incidence ϑ and:

1. The fundamental wave, $E_1(\omega)$, and the harmonic wave, $E_2(2\omega)$, are polarized perpendicularly to each other.
2. An ordinary ray of light of the fundamental frequency, ω, generates an extraordinary ray of the double frequency, 2ω.

With respect to the refractive indexes, that of the ordinary ray $n_o(\omega)$ is not dependent upon ϑ whilst the refractive index of the extraordinary ray, $n_e(2\omega)$, depends strongly upon ϑ and, due to the larger frequency for $\vartheta = 0°$, is also greater than $n_o(\omega)$. Figure 1 illustrates these relationships for a KDP crystal. $n_e(2\omega)$ describes an ellipse in the z-y plane with the maximum value in the z direction. But $n_o(\omega)$ is a circle with $n_o(\omega) =$ constant. We see immediately that there is a quite specific angle of incidence, ϑ_o, to the optic axis, z, for which the refractive index, $n_o(\omega)$, of the ordinary ray, the fundamental wave, is equal to the refractive index, $n_e(2\omega)$ of the extraordinary, harmonic wave: $n_o(\omega) = n_e(2\omega)$.

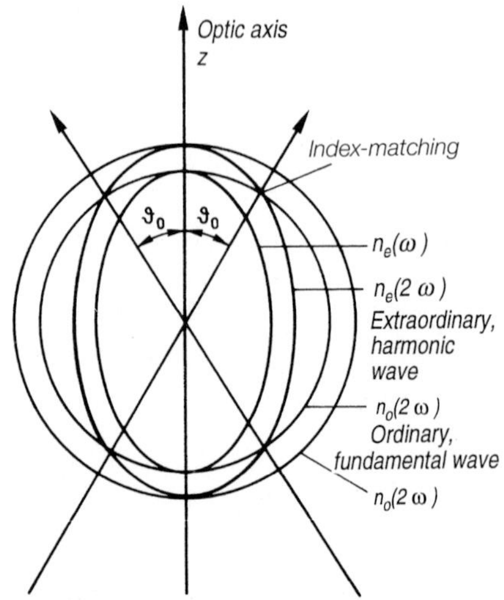

Frequency doubling. Fig. 1. Index ellipsoid for index matching; KDP crystal.

There is a matching of the refractive indexes. If the beam of light propagates at this angle, ϑ_o to the optical axis of the KDP crystal, the velocity is the same for the fundamental, polarization and harmonic waves. The polarization and harmonic waves, in particular, do not dephase. All harmonic wave contributions overlap in phase and the intensity of the harmonic wave increases with increasing crystal thickness until all the intensity of the fundamental wave has been converted into light of the harmonic frequency. The crystal thickness required for this depends upon the intensity of the light and the crystal parameters.
In terms of photons, frequency doubling occurs when two photons of frequency, ω, or energy, $\hbar \cdot \omega$, are converted into a photon of the double frequency, 2ω or energy, $\hbar \cdot 2\omega$:

$$\hbar\omega + \hbar\omega = \hbar 2\omega. \ (\hbar = h/2\pi)$$

This obeys the law of conservation of energy. From the equality of the refractive indexes and the resulting equal velocities of the light beams, $c_1(\omega) = c_2(2\omega)$ it follows for the wavelengths of the fundamental and harmonic waves that $\lambda_1 = 2\lambda_2$ or synonymously, $1/\lambda_2 = 2/\lambda_1 = 1/\lambda_1 + 1/\lambda_1$. If we introduce the wave vector, $|\vec{k}| = 2\pi/\lambda$ or the → photon momentum, $p = h/\lambda$, the index matching equation is:

$$\frac{h}{\lambda_2} = \frac{h}{\lambda_1} + \frac{h}{\lambda_1} \to P_2 = P_1 + P_1$$

or

$$\frac{h}{2\pi} \cdot k_2 = \frac{h}{2\pi} k_1 + \frac{h}{2\pi} k_1 \to k_2 = k_1 + k_1.$$

Thus, this formulation corresponds to the law of conservation of momentum in physics.

Ref.: J. Wilson, J.F.B. Hawkes, *Optoelectronics: An Introduction*, Prentice Hall International, Englewood Cliffs (NJ), London, Sydney, Toronto, **1983**.

Frequency-sweep method, <*Frequenz-Sweep-Methode*> → field-sweep method.

Fresnel rhomb or parallelepiped, <*Fresnelsches Parallelepiped*>, a device made of crown glass, developed by Fresnel in 1823, for the conversion of linearly to circularly polarized light. The cross section of a Fresnel rhomb is a parallelogram (see Figure 1). The acute angle at *A* and *C* is 54°37'. Linearly polarized light incident normally upon the front face, *AD*, is reflected twice at angle, 54°37' at the lateral faces, *AB* and *CD* and leaves the crown glass parallelepiped again, normally, through the opposite face, *BC*. The emerging light is circularly polarized if the vibrational plane of the incident, linearly polarized light

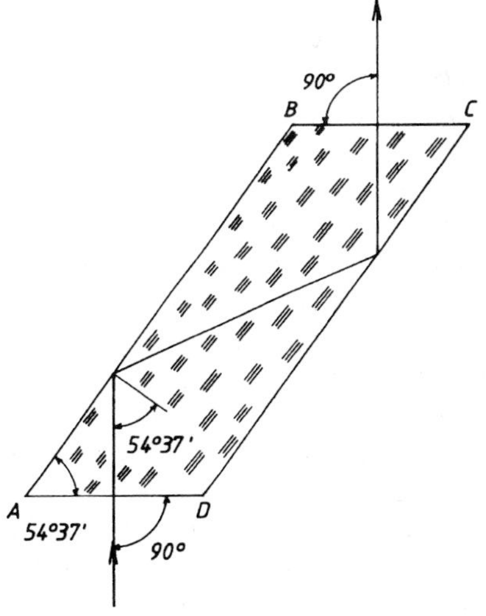

Fresnel rhomb. Fig. 1. Cross section of a Fresnel rhomb

forms an angle of 45° with the plane of incidence upon the rhomb. A phase shift, $\Delta = \pi/2$, takes place as a consequence of the two total reflections, i.e. the phase difference between the parallel (||) and perpendicularly (⊥) polarized light is $\Delta = \delta_{||} - \delta_{\perp} = 90°$ and this corresponds exactly to circularly polarized light. According to Fresnel's formula, the phase shift is given by:

$$\tan\left(\frac{\Delta}{2}\right) = \frac{\sqrt{(1 - \sin^2\alpha)(\sin^2\alpha - 1/n^2)}}{\sin^2\alpha};$$

where α is the angle at which the total reflection occurs (see Figure 1) and n is the refractive index of the material from which the Fresnel rhomb is constructed (the optically denser material). Since the above formula contains the refractive index, the phase shift depends upon the wavelength. For quartz with the angle, $\alpha = 52°$, Δ decreases from 94.7° to 84.1° in the region of 214 to 600 nm. The value $\pi/2$ is obtained at $\lambda = 275$ nm. The use of the Fresnel rhomb is somewhat impractical because the paths of the two beams are shifted parallel to each other. It is important that a Fresnel rhomb is manufactured from non-doubly refracting material, e.g. fused quartz. The Fresnel rhomb was formerly used in → CD instruments to generate circularly polarized light.
If two Fresnel rhombs of suitable material are bonded in optical contact the fourfold → total reflection produces a total phase difference of π = 180° between light beams polarized parallel and perpendicular to the plane of incidence. Thus, the device rotates the plane of linearly polarized light through 90°.

FT-IR spectrometer, <*FT-IR-Spektrometer*> → Fourier-transform infrared spectrometer, (FT-IR).

FT-NMR spectroscopy, <*FT-NMR-Spektroskopie*> → Fourier-transform NMR spectroscopy, (FT-NMR).

FT-photoluminescence spectrometer, <*FT-Photolumineszenz-Spektrometer*> → Fourier-transform photoluminescence spectrometer (FT-photoluminescence).

FT-Raman spectrometer, <*FT-Raman-Spektrometer*> → Fourier-transform Raman spectrometer, (FT-Raman).

FT-UV-VIS spectrometer, <*FT-UV-VIS-Spektrometer*> → Fourier-transform UV-VIS spectrometer, (FT-UV-VIS).

Fundamental series, <*Fundamentalserie, Bergmann-Serie*>, the line series in atomic spectra characterized by the running term *nF* which, in the spectrum of the hydrogen atom, corresponds to the Bergmann series. The name *fundamental series* instead of *Bergmann series*, is common, particularly in the English and American literature, and is the reason for the symbol, *nF.* The name originates in the similarity of the Bergmann series to the spectrum of hydrogen, although this similarity is not a decisive feature of the former. The names, *nD* and *nS*, also refer to features such as diffuse (*D*) and sharp (*S*) which do not always apply.
Ref.: A. Sommerfeld, *Atomic Structure and Spectral Lines*, translated by H.L. Brose, E.P. Dutton, New York, **1934**.

Fundamental vibration, fundamental band, <*Grundschwingung*>, the vibration of a molecule which arises from the excitation of a vibration from the vibrational ground state, $v'' = 0$, to the first excited state, $v' = 1$. It is the most intense band (→ overtone) and leads to a pronounced absorption maximum in the → IR spectrum, if the vibration is → IR active. All the → overtones can be constructed as approximately whole-number multiples of the fundamental vibrations.

G

Galvanometer grating drive, <*Galvanometerantrieb*>, an electromagnetic grating drive in which the grating is rotated by means of a controllable magnetic field. The drive is not subject to wear and offers the best line position reproducibility on account of the very small of steps. It is uniquely fast (1000 nm/s).

Gamma radiation, <*Gammastrahlung, γ-Strahlung*>, a very short-wavelength and hence extremely penetrating (hard), electromagnetic radiation. The shortest wavelength so far measured is found in the gamma radiation of thorium, *C″*. It is $4.66 \cdot 10^{-11}$ cm, which corresponds to a frequency, ν, of $6.43 \cdot 10^{20}$ s^{-1} and a photon energy, $h\nu$, of 2.66 MeV. The wavelengths are generally considerably longer and the energies correspondingly smaller, i.e. they are then X-rays. In some cases, gamma radiation occurs in radioactive decay.

Gas-discharge lamp, <*Gasentladungslampe, Entladungslampe*>. In a gas discharge, free electrons move through the gas between two electrodes. If enough free electrons are initially available, e.g. by emission from the cathode, and if they are sufficiently accelerated in the electric field between the electrodes, inelastic collisions occur between the accelerated electrons and the neutral gas atoms. Depending upon the energy transferred in the collisions, this leads to excitation or ionization of the gas atoms. The return transition from the excited state to the ground state generally occurs spontaneously and the energy difference is emitted as radiation. In order to maintain the gas discharge, as many ionizing collisions as possible, in which the electrons released are accelerated again in the electric field and can ionize further atoms, must occur. In order to obtain a high radiation flux from the gas discharge, the greatest possible number of atoms must be transferred to the excited state by the collisions. We can conclude from this that primarily only discrete spectral lines of a characteristic emission spectrum are to be expected. For that reason, these lamps are also called spectral lamps. → Metal-vapor discharge lamps are of this type. The most important gas- or vapor-discharge lamps in the UV-VIS spectral region are → xenon lamps, → hydrogen lamps and → mercury-vapor lamps.

Gas-dynamic laser, <*gasdynamischer Laser*>, a molecular laser in which the inversion state (→ laser) of the active medium is established by the supply of thermal energy. At a temperature of ~ 3000 K, the thermal energy, $R \cdot T$ is approximately 12,500 J mol^{-1} which corresponds to an excitation energy of ca. 2000 cm^{-1}. We see immediately that the achievable excited states of the working gas are of the order of magnitude of vibrational excited states. However, we must take account of the fact that, according to → Boltzmann statistics, the supply of thermal energy alone is not sufficient to reach an → inversion state. There is a second effect due to the varying lifetime of the thermally occupied vibrational levels which achieves this end better. Figure 1 illustrates these relationships

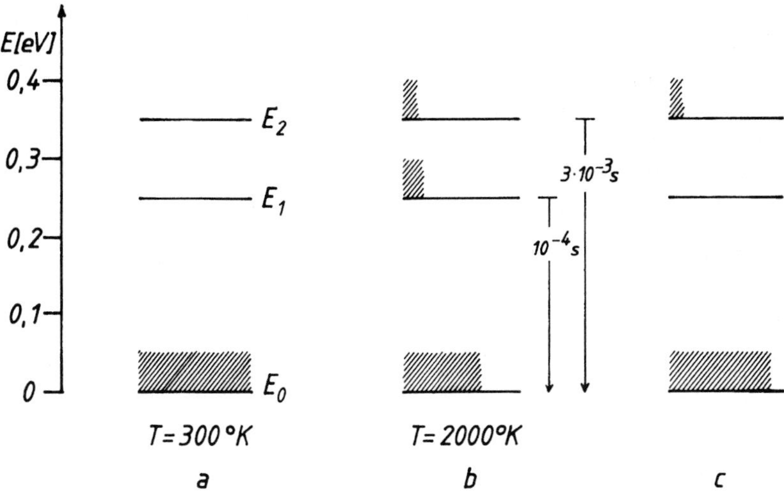

Gas-dynamic laser. Fig. 1. Energy-level scheme for a gas-dynamic laser, CO_2 as example

for CO_2 using a simplified energy-level scheme. The level E_2 corresponds to the excited vibrational state, $v' = 1$, of the asymmetric stretching vibration, ν_3. E_1 corresponds to the excited symmetric stretching vibration, ν_1, and the doubly excited state of the deformation vibration, ν_2, which is in → Fermi resonance with it. E_1 and E_2 are the two levels between which laser emission occurs. At room temperature, 300 K, all molecules are in the ground state, E_o, (Figure 1a). If the gas is heated to a temperature of 2000–3000 K in a furnace, the levels, E_1 and E_2, are significantly occupied (20–40% according to Boltzmann statistics). If the gas is then expanded through a jet it cools down to room temperature and, at thermal equilibrium, all molecules are in the ground state, E_o. If the expansion occurs very rapidly, a certain time elapses before the equilibrium state is established. In the case of the CO_2 laser, the decay times of the two laser levels are very different. Whilst the lower level, E_1, is emptied in 10^{-4} s the upper level requires $3 \cdot 10^{-3}$ s, i.e. the lower level is effectively empty after a few tenthousandths of a second whilst the upper level, E_2, is still occupied (Figure 1c). Thus, a state of inversion is present and light of the frequency $\nu = (E_2 - E_1)/h$ can be amplified. Since the gas-dynamic CO_2 laser has the same energy-level scheme as the → CO_2 laser, we are dealing with the same laser emission lines at 10.6 μm and 9.6 μm. We see two lines because the lower laser level consists of two neighboring levels, see above. Figure 2 shows a gas-dynamic

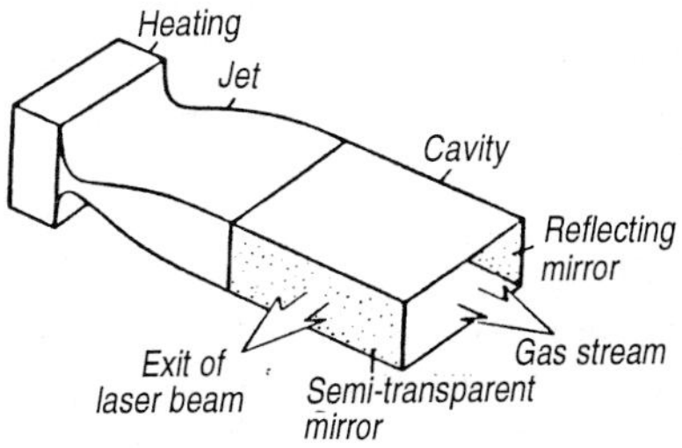

Gas-dynamic laser. Fig. 2. The basic construction of a gas-dynamic laser

laser schematically. Further details are given on the figure.

Ref.: J.D. Anderson, *Gas-Dynamic Lasers: An Introduction*, Acad. Press, New York, **1976**; S.A. Losev, *Gas-Dynamic Lasers*, Springer Series in Chemical Physics, vol. 12, Springer Verlag, Berlin, Heidelberg, New York, **1981**.

Gas-filter correlation, <*Gasfilterkorrelation*>, a method of measurement frequently used in → on-line photometers. The probe light, which is usually limited to a narrow spectral region by an → interference filter, passes through a gas cell which contains a very high concentration of the gas under investigation so that all wavelengths capable of being absorbed are completely removed. Only the nonabsorbed light then passes through the sample cell and generates a reference signal. If the probe light now passes directly through the sample cell, a fraction is absorbed by the substance (see above), but the reference beam reaches the receiver without loss as before. In this technique, the absorption usually depends linearly upon the pressure (concentration) of the gas to be analyzed. Alternatively, a correlation between the instrumental signal and the pressure or concentration can be established with a calibration. This substance-specific method has been implemented in instruments such as the → RADAS IG, → UVAS 247, → URAS, → Spectran etc.

Gas-flow counter, <*Gasdurchflußzähler*>, proportional meters used as detectors in wavelength-dispersive → X-ray spectrometers. They generate voltage pulses with amplitudes which are proportional to the energies of the incident X-ray quanta and thus provide a spectrally resolved measurement of the radiation. A gas-flow counter consists of a metal tube of diameter ca. 3 cm with a ca. 100 μm diameter wire along its longitudinal axis. The radiation passes through a side window and ionizes the gas in the tube. If a high voltage (1.5 kV) is applied between the wire and the metal tube, an electron avalanche is generated which passes to the wire anode and produces a negative voltage pulse. Its amplitude is a measure of the energy of the quantum. Argon and neon are frequently used as gases for counter tubes ($\varepsilon \approx 27$ eV effectively); 10% methane is added as a quencher. Thin, ca. 1 μm thick supported foils are used as windows.

Gas laser, <*Gaslaser*>, a type of laser, in which the → active medium is present in the gas or vapor phase. The number of elements in which laser activity has been observed in this state is extremely large. Figure 1 lists some examples.

A gas discharge is used to produce the → population inversion in the gaseous or vapor-like laser medium. Neutral atoms are ionized, i.e. positive ions and electrons are formed. The electrons are accelerated by the electric field prevailing in the discharge and collide, after a short acceleration time, with atoms or ions to which they transfer their kinetic energy in the form of excitation energy. The latter can be used directly, or indirectly through collisions, for the population of the upper laser levels (see the → Franck-Hertz experiment for the principles of excitation by an electron collision).

In comparison with the → ruby or → neodymium lasers, which are typical

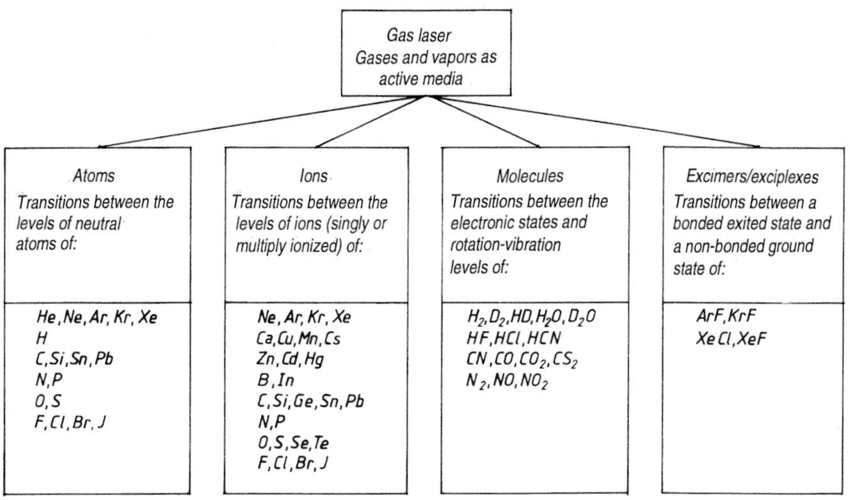

Gas laser. Fig. 1. Examples of gas lasers

→ solid-state lasers, the pump energy of most gas lasers is very much greater than the photon energy which is obtained in the laser emission. Consequently, the energy losses are also quite considerable. Figure 2 shows schematically a summary of these losses for four different gas lasers. Despite these energy losses, some gas lasers have proved to be outstanding for applications in spectroscopy and other fields. The most important ones are the → helium-neon laser, the → argon-ion laser, the krypton-ion laser

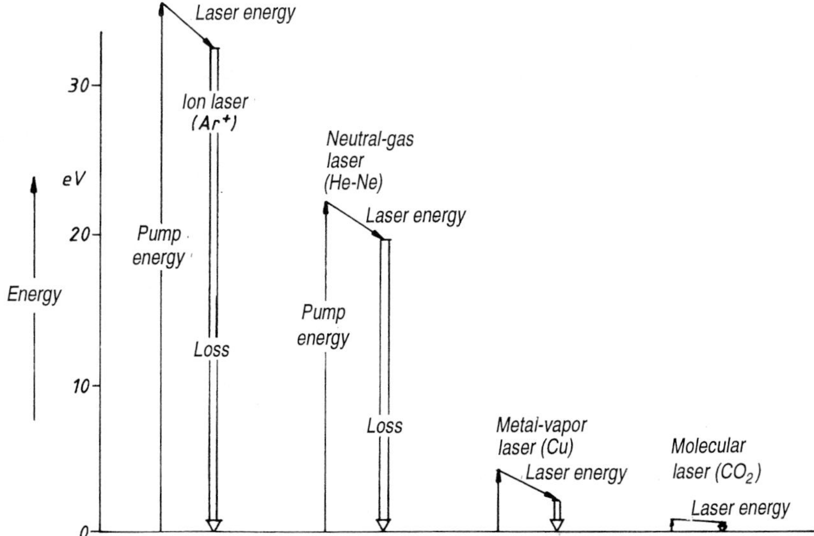

Gas laser. Fig. 2. Energy losses in four different gas lasers

(→ noble-gas ion laser), the → CO_2 laser, the → nitrogen laser and the → excimer laser.
Ref.: D. Eastham, *Atomic Physics of Lasers*, Taylor and Francis, London and Philadelphia, **1989**.

Gas pumping, *<Gaspumpen>* → noble-gas ion laser.

Gaussian band shape, *<Gauß-Profil>* → line shape, → band shape.

g factor, Landé g factor, *<Landé-Faktor, g>*, the factor which takes account of the varying magnitudes of the magnetic moments due to pure orbital angular momentum ($S = 0$, $L \neq 0$) or pure spin angular momentum ($L = 0$, $S \neq 0$). The magnetic moment, $\overrightarrow{\mu_D}$, of an atom, which is due to the external electrons, is given in general by:

$$\overrightarrow{\mu_D} = -g\mu_B\sqrt{D(D+1)}$$

where μ_B is the → Bohr magneton and D is a general symbol for the angular momentum quantum numbers, S, L and J or s, l and j. g is the Landé factor. For the two limiting cases we have:

$g = 1$ for $S = 0$ but $L = 1, 2, 3$ ... and
$g = 2$ for $L = 0$ but $S = 1/2, 2/2, 3/2$ ...

The energy of the atom in a magnetic field, $\overrightarrow{H}$, is:

$$E_{\text{mag}} = E_o - \mu_o\overrightarrow{H}\cdot\overrightarrow{\mu_D}$$

where E_o is the energy in the field-free case, μ_o is the vacuum permeability and $\overrightarrow{H}\cdot\overrightarrow{\mu_D}$ is the scalar product of $\overrightarrow{H}$ and $\overrightarrow{\mu_D}$. Thus, if ($M_D = M_S, M_L, M_J$) is the component of the angular momentum vector in the direction of the magnetic field, then:

$$E_{\text{mag}} = E_o + g\mu_B\mu_o HM_D$$

In the case of → Russell-Saunders coupling ($J = L + S$), g is a rational number which depends in a simple way upon L, J and S. Landé derived the formula in 1921:

$$g = 1 + \frac{J(J+1) + S(S+1) - L(L+1)}{2J(J+1)}$$

We see immediately that both limiting cases can be derived directly from this formula (see G. Herzberg, *Atomic Spectra and Atomic Structure*, Dover Publications, New York, **1945**, p. 106 ff, for the derivation of Landé's formula). A complete explanation of the anomalous → Zeeman effect is possible by calculating the g value required for each term.
As stated above, $g = 2$ for an individual, completely free electron. The exact value is $g = 2.002319$. This value is of importance in → magnetic resonance spectroscopy and in the relativistic theory of the atom.

Glan prism, *<Glan-Prisma>*, a → polarizer for the UV region developed by P. Glan in the last century. The Glan prism is produced from UV transparent → calcite which has almost no streaks or bubbles. In contrast to the → Nicol prism and the → Glan-Thompson prism, the two halves of the prism are not cemented together with Canada balsam but are fixed in a mounting with an air gap of 0.07 mm. This eliminates the interfering UV absorption in the cement layer. Figure 1 shows a cross section of a Glan prism. On entry into the prism, the light is split into an ordinary (o) and an extraordinary (e) ray which propagate with different velocities due

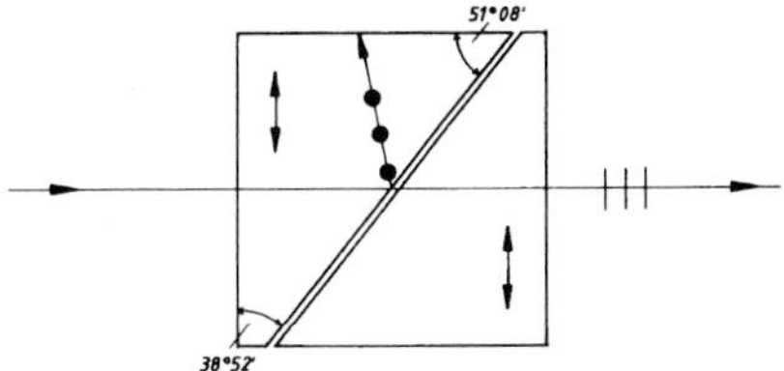

Glan prism. Fig. 1. Cross section of a Glan prism

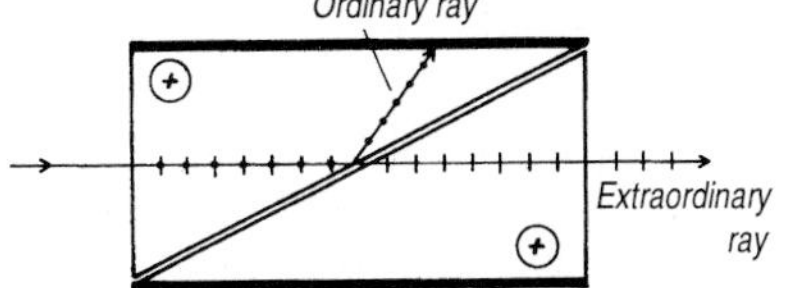

Glan-Thompson prism. Fig. 1. Cross section of a Glan-Thompson prism

to the different refractive indexes. For that reason also, there are two critical angles for → total reflection α_o = 37°05', α_e = 42°17'. Within these critical angles, the transmitted light is polarized. The ordinary ray is totally reflected at the air gap whilst the extraordinary ray exits as linearly polarized light in the direction of the incident light. Outside the range of these angles, both rays are either totally reflected by the prism or are allowed through completely unpolarized. Thus, the faces at the junction of the two components must lie at quite specific angles of 38°52′ and 51°8′ to the front and exit faces as shown in Figure 1. The ratio of the height to the length is 1:1.25. These dimensions provide a symmetrical field of vision of ±3° in the region of 215 to 700 nm which cannot be exceeded.

Glan-Thompson prism, *<Glan-Thompson-Prisma>*, a prism made from → calcite which is used today mainly as a → polarizer or → analyzer. Compared to the → Nicol prism, its main advantages are that the end faces are perpendicular to the longitudinal direction and a relatively large aperture. Figure 1 shows that unpolarized incident light is split into an ordinary (*o*) and an extraordinary (*e*) ray due to the birefringence. The ordinary ray is completely reflected by the cement layer and absorbed by the black lacquered layer on the lateral surfaces. There are two versions of the Glan-Thompson prism:

a) Normal version – 73° angle between the entrance and internal surface, an evenly polarized field of vision 28° in air, cross section to length ratio 1:3;
b) shortened version – 67°30' angle between the entrance and internal surface, an evenly polarized field of vision of 13° in air, cross section to length ratio 1:2.

Glass, optical, *<Glas, optisches Glas>*. Although glass is one of the oldest materials, most types of glass suitable for optical purposes were only developed from 1880 onwards. This development was based on a fruitful cooperation between O. Schott (1851–1935) and E. Abbe (1840–1905). Prior to 1880, optical systems were constructed from simple crown glasses (sodium-calcium-silicate glasses) with a relatively low dispersion and simple flint glasses (lead-alkali-silicate glasses) with a relatively high dispersion. The mean dispersion is generally used as the measure of the → dispersion. It is defined as the difference between the refractive indexes, $n_F - n_C$. The letters, *F* and *C*, refer to the measurement of the refractive index at the blue (486.1 nm) and the red (656.3

nm) hydrogen (→ Fraunhofer) lines. The relationship, $(n_F - n_C)/(n_d - 1)$ is the relative dispersion, with n_d, the refractive index measured at the yellow helium line, $\lambda = 546.1$ nm. The reciprocal of this expression $v_d = (n_d - 1)/(n_F - n_C)$ is called the Abbe number, v_d. A low Abbe number, v_d, denotes a glass with high dispersion relative to its refractive index, n_d. Such glasses are called flint glasses. In contrast, crown glasses have relatively large Abbe numbers relative to their refractive indexes, n_d. If we plot the refractive index, n_d, for the different optical glasses on the ordinate against the associated Abbe number, v_d, on the abscissa, we obtain a diagram such that as shown in Figure 1 which is used to characterize optical glasses. Every point on this diagram is assigned to a specific glass. The different glasses have been traditionally denoted by the letters *F* for flint and *K* for crown (German *Kron*). The prefixed letters, *L* and *S* are given for light and heavy (German *schwer*) and characterize a glass with a low (*L*) or high (*S*) refractive index, e.g. *LF* = light flint; *SK* = heavy crown; *SF* = heavy flint. In addition, the notation *B* = boron, *Ba* = barium, *La* = lanthanum, *FK* = crown fluoride and *P* = phosphate, has been used to indicate the different chemical compositions of the melts. Furtheremore, *LL* means double light and *SS* double heavy; see the corresponding regions for *SSK* and *LLF* as well as *Ti* for deep (German *tief*) in the diagram.

The Abbe numbers, v_d, of optical glasses lie between 20 and 90; the border between crown and flint glasses lies at $v_d = 50$ by definition. According to the DIN standard 58295 T.2, the refractive index, n_e, for the wavelength, λ_e =

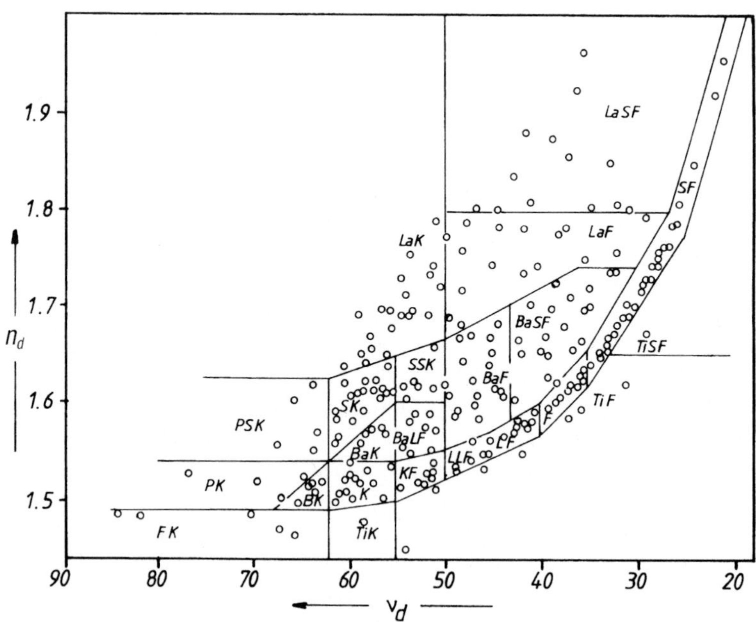

Glass, optical. Fig. 1. Diagram of n_d against (v_d) to characterize optical glasses

546.1 nm (see above) should be given and the Abbe number, $\nu_e = (n_e - 1)/(n_{F'} - n_{C'})$ determined with the lines F': $\lambda = 480$ nm and C': $\lambda = 643.8$ nm. However, the above classification with n_d and ν_d is still the most commonly used. A six-digit code number is formed from these values for each glass type. The Schott crown glass BK7, one of the most frequently used commercial glasses, is denotated by BK7–517642. The number 517 gives the refractive index, $n_d = 1.517$, i.e. $(n_d - 1) \cdot 10^3 = 517$, and 642 gives $\nu_d = 64.2$ in the form $10\nu_d = 642$. The digit 7 is the type number of the BK glass.
Ref.: G.W. Morey, *The Properties of Glass*, 2nd ed., Reinholt, New York, **1954**; W.L. Wolfe, *Properties of Optical Materials*, sec. 7, *Handbook of Optics*, Optical Society of America, **1978**.

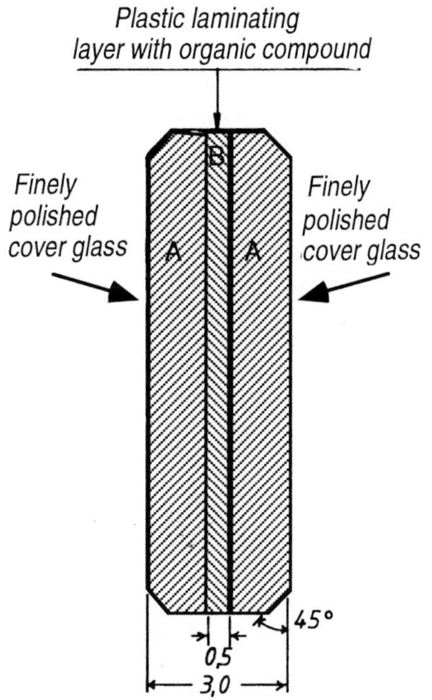

Glass-plastic laminated filter. Fig. 1. Construction of a glass-plastic laminated filter

Glass-plastic laminated filter, <*Kunststoffverbundfilter, KV-Filter Glas-Kunststoff-Verbundfilter*>, a Schott filter which, among other applications, is used as an UV blocking filter. On account of its steep edges it is an almost ideal → edge filter. The required properties reside in the plastic layer, i.e. it contains the absorbing organic compounds dissolved as a molecular dispersion. Figure 1 illustrates the construction of a plastic laminated filter. Two 1.25 mm thick glass plates are bound together by a 0.5 mm plastic layer, *B*, so that the total filter is 3 mm thick. The plastic layer, like the dissolved substances, must be chemically inert and it must also adhere extremely strongly to the glass plates which are finely polished. The polish meets the high demands made upon photographic filters and these filters can therefore be used not only for scientific but also for photographic purposes. The fluorescence of plastic laminated filters is extremely weak and they are therefore used as blocking filters in → fluorimetry.
Figure 2 shows the transmission on a log(log) scale as a function of the wavelength (→ diabatic scale).
Ref.: H. Dislich, K.H. Wiesner, *Neue Glas-Kunststoff-Verbundfilter,* Optik, **1970**, *30*, 340.

Global radiation, <*Globalstrahlung*>, the sum of spectrally undispersed direct solar radiation and scattered solar radiation (sky radiation) which reaches a horizontal surface on the earth. Since specific parts of the extraterrestrial solar radiation are scattered by molecules and aerosols, reflected

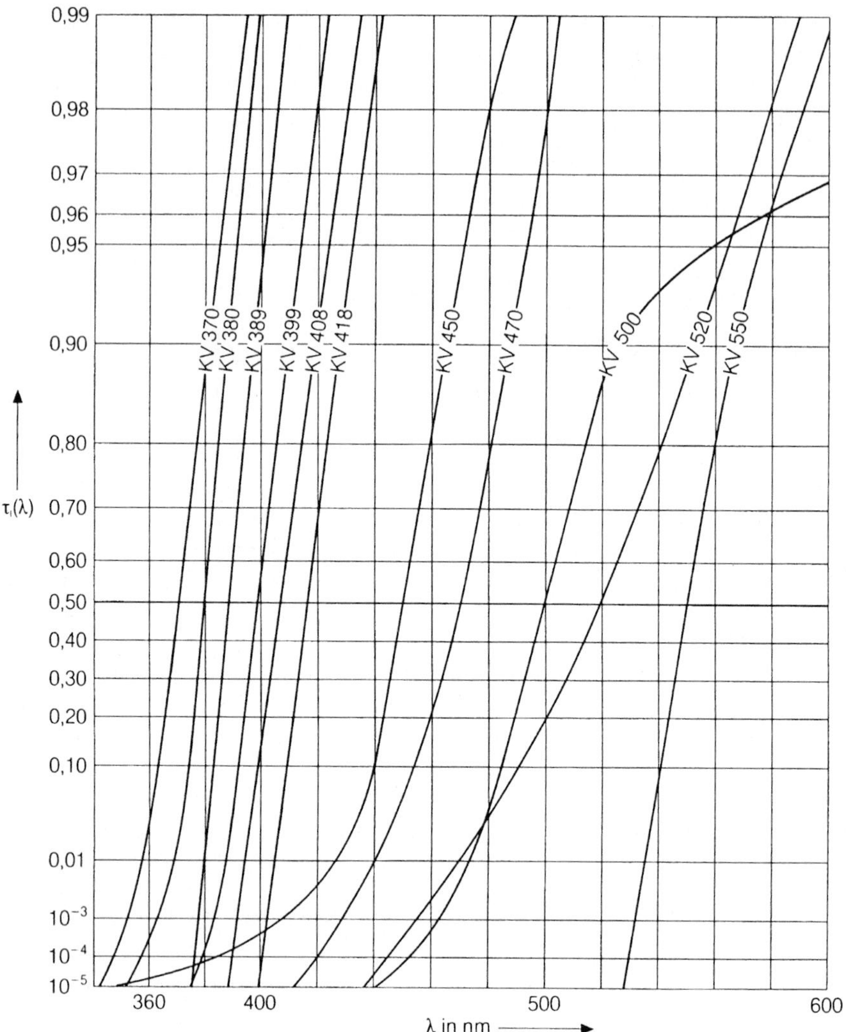

Glass-plastic laminated filter. Fig. 2. Transmission curves for glass-plastic laminated filters

by the clouds and absorbed by water vapor, ozone, oxygen and aerosols, the global radiation depends to a large extent upon the consistency of the earth's atmosphere, the angle of incidence of the sun and the position of the observer (high mountain, flat land, large city, industrial site). The intensity of the extraterrestrial solar radiation has been determined by various measurements and found to be 136 mW cm^{-2}. This is the radiation intensity of the sun outside the earth's atmosphere onto a surface perpendicular to the irradiation. It is known as the extraterrestrial radiation intensity or solar constant, I_o.

Ref: G. Abetti, *The Sun*, 2nd English ed., Translator J.B. Sidgwick, Faber and Faber, London, **1957**.

Globar, <*Globar*>, a → thermal radiator, which is frequently used in IR spectroscopy, in addition to the → Nernst glower. The globar consists of a solid or hollow cylinder of silicon carbide. The part, which actually glows, is 6 to 8 mm thick and several centimeters long whilst the ends are generally kept cold by an enlarged cross section, i.e. a reduced resistance. In contrast to the → Nernst glower, the globar can be lit by simply applying a voltage since silicon carbide also conducts electricity in the cold state. A low voltage and a very high current are required for the operation, thus, a low-voltage transformer or DC source with a high power capacity is necessary. The maximum utilizable temperature is 1400 K and the emission reaches ca. 75% of that of a → black body at the same temperature. The life and mechanical stability of the globar are superior to those of the Nernst glower.

Glow curve, <*Glow-Kurve*>, a curve, which represents the dependence of the intensity of → phosphorescence upon the temperature of an excited, phosphorescing solid. Analogously, an increase of the dark conductivity with increasing temperature can be observed in inorganic and organic solids which have been excited at a low temperature. Characteristic glow peaks frequently occur at specific temperatures.

Golay detector, <*Golay-Detektor*>, a detector of the → pneumatic receiver type used together with the → thermocouple and → bolometer in IR spectroscopy. In the Golay detector, the heating caused by the absorption of radiation is measured by the expansion of a gas enclosed in a small cell. The expansion is detected optically. Figure 1 shows the construction of a Golay detector. A metal foil, which is usually blackened, is the radiation absorber and also forms a rigid wall of the cell which encloses the gas. The opposite side of this cell is a metallized, flexible wall. The heat developed in the radiation absorber is transferred to the gas contained in the cell and causes a pressure increase. If the radiation to be measured is modulated, a periodically varying pressure results and this is followed by the rear wall of the cell which is formed as a flexible mirror. The mirror is part of an optical imaging system which also

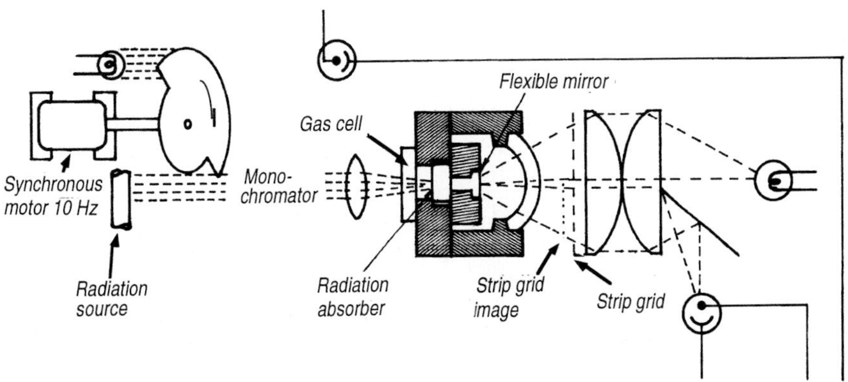

Golay detector. Fig. 1. Cross section of a Golay detector

includes an incadescent lamp, a light condenser and a strip grid. The grid is positioned in the beam path of this optical system in such a way that its real image normally covers it, bar on gap and gap on bar, so that the light of the lamp above the angled plane mirror does not reach the laterally mounted photocell. If the flexible mirror is deflected by the pressure increase in the gas cell due to radiation absorption, the image of the grid changes its position relative to the grid itself in such a manner that the gaps in both image and grid more or less coincide. This permits the light to reach the photocell and the change of intensity provides a measure of the radiation absorbed. The varying photocurrent is amplified and directed to a display or recording instrument. The reaction time of the Golay detector can be set electronically between 1 and 30 ms, the modulation frequency lies at 10 Hz and the detection threshold is quoted as better than $6 \cdot 10^{-11}$ W.
Ref.: G.K.T. Conn, D.G. Avery, *Infrared Methods*, Academic Press, New York, London, **1960**, pp. 84–87

Graded filter, linear variable interference filter, <*Verlauffilter*>, an → interference filter the spectral effect of which, in general, depends linearly upon a coordinate. The path length of linear graded filters rises wedge-like in the longitudinal direction. Thus, the wavelength of maximum transmission becomes a linear function of the longitudinal coordinate of the filter. The filter can be used, for example, for the construction of a compact, light-intense monochromator. Such filters are usually manufactured for the visible spectral region (400–700 nm) and they frequently have three mirrors and two distance pieces and are known as S3 systems (a type of the MDI-II-S3 filter). They have typical half-width values (→ band-pass filter) of ca. 12 nm with a maximum transmission, $\tau_{max} \approx 30\%$. Schott supplies graded filters S 60, S 200 and X 160 with the name VERIL for the spectral region of 400 to 700 nm (see Table). Schott also produces graded filters, VERIL B60 and VERIL BL200, with half-width values of ca. 23 nm for the spectral regions 400 to 700 nm and 400 to 1000 nm respectively.

Graphite furnace technique, <*Graphitrohrtechnik*>, a technique in → atomic absorption spectroscopy in which the sample atomization is made in a graphite tube furnace, ca. 30 mm long with an internal diameter of ca. 6 mm. The furnace is mounted in the beam of the → atomic absorption spectrometer. It is held by graphite contacts in a water-cooled housing and is flushed internally and externally with an inert gas such as argon or nitrogen to prevent combustion of the graphite (see Figure 1).
After introduction of the sample through the inlet orifice in the graphite tube, the tube is heated electrically in programmed steps with a variable voltage up to ca. 8 V and a current of ca. 400 A. The various programmable temperature steps are matched to the sequential processes of → atomization. The free atoms finally formed remain in the graphite tube, i.e. in the measuring light beam, for up to 1 s. This is between 100 and 1000 times longer than in the → flame technique where the atoms are transported through the probe beam in a few ms by the gas flow in the flame. Thus, many more atoms are available for light

Graded filter. Table. Schott linear variable interference filters

Kind of filter	Linear variable line filter	Linear variable line filter	Linear variable line filter
Spectral range in nm	400–700	400–700	400–700
Corresponding kind of homogeneous filter	Line filter	Line filter	Line filter
Construction	MDI-II-S3	MDI-II-S3	MDI-II-S3
Filter type	VERIL S 60	VERIL S 200	VERIL X 160
Filter length in mm	60 + 0 – 0.3	200 + 0 – 0.3	160 + 0 – 0.3
Filter width in mm	25 + 0 – 0.3	25 + 0 – 0.3	25 + 0 – 0.3
Filter thickness in mm	max. 5	max. 6	max. 7
Spectrum length in mm	37–51	111–142	111–142
Dispersion in mm/nm	0.12–0.17	0.37–0.47	0.37–0.47
HW (450 nm) in nm	10–16	10–16	10–16
HW (550 nm) in nm	10–14	10–14	10–14
HW (650 nm) in nm	10–18	10–18	10–18
τ_{max} (450 nm)	≥0.25	≥0.25	≥0.25
τ_{max} (550 nm)	≥0.30	≥0.30	≥0.30
τ_{max} (650 nm)	≥0.25	≥0.25	≥0.25
$\frac{\text{Tenth-width}}{\text{Half-width}}$	ca. 1.8	ca. 1.8	ca 1.8
$\frac{\text{Thousandth-width}}{\text{Half-width}}$	ca. 7.0	ca. 7.0	ca. 7.0
Blocking range in mm	up to 750	up to 750	unlimited
Maximum degree of transmission in blocking range	10^{-3}	10^{-3}	10^{-3}
Number of mechanical touching points	0	0	0

absorption in the graphite furnace technique which makes it possible to use very small quantities of sample and to detect the minutest absolute quantities of trace elements. The total sample introduced into the furnace, independent of volume or quantity, is atomized in a time characteristic for the element and the selected temperature. Thus, the magnitude of the signal is proportional to the mass, and not the concentration, of the element to be determined. Therefore, in flameless atomic absorption spectroscopy there is an interchangeability between volume and concentration, and by using larger volumes correspondingly higher signals can be obtained and the lowest concentrations measured. The greatest advantage of the graphite furnace technique is its unchallenged sensitivity of detection for the majority of

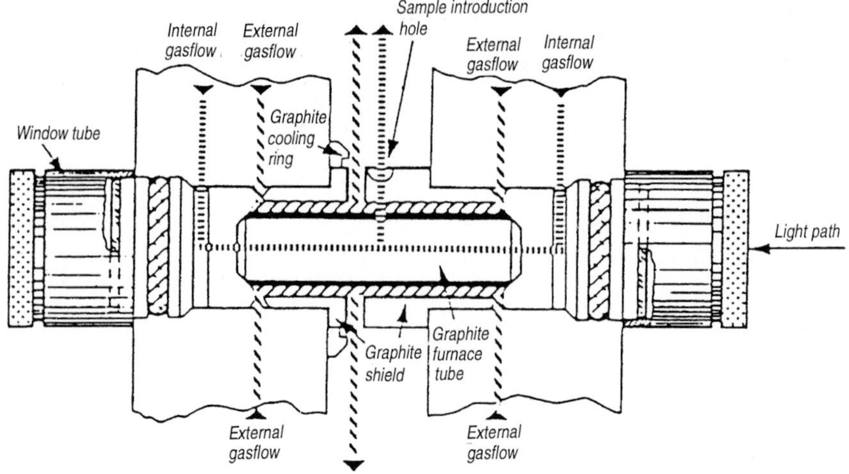

Graphite furnace technique. Fig. 1. Cross section through a graphite furnace

the elements which can be determined using atomic absorption spectroscopy, combined with the fact that only very small quantities of sample are required. The graphite furnace technique is especially suited to micro- and trace analysis.

Grating, <*Gitter*>, a planar arrangement of *p* parallel slits of width, *b* (→ diffraction). The grating and the → prism are extremely important → dispersing elements. The basis of → dispersion by a grating is the Huygens wave theory of light according to which light can be diffracted through an angle α by a row of many very fine slits, i.e. a grating (Fraunhofer 1821). The light rays diffracted in the same direction are coherent, i.e. they oscillate synchronously. This coherence is required if → interference of the waves is to take place. The result of the interference is that, where wave crest meets wave trough the light is extinguished, but where crest meets crest it is intensified. Therefore, if two neighboring waves differ in their path lengths by an odd integral multiple of $\lambda/2$ they are extinguished, but if they differ by an integral multiple of λ they are intensified. The diffraction angle, α_K, is given by the simple equation

$$\sin\alpha_K = \frac{K \cdot \lambda}{a};$$

where $K = 1, 2, 3 \ldots$ is the order of the diffraction, λ the wavelength and a the grating constant which is the distance between the slits measured from center to center. Therefore, the angle of diffraction, α, depends upon the wavelength, λ, i.e. red light is diffracted more strongly than blue. This is the exact opposite of the dispersion by a prism. If white light falls upon the grating, the diffraction produces the spectrum of the spectral colors (→ normal spectrum). The spectra in higher orders (larger K) show a greater dispersion which is directly proportional to the order number K. However, since spectra of different orders overlap, when using gratings special measures must be taken to remove unwanted orders. Fraunhofer introduced transmission gratings in which the spectrally dispersed light was produced following passage through the grating. For spectroscopic measurements these have been replaced by reflection gratings; mirrors upon which fine grooves separated by the grating constant, a, have been ruled. There can be thousands of these grooves per mm. The diffraction of the light, in all directions, takes place on the remaining reflecting lines between the grooves, as indicated for the incident ray 5 in Figure 1. Diffraction at the grating produces a parallel beam of light in a particular direction (characterized by α_r) for a particular order, K, as shown in Figure 1. In this case also, the angle of diffraction is given by the equation above, i.e. there is a different diffraction angle, α, for every wavelength, λ. If the parallel beam is focused then the images of the various wavelengths appear side by side and form a spectrum. Both transmission and reflection gratings diffract the incident light in two directions, to the left and to the right of the normal to the grating, for every order. In the first order ($K = 1$), these two directions are known as the +1st and –1st order. Since, in practical applications of gratings (→ grating monochromator), the dispersed light is observed in only one direction, one half of the intensity in the first order is lost. The intensity distributed in the higher orders is also lost, though this is usually small. The available radiation flux therefore depends largely upon how the light is distributed in the two directions of the first order. To optimize this situation blazed or echelette gratings, which have a very unsymmetrical groove profile (see Figure 1), are used. These gratings concentrate the radiation of the required wavelengths in one of the first order directions. The concentration of the radiation is optimal at a specific angle of reflection, α_B, i.e. at a specific wavelength, λ_B. This optimum angle and wavelength are called the blaze angle and blaze wavelength of the grating. The further from this wavelength the less efficient is the grating. A blaze wavelength between 210 and 250 nm is normally chosen for the UV region, and for the visible region, gratings are blazed at ca. 500 nm. In practice, a classically ruled reflection grating can be produced for any chosen blaze angle from the UV to the IR. The rule of thumb for the most effective working range is:

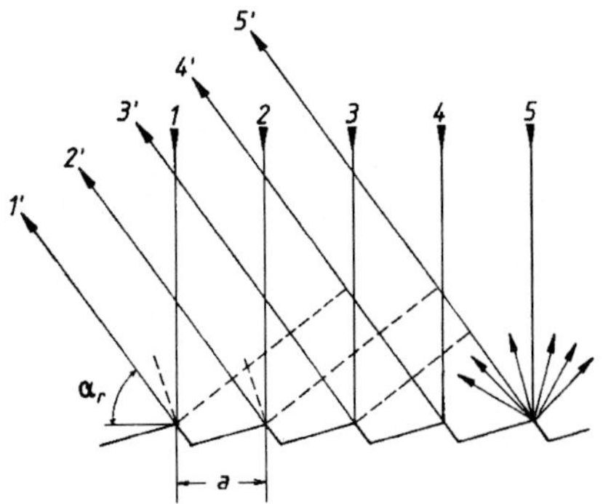

Grating. Fig. 1. A reflection grating

$$\lambda_B - \lambda_B/2 \rightarrow \lambda_B + \lambda_B/2.$$

Every grating produces a certain amount of scattered light. The effects of this scattered light are known as → grating ghosts. The smaller the scattering the more efficient is the optical system. In mechanically ruled gratings, the scattering arises mainly from roughness in the grooves, which should ideally be smooth, and from periodic and nonperiodic irregularities in the distance between the grooves. Imperfections are much reduced in the holographic gratings prepared by photographic methods. Such gratings therefore show much less scattered light and are finding increasing use in precision spectrometers. In general one can say that the reflection grating has largely replaced the prism as a dispersing element.

Grating ghosts, <*Gittergeister*>, stray light found with mechanically ruled reflection gratings which arises because of unavoidable imperfections in their manufacture. These imperfections can arise, for example, from the form and depth of the grooves or from systematic errors in their spacing. They cause phase shifts in the diffracted light which can lead to losses in the → resolving power or to the appearance of false lines (ghosts) in the spectrum.

Grating monochromator, <*Gittermonochromator*>. A monochromator consists of the following characteristic components: entrance slit, collimator, dispersive element, collector and exit slit. When a → grating is used as the dispersive element, the collimator and collector are always concave mirrors. Various solutions to the technical problem of the construction of a grating monochromator are known. The most important are the simpler → Littrow monochromator and the somewhat more costly → Ebert which is often constructed as a modified → Czerny-Turner monochromator. The concave mirror collimator directs a parallel beam of the light from the entrance slit onto the grating. The parallel beam is spectrally dispersed by the grating and imaged as a spectrum in the focal plane of the monochromator by the concave mirror collector. The exit slit, which normally has the same dimensions as the entrance slit, also lies in the focal plane. The various wavelengths of the spectrum can be moved across the fixed exit slit by rotating the grating on a precision spindle either manually or by means of a computer-controlled stepper motor. The following quantities are important for the characterization of the performance of a monochromator:

1. Resolving power;
2. reciprocal linear dispersion;
3. spectral slit width (band width);
4. geometrical slit width;
5. stray light;
6. reflectance of the integral mirrors.

The → resolving power (R) of a grating monochromator is given by:

$$R = \frac{\lambda}{\Delta\lambda} = n \cdot K;$$

where n is the total number of lines on the grating and K the order.
The reciprocal linear dispersion, $1/(DL)$, (→ monochromator) is given by:

$$\frac{1}{DL} = \frac{a \cdot \cos\alpha}{f \cdot K} = \frac{\text{spectral slit width}}{\text{geometrical slit width}} \left[\frac{\text{nm}}{\text{mm}}\right].$$

Here a is the grating constant, i.e. the distance between the grooves in mm, K the order, f the focal length of the monochromator in mm and α the angle of diffraction.
The spectral → slit width, which is also known as the band width, is the sec-

tion of the spectrum in nm which, for a given geometrical slit width, fills the exit slit with light.
The stray light is a result of undirected reflection and scattering by dust particles as well as imperfections in the grating grooves (→ grating ghosts). It overlies the dispersed light at the exit slit and can therefore lead to errors in measurements. The influence of stray light can be considerably reduced by the use of holographic gratings and coated mirrors.
The reflectance of the mirrors used has a decisive effect upon the amount of light emerging from the exit slit. Thus, the aluminium-coated mirrors and gratings are protected with a SiO film in modern instruments.

Ground term, ground state <*Grundterm*>, the term which describes the ground state of an atom or molecule, i.e. that state in which the total electronic energy of the atom (molecule) is as low as possible. The → Pauli principle and → Hund's rules must be observed when determining this term. For the N atom with the electron configuration $1s^2\ 2s^2\ 2p^3$ for example:

a) The two $1s$ and two $2s$ electrons do not contribute to the ground term (closed shells, i.e. $L = 0$ and $S = 0$ are not considered).
b) The Hund's first rule gives $\sum s_i = S = 3/2$ and therefore the maximum multiplicity is $2S + 1 = 4$.
c) According to the Hund's second rule, the maximum component of the resultant orbital angular momentum is given by: $M_{L,\max} = \sum m_{\ell i}$. Since, according to the first rule, the three electrons are in the three degenerate p orbitals with $m_\ell = +1, 0, -1$, $\sum m_{\ell i} = 0$. If $M_{L,\max}$ is the maximum component of the resultant orbital angular momentum then $L = 0$ because $M_L = 0$, i.e. we have an S term.
d) The coupling of $L + S$ gives $J = 3/2$ so that the term symbol for the ground state of the N atom is $^4S_{3/2}$.

Further, Hund's third rule also applies when determining the ground state of atoms. This rule states that the state with the lowest J value is the ground state until all degenerate orbitals are half full after which the state with the highest J value becomes the ground state and remains so during the further filling.
The electronically and vibronically nonexcited state is always regarded as the ground state in molecules. With regard to rotational states, it must be noted that some molecules may be in excited rotational states at room temperature according to → Boltzmann statistics. This applies also for very small vibrational quanta and the result is → hot bands.

Group frequency, <*Gruppenfrequenz*>, a molecule of N atoms has $3N$–6, or in the case of a linear molecule $3N$–5, → normal vibrations. However, theory shows that not all of them need be → IR or → Raman active. In the case of larger molecules, IR absorption spectra (→ IR spectrum) are usually complicated because of the vibrational interaction of all the atoms in the molecule. This leads to coupling phenomena between the possible vibrations of the molecule and makes the main concern of IR spectroscopy, i.e. to draw conclusions concerning specific parts of a molecular structure from the IR bands, more difficult. However, theory also shows that such structural units can enter into a strong vibrational interaction only if their masses and force constants are of the same or similar magnitude. Therefore, when mutual interactions are small, the IR absorption maxima of the structural units are limited to relatively narrow and characteristic wavenumber regions, known as group frequencies, to which the structural units of a molecule may be empirically assigned. The region, in which a group frequency occurs is called its → expectation region. Examples of important structural units are the: -C-C- single bond, >C=C< double bond, -C≡C- triple bond and the groups: -OH, -SH, $-NH_2$, C=O, C-X (X: F, Cl, Br, I), $-NO_2$, -NO, $-CH_3$, –C≡N and many more including, of course, those in inorganic compounds. Since the absorption band(s) in their specific expectation region(s) are characteristic for a particular structural unit they are also called characteristic frequencies. On the basis of this empirical study of IR spectra, extremely comprehensive assignment tables have been created for the group frequencies which are of great value in the practical application of IR spectroscopy in structural analysis.

Ref.: L. J. Bellamy, *The Infrared Spectra of Complex Molecules*, 3rd ed., Chapman and Hall, London, **1975**; K. Nakamoto, *Infrared and Raman Spectra of Inorganic and Coordination Compounds*, 3rd ed., J. Wiley and Sons Ltd., London, New York, **1978**.

Group velocity, <*Gruppengeschwindigkeit*>, the velocity, v_g, of a wave group or a wave packet, i.e. a group of mutually superimposed waves, propagating in the same direction, the frequencies, and consequently the wavelengths, of which are slightly different. If they have varying phase velocities, v_p, the amplitude maximum resulting from the superimposition of the individual waves of the wave group or its square, which gives the intensity maximum, propagates with the group velocity, v_g.

Gudden-Pohl effect, <*Gudden-Pohl-Effekt*> → electrophotoluminescence.

H

Half-wave plate, <*Halbwellenplättchen*>. If unpolarized, monochromatic light falls vertically onto a plane-parallel, doubly refracting (birefringent) crystal plate, in which the plate normal coincides with a symmetry direction of the crystal then, on entering the plate, the light is split into two beams, polarized orthogonally to each other ($E_{\parallel}$ and $E_{\perp}$). These two beams propagate through the crystal with different velocities which correspond to the principal refractive indexes of the crystal, n_1 and n_2. Due to their different velocities, the two waves traversing the crystal plate leave the plate with a phase difference determined by the wavelength, λ, and the plate thickness, d. For a path difference, D, and a phase difference, δ, we have:

$$D = d(n_1 - n_2);$$

$$\delta = 2\pi D/\lambda.$$

This means, in practice, that a doubly refracting plate of suitable thickness, orientated diagonally ($\Psi = 45°$) between two parallel → Nicol prisms, can eliminate light of a specific wavelength. If the Nicol prisms are crossed, the light passes unattenuated through the → analyzer from which we can conclude that the plate has obviously rotated the plane of polarization by 90° (or −90°). The following general the equation holds:

$$d(n_1 - n_2) = (2K + 1)\lambda/2.$$

The path difference in this case is $\lambda/2$ ($K = 0$), or an uneven multiple of half the wavelength; the phase difference, δ, is 180°. This also applies to all other relative orientations of plate and polarizer. Crystal plates with the path difference $\lambda/2$ are therefore suitable for setting any plane of polarization for linearly polarized light. Plates with this specific path difference are frequently used and are generally called half-wave plates. They are usually produced by cleaving gypsum or mica, but they may be cut from other suitable crystal materials. The table shows the approximate minimum thicknesses (for $K = 0$) for different half-wave plates.

The above equation shows immediately that plates of three, five, seven ... times the thickness induce the same effect. This is particularly important for $\lambda/2$ or $\lambda/4$ plates made of quartz because very thin plates are difficult to produce. Halving the values in the table gives the thickness required for $\lambda/4$ plates which are also frequently used, usually to generate a phase difference of $\delta = 90°$ between two beams polarized perpendicularly to each other. A difference of 90° produces

Half-wave plates. Table.

λ [nm]	656	600	589	527	486	431	400	300	200
Gypsum, d [μm]	33	30	27	24	22				
Mica, d [μm]	77	70	62	57	51				
Quartz, d [μm]		32					20	14	7.4

elliptically polarized light. If the amplitudes of the two beams are the same, i.e. diagonal orientation of the $\lambda/4$ plate, → circularly polarized light (→ polarized light) is produced. $\lambda/4$ plates can therefore be used for the generation and analysis of circularly and any elliptically polarized light.
Ref: R.W. Ditchburn, *Light*, Academic Press, London, New York, San Francisco, **1976**, vol. 1, chapt. 12.

Half-width, <*Halbwertsbreite*>, the difference between the wavenumbers, $\Delta\tilde{\nu}_{1/2}$, or wavelengths, $\Delta\lambda_{1/2}$ at which the intensity (height) of a peak has been reduced to 50% of its maximum value (ε_{max}) as shown in Figure 1. To avoid confusion with other terminology, this definition of the half-width is sometimes termed the full width at half-height (FWHH). The half-width, $\Delta\tilde{\nu}_{1/2}$, is of practical importance in the interpretation of the shape of an absorption band and for → band analysis. With the Lorentzian band shape, for example, the following correlation

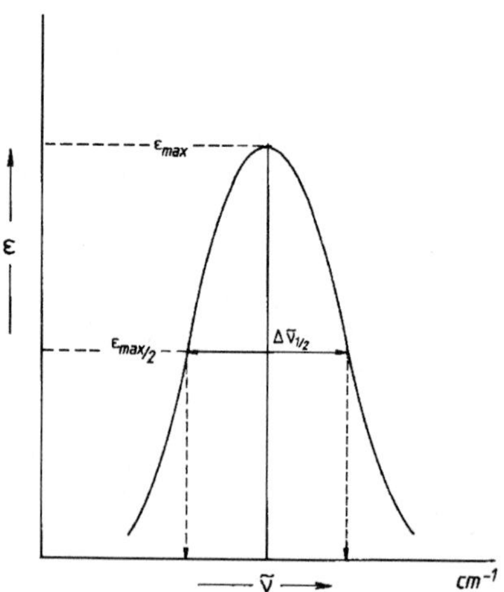

Half-width. Fig. 1. The definition of the half-width

exists between the molar decadic → extinction coefficient, ε, the wavenumber, $\tilde{\nu}$ and the half-width:

$$\varepsilon = K \cdot \frac{\Delta\tilde{\nu}_{1/2}}{4(\tilde{\nu}_o - \tilde{\nu})^2 + \Delta\tilde{\nu}_{1/2}};$$

$$K = \frac{2N_L e_o^2}{2.303 \cdot 1000 m_o c_o^2}$$

K is a constant, $\tilde{\nu}_o$ the wavenumber of the maximum and $\tilde{\nu}$ the variable wavenumber in the region of $\tilde{\nu}_o$. The half-width can also be used for an approximate determination of the → integral absorption.

Halogen lamp, <*Halogenlampe, Halogen-Glühlampe*>, a lamp without the disadvantage of the normal → tungsten lamp, i.e. the blackening of the glass envelope by the evaporating tungsten which considerably reduces the service life. The blackening of the envelope can be prevented, for example, by iodine vapor. The iodine reacts with the tungsten on the glass and forms a volatile compound which decomposes into tungsten and iodine on the hot surface of the filament. A continuous cycle is established in the lamp which ensures the transparency and prolongs its life. Furthermore, without reduction of its life the filament can be operated at considerably higher temperatures, which gives an increase in the radiation intensity in the VIS and near UV regions. Since iodine vapor absorbs a small proportion of the radiation in the VIS region, the light of the halogen lamp has a slightly purple glow. Such a lamp can also be used as a calibration standard when fitted with a quartz envelope.
Ref.: R. Stair, W.E. Schneider, J.K. Jackson, *Appl. Opt.*, **1963**, *2*, 1151.

Harmonic oscillator, <*harmonischer Oszillator*> → vibrational spectrum.

HBO lamp, <*HBO-Lampe*>, a trade name for high-pressure mercury lamps (→ mercury-vapor discharge lamp). An additional number indicates the power of the lamps, e.g. HBO 100, HBO 250 ...
Ref.: see the Osram catalogue, '93/94.

Heat filter, <*Wärmeschutzfilter, Wärmefilter*>, a → filter which is opaque to infrared thermal radiation. Liquid dye filters and glass filters are suitable for this purpose. The simplest filter is a cuvette with a path length of 5–10 cm filled with distilled water. Heat filters are used particularly in conjunction with → xenon lamps in order to block the high radiation intensity in the NIR region. Schott glass heat filters carry the type description KG.

Heavy atom effect, <*Schweratomeffekt*>. In principle, singlet-triplet, electronic transitions are forbidden. However, for atoms this prohibition is progressively relaxed with increasing atomic weight. Mercury provides a very good example; → mercury spectrum.
Singlet-triplet transitions in organic molecules can be made more probable in two ways:

1. A heavy atom may be chemically substituted into the molecule concerned; internal heavy atom effect.
2. The molecule may be dissolved in a solvent containing heavy atoms e.g. C_2H_5Br or C_2H_5I; external heavy atom effect.

Hefner candle, <*Hefner-Kerze, HK*>, an outdated unit of light which was used, in conjuction with a combustion lamp with a self-luminous flame, as a light-intensity standard for liquid fuels. It was valid in Germany and Austria until 1948 and was represented by the light intensity of the → Hefner lamp perpendicular to the flame. It was replaced by the → candela, which is a unit ca. 10% greater than the Hefner candle.

Hefner lamp, <*Hefner-Lampe*>, a lamp of a prescribed dimension the wick of which is fueled with *iso*-amyl acetate. It was constructed by Hefner-Altenbeck and was previously used as a light-intensity standard to represent the old light-intensity unit, the → Hefner candle. This lamp can also be used as a total radiation standard if high accuracy is not required. The wick is 8 mm thick and fitted into a nickel-silver tube with a wall thickness of 0.15 mm. The flame is set to a height of exactly 40 mm by adjusting the wick height with an optical sight. In this standard state, the lamp has the light intensity of one Hefner candle in the horizontal direction, i.e. a radiation intensity of 0.94 W/sr. A surface irradiated by it at 1 m distance receives a radiation flux of $94 \cdot 10^{-6}$ W/cm^2. Since the international introduction of the candela (→ photometric units), which is based on the black-body radiator, the Hefner lamp is no longer of great importance.

Heisenberg uncertainty principle, uncertainty principle, <*Heisenbergsche Unschärferelation, Unbestimmtheitsrelation*>, the equation proposed by W. Heisenberg in his paper: *Concerning the descriptive content of quantum-theoretical kinematics and dynamics* (*Zeitschr. f. Physik*, **1927**, *43*, 172):

$$\Delta x \cdot \Delta p_x \geq h/2\pi \equiv \hbar;$$

Δx is the uncertainty in the determination of the position and Δp_x the uncertainty in the momentum, mv_x. The Heisenberg uncertainty principle states that the position and momentum cannot be measured exactly at the same time. If the momentum is measured exactly, then the associated position is completely uncertain, and vice versa. Heisenberg also proved that this principle is correct for any pair of canonically conjugate quantities and not just for the pair, x and p_x. The uncertainty principle is particularly valuable since it removes all the contradictions which could arise from the dual character (wave/particle) of matter and light.

The most important uncertainty relationships may be formulated as follows:

Position – momentum	$\Delta x \cdot \Delta p \geq h/2\pi$
Frequency – time	$\Delta \nu \cdot \Delta t \geq 1/2\pi$
Energy – time	$\Delta E \cdot \Delta t \geq h/2\pi$

The minimum uncertainty, i.e. the equality sign, is always achieved when the quantity concerned has a Gaussian probability distribution. A number of spectroscopic statements can be justified with the help of the uncertainty relationship. For example, the → zero-point energy, $h\nu/2$, is a direct consequence of the relationship. The formulation of the energy-time relationship, $\Delta E \cdot \Delta t \geq h/2\pi$, gives expressions for line widths, i.e. the uncertainty in the frequency of a transition. For example, the energy, E_i, of an excited state of average lifetime, τ_i, can only be determined with the minimal uncertainty, $\Delta E_i \cdot \tau_i = h/2\pi$ or $\Delta E_i = h/(\tau_i 2\pi)$. Therefore, the frequency, $\nu_{i,k} = (E_i - E_k)/h$, of a transition to the stable ground state has an uncertainty given by:

$$\delta \nu_{i,k} = \Delta E_i/h = (h/\tau_i 2\pi) \cdot (1/h) = 1/(2\pi\tau_i)$$

or in wavenumbers

$$\delta \tilde{\nu}_{i,k} = \delta \tilde{\nu}_{i,k}/c = 1/(2\pi c \tau_i)$$

(see → line width, natural)

Helium-cadmium laser, He-Cd$^+$ laser, *<Helium-Cadmium-Laser, He-Cd$^+$-Laser>*, a gas laser. Older models of this laser use the positive column of a gas discharge for excitation, as is the case with most gas lasers. In a newer version, the hollow cathode He-Cd$^+$ laser, the region of negative glow discharge is used to excite the laser. For this purpose, the cathode is constructed as a tube and the negative glow discharge is concentrated inside this tube. This produces a highly ionized plasma with an optimal plasma density in the center of the tube along the optic axis. Figure 1 shows the construction of the laser schematically.

The shading of the diagram shows the hollow cathode which forms part of the laser tube. It contains several lateral holes for the generation of the negative discharge via the anodes, A_1 – A_5. Below each hole there is a cadmium reservoir. The cathode tube is connected to the supply at K. The additional anodes, $SA1$ and $SA2$, prevent the condensation of cadmium vapor onto the Brewster windows, $B1$ and $B2$, of the laser tube. The helium-gas pressure required for the laser is set to the optimal value by means of a pressure sensor, pressure controller and helium-gas supply. A getter, G, continuously removes impurities from the gas. The Cd reservoirs are placed so that a particular Cd-vapor pressure can be set by means of externally controlled heating.

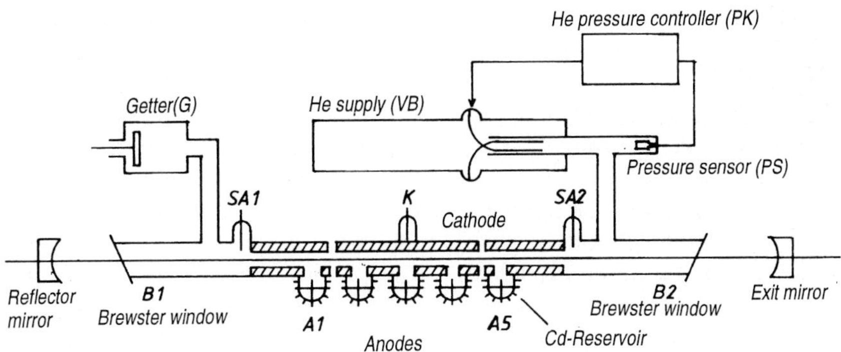

He-Cd⁺ laser. Fig. 1. He-Cd⁺ laser; cross section of the laser tube

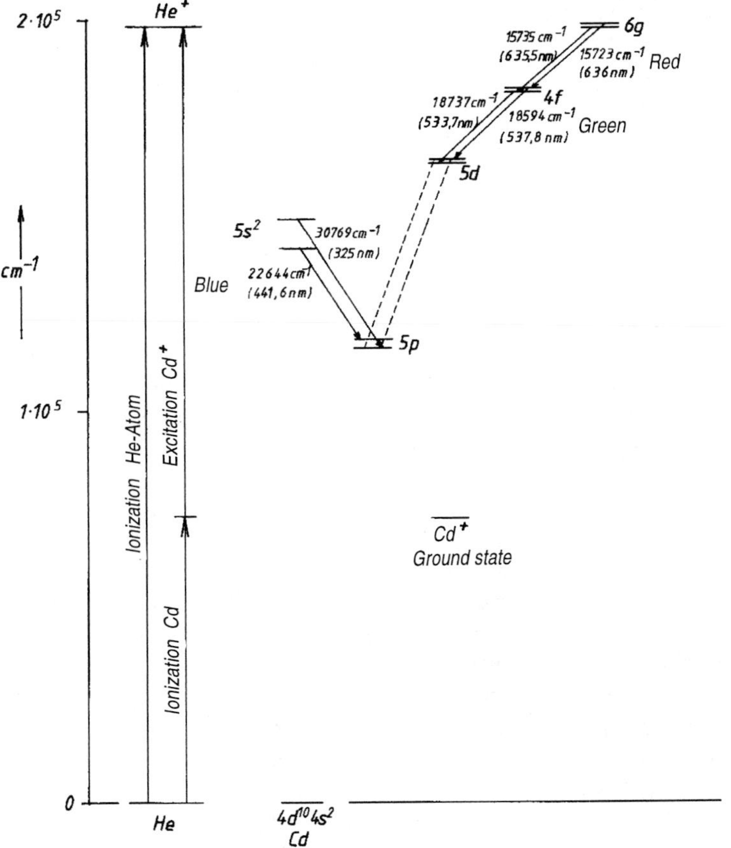

He-Cd⁺ laser. Fig. 2. Energy-level scheme for the Cd⁺ ion and the laser transitions

The mechanism of the He-Cd^+ laser can be explained in two steps:

1. Ionization of the cadmium atoms which requires 8.96 eV.
2. Excitation of the Cd^+ ions.

Because of the high ionization energy of the helium atom (24.6 eV), electrons with high energies are produced which pump the cadmium ions up the level *6g*. From this state, laser emission to the *4f* state takes place, and also from the *4f* to the *5d*. In the first case, a red, and in the second case a green, doublet is emitted. A further laser emission can arise from the state $5s^2$ which is excited directly by collisions with the electrons. The transition, $5s^2 \rightarrow 5p$ consists of two lines which, because of the relatively large doublet splitting of the $5s^2$ and $5p$, lie in the blue and ultraviolet regions of the spectrum. The situation is illustrated in a simplified energy-level scheme in Figure 2. Because of the almost simultaneous emission of red (635.5 and 636.0 nm), green (533.7 and 537.8 nm) and blue (441.6 nm) lines, this laser is marketed by manufacturers as an RGB laser.
Supplier: Oriel Corporation, 250 Long Beach Blvd., PO Box 872, Stratford (CT), USA 06497.

Helium-neon laser, He-Ne laser, <*Helium-Neon-Laser, He-Ne-Laser*>, a neutral gas → gas laser. Developed in 1961, it was the first continuously operating laser. A gas discharge is created in the gaseous laser medium to produce a → population inversion. In the discharge, the neutral gas partially dissociates into electrons and ions. The electrons are accelerated by the field prevailing in the discharge and, after a short acceleration time, they can transfer their kinetic energy to a collision partner here, for example, to a neon atom. The maximum of the electron-energy distribution (3–10 eV) generally lies well below the laser level (16–21 eV) to be excited. Under these conditions, only those electrons from the tail of the Maxwell-Boltzmann energy distribution (→ Boltzmann statistics) have a sufficiently high kinetic energy to excite the laser levels by a collision. The rate of population of the laser levels is therefore small and a population inversion in the stationary state is not achievable since the total system is in a thermal equilibrium. The population inversion required for the laser activity can only be achieved by a trick. Helium as a pump gas is added to the active medium, neon in this case. The He atom has suitably positioned metastable levels such as the 2^1S and 2^3S state, from which the upper laser levels of the Ne atom can be excited. The population of the metastable 2^1S and 2^3S state of the He atom occurs partly by direct excitation but mainly by radiative transitions from higher levels. Due to the long lifetime of the metastable levels, 2^1S and 2^3S, the excitation energy is stored over a long period of time and represents a monoenergetic energy store in the stationary state. If a helium pump atom, the metastable level of which is excited, collides with a neon atom then, depending upon the transition probability, the excitation energy can be transferred to the neon atom in the collision. Very high values of population inversion can be achieved in the laser gas by this mechanism. Figure 1 shows this by means of a combined energy-level scheme. The upper laser levels, *4s* and *5s* of the neon atoms are

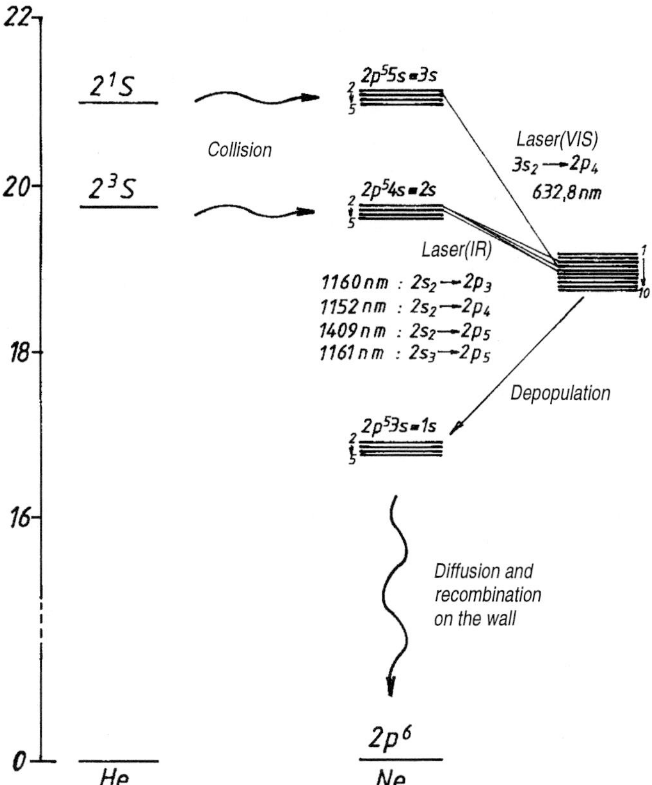

Helium-neon laser. Fig. 1. Energy-level scheme and laser transitions in the He-Ne laser

populated by collisions between excited helium atoms and neon atoms. The lower laser level, which is empty, is the 3*p* level. Therefore, an → inversion of population exists between the 4*s* or 5*s* and the 3*p* level. The numbers 2–5 or 1–10, given for the *s* states and the 3*p* state refer to the → Paschen series of the neon atom. This scheme shows that several laser transitions are possible on account of the numerous energy levels. The most important lines of the He-Ne laser are: $\lambda = 632.8$ nm, $\lambda = 1152.3$ nm and $\lambda = 3391.3$ nm. The last arises from the transition 5*s* → 4*p*, not shown here. As shown in Figure 2, a typical discharge tube is closed by windows which are inclined

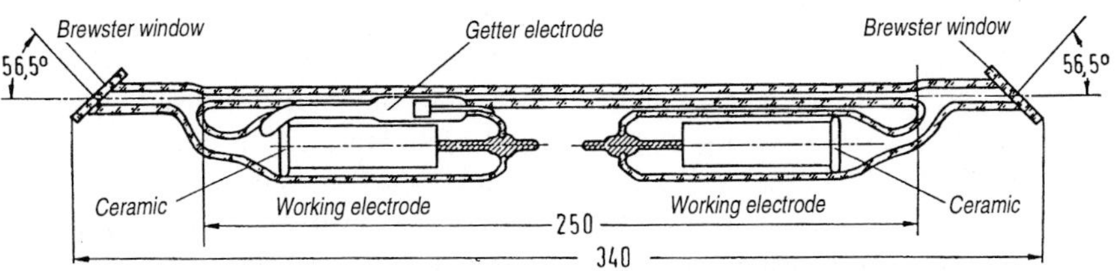

Helium-neon laser. Fig. 2. Cross section of the He-Ne laser tube

towards the tube axis at the Brewster angle (56.8° for crown glass, → Brewster's law). Light polarized in the plane, defined by the tube axis and the normal to the window (the plane of the diagram), can pass through the windows without incurring a loss by reflection.
Ref.: D. Eastham, *Atomic Physics of Lasers*, Taylor and Francis, London, Philadelphia, **1989**, section 6.2.

Herapathite, <*Herapathit*>, dichroic, single crystals of artificially grown quinine sulfate periodide plates which are cemented between glass plates for use as polarization filters (→ polarized light). They are named after J. Herapath (→ dichroism → anisotropy of light absorption).

Hertz, Hz, <*Hertz, Hz*>, the special unit commonly used for the → frequency of an oscillatory process having a 1 s period. It is named after Heinrich Hertz (1857–1894) the discoverer of electromagnetic waves.

Hertz vector, <*Hertzscher Vektor*>, a vectorial quantity frequently used in the electromagnetic theory of light and in general more widely in the theory of electromagnetic waves. By its use, the six components of the electromagnetic field, i.e. the two mutually independent vectors representing the electric field, E, and the magnetic field, H, both of which must satisfy Maxwell's equations, may be represented as a single vector, $\vec{Z}$.
For harmonic waves $\vec{Z} = z \cdot \exp[-\omega t]$ with $\omega = 2\pi\nu$, the circular frequency. $\vec{Z}$ satisfies the time-independent wave equation or vibration equation, $\Delta^2\vec{Z} + k^2\vec{Z} = 0$ where $k = 2\pi/\lambda$, the → wave vector.

Heterochromous, <*heterochrom*>, multicolored.

HFS, <*HFS*> → hyperfine structure of atomic spectra.

High-current arc, <*Hochstrombogen*>, a gas discharge (→ arc discharge) which is formed with discharge currents above 80 A. Finkelnburg named it but the high-current arc was discovered much earlier by Beck. The Beck arc is a special form of the high-current arc which is very important for lighting technology. The high-current arc is characterized by a particular anodic mechanism and by the specific formation of the arc column. A turbulent evaporation of the anode material takes place at the typical anode loads of 100 $A \cdot cm^{-2}$ to 150 $A \cdot cm^{-2}$ because the power supplied to the anode by the impinging electrons can no longer be dissipated by convection and thermal conduction. The discharge is positive in the high-current region on account of the strongly positive characteristics of the abnormal anode voltage drop. The material evaporating from the anode flows out of the crater at high velocity as a thermally glowing anode flame, also known as the positive flame or Beck flame. The form of the anode flame depends upon the load and the anode material. The Beck arc is a high-current arc the carbon electrode of which consists of a wick (with up to 60% cerium fluoride as emissivity additive) and a sheath. A deep crater usually forms in the electrode because the wick evaporates faster than the sheath. Within the crater there is a very high light intensity. For this reason, the Beck arc was found to be an excellent, intense light source for technical purposes. Above 100 to 130 A, the arc discharge is strongly constricted and becomes a darting negative flame (cathode flame) which blows the anode flame aside.

High-frequency discharge, <*Hochfrequenzentladung*>, a gas discharge under a high-frequency voltage. In discharge tubes with oxide electrodes, the difference between the reignition and operating voltage disappears at frequencies of approximately 500 Hz. The current intensity is proportional to the operating voltage above 5000 Hz because the conductivity of the gas cannot follow the rapid current variations. The phenomena in a high-frequency discharge are completely different from those of a DC discharge. The spatial charges, light and dark sections (glow discharge) characteristic of the latter are no longer formed, even transiently, since they cannot follow the frequency. Instead, a stationary state is established which does not show the characteristic phenomena of DC operation. The operating voltage of a high-frequency discharge is very low because the electrons effectively shuttle to and fro in the high-frequency field and generate sufficient carriers during their movement so that the cathode need not supply additional electrons.
For that reason, discharges can be operated at high frequencies (larger than 10^6 Hz), where the shuttle period of the electrons is comparable with the frequency, with higher current intensities and very low pressures. Such discharges cannot be kept stable with DC or alternating voltages of low frequency. One characteristic feature of the high-frequency discharge is that the current need not be supplied to the discharge through electrodes fitted in the discharge chamber. Instead, the current can be induced capacitively, through the glass wall, since no electrons need to be generated at the cathode. For this electrodeless discharge (electrodeless → discharge lamp), the electrodes take the form of metal layers on the outside of the discharge tube. Closed ring electrodeless discharges can be formed in high-frequency magnetic fields in vessels of a suitable form.

High-frequency spectroscopy, <*Hochfrequenzspektroskopie*>, a comprehensive term for all spectroscopic methods which lie adjacent to → optical spectroscopy in the IR region on the side of lower frequencies, longer wavelengths or lower wavenumbers. This includes methods such as → microwave spectroscopy (MW), → magnetic resonance spectroscopy (ESR) and → NMR spectroscopy which are very important in their application. Their frequency ranges can be delimited as follows:

MW	ν:	$1 \cdot 10^9 - 10^3 \cdot 10^9$ Hz (1GHz - 10^3GHz)
	λ:	30 cm - 0,03 cm
	$\tilde{\nu}$:	0.033 cm^{-1} - 33.3 cm^{-1}
ESR	ν:	$\sim 3 \cdot 10^{10}$ Hz (30000 MHz = 30 GHz)
	λ:	$\sim$ 1 cm
	$\tilde{\nu}$:	~ 1 cm^{-1}
NMR	ν:	$\sim 3 \cdot 10^6$ Hz (30 MHz = 0,03 GHz)
	λ:	$\sim 10^3$ cm
	$\tilde{\nu}$:	$\sim 10^{-3}$ cm^{-1}

In contrast to MW spectroscopy, the strength of the applied magnetic field also determines the resonance frequencies in ESR and NMR spectroscopy.

The grouping of these methods under the comprehensive term *high-frequency spectroscopy* is based on the fact that the techniques of high-frequency physics are used to produce the required electromagnetic radiation.

High-pressure discharge lamp, high-pressure lamp, high-pressure gas discharge lamp, <*Hochdruckgasentladungslampe, Hochdrucklampe*> → metal-vapor discharge lamp.

High-pressure gas discharge, high-pressure discharge <*Hochdruckgasentladung, Hochdruckentladung*>, a gas discharge at higher pressures where the temperature of the electrons has reached the temperature of the gas. The ionization of the → plasma occurs in an essentially thermal manner. An intense, continuous spectrum in the visible and ultraviolet spectral region occurs during a high-pressure discharge on account of the bremsstrahlung of the electrons and the recombination of the ions. This discharge finds practical application in xenon and mercury high-pressure discharge lamps (→ metal-vapor discharge lamp).

High-pressure mercury lamp, <*Höchstdrucklampe, Superhochdrucklampe*>, a → metal-vapor discharge lamp which reaches light densities up to $18 \cdot 10^8$ cd m^{-2} at an Hg vapor pressure of between 50 atm and 200 atm. If the discharge tube is a quartz capillary with a diameter of ca. 1 mm to 3 mm, the tube must be cooled with water. This cooling is not required for the Rompe and Thouret high-pressure lamp because it has a spherical discharge chamber. Such mercury lamps are frequently used for photochemical investigations on the technical scale. The spectrum consists of very strongly broadened lines and an intense, continuous background which is particularly pronounced in the infrared spectral region; → mercury-vapor (discharge) lamp.

High-temperature phosphorescence, <*Hochtemperaturphosphoreszenz*>, a type of → phosphorescence which occurs in organic molecules when the first excited singlet state, S_1, and the lowest triplet state, T_1, lie energetically close together. An → intersystem crossing, $T_1 \rightarrow S_1$, which results from the thermal excitation of vibrations, can take place during the life of T_1 and a radiative deactivation from S_1 is then seen as fluorescence. Thus, the spectral distribution of high-temperature phosphorescence corresponds to the fluorescence spectrum. The characteristic feature of this emission is that it is chronologically very delayed on account of the long → fluorescence lifetime of the T_1 state; therefore, we speak of → delayed fluorescence. This type of luminenscence is particularly observed in those dyes where the triplet state, T_1, lies close to the S_1 state. High-temperature phosphorescence can be explained simply by means of Figure 1 of the → Jablonski diagram; see path *a*, *m*, *n*, *e*, involving T_2 in this case.

HMO energy-level scheme, Hückel molecular orbital energy-level scheme, <*HMO-Termschema, Hückel-Molekular-Orbital-Termschema*>, an energy-level scheme based on the Hückel molecular orbital (HMO) theory in which only the p_z orbitals of the sp^2 hybridized C atoms are combined, in contrast to the general LCAO-MO theory in which all atomic valence orbitals are used. The orbitals of the σ bonds of the participating C atoms generally lie much lower energetically than the π orbitals which are formed from the p_z orbitals. Furthermore, most organic molecules whose σ bonded skeletons are formed from sp^2 hybridized carbon atoms are planar, and the → symmetry species of the molecular orbitals formed from the p_z and sp^2 atomic functions are different for such molecules. This provides another reason for treating the π and σ electron systems separately. Figure 1 shows a generalized MO energy-level scheme which takes account of the relative energies of π and σ levels. Important concepts for unterstanding the spectra of organic compounds can be derived from this diagram. $\pi \rightarrow \pi^*$ transitions have the smallest energy difference, i.e. the smallest excitation energy. Therefore, the absorption or fluorescence spectra measured in the UV-VIS region should generally be classified as $\pi \rightarrow \pi^*$ transitions.

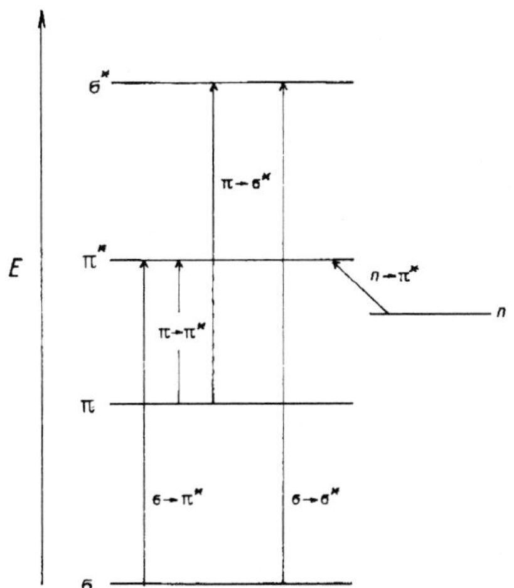

HMO energy-level scheme. Fig. 1. A general MO energy-level diagram

$\sigma \rightarrow \pi^*$, $\sigma \rightarrow \sigma^*$, $\pi \rightarrow \sigma^*$ transitions require a considerably higher excitation energy and they are generally observed at wavelengths < 200 nm or wavenumbers > 50,000 cm^{-1}.

If a molecule has an atom, or several atoms, with a nonbonding electron pair, such as a carbonyl group or an N atom, the nonbonding level (n), which corresponds to this electron pair, lies energetically between the π and π^* state. As illustrated in Figure 1, an $n \rightarrow \pi^*$ transition requires a considerably lower excitation energy than a $\pi \rightarrow \pi^*$ transition and is thus shifted to the red vis-a-vis the latter transition. An energy-level diagram, constructed according to the HMO theory, takes account only of the π and π^* levels, the energies of which are given as $E_i = \alpha + x_i\beta$. α is the Coulomb integral and β the resonance integral ($\beta < 0!$); the x_i are the solutions of the secular equations (determinant). Figure 2 shows the HMO energy-level diagrams corresponding to the polyenes, C_mH_{m+2}

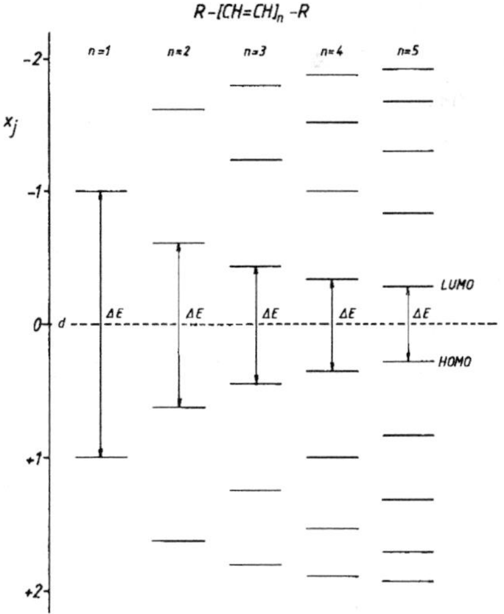

HMO energy-level scheme. Fig. 2. An HMO energy-level diagram for the polyenes

(m even), up to $m = 10$. If we compare the corresponding excitation energies, ΔE, we see that they decrease with increasing conjugation which is in agreement with experiment.

It is usual in semiempirical MO theory to denote the lowest unoccupied MO with LUMO and the highest occupied MO with HOMO.

Hole burning, <*Hole-Burning, Lochbrennen*>, a variant of → saturation spectroscopy which is based upon the principle of selective saturation or selective photochemistry. Bloembergen, Purcell and Pound described selective saturation, in an NMR experiment, for the first time in 1948. Figure 1 shows the general principle of the method schematically. Consider, for example, five molecules in a matrix which have different ground states on account of their varying molecular environments (→ matrix effect). A molecule in its matrix environment is generally said to be at a *site*. For the multiplicity of molecules, which are present in every macroscopic sample, a spectrum with a broad Gaussian (i.e. inhomogeneous) band results. The parameters of interest, such as the phonon and vibrational energies, cannot be determined from this band. If we irradiate at the mean frequency, ν_o, with monochromatic radiation of such a high intensity that this transition is selectively saturated, the same number of molecules can be found in the ground and excited states at the irradiation frequency (Figure 1a). Consequently, the sample is transparent, i.e. the absorption curve has a hole at that site; as crudely illustrated for a Gaussian band in Figure 1c.

The performance of such saturation experiments causes certain experimental problems. Firstly, the high laser power required can lead to the destruction of the sample. Secondly, the optical spectrum must be scanned with a variable light source within the lifetime, T_1, of the optically excited state, immediately after the saturation. This presents no great problem in NMR experiments, for which the method was originally developed, since typical T_1 times can be of the order of seconds, minutes or even

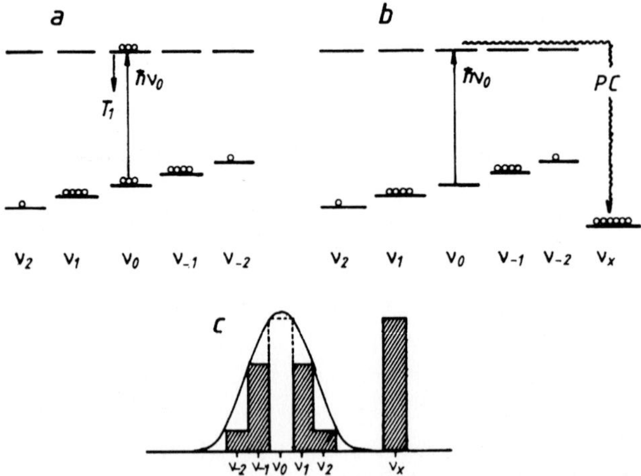

Hole burning. Fig. 1. The principle of the method of hole burning

hours. But the T_1 times, i.e. the lifetime of the excited states, for optical excitation are of the order of 10^{-8} to 10^{-10} seconds. Thus the experiment, though simple in principle, becomes almost impossible. However, it can be relatively easily executed with the help of a small variation. Figure 1b illustrates a situation in which the excited state, produced by irradiation at ν_o, can convert into a photochemically or photophysically changed state, X, which differs spectroscopically from the molecule originally excited. Monochromatic radiation of high intensity is not required; a long period of monochromatic irradiation at low intensity is sufficient to drive a photochemical process forward, depending upon its quantum yield. Furthermore, the spectrum can be scanned after any time delay if the new state, X, is stable. This procedure is also called photochemical hole burning. Figure 2 illustrates an experiment of this type using the results for the long-wavelength band of phthalocyanine in polystyrene at $T = 4$ K from D. Haarer and J. Friedrich. A laser of wavelength 688.0 nm was used for the irradiation. The sharp cut, i.e. the hole in the absorption band, can be seen quite clearly and also the phonon hole (phonon side band) at longer wavelength.

Photophysical hole burning usually involves rearrangement processes which change the interaction between the molecule and the matrix. Photochemical and photophysical hole burning are distinguished by the fact that the photoproduct lies relatively close to the energy selected by the laser in the latter case, whilst in the former the spectrum of the product may well lie outside the inhomogeneous band. Figure 3 illustrates the differences.

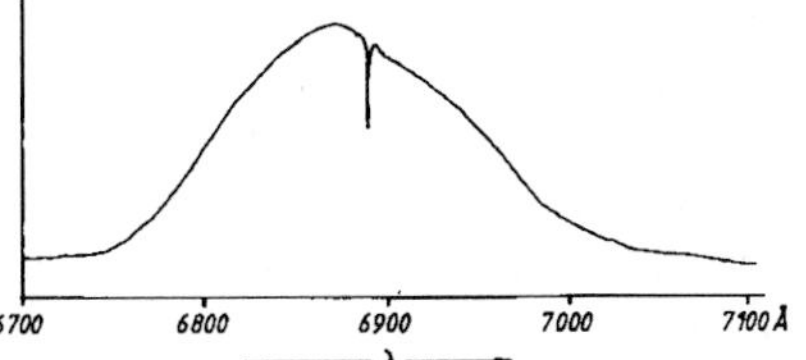

Hole burning. Fig. 2. Photochemical hole burning; example of phthalocyanine

This method is applied primarily to host/guest systems, e.g. in the spectroscopy of glasses and polymers where dye molecules, introduced in small concentrations, serve as probes. The extraordinarily narrow holes, which can be burnt into the absorption bands of dye molecules using laser light, make it possible to obtain information about the guest/host system. The high resolution of these methods provides new information about the parameters which were previously hidden by the effects of inhomogenity. This also includes investigations of the influence of small perturbations upon disordered lattices such as the in-

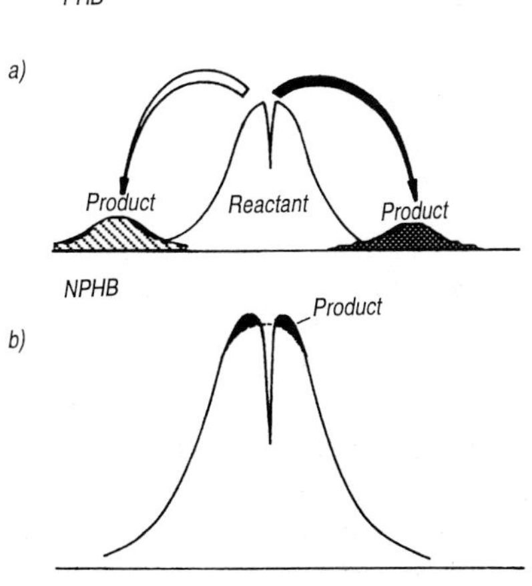

Hole burning. Fig. 3.
a. Photochemical hole burning;
b. photophysical hole burning

fluence of external electric fields (→ Stark effect) or magnetic fields (→ Zeeman effect). The method has been applied successfully to the study of antenna (light gathering) pigments. The development of optical storage devices is of particular interest for technical applications of hole burning.
Ref.: J. Friedrich, D. Haarer, *Angewandte Chem. Int. Ed. Eng.*, **1984**, *23*, 113–140;

Hollow cathode lamp, <*Hohlkathodenlampe*>, a lamp which is now normally used in → atomic absorption spectroscopy. It consists of a glass cylinder, which may have quartz windows, filled with a few Torr of noble gas (neon or argon) and into which a cathode and anode are fused. The cathode has the form of a hollow cylinder and is either made of the metal of interest or filled with that metal. The anode is a thick wire, usually made of tungsten or nickel. When a voltage of several hundred volts is applied to the electrodes a glow discharge develops. If the cathode consists of a hollow cylinder then, under certain conditions, the discharge is almost entirely confined within the cathode where two processes take place. The stream of positive gas ions, which strike the cathode, eject metal atoms from its surface. These enter the region of intense discharge where they meet a concentrated stream of gas ions and excited noble-gas atoms and are excited to emission by them. Because the greater part of the emission comes from the inside of the hollow cylinder the emission is relatively well concentrated. The disadvantage of the hollow cathode lamp is that a separate lamp is required for each metal. To overcome this, multielement hollow cathode lamps have been constructed in which, for example, the cathode is made of various powdered metals, mixed in specific proportions, compressed and sintered. However, with these lamps one must always guard against → spectral interferences.
Ref.: B. Welz, *Atomic Absorption Spectrometry*, 2nd edtn., Verlag Chemie, Weinheim, Deerfield Beach, Florida, Basel, **1986**.

Hollow cathode laser, <*Hohlkathodenlaser*> → He-Cd^+ laser.

Homologous pair of lines, <*homologes Linienpaar*>, a pair of lines, used in quantitative atomic emission spectroscopy, consisting of one from the element to be determined and one from an additional element. The lines should not be too distant from each other. The equality of intensity (equal line density) of the two lines, under specified discharge conditions, is a criterion for a particular relative concentration; (→ spectral analysis).

Hooke's law, <*Hooke'sches Gesetz*>, describes the relationship between the extension, x, of a spring and the associated restoring force, F. For small extensions (e.g. → harmonic oscillator) the relationship is $F = -kx$. The quantity, k, defines the force constant of the spring.

Hot band, <*Hot-Band*>. In the electronic excitation of molecules (UV-VIS, NIR) it is assumed, according to the Boltzmann energy distribution (→ Boltzmann statistics), that the transitions all originate in the vibrational ground state, $v'' = 0$, of the electronic ground state. This assumption only holds if the vibrational excitation

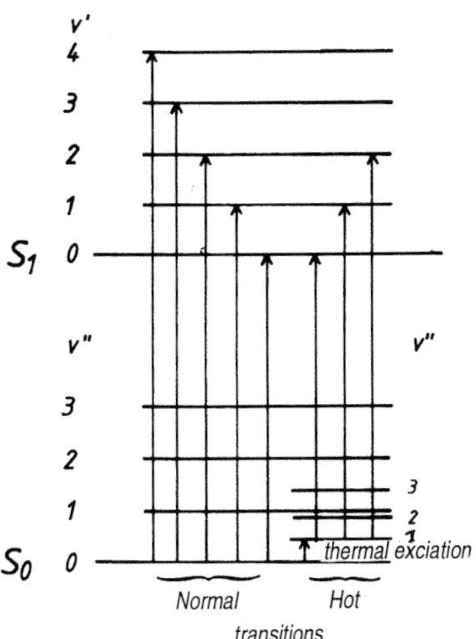

Hot band. Fig. 1. Energy-level scheme illustrating hot bands

energy is fairly large (≥ 1000 cm^{-1}). For vibrations, with an excitation energy in the region of 200 cm^{-1}, approximately 27% of all molecules are in the first excited state, $v'' = 1$, at room temperature. Thus, a relatively large number of molecules can undergo transitions to the electronic excited state from this vibrational state. The associated bands are shifted to lower wavenumbers. Because these transitions originate from thermally excited or *hot* molecules the bands are known as hot bands.

Figure 1 shows an energy-level scheme including such bands. Since the occupation of the vibrational energy levels depends upon temperature, according to the Boltzmann distribution, the intensity of hot bands is also very temperature-dependent. An increase of temperature leads to an increase in intensity; a decrease in temperature to a loss of intensity. Such bands have been identified in the UV spectrum of benzene.

Hückel molecular orbital energy-level scheme, <*Hückel-Molekular-Orbital-Termschema*> → HMO energy-level scheme.

Hue, <*Farbton*>, in everyday speech the sensations of green, yellow, purple etc. would be described as colors; but not black, gray or white. In color science, however, these last three sensations are regarded as color sensations in their own right. But they are differentiated from the true colors by saying that they do not possess the attribute of hue. They are therefore achromatic sensations. Hue is the most important variable of perceived color and is the only term ever used in color science to describe this attribute. Though it is not common in everyday speech, when hue is used it has exactly the same meaning.

Hund's coupling cases, <*Hundsche Kopplungsfälle*>, the description of the different possibilities for the coupling of rotational and electronic motion in the electronic band spectra of diatomic molecules. The electron spin, *S*, the electron orbital angular momentum, Δ, and the angular momentum due to the rotation of the nuclei, *N*, combine to form a resultant which is denoted by *J*. If *S* and Δ for the electrons both equal zero, we have a $^1\Sigma$ state and *J* is identical with the angular momentum due to the nuclear rotation, *N*, giving a simple rotator (→ rotation-vibration spectrum of linear molecules). In all other cases, we must distinguish between the different possibilities for the coupling of the angular momenta which Hund first discussed in detail. Of the total of five possibilities, only two are of general

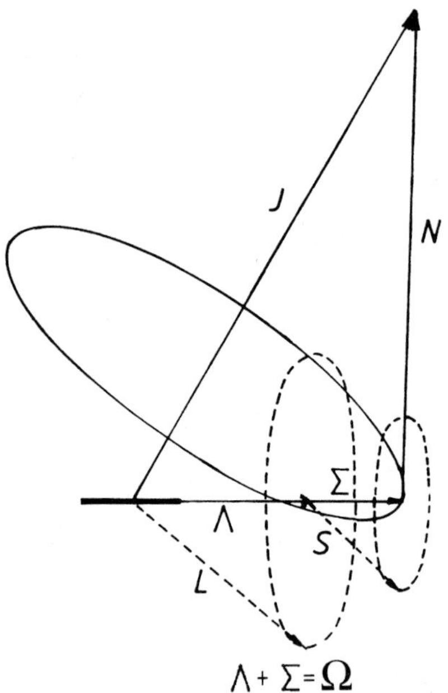

Hund's coupling cases. Fig. 1. Vector diagram of Hund's case a

practical importance, i.e. Hund's coupling cases a and b.

Hund's case a: It is assumed that the interaction between the nuclear rotation, N, and the electronic motion (both spin and orbital) is very weak whilst the electronic motion itself is rigidly coupled to the internuclear axis. Then, $\Omega = |\Lambda + \Sigma|$ (Λ and Σ are respectively the components of the orbital angular momentum, L, and the spin, S, along the internuclear axis) is also well defined in the rotating molecule and together with the nuclear rotational angular momentum, N, forms the resultant, J. This corresponds exactly to the situation discussed in the case of the → symmetric top molecule. Figure 1 illustrates the vectors involved. J is a vector constant in space and time. Ω and N rotate about this vector (nutation). There is a simultaneous precession of L and S about the internuclear axis which is assumed to be much more rapid in case a than the nutation of the → figure axis. The approximate frequency of the latter is given by $\nu_{rot} \approx c2BJ$.

As a first approximation to the rotational energy we have:

$$F(J) = B_v[J(J + 1) - \Omega^2].$$

This expression is quite similar to the formula which applies to the symmetric top $F(J) = B_v J(J + 1) + (A - B_v)\Lambda^2$. However, the term containing A has been omitted since it is a constant for a given electronic state and can therefore be included in the electronic energy. B_v has the same value for all the components of a specific multiplet term, except in the case of a very large multiplet splitting. Since Ω is constant for a specific multiplet component, the rotational energy, according to the above equation, is practically the same as for the vibrating rotator, but with the exception that some of the first rotational levels drop out because J cannot be smaller than Ω (see Figure 1). In fact:

$$J = \Omega, \Omega + 1, \Omega + 2, \ldots$$

It follows that J is half-integral if Ω is half-integral, i.e. for an odd number of electrons. The electronic energy of the multiplet component is given to a first approximation by $T_e = T_o + A \cdot \Lambda \cdot \Sigma$. A is the coupling constant which is a measure of the strength of the interaction between the orbital and spin angular momenta. Figure 2 shows the example of the rotational terms of a $^2\Pi$ in case a. The levels which drop out are indicated by dashed lines.

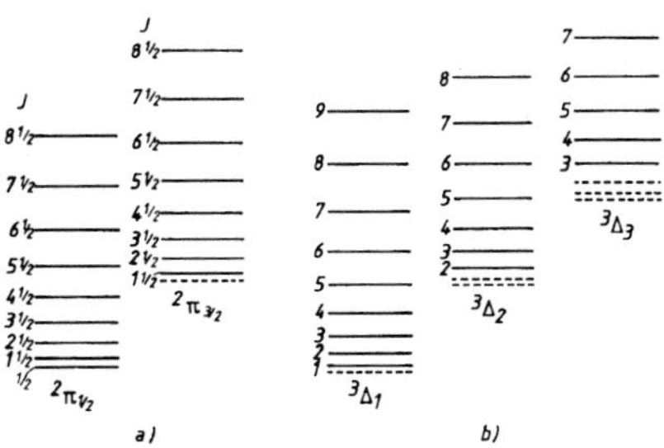

Hund's coupling cases. Fig. 2. Rotational levels for Hund's case a

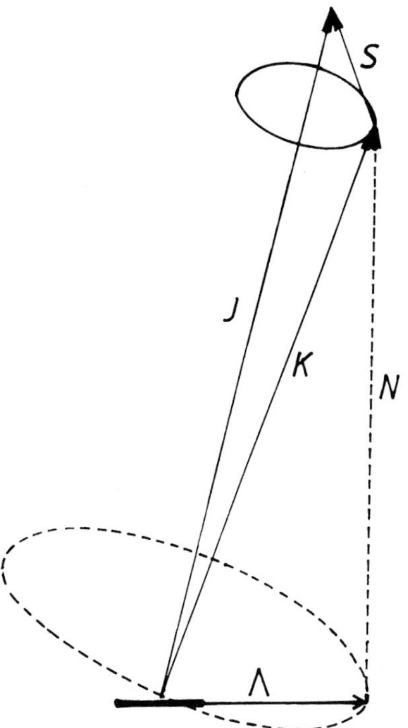

Hund's coupling cases. Fig. 3. Vector diagram of Hund's case b

Hund's case b: If $\Lambda = 0$ and $S \neq 0$, S is not coupled at all with the internuclear axis. That means that Ω is not defined and Hund's case a cannot be present. With light molecules, under certain circumstances S can be coupled extremely weakly to the internuclear axis even if $\Lambda \neq 0$. This is the characteristic feature of case b. Here, Λ (if it differs from 0) forms a resultant with N (in the same manner as Ω and N in case a) which is denoted by K (see Figure 3). The associated quantum number, K, can take the integer values:

$$K = \Lambda, \Lambda + 1, \Lambda + 2, ...$$

K is the total angular momentum, apart from the spin. (If $\Lambda = 0$, $K \equiv N$ and K can assume all integral values from 0 upwards). K and S then combine to form a resultant, J, the total angular momentum, including the spin (see Figure 3). The possible values of J for a specific K are, according to the principles of vector addition, $J = (K + S), (K + S - 1), (K + S - 2) \ldots, |K - S|$. Thus, each level with a specific K consists in general (except when $K < S$) of $(2S + 1)$ components, i.e. as many components as are given by the multiplicity. These components each have a slightly different energy on account of the coupling between K and S, although it is very small. The splitting increases with increasing K. J is again half-integral for an odd number of electrons and integral for an even number. In practice, only the cases with $S = 1/2$ (doublets) and $S = 1$ (triplets) have been observed to date. Figure 4 shows the energy-level scheme for a $^2\Sigma$ term. The rotational energy in this case is given by:

$$F_1(K) = B_v K(K+1) + \frac{\gamma}{2}K;$$

$$F_2(K) = B_v K(K+1) - \frac{\gamma}{2}(K+1),$$

where $F_1(K)$ refers to the components with $J = (K + 1/2)$ and $F_2(K)$ to those

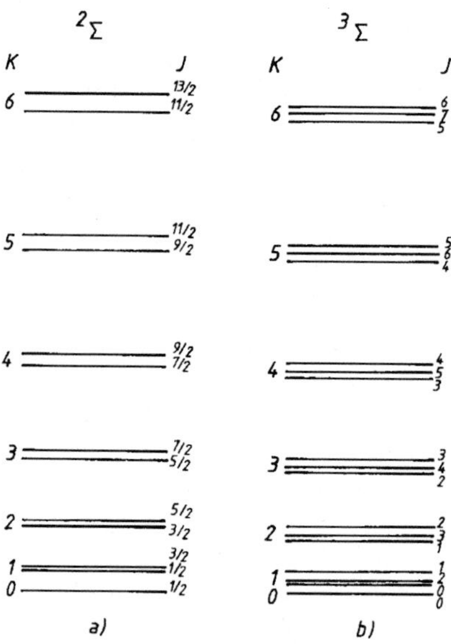

Hund's coupling cases. Fig. 4. Rotational levels for Hund's case b

with $J = (K - 1/2)$. γ is a constant which is very small compared with B_v. These are essentially the energy levels of the rotator, but each level is split into two components and the splitting increases linearly with K (see Figure 4a).

The formulas which apply to the $^3\Sigma$ terms are complicated. Of course, the distinction between case a and b does not exist for singlet terms; for them $\Lambda = \Omega$ and $K = J$.

Ref.: G. Herzberg, *Molecular Spectra and Molecular Structure I, Spectra of Diatomic Molecules*, 2nd ed., D. Van Nostrand Co. Inc., Princeton, New York, London, **1950**.

Hund's rules, <*Hundsche Regeln*>, the three rules, derived by F. Hund, which are used for the determination of the → ground term (state) of an atom.

Hund's first rule: If several electrons are to be placed in $(2\ell + 1)$ degenerate atomic orbitals then, whilst observing the → Pauli principle, they must be distributed in such a way that the maximum spin, $S_{\max}$, results, i.e. they are initially distributed to the degenerate orbitals with the same (parallel) spin component, m_s. Since the total spin determines the multiplicity $(2S + 1)$ of a term, this is also called the rule of maximum multiplicity.

Hund's second rule: The population of the degenerate orbitals must be such that the sum of the m_ℓ values of the populated orbitals assumes a maximum value: $(\Sigma m_\ell)_{\max} = M_{L,\max}$. This maximum value corresponds to the z component, M_L, of the resulting total orbital angular momentum, L. This means that $M_{L,\max}$ determines the ground state.

Hund's third rule: According to the → Russell-Saunders coupling scheme, the total angular momentum, J, is given by $J = L + S$. The associated quantum number determines the → multiplet structure of a term. Hund's third rule states that the term having the lowest energy is the one with the smallest J value, until the degenerate orbitals are half full, $\leq (2\ell + 1)$. Upon further population of the orbitals, up to $\leq 2(2\ell + 1)$, the lowest term is the one with largest J value. Hund's third rule is sometimes divided into a third and a fourth rule.

Hundreth-width → one hundreth-width.

Hybrid band, <*Hybridbande*>. Using the selection rules for the change of the rotational quantum number, J, and the change of the quantum number, K, which takes account of the additional angular momentum in a symmetric top molecule, we can distinguish between specific band types in the → rotation-vibration spectra of nonlinear molecules. The direction, with respect to the three principal axes of inertia, of the dipole moment change during the vibration is decisive here. In a symmetric top ($I_A \neq I_B = I_C$) therefore, we distinguish initially between → parallel and → perpendicular bands, for which the following selection rules apply:

Parallel bands: $\Delta K = 0$, $\Delta J = 0, \pm 1$ $\quad K \neq 0$

$\Delta K = 0$, $\Delta J \pm 1$, $\quad K = 0$

Perpendicular bands: $\Delta K = \pm 1$, $\Delta J = 0, \pm 1$

Vibrations, in which the change of dipole moment occurs not only in the direction of (parallel) but also perpendicular to the top axis, show the characteristics of both parallel and perpendicular bands. The selection rules given above apply simultaneously in this case. Such bands are termed hybrid bands; they are sometimes also called mixed or bastard bands. Hybrid bands which can be assigned to mixtures of type A, type B and type C bands occur in the asymmetric top; especially in molecules of low symmetry (→ rotation-vibration spectrum of nonlinear molecules).

Hydride technique, *<Hydridtechnik>*. The basis of the hydride technique in → atomic absorption spectroscopy is the fact that antimony, arsenic, bismuth, selenium, tellurium, and tin and, with certain limitations, also germanium, lead and phosphorus are capable of forming gaseous hydrides. These hydrides are relatively stable at room temperature, but they dissociate rapidly into atoms at temperatures of 800–1000°C. Figure 1 shows the functional principles of equipment for the hydride technique. The reducing agent, usually a 3% solution of sodium borohydride in 1% sodium hydroxide, is contained in a reservoir. Prior to the reduction an inert gas, usually nitrogen, flows

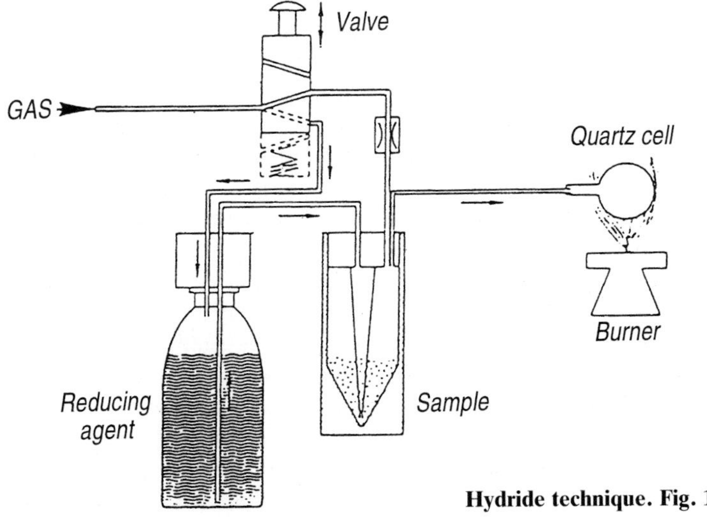

Hydride technique. Fig. 1. The apparatus

through the reaction vessel to purge the air from the entire system, including the quartz cell. By means of a valve, the major portion of the gas stream is diverted to the reducing agent reservoir where it pushes the sodium borohydride solution into the reaction vessel. There, hydrogen and the covalent, gaseous hydrides are formed in a rapid reaction and they flow, together with an additional inert gas stream, into the heated quartz cell where the hydrides are thermally dissociated into atoms of the element of interest (and hydrogen). The cell is aligned in the beam of an → atomic absorption spectrometer. Depending upon the element under investigation, the quartz cell is heated to the usual 800–1000°C by means of the air/acetylene flame of the AAS instrument or electrically.

For the elements mentioned, considerably improved relative limits of detection can be achieved with the hydride technique as compared with the → graphite furnace. A further advantage is the selective volatilization of the elements to be analyzed which separates them from accompanying substances. For this reason, matrix interferences occur only rarely with the hydride technique and the use of background compensation is usually not required.

Hydrogen anomaly, <*Wasserstoffanomalie*>, a phenomenon the origin of which lies in the different populations of rotational states. It is directly related to the → intensity alternation in the → rotational Raman spectrum of the H_2 molecule, for which Figure 1 shows the relevant energy-level scheme. The rotational states are symmetric for J = 0, 2, 4 ... and antisymmetric for J = 1, 3, 5 ... If we take

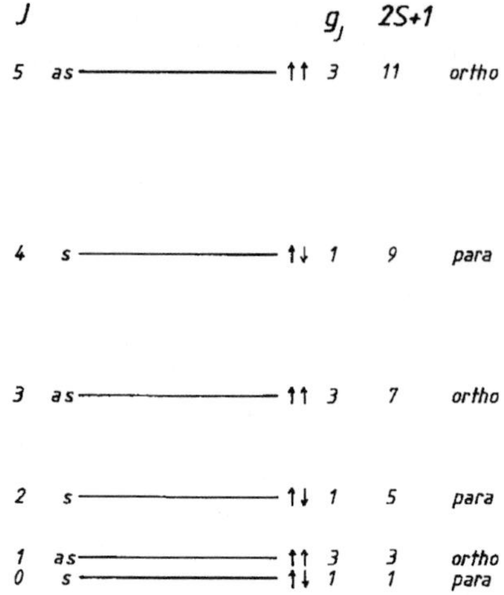

Hydrogen anomaly. Fig. 1. Rotational energy-level scheme for the hydrogen molecule

account of the nuclear spin of the H atom with $I(^1H) = 1/2$ then, according to Fermi-Dirac statistics, symmetric states have the statistical weight, g_J = 1 and the antisymmetric states the statistical weight, g_J = 3. Thus the states having J = 1, 3, 5 ... have the greater population in the ratio 3:1. The antisymmetric states have a total spin, I_{tot} = 1, whilst the symmetric states have a total spin, I_{tot} = 0. For the sake of completeness, the degeneracy in J has been added in Figure 1. The → selection rule antisymmetric $\nleftrightarrow$ symmetric, shows that intercombination transitions are strictly forbidden. Therefore, transitions between adjacent J states (ΔJ = ±1) are not allowed and, in principle, we have for the H_2 molecule two energy-level systems which do not combine with each other. The existence of these two sets of levels, in which the distances between the rotational states are particularly large because of the large rotational constant (B_e = 30.43 cm^{-1}), leads to the hydrogen anomaly. If the temperature is continuously decreased, not all of the hydrogen molecules can move smoothly into the rotationless state, $J = 0$. Only those molecules with antiparallel nuclear spins ($I_{tot} = 0$; $g_J = 1$) can occupy this state whilst those with parallel nuclear spins ($I_{tot} = 1$; $g_J = 3$) must remain in the state $J = 1$, with no connection to $J = 0$; no matter how low the temperature to which the hydrogen is cooled. Fundamentally, this is due to the → Pauli principle and it is expressed in the anomaly in the specific heat of the hydrogen at low temperatures. Molecular hydrogen behaves as if it consists of two different modifications (see Figure 1) to which Bonhoeffer and Harteck gave the names ortho- and para-hydrogen in analogy with ortho- and para-helium (K. Bonhoeffer, P. Harteck, *Z. Phys. Chem.*, **1929**, *B4*, 113). Analogous effects must be taken into consideration for D_2 and N_2 for which we can also distinguish between ortho- and para-modifications. The modification with the greater statistical weight is always called the ortho-modification and the one with the smaller weight the para-modification. If the gas is kept at a low temperature for several days or weeks, all molecules finally move into the $J = 0$ state.

Hydrogen lamp, <*Wasserstofflampe*>, the usual radiation source in → dispersive spectrometers below 350 nm (above 28,500 cm^{-1}) because a discharge in hydrogen provides a continuous spectrum between 160 and 400 nm (62,500–25,000 cm^{-1}). Figure 1 shows the construction of a hydrogen lamp. A tungsten helix is mounted as an anode on the axis of a cylindrical discharge tube made of quartz. An

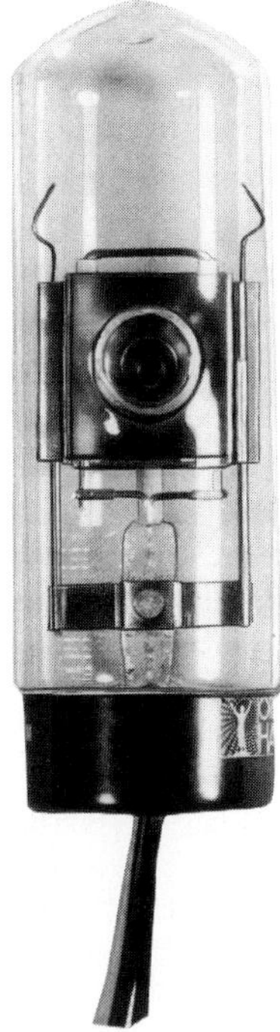

Hydrogen lamp. Fig. 1. Construction of a hydrogen or deuterium discharge lamp

activated tungsten double spiral is generally fitted at the side as cathode. The lamp is filled with hydrogen or deuterium at approximately 10 Torr. A heating voltage is usually applied to the cathode coil so that its favorable thermal emission properties keep the ignition voltage low (200–400 V). Hydrogen, being the lightest gas, has a very high diffusion velocity. Consequently, the energy loss by heat conduction is very large and the radiation yield correspondingly low. Therefore, deuterium is used in modern lamps which improves the radiation yield by ca. 30% because of the doubling of the molecular weight.

In order to generate the highest possible radiation densities which are required in optical instruments, it is

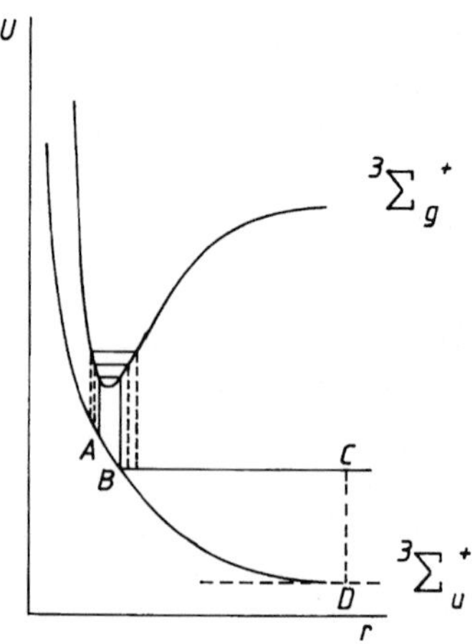

Hydrogen lamp. Fig. 2. Potential curves to explain the hydrogen continuum, see text

necessary to restrict the discharge between the cathode and anode by means of an aperture with a small cross section (approximately 1 mm^2) in a refractory metal (molybdenum).
The continuum between 160 and ca. 400 nm derives from the transition between a stable, excited state; the lowest triplet state, $^3\Sigma_g^+$ and a repulsive state, $^3\Sigma_u^+$, of the H_2 molecule (cf. Figure 2). When the molecule falls to the lower state, a simultaneous dissociation occurs and the emitted continuous spectrum is therefore also called → decay luminescence. The hydrogen or deuterium atoms formed recombine on the cold surfaces of the discharge tube giving H_2 or D_2 molecules. In addition to the continuum, the Balmer series (→ Balmer's formula) of the H or D atoms occurs in the visible spectral region so that the corresponding H_α, D_α ($n = 3 \rightarrow n = 2$) or H_β, D_β ($n = 4 \rightarrow n = 2$) lines can be used to calibrate the wavelength scale of the UV-VIS spectrometer. If the lamp is filled with a mixture of hydrogen and deuterium, the Balmer series of the two gases are emitted. The H_β line lies at 486.12 nm and the D_β line at 485.99 nm. The difference of $\Delta\lambda$ = 0.13 nm or $\Delta\tilde{\nu} = 5.5$ cm^{-1} can be conveniently used to check the → resolving power of a → spectrometer.

Hydrogen-like ions, <*wasserstoffähnliche Ionen*>, ions which, after a single or multiple → ionization, have the electronic configuration of the H atom, $1s^1$. Examples of such ions are: He^{1+}, Li^{2+}, Be^{3+}, B^{4+} ...
One might expect, therefore, that the associated → spark spectra HeII, LiIII, BeIV, BV ..., would be completely hydrogen-like if account is taken of the effective nuclear charge $Z - p$. p is the number of remaining core electrons and in this case $p = 0$ and $Z - p = Z$ which takes the values 2, 3, 4, 5 ... in the series He^{1+}, Li^{2+}, Be^{3+}, B^{4+} .... The wavelengths of the lines and the term values should therefore differ by the constant factor Z^2. I.e. if the term value is divided by Z^2 we should obtain the same numerical result for all the spectra in a series. Apart from a very small effect due to the change of nuclear mass (→ Rydberg constant) this expectation is fulfilled. Thus the spectra of all the hydrogen-like ions can be represented with the same energy-level diagram, practically the only difference being a change in the energy scale. Cf. → alkali-like ions.

Hyperchromism, <*Hyperchromie*>, an observed increase in intensity in conjunction with the → bathochromic shift of the characteristic bands of an UV-VIS absorption spectrum. When the intensity decreases it is known as hypochromism.

Hyperfine structure of atomic spectra, HFS, <*Hyperfeinstruktur des Atomspektrums, HFS*>. The fine structure

(FS) of a line spectrum is generally described in terms of (*LS*) coupling, → multiplet structure. If the individual components of a multiplet are measured with instruments having a very high resolving power, such as → interference spectroscopes or gratings in high orders, many atomic spectra show that they have been split further into a number of extremely closely spaced components called hyperfine structure (HFS). The total splitting, i.e. the distance between the two outer components, is of the order of 2–3 cm^{-1} at most and is frequently below 1 cm^{-1}. It is obvious, therefore, that separations of $\Delta\tilde{\nu} < 0.1$ cm^{-1} can occur between adjacent lines in hyperfine structure. The properties of the atomic nucleus must be considered when explaining hyperfine structure; and we recognize two different phenomena:

a) The → isotope effect and b) the nuclear spin.

a. Because both particles, the nucleus and the electron, move around their common center of gravity, the → Rydberg constant, *R*, depends upon the nuclear mass. This is taken into account by use of the → reduced mass, $\mu = mM/(M + m)$; where *m* is the mass of the electron and *M* the mass of the nucleus. It follows from the series formulas that a heavy isotope of an element must show a shift of the spectral lines vis-a-vis a lighter isotope. Urey *et al.* were the first (1932) to find a very weak spectral line on the short-wavelength side of each of the Balmer lines, H_α (656.279 nm), H_β (486.133 nm), H_γ (430.447 nm) and H_δ (410.174 nm) at separations of 1.79 Å (H_α), 1.33 Å (H_β), 1.19 Å (H_γ) and 1.12 Å (H_δ). The wavelengths of the additional lines agree completely, within the experimental error, with the values calculated using → Balmer's formula and the mass $M = 2$, instead of $M = 1$, in the Rydberg constant. This also provided proof of the existence of the hydrogen isotope with mass $M = 2$; deuterium. The effect has also been observed in heavier atoms such as ^{6}Li and ^{7}Li, ^{20}Ne and ^{22}Ne and in the Zn isotopes, ^{64}Zn, ^{66}Zn and ^{68}Zn. The intensities of the lines correspond to the abundance of the isotopes and therefore also provide evidence that the observed HFS is due to an isotope effect. With increasing atomic number and weight, we may expect that the movement of the nucleus will decrease in importance, although the example of the Zn isotopes shows that an isotope effect can still be observed. An estimate of the splitting, $\Delta\tilde{\nu}$, in wavenumbers, caused by the isotope effect at the mean wavenumber, $\tilde{\nu}$, can be obtained by using the relationship:

$$\Delta\tilde{\nu} = \tilde{\nu} \cdot \frac{m_e \Delta M}{M^2};$$

where m_e is the mass of the electron, *M* the mean atomic mass and ΔM the mass difference of the two isotopes. The lead isotopes, ^{204}Pb and ^{206}Pb ($\Delta M = 2$) would show a splitting of ca. 0.0008 cm^{-1} at $\tilde{\nu} \sim 30{,}000$ cm^{-1} which would not be observed.

b. The isotope effect is not sufficient to explain the HFS in very many cases. The number of components is frequently much greater than the number of isotopes. In particular, elements such as ^{209}Bi and ^{141}Pr, which have only one isotope, show HFS splittings. However, their HFS can be explained if we remember that the atomic nucleus, like the electron, has a spin angular momentum with an associated magnetic moment (→ NMR spectroscopy). The magnetic moment of a nucleus, μ, is given by:

$$\mu = g_I \cdot \vec{\mu}_N \cdot \sqrt{I(I+1)};$$

where g_I is the nuclear g factor, μ_N the → nuclear magneton and I the nuclear spin quantum number which assumes integer or half-integer values between 0 and 9/2, depending upon the nucleus.

A magnetic interaction occurs between the nucleus and the electron shell which is very small because the nuclear magneton is only 1/1836 of the → Bohr magneton. Nevertheless, it is this magnetic interaction which leads to the very small HFS splittings of the terms of an atom. In a manner similar to (LS) coupling (→ Russell-Saunders coupling) where a total angular momentum, J, results for the outer electrons, we obtain the total angular momentum, F, of the whole atom, including the nuclear spin, by coupling J and I. The associated quantum number can assume the values, $J + I$, $J + I - 1$, $J + I - 2$ ... $|J - I|$; this gives a total of $(2J + 1)$ values for $J < I$ and $(2I + 1)$ values for $J > I$, quite analogous to the coupling between L and S in the electronic structure. The selection rule for transitions between the HFS terms is also analogous:

$$\Delta F = 0, \pm 1.$$

The energy-level diagram, Figure 1, shows the origin of the HFS of the BiI line at 24,260 cm^{-1} (4122 nm). The nuclear spin quantum number of the Bi atom is $I = 9/2$. J has the value 1/2 in both states so that two values result for F (because $J < I$), i.e. $F = 5$ and 4. Taking account of the selection rule, $\Delta F = 0$ and ± 1, a total of four lines is obtained as shown in Figure 1. Furthermore, we see that the energy differences of the upper two and lower two states each occur twice, which is in complete agreement with experiment. In an external magnetic field the HFS terms split into $(2F + 1)$ components (→ Zeeman effect) which are defined by the quantum number, M_F, for the components of the total angular momentum, F, in the direction of the field (space quantization of F): $M_F = F, F - 1, F - 2 \ldots - F$. Here also, in stronger magnetic fields the Zeeman effect gives way to the → Paschen-Back effect where J and I are independently space-quantized with components, M_J and M_I. This occurs at considerably smaller magnetic field strengths in HFS due to the weakness of the coupling. The directional quantization of J results in the usual Zeeman effect with line separations which are considerably greater, at sufficient field strength, than those of the HFS components without a field. However, each term with a specific M_J is split further into $(2I + 1)$ components, because M_I can assume the values, I, $I-1$, $I-2$ ... $-I + 1$, $-I$, i.e. $(2I + 1)$ values. This number of components is the same for all the terms of an atom since I is constant for an atom. Figure 2 shows these relationships for $J = 1/2$ and $I = 3/2$. It follows from the selection rule, $\Delta M_I = 0$, and the fact that the magnitude of the splitting varies for different states, that each of the simple anomalous Zeeman components (not taking account of the nuclear spin) consists of $2I + 1$ very closely spaced, equidistant lines. From measurements in a sufficiently strong magnetic field and with adequate resolving power it is possible to determine the nuclear spin quantum number, I, simply by counting the components of the line.

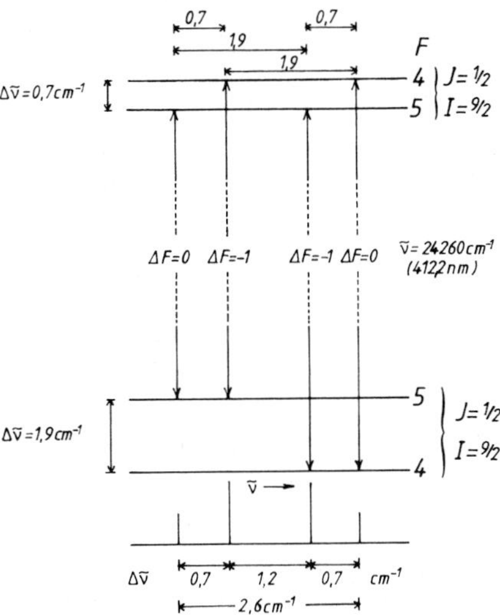

Hyperfine structure. Fig. 1. HFS; the example of the BiI line at 24,260 cm^{-1}

Ref.: G. Herzberg, *Atomic Spectra and Atomic Structure*, Dover Publications, New York, **1944**,

Hypochromism, *<Hypochromie>* → hyperchromism.

Hypsochromic shift, blue shift, *<Hypsochromie, Blauverschiebung>*, a term which describes the shift of the UV-VIS absorption spectrum of a → chromophore to shorter wavelengths or greater wavenumbers under the influence of substituents or solvents. The opposite effect is called a → bathochromic shift. We observe a pronounced hypsochromic shift especially with $n \rightarrow \pi^*$ transitions on changing from a nonpolar solvent (e.g. n-hexane) to a polar solvent (e.g. ethanol); $\pi \rightarrow \pi^*$ transitions generally show a bathochromic shift in this case. For further information on $n \rightarrow \pi^*$ and $\pi \rightarrow \pi^*$ transitions → HMO energy-level scheme.

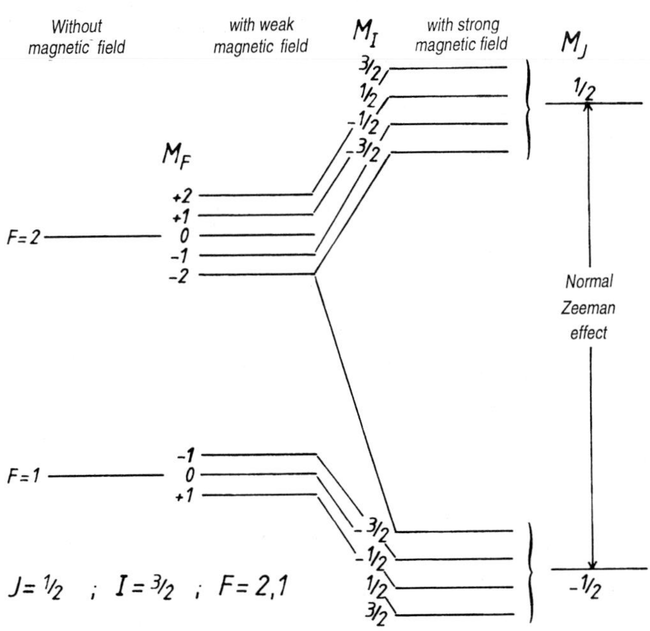

Hyperfine structure. Fig. 2. HFS splitting of terms in weak and strong magnetic fields, see text

I

ICP, <*ICP*> → inductively coupled plasma.

Image error, <*Abbildungsfehler*>, errors which are to some extent caused by the optical properties of the lens material and which can cause problems in the construction of spectroscopic instruments. The most important errors are → chromatic and → spherical aberration, → coma, → astigmatism and curvature of the field of vision. These image errors can be eliminated or partially reduced by the use of compound lenses.

Incoherent light source, <*inkohärente Lichtquelle*>, a conventional light source such as is used, for example, in optical spectroscopy. → Thermal radiators (incadescent lamp, Nernst glower, globar) and → gas or metal-vapor discharge lamps are examples. The incoherence of conventional light sources derives from the mechanism by which the light emitted by them is generated. The centers of a light source, which actually emit the light, are atoms or molecules the outer electrons of which are moved into excited states by the supply of energy (→ excitation). After a specific lifetime of ca. 10^{-9} to 10^{-7} s, they return to the ground state with the emission of light. The emission has the form of a damped wave which can be regarded, in approximation, as a wave train of a limited length. In compounds containing many atoms or molecules, the individual acts of light emission at the different centers occur statistically so that the individual wave trains have changing phase constants. Therefore, the light waves radiated from the different points or centers of a light source are incoherent with each other. The same applies to different light sources.

Incoherent(ly) scattered radiation, <*inkohärente Streustrahlung*>, scattered radiation in which, in contrast to coherent scattered radiation, no fixed phase relationship exists between the phases of the scattered radiation. This is the case, for example, if a change of frequency is associated with the scattering as in scattering by free, or only loosely bound, electrons (→ Compton effect). The scattered radiation is also incoherent, if unspecific, i.e. varying, time intervals are present between the emission of the scattered radiation and the impact on the incident radiation on the scattering particles. A change of frequency (→ Raman effect) almost always occurs with incoherent scattering. The scattered radiation in the purest media (→ Tyndall effect) is also incoherent.

Index ellipsoid, <*Indexellipsoid*>. In anisotropic materials the dielectric polarization, *P*, does not generally lie in the direction of the electric field, *E*. Each component of the dielectric polarization depends upon the *x*, *y* and *z* components of the field:

$$\begin{aligned} \vec{P}_x &= \varepsilon_0(\chi_{xx}\vec{E}_x + \chi_{xy}\vec{E}_y + \chi_{xz}\vec{E}_z) \\ \vec{P}_y &= \varepsilon_0(\chi_{yx}\vec{E}_x + \chi_{yy}\vec{E}_y + \chi_{yz}\vec{E}_z) \quad (1) \\ \vec{P}_z &= \varepsilon_0(\chi_{zx}\vec{E}_x + \chi_{zy}\vec{E}_y + \chi_{zz}\vec{E}_z) \end{aligned}$$

(where χ is the electric susceptibility and ε_0 the vacuum permittivity.) Similarly, for the electric displacement vector, *D*, we have:

$$\vec{D}_x = \varepsilon_{xx}\vec{E}_x + \varepsilon_{xy}\vec{E}_y + \varepsilon_{xz}\vec{E}_z$$

$$\vec{D}_y = \varepsilon_{yx}\vec{E}_x + \varepsilon_{yy}\vec{E}_y + \varepsilon_{yz}\vec{E}_z \quad (2)$$

$$\vec{D}_z = \varepsilon_{zx}\vec{E}_x + \varepsilon_{zy}\vec{E}_y + \varepsilon_{zz}\vec{E}_z;$$

where $\varepsilon_{ik} = \varepsilon_0\chi_{ik}$ for $i \neq k$ and $\varepsilon_{ik} = \varepsilon_0(\chi_{ik} + 1)$ if $i = k$ $(i, k = x, y, z)$. Thus the fields, $\vec{D}$ and $\vec{E}$, are connected by a tensor, $\underline{\varepsilon}$, the components of which can be determined with equation (2).

It can be shown, using Maxwell's equations, that the tensor is symmetric which means that this tensor and its reciprocal tensor, $\underline{\varepsilon}^{-1}$, can be represented by ellipsoids. This is analogous to the representation of the inertial tensors in mechanics.

The tensor $(\varepsilon_0\underline{\varepsilon})^{-1}$ is called the Fresnel ellipsoid and the tensor, $\varepsilon_0\underline{\varepsilon}^{-1}$, the index ellipsoid.

The relationships between $\vec{E}$ and $\vec{D}$ as well as between the diffusion of energy in the direction of the vector $\vec{w}$, ($\vec{w}$ is perpendicular to $\vec{E}$) and the propagation of the wave surfaces in the direction of the unit vector $\vec{s}$, ($\vec{s}$ is perpendicular to $\vec{D}$) are important in crystal optics.

Index matching, <*Index-Matching*>, the matching of refractive indexes → frequency doubling.

Indicator, <*Indikator*>, a substance which can show the equivalence point of a reaction visually. This may result from a change of color, fluorescence or turbidity or from an increase of color due to absorption. Indicators are used internally in most cases, i.e. a few drops of an indicator solution are added to the reaction mixture under investigation. In specific cases, as determined by the system, the use of an external indicator is preferred. The best-known external indicators are the pH indicator papers, reagent papers, test sticks, etc. The most important and most common indicators are the acid-base indicators for aqueous systems which, in general, are used internally. They are acids or bases and change color upon loss or gain of a proton. Particular examples are azo dyes, nitrophenols, phthaleins, sulfophthaleins and triphenyl methane dyes which change their π electron systems upon protonation or deprotonation in such a way that the position of the longest wavelength absorption maximum is significantly shifted. If this shift occurs within the VIS spectral region a change of color occurs, i.e. two-color indicator such as bromocresol green and methyl red. If the absorption spectrum is shifted hypsochromically (→ hypsochromic shift) out of the VIS region or bathochromically (→ bathochromic shift) into it, a decoloration or coloration of the previously colored or colorless solution occurs, i.e. a single color indicator such as phenolphthalein. Since the change of color is brought about by protonation or deprotonation, the color change is dependent upon the hydrogen-ion concentration i.e. the pH value. The turning point of the indicator is determined by that hydrogen-ion concentration at which the concentrations of the alkaline and acid form of the indicator are equal. This point lies at a different hydrogen-ion concentration for each dye. Thus, it is possible to show each such concentration or pH value by the turning point of a specific indicator.

However, in practice indicators do not show a sharp turning point with a sudden change of color. Instead they have a pH region of varying extent which is characterized by a gradual color transition via a mixed color. This is caused by the different positions of the absorption maxima of the two molecular forms in the pH region of their interchange. For the most exact visual recognition of the equivalence point of the acid-base pair under investigation, two things are therefore required: the largest possible jump in the pH value around the equivalence point and the use of a small quantity of the indicator, since it must also be titrated. By adding a pH neutral dye such as methylene blue to methyl red or by mixing several indicators, the recognition of the color change can be improved considerably. The first type of mixture is known as a screened indicator and the second as a mixed indicator.

Due to hydrolytic processes, which are a consequence of the different degrees of dissociation of the individual acids and bases, the equivalence and neutral point do not always coincide. Thus, the determination of the equivalence point requires the selection of an indicator the turning range of which covers the pH value of the equivalence point. Therefore, for every acid-base system there is an optimum indicator which depends upon the pH value at the equivalence point.

The error, which can occur in the usual visual and thus subjective acid-base titration with an indicator, can be largely eliminated by the application of an objective method of detection. The combination of an automatic piston burette with a → photometer makes it possible to plot the whole titration curve, in which the inflection point of the S-shaped curve can be recognized accurately, from the first derivative if necessary. Such a method can also be extended in process control. In addition to acid-base indicators for aqueous solutions, there are a number of other indicators whose application is based upon a change of their spectroscopic properties.

Chemiluminescence indicators are organic compounds which either show chemiluminescence or quench it at or near the equivalence point of a titration. The energy for this → chemiluminescence is normally generated by an oxidation-reduction (redox) process. The emitted, cold light can be measured with a fluorescence spectrometer. On account of their reaction mechanism, chemiluminescence indicators are used as redox indicators.

Since the redox potential depends upon the pH value, they can also be used as acid-base indicators.

Extraction indicators are compounds which move suddenly from one liquid phase to the other at or near the equivalence point of a two-phase titration. The indicator must show no change of its absorption spectrum in this process.

Fluorescence indicators change their fluorescence properties at or near the equivalence point of a titration. A few such indicators change their fluorescence spectrum, others are converted to a fluorescing form and show fluorescence for the first time at the equivalence point. The fluorescence of these organic compounds can be influenced by various parameters such as pH value, oxidation potential or the presence of metal ions. Thus, they can be used to determine these parameters. However, in most cases fluorescence indicators are used for the determination of the pH value of solutions in which the common pH indicators show only a blurred end point, e.g. drinks, synthetic resins, oils, etc.

Metal-fluorescence indicators are a special form of fluorescence indicators. Under suitable irradiation they change their fluorescence spectrum in the presence of specific metals.

The metallochromic indicators required for complexo-metric titrations form another very important group. They are themselves complexing agents for the metal ion to be titrated and frequently show different absorption spectra in the free and complexed states.

Redox indicators are easily reversible, oxidizable and reducible organic compounds whose absorption spectra change at or near the equivalence point of an oxidation or reduction. Their action does not depend upon the character of the agent to be oxidized or reduced but only upon the relative positions of the oxidation potentials of the system to be titrated and the indicator. Therefore, the turnover potential of the indicator and the redox potential at the equivalence point of the titration must agree as closely as possible. If, in addition to the movement of electrons, the color change of a redox indicator is connected with a movement of protons, then the turnover potential also depends upon the pH. In such a case, this particular pH value must be present at the equivalence point of the titration for an error-free end point determination.

Ref.: J.G. Dick, *Analytical Chemistry*, Robert E. Krieger Publishing Co., Huntington, New York, **1978**.

Induced emission, *<induzierte Emission>* → laser.

Inductively coupled plasma, ICP, *<ICP, induktiv gekoppeltes Plasma>*, an electrodeless → plasma generated by induction heating in the coil of a high-frequency (HF) generator. A torch consisting of three concentric quartz tubes is mounted in the coil of the HF generator. A gas such as argon flows through the tubes. If the HF field is switched on and ionization is initiated by a spark of high energy density (Tesla spark) the HF field transfers electrical energy to the gas and heats it. The transfer is made in such a way that the HF field generates a magnetic field which, in turn, produces an electric field. If a conducting material (ionized gas) is present in this electric field, the field induces a high-frequency current in it. The electrical energy absorbed by the charge carriers is converted into kinetic energy. Heat energy is released by the collisions of the charge carriers with the other gas particles causing evaporation, dissociation, excitation and ionization of the introduced sample aerosol. In the ICP, the kinetic temperature is ca. 3000 K, the excitation temperature ca. 5000 K and the ionization temperature ca. 8000 K.

Figure 1 shows the characteristic temperature profile of an ICP. These temperatures are sufficient for the formation of free atoms (ca. 4000 K) and also ions (ca. 7000 K) and the ICP therefore provides a spectrum rich in atomic and ionic lines.

By the selection of specific conditions (frequency, torch construction and gas velocity), the plasma is brought to a toroidal form in which the axial zone is cooler than the surrounding ring (see Figure 1). If a gas stream with a diameter of 1–2 mm is directed at the center of the plasma it drills a tunnel in the plasma without destroying its stability. This provides the possibility

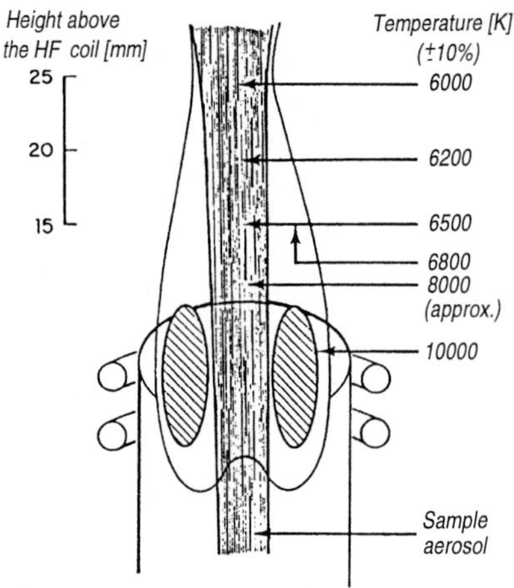

Inductively coupled plasma. Fig. 1. Cross section of a plasma torch and its temperature profile

of introducing an aerosol efficiently into the plasma. The toroidal structure of the ICP is extremely effective for introducing the aerosol into the plasma which is an important reason for the high sensitivity of the ICP. The high temperatures in the tunnel (ca. 6000 to 8000 K, see Figure 1) and the long dwell time of the sample (ca. 1 ms) are decisive for the effectiveness of the energy transfer from the ionized gas to the sample and thus for its → atomization, → ionization and excitation. The sample is transported via the tunnel to the zones lying above the HF coil. There, the toroidal structure of the plasma gradually disappears and the plasma takes the form of a flame. An observation height of approximately 20 mm above the HF coil has proved to be the most favorable position for emission measurements. Here, the temperatures lie between 6000 and 6500 K.

In addition to the temperature and the long dwell time, the chemical environment is also important. Since the plasma consists essentially of argon, and oxygen is only formed in small quantities by the dissociation of the water introduced with the sample, a low partial pressure of oxygen is present and the formation of oxides is quite improbable. Therefore, elements with a high affinity for oxygen can be atomized very effectively in the ICP.

Infrared, <*Infrarot-...*> → IR ...

Infrared polarizer, <*Infrarotpolarisator*>. The phenomenon of polarization by → reflection at a dielectric medium is primarily used for the generation of polarized light in the infrared spectral region. If the angle of incidence is selected in such a way that the refracted beam is perpendicular to the reflected ray then, in the ideal case, the reflected light is completely polarized (Brewster's law). Its electric vector is perpendicular to the plane of incidence. The transmitted beam is also polarized, though to a lesser degree, with its electric vector parallel to the plane of incidence.

The angle of incidence, α, for which both beams are perpendicular to each other, depends upon the ratio of the refractive indexes of the two media at the interface where the reflection takes place. Normally, this occurs only at the interface between air and another substance with the refractive index, n. The Brewster angle is then given by the simple equation:

$$\tan\alpha = h$$

This is the principle of almost all infrared polarizers and use can be made of the reflected radiation, following one to two reflections, as well as the transmitted radiation. On account of the low degree of polarization, several passes are required in the latter case. The previously more common reflection polarizers used selenium mirrors with an angle of incidence of 71°, and also devices with germanium and PbS. Although a very high degree of polarization is possible, almost all these polarizers have the disadvantage that the deflection of the beam caused by reflection must be compensated by additional mirrors. Transmission polarizers do not require such deflecting mirrors. Several thin plates of an electrically nonconducting material, which is as transparent as possible in the whole IR region, are placed in the beam at the appropriate Brewster angle. Selenium ($\alpha = 68°$) can be used as the material. Thin layers can be easily prepared by evaporation onto a substrate which can be removed afterwards with a solvent. Silver chloride ($\alpha = 63°$), KRS-5 ($\alpha = 67°$), tellurium ($\alpha = 78°$), germanium ($\alpha = 76°$) and, especially for the long-wavelength region, polyethylene ($\alpha = 55.5°$) plates are frequently used. In all these polarizers the plates are mounted parallel to each other and in such a way that the beam is incident on the reflecting surface at the Brewster angle. The complete polarizer can also be rotated in both senses around the direction of the transmitted beam. In the case of polarizers of the type described, the degree of polarization of the transmitted radiation depends primarily upon the number of plates, provided that the materials are optically isotropic. The degree of polarization lies between 85 and 99 % for most polarizers.

For the sake of completeness, a number of other polarizers should be mentioned. However, they are only suitable for specific purposes on account of their limited transmittance. Dichroically orientated foils of polyvinyl alcohols are available for the short-wavelength infrared where they can be used, with limitations, up to ca. 6.5 μm. Very high degrees of polarization (> 99 %) can be achieved with them. Polarizers made of doubly refracting materials such as quartz and calcite are also suitable in the short-wavelength infrared. If calcite is cut into thin plates at specific angles to the axes, very good polarizers are obtained for the region between 2.5 and 16 μm. A transmission polarizer, which has very good polarization properties, can be produced for the long-wavelength region, from 10 to 650 μm, from pyrolitic graphite.

Apart from those described last, all other IR polarizers have the great disadvantage that, due to their size, they can only be placed in the light beam in the sample compartment of the IR instrument; i.e. the most favorable position, behind the sample and immediately in front of the entrance slit of the monochromator, can only be achieved with difficulty. The → wire grid polarizer is a new type of polarizer.

Infrared source, *<Infrarotstrahler, IR-Strahler>*, a radiation source with a strong emission in the IR spectral region from 0.8 to ca. 10^3 μm. Continuous IR sources are of particular importance for technical applications and especially for use in → IR spectrometers. They are usually → thermal radiators.

The most common thermal radiators are:

1. Metal-filament incandescent lamps;
2. metal coils made of noble metals on or in ceramic bodies;
3. Nernst glowers;
4. globars;
5. zirconium point lamps;
6. mercury-vapor discharge lamps.

In recent years, the importance of the Nernst glower and globar, as radiation sources in the near and mid IR region, has been greatly reduced because of the use of the sources mentioned under 2.

Instrumental line width, <*Apparatebreite*>, the contribution to the width of a spectral line which depends upon the → spectroscopic instrument.

Integral absorption, integrated absorption, <*integrale Absorption, integrale Intensität*>, the integral over an absorption band. If the absorption band is represented in general by $\varepsilon_{\tilde{\nu}} = f(\tilde{\nu})$:

$$A_g = 2.303 \int_{Band} \varepsilon_{\tilde{\nu}} d\tilde{\nu},$$

where A_g is the integral absorption.
If $\varepsilon_{\tilde{\nu}}$ is substituted by the measured absorbance, A, according to the → Bouguer-Lambert-Beer law, we obtain:

$$A_g = \frac{2.303}{c \cdot d} \int_{Band} A_{\tilde{\nu}} d\tilde{\nu}$$

(c = the concentration in $\mathrm{mol \cdot l^{-1}}$; d = the path length in cm).
The integral represents the area beneath the absorption band. Assuming that the absorption band is structureless and corresponds approximately to a Lorentzian function, the integrated absorption can be approximated by the expression:

$$2.303 \left[\frac{\pi}{2} \varepsilon_{max} \Delta\tilde{\nu}_{1/2} \right]$$

ε_{max} is the molar decadic → extinction coefficient at the absorption maximum and $\Delta\tilde{\nu}_{1/2}$ the → half-width of the absorption band. The units of A_g are then $\mathrm{l\ mol^{-1} \cdot cm^{-2}}$ or $\mathrm{1000\ mol^{-1} \cdot cm}$.
From the integral absorption it is possible to calculate the → oscillator strength, f, which, in turn, is related to the → transition dipole moment based on quantum mechanics. Apart from the case of vibrational structure, where there are superimposed absorption bands a → band analysis, i.e. a deconvolution, must first be carried out before determining the integral absorption. A_g can be determined in three different ways, i.e.:

1. Planimetric determination of the area;
2. cutting out and weighing the area;
3. application of numerical integration rules such as the trapezium rule etc.

In → IR spectroscopy, most bands usually have a small half-width so that the distortion of the extinction coefficients by the finite slit width may be noticeable. In such cases, the measured integral absorption values must be corrected. This is particularly important in the photometric evaluation of IR measurements for the determination of concentration. Various authors have proposed procedures for this purpose.

Ref.: V.J.I. Zichy in *Laboratory Methods in Infrared Spectroscopy*, 2nd ed. (Eds: R.G.J. Miller and B.C. Stace), Heyden and Son Ltd., London, Philadelphia, Rheine, **1979**, chapt. 5.

Integrating sphere, reflectance attachment, <*Reflexionszusätze, Remissionszusätze*>, an attachment used with conventional UV-VIS spectrometers to measure diffuse reflectance. In the practical application of → reflection spectroscopy, the relative diffuse reflectance, $R'_\infty = R_\infty / R'_{\text{standard}}$ is defined. This definition has a formal similarity with the definition of the transmittance in normal absorption measurements. There, the attenuated intensity, I, of the light transmitted through the sample is related to the unattenuated intensity, I_o, transmitted through a reference cuvette. But in reflection spectroscopy, the diffuse reflectance is always related to that of a white standard. This is the principle upon which the construction of reflectance attachments for most spectrometers and the experimental method itself is based. A photometric sphere is an essential component of a reflectance attachment. It is coated internally with MgO or $BaSO_4$ and its purpose is to integrate the diffusely reflected light from the sample and standard. This is the reason for the name *integrating sphere*.

Figure 1 shows an integrating sphere for single-beam operation schematically. The light leaving the exit slit of the monochromator is imaged via a lens and deflecting mirror onto the sample. The light diffusely reflected by the sample is collected in the sphere and falls upon to the photomultiplier. The measured signal is proportional to the diffuse reflectance of the sample. If the mirror is then rotated by a small angle (dashed line), the light falls onto a position on the internal wall of the integrating sphere which is also used as the → white standard. A measurement of the diffusely reflected light then provides the signal of the standard. By forming the ratio of the two signals the relative diffuse reflectance, R'_∞, of the sample at the set wavelength, λ is immediately obtained. This experimental geometry is called ${}_oR_d$ where, the left subscript indicates the angle of incidence of the radiation and the right subscript stands for measured diffuse reflection.

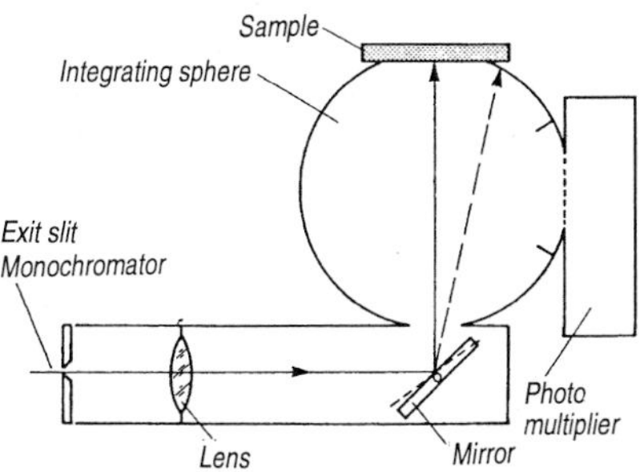

Integrating sphere. Fig. 1. Attachment for a single-beam instrument

It is frequently preferable to work with a movable sample holder which makes it possible to place the sample and the standard, one after the other, into the same position. This requires no adjustment of the mirror. The experimental configuration shown schematically in Figure 1 can be reversed, i.e. irradiate with diffuse white light and image the light reflected by the sample onto the entrance slit of the monochromator. This corresponds to the experimental arrangement, ${}_dR_o$. This is easily achieved by replacing the photomultiplier shown in Figure 1 with a continuum light source. Screening apertures (stops) must be used to ensure that no light falls directly upon the sample and standard. Figure 2 illustrates schematically the light path of a dual-beam spectrometer with an integration sphere (a simplified representation of the beam path in an attachment for

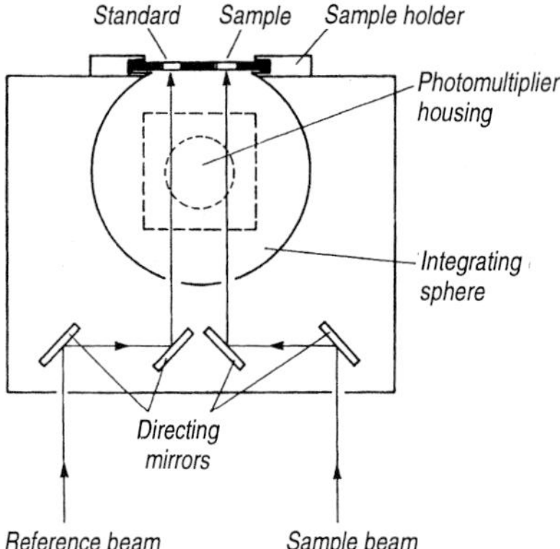

Integrating sphere. Fig. 2. Attachment for the Perkin-Elmer, model 55X dual-beam instrument

the Perkin-Elmer series 55X spectrometers). This attachment is equipped with a holder for both solid samples and cuvettes. The solid sample holder accepts samples with minimum dimensions of 12 mm × 22 mm up to a maximum of 40 mm × 40 mm and with a thickness of 6 mm. The experimental geometry, $_oR_d$, is also available with this attachment. The photomultiplier (dashed line) stands on the spherical surface vertically above the light beams. With this attachment and other auxiliary equipment it is possible to measure transmission spectra of turbid solutions and transparent solids. Using fiber bundles or → light pipes it is possible to measure with a photometric sphere outside the spectrometer.

Intensity, <*Intensität*>.

1. A property of electromagnetic radiation which is proportional to the square of its vibrational amplitude;
2. → integral absorption.

Intensity alternation, <*Intensitätswechsel*>, a phenomenon which is observed in → pure rotation spectra, the rotational structure of → rotation-vibration spectra and → electronic band spectra when measured in the gaseous state. For homonuclear diatomic molecules (O_2, N_2, H_2, etc.) and linear molecules of the point group $D_{\infty h}$, successive rotational states are alternately symmetric or antisymmetric. Thus, the states with even J (0, 2, 4, 6 ...) are symmetric and those with odd J (1, 3, 5, 7 ...) antisymmetric. An explanation of this intensity alternation lies in the nuclear spin, I_i, which determines the statistical weight, g_J, of a rotational state and hence its population. Because of this fact, successive rotational states can have varying statistical weights. These depend upon whether J is even (Ψ_{rot} = symmetric) or uneven (Ψ_{rot} = antisymmetric) and whether the nuclear spin, I_i, is integral or half-integral. Nuclei with spin $I_i = 0, 1, 2$ ... are described by → Bose-Einstein statistics (B.E.) and those with I_i = 1/2, 3/2, 5/2 ... by Fermi-Dirac statistics (F.D.) → quantum statistics. The statistical weights, g_J, are given for linear molecules with a center of symmetry ($D_{\infty h}$) by:

$$g_J = \frac{1}{2}\left[\Pi_i(2I_i + 1)^2 + \Pi_i(2I_i + 1)\right]$$

B.E.: for $J = 0, 2, 4, 6$ ...
F.D.: for $J = 1, 3, 5, 7$ ...
or

$$g_J = \frac{1}{2}\left[\Pi_i(2I_i + 1)^2 - \Pi_i(2I_i + 1)\right]$$

B.E.: for $J = 1, 3, 5, 7$ ...
F.D.: for $J = 0, 2, 4, 6$ ...

The products Π_i must only be extended over those nuclei which do

Intensity alternation. Table 1.

Molecule	I_i	I_{total}	Statistics	g_J	
				Ψ_s;J = 0,2,4,...	Ψ_a;J = 1,3,5,...
$^1H_2$	$I_{1H} = 1/2$	1/2	F.D.	1	3
O_2	$I_{16O} = 0$	0	B.E.	1	0
N_2	$I_N = 1$	1	B.E.	6	3
CO_2	$I_{12C} = I_{16O} = 0$	0	B.E.	1	0
CS_2	$I_{12C} = I_{32S} = 0$	0	B.E.	1	0
$^{12}C_2^1H_2$	$I_{12C} = 0; I_{1H} = 1/2$	1/2	F.D.	1	3
$^{12}C_2^2D_2$	$I_{12C} = 0; I_{2D} = 1$	1	B.E.	6	3
$^{13}C_2^1H_2$	$I_{13C} = 1/2; I_{1H} = 1/2$	1	B.E.	10	6
$^{13}C_2^2D_2$	$I_{13C} = 1/2; I_{2D} = 1$	3/2	F.D.	15	21

not lie at the center of symmetry. Furthermore, nuclei which exchange their positions upon inversion in the center of symmetry may only be counted once. The table shows the statistical weights, g_J, of a few diatomic and linear molecules ($D_{\infty h}$).

It follows for the H_2 molecule that the antisymmetric rotational states (J = 1, 3, 5 ...) have a threefold statistical weight (g_J = 3) while the symmetric (J = 0, 2, 4, 6 ...) states have g_J = 1; i.e. an intensity alternation occurs in the Raman spectrum. The unequal populations of the states J = 0 and J = 1, together with the applicable selection rules, results in the → hydrogen anomaly. Since g_J = 0 for the antisymmetric rotational states of O_2, CO_2 and CS_2, every second line is missing in their rotational Raman spectra. All the other molecules in table 1 show an intensity alternation, corresponding to the ratio of their given statistical weights, in their rotational Raman spectra or rotation-vibration spectra. No distinction between symmetric and antisymmetric rotational states can be made for linear molecules of the point group $C_{\infty v}$ such as HX(X: F, C1, Br, J), HCN, HC≡C-Cl, COS etc. since they have no center of symmetry. Therefore no intensity alternation occurs. For this reason, conclusions about the structure of a molecule can be drawn from its rotational structure. The molecule N_2O is the classical example. Its structure is N-N-O and not N-O-N since no intensity alternation is observed in its rotational spectrum.

The relations become rapidly more complicated in → top molecules. Table 2 shows the alternation of the statistical weights of the totally symmetric vibrational states of molecules with a p-fold principal axis of rotation, i.e. → symmetric top molecules. They are related to the intensity alternation of the K level in the sequence $K = p$, $p + 1 \ldots 2p$.

Ref.: G. Herzberg, *Molecular Spectra and Molecular Structure*, Vol. II and III, D. Van Nostrand, Princeton, New Jersey, Toronto, New York, London, **1966**.

Intensity-modulated light, <*Wechsellicht, frequenzmoduliertes Licht*>, light which is generated in → double-beam spectrometers by rotating sector

Intensity alternation. Table 2.

Molecule	Point group	Nuclear spin of identical atoms	Alternation in g_K	Alternation in J for $K = 0$
XY_3XYZ_3	C_{3v}	$I = 0$	1:0:0:1	–
NH_3;$HCCl_3$		$I = 1/2$	4:2:2:4	–
CH_3-X		$I = 1$	11:8:8:11	–
		$I = 3/2$	24:20:20:24	–
XY_3,X_3	D_{3h}	$I = 0$	1:0:0:1	1:0
		$I = 1/2$	4:2:2:4	0:4
		$I = 1$	11:8:8:11	10:1
		$I = 3/2$	24:20:20:24	4:20
cyclo-C_3H_6	D_{3h}	$I = 1/2$	24:20:20:24	8:16
XY_4	C_{4v}	$I = 0$	1:0:0:0:1	–
		$I = 1/2$	4:3:6:3:4	–
		$I = 1$	24:18:21:18:24	–
XY_4	D_{4h}	$I = 0$,	1:0:0:0:1	1:0
		$I = 1/2$	4:3:6:3:4	1:3
		$I = 1$	24:18:21:18:24	21:3
X_2Y_6(C_2H_6)s	D_{3d}	$I_x = 0$, $I_y = 1/2$	24:20:20:24	8:16
		$I_x = 0$, $I_y = 1$	249:240:240:249	138:111
C_6H_6	D_{6h}	$I = 1/2$	10:11:9:14:9:11:10	7:3
C_3H_4 (Allene)	D_{2d}	$I = 1/2$	10:6:10:6:10	7:3

mirrors or by → choppers. When it falls on a photoelectric detector it generates an alternating current which can be measured with the aid of an AC amplifier or a frequency-selective, lock-in amplifier; → light modulation.

Intensity standard (lamp), <*Lichtstärkenormal*> → light intensity standard.

Intercombination prohibition, intercombination ban, <*Interkombinationsverbot*> → intercombination transition.

Intercombination transition, <*Interkombinationsübergang*>, a transition between states of different → multiplicity in which the selection rule $\Delta S = 0$ is contravened, is, in principle, forbidden (intercombination prohibition). The theoretical explanation of the appearance of such transitions is based upon the strength of the coupling between the spin and orbital angular momenta. This coupling is large in elements with a high atomic number so that the intercombination ban applies less strictly with increasing atomic number. Thus, though it applies strictly to He, the intercombination line of Hg at 39,417 cm^{-1} or 253.7 nm is actually the most intense line in the spectrum of the Hg atom. It corresponds to the intercombination transition, $6^1S_0 \rightarrow 6^3P$ (→ Franck and Hertz experiment). Singlet and triplet energy levels are also present in organic molecules, especially in unsaturated hydrocarbons and their derivatives. Here the intercombination transitions are also forbidden which is manifested experimentally in the extremely small molar decadic → extinction coefficients for the $S_0 \rightarrow T_1$ transition which is about ~ 10^{-4} $l \cdot mol^{-1} \cdot cm^{-1}$. The $S_0 \rightarrow S_n$ transitions have extinctions of the order of $10^3 \leq \varepsilon \leq 10^5\ l \cdot mol^{-1} \cdot cm^{-1}$.

The → phosphorescence of these compounds corresponds to the transition, $T_1 \rightarrow S_0$ where T_1 is not populated by direct excitation but by → intersystem crossing. The long life of the T_1 state, in comparison to the S_1 state, is in agreement with the fact that this transition is also subject to the intercombination ban. Intercombination transitions in molecules are promoted by the incorporation of heavy atoms such as I or Br (internal → heavy atom effect) or by the interaction with solvents which contain heavy atoms (external heavy atom effect).

Interference at thin films, *<Interferenz an dünnen Blättchen>*, a phenomenon (→ interference of light waves) in which two surfaces lying close together produce the path difference of the interfering rays; e.g. an oil film on water, thin mica plates etc. Interferences at thin films are generally caused when a parallel pencil of rays falls onto a thin film. Figure 1 shows that the beam is split into a reflected and a transmitted component at both the front and the back surfaces of the film. Both the light passing through the film and those components reflected at the front and back surfaces can interfere. When viewed perpendicularly, the interference fringes lie in the upper film surface. In daylight or lamplight, the human eye sees the fringes in the usual polychromatic colors (→ interference colors) and we speak of the colors of thin films. When the fringes form circular patterns, such as occur at the point of contact between a lens with a large radius of curvature and a plane plate, they are known as Newton's rings.

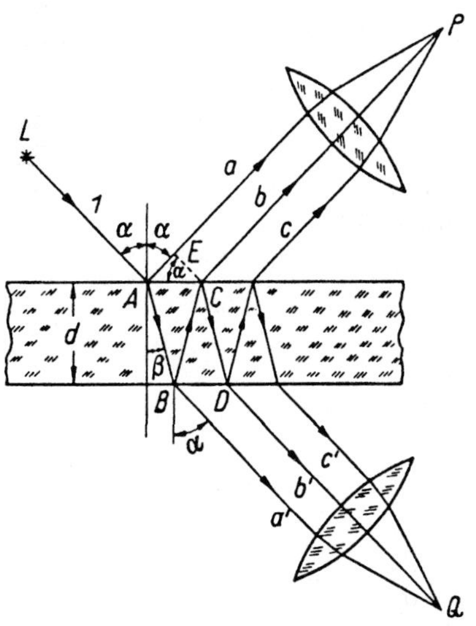

Interference at thin films. Fig. 1. Interference at a plate of thickness *d*; division into a reflected (*a*, *b*, *c*) and a transmitted component (*a'*, *b'*, *c'*)

Interference colors, *<Interferenzfarben>*, the various colors which generally occur in interference phenomena with polychromatic light. They are caused by the fact that some of the wavelengths contained in the incident light are eliminated by interference if the path difference, at the position where the interference phenomenon is observed, equals 1/2, 3/2, 5/2 ... of this wavelength. The remaining wavelengths, which are not attenuated or only partially so, appear as the superimposed interference colors. Interference colors are really only important in white light. In natural light, interference colors are seen with thin transparent plates, layers, oil films, annealing layers (annealing colors) when observed in transmitted or reflected light. They are called colors due to thin films, → interference at thin films. Interference colors can be observed in every interference device, such as Newton's rings, if the path difference can be made sufficiently small.

Interference filter, <*Bandfilter, Interferenzfilter*>, filters which utilize interference in order to transmit or reflect specific spectral regions. They consist of numerous thin layers the optical thickness of which is usually a quarter of the central wavelength or a multiple of it. Depending upon the type, number, thickness and arrangement, a vast range of band widths with high transmission or high reflection can be produced. With respect to the construction, we generally distinguish between → MDI filters, → ADI filters and filters with induced transmission which are a combination of the MDI and ADI filters. Figure 1 shows the construction of an MDI filter schematically. N is a transparent distance layer of thickness, d, of MgF_2 for example (→ dielectric layers), mounted between two semitransparent silver films, M_1 and M_2 on a support, G. A cover plate of colored glass, D, is used as protection and for → blocking undesirable radiation. If a parallel beam of rays is incident perpendicularly, the beam is split into a transmitted and a reflected component at each of the mirror surfaces, M_1 and M_2. Interference occurs on account of the multiple reflection. Only radiation of that wavelength, λ, is increased for which the path length, d, of the spacer (more exactly $n \cdot d$) is an integral multiple k of $\lambda/2$, i.e. the wavelength for which $n \cdot d = k\lambda/2$. All other wavelengths are attenuated or eliminated by interference. The transparency of such a filter is given by:

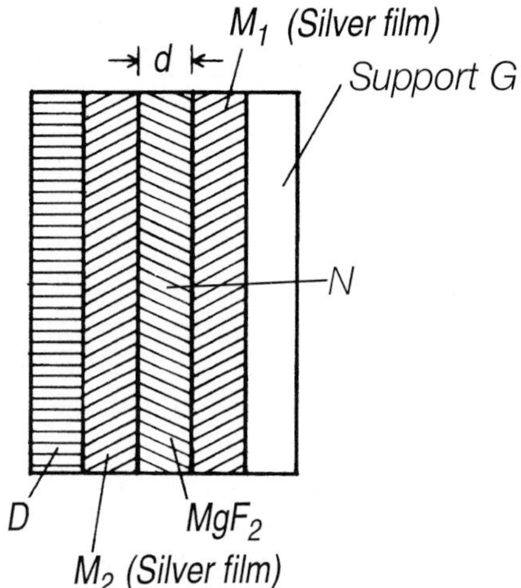

Interference filter. Fig. 1. Schematic construction of an interference filter. The layers are shown much enlarged

$$\tau_F = \frac{\tau_{Ag}^2}{(1-R)^2 + 4R\sin^2(2nd/\lambda)},$$

where τ_{Ag} is the transparency and R the reflectivity of the silver foil ($0 \le R \le 1$); n is the effective refractive index of the system.

For reasons of simplicity, phase shifts at the metal/dielectric medium interface are not taken into consideration here. When $n \cdot d$ is an integral multiple, k, of $\lambda/2$ there is maximum transparency given by:

$$\tau_{F,max} = \frac{\tau_{Ag}^2}{(1-R)^2}.$$

$k = 1$ gives a first order maximum of transparency, τ_1 at λ_1; $k = 2$ and $k = 3$ correspond to the second and third order transparency maxima at $\lambda_1/2$ and $\lambda_1/3$, and so on. The changes in the optical parameters of an interference filter with increasing angles of incidence or aperture depend upon the transmitted wavelength, the polarization, the layer material and the mode of construction of the device as a whole. The last is not constant, even for a single filter type.

The region of transmittance of a filter is always shifted towards shorter wavelengths when the filter is inclined to a parallel beam. To a good approximation, for angles in the range $\alpha = 0$ to ca. 30°, the shift is given by:

$$\Delta\lambda = \lambda_{m,\alpha=0} - \lambda_{m,\alpha} \approx K \cdot \sin^2\alpha$$

K is a constant for a specific filter. On request, Schott will provide λ_m values for $\alpha = 0$ and, for example, for $\alpha = 30°$. The intermediate values can generally be calculated with sufficient accuracy using the above formula. The exact spectral position required can be obtained by using an interference filter with a plus tolerance for the required wavelength and then inclining the filter. The shift for $\alpha = 20°$ is usually approximately 7 nm in a MDI filter for 550 nm.

The position of transmittance is also shifted towards shorter wavelengths if the beam is incident perpendicularly but is also convergent. In the case of smaller apertures, the value of the shift is approximately half as large as that for the maximum possible angle. The effective filter function (transmission against wavelength) is also broadened and reduced. This effect is greater in filters with a narrower band width than in band-pass filters with relatively larger half-widths. The construction of a Schott MDI (M) or ADI (A) interference filter is usually denoted with the letters Z or D after the M or A as described in table 2. A mirror is generally made from a thin metal layer or a series of alternating dielectric layers. The absorption-free distance piece between two mirrors is called a cavity. A 2-cavity filter has 3 mirrors and 2 cavities and is denoted by Z; a 3-cavity filter by D. If more mirrors are used, the shape of the transmittance curve, among other things, is changed from triangular towards rectangular. The ratio Q:*TW*/*HW*, i.e. the ratio → one tenth-width to → half-width, is used as a measure of the steepness of the edge. The following are approximate Q values:

System	S2	S3	S4	S6	S7	S14
Q	3.00	1.75	1.40	1.25	1.15	1.10

Thus, UV – DAD 8–0.5 has a center wavelength below 400 nm and additional short- and long-wave blocking. It is of ADI type with 3 cavities and an approximate half-width of 8 nm. Depending upon the type of filter, we distinguish between, line, double-line, band, double-band, broad-band and graded filters. Manufacturers will produce these filters to the requirements of the user. The table lists some of the Schott filters available for the UV and NIR region. (→ Half-width, and for definitions of τ_{max}, τ_s, Q and q → band-pass filter)

Interference of light waves, <*Interferenz von Lichtwellen*>, a phenomenon which occurs when two or more wave systems coincide. The resulting

Interference filters. Table. A Selection of Schott interference filters and their characteristic properties

Kind of filter	Type[1] designation	λ_m range [nm]	HW[2] [nm] approx.	τ_{max}[2] approx.	Q[2] approx.	Blocking range [nm]	τ_{SH}[3]
UV line filter	UV-KMD 12-1	195– 333	12	0.18	1.85	unlimited	$\leq 10^{-5}$
UV band filter	UV-KMZ 20-3	195– 333	20	0.2	2.0	unlimited	$\leq 10^{-4}$
UV line filter	UV-MAZ 8-1	220– 333	8	0.15	1.75	unlimited	$\leq 10^{-5}$
UV line filter	UV-DAD 8-1	334– 399	8	0.3	1.5	up to 1200	$\leq 10^{-5}$
UV band filter	UV-DAD 15-3	334– 399	15	0.3	1.5	up to 1200	$\leq 10^{-5}$
Line filter	MAZ 3-0.3	400–1100	3	0.4	1.75	unlimited	$\leq 10^{-5}$
Line filter	MAD 8-2	400–1100	8	0.45	1.5	unlimited	$\leq 10^{-5}$
Line filter	DAD 8-1	400–1100	8	0.65	1.5	up to 1200	$\leq 10^{-5}$
Line filter	DMZ 12-0.5	400– 599	12	0.4	1.8	unlimited	$\leq 10^{-5}$
Line filter	KMZ 12-0.5	600–1100	12	0.35	1.8	up to $2 \cdot \lambda_m$	$\leq 10^{-5}$
Band filter	DMZ 20-2	400– 599	20	0.55	1.8	unlimited	$\leq 10^{-5}$
Band filter	KMZ 20-1	600–1100	20	0.5	1.8	up to $2 \cdot \lambda_m$	$\leq 10^{-5}$
Broadband filter	KMZ 50-2	400–1400	50	0.45	1.8	unlimited	$\leq 10^{-4}$

[1]) **Type designation scheme:**

UV – D A D 8 – 1
1. 2. 3. 4. 5. 6.

Explanation of symbols

1. *Spectral position of center wavelength* λ_m
   No indication: $\lambda_m \geq 400$ nm
   UV-: $\lambda_m < 400$ nm
2. *Type of additional blocking*
   D: Blocking in short- and long-wave blocking range via ADI filter and/or optical glass filters
   M: Blocking in short- and long-wave blocking range via MDI filters and optical glass filters and/or ADI filters
   K: Blocking in short-wave blocking range via optical glass filters and/or ADI filters
3. *Type of bandpass system*
   A: ADI design of bandpass system
   M: MDI design of bandpass system
4. *Cavity number of bandpass system*
   Z: 2 cavities
   D: 3 cavities
5. *Half-width HW*
   Numerical value: Approximate value of HW in nm
6. *Tolerance of center wavelength* λ_m
   Numerical value: Total tolerance of λ_m in % of λ_m

[2]) → Band-pass filter

[3]) Mean optical transmittance in blocking range

process can be approximately constructed by overlapping the two unperturbed wave systems, i.e. each system behaves as if the other was not present. The total field resulting from the overlapping of two light waves with the electric field strengths E_1 and E_2 can be determined at every point by adding the primary fields, E_1 and E_2, vectorially. The addition of the field strengths means that the intensities, I_1 and I_2, which are given by the squares of the field strengths, are not simply additive. We describe any deviation from the additivity of the intensity when waves overlap as interference, especially in the case of light where only the intensities are observed. For two waves this can be represented mathematically by:

$$I = I_1 + I_2 + [2\sqrt{I_1 I_2} \cos 2\pi(\frac{r_2 - r_1}{\lambda} + \delta_2 - \delta_1)].$$

I is the total intensity, I_1 and I_2 are the intensities of the two light waves and r_1, r_2 are the distances which the two waves covered from their starting point to the point where they overlap. Thus, $r_1 - r_2$ is a path difference. δ_1, δ_2 are the phase constants of the two light waves and λ is the wavelength of the (monochromatic) light.
Thus, the total intensity, I, is generally not equal to the sum of the individual intensities, I_1 and I_2, because the interference term, which is emphasized in the above equation by square brackets, must also be included. The result of this term is that the intensity at different points oscillates about the mean value, $I_1 + I_2$. The maximum intensity occurs for the case $\delta_1 = \delta_2$ when the path difference between the two waves at their meeting point is an integral multiple of the wavelength, λ:

$$r_2 - r_1 = m \cdot \lambda$$
$$(\cos 2\pi \cdot m = 1).$$

A minimum occurs if the path difference is a half-odd-integral multiple of the wavelength:

$$r_2 - r_1 = [(2m + 1)2] \cdot \lambda;$$
$$(\cos 2\pi \frac{2m + 1}{2} = -1).$$

For $I_1 = I_2$ and $\delta_1 = \delta_2$, the maximum intensity is $I = 4I_1$ and the minimum intensity is $I = 0$. Figure 1 shows the intensity as a function of the path difference for this case.
If $\delta_1 \neq \delta_2$ the curve shown in Figure 1 is shifted by $\lambda(\delta_2 - \delta_1)$. Thus, the difference of the phase constants $\delta_2 - \delta_1$ acts like an additional path difference, and generally we have for phase differences:

$$2\pi \frac{r_2 - r_1 + \lambda(\delta_2 - \delta_1)}{\lambda} = 2\pi m,$$

where $m = 0$ or 1, 2, 3 ... for intensity maxima, and:

$$2\pi \frac{r_2 - r_1 + \lambda(\delta_2 - \delta_1)}{\lambda} = 2\pi \frac{(2m + 1)}{2},$$

where $m = 0$ or 1, 2, 3 ... for intensity minima.

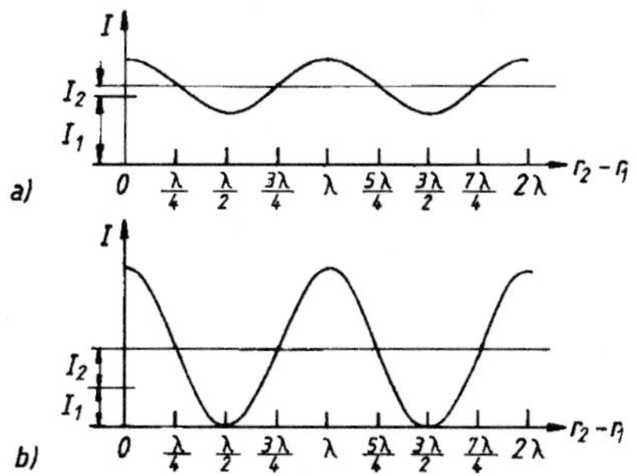

Interference of light waves. Fig. 1. The intensity due to the interference of two light waves as a function of their path difference

However, in optics we generally observe that, contrary to the above discussion, the intensities are additive, i.e. no interference takes place between two light waves which originate from different light sources. The reason for this is the fact that the difference between the phase constants of the light waves, $\delta_2 - \delta_1$, is generally not constant in time and during the period of observation it can take any positive or negative value. Therefore, the above equation must be averaged over the difference between the phase constants which eliminates the interference term (see square brackets) giving a total intensity equal to the sum of the individual intensities. Two wave systems with a constant $\delta_2 - \delta_1$ are called coherent to express the fact that a fixed phase relationship exists between the two wave systems. If the two phase constants, δ_2 and δ_1, change in an unconnected way the waves are called incoherent. Thus, interference can only be observed if the overlapping wave systems are coherent. The terms *coherence* and *incoherence* are also applied to the sources which send out light waves. The conventional light sources in spectroscopy are → incoherent light sources. The development of → lasers made a → coherent light source available to spectroscopy. Although the usual light sources in optics are incoherent radiators, interference still has a practical importance. The phenomenon of the → diffraction of light at slits and gratings especially must be described in terms of interference.

Ref.: R.W. Ditchburn, *Light*, Academic Press, London, New York, San Francisco, **1976**, vol. 1.

Interference spectroscopy, <*Interferenzspektroskopie*>, a method in atomic spectroscopy which utilizes the fact that conclusions can be drawn about the structure of spectral lines from observations of interference. A. Fabry and Ch. Perot (1897) were the first to describe the principles of interference spectroscopy using the interferometer which they developed (→ Fabry-Perot interferometer). The method uses the multiple-beam interference between two semitransparent, silver-plated, plane-parallel glass plates separated by an air gap of thickness, d. The resolving power, A, of an interference spectrometer of this type is given by:

$$A = \frac{\lambda}{\delta\lambda} = \frac{4\pi nd}{(1-R)\lambda}.$$

For a reflectivity $R = 0.95$, $d = 10^{-2}$ m, $n \sim 1$ and $\lambda = 400$ nm, $A = 6 \cdot 10^6$; resolution which cannot be achieved with → prism or grating spectrometers. Two lines with a separation $\delta\lambda = 0.0004$ can still be resolved at 400 nm. However, the advantge of high resolving power is counteracted by an extremely small → free spectral range, $\Delta\lambda$, the value of which is given by $\Delta\lambda = \lambda^2/(2d \cdot n)$. $\Delta\lambda \sim 8 \cdot 10^{-3}$ nm at $\lambda = 400$ nm, $d = 10^{-2}$ m and $n \sim 1$ (air). i.e. twenty times the resolving power. Thus, the Fabry-Perot interferometer can only be used if the radiation is restricted to the narrow wavelength region, $\Delta\lambda$, which can be achieved, for example, by a preselection. → Interference filters, → prism or → grating spectrometers are used for this purpose. The preselection can also be made using a second Fabry-Perot interferometer, or duplex interferometer, since the resolving power, A, and the free spectral range, $\Delta\lambda$, depend upon upon the plate separation of the interferometer. A is directly proportional to d whilst $\Delta\lambda$ is inversely proportional to d. For plate separations, d, of 10^{-1} m, 10^{-2} m and 10^{-3} m and the above values of λ, R and n, we obtain the following data:

Interference spectroscopy. Table.

d [m]	A	$\Delta\lambda$ [nm]
10^{-1}	$6 \cdot 10^{7}$	0.0008
10^{-2}	$6 \cdot 10^{6}$	0.008
10^{-3}	$6 \cdot 10^{5}$	0.08

An interferometer with a smaller plate separation, d has a larger free spectral range, $\Delta\lambda$, but a smaller resolving power, A. H. Krause and K. Krebs (*Optik*, **1963**, *20*, 471) have demonstrated the operation of a duplex interferometer using the green mercury line. When constructing a duplex interferometer the plate separations, d_1 and d_2, of the two interferometers must have an integral relationship to each other. These spectrometers are preferred for the investigation of the → hyperfine structure of spectral lines on account of their high resolving power. In principle, a cuvette is also a Fabry-Perot interferometer. Multiple interference can also occur between the two cuvette windows which may be observed in transmission or reflection.

In transmission, the interference maxima obey the relationship: $2nd = m\lambda$ ($m = 1, 2, 3 \ldots$ is the order of interference) and, taking account of the phase change on reflection:

$$(\lambda/2) \cdot 2nd + \frac{\lambda}{2} = m\lambda.$$

With polychromatic light, the interference maxima and minima form a characteristic interference pattern within the spectral region under investigation which is also called an interference spectrum. If we observe interference maxima at wavelengths, λ_1 and λ_2, which are separated by an integral number of maxima, δm, then in both cases we have:

$$2nd = \frac{\lambda_1\lambda_2}{\lambda_1 - \lambda_2}\delta m \ (\delta m = 1, 2, 3 \ldots).$$

This well-known relationship is used for the determination of the path length of thin films in reflection or of cuvettes in transmission, and especially for the small path lengths of IR cells. However, it is a precondition for its use that the refractive index, n, is constant or corresponds to that of air ($n \sim 1$) in the case of empty cuvettes. Alternatively, if the path length, d, and the order, m, are known, the → refractive index of a solution in the cuvette can be calculated from the interference spectrum. The dependence of the refractive index upon wavelength, i.e. the dispersion curve, can be determined in this way, (→ dispersion spectroscopy). The interference spectrum can also be used to calibrate the wavelength scale of a spectrometer. The emission spectrum of a line source, the wavelengths of which are known accurately, is used for this purpose, e.g. the → mercury spectrum. If the interference spectrum is also measured, an exactly known wavelength, λ_m, of a mercury line can be assigned to the nearest interference extremum of order m. The value of m can be determined by using the adjacent interference extremum of order $m + 1$ with wavelength, λ_{m+1}. In the case of a transmission measurement, using the interference conditions we have $m = \lambda_{m+1}/(\lambda_m - \lambda_{m+1})$. The value

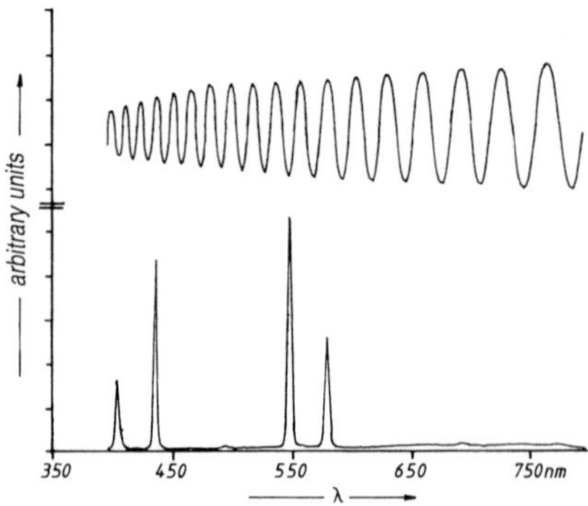

Interference spectroscopy. Fig. 1. Wavelength calibration using an interference spectrum, see text

for m generally contains an error, it is therefore rounded to an integral value (interference maximum) or to a half-integral value (interference minimum). The order numbers, n, of all other maxima or minima can then be obtained from the values for m and λ_m by simple counting. The corresponding values of the wavelengths are then given by:

$$\lambda_n = \frac{m}{n}\lambda_m.$$

Figure 1 illustrates the method of calibrating the wavelength of a spectrometer in the visible spectral region. The interference spectrum shows a total of 36 extrema which provide 36 points on the wavelength scale. If the mercury spectrum were used it would only give four calibration points. Figure 2 shows schematically a modern apparatus for the measurement of the interference spectra of thin films in reflection. It is also the preferred equipment for the determination of film thickness when $d \leq 50$ μm. The light source and → multichannel spectrometer are connected to the sample surface via a *Y*-form light pipe which is also shown in Figure 2.

Ref.: A.P. Thorne, *Spectrophysics*, Chapman and Hall, London, **1974**; H.H. Schlemmer, H. Mächler, *J. Phys. E: Sci. Instrum.*, **1958**, *18*, 914.

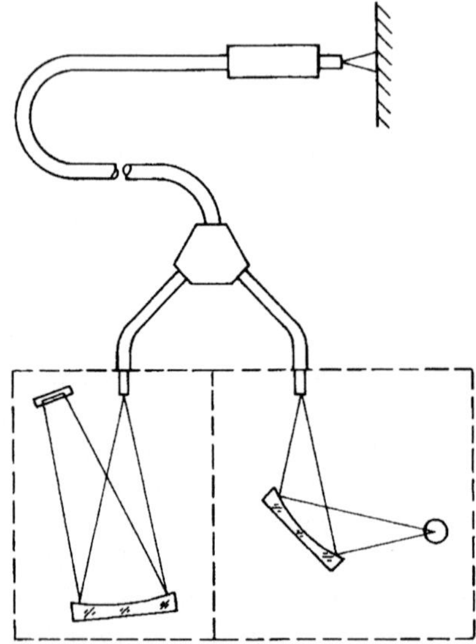

Interference spectroscopy. Fig. 2. A multichannel spectrometer for the measurement of interference spectra, see text

Interference spectrum, *<Interferenzspektrum>* → interference spectroscopy.

Interferometer, *<Interferometer>*, an optical instrument which utilizes the interference phenomena which arise when light waves overlap for the purposes of measurement and observation. Therefore, interferometers have components which split a light wave into two or more coherent wave systems which are reunited after traversing different optical path lengths, $n \cdot d$. Figure 1 shows the basic principle of all interferometers schematically. The incident light with the intensity, $I_0 \sim A_0^2$ (A_0 = amplitude of the incident

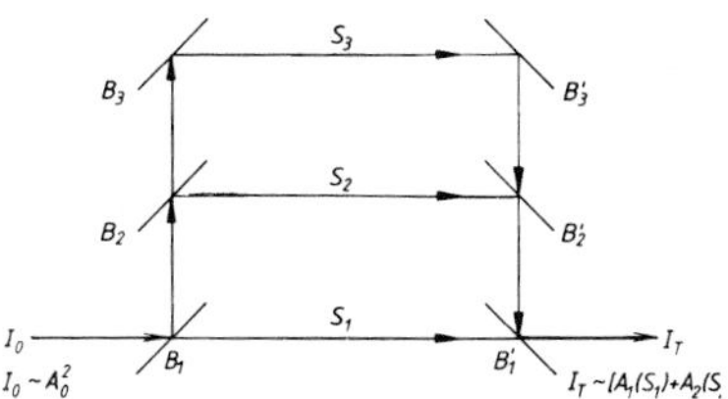

Interferometer. Fig. 1. The basic principle of the interferometer, see text

light) is directed by the → beam splitters, B_1, B_2, B_3 along several optical path lengths, S_1, S_2, S_3 ... In general, amplitude $= A_k$, and $S_k = nx_k$ (n = refractive index; x_k = geometric path length of the partial beam). The beams are reunited at the interferometer exit. Since all the partial beams originate from the same light source they are coherent so long as the maximum path difference is not longer than the coherence length of the light. The total amplitude, $\sum_k A_k$, of the transmitted wave, i.e. the superposition of all the partial waves, depends upon the individual amplitudes, A_k, and the phases, $\Psi_k = \Psi_0 + 2\pi S_k/\lambda$ of the partial waves. Therefore, a sensitive dependence upon the wavelength, λ, results. The maximum transmitted intensity arises when all the partial waves give constructive interference, i.e. overlap in phase and reinforce each other. This gives the condition for the optical path difference, $\Delta S_{ik} = S_i - S_k = m\lambda$ with $m = 1, 2, 3$ ... The transmitted intensity, I, is proportional to the square of the total amplitude: $I = [\sum_k A_k]^2$.

The requirement, $\Delta S_{ik} = m\lambda$ for the maximum transmission of an interferometer applies not only to the individual wavelength, λ, but also to the wavelength, λ_m, for which $\lambda_m = \Delta S/m$. The wavelength interval, $\delta\lambda = \Delta S/m - \Delta S/(m + 1) = \Delta S/(m^2 + m)$ is called the → free spectral range of the interferometer. It is more appropriate to express this as a frequency. With $\nu = c/\lambda$, $\Delta S = mc/\nu$ and the free spectral range becomes $\delta\nu = c/\Delta S$, which is independent of m.

Since the interference depends upon the ratio of the path difference $\Delta(nd)$ to the wavelength, λ, interferometers can be used for the determination of the quantities: path length, d (measurement of length, investigation of surfaces), → refractive index, n (refractometry) and → wavelength, λ (spectroscopy), or to investigate changes in these quantities.

A distinction is made between double-beam and multibeam interferometers. The → Michelson interferometer, which is of great importance in spectroscopy, is of the former type. The → Fabry-Perot interferometer, which is an extremely important optical instrument in → interference and → laser spectroscopy, belongs to the latter. Other double-beam interferometers of importance for spectroscopic purposes are the → Mach-Zehnder and the → Jamin interferometers. They all operate by means of a physical separation, i.e. a partitioning of the amplitude. Interferometers which effect a geometric separation or wave front separation are rarely used in spectroscopy.

Internal conversion, *<innere Umwandlung, innere Konversion>*, the radiationless deactivation of excited electronic states. As the name implies, this involves an internal conversion of the excitation energy which is then released as heat. This occurs via molecular vibrations and collisions with the surrounding molecules (those of the same species and solvent molecules) and corresponds to the transitions *i*, *k* and *l* in Figure 1 of the → Jablonski diagram. Whilst the transitions, *i*, are completed in 10^{-12} to 10^{-13} s the transition, *k* ($S_1 \rightarrow S_0$) occurs only after an average life of 10^{-9} to 10^{-8} s and the transition, *l* ($T_1 \rightarrow S_0$) only after 10^{-4} s. Both radiationless processes, *l* and *k*, compete with the radiative deactivation, i.e. with → fluorescence and → phosphorescence which, in agreement with → Kasha's rule, start from S_1 or T_1. A → fluorescence quantum yield of 1 can therefore only occur if no radiationless deactivation takes place. The radiationless deactivation, i.e. the internal conversion of S_P ($p > 1$) to S_1 takes place with a quantum yield of 1 according to → Vavilov's law.

The conversion of electronic excitation energy into heat by internal conversion is an extremely important precondition for → photoacoustic spectroscopy and → thermal lensing spectroscopy.

Internal filter effect, <*innerer Filtereffekt*>, an interfering effect in quantitative fluorescence measurements which arises because the fluorescence, generated by excitation of the dissolved molecules within a fluorescence cuvette, traverses a finite path length through the solution on its way out of the cuvette and suffers a reabsorption or → self-absorption. Thus, a solution capable of fluorescing acts like a filter; hence the name *internal filter effect*. It depends upon three factors:

1. The greater the overlapping of the fluorescence and the absorption bands the larger is the reabsorption.
2. The higher the concentration the greater is the effect.
3. The path length of the fluorescence cuvette and the geometric position in the beam path of the fluorescence spectrometer are of critical importance.

A simple way to reduce reabsorption is to tilt or rotate the fluorescence cuvette through an angle of 20°–30° towards the beam of the exciting light. Then, only the fluorescence light which is excited within a narrow zone behind the cuvette window will be detected and the reabsorption is low because of the reduced exit path length. In addition to this simple procedure, numerous methods which deal with the reduction or elimination of the internal filter effect are described in the literature.

Ref.: J.N. Miller, *Standards of Fluorescence Spectrometry*, Chapman and Hall, London, New York, **1981**, chapt. 5.

Internal reflectance spectroscopy, internal reflection spectroscopy, <*innere Reflexionspektroskopie, ITR, TIR*> → attenuated total reflection.

Interruption arc, <*Abreißbogen*>, an undesired → arc discharge when opening a switch in an electric circuit. It is extinguished at a specific length and voltage. Inductance in the circuit increases the duration of the interruption whilst a capacitor wired in parallel to the contact accelerates the extinction of the interruption arc. The Pfeilsticker interrupted arc is a periodically interrupted arc (→ arc discharge) which was used particularly as a light source for the detection of trace elements (trace analysis) in qualitative → spectral analysis. Vis-a-vis the continuous arc, it has an advantage in the very clear background of its spectra. In addition, it is characterized by the low intensity of interfering bands. The Pfeilsticker interrupted arc is controlled by a circuit breaker and a high frequency spark of low intensity to ignite the arc after the interruption.

Intersystem crossing, *<Intersystem-Crossing>*, a radiationless transition which takes place with a change of multiplicity. These are the transitions, $S_1 \rightarrow T_1$, $S_1 \rightarrow T_2$, $S_2 \rightarrow T_3$ and $T_2 \rightarrow S_1$ in the → Jablonski diagram. As detailed study shows, these occur isoenergetically. Thus, for the transitions $S_1 \rightarrow T_1$ or $T_1 \rightarrow S_0$ (Figure 1) the energy of the S_1 (T_1) level corresponds to a vibrationally excited state of the $T_1(S_0)$ level.

Inversion doubling, *<Inversionsverdopplung>* → inversion vibration.

Inversion vibration, *<Inversionsschwingung>*, a vibration which

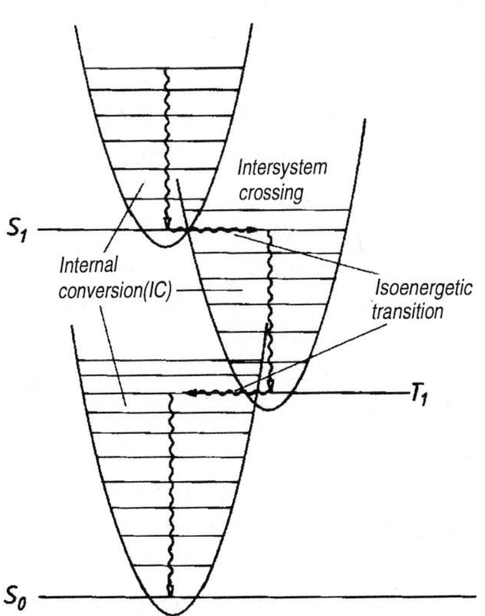

Intersystem crossing. Fig. 1. Energy-level diagram with potential curves to illustrate intersystem crossing, see text

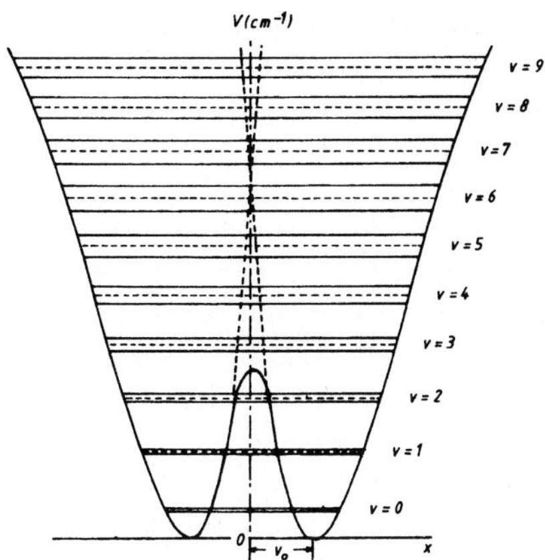

Inversion vibration. Fig. 1. Potential curve for the NH_3 molecule, see text

occurs in some molecules of the XY_3 type. These molecules have a pyramidal structure and belong to the → symmetry point group C_{3V}. The best-known and most frequently studied example is the NH_3 molecule where the N atom is found at the top of the pyramid and can vibrate through the plane defined by the three H atoms. The N atom then adopts an equivalent position on the other side of the plane. This is the inverted configuration. It is important that the two configurations cannot be interconverted by a rotation of the molecule. Thus, there are two modifications of this molecule which we can describe as a left-handed and right-handed form, similar to an optical isomer. A potential barrier, the height of which depends upon the geometrical structure of the molecule concerned, must be overcome during the transition to the inverted configuration. Figure 1 shows the potential energy as a function of the distance of the N from the H_3 plane. The dashed horizontal lines correspond to the energy levels which would be obtained if two identical potential curves (dashed), independent of each other and not connected by a potential barrier, were present.

On account of a resonance between the two modifications of the molecule and in conjunction with a perturbation, which here corresponds to the deviation of the actual potential curve from the dashed lines, each pair of degenerate levels splits into two levels which are drawn as continuous horizontal lines in Figure 1. This splitting, known as inversion doubling, increases strongly with increasing vibrational quantum number, v. Of the six normal vibrations of the NH_3 molecule, ν_1 and ν_2 of species, A_1 plus ν_3 and ν_4 of species E (doubly degenerate), the vibration ν_2 shows a marked inversion doubling. For $v_2 = 0 \rightarrow v_2 = 1$ the splitting is 35.7 cm^{-1} and for $v_2 = 0 \rightarrow v_2 = 2$ it is already 312.5 cm^{-1}. Vibration ν_2 which is illustrated in Figure 2, is characterized by the fact that the N atom vibrates perpendicularly to the H_3 plane, as was assumed for the potential curve in Figure 1. An exact evaluation of the IR spectrum, including inversion doubling, gives a value of 2076 cm^{-1} for the potential barrier in NH_3; this corresponds to the excitation of $2\nu_2 \sim 1900$ cm^{-1} ($\nu_2 = 950$ cm^{-1}). In vibration ν_1 the N atom also vibrates perpendicularly to the H_3 plane, but away from it. In contrast to vibration, ν_2, this makes inversion doubling more difficult. For the transition $v_1 = 0 \rightarrow v_1 = 1$ the splitting is 1 cm^{-1}. In the case of the degenerate vibrations ν_3 and ν_4, the N atom moves parallel to the H_3 plane, therefore, the conditions for inversion are considerably less favorable than for ν_1. The splitting of these vibrations is less than 1 cm^{-1}. The fact that a transition between the two configurations is observed in the ground state, $v_2 = 0$, and in the first excited state, $v_2 = 1$, despite the high potential barrier, can be explained with reference to the → tunnel effect. A splitting of 0.66 cm^{-1} is found for the ground state, $v_1 = v_2 = v_3 = v_4 = 0$, using → microwave spectroscopy. In the pure rotation spectrum of NH_3, every individual rotational line shows an inversion doublet. In addition to NH_3, ND_3, ND_2H, NDH_2 and PH_3 also show this effect.

Ref.: G. Herzberg, *Molecular Spectra and Molecular Structure II. Infrared and Raman Spectra of Polyatomic Molecules*, D. Van Nostrand Co. Inc., Princeton, New Jersey, **1966**.

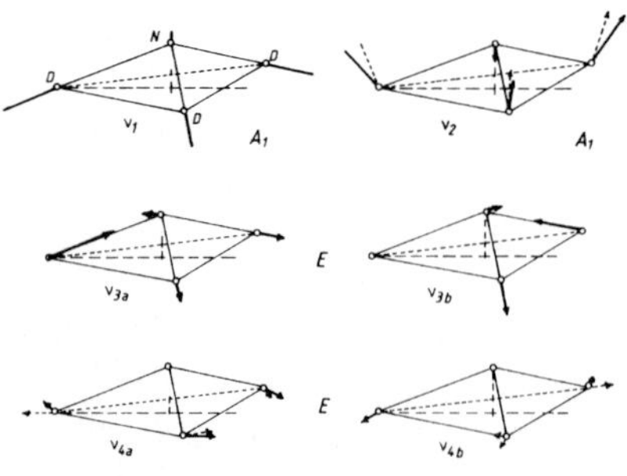

Inversion vibration. Fig. 2. Vibrations of the NH_3 molecule

Ion-bombardment ion source, *<Ionenbeschußionenquelle>*, a source in which the samples under investigation are bombarded with primary ions, usually argon or oxygen ions. Ion bombardment forms positive ions of the sample, i.e. secondary ions; these sources are therefore also called secondary ion sources. Gas-discharge ion sources or electron-impact ion sources are used to generate primary ions such as Ar^+. The primary ion beam is directed with an energy of 1 to 10 KeV onto the surface of the target by means of an accelerating and focusing system. From the magnitude of the primary current it is possible to determine whether one atom, several atoms or molecular layers are removed from the surface. It is specifically the surface which is the subject of the investigation. The mass spectroscopic technique based on this method, → SIMS, finds its applications in semiconductor technology and in the investigation of catalysts, corrosion, diffusion phenomena and surface reactions.

Ionization, *<Ionisation, Ionisierung>*, the conversion of an electrically neutral particle (atom or molecule) into a singly or multiply positively or negatively charged particle. The formation of a singly positive atom or molecular ion, M^+, is the commonest process. In order to achieve ionization, the neutral particle must be provided with a specific minimum energy which is called the → ionization energy, *I*. The ionization of atoms and molecules is of particular importance to several spectroscopic methods such as → spark spectra, and generally in → atomic emission spectroscopy, → photoelectron spectroscopy and → mass spectrometry. A number of methods are available for the generation of positive ions, M^+: → electron-impact ionization, → thermal ionization, → photoionization, → spark ionization, → fast atom bombardment, → chemical ionization, → field ionization and laser ionization. Most of these methods are used in the → ion sources of → mass spectrometers. Spark ionization is used for the generation of the → spark spectra of atoms and → photoionization in → photoelectron spectroscopy.

Ionization energy, *<Ionisierungsenergie>*, an amount of energy which must be supplied to convert a neutral free atom or molecule into a singly positively charged ion. Since multiple → ionization is also possible, the energy required for the generation of the singly positive ion is called the first ionization energy while the others are described as the higher ionization energies. This applies especially to atoms in which fivefold and higher ionizations are observed.

The first ionization energy of the elements is obtained from their → atomic spectra, i.e. from the series limit of the principal series. The first ionization energy can be determined for simple molecules (e.g. diatomics) from the series limit of the Rydberg series. → Photoelectron spectroscopy can be applied generally to molecules. → Electron-impact ionization is another frequently used method. It is used to determine the ionization energy by means of the → appearance potential in mass spectrometry, for example. From the analysis of electron donor-acceptor (EDA) interactions, which give rise to → charge-transfer complexes, it is possible to determine the ionization energies of the electron donors, which are usually aromatic substances and other π-electron systems.

Ion source, *<Ionenquelle, Ionisierungsquelle>*, the component of a → mass spectrometer where the sample is ionized before being extracted, in the direction of an extraction electrode, by an accelerating voltage, U_B, of between 1 kV and 10 kV. Ionization sources are classified according the type of ionization. The → electron-impact ion source utilizes → electron-impact ionization and is the most common source. → Field ionization, and the variant → field desorption, are utilized in thermal surface-ion sources. Chemical ion sources are based on → chemical ionization, the vacuum-discharge ion source uses → spark ionization and the photoion source → photoionization. In the case of bombardment ion sources, we distinguish between → fast atom bombardment, → ion bombardment and → electron bombardment ion sources. They are preferred in → mass spectrometry for the investigation of solids, especially of surfaces where → laser ion sources are also used.

IR active, *<IR-aktiv, infrarot-aktiv>*. The excitation of a molecular vibration in the → IR spectrum can only occur without simultaneous electronic excitation if a change of the dipole moment, or a component of that moment, is associated with the vibration. Vibrations, which meet this precondition, are called IR active and those which do not meet it, IR inactive. For diatomic molecules, this means that they must have a permanent dipole moment, i.e. only heteronuclear diatomic molecules can be excited, but not homonuclear diatomics such as H_2, N_2, O_2, Cl_2 etc. In small molecules, $N \geq 3$, it is easy to say which of the $3N$-5 or $3N$-6 vibrations are IR active or IR inactive on account of the small number of → vibrational modes. For larger molecules, if the vibrational → symmetry species are known, we can use the → character tables to determine which are the symmetry species of the IR active vibrations. The character table lists all the data required to calculate the number of vibrations which belong to a particular symmetry species.

IR diode laser, *<IR-Diodenlaser>* → diode laser.

IR gas analyzer (Ultramat), *<Ultramat>* a → nondispersive infrared instrument made by Siemens for the analysis of gases. In contrast to the → URAS, the Ultramat is a single-beam instrument based upon the first description of such an apparatus by K.F. Luft in 1958. It is illustrated schematically in Figure 1. Infrared radiation from a radiating filament (3), heated to 600 °C, is modulated at 8.33 Hz by a

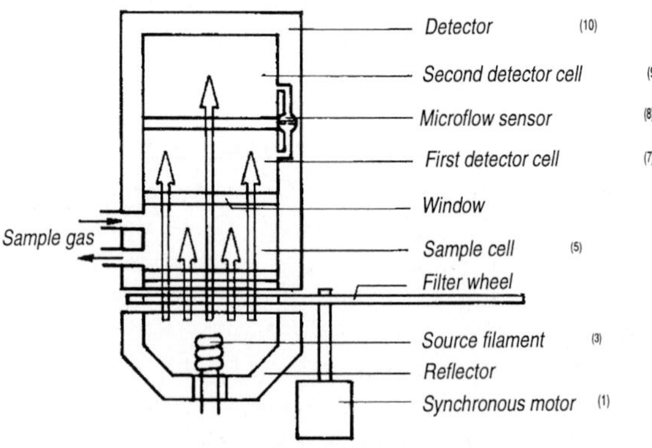

Ultramat, Fig. 1. Schematic of the construction of the Siemens Ultramat

→ chopper (4) driven by a synchronous motor. After passing through the measurement cell (5), the radiation is measured by a double-chamber detector (10). The detector consists of two gas cells (7,9) of different path lengths in series which are connected via a → pneumatic detector (8). The unabsorbed IR radiation leaving the measurement cell (5) is absorbed in the two detector cells (7,9), one after the other. This leads to a differential warming in the two cells and therefore to a pressure difference between them, which is modulated at the chopper frequency. The pneumatic detector, the microflow sensor (8), converts these periodic pressure changes into an alternating electrical signal which is amplified and indicated. The detector is optimized so that only those IR wavelengths are absorbed which are specific for the gas to be measured, i.e. the two detector cells are themselves filled with the gas to be analyzed, which is the principle of → positive filtering. The length of the detector cells (7,9) is chosen such that if there is no absorption in the measurement cell (5), the absorption and hence the pressure increase, is approximately the same in both. Therefore, the second cell (9) is always longer than the first (7).

The instrument is calibrated before delivery when purchased commercially. It is preferred for the on-line analysis of CO, CO_2, SO_2 and CH_4.

IR gas analyzer (URAS), *<Ultrarotabsorptionsschreiber, URAS>*, an analytical instrument developed in 1938 which is a → nondispersive spectrometer and operates according to the principle of positive filtering. Figure 1 shows its construction. Two nickel-chromium coils heated electrically to ca. 1000 K are used as radiation sources. The radiation is modulated by a → chopper which is formed like an aperture. A → pneumatic detector is used to detect the radiation and its two chambers are connected by a diaphragm capacitor. The chambers are filled with the gas to be measured, usually diluted with N_2. The sample cell is flushed with this gas whilst the reference cell is filled with N_2. If there is absorption of radiation in the sample cell, the gas pressure in the sample chamber of the detector varies and the capacitance of the diaphragm capacitor also changes at the modulation fre-

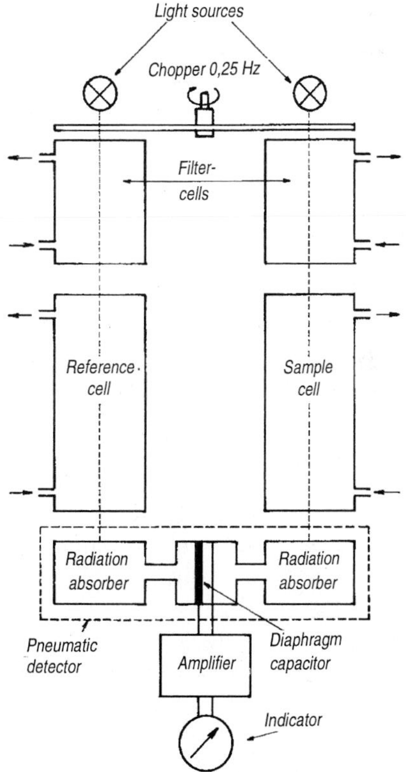

IR gas analyzer (URAS). Fig. 1. The constructional principles of a URAS

quency. The resulting AC voltage is amplified, rectified and recorded. The energy in the two beams can be made equal by means of an adjustable aperture. The sensitivity can be modified by varying the DC voltage on the capacitor. The range of measurement is determined by the length of the sample cell in any particular case. The interference of accompanying gases is eliminated by filling the filter cells with them. This specifically removes the radiation which they absorb from the two beams. The most important gases and vapors which can be measured with the IR gas analyzer, include CO, CO_2, methane, acetylene, ethane, propane, NO, N_2O, ethylene oxide, butadiene, acetone, ethyl alcohol, benzene, dimethyl ether etc.

Ref.: J.C. Wright in *Laboratory Methods in Infrared Spectroscopy*, 2nd ed. (Eds.: R.G.J. Miller and B.C. Stace), Heyden and Son Ltd., London, Philadelphia, Rheine, **1979**, chapt. 6.

IR inactive, <*IR-inaktiv*> → IR active.

IR radiation, infrared radiation, <*IR-Strahlung, Infrarotstrahlung, Ultrarotstrahlung*>. William Herschel (1800) provided the proof of infrared radiation in the following experiment. Sunlight split into its spectral colors by a prism fell onto a table in the laboratory where several mercury thermometers with blackened bulbs were mounted to investigate the heat distribution of the dispersed radiation. The surprising result of this experiment was that Herschel found that the temperature maximum was not in the region of the greatest physiological perception of brightness, i.e. yellow-green, but near red in the invisible spectral region. In further investigations he was able to show that this new radiation also obeyed the laws of optics, just like visible light. The IR region adjacent to the visible spectral region extends from approximately 800 nm to ca. 1000 μm or 12,400–10 cm^{-1} where the upper and lower limits depend upon the current state of the methods for the generation and detection of IR radiation. The IR region is divided as follows:

Near IR: NIR 12,500–4000 cm^{-1}
Mid IR: MIR 4000–200 cm^{-1}
Far IR: FIR < 200 cm^{-1}

→ IR spectroscopy is routinely applied in the spectral region of 800 nm to 50 μm or 12,500 to 200 cm^{-1}, i.e. in the MIR and NIR; → near infrared spectroscopy.

The fact that Herschel discovered IR radiation using a temperature measurement and that a physiological perception of heat was connected with the action of this radiation led to the name *heat radiation*.

IR spectrometer, infrared spectrometer, <*IR-Spektrometer, IR-Spektralphotometer, Infrarotspektralphotometer*>. The construction of a dispersive IR spectrometer (→ Fourier-transform infrared spectrometer) differs from that of the common → spectrometers which are used in the UV-VIS and NIR region because the → monochromator and → sample chamber (cells) are arranged in a different way. The most commonly used double-beam instruments also divide and recombine the beams differently. The splitting into two identical beams is made immediately in the source unit, behind which the sample compartment is mounted. The reference and sample beam are recombined before reaching the monochromator into which they then pass. The detector is mounted behind the monochromator exit slit. A → Nernst glower, → globar or a tungsten coil wound in or on a ceramic tube, which also acts as a thermal radiator after heating, are used as light sources. NaCl, KBr and KI prisms (→ prism materials) were previously used as dispersive elements. However, → gratings are employed almost exclusively today. Compared to the UV-VIS region, where 1200–2400 lines/mm are needed, good resolution can be achieved with ≤ 240 lines/mm because of the longer wavelengths. For many years, the thermopile was used as a detector, but it has been increasingly replaced by → pyroelectric detectors. Their introduction has made it possible to replace the optical null principle, used with thermopiles, with the method of → ratio recording. The optical null principle required that the photometer component was placed before the monochromator and thus that the sample compartment of IR spectrometers be mounted before the monochromator.

Figure 1 shows the optical system of a modern, dispersive IR spectrometer; Perkin-Elmer model 882.

A ceramic tube with an internal heating element (1100°C) is the radiation source. The light from this source is divided by the toroidal mirrors, M_1 and M_2, into two symmetrical parallel beams, the reference and sample beams, and focused in the sample compartment at the positions of the sample and reference cells. The two beams are recombined by the plane mirrors, M_3 and M_4 and a sector mirror (chopper). The sample beam is reflected by the sector mirror to the toroidal mirror M_5 via mirror M_3, whilst the reference beam passes to the sector mirror via mirror M_4 and is then reflected at mirror M_5. The sector mirror has another position, where neither the reference nor sample beam

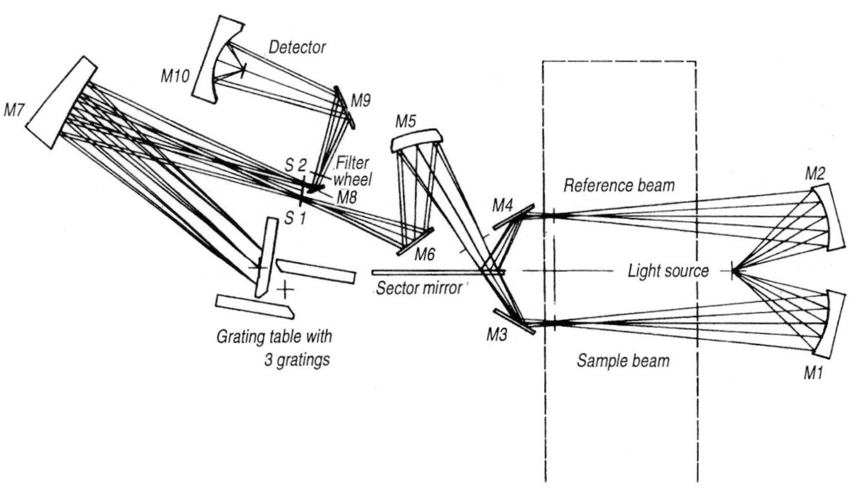

IR spectrometer. Fig. 1. Beam path in the dispersive Perkin-Elmer IR grating spectrometer 882, see text

is transmitted, which provides a pure reference pulse. The sector mirror fulfills the following functions twice for every rotation: reference beam pulse (mirror position), sample beam pulse (transmission, free sector) and pure reference pulse (both beams blocked). The recombined beam is then focused onto the entrance slit, S_1 via mirror M_5 and the plane mirror M_6 and passes to one of the three gratings via the parabolic mirror M_7. The gratings cover the following ranges:

Grating 1: 4000–2000 cm^{-1}; 240 lines/mm
Grating 2: 2000–600 cm^{-1}; 90 lines/mm
Grating 3: 600–200 cm^{-1}; 25 lines/mm

The dispersed light then passes to the elliptical mirror M_{10}, via mirror M_7, exit slit S_2, and the plane mirrors M_8 and M_9. The detector is mounted at the focus of mirror M_{10}. It is pyroelectric and consists of triglycine sulfate (TGS) doped with L-alanine. The detector has a 0.5 mm × 2.0 mm surface (effective surface 0.38 mm × 1.5 mm).

As determined by the operational mode of the sector mirror, the detector receives reference and sample beams alternately, separated by the blank beam as a pure reference signal. The intensity ratio is formed mathematically. The resulting signal is transmitted to an analog-to-digital converter via a preamplifier. To eliminate the higher orders of the IR radiation, a set of eight optical filters is mounted between mirrors M_8 and M_9.

The instrument can be flushed completely with purified nitrogen to eliminate the interfering absorption of water vapor and CO_2. The whole spectrometer is microprocessor-controlled. The course of the measurements can be followed on a screen which simultaneously displays all data which can be printed out by a printer/plotter. It is also possible to store the spectra and to obtain difference spectra.

Ref.: *Instrument Handbook*, Perkin-Elmer Corp., Instrument Division, Norwalk, CT, USA.

IR spectroscopy, infrared spectroscopy, vibrational spectroscopy, <*IR-Spektroskopie, Infrarotspektroskopie, Schwingungsspektroskopie*>, a part of → molecular spectroscopy in which the excitation of the discrete (quantized) vibrational and rotational states of molecules is measured by the absorption of electromagnetic radiation. Conventional dispersive → IR spectrometers and → Fourier-transform infrared spectrometers cover the range 4000 cm^{-1} to ca. 200 cm^{-1} of → IR radiation in which most → normal vibrations of molecules are found. Because of the size of the molecules it is frequently not possible to resolve the overlapping of the vibrational and rotational excitation. Thus, the bands in the → IR spectrum appear broadened. They are not the sharp lines of individual vibrational excitations but are extensively overlapped by unresolvable rotational excitations. IR spectroscopy, together with → Raman spectroscopy, is also called vibrational spectroscopy and an IR spectrum a → vibrational spectrum because the individual bands in the spectrum can be assigned to the normal vibrations and combination vibrations of the measured molecules. IR spectroscopy has become very important for the determination of molecular structure using → expectation regions or → group frequencies. Quantitative measurements can be carried out using the → Bouguer-Lambert-Beer law. In addition to routine measurements in solution or using KBr discs and nujol mulls (→ nujol technique and → KBr technique), IR spectroscopy is used for studies of gases, solids and surfaces, in conjunction with gas chromatography in analytical chemistry and for the investigation of intermolecular interactions.

IR spectrum, infrared spectrum, *<IR-Spektrum, Infrarotspektrum>*, a spectrum which is obtained by the measurement of light absorption in the wavenumber region between 4000 cm^{-1} and 200 cm^{-1} (2.5 μm to 50 μm) which is accessible to most → IR spectrometers. The excitation energies involved here are small when compared to UV-VIS spectroscopy, so that only that absorption is measured which is due to the excitation of the vibrations and rotations of molecules. Thus, the IR spectrum represents the → rotation-vibration spectrum of a molecule. Since in most cases, the measurements are carried out on solids or solutions, the rotational structure is strongly disturbed by interactions with the surrounding molecules and cannot be resolved. Therefore, the IR spectra of molecules in solution have numerous bands which are for the most part strongly broadened and the maxima of these bands correspond to the → vibrational spectrum. The rotation-vibration spectrum can be resolved in the gas phase, especially for smaller molecules, if the IR instrument has good resolution. In the case of larger molecules which can be vaporized without decomposition, the resolution of most IR spectrometers is too low for the analysis of the rotational structure so that we recognize only the envelope of this structure which can be very characteristic. The rotations are effectively frozen in a crystalline solid so that relatively sharp bands which correspond to the → normal and lattice vibrations occur.
In contrast to a → UV-VIS spectrum, an IR spectrum is normally plotted or recorded directly in the form of → transmittance ($T = I/I_0$) in % against the wavenumber, $\tilde{\nu}$. The reason for this is that in many measurements the concentration and path length are not sufficiently accurately known to permit a calculation of the → extinction coefficient, ε, using the measured → absorbance. However, for quantitative applications the extinction coefficient must be calculated or the path length must be reproducible and set accurately in order to work with a → standard calibration method.
Figure 1 shows a typical IR spectrum which can be used as a model to explain some significant details:

1. Base line: The curve which the instrument records without a sample in the beam. In most cases, it is not displayed and in computer-controlled IR instruments it is stored and used for background correction.
2. Area between the base line and the 100% T line, also called the background.
3. Band, absorption maximum or transmittance minimum. These terms are used to characterize the positions of maximum absorption, $\tilde{\nu}_{max}$, by the molecule.
4. Shoulder: two unresolved bands.
5. Switch-over point: Most IR spectrometers have a change of grating,

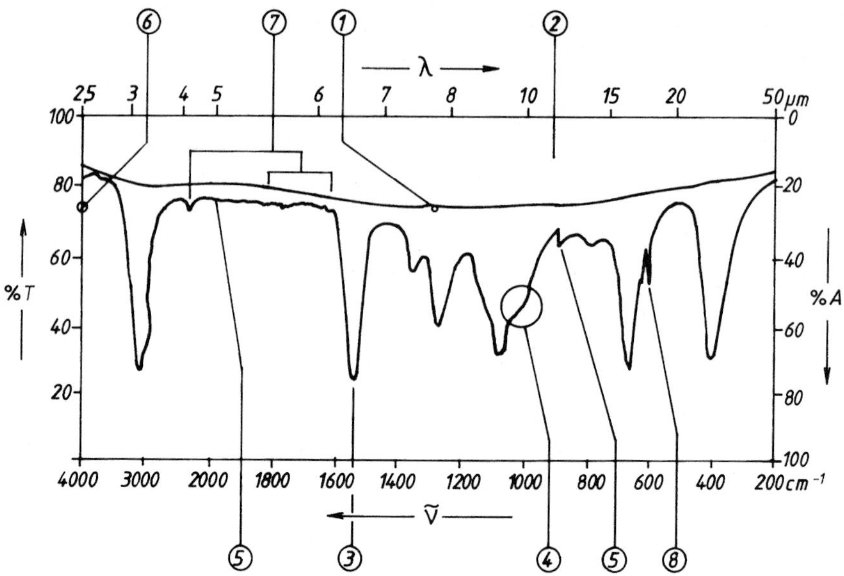

IR spectrum. Fig. 1. Example of an IR spectrum, see text

filter or scale at a specific wavenumber, $\tilde{\nu}$ (frequently $\tilde{\nu}$ = 2000 cm^{-1}) where the paper feed is halted during the switch-over process. The position on the abscissa of this change is specific to the instrument and must agree with the preprinted wavelength/wavenumber scale.

6. Initial mark to indicate the start of recording.
7. Absorption by atmospheric CO_2 and H_2O.
8. Spikes which can be recognized by their small half-width. These are recorder deflections caused by uncontrolled voltage pulses.

When measuring an IR spectrum, the sample concentration should be such that a transmittance of T = 75–80 % is shown initially at approximately 4000 cm^{-1} (recorder deflection, digital display). Generally, good IR spectra are obtained if the scale is fully used with a 95 % transmittance at the position of the greatest transmittance and a ~ 5 % transmittance at the position of the strongest bands. The name and origin of the sample, sampling method (film, solution, halide disc, matrix), purity (GC, distillation fraction), date and operator's name should also be recorded for every spectrum. The selection of the → solvent is of the greatest importance when measuring IR spectra. In contrast to → UV-VIS spectroscopy, there is practically no solvent which is transparent throughout the whole IR region. Hence, solvents are selected which are transparent over as large a wavenumber range as possible, e.g. carbon tetrachloride and carbon disulfide, which complement each other well with respect to their ranges of transparency. The absorption of the solvents themselves must be carefully compensated in such measurements. It is important that the amount of the solvent present in the reference beam is equivalent to that in the sample beam, i.e. the path length of the cell in the reference beam is always smaller than that in the sample beam. And the more concentrated the solution to be measured, the greater the difference. Problems with solubility occur frequently on account of the restricted choice of suitable solvents so that other sampling methods such as the → KBr disc, nujol mull or capillary film must be used (→ KBr technique, → nujol technique).

Isochromatic curve, *<Strahlungsisochromat, Strahlungsisotherme>*, the curve which is obtained if the energy of a black-body radiator (→ thermal radiator) is plotted as a function of the temperature at a constant wavelength in accordance with → Planck's radiation law. Correspondingly, the graph of the radiant energy as a function of the wavelength at constant temperature is described as an isotherm, as is the usual practice for a measurement at constant *T*. This representation is also called the spectral energy distribution.

Isosbestic point, *<isosbestischer Punkt>*. If the spectrum of a reacting mixture is repeatedly scanned over a broad wavelength range then the → reaction spectra obtained are found to intersect at one or more wavelengths. These points of intersection are the isosbestic points. They occur not only in the investigation of reaction kinetics, but also in spectroscopic studies of equilibria such as of protolytic equilibria where the absorption spectrum depends upon the pH value. The variable in this case is the concentration rather than the time. Figure 1 shows the pH dependence of the UV-VIS absorption spectrum of p-nitrophenol with three isosbestic points. The occurrence of isosbestic points is an indication of the uniformity of the reaction or equilibrium.

Ref.: J. Polster and H. Lachmann, *Spectrometric Titrations*, VCH Publishers, Weinheim, New York, **1989**.

Isomer shift, chemical isomer shift, *<Isomerieverschiebung>* → Mössbauer spectroscopy

Isotope effect, *<Isotopieeffekt>*, the change of a spectroscopic observable as a consequence of the replacement of a light isotope by a heavy one. This effect is most clearly observed in the excitation of molecular vibration and rotation. The molecular properties due to the bonding electrons are not influenced by the introduction of the heavy isotope, i.e. the bond strength, force constant and bond length do not change. But it is different with the properties into which the masses enter. Using a diatomic molecule as an example, these are the frequency of the → harmonic oscillator and the → rotational constant.

The frequency of the harmonic oscillator is given by; $\nu = (1/2\pi)\sqrt{(k/\mu)}$, and therefore depends upon the → reduced mass, μ. For an isotopic molecule, index *i*, we have; $\nu_i = (1/2\pi)\sqrt{(k/\mu_i)}$. Since the force constant, *k* does not change we obtain the ratio, $\nu_i/\nu = \sqrt{(\mu/\mu_i)} = \varrho$, with $\varrho < 1$ since $\mu_i > \mu$. From this it follows that: $\nu_i = \varrho\nu$. For the rotational constant, $B = h/8\pi^2c\mu r^2$ or $B_i = h/8\pi^2c\mu_i r^2$ it follows analogously that: $B_i/B = \mu/\mu_i = \varrho^2$ or $B_i = \varrho^2 B$. The energy levels of

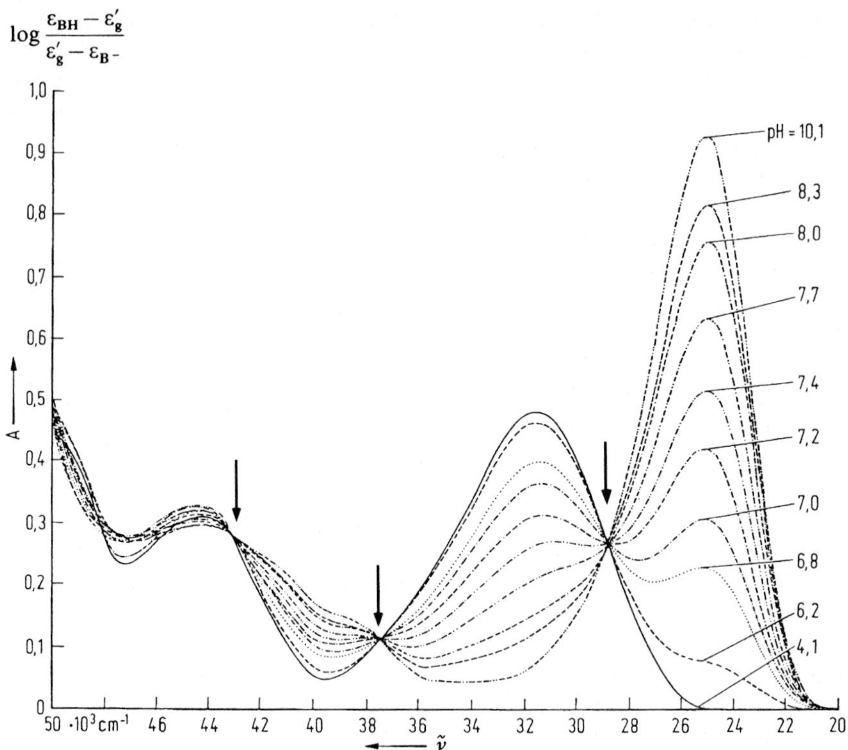

Isosbestic point. Fig. 1. pH dependence of the UV-VIS spectrum of p-nitrophenol with 3 clear isosbestic points

the heavier molecule are therefore given by:
Vibration: $G^i(v) = \varrho\tilde{\nu}(v + ½)$;
Rotation: $F^i(J) = \varrho^2 BJ(J + 1)$.

An approximation is involved in using $\tilde{\nu}$ rather than ω in the expression for the vibrational energy. But the error involved is small, see Herzberg (reference below), p. 228.

Since $\varrho < 1$, the energy levels of the molecules with the heavier isotope always lie lower than those of the lighter molecule. This expected isotope effect was found initially in the → rotation-vibration spectrum of the HCl molecule. HCl consists of the two isotopic molecules $H^{35}Cl$ and $H^{37}Cl$. From the reduced masses of the molecules $\varrho = 0.99925$ and $\varrho^2 = 0.9985$. For the vibrational transition $v'' = 0 \rightarrow v' = 1$, $\tilde{\nu} = 2885.9$ cm^{-1} and thus $\tilde{\nu}_i = 2883.6$ cm^{-1}. The isotope effect causes a shift to smaller wavenumbers of $\Delta\tilde{\nu}_i = 2.3$ cm^{-1}. Therefore, in the IR spectrum of natural HCl we observe a doubling of each rotational line in the rotation-vibration spectrum. The lines of $H^{37}Cl$ appear shifted to longer wavelengths with a separation of 2.3 cm^{-1} from the lines of the $H^{35}Cl$ molecule. The isotope effect gives a shift of $\Delta\tilde{\nu}_i = 0.03$ cm^{-1} in the pure rotational spectrum. This requires a spectrometer of high resolving power in the far IR region. -X-H groups in which the hydrogen atom ($m = 1$) is exchanged for deuterium ($m = 2$) show the greatest isotope effects. For groups such as the -OH group $\varrho = 0.7276$. Therefore, for $\tilde{\nu}_{OH} = 3500$ cm^{-1}, $\tilde{\nu}_{OD} = 2500$ cm^{-1}, i.e. a very large shift towards longer wavelengths. This large isotope effect can be utilized in the analysis of the vibrational spectra of larger molecules. All vibrations in which → valence or → bending vibrations of the XH or XH_2 groups are involved show a large isope effect in an H/D exchange, whilst others such as the skeletal vibrations are almost unaffected. The fact, that the force constant, k, does not change in an isotope exchange is of critical importance for the vibrational analysis of complex spectra. Here, use is made of the product rule established by Teller and Redlich (see G. Herzberg, *Molecular Spectra and Molecular Structure II. Infrared and Raman Spectra of Polyatomic Molecules*, D. Van Nostrand Co. Inc., Princeton, New Jersey, **1966**, p. 231 ff.)

A careful study of the isotope effect can also be used to obtain an exact value for the mass ratio of two isotopes by determining $\varrho = \sqrt{(\mu/\mu_i)}$. The isotope effect in → electronic band spectra has even led to the discovery of new isotopes. It is important to → mass spectroscopy that most elements occur as isotopic mixtures. This has the result that molecular ions do not give just one peak in the → mass spectrum. Whether or not an isotopic peak can be observed in a mass spectrum depends upon the magnitude of the natural isotopic content. The values lie well below 1% for ^{2}H, ^{15}N, ^{17}O and ^{18}O, so that their isotopic peaks are generally not observed. For carbon with a natural content of 1.1% of ^{13}C, in addition to ^{12}C, there is sufficient of the former to produce a satellite peak which lies one mass unit higher. The intensity of this peak for a molecular ion which contains n C atoms is $\sim n \cdot (1.1\%)$ of the ^{12}C peak. Therefore, for a molecule with 10 C atoms, e.g. naphthalene, the intensity of the satellite peak is 11% of that of the ^{12}C peak. In sulfur, chlorine and bromine, the natural content of the isotope, which is heavier by two mass units, is relatively large so that the corresponding peaks are easily recognizable. An isotope effect must also be taken into consideration in → NMR spectroscopy. It is characterized by the fact that the isotopic atom types have different nuclear magnetic moments which leads to very widely varying resonance frequencies.

ITR, internal total reflectance, *<ITR>* → attenuated total reflection, ATR.

IVR, intramolecular vibrational redistribution, *<IVR>*. In this process excitation energy is distributed over the internal degrees of freedom of a molecule without interaction with adjacent molecules. It differs from a → radiationless transition in which the energy is transferred to the surrounding medium by collisions etc. The process determines the reactivity of molecules in the excited electronic state.

J

Jablonski diagram, <*Jablonski-Termschema*>, the → energy-level diagram for the electronic states of molecules which includes both singlet and triplet states. In Figure 1, a Jablonski diagram illustrating the primary photophysical processes is shown without including the superimposed vibrational levels. The individual processes shown are:

a) Singlet-singlet absorption:
$S_0 \rightarrow S_1$, $S_0 \rightarrow S_2$ ...; spin-allowed

b) Singlet-triplet absorption:
$S_0 \rightarrow T_1$, $S_0 \rightarrow T_2$ ...; spin-forbidden

c) Singlet-singlet absorption:
$S_1 \rightarrow S_2$, $S_1 \rightarrow S_3$ ...; spin-allowed

d) Triplet-triplet absorption:
$T_1 \rightarrow T_2$, $T_1 \rightarrow T_3$ ...; spin-allowed

e) Fluorescence:
$S_1 \rightarrow S_0$; spin-allowed

f) Phosphorescence:
$T_1 \rightarrow S_0$; spin-forbidden

g) Resonance fluorescence:
$S_p \rightarrow S_0$ ($p \geq 2$); spin-allowed

h) Triplet fluorescence:
$T_p \rightarrow T_1$ ($p \geq 2$); spin-allowed

i) Internal conversion:
$S_p \rightarrow S_1$ ($p \geq 2$); $T_p \rightarrow T_1$ ($p \geq 2$); deactivation

k,l) Internal conversion:
$S_1 \rightarrow S_0$; $T_1 \rightarrow S_o$ in competition with e) and f)

m) Intersystem crossing:
$S_1 \rightarrow T_1$; $S_1 \rightarrow T_2$; $S_1 \rightarrow T_3$

n) Intersystem crossing:
$T_2 \rightarrow S_1$

a)–h) are the radiative processes (continuous lines) which take place with the emission or absorption of electromagnetic radiation. i)–n) are the radiationless deactivation steps (dashed lines) which are called → internal conversion when the → multiplicity is unchanged and → intersystem crossing when the multiplicity changes ($S \rightarrow T$). The absorption processes c and d begin from states S_1 and T_1 respectively. They are two-photon transitions because in each case a photon is first required to excite the system to the states S_1 or T_1.

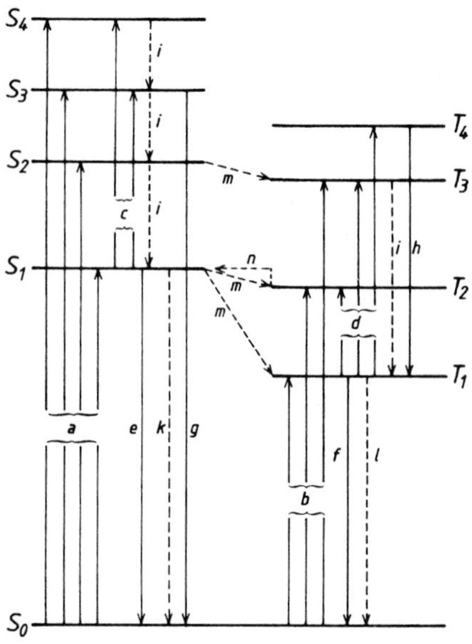

Jablonski diagram. Fig. 1. A general Jablonski diagram showing the possible transitions, see text

Jamin interferometer, <*Jaminsches Interferometer*>, a double beam interferometer in which a beam splitter and a mirror are combined in one optical element. As shown in Figure 1, it consists of two glass plates, P_1 and P_2, 3 to 5 cm thick and mirrored on the back in order to increase their reflectivity and hence the intensity of the emerging light. The incident light is partially

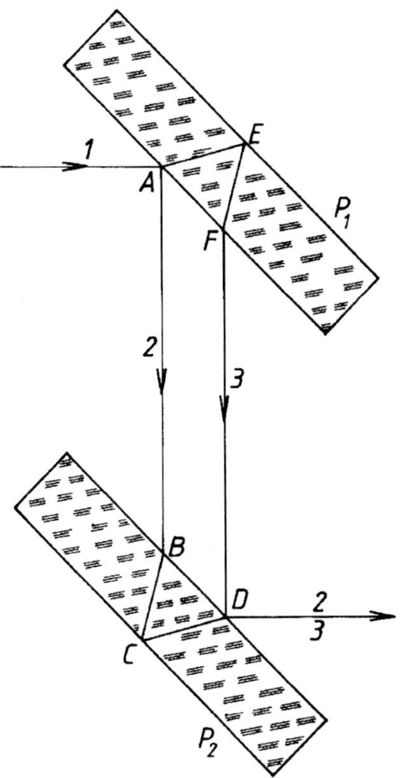

Jamin interferometer. Fig. 1. Ray paths in a Jamin interferometer, see text

reflected at P_1 as ray 2. The other part, ray 3, is refracted, reflected on the rear side of P_1 and refracted again as it leaves P_1. Similarly, ray 2 is refracted on entering P_2, reflected on the back surface and refracted again upon exiting. If we consider only that part of ray 3 which is reflected at the surface of P_2, then rays 2 and 3 are superimposed. The optical path difference between the two rays $\Delta S_{2,3} = n\,(ABCD-AEFD)$, where n is the refractive index. If the plates P_1 and P_2 are exactly parallel to each other, $\Delta S_{2,3} = 0$. A screen illuminated by rays 2 and 3 is then bright, and this is true for every angle of incidence. A path difference can be generated if P_2 is rotated with respect to P_1 about an axis either perpendicular to the plane of the drawing or lying in the plane of the drawing. Since the two beams, 2 and 3, are rather well separated from each other (they have the greatest separation for an angle of incidence of 49°), samples having the same path length, L, but different refractive indexes, may be placed in the beam, as with the → Mach-Zehnder interferometer. The optical path difference is then given by $\Delta S = (n_2 - n_1)L$. The sensitivity of this arrangement makes it possible to determine very small changes in refractive index with very high precision. For this reason, the Jamin interferometer is also called the Jamin interference refractometer.

Ref.: T.B. Brown in *The Lloyd William Taylor Manual of Advanced Undergraduate Experiments in Physics*, (Ed.: T.B. Brown), Addison-Wesley Publishing Co. Inc., Reading (MA), London, **1959**, p. 235.

K

Kasha's rule, <*Kasha-Regel*>, a rule, discovered by M. Kasha in 1950, according to which emission (fluorescence or phosphorescence) in larger molecules (e.g. aromatic hydrocarbons), always comes from the lowest excited state of the appropriate multiplicity, S_1 or T_1, → Jablonski diagram. For molecules in solution, especially unsaturated organics, the fluorescent state is always the first excited singlet, apart from a few exceptions. The energy of this state is usually lowered further during its lifetime of ~ $10^{-9}-10^{-8}$ s as a result of interaction with the solvent. Higher excited singlet states undergo rapid ($10^{-13}-10^{-12}$ s) radiationless deactivation to the lowest excited singlet. Kasha's rule is general, it also applies to triplet states.

KBr technique, <*KBr-Technik*> → potassium bromide technique.

KDP crystal, <*KDP-Kristall*>. Potassium dihydrogen phosphate (KH_2PO_4) is an optically uniaxial crystal which is preferred as a → frequency doubling crystal for lasers. It is also used in → Lyot filters and in → Pockels cells.

Kerr effect, <*Kerr-Effekt*>. The phenomenon, discovered by J. Kerr in 1875, whereby electric fields render isotropic materials, in all states of aggregation, doubly refracting or birefringent. The Kerr effect is particularly strong in liquids, especially in nitrobenzene and p-nitrotoluene. The electric field aligns the molecules along the lines of force and the liquid becomes anisotropic. This is the orientational Kerr effect. The refractive index of light polarized parallel to the field direction, n_1, is different from that of light polarized perpendicularly to the applied field, n_2. The liquid has become doubly refracting and behaves like a uniaxial, doubly refracting crystal. The difference between the two refractive indexes, Δn, is found experimentally to be given by the equation:

$$\Delta n = n_1 - n_2 = K \cdot \lambda \cdot E^2.$$

K is the Kerr constant, λ (in m) the wavelength of the light used and E (in V/m) the strength of the electric field. The Kerr effect is also known as a quadratic electrooptical effect because of the above relation. A Kerr cell is a glass container with plane-parallel windows and an integral electrical capacitor, filled with a liquid which has a large Kerr effect, usually nitrobenzene or nitrotoluene.

It is characteristic of the Kerr effect that it follows a rapidly changing field effectively without lag. Since the double refraction vanishes at $E = 0$, the combination of a polarizer, Kerr cell and crossed analyzer blocks light transmission once every half-period of a high-frequency field. Thus, a Kerr cell makes an excellent, highly resolving (in time) shutter. A suitable experimental arrangement, with a parallel analyzer, is illustrated schematically in Figure 1. Polarizer, I, produces light

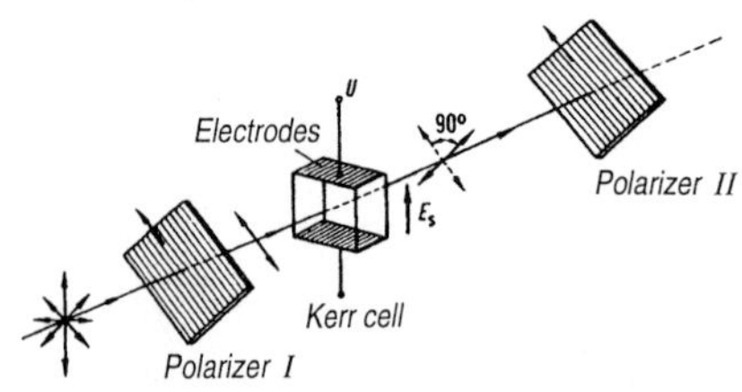

Kerr effect. Fig. 1. A Kerr cell as part of a light shutter

polarized at an angle of 45° to the field direction, E_s. The Kerr cell rotates the plane of polarization by 90°, and polarizer II blocks the light. When the voltage is switched off, no rotation takes place and the light passes through. Switching the voltage on and off changes the transmission of the unit from 0 to 100%.
If polarizer II is replaced by a mirror, an optical element of variable reflectivity is obtained. Since the Kerr liquid is usually nitrobenzene the Kerr cell is only usable with visible light.

Kinetic degrees of freedom, degrees of freedom, *<Bewegungsfreiheitsgrade>*. The degree of freedom in the description of the position of a point mass in three-dimensional space is 3. For two point masses 6, for three 9 and for N point masses $3N$ coordinates are necessary and, clearly, with $3N$ coordinates all the possible motions of the N point masses in three-dimensional space can be described. This concept may be expressed by saying that a system of N point masses has $3N$ kinetic degrees of freedom. However, in the case of a molecule with N atoms (point masses), the degrees of freedom are not independent of each other but are related by the molecular bonding. The translational motion in the three spatial directions can be referred to the center of gravity so that only 3 degrees of freedom are required for the translation of the $3N$ masses. For rotation of the molecule, also, it is clearly sufficient to refer the motion to three, mutually perpendicular inertial axes through the center of gravity so that, again, only 3 degrees of freedom are required for the rotational motion. For linear molecules only two mutually perpendicular axes of rotation can be specified since the third axis passes through all the atoms and a rotation about this axis produces no change of the positions of the atoms in space. The third possibility of motion for a system of N atoms is the vibration of the masses with respect to one another. Since the total number of degrees of freedom, translation, rotation and vibration is $3N$, the number of degrees of vibrational freedom must be:
$3N-5$ for linear molecules and
$3N-6$ for nonlinear molecules

Kinetic energy, *<Bewegungsenergie>*, the break-down of the internal energy, U, which is acquired when thermal energy is supplied to a molecule according to thermodynamics, into translational (E_{trans}), rotational (E_{rot}) and vibrational (E_{vib}) energy. The balance may be written:

$$U = E_{tot} = E_{trans} + E_{rot} + E_{vib}$$

From quantum mechanics we know that the rotational and vibrational energies are quantized and that the magnitudes of the quanta are of different orders of magnitude (→ spectral region). Translational motion is also quantized when boundary conditions are applied (e.g. the particle in a box), but the translational quanta are very small and the states so close together that they can be regarded as a continuum and thus unquantized. At sufficiently high temperatures, the electronic energy (E_{elec}) must also be considered and we have the following series of individual energies:

$$E_{trans} < E_{rot} < E_{vib} < E_{elec}$$

The → excitation energies of the corresponding quantized states follow the steps of this energy sequence and it is always true that higher excitation energies are able to include the smaller excitations, e.g. → rotation-vibration spectra and → electronic band spectra. The connection between the macroscopic thermodynamic quantities and quantized energy states is made through the → partition function and the statistical theory of matter. IR and Raman spectroscopy provide excellent examples of this.

Kinetics of chemical reactions, *<Kinetik chemischer Reaktionen>*. A chemical reaction is always easy to follow spectrophotometrically if the spectra of reactants and products are different. Since kinetics involves the observation of the changing of the concentration of the components with time, there is always a connection to the → Bouguer-Lambert-Beer law. For a simple first order reaction with a stoichiometry coefficient $\nu = 1$ and a rate constant k_1

$$a(\text{reactant}) \xrightarrow{k_1} b(\text{product})$$

then, at time t the absorbance measured at a particular wavelength λ is:

$$A_{\lambda t} = [a]_t \varepsilon_{a,\lambda} d + [b]_t \varepsilon_{b,\lambda} d.$$

Using the transformation variable, $x(t)$, and with $[a]_t = [a]_o$ and $[b]_t = 0$ at time $t = 0$

$$A_{\lambda,t} = ([a]_o - x)\varepsilon_{a,\lambda} d + x\varepsilon_{b,\lambda} d.$$

with $A_{\lambda,o} = [a]_o \cdot \varepsilon_{a,\lambda} \cdot d$ it follows that

$$A_{\lambda,t} = A_{\lambda,o} + x \cdot (\varepsilon_{b,\lambda} - \varepsilon_{a,\lambda}) \cdot d$$

or

$$A_{\lambda,t} - A_{\lambda,o} = q_\lambda \cdot x,$$

where $q_\lambda = (\varepsilon_{b,\lambda} - \varepsilon_{a,\lambda}) \cdot d$. Thus, we obtain:

$$x = \frac{A_{\lambda,t} - A_{\lambda,o}}{q_\lambda}$$

and

$$\frac{dx}{dt} = \frac{1}{q_\lambda} \cdot \frac{dA_\lambda}{dt}.$$

For a first order reaction the rate law, with the above definition of x, is:

$$\frac{dx}{dt} = k_1([a]_o - x)$$

$$= k_1\left([a]_o - \frac{A_\lambda - A_{\lambda o}}{q_\lambda}\right)$$

or

$$\frac{dx}{dt} = \frac{1}{q_\lambda}\frac{dA_\lambda}{dt}$$

$$= k_1\frac{\left([a]_o q_\lambda - A_\lambda + A_{\lambda,o}\right)}{q_\lambda}$$

and

$$\frac{dA_\lambda}{dt} = k_1(A_{\lambda,\infty} - A_\lambda); \qquad (a)$$

$A_{\lambda,\infty} = [a]_o \varepsilon_{b,\lambda} d$, for $t \to \infty$ $[b]_\infty = [a]_o$, i.e. there is complete transformation. Integration of equation (a) gives:

$$\ln\frac{A_{\lambda,\infty} - A_\lambda}{A_{\lambda,\infty} - A_{\lambda,o}} = -k_1 t \qquad (b)$$

Special cases:
1. Product b does not absorb at the wavelength λ:

$$\varepsilon_{b,\lambda} = 0,$$

i.e.

$$A_{\lambda,\infty} = 0; A_\lambda = A_{\lambda,o} e^{-k_1 t}. \qquad (c)$$

2. Reactant a does not absorb at wavelength λ:

$$\varepsilon_{a,\lambda} = 0,$$

i.e.

$$A_{\lambda,\infty} = 0; A_{\lambda} = A_{\lambda,\infty}(1 - \mathrm{e}^{-k_1 t}). \quad (d)$$

Absorbance-time measurements can be immediately evaluated using equations (b) and (c).
The rate laws for reactions of higher orders can be written in terms of absorbance in an analogous manner.
Ref.: H.H. Perkampus, *UV-VIS Spectroscopy and its Applications*, Springer Verlag, Berlin, Heidelberg, New York, London, **1992**, chapt 7.

Kinetics of photochemical reactions, <*Kinetik photochemischer Reaktionen*>. The basis for establishing rate laws for photochemical reactions is the equation which defines the true, differential → quantum yield:

$$\phi_b^a = \frac{\mathrm{d}c_b(t)/\mathrm{d}t}{I_a} = \frac{\dot{c}_b}{I_a}$$

$$= -\frac{\mathrm{d}c_a(t)/\mathrm{d}t}{I_a} = -\frac{\dot{c}_a}{I_a}.$$

from which it follows that:

$$\frac{\mathrm{d}c_b(t)}{\mathrm{d}t} = \phi_b^a I_a \text{ and } \frac{\mathrm{d}c_a(t)}{\mathrm{d}t} = -\phi_b^a \cdot I_a.$$

Here we are dealing with a simple, uniform photoreaction in which a is irradiated and b is formed: $a \rightarrow b$. Although these equations are straightforward, a simple integration is not possible because the quantity of light absorbed, I_a, depends, in a complex manner, on where within the reacting medium the reaction takes place and the duration of the irradiation. If certain preconditions are observed during the photochemical reaction, the following equation is obtained for $\dot{c}_a$:

$$\dot{c}_a = -1000 \cdot \phi_b^a \varepsilon'_{a,\lambda e} c_a I_o \frac{1 - \mathrm{e}^{-A'_{\lambda e}}}{A'_{\lambda e}}.$$

$\varepsilon'_{a,\lambda e}$ and $A'_{\lambda e}$ are the molar natural → extinction coefficient and the natural absorbance measured at the wavelength of the irradiating light λ_e. The natural absorbance, A', is given by:

$$A' = ln(I_o/I);\ A' = \sum_i \varepsilon'_i c_i \cdot d;$$

I_o is the intensity of the radiation at the surface in Einstein $\cdot$ cm^{-2} $\cdot$ sec^{-1}. The term (1-e$^{A'}$)/A' is the → photokinetic factor which must be taken into account in all photokinetic equations if the quantum yield is independent of concentration and intensity. The determination of the quantum yield, ϕ_b^a, by means of the above equation assumes that:

1. The reaction mechanism is known.
2. The extinction coefficients, ε_i', at the wavelength of irradiation, λ_e, are measurable. (Note that ε_i' = 2.303ε_i)
3. The change in concentration of the substances formed or consumed can be followed.
4. It is possible to determine I_o, the intensity in Einstein $\cdot$ cm^{-2} $\cdot$ sec^{-1} of the light source at the surface, which is required to calculate the amount of light absorbed, I_a, at wavelength λ_e. This can be done with suitable physical → detectors or better, with → actinometers.

For a spectrophotometric study of a reaction of the above type, $a \rightarrow b$, the concentration $c_a(t)$ can be written, with the help of the → Bouguer-Lambert-Beer law, in terms of the transformation variable $x(t)$ (→ kinetics of chemical reactions). The rate law or photokinetic equation is:

$$\frac{dA_\lambda}{dt} = 1000\phi_b^a \varepsilon'_{a,\lambda_e} I_o (A_{\lambda,\infty} - A_\lambda)\frac{1-e^{-A'_{\lambda e}}}{A'_{\lambda_e}} \quad (a)$$

in which $A_{\lambda,\infty} = c_a(0) \cdot \varepsilon_b \cdot d$; $c_a(0) = c_{b\infty}$; $c_{a\infty} = 0$. Rearranging (*a*) we obtain:

$$\frac{dA_\lambda}{dt} = (Z_{1,\lambda} + Z_{2,\lambda} \cdot A_\lambda) \cdot \frac{1-e^{-A'_{\lambda e}}}{A'_{\lambda_e}} \quad (b)$$

where

$$Z_{1,\lambda} = 1000\phi_b^a \varepsilon'_{a,\lambda_e} \cdot I_o \cdot A_{\lambda,\infty};$$
$$Z_{2,\lambda} = -1000\phi_b^a \varepsilon'_{a,\lambda_e} \cdot I_o.$$

Equation (*b*) can be written in the exponential form with the basis 10 as:

$$\frac{dA_\lambda}{dt} = (Z'_{1,\lambda} + Z'_{2,\lambda} \cdot A_\lambda)\frac{1-10^{-A_{\lambda_e}}}{A_{\lambda_e}} \quad (b')$$

which can be rearranged to give:

$$\frac{dA_\lambda/dt}{1-10^{-A_\lambda}} = Z'_{1,\lambda}\frac{1}{A_\lambda} + Z'_{2,\lambda} \quad (c)$$

($\lambda = \lambda_e$, i.e. the wavelength of irradiation, λ_e, is the same as that of observation, λ):

$$Z'_{1,\lambda} = 2303\phi_b^a \varepsilon_{a,\lambda} \cdot I_o A_{\lambda,\infty}$$

$$Z'_{2,\lambda} = 2303\phi_b^a \varepsilon_{a,\lambda} I_o.$$

$Z'_{1,\lambda}$ and $Z'_{2,\lambda}$ can be obtained from a graph of the left-hand side of equation (c) against A_λ^{-1}. If $\varepsilon_{a\lambda}$ and I_o are known, $\phi_b{}^a$ can be calculated from $Z'_{2,\lambda}$. To reduce the effect of errors in the measurements, the formal integration method of Mauser (*Z. Naturforschg.*, **1971**, *26b*, 203) is recommended for the evaluation of the experimental data. With $A_{\lambda e} = A_\lambda$, ($b'$) becomes:

$$A_\lambda(t'') - A_\lambda(t') = Z'_{1,\lambda}\int_{t'}^{t''} \frac{1-10^{-A_\lambda}}{A_\lambda}dt + Z'_{2,\lambda} \cdot \int_{t'}^{t''} (1-10^{-A_\lambda})dt.$$

Rearranged for data evaluation, the equation becomes:

$$\frac{\Delta A_\lambda(t)}{(1-10^{-A_\lambda})dt} = Z'_{1,\lambda}\frac{\int \frac{1-10^{-A_\lambda}}{A_\lambda}dt}{\int (1-10^{-A_\lambda})dt} + Z'_{2,\lambda}.$$

If the left-hand side is plotted against the quotient of the integrals on the right for various time intervals, Δt, a straight line is obtained from which $Z'_{1,\lambda}$ and $Z'_{2,\lambda}$ can be found, and ϕ_b^a can be determined from $Z'_{2,\lambda}$ as indicated above. For complicated, uniform photoreactions the mechanism must be known and the partial quantum yields used.

Ref.: K.J. Laidler, *The Chemical Kinetics of Excited States*, Oxford University Press, Oxford, New York, Toronto, Melbourne, **1955**.

Kirchhoff's law of thermal radiation, <*Kirchhoffsches Strahlungsgesetz*>, the radiation law for → thermal radiators proposed and established by G. Kirchhoff in 1861. It states that, for a given temperature, T, and wavelength, λ, the spectral emissivity, $\varepsilon(\lambda,T)$, of any body is equal to its spectral absorptivity $\alpha(\lambda, T)$. Thus $\varepsilon(\lambda,T) = \alpha(\lambda,T)$, or alternatively, $\varepsilon/\alpha = f(\lambda,T)$. The function $f(\lambda,T)$ depends only upon temperature and wavelength and is the same for all bodies. Thus, at a given temperature, T, the body which has the greatest emissivity, ε, is that which has an absorptivity $\alpha = 1$ and which absorbs completely all radiation which falls upon it, i.e. a → black body. Numerically, α and ε are equal.

KRS-5, *<KRS-5>* is a mixed crystal of thallium bromide and iodide which has good optical properties in the IR. Therefore, prisms of this material are used in the region 24 μm to 40 μm, i.e. 400–250 cm^{-1}. Because of its thallium content this material is very toxic.

Krypton-ion laser, *<Kryptonionen-Laser>* → noble gas ion lasers.

K shell, *<K-Schale>*, the innermost shell in the electronic structure of an atom. It is closest to the atomic nucleus and, in accordance with the Pauli principle, can only contain two electrons. Energetically, it is the shell with the principal quantum number $n = 1$. The designation, *K* shell, is due to Barkla who also introduced the nomenclature: L ($n = 2$), M ($n = 3$), N ($n = 4$), O ($n = 5$), etc. for further shells.

L

L-alanine-triglycinesulfate detector, L-ATGS, *<L-Alanin-Triglycinsulfat-Detektor>* → pyroelectric detector.

Lambda plate, λ-plate, *<Lambda-Plättchen, λ-Plättchen>*, a phase plate made from an optically anisotropic crystal, frequently mica or gypsum because of their good cleavage properties. The thickness of a lambda plate is chosen such that the two beams which, because of the → birefringence, emerge from the plate linearly polarized in mutually perpendicular planes, have a path difference of one wavelength. If the optical path difference, $\Delta\lambda$, is about 550 nm, then, when white light is used, the crystal plate (lambda plate red: first order) viewed through an analyzer crossed with a polarizer appears in a wine red → interference color. This is said to be a sensitive color since, when a slightly thicker or thinner plate is used, the wine red is replaced by markedly changed mixed colors.

Lambert-Beer law, *<Lambert-Beersches Gesetz>* → Bouguer-Lambert-Beer law.

Lamb formula, *<Lambsche Formel>*. A formula by means of which the → shielding constant, σ, of an individual atom can be calculated. σ is important in → NMR spectroscopy because it gives the → chemical shift which would be observed if the electron distribution around the atom were spherically symmetrical and the electrons were able to orbit the atom unhindered. The magnetic field produced by this motion is proportional to the applied field, B_o, (B_o = the magnetic flux density in Tesla) and in the opposed direction. The field at the atomic nucleus is reduced by this effect giving a local field, B_{loc}, according to the equation:

$$B_{loc} = B_o(l - \sigma).$$

Lamb's formula for σ (W.E. Lamb, *Phys. Rev.*, **1941**, *60*, 817) is:

$$\sigma = \frac{4\pi e^2}{2mc^2} \int_o^\infty r\varrho(r)\mathrm{d}r.$$

Here e is the electronic charge, m the mass of the electron, c the velocity of light and $\varrho(r)$ the electron density at a distance r from the nucleus in a spherically symmetrical S state. For neutral atoms, B_{loc} can be determined with the aid of σ values calculated with this formula (W.C. Dickinson, *Phys. Rev.*, **1950**, *80*, 563). E.g. for $B_o = 1\,\mathrm{T} \approx H_o = 10^4$ Gauss, we find for the proton $B_{loc} = 0.999982$ T, for carbon $B_{loc} = 0.999739$ T, for iodine $B_{loc} = 0.9950$ T and for lead $B_{loc} = 0.9900$ T. The more electrons there are in the atomic electron cloud, the greater is the opposing magnetic field which they induce and the weaker is the strength of the applied field at the nucleus. The situation is appreciably more complicated in the case of molecules. In practical applications of NMR spectroscopy, therefore, the → chemical shifts are always referred to a reference substance.

Laminar grating, *<Laminargitter>*. A reflection, diffraction grating (→ grating) with grooves having a rectangular cross section, Figure 1. It is preferred

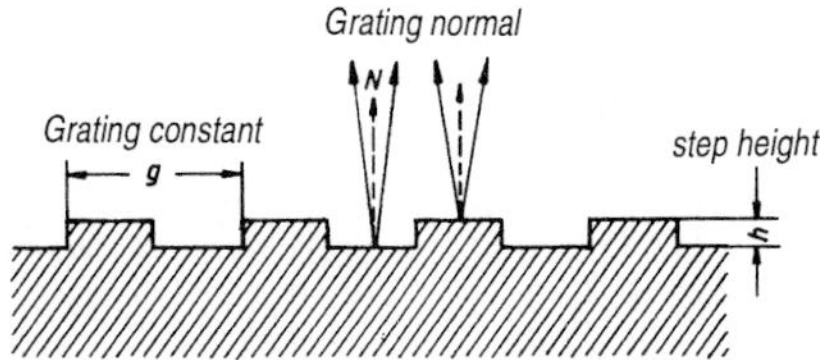

Laminar grating. Fig. 1. Section through a laminar grating

for use in the infrared region above 30 μm. A laminar grating can be produced by etching the desired grooves on a glass plate followed by a silvering of the whole surface. In this case, metallic reflection takes place both from the grooves and from the higher areas between them. The maximum intensity is obtained when $\lambda_o = 4h$ (h is the step height) because then, for normal incidence, the path difference between rays reflected by the tops and bottoms of the grooves in zero order is just $\lambda/2$ and no energy is lost. The depth of the grooves is therefore chosen such that this condition is fulfilled for a wavelength in the middle of the region of interest. The advantages of the laminar grating make themselves felt in the range $2h < \lambda < 4h$. Normally, gratings in which the width of the grooves is equal to half the grating constant are used. Then all the even orders disappear. If, in addition, the angles of incidence and refraction are equal and very small, a condition which is largely fulfilled in the usual instrumental designs, the intensities, J, in the odd orders $m = 1, 3, 5 \ldots$ are:

$$J \sim \frac{1}{m^2}(1 - \cos\frac{4\pi h}{\lambda}),$$

and for $m = 0$.

$$J \sim 1 + \cos\frac{4\pi h}{\lambda}$$

If $\lambda = \lambda_o = 4h$, the intensity in the *zero*th order is zero and the intensity in the useful orders is enhanced at the cost of the zero order intensity.

LAMMA, laser microprobe mass analyzer, <*LAMMA, Laser-Mikrosondenmassenanalysator*>. An apparatus originally developed for the elemental microanalysis of biological materials. It is based upon the laser generation of a microplasma followed by identification of the positive and negative ions in the plasma by means of a → time-of-flight mass spectrometer. Apart from the original objectives, which were biological and medical problems, LAMMA has found increasing application in surface analysis. The reason for this is that the laser pulse evaporates and ionizes a chosen sample volume from a precisely defined spatial region: probe area < 1 μm diameter, probe volume 10^{-13} cm^3. The applications are many and varied: organic and inorganic materials science, metallurgy, geology, mineralogy, biology, production control, forensic science, microelectronics etc. Since the probe can be moved exactly in the x, y or z directions, while its position is monitored with a microscope or a video camera and screen, it is possible to scan the surface of a sample. This is especially important in surface analysis. A LAMMA instrument made by Leybold-Heraeus (LAMMA 1000) is shown schematically in Figure 1. The laser is a Nd:YAG, frequency quadrupuled to give $\lambda = 265$ nm. The annotation of the diagram gives further details of the components. Since laser-induced ionization also gives large numbers of negative ions, all potentials on the spectrometer electrodes are reversible.

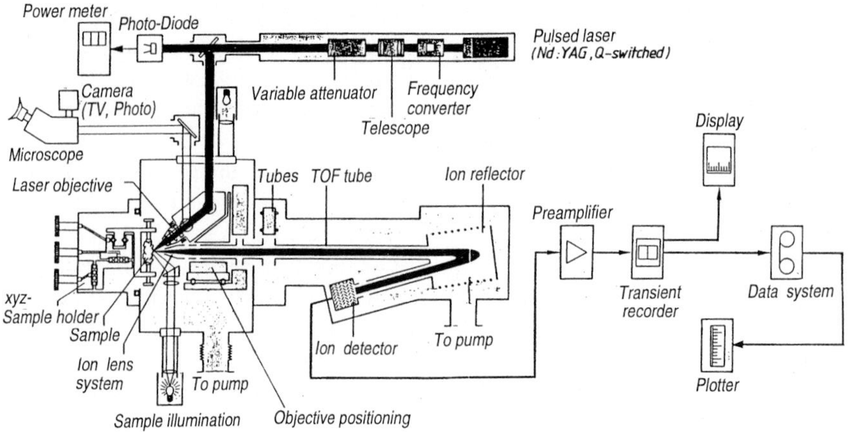

LAMMA. Fig. 1. The Leybold-Heraeus LAMMA 1000

Ref.: H.J. Heinen, R. Holm, *Scanning Electron Microscopy*, **1984**, III, SEM Inc., AMF O'Hare, Chicago, pp. 1129–1138.

Landé g factor, <*Landé-Faktor*> → g factor.

Laporte rule, <*Laporte-Regel*>, a → selection rule for atomic spectral transitions proposed by O. Laporte in 1924. All atomic wave functions must be either symmetric *g* (from the German *gerade* = even) or antisymmetric *u* (*ungerade* = uneven) with respect to inversion in the nucleus which is the center of symmetry of the atom (→ symmetry element). *g* states (orbitals) are said to be of even parity, examples are *S*, *D*, *G*; *u* states (orbitals) are of odd parity, e.g. *P*, *F*, *H*. Laporte's rule states that all atomic spectral lines due to electric-dipole radiation arise from transitions between states of opposite parity. Thus $u \to g$ and $g \to u$ are allowed but $u \to u$ and $g \to g$ are forbidden.

Laser, light amplification by stimulated emission of radiation, <*Laser*>. An excited atom or molecule falls to a lower state stimulated by an incident photon, (light quantum). The energy thereby released is emitted as a photon. In contrast to spontaneous emission (→ fluorescence), this induced emission does not occur in any direction, but in the direction of the incident photon which induced the emission process. Both photons have the same direction and frequency. If the light falls simultaneously on very many excited atoms, then many atoms can be induced to emit and the photons produced all have the same direction and frequency. It is not necessary that the atoms are all regularly ordered. The incident light wave that induces all the atoms to emit also induces them to emit in phase, no matter how they are distributed in space.
It is important for laser emission that the population of the excited state (lasing state) is greater, in general very much greater, than the ground state population. Only then do the induced

emission processes dominate and intensify the stimulating light. This situation is known as a → population inversion. The inversion must always be brought about by some external action so that the extra radiated energy is drawn from an external energy source. The → active medium in which a population inversion can be created is important for a laser. Lasers are classified as → solid-state, → semiconductor (→ diode), → gas, → excimer and → dye according to the nature of this medium. The active medium may be contained in an → optical resonator, the purpose of which is to bring about a further increase in the intensity of the laser light. This is always done when the active medium shows too little amplification. The advantages of the laser over conventional light sources are the coherence and good monochromacity of its light, properties which make the laser an excellent light source for spectroscopy.

Ref.: D. L. Andrews, *Lasers in Chemistry*, 2nd edtn., Springer Verlag, Berlin, Heidelberg, New York, **1990**.

DCM

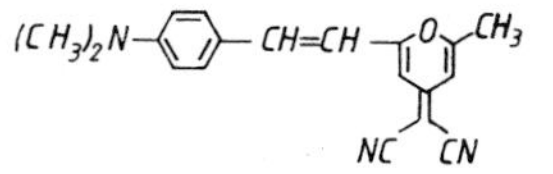

LD 700 — Rhodamine 700

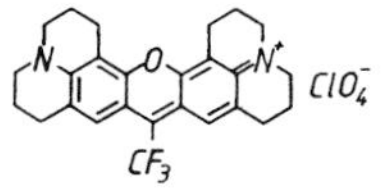

PYR 1 — Pyridine 1

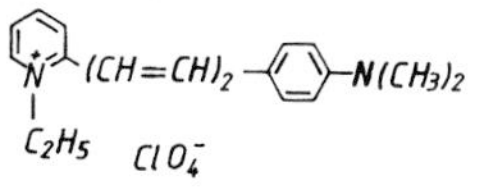

OX 750 — Oxazine 750

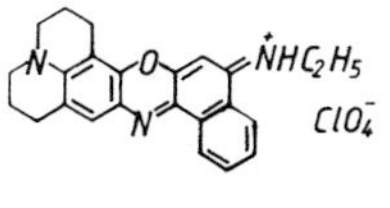

Sty 9 — Styryl 9

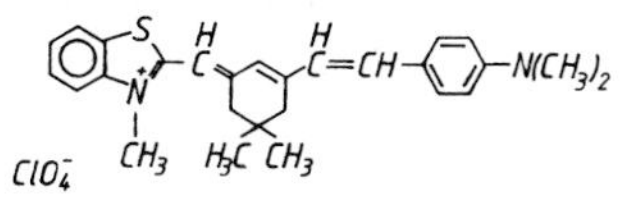

HITCI

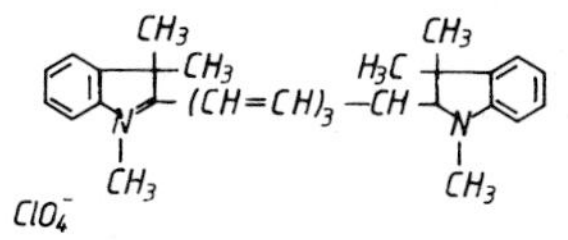

Laser dyes. Fig. 1. The structural formulas of some laser dyes

S1 Stilbene 1

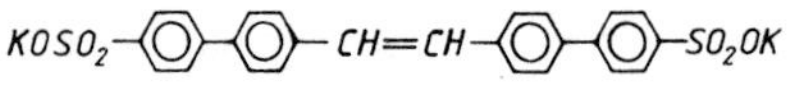

S3 Stilbene 3

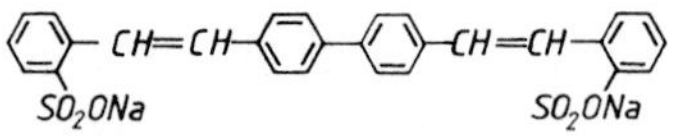

C102 Coumarin 102

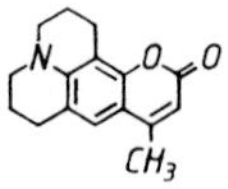

C7 Coumarin 7

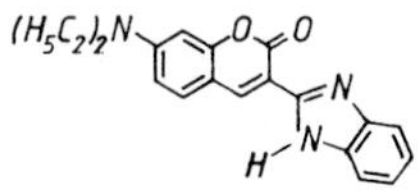

C6 Coumarin 6

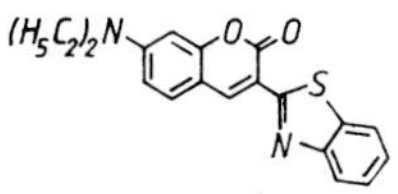

R110 Rhodamine 110

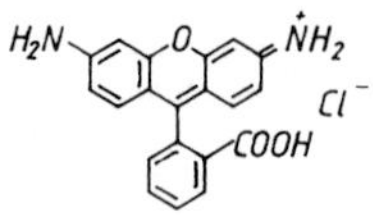

R6G Rhodamine 6G

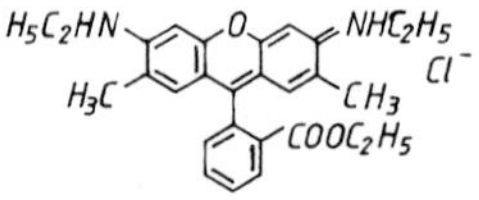

IR 140

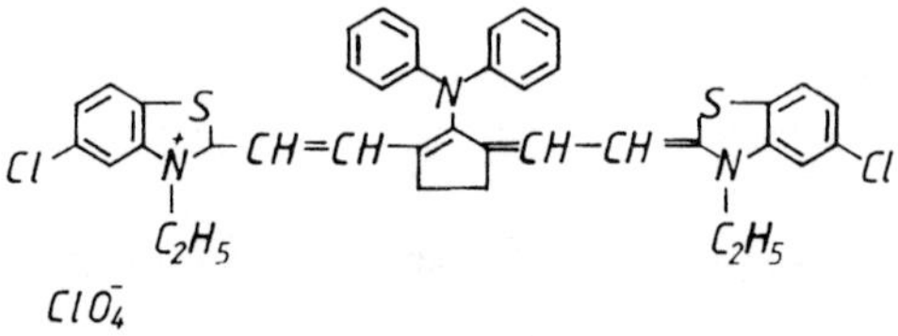

Laser dyes. Fig. 1. Continued

Laser dye, <*Laserfarbstoff*>. Fluorescent organic compounds which, in suitable solvents, form the active media of pulsed or continuous → dye lasers. Since it is the → fluorescence of these compounds which is exploited for the laser emission, the → fluorescence quantum yield should be as large as possible so that the competing processes such as → intersystem crossing ($S_1 \rightarrow T_1$) and radiationless deactivation play only a minor role. Normally, the $S_o \rightarrow S_1$ transition, which should have a high → extinction coefficient ($2 \cdot 10^4 \leq \varepsilon \leq 10^5$ l·mol^{-1} cm^{-1}), is excited with a suitable pump laser (→ optical pumping). Because the $S_o \rightarrow S_1$ excitation energy determines the position of the long-wavelength absorption band and that of the red-shifted, mirror-image fluorescence, the laser emission is confined to a specific, narrow range in the VIS or NIR region. Thus, a range of dyes is required to cover the whole region from 400 to 1000 nm. In addition to the above factors, the choice of suitable laser dyes is limited by the requirement that they be photochemically stable. Approximately 100 compounds suitable for use as laser dyes are now known and are available commercially. The structural formulas and trivial names of a selection of 14 compounds are shown in Figure 1.

Ref.: U. Brackmann, *Lambdachrome® Laser Dyes*, 2nd edtn., Lambda Physik GmbH, Göttingen, **1994**.

Laser ion source, *<Laserionenquelle>*, an → ion source in which the surface of a solid (the target) is irradiated with a focused laser beam. Very high temperatures ($7 \cdot 10^3 - 18 \cdot 10^3$ K) are generated at the sample surface and in the laser microplasma. A laser ion source can evaporate and ionize all the elements present in a sample. The principle of the device is illustrated in Figure 1a. Thermal ions produced in the laser microplasma are drawn out of the region of ionization by an accelerating field, U_B. Apart from this direct method, laser ion sources may also be combined with a low-voltage discharge (Figure 1b → spark ionization) or → electron-impact ionization (Figure 1c). Pulsed → solid-state

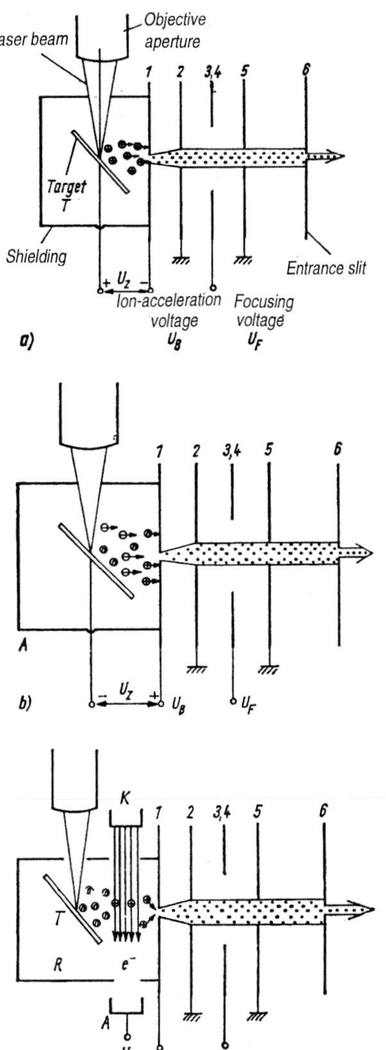

Laser ion source. Fig. 1.
a) The principle of a laser ion source
b) Combination with a low-voltage discharge
c) Combination with electron-impact ionization

lasers are used, e.g. → ruby or → neodymium lasers, with powers of 10^7 to 10^{13} W cm^{-2} and pulse durations of 1 μs to 100 ns, depending upon the mode

of action. Very high radiation intensities result from the focusing so that, in effect, all materials can be vaporized and ionized by these methods.
Ref.: *Lasers and Mass Spectrometry*, Ed.: D.M. Lubman, Oxford University Press, Oxford, New York, **1990**.

Laser spectroscopy, <*Laserspektroskopie*>, a general term for the spectroscopic methods which use a laser as light source. Rationally, one could also speak of spectroscopy with conventional light sources as tungsten lamp, deuterium lamp, Nernst glower and hollow cathode lamp spectroscopy etc. The reason for the separate name for laser spectroscopy is the special properties of this light source which conventional light sources do not have. Thus, some newer methods such as, for example, → multiphoton spectroscopy, → hole burning and → frequency doubling first became possible following the development of lasers.
Ref.: W. Demtröder, *Laser Spectroscopy*, Springer Series in Chemical Physics, Vol 5., Springer Verlag, Berlin, Heidelberg, New York, **1983**.

Laue relationship, <*Laue-Beziehung*> → X-ray diffraction.

Ligand field theory, <*Ligandfeldtheorie*>, a development of the crystal field theory (CFT) proposed by H. Bethe in 1929 and first applied to transition-metal ions by H.H. Van Vleck in 1932. CFT provides a simple way of describing the energies and occupation of the *nd* atomic orbitals of transition-metal atoms or ions (3*d*: Ti–Zn; 4*d*: Zr–Cd; 5*d*: Hf–Hg) in the complexes which they form with ions and neutral molecules. Typical examples of such complexes are: $[CrCl_6]^{2+}$, $[Mn(OH_2)_6]^{2+}$ and $[Co(NCS)_4]^{2-}$. The surrounding molecules or ions, Cl^-, H_2O and $(NCS)^-$ in the above examples, are known collectively as ligands. Transition-metal complexes are characterized by a high degree of symmetry; e.g. the first two examples above are octahedral while the last is tetrahedral. The high symmetry made the use of → symmetry point group theory very effective and the development of the CFT paralleled the growing use of symmetry theory in chemistry. Since their electronic spectra and magnetism are primarily determined by the energy and occupation of the metal *d*-orbitals, a knowledge of these factors is required to interpret the most striking physical properties of transition-metal complexes.
Water and ammonia played an important role in early experimental work of transition-metal complexes and these two ligands are bound to the central metal through their electron-rich oxygen and nitrogen atoms respectively. Thus, the CFT, which is concerned only with electrostatic interactions, envisages a repulsive interaction between the metal *d*-electrons and the surrounding ligands. The *d*-orbitals are orientated in space as illustrated in Figure 1 and the electrons occupying these orbitals will interact with the complexing ligands to different extents, depending upon the positions of the ligands and the orientation of the *d*-orbitals. The regions of maximum electron density of the $d(z^2)$ orbital are directed along the $\pm z$-axis and the $d(x^2\text{-}y^2)$ has its maxima along the $\pm x$- and $\pm y$-axes. The remaining orbitals are directed between the Cartesian axes. It is important to note that these angular properties of the *d*-orbitals apply for all principal → quan-

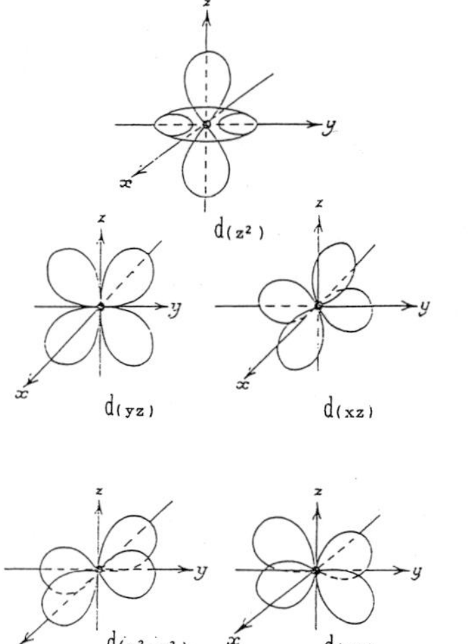

Ligand field theory. Fig. 1. The forms of the five *nd*-orbitals

tum numbers, *n*, and for any number of electrons. As the value of *n* increases on going to the heavier transition metals, the orbitals reach out further from the atomic nucleus, but their angular disposition in space is not changed.

If we consider the effect upon the 3*d*-orbitals of the Mn^{2+} ion of placing six water molecules around it, each on a Cartesian axis in an octahedral array, then the $d(z^2)$ and $d(x^2\text{-}y^2)$ orbitals will be most strongly repelled by the electron density on the oxygen atoms of the water ligands and the *d*(*xz*), *d*(*yz*) and *d*(*xy*), which point between the water molecules, will be repelled less strongly. Thus, the five 3*d*-orbitals, which were degenerate in the free Mn^{2+} ion, will not remain degenerate in the $[Mn(O_2H)_6]^{2+}$ complex. It is clear that the *d*(*xz*), *d*(*yz*) and *d*(*xy*) will have the same energy but it is not obvious what the relative energies of the $d(z^2)$ and $d(x^2\text{-}y^2)$ orbitals will be. However, group theory shows that they have the same energy and for an octahedral complex (→ symmetry point group O_h) the two sets of degenerate orbitals belong to the symmetry species e_g and t_{2g}. For a regular tetrahedral (T_d) complex the five *d*-orbitals split into the same two degenerate sets, but in this case the $d(z^2)$ and $d(x^2\text{-}y^2)$ orbitals have the higher energy and the *d*(*xz*), *d*(*yz*) and *d*(*xy*) the lower energy. The two sets of orbitals span the symmetry species *e* and t_2 respectively.

The *d*-orbital splitting patterns for the more symmetric complex geometries are illustrated with the energy-level scheme in Figure 2. Less symmetrical

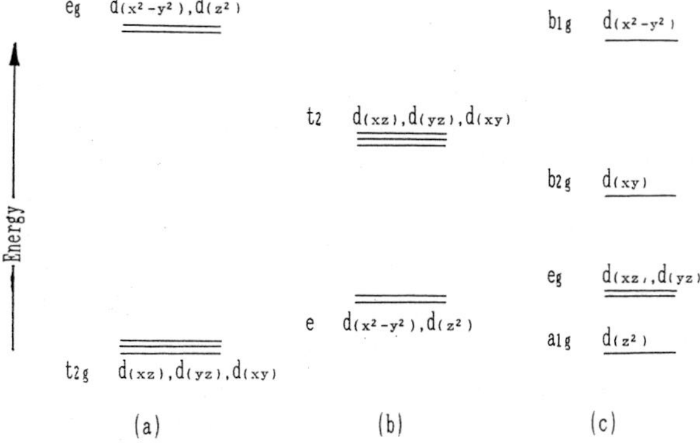

Ligand field theory. Fig. 2. Energy-level schemes for *nd*-orbitals in a) octahedral [O_h], b) tetrahedral [T_d] and c) square planar [D_{4h}] coordination

structures are normally described with reference to the symmetrical cases. Having obtained the energies of the *d*-orbitals in a complex it is next necessary to determine how they are occupied by the available electrons, and here two factors have to be considered. The aufbau principle requires that the electrons fill the orbitals, two at a time (→ Pauli exclusion principle), from the lowest in energy upwards. But interelectronic repulsion also has a role to play and if, for example, the splitting between the e_g and t_{2g} sets of orbitals in an octahedral complex is small, it will be energetically more favorable for the electrons to occupy both sets of orbitals singly, with parallel spins (→ Hund's rule), than to occupy the t_{2g} orbitals in pairs with opposed spins. Complexes with unpaired, parallel spins are termed high-spin complexes, those with paired electrons are low-spin complexes. The magnetic properties of a complex are directly determined by the number of unpaired spins (→ Bohr magneton). Thus, the → multiplicity, magnetism, paramagnetic resonance (→ ESR) and electronic spectroscopy of a transition-metal complex depend critically upon a delicate balance between the splitting of the *d*-orbitals by the crystal field and the effects of electron repulsion. A high crystal field leads to a large splitting and to pairing of electrons in the lower orbitals, i.e. high field ⇨ low spin ⇨ diamagnetism. Conversely, a low field leads to high spin and paramagnetism. For octahedral complexes the field strength is normally given the symbol Δ (or 10 *Dq*) and ligands are classified according to the strength of their field in the spectrochemical series:

$I^- < Br^- < S^{2-} < SCN^- < Cl^- < NO_3^- < F^- < OH^- < (oxalate)^{2-} < H_2O < NCS^- < CH_3CN$, $NH_3 <$ ethylenediamine < bipryridine < o-phenanthroline $< NO_2^- < CN^- < CO$

Thus, the six 3*d*-electrons in the complex ion $[Fe(CN)_6]^{4-}$ with the strong-field ligand CN^- are paired in the three t_{2g} orbitals; the ground state of the complex is a singlet and it is low-spin and diamagnetic. In $[Fe(H_2O)_6]^{2+}$ with the much weaker H_2O ligand the electrons are distributed over all five 3*d*-orbitals so that four are unpaired and the ground state of the complex is a quintet. It is high-spin and shows a large magnetic moment. The relative strength of the two ligand fields is revealed by the fact that the lowest energy $d(t_{2g}) \rightarrow d(e_g)$ transition of the high-spin complex lies at 10,400 cm^{-1} while for the low-spin complex it is found at 31,000 cm^{-1}. It should be noted that these *d-d* transitions are formally forbidden by the → Laporte rule.

The balance of crystal field and electron replusion, plus the fact that the energy of an electronic transition is not simply related to the difference in energy between the orbital which the electron leaves and that which it enters, makes the relationship between the observed absorption bands of a complex and the various theoretical parameters rather complicated. L.E. Orgel derived diagrams (Orgel diagrams) relating the experimental data and the theoretical parameters. Y. Tanabe and S. Sugano produced similar but more elaborate diagrams.

Though the succcesses of the CFT in explaining the stability, magnetism and electronic spectroscopy of transition-metal complexes were many, it was clear that a purely electrostatic theory could not fully represent the bonding of the ligands to the central metal ion. Meanwhile, the use of molecular orbital (MO) theory to interpret chemical bonding and structure had made great advances and in 1952 M. Wolfsberg and L. Helmholz reported MO calculations on the ions MnO_4^-, CrO_4^{2-} and ClO_4^-. In their calculations they made extensive use of symmetry theory and confirmed the view that the high symmetry of transition-metal complexes plays a central role in determining their properties. Thus, the results of the MO calculations are qualitatively very similar to those of the CFT, which goes a long way to account for the latter's success. In the last 40 years theoretical models of transition-metal complexes embracing both CFT and MO concepts have been developed. The whole gradation of these theories, from the point-charge CFT to the pure MO method, is generally known as ligand field theory and, usually, no specific form of this wide range of theoretical models is implied by the term. Nowadays, a variety of semiempirical molecular orbital computer packages are also available for the calculation of the electronic structures of transition-metal complexes. *Ab initio* methods have been used in some cases but, at the present time, the large number of atoms in a typical transition-metal complex frequently makes the use of semiempirical methods the only realistic possibility.

Ref.: H.L. Schläfer, G. Gliemann, *Basic Principles of Ligand Field Theory*, J. Wiley and Sons Ltd., London, New York, Sydney, Toronto, **1969**; J.S. Griffith, *The Theory of the Transition-Metal Ions*, Cambridge University Press, Cambridge, London, New York, **1961**; A.B.P. Lever, *Inorganic Electronic Spectroscopy*, 2nd ed., Elsevier, Amsterdam, Oxford, New York, Tokyo, **1984**.

Light, electromagnetic radiation, *<Licht, elektromagnetische Strahlung>*, a transverse electromagnetic wave phenomenon. Light is produced by emission from excited atoms or molecules in which the excitation can be the result of provision of thermal energy (→ thermal radiator), electrical energy (→ gas-discharge lamp) or of a resonance process (→ fluorescence, → phosphorescence). In the excited state, there is usually a charge separation which leads to an electric dipole which attempts to return to the ground state and charge neutrality by means of damped relaxation vibrations. The vibrating dipole is a Hertz oscillator which sends out electromagnetic waves, as illustrated for a single oscillator in Figure 1. The emitted wave consists of an oscillating electric and magnetic field, the directions of which depend upon the polarization of the oscillating dipole. The electric field, E, the magnetic field, H, and the direction in which the wave propagates, Z, form a right-handed coordinate system. The electric field vibrates in

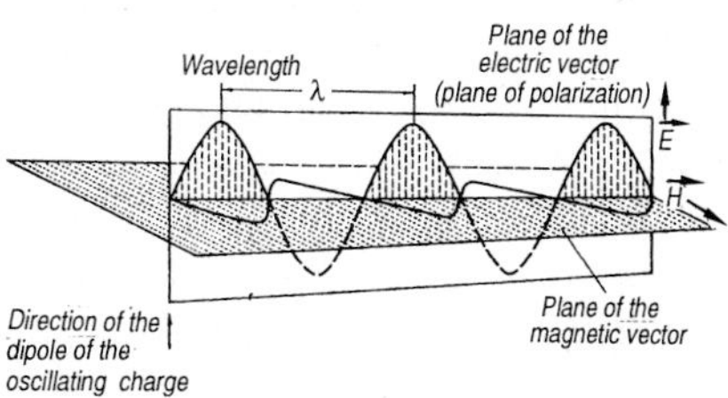

Light, electromagnetic radiation. Fig. 1. The electromagnetic wave of an individual Hertz oscillator

the same plane as the oscillating charge distribution which is known as the plane of vibration or polarization. Natural light shows no preference for a particular plane of polarization and is completely axially symmetric. This is because the extremely large number of oscillators in a light source are statistically distributed, i.e. the vibrational directions of the independent oscillators are distributed completely uniformly over all directions. However, if a light source is placed in a magnetic field, the statistically uniform distribution of the vibration directions is removed and emission of polarized radiation is observed (→ Zeeman effect). In natural light the electric field vector, $\vec{E}$, vibrates in all directions perpendicular to the direction of propagation, Z. With linearly polarized light, however, the E vector vibrates in a particular plane called the vibrational plane or the plane of polarization. The spreading of an electromagnetic wave is characterized by the propagation velocity, c, and the wavelength, λ, which are related by the → frequency, ν, in $c = \lambda \cdot \nu$. When an electromagnetic wave travels through media with different refractive indexes, n_i, then, because ν is constant, $c_i = c_o/n_i$ and therefore, $\lambda_i = \lambda_o/n_i$. c_o is the velocity of electromagnetic radiation in a vacuum ($c_o = 3 \cdot 10^8$ ms^{-1}), i.e. the velocity of light. Alongside the above wave model of electromagnetic radiation as spreading vibrations or waves, the particle model (Einstein, 1905) of light as a stream of particles must also be considered. Here, with the introduction of quantum mechanics, the wave-particle duality of light comes to our attention. For example, the particle picture (light quanta, → photons) explains the momentum, the → photoelectric effect and the → Compton effect while the wave description finds application in the interpretation of diffraction and interference phenomena. The connection between the particle and wave models is given by the de Broglie relationship, $\lambda = h/(m \cdot v)$, in which h (= $6.63 \cdot 10^{-34}$ J s) is → Planck's constant and $m \cdot v$ is the momentum of the particle (m = mass in kg and v = velocity in $m \cdot s^{-1}$). Depending upon the frequency and the application of the radiation, either the particle or the wave properties come to the fore.

Light attenuation, *<Lichtschwächung>*, a technique in → photometry in which two beams of light are used (double-beam technique → spectrophotometer). One of the beams of white or monochromatic light is attenuated when it passes through a homogeneous, isotropic, absorbing medium as required by the → Bouguer-Lambert-Beer law. The intensity of the other beam, the reference beam, is reduced by mechanical attenuation until it is equal to that of the first beam, i.e. they each give the same signal at a detector. This → optical null principle was used in IR spectrometers until recently. Originally, it was also used in UV-VIS instruments, but there it was replaced by → ratio recording at an earlier date.

A specific degree of light attenuation, which is desirable for the study of some phenomena, can be achieved by placing → rotating sectors, → neutral density filters or stops in the beam. Grids made of thin wires are very effective and they can be obtained in sets covering a range of attenuations from most suppliers of spectroscopic accessories. They are also quite simple to construct in the laboratory.

Light barrier, <*Lichtschranke*>, a relay switch employing photoelectric detectors which is almost without lag (→ photoelectric measurement methods).

Light beam, <*Lichtstrahl*>, in the wave theory of light, the normal to the wave front. In everyday speech, a light beam is a ray of light which is obtained when light passes through a small orifice. The cross section of the ray is disregarded and it is treated as if it were a geometrical straight line. A laser provides a very narrow beam of almost constant width. The direction of a light beam is the direction in which the light energy flows.

Light conductance, <*Lichtleitwert*>, a measure introduced by Hansen to quantify optical apparatus. It is defined as follows:

$$L = \frac{n^2 F_1 F_2}{a^2}.$$

F_1 and F_2 are the areas of the first aperture (entrance slit) and the second aperture (exit slit), respectively. a is the distance between the two apertures (slits). The factor n^2 takes account of the passage of the light through a medium of higher refractive index (prism or solution). The performance of the optical components between the source and the detector characterizes the effectiveness of an optical setup. The best use of the light is always obtained when every slit is equipped with a condensing lens which forms a sharp image of the foregoing slit upon the next in such a way that the size of the image agrees exactly with the aperture of the next slit. The slit is then said to be filled. For a monochromator, this means that the image of the entrance slit must be exactly superimposed upon the exit slit. The light conductance value determines the performance of the whole optical arrangement and can therefore be used to compare one setup with another.
Ref.: G. Hansen, E. Mohr, *Spectrochim. Acta*, **1949**, *3*, 584.

Light conductor, light pipe, light conduit, <*Lichtleiter*>, a transparent body along which light is conducted by multiple reflection at the surface cladding. Glass, quartz, acrylic glasses and polystyrene are suitable materials. Use is exclusively made of the → total reflection which always takes place in the optically more dense medium. Thus, rods of glass, quartz or plexiglas (perspex) make good light pipes in air. However, they are not optically isolated, i.e. the total reflection can be disrupted, with a resulting loss of light and scattering, by supporting devices or by close contact with a neighboring light pipe. Therefore, the basic form of a light pipe screened from external influences is a straight rod with refractive index, n_1, covered with a cladding with refractive index, $n_2 < n_1$. For light rays propagating in a plane, which contains the axis of the pipe, the aperture, NA, of a light pipe is given by the same equation as for an optical fiber (→ fiber optics): $NA = n_o \sin\alpha_o = (n_1^2 - n_2^2)^{1/2}$, since the same model applies to both. The individual optical fiber and the light conductor differ only in the fact that the former is flexible while the latter is rigid. With air as the cladding, $n_o = n_2 = 1$, $n_1 \geq \sqrt{2}$, and the above equation gives; $\sin\alpha_o = 1$ i.e. the angle $\alpha_o = 90°$. Thus, the light conductor can accept light from

the whole hemisphere and conduct it further by total reflection. The requirement, $n_1 \geq \sqrt{2}$, is fulfilled by all the usable glasses and plastics.
Ref.: H.-G. Unger, *Planar Optical Waveguides and Fibres*, Oxford University Press, **1977**.

Light-emitting capacitor, <*Leuchtkondensator*>, a capacitor in which the dielectric is a light-emitting material which is excited by an alternating electric field. The light-emitting dielectric is sandwiched between two glass plates the inner surfaces of which carry electrodes formed from conducting layers. One of the electrode layers must be transparent. As a result of the applied alternating voltage, e.g. 220 V, an alternating electrical field with high local field strengths and high electron acceleration is formed in the dielectric. As the applied voltage passes through zero, the field changes direction and the accelerated electrons in the dielectric give out their excess energy as radiation. The radiation is emitted through the transparent electrode; its intensity depends upon both the frequency and the magnitude of the applied voltage. Figure 1 shows the construction of a light-emitting capacitor schematically. The spectral energy distribution depends upon the dielectric; an example is given for a green-emitting material in Figure 2. The light intensity at 100 Hz is 3 times and at 600 Hz, 17 times that at 50 Hz, which is said to be about 0.001 cd/cm^2. The light yield is currently approximately 7.5 lm/W for green emission. Light-emitting capacitors could find uses as large area ceiling and wall illumination panels with very low power consumption. Their present service life is said to be 2000 hours. The construction and application of these devices is still at the development stage.
Ref.: H.K. Henisch, *Electroluminescence*, International Series of Monographs on Semiconductors, Vol. 5, Pergamon Press, Oxford, London, New York, Paris, **1962**.

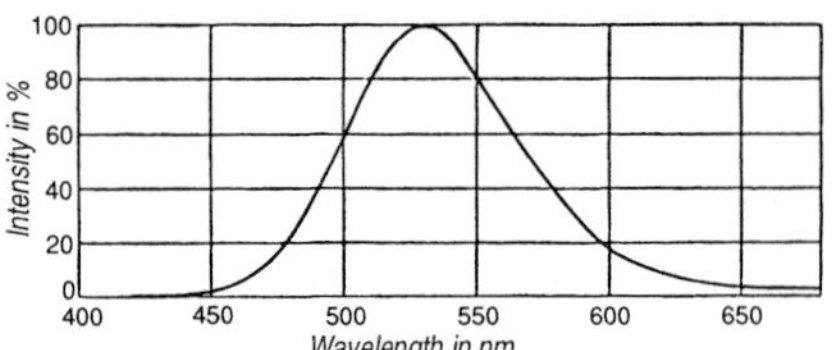

Light-emitting capacitor. Fig.2. Emission curve for a green emitter

Light-emitting diode, LED, <*Lumineszenzdiode, lichtemittierende Diode*>, a diode which emits radiation in the VIS or NIR spectral regions without passing through the diversion

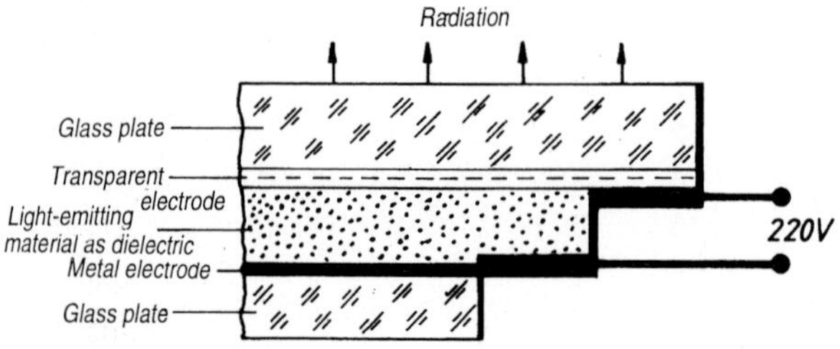

Light-emitting capacitor. Fig. 1. Section showing the construction

of a thermal energy step (→ thermal radiator). The action is the reverse of that which takes place in a → photodiode. An electric field applied to a *pn* semiconductor junction induces charge separation; when the charges recombine energy is emitted in the form of electromagnetic radiation. The voltage must lie in the conducting direction and the current should be limited by a preceding resistor or a constant current source.

Particularly suitable semiconductor materials are gallium phosphide (GaP) which emits in the green (~ 560 nm), gallium arsenide (GaAs) which emits in the NIR (~ 950 nm) and mixed crystals ($GaAs_{1-x}P_x$) for the intervening, red-green, region e.g. $GaAs_{0.6}$ $P_{0.4}$ for red (~ 645 nm). In general, the emission is narrow banded with a half-width of the order of 30 nm. The maximum of the emission increases with temperature; a temperature increase of 3 K gives a shift of approximately 1 nm. Naturally, the temperature of the *pn* layer is important. The upper limit of the conduction current for continuous operation is 25–30 mA. Higher current pulses are possible. For a maximum pulse length of 1 μs, current pulses of between 1 and 5 A, depending on the particular diode, are allowable. At higher current densities and with a suitable construction (use of a resonator) the LED becomes a → diode laser.

In principle, LEDs are very cheap, almost monochromatic light sources. In the NIR, where the → spectral energy distribution of the LED is well matched to the → spectral sensitivity of the photodiode detector, the conditions for applications are particularly favorable.

Ref.: J. Wilson, J.F.B. Hawkes, *Optoelectronics: An Introduction*, Prentice Hall International Inc., London, Englewood Cliffs (NJ), **1983**.

Light-emitting materials, <*Leuchtstoffe*>, materials capable of luminescence in the visible spectral region (luminophores).

Light flash, <*Lichtblitz*>, a very short emission of light. To produce such emissions electronic flash discharge lamps (→ flashlamps) with a flash duration of $10^{-4}-10^{-6}$ s and a repetition frequency of 10^5 s^{-1} are used. Shorter flash times ($10^{-8}-10^{-9}$ s) are possible with pulsed → nitrogen-discharge lamps, sometimes called nanosecond lamps, also with nitrogen lasers and generally with pulsed lasers. The applications of light flashes are found in → flash photolysis for the determination of the lifetimes and in following the kinetics of excited states.

Light intensity, <*Lichtstärke*>, the luminous intensity of a light source per unit solid angle; symbol *I*. The unit is the → candela (→ photometric units).

Light intensity, unit of <*Lichteinheit*> → candela, → photometric units.

Light-intensity standard, <*Lichtstärkenormal*>, a lamp which emits light of a definite intensity. Nowadays, lamps with ribbon incandescent filaments arranged in a plane (photometer lamps) are used. The intensity given out perpendicular to the emitting elements is then well defined, provided that there are no imperfections in that part of the glass envelope of the lamp through which the measured light emerges. The envelope has

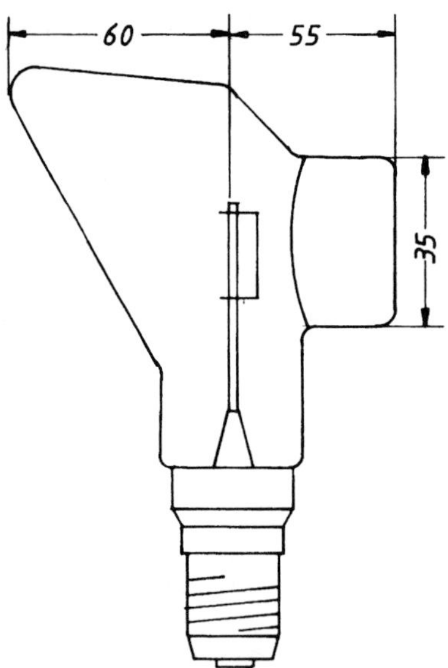

Light-intensity standard. Fig. 1. Construction of a tungsten lamp, Osram Wi 17/G

a conical shape so that light reflected from the inner wall does not emerge in the direction of measurement, see Figure 1. To guard further against the disturbing influence of reflected light from the inside of the lamp, the side of the envelope directed towards the photometer is covered, apart from the window, by a light-tight screen.

Light modulation, *<Lichtmodulation>*, the conversion of light of a constant intensity to light alternating in intensity with frequency ν s^{-1}, by means of electrical or mechanical devices. For this purpose rotating sectors (→ chopper), an electrooptical effect such as the → Kerr or → Pockels effect, pulsed → noble-gas discharge lamps and → lasers can be used.

Light path, beam path, *<Strahlengang>*, the path of the light (beam, ray) through any optical device. In particular, the light path is used in conjunction with the disposition of the optical components, to describe the functional principles of spectrometers; e.g. → atomic absorption, → UV-VIS, → fluorescence and → IR spectrometers.

Light quantum, *<Lichtquant>* → photon.

Light reaction, *<Lichtreaktion>*, a chemical reaction in which light is involved; a → photochemical reaction. In general, the course of such reactions is different from those which do not depend upon light and are described in terms of the usual chemical kinetics.

Light scattering, *<Lichtstreuung>*, the scattering of photons by an assembly of particles such as atoms, molecules, crystallites, colloidal or dust particles, or even variations of density or concentration. In general, the term *scattering* is used when the dimensions of the particles are comparable with the wavelength of the incident electromagnetic radiation or even smaller. A distinction may be made between simple and multiple scattering. The first occurs where the particles scatter independently of each other. An estimate shows that this is the case when the mean distance between the particles is at least twice as large as their diameter. In this situation, it can be assumed that the amplitudes of the waves scattered by the individual particles in all directions may be simply summed without considering the → phase, i.e. the scattering can be considered to be → incoherent. The apparent absorption (→ absorbance, apparent) measured in transmission is a criterion for simple scattering. Analogously to the → Bouguer-Lambert-Beer law, this is given by:

$$I' = I_o \exp[-S' \cdot d]$$

or

$$\ln (I_o/I') = S' \cdot d$$

I_o is the incident and I' the transmitted light intensity, d is the path length. S' is the scattering coefficient or → turbidity and is more accurately defined by S' (in cm^{-1}) $= \sigma n_o$, in which σ is the scattering cross section in cm^2 and n_o the number of particles per cm^3. If $S' < 0.1$ the scattering is essentially simple, if $S' > 0.3$ multiple scattering is dominant. This is always the case when the distance between the particles decreases, i.e. when the particle density, n_o, increases as, for example, in clouds, frosted glass, concentrated sols, polymer solutions, crystal powders and pigments. In opaque samples, where the incident radiation is effectively entirely reflected, e.g. a thick fog, multiple scattering dominates.

Rigorous theoretical treatments are avaivable only for simple scattering by molecules which are small compared to λ (→ Rayleigh scattering) and for isotropic, spherical particles of any size (→ Mie scattering). Rayleigh scattering is a limiting case of Mie scattering. In the photon model, we distinguish between elastic and inelastic scattering in the interaction between a photon with energy $h\nu_o$ and a scattering particle. In the first case, there is no change in the frequency of the photon, i.e. the incident and the scattered light have the same photon energy, $h\nu_o$. Rayleigh and Mie scattering are of this type. In inelastic scattering there is a change of frequency and the incident and scattered light do not have the same photon energy, $h\nu_o \neq h\nu_{scat}$. The → Compton and → Raman effects are examples of inelastic scattering.

Light-scattering photometer, <*Lichtstreuphotometer*>, a → photometer which is, in principle, a → turbidity meter. Although these instruments are usually fitted with a tungsten lamp, lasers are used as sources in modern light-scattering photometers. They make the technique of measuring the → Mie scattering in the forward direction easier because the extremely narrow laser beam can be separated from the forward scattered light by means of suitable apertures (stops). In tune with modern technical developments, light-scattering photometers are microprocessor-controlled and, in combination with a computer, offer every possibility for the evaluation of the experimental data.

Light sensitivity, photosensitivity, <*Lichtempfindlichkeit*>, the property whereby a substance undergoes a chemical change as a result of the action of light, (→ photochemistry). Thus, light sensitivity is confined to particular spectral ranges. Photography depends upon the light sensitivity of the silver halides. For the light sensitivity of photographic materials → characteristic curve.

Light source, <*Lichtquelle*>, a body which sends out light. A light source consists of many small transmitters each of which sends out wave trains (the photons) independently. Every individual wave train is coherent (→ coherent light) and polarized, i.e. the electric vector oscillates in the same plane along the whole wave train, (→ polarized light). The magnetic vector oscillates at right angles to this vector and to the direction of propagation. The velocity of light in a vacuum is $3 \cdot 10^8$ m/s. Since a wave train is some meters in length (→ coherence length), it persists for about 10^{-8} s at any fixed point which it passes. The wave trains emitted from the different individual points of the light source are totally independent of each other and, therefore incoherent. Natural light is of this type; incoherent wave trains spreading out from the source in straight lines in all directions.

Light sources are the most important primary components in → optical spectroscopy. Continuum or line sources are used depending upon the requirements of the various methods. The first are generally → thermal radiators, they are used in the UV, VIS, NIR and IR regions. However, in the near and central UV region, the intensity of the thermal radiator is very low and → hydrogen lamps, which produce a continuum between 160 and 400 nm, are used for absorption measurements. The line sources are usually → metal-vapor discharge lamps or → gas-discharge lamps. The widely used → xenon lamps also show a broad continuum in the UV-VIS region and, because they have a high intensity, they are preferred for application in → fluorescence and → photoacoustic spectroscopy. In → Raman spectroscopy → lasers are now effectively the only source and in → atomic absorption spectroscopy, the hollow cathode lamp and the electrodeless discharge lamp (ELD) have increasingly proved their worth.

The table gives an overview of the sources used in optical spectroscopy. Point light sources in which the dimensions of the source can be neglected in comparison with all other relevant distances and which radiate equally in all directions are ideal sources. The → tungsten point lamp, the crater in the electrode of a small arc lamp and the tiny arc of a small high-pressure mercury lamp are examples of point light sources. A circular orifice illuminated from behind can also be used as a point source, though only in one half of the surrounding space.

Light source. Table.

Spectroscopy	Light source
Vacuum UV	noble-gas low-pressure discharge lamps
UV	hydrogen lamp
VIS	tungsten/tungsten-halogen lamp
NIR	tungsten/tungsten-halogen lamp
IR	Nernst glower, globar, thermal radiator
Luminescence	xenon lamp, mercury lamp
Raman	mercury lamp, laser
AAS	hollow cathode lamp, discharge lamp
AFS	hollow cathode lamp
ORD	xenon lamp
PA	xenon lamp, laser

Light throttle, <*Lichtdrossel*>, a set of filters in a → photocell which limits its spectral sensitivity to a particular wavelength interval.

Light yield, <*Lichtausbeute*>, the ratio of the light given out to the power taken up by a light source; (c.f. the situation with → lasers). The unit is lumen/Watt. In luminescence processes the light yield is the → quantum yield of the luminescence.

Limiting term, fixed term, <*Grenzterm*>, the term which represents the series limit in → atomic spectra. For the principal series this is the ionization energy of the atom.

Linear optics, <*lineare Optik*>, a classical area of physics which has long been regarded as complete. It is characterized by the concept that the propagation of a light beam in a medium is determined by the optical constants themselves, i.e. by the refractive index, n, and the absorption coefficients $\alpha(\kappa,\beta)$ and that the magnitudes of these constants are independent of the intensity of the light. Thus, reflection, refraction, diffraction, interference and absorption are independent of light intensity.

Linear variable interference filter, <*Verlauffilter*> → graded filter.

Line curvature, <*Linienkrümmung*>, a characteristic common to all spectroscopic devices containing prisms, (→ prism spectrograph, → prism spectrometer, → prism monochromator). The origin of the phenomenon lies in the fact that the rays emanating from each individual point along the entrance slit pass through the principal section (→ prism) of the prism at different angles. The spectral line is approximately a parabola with a radius of curvature, r, at the apex given by:

$$r = \frac{n^2 - f}{2(n^2 - 1)} \cot \alpha$$

Here, n is the refractive index of the prism material, f the focal length of the camera lens and α the angle of incidence. The line curvature becomes greater, i.e. r becomes smaller, as the wavelength (increase of n) and f become smaller. The curvature is concave when viewed from the UV side, i.e. the apex of the parabola lies to the red. Allowance for line curvature can be made by the use of nonlinear exit slits. The spectral lines from → diffraction gratings are not curved.

Line reversal, <*Linienumkehr, Selbstumkehr*>, the phenomenon that an atom is itself able to absorb many of the lines which it emits. This was first demonstrated in Bunsen and Kirchhoff's famous experiment (1860) on the resonance line of sodium. Line reversal occurs because every emitting atomic vapor is surrounded in its outer, cooler regions by a cloud of unexcited atoms in their ground states. These ground state atoms absorb the radiation emitted from the hot inner zone of the flame, which is the emission spectrum of the element in question, in a resonance process. The capacity for absorbing the first line of the principal series is very large (→ resonance line), thus the effect is especially clear with these lines. There is the additional fact that the width of the spectral lines in absorption and emission are not the same, the emission lines having a greater width than the absorption profiles of the atoms in their ground states. Thus, the atoms in the outer, cooler region absorb the center out of the emission lines and a sharp absorption line is observed on the background of a broadened emission line. This is illustrated schematically in Figure 1.

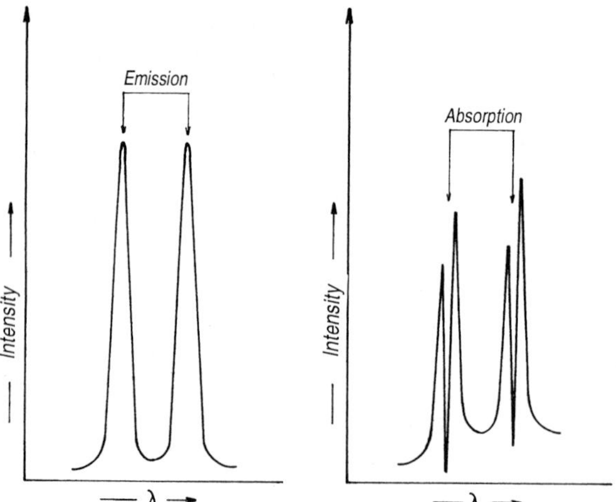

Line reversal. Fig. 1. Line reversal of emission lines in absorption (reabsorption)

Line shape, line profile, <*Linienprofil*>, the general term for the function $I(\nu)$ which describes the spectral distribution of the absorbed or emitted light about the central frequency, ν_o. $\nu_o = (E_i - E_k)/h$ corresponds to a resonance transition between the states i and k with the energy difference $\Delta E = E_i - E_k$. Gaussian or Lorentzian functions are frequently used to describe $I(\nu)$ analytically and we speak of the Lorentzian or Gaussian profile of a line or band.

As an example, the molar decadic → extinction coefficient can be expressed as a Lorentzian profile as follows:

$$\varepsilon = \frac{\varepsilon_{max} \cdot \Delta\tilde{\nu}_{1/2}}{4(\tilde{\nu}_o - \tilde{\nu})^2 + \Delta\tilde{\nu}^2_{1/2}}.$$

ε_{max} is the maximum value of the extinction coefficient at the absorption maximum $\tilde{\nu}_o$, $\Delta\tilde{\nu}_{1/2}$ the half-width of the line or band and $\tilde{\nu}$ the running wavenumber. The equation is also suitable for the description of the shape of the electronic absorption bands of substances in solution. The Gaussian profile derives from the Gaussian error function. In this form, the extinction coefficient may be expressed by the following equation:

$$\varepsilon = \varepsilon_{max} \exp\{-k(\tilde{\nu}_o - \tilde{\nu})^2\}.$$

Converting to logarithms, the equation of a parabola is obtained:

$$\ln \frac{\varepsilon_{max}}{\varepsilon} = k'(\tilde{\nu}_o - \tilde{\nu})^2;$$

$$k' = 4 \ln 2/\Delta\tilde{\nu}^2_{1/2}$$

Lorentzian and Gaussian profiles are shown together in Figure 1. In most cases, the Lorentzian shape provides a better approximation to the profile of a line or band. Both functions can also be used for → band analysis or deconvolution.

In the case of → Doppler broadening, more detailed considerations show that the intensity profile $I(\nu)$ can be represented by a convolution of the Gaussian and Lorentzian functions. This is called a Voigt profile and is considerably broader and flatter.

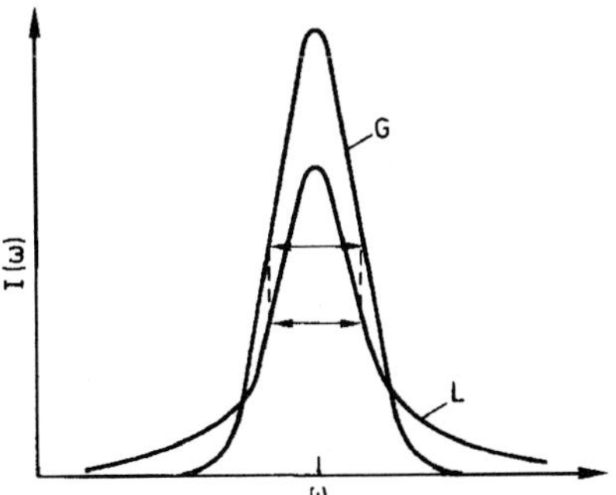

Line shape. Fig. 1. Comparison of Gaussian and Loretzian line shapes

If the transition probability for the absorption or emission of electromagnetic radiation of frequency ν_o corresponding to the transition $i \rightarrow k$ is the same for all molecules in the sample, which are in the same original energy level i, then the spectral line profile of this transition is said to be homogeneously broadened. This natural line broadening is an example of a homogeneous line profile which can be described by a normalized → Lorentzian band shape with the central frequency ν_o. The assumption that all molecules are excited from one and the same state is not always justified. In such cases, inhomogeneous line broadening is observed, the standard example of which is → Doppler broadening. In this case, the probability of the absorption or emission of monochromatic light $E(\nu)$ in the gas phase is not the same for all molecules but depends upon their velocity, υ. And υ is determined by the Maxwellian distribution. Therefore, every single component of velocity corresponds to a definite transition, all of which are hidden under the Doppler broadened band. This is generally the case for gas phase spectra so that → Doppler-free spectroscopy is required for the determination of exact spectroscopic data.

For a crystalline solid in which all atoms or molecules occupied absolutely equivalent positions on an ideal lattice, the total emission or absorption for a transition $i \rightarrow k$ of all the atoms or molecules would be a line homogeneously broadened by the lattice vibrations or phonons. In reality however, the various atoms or molecules occupy nonequivalent lattice positions. This is especially true of amorphous solids and supercooled liquids such as glasses and rigid solvents (solid, isotropic → solvents) which have no regular lattice structure. A guest or probe molecule, e.g. a dye molecule, incorporated into such host lattices experiences very different molecular environments which influence the ground and excited states of the guest through intermolecular interactions. As a result, the differently positioned guest molecules in the host lattice have different ground state energies and therefore different excitation energies, frequencies or wavenumbers. Every individual, discrete transition has a homogeneously broadened line shape. The total gives an inhomogeneously broadened line profile under which the multitude of individual lines is hidden.

The entity consisting of the guest molecule and its surrounding matrix in a host/guest system is called a site. In Figure 2 three sites, the individual transitions and their overlapping to form an inhomogeneous line profile are shown schematically. The result is totally analogous to the Doppler broadening in gases, but with the difference that for host/guest systems the line widths are some 100 cm^{-1} greater

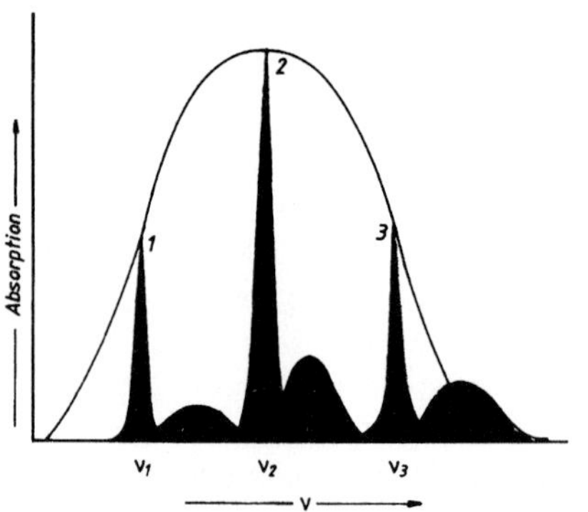

Line shape. Fig. 2. An inhomogeneous line shape resulting from the overlap of homogeneously broadened lines from three sites

than for Doppler broadening where line widths are < 1 cm^{-1}. Experimentally therefore, strongly broadened bands are observed beneath which much spectroscopic information is hidden. Further data can be extracted with modern laser spectroscopic techniques such as saturation spectroscopy (→ saturation) and → hole burning.
Ref.: W. Demtröder, *Laser Spectroscopy*, Springer Series in Chemical Physics, Vol. 5, Springer Verlag, Berlin, Heidelberg, New York, **1981**, p. 78 ff.

Line source, *<Linienstrahler>*, the general description of a radiation source which emits a spectrum of individual spectral lines rather than a continuum. Examples are the → metal-vapor discharge lamps, of which the various forms of the → mercury-vapor discharge lamp are the best-known representatives. Sources of element-specific lines are known as spectral lamps. The → hollow cathode lamps used in → atomic absorption spectroscopy are line sources.

Line spectrum, *<Linienspektrum>*, the total spectrum of the light emitted by excited atoms or ions after spectral dispersion (→ atomic spectra, → alkali spectra, → Balmer formula, → atomic emission spectroscopy, → spectroscopic analysis).

Littrow mounting, *<Littrow-Aufstellung>*, a proven method for the mounting of a → prism in a → monochromator. The rear side of a 30° prism (half-prism) is mirrored so that the light returns through the prism having been totally reflected. Thus, the effect is that of a 60° prism with the corresponding base length. This type of → autocollimation prism gives a constant deviation of 180° which can

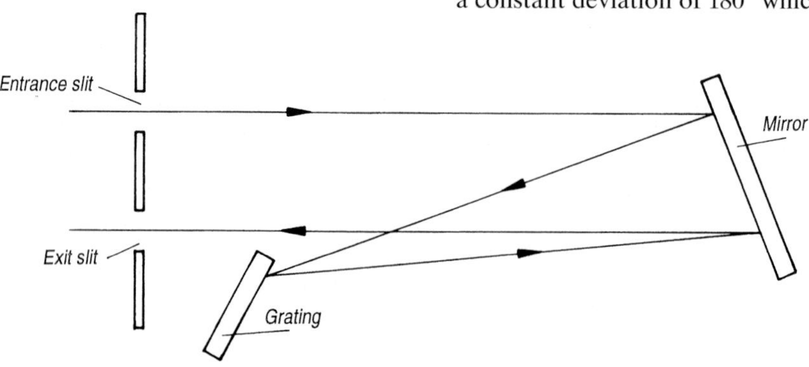

Littrow mounting. Fig. 1. A grating monochromator in a Littrow mounting

be set for every wavelength by rotating the prism about an axis parallel to the prism angle, (→ prism monochromator). In a → grating monochromator, a Littrow mounting uses only one concave mirror which serves as both collimator and collector; Figure 1. The entrance and exit slits lie alongside each other, but they can be placed at the side if an additional, planar, diverting mirror is used. This is often advantageous in the construction of spectrometers.

Log(log) scale → diabatic scale.

Long-wavelength limit, <*langwellige Grenze*> → photoelectric effect, external and internal.

Long-wavelength tail, <*langwelliger Ausläufer*>, that part of the → spectral sensitivity curve of a photoelectric radiation detector which lies at the very longest wavelengths. The expression is also used to describe the region of very low absorption on the long-wavelength flank of a broad UV-VIS absorption band.

Long-wave pass filter, <*Langpaßfilter*>, an → edge filter which has a region of higher transmission in the long-wavelength spectral region which adjoins a region of low transmission, the blocked region, at shorter wavelengths. Typical examples of long-wave pass filters are the → colored glass filters such as → Schott glass filters of types OG, GG, and RG and most → glass-plastic laminated filters.

Lorentzian band shape, <*Lorentz-Profil*> → band shape.

Low-pressure gas discharge, <*Niederdruckgasentladung*>, a gas discharge in which the electron temperature is much higher than the gas temperature. Pressures in low-pressure gas discharges are of the order of a Torr. The gas temperature is a few hundred degrees at the most and the emission is due mainly to collisions and is therefore essentially a line spectrum. Thus, a low-pressure discharge is characterized by sharp lines (the → hyperfine structure of the lines may be resolvable) and a very weak continuum.

Low-pressure lamp, <*Niederdrucklampe*>, usually a mercury lamp operating under conditions of a → low-pressure gas discharge, e.g. type Nk 4, from Heraeus.

L shell, <*L-Schale*>, the second electron shell of an atom. It corresponds to a principal quantum number $n = 2$. Since $l = 0$, 1 and $m_s = \pm 1/2$, it can accommodate 8 electrons in accordance with the → Pauli principle.

Luminescence, <*Lumineszenz*>, the emission of cold light, i.e. light which is not produced by a raising of temperature (→ thermal radiator). The most common, and for spectroscopy the most important, luminescence phenomena are → fluorescence and → phosphorescence which are excited by the absorption of radiation and therefore also called → photoluminescence. Three further categories, → electroluminescence, → chemiluminescence and → bioluminescence are distinguished by the origin of the luminescence.

Luminescence analysis, <*Lumineszenzanalyse*>, a method which uses the intensity of luminescence (→ fluorescence or → phosphorescence) analytically, e.g. its dependence upon the concentration of a component in a system or the change with time in a system to be analyzed.

Luminescence spectrometer, <*Lumineszenzspektrometer*> → fluorescence spectrometer.

Lux, <*Lux*>, unit of illumination, symbol lx, → photometric units.

Lyman series, <*Lyman-Serie*>, the principal series of the hydrogen atom spectrum (→ Balmer formula). It corresponds to the transitions $1S \rightarrow n$P ($n \geq 2$) and lies in the → vacuum UV spectrum.

Lyot filter, <*Lyot-Filter*>, a polarization-interference filter developed by B. Lyot in 1933. It consists of a doubly refracting crystal, e.g. potassium dihydrogen phosphate (KDP), sandwiched between two linear polarizers (P-KDP-P), Figure 1a. The intensity of the light transmitted by this optical unit is given by:

$$I = I_o[1 - \sin^2(\delta/2)] \cdot \sin^2(2\alpha).$$

Here, δ is the phase difference in a light wave brought about by the difference in the refractive indexes for the ordinary, n_o, and the extraordinary, n_e, rays as they emerge from the KDP crystal and α is the angle between the KDP crystal and the polarizers which are aligned parallel to each other. Using the thickness, L, of the KDP plate, δ can be written:

$$\delta = (2\pi/\lambda)(n_o - n_e) \cdot L.$$

For the arrangement in which $\alpha = 45°$, the intensity is:

$$I = I_o \mid 1 - \sin^2(\delta/2) \mid$$

$$= I_o \cdot \cos^2 \frac{\pi(n_o - n_e)L}{\lambda};$$

from which the transmission is found to be:

$$T(\lambda) = \frac{I}{I_o} = T_o \cos^2 \frac{\pi \cdot L(n_o - n_e)}{\lambda},$$

where T_o is the maximum transmission, allowing for absorption and reflection, i.e. $T_o/T_{o(\max)} < 1$.
The intensity of light passing through this device will be modulated by the function $\cos^2\delta$. Maximum transmission occurs when $\delta = k \cdot \pi$, with $k = 1, 2 \ldots$, i.e. for $\lambda = L \cdot (n_o - n_e)/k$. Dark bands, known as Müller's bands,

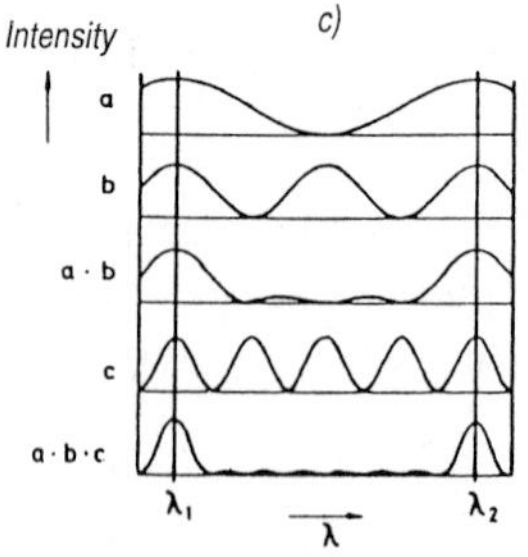

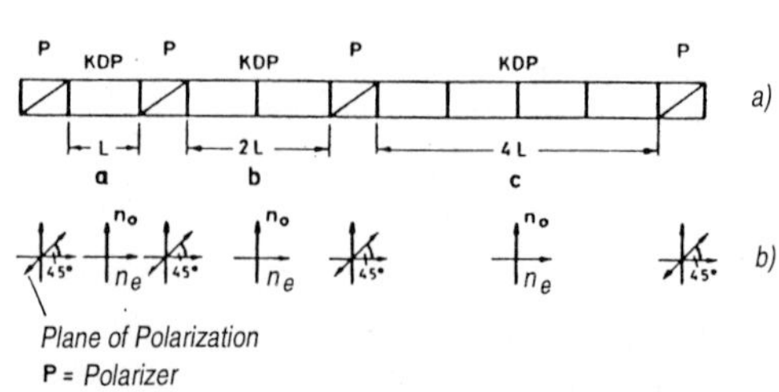

Lyot filter. Fig. 1. a) The construction of a Lyot filter; b) the arrangement of KDP crystals and parallel-orientated polarizers; c) the intensity distribution in a Lyot filter constructed as in a.

with extinctions at $\lambda = 2L(n_o - n_e)/(2k + 1)$, lie between these maxima. A monochromator of high light throughput can now be obtained if further KDP-P pairs, in the same orientation and with the thicknesses, L_m, of successive KDP crystals increasing in the ratio 1:2:4:8, are added to the first P-KDP-P unit. For this series of N elements of thickness L_m the transmission is given by:

$$T(\lambda) = \prod_{m=1}^{N} (T_{om}\cos^2[\pi(n_o - n_e)L_m/\lambda])$$

The transmission is determined only by the product $(\cos^2\gamma \cdot \cos^2 2\gamma \cdot \cos^2 4\gamma \ldots)$ where $\gamma = \pi(n_o - n_e) \cdot L_m/\lambda$, with $L_m = 1L, 2L, 4L, 8L \ldots$ Only the regions around the positions $\gamma = k \cdot \pi$ remain as maxima, while the areas of zero intensity, which are spread evenly over the spectrum, double with every additional KDP plate and squeeze the maxima together from both sides so that they rise steeply out of the background. This is illustrated in Figure 1c for L, $2L$ and $4L$. The free spectral range of a Lyot filter, $\delta\nu$ or $\delta\tilde{\nu}$, is given by:

$$\delta\nu = c/[(n_o - n_e) \cdot L]$$

or in wavenumbers:

$$\delta\tilde{\nu} = 1/[(n_o - n_e) \cdot L]$$

For a KDP crystal with $n_o = 1.51$ and $n_e = 1.47$, $n_o - n_e = 0.04$ at 530 nm. If $L = 2$ cm and $c = 3 \cdot 10^{10}$ cm s^{-1} then $\delta\nu = 3.75 \cdot 10^{11}$ s^{-1} i.e. $\delta\tilde{\nu} = 12.5$ cm^{-1}. It is possible to tune through the frequency or wavelength of maximum transmittance with the help of the linear electrooptical effect. Optically uniaxial crystals like KDP become optically biaxial under the influence of an external electric field along the optic axis (z axis). In addition to the natural double refraction of these crystals, the field induces a double refraction which is approximately proportional to the field strength (→ Pockels effect). This additional component in the z direction is given by:

$$\Delta n(E_z) = \Delta n_{E1} = \frac{1}{2} n_1^3 d_{36} E_z,$$

where d_{36} is the electrooptical coefficient of KDP which has the value of $-1.01 \cdot 10^{-11}$ m/V. The maximum of the transmission is now given by $(\Delta n + \Delta n_{E1})L/\lambda = k$, where $k = 0, 1, 2, 3 \ldots$ and $\Delta n = n_o - n_e$. Thus, λ is given as a function of the applied electric field by:

$$\lambda = (\Delta n + 0.5 \cdot d_{36} E_z)L/k.$$

Electrooptical tuning of a Lyot filter allows a rapid changing of the position of maximum transmission, but in many cases mechanical tuning is the easier method. For this purpose an inclined doubly refracting plate is placed inside the resonator of a cw laser to restrict the spectral width of the laser beam. If the angle of inclination is equal to the Brewster angle (→ Brewster's law), then the plate causes a shift δ as well as a polarization, i.e. additional polarizers are not required as they are with the Lyot filter described above. When the beam enters the doubly refracting plate, which has its optic axis (z axis) at right angles to the plane of the diagram, Figure 2a, it is split into an ordinary and an extraordinary ray. The refractive index, n_o, for the ordinary ray in a uniaxial crystal is constant for all angles with respect to the optic axis, but the refractive index n_e depends on the

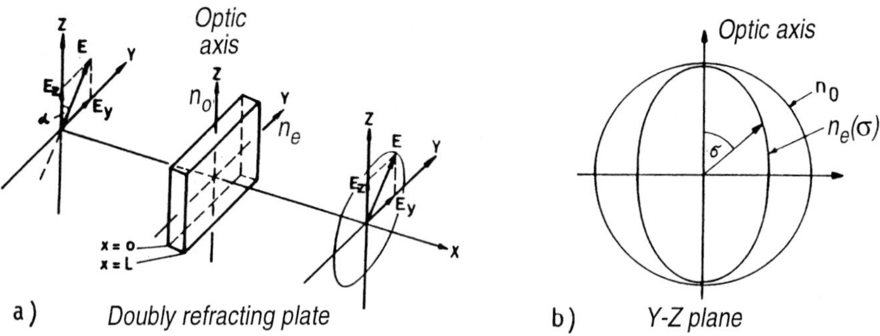

Lyot filter. Fig. 2. a) Mechanical tuning of a Lyot filter by means of doubly refracting plate; b) index ellipsoid for n_o and n_e

angle σ. This is clarified in Figure 2b which shows the → index ellipsoid and how a change of σ changes the difference $\Delta n = n_o - n_e(\sigma)$. This change of σ can be made by rotating the crystal about the x axis which is perpendicular to the yz plane. The result is a continuous change of Δn and a tuning of the wavelength through the maximum transmission.

Following the development of → laser spectroscopy, Lyot filters of the type described here have become very important optical elements. They are usually built directly into the resonator.

Ref.: W. Demtröder, *Laser Spectroscopy*, Springer Series in Chemical Physics, Vol. 5, Springer Verlag, Berlin, Heidelberg, New York, **1987**, Section 4.2.11, p. 177 ff.

M

Mach-Zehnder interferometer, <*Mach-Zehnder-Interferometer*>, a double-beam interferometer with amplitude division of the incoming beam. The construction is illustrated in Figure 1a. The two paths $B_1M_1B_2$ and $B_1M_2B_2$ are absolutely equal. When the beam splitters, B_1 and B_2, and the mirrors, M_1 and M_2, are all exactly parallel, the path difference between the two separated beams is independent of the angle, α, of the incident beam. With this symmetrical construction, i.e. without the sample (length L) in the path M_1B_2, the path difference is zero. With a sample between M_1 and B_2 there is a path difference Δp which is given by $\Delta p = n_L \cdot L$ (where $n_L = n_s - 1$; $n = 1$ for air and n_s is the refractive index of the sample). It follows from the general relationships for interference (→ interferometer) that when $\Delta p = m\lambda$, with $m = 1, 2, 3 \ldots$, there is constructive interference in the plane of observation behind the interferometer. The wavelengths of maximum transmission are given by:

$$\lambda_m = \Delta p/m = (n_s{-}1)L/m$$

This simple view of the Mach-Zehnder interferometer shows that it is very well suited for the exact determination of refractive indexes. In combination with a spectrograph this interferometer can be used to measure the refractive index of an atomic vapor in

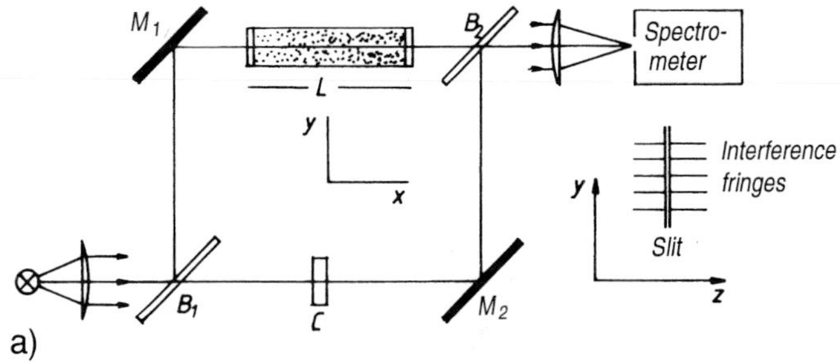

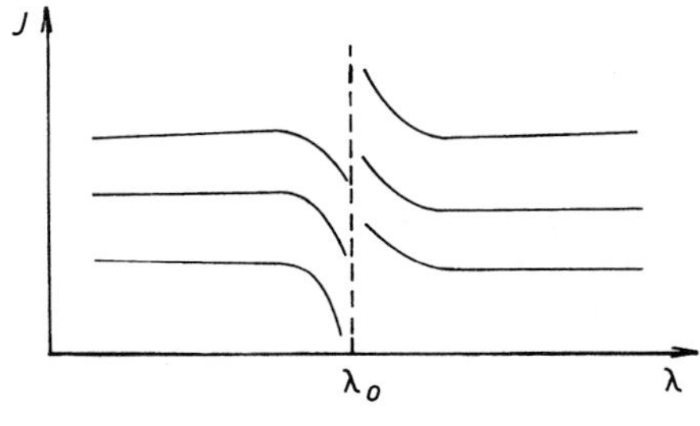

Mach-Zehnder interferometer. Fig. 1. a) The construction of a Mach-Zehnder interferometer; b) the behavior of the refractive index on passing through an atomic spectral line

the immediate neighborhood of a spectral line. See the example in Figure 1b.
Ref.: W. Demtröder, *Laser Spectroscopy*, Springer Series in Chemical Physics, Vol. 5, Springer Verlag, Berlin, Heidelberg, New York, **1987**, sec. 4.2.4, p. 149 ff.

Magnetic anisotropy, <*magnetische Anisotropie*>. Consider a linear molecule, *H-X*, in which an electric current has been induced on atom *X* by an external magnetic field, B_o. The shielding of the proton from the magnetic field, by this current, depends upon the angle between the internuclear axis and the direction of the field. If the two are parallel, Figure 1a, the field at the proton is reduced or the shielding increased. On the other hand, if the field is perpendicular to the internuclear axis, Figure 1b, the current on atom *X* increases the field at the proton and diminishes the shielding. These two opposing effects may cancel each other, or one may dominate, depending upon whether the susceptibility of the *X* atom is the same in all directions (isotropic) or not (anisotropic). If the diamagnetic susceptibility in the direction of the *X-H* axis is greater than perpendicular to it, as is often the case in linear molecules, the shielding of the proton is increased and the → NMR resonance signal appears at higher field than would have been expected on the basis of the electron density.
An interesting example of this effect is acetylene (ethyne) for which the susceptibilities along the line of the atoms and at right angles to it are different. Thus, the shielding of the protons depends upon their position with respect to the C≡C triple bond. The regions of positive and negative shielding are shown in Figure 2a. In linear acetylene, the protons lie in regions of positive shielding and, consequently, their resonance signals appear at low values of the → chemical shift, δ. Similarly, the small shielding of aldehydic protons can be explained with reference to the anisotropy of the carbonyl group. Experience shows that there are regions of higher and lower shielding around the C=O bond. They are illustrated in Figure 2b where it can be seen that the protons lie in a region of weak shielding so that their signals lie at higher δ values. Magnetic anisotropy can be used to explain that part of the → shielding constant which depends upon neighboring atoms and bonds.

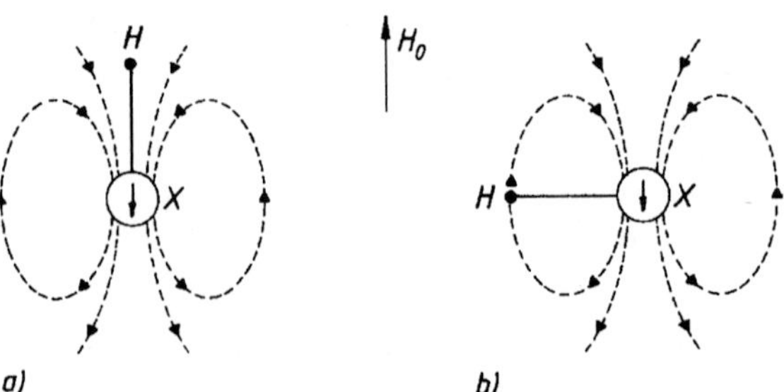

Magnetic anisotropy. Fig. 1. Magnetic anisotropy in a linear molecule, *H-X*

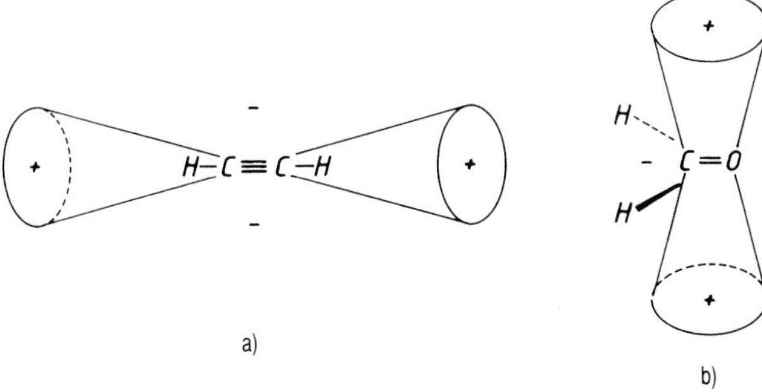

Magnetic anisotropy. Fig. 2. Magnetic anisotropy in a) acetylene and b) formaldehyde

Magnetic circular dichroism, MCD, *<Magnetocirculardichroismus, magnetischer Zirkulardichroismus, MCD>*, a phenomenon which, like → magnetooptical rotation, is a consequence of the → Faraday effect. A magnetic field induces an optical rotatory power. This is a general phenomenon which can be observed with every dielectric material, provided that it is transparent. As with the → circular dichroism of naturally optically active substances and solutions, magnetic circular dichroism derives from the differential absorption of left- and right-circularly polarized light. Thus, for a transparent sample, the behavior outside the regions of absorption is described by the normal → magnetooptical rotatory dispersion (→ magnetooptical rotation) and the behavior in the region of absorption bands by magnetic circular dichroism. The origin of the phenomenon lies in the → Zeeman effect, the splitting of the degenerate energy levels of atoms and molecules and the mixing of their electronic states by a magnetic field.

The simplest interpretation of magnetic circular dichroism is obtained by considering the states of an atom. Consider the transition ${}^1S_o \rightarrow {}^1P_1$. The magnetic field splits the ${}^1P_1$ term into three components which are distinguished by their M_J values of +1, 0, –1. The selection rules for ΔM_J are $\Delta M_J = \pm 1, 0$. The transitions with $\Delta M_J = \pm 1$ are polarized perpendicular (σ) to the magnetic field, while the transition $\Delta M_J = 0$ is polarized parallel (π) to the field. More exactly, $\Delta M_J = +1$ is left-circularly polarized (LCP) and $\Delta M_J = -1$ is right-circularly polarized (RCP). These two transitions are important for the discussion of the Faraday effect and hence for magnetic circular dichroism. The situation described above is illustrated schematically in a simplified manner in Figure 1. For case I, a) shows the levels with and without the magnetic field and the transitions which are distinguished by σ_L and σ_R and b) shows the absorption spectra for unpolarized light without a magnetic field ($H = 0$) and for left- and right-circularly polarized light with $H > 0$. The two separated absorption bands have equal intensities. The MCD spectrum, $\Delta\varepsilon = f(\tilde{\nu})$, in c) crosses the zero at $\tilde{\nu}_o$. This form of MCD spectrum is described as a Faraday *A* term.

Case II in Figure 1 assumes that the ground state is degenerate and split by

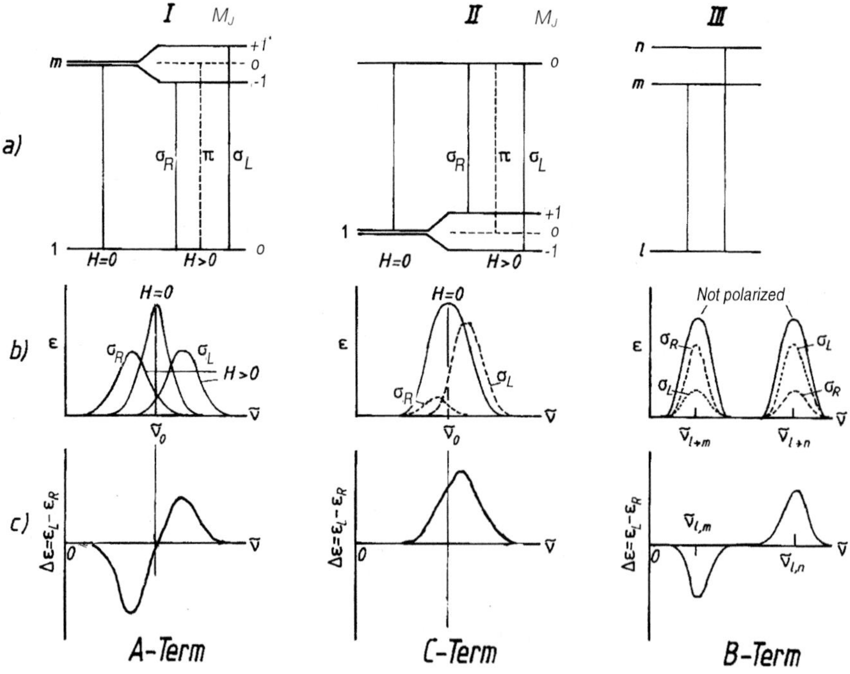

Magnetic circular dichroism. Fig. 1. Energy-level schemes and spectra to explain the Faraday A, B and C terms in MCD spectroscopy; see text

the magnetic field while the excited state is not split. However, the transitions $+1 \rightarrow 0$ and $-1 \rightarrow 0$ (II, a) do not have the same intensity because +1 has a smaller population than –1 on account of the Boltzmann distribution (→ Boltzmann statistics). Therefore, the absorption spectra measured with left- and right-circularly polarized light have very different extinction coefficients and $\varepsilon_L > \varepsilon_R$; so that $\Delta\varepsilon = \varepsilon_L - \varepsilon_R$ is positive, as can be seen from the spectrum. This type of MCD feature is known as a Faraday C term.

Case III, the Faraday B term, shows no Zeeman effect. It arises as a result of the mixing of states by the magnetic field and there must be a magnetic dipole transition moment between the states to be mixed. Following the original work of A.D. Buckingham and P.J. Stephens, there have been several different theoretical formulations of MCD as the sum of the A, B and C terms and some confusion over signs and notation has arisen. A standard form of all relevant equations has been suggested by S.B. Piepho and P.N. Schatz and is detailed in Section A.4 of their book, see literature. Their equation for the molar ellipticity, $[\Theta]_M$ is:

$$\frac{[\Theta]_\mathrm{M}}{\mathrm{E}} = 5.028 \times 10^5 B \left\{ \mathcal{A}_1 \left(\frac{-\partial f}{\partial \mathrm{E}} \right) + \left(\mathcal{B}_0 + \frac{\mathcal{C}_0}{\mathrm{kT}} \right) f \right\}$$

$E = h\nu$, the energy of the transition and all energies are in cm^{-1}, the field,

B, is in Tesla, magnetic dipole moments are in → Bohr magnetons and electric dipole moments in Debyes. The exact expressions for $\mathscr{A}_1$, $\mathscr{B}_0$ and $\mathscr{C}_0$ can be also found in Piepho and Schatz. f is an assumed → line shape. The results of detailed theoretical studies may be summarized as follows.

a) A terms change sign at the position of the absorption maximum (case I, Figure 1c) while B and C terms show extrema at the absorption maximum, (cases II and III, Figure 1c).
b) C terms are inversely proportional to the absolute temperature.
c) A terms are only possible when either the ground or excited state, connected by a transition, is degenerate. This is also a criterion for a Zeeman effect.
d) C terms are only possible when the ground state (strictly, all thermally populated states) is degenerate.
e) In general, in molecules B terms are present for all transitions.
f) In contrast to the extinction coefficient, the Faraday parameters can change their signs.

MCD spectra have proved to be a valuable adjunct to the interpretation of the electronic spectra of metal complexes and complex organic molecules.

Ref.: S.B. Piepho and P.N. Schatz, *Group Theory in Spectroscopy, with Applications to Magnetic Circular Dichroism*, J. Wiley, New York, Chichester, Brisbane, Toronto, **1983**.

Magnetic resonance spectroscopy, electron paramagnetic resonance, nuclear magnetic resonance, <*magnetische Resonanzspektroskopie, ESR-Spektroskopie, MKR- bzw. NMR-Spektroskopie*>, a form of spectroscopy which depends upon the magnetic properties of the electron or the atomic nucleus. In the case of unpaired electrons with spin angular momentum S, we have electron spin resonance (ESR) or electron paramagnetic resonance (EPR). The terminology *EPR* is more appropriate where orbital angular momentum, and not just spin angular momentum (→ electronic angular momentum), make a significant contribution to the observed spectrum. For nuclei with nuclear spin I, we have nuclear magnetic resonance (NMR). The basic theoretical principles of NMR and ESR are the same. Most nuclei, including the various isotopes of an element, and the electron possess spin angular momentum. Quantum mechanics shows that the spin is quantized according to the equations, for nuclei:

$$P_I = \frac{h}{2\pi}\sqrt{I(I+1)}$$

and for electrons:

$$P_S = \frac{h}{2\pi}\sqrt{S(S+1)}$$

h = Planck's constant = $6.63 \cdot 10^{-34}$ J s.
I is the nuclear spin quantum number, $I = 0, 1/2, 1, 3/2, 2 \ldots$
S is the electron spin quantum number, $S = 1/2$.

In what follows the use of I or S will always indicate whether the quantity in question refers to a nucleus or the electron. The spin is always associated with a magnetic moment, for

nuclei: $\mu_I = g_N \cdot \mu_N \sqrt{I(I+1)}$

electrons: $\mu_S = -g_e \cdot \mu_B \sqrt{S(S+1)}$

g_N the nuclear g factor and g_e the → Landé g factor, or simply g factor, are dimensionless numbers. For the proton $g_H = 5.585$ and for the free electron $g_e = 2.0023$.
μ_N and μ_B are the → nuclear magneton and → Bohr magneton respectively. Their numerical values are:

$$\mu_N = 5.051 \cdot 10^{-27}\ JT^{-1}$$

and

$$\mu_B = 9.27 \cdot 10^{-24}\ JT^{-1}$$

The opposite sign in the expressions for the magnetic moment is due to the fact that the nucleus carries a positive charge and the electron a negative charge. We can draw the important conclusion that the magnetic moment is proportional to the spin and the two vector quantities have the same direction in the case of the proton and opposite directions in the case of the electron. On account of the large mass of the proton, compared to that of the electron, and taking account of the different g factors, the magnetic moment of the electron is about 650 times that of the proton. Systems with a spin $\vec{P}$ are degenerate and the number of degenerate states is determined by the spin quantum number. There are $2I + 1$ degenerate states for a nuclear spin quantum number I and $2S + 1$ for an electron spin S. If such a system is brought into a magnetic field there is a → space quantization of the spin angular momentum, i.e. the → Zeeman effect comes into play and $2I + 1$ (or $2S + 1$) states of different energies result. These states differ in the values of their magnetic quantum numbers m_I (or m_S).
A particular I or S value has the following m_I or m_S values.

$$m_I = I, I-1, \ldots -I+1, -I;$$

$$m_S = S, S-1, \ldots -S+1, -S.$$

$m_I(h/2\pi)$ and $m_S(h/2\pi)$ give the components of the spin in the direction of the applied magnetic field which is generally chosen to be the z direction. The energy of these states in a magnetic field of flux density B is given by:

$$E_I = -g_N \cdot \mu_N \cdot m_I \cdot B$$

or

$$E_S = g_e \cdot \mu_B \cdot m_s \cdot B$$

Using the selection rule $\Delta m = \pm 1$, we obtain the energy differences for the resonance which are independent of m.

$$\Delta E = g_N \cdot \mu_N \cdot B$$

or

$$\Delta E = g_e \cdot \mu_B \cdot B$$

And using $\Delta E = h\nu$ the resonance frequencies or Larmor frequencies:

$$\nu_N = g_N \cdot \mu_N \cdot B/h$$

or

$$\nu_e = g_e \cdot \mu_B \cdot B/h$$

In NMR it is usual to use the angular frequency, $\omega = 2\pi\nu$ in the resonance equation which gives:

$$\omega = \frac{g_N \mu_N}{(h/2\pi)} \cdot B = \frac{[\mu_I]}{[P_I]} \cdot B = \gamma \cdot B.$$

In the last equation we have the quantity γ, the → magnetogyric ratio, which gives the relationship between the nuclear magnetic moment, μ_I, and the spin angular momentum, P_I.

γ is a constant for a given nucleus and it is defined with respect to the bare nucleus. All frequency differences, which are measured in the nuclear resonance of real molecules, are described by introducing a new quantity, the → chemical shift. A correspondingly defined γ for the electron is, strictly speaking, not a constant. Orbital angular momentum, introduced by spin-orbit coupling, may modify the g factor of an electron which therefore depends upon the electronic structure of the molecule or atom; → g factor.

Resonance can be obtained not only by varying the frequency, ν, with a fixed magnetic field (frequency-sweep method), but also by varying the field, B, at a fixed frequency (→ field-sweep method). Generally, NMR measurements are made with magnetic fields of 1–8 Tesla which, for protons, corresponds to frequencies of 60–360 MHz, wavelengths in the radio wave region, i.e. $5\text{m} \geq \lambda \geq 0.8\text{m}$ and wavenumbers from $2 \cdot 10^{-3}$ cm^{-1} to $1.25 \cdot 10^{-2}$ cm^{-1}. However, there are many advantages in making NMR measurements at very high fields and instruments operating at 17.63 Tesla ≈ 750 MHz for protons are now offered by at least two manufacturers.

For electrons with the usual g factor of 2 and a magnetic field strength of 0.3–0.5 Tesla, the resonance frequency is approximately 9 GHz ($9 \cdot 10^9$ Hz) which corresponds to wavelengths in the microwave region. Because of this large difference in wavelengths, the practical aspects of NMR and ESR spectroscopy are very different.

Magnetogyric ratio, <*magnetogyrisches Verhältnis*>, a constant specific to a bare (all electrons removed) atomic nucleus. The resonance frequency in → NMR spectroscopy is given by:

$$\nu_o = g_N \cdot \mu_N \cdot B_o / h$$

where g_N is the nuclear g factor, μ_N the → nuclear magneton, h Planck's constant and B_o the magnetic field. If ν_o is replaced by the angular frequency, $\omega_o = 2\pi\nu_o$, we obtain:

$$\omega_o = \frac{g_N \mu_N}{h/2\pi} \cdot B_o = \gamma_I \cdot B_o$$

in which ω_o is the Larmor frequency. γ_I is the magnetogyric ratio. It relates the magnetic moment $\vec{\mu}_I$ to the nuclear spin $\vec{P}_I$ as can be readily seen by multiplying the top and bottom of the above equation by $\sqrt{I(I+1)}$

$$\gamma_I = \frac{g_N \mu_N \sqrt{I(I+1)}}{(h/2\pi)\sqrt{I(I+1)}} = \frac{|\vec{\mu}_I|}{|\vec{P}_I|}.$$

Magnetooptical rotation, <*magnetooptische Drehung*> → Faraday effect.

Mass defect, <*Massendefekt*>, the phenomenon that the mass of a nucleus is always less than the sum of the masses of its component particles. If we compare the atomic weight of a nucleus, A_K, with the sum of the masses of the protons, A_P, and the neutrons, A_N, which it contains, the mass defect ΔM appears to be in conflict with the law of conservation of mass, a principle which is a foundation stone of chemistry. ΔM is given by:

$$\Delta M = Z \cdot A_P + (A - Z)A_N - A_K$$

where Z is the atomic number, A the mass number, and A_K the atomic mass of the nucleus; the difference between the mass of the isotope, A_A, and the mass of the outer electrons, i.e. $A_K = A_A - m_e \cdot Z$. All masses are measured in units of $^{12}C/12$ so that M is also obtained in these units. For the α particle, He, with mass number $A = 4$ and atomic number $Z = 2$, the known values of A_P, A_N and A_K (or A_A) inserted in the above equation give:

$$\Delta M = 0.030378 \text{ amu (atomic mass units, } ^{12}\text{C}/12)$$

The apparent contradiction of the law of conservation of mass can be explained with reference to the equivalence of mass and energy, $E = mc^2$, where c is the velocity of light. When an atom is formed from protons and neutrons a stable nucleus must result and the bonding energy must be set free. The mass equivalent of this bonding energy is the mass defect. For ^4He, the ΔM value of 0.0304 amu, implies a bonding energy of 28.3 MeV ($28.3 \cdot 10^6$ eV). Nuclear bonding energies are of the order of 10^6 times greater than the normal chemical bonding energies of molecules. If the bonding energies of molecules are converted into masses then mass defects are found which lie far below the accuracy of mass measurement and cannot therefore be detected. But the mass defect in the form of an observable mass difference is of great importance in → mass spectroscopy. The practical consequence of the mass defect is that there is a measurable mass difference between combinations of elements of nominally equal mass. Thus, $^{12}C^{16}O$, $^{14}N_2$ and $^{12}C_2{}^1H_4$ have the same nominal mass, $M = 28$ amu. But the mass difference between CO and N_2 is $11.23 \cdot 10^{-3}$ amu and this can be measured very accurately in a double-focusing → mass spectrometer because of the good → resolution of such instruments. For a resolution, $R = 20{,}000$, the equation $R = M/\Delta M$ gives $\Delta M = 1.4 \cdot 10^{-3}$ amu at $M = 28$. Since the resolving power of modern mass spectrometers is greater than 20,000 the mass differences which arise as a result of the mass defect are, in general, readily measurable. This is of importance for an exact mass determination in conjunction with an elemental analysis. In the table, some mass differences for various combinations of elements are given in units of 10^{-3} amu.

Ref.: R.M. Eisberg, *Fundamentals of Modern Physics*, J. Wiley and Sons, Inc., New York, London, Sydney, **1961**.

Mass defect. Table.

Nominal mass	Element combination		$\Delta M \cdot 10^3$ [amu]
14	$^{12}C^1H_2$	^{14}N	12.58
16	$^{12}C^1H_4$	^{16}O	36.39
18	$^{15}N^1H_3$	^{18}O	24.42
28	$^{12}C^{16}O$	$^{14}N_2$	11.23
30	$^{12}C_2{}^1H_6$	$^{14}N^{16}O$	48.86
44	$^{12}C_3{}^1H_8$	$^{14}N_2{}^{16}O$	61.54

Mass difference, <*Massendifferenzen*> → mass defect.

Mass filter, <*Massenfilter*> → quadrupole mass spectrometer.

Mass spectrometer, <*Massenspektrometer*>, an instrument for the separation of ions according to their masses, especially for separation on the basis of *m/e* values. The recording of the ions by a mass spectrometer is electrical, instruments using photographic plates are known as mass spectrographs. The first apparatus, built by Aston in 1919, used a photographic plate and was therefore a mass spectrograph, but the majority of modern instruments are mass spectrometers. They consist essentially of four components (see Figure 1):

1. An inlet system for introducing the sample;
2. an ion source where the ions are produced;
3. an analyzer to separate the ions according to their *m/e* values and
4. a detector and amplifier to detect and record the ions in the form of a → mass spectrum.

Commonly, three techniques are used to introduce the sample; direct injection, indirect injection and coupling with a gas chromatograph. In the ion source (see Figure 2), the gaseous molecules are generally ionized by electron impact. The electrons from a heated filament, *HF*, are accelerated across the sample space towards an anode, *A*, by a potential difference of ~ 70V and acquire an energy of ~ 70 eV. The most important process is the formation of positive ions, M^+; the probability of forming negative ions is approximately 1000 times smaller. Apart from electron-impact ionization, other methods are used, mostly for special purposes. They include → fast atom bombardment, → photo-, → field → thermal, → chemical and laser ionization with → laser ion sources and → spark ionization in vacuum discharges. Following ionization, the ions with charge $+e$ and mass, m, are drawn towards an electrode, Z, by a potential gradient of 1 to 10 kV and are focused onto an entrance slit, S_1, by means of a focusing electrode, *F*, (see Figure 2). On leaving the ion source, the ions have a velocity, v, which is obtained by conversion of the potential energy, eU, into kinetic energy, $m \cdot v^2$, i.e. $v = (2eU/m)^{1/2}$. This dependence of velocity upon mass is exploited in the → time-of-flight mass spectrometer.

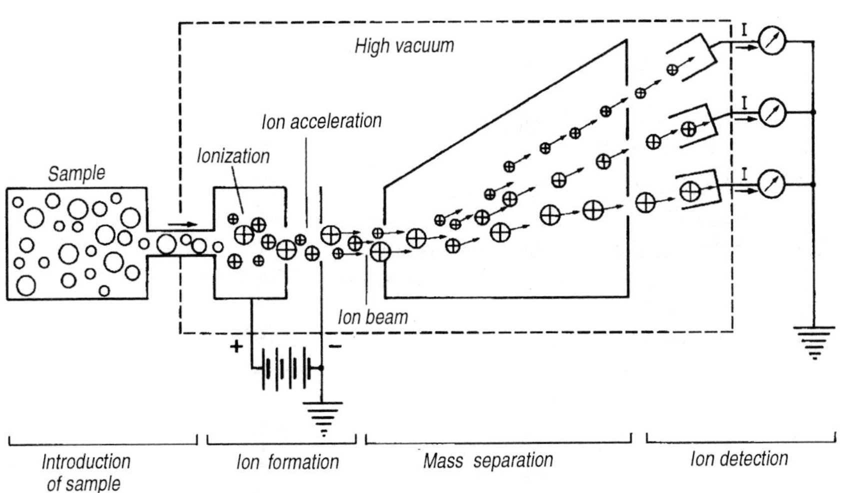

Mass spectrometer. Fig. 1. The principle of the mass spectrometer

The ions pass with velocity, v, through the entrance slit, S_1, into the analyzer. The commonest mass spectrometers use either a magnetic field alone, or a combination of magnetic and electric fields, as analyzer. The first type is known as a single-focusing and the second as a double-focusing mass spectrometer. Another method of analyzing is realized in the → quadrupole mass spectrometer. The effect of the magnet in separating the ions depends upon their deviation as they pass through the field. The particular radius, r, of an ion's path results from the combination of the Lorentzian and centrifugal forces and is given by the spectrometer equation:

$$r = \frac{m \cdot v}{eH}$$

or, substituting the above equation for v,

$$r = \frac{1}{H}\sqrt{\frac{2mU}{e}}.$$

For a constant magnetic field, H, and constant voltage, U, ions having different m/e ratios have different path radii. Thus, the magnetic field has a dispersing effect with respect to the masses (m/e ratios) and energies (velocities) of the particles passing through it. In addition, a magnetic field has a focusing effect upon a divergent entering beam of ions having the same mass and the same energy. This is known as direction focusing (see Figure 1). By varying the magnetic field, H, at constant accelerating voltage, U, or vice versa, ions with different m/e ratios can be brought, one after the other, to the fixed exit slit of the analyzer. The magnetic fields are designed as sector fields with which various angles of deviation, e.g. 30°, 60° or 90°, are possible, depending on the constructional details. Figure 2 shows a 60° sector field. Ions which have left the analyzer and passed through the exit slit (also known as the collector slit) reach a detector (see Figure 2) which

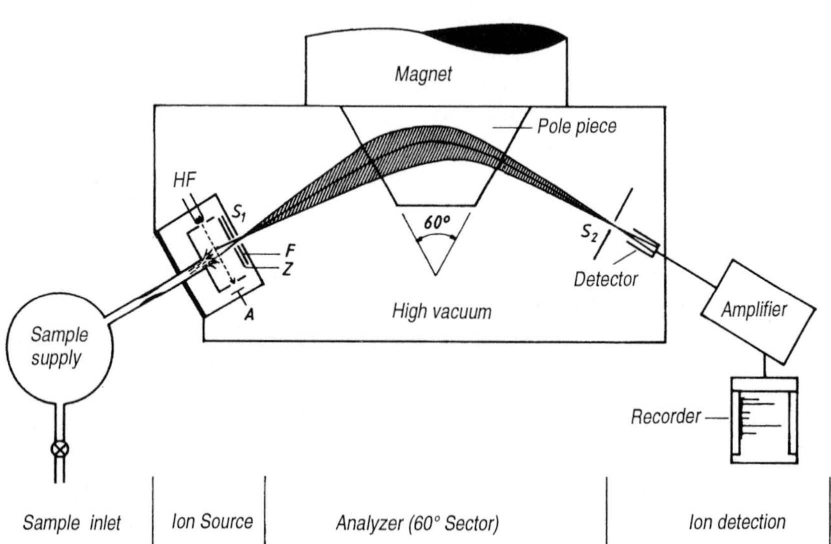

Mass spectrometer. Fig. 2. A 60° sector instrument, see text

may be a → Faraday cup, an → electron multiplier or a photographic plate. In the case of the first two, the discharge current is amplified and immediately plotted as a function of H to give the → mass spectrum. When a photographic plate is used, the various masses having different path radii are simultaneously registered, side by side, on the plate in the exit plane of the analyzer. Analysis of the developed plate then provides a line spectrum. In the single-focusing mass spectrometer illustrated in Figure 2, the range of energies (velocities) of the accelerated ions leads to a broadening of the ion beam and, consequently, to a limitation of the resolving power. This dispersion of energy can be compensated by velocity or energy focusing by means of a foregoing electrostatic sector field. Mass spectrometers of this type are known as double-focusing instruments. They are capable of a resolution of $R = 60{,}000$ where single-focusing instruments generally have resolutions in the range $1000 \leq R \leq 2000$, which can be raised to 10,000.

Ref.: M.E. Rose and R.A.W. Johnstone, *Mass Spectrometry for Chemists and Biochemists*, Cambridge University Press, Cambridge, London, Sydney, **1982**.

Mass spectroscopy, mass spectrometry, *<Massenspektroskopie, Massenspektrometrie>*, a spectroscopic method which is concerned with the separation and determination of atomic and molecular masses. The molecules or atoms are brought into the gas phase and introduced into a → mass spectrometer where they are ionized, separated according to their masses, and finally detected. The result is a → mass spectrum.

The component which separates, or disperses, the masses is a magnetic and/or electric field or an electric quadrupole field. Thus, this component corresponds formally to the dispersive element in optical spectroscopy i.e. the → prism, → grating or → filter.

Mass spectrum, *<Massenspektrum>*. At the exit of a → mass spectrometer, ions with different m/e ratios are detected, one after the other, by variation of the magnetic field. Electrical recording provides an ion current which is proportional to the number of ions and the plot of the current against m/e gives the mass spectrum. In the simplest case, this permits the analysis of a gas mixture with various atomic or molecular masses which are separated according to their m/e ratios. But the particular significance of mass spectroscopy lies in the fact that even pure samples of a single molecular species give a mass spectrum and not, as might have been expected, a single peak at the molecular mass, M^+. There is, in general, a → fragmentation due to the fact that the → ionization is brought about by collisions with electrons which have energies which are relatively high in comparison with the bonding energies of molecules. The molecules therefore break up into fragments which are, in part, characteristic of the molecular structure. The use of mass spectroscopy in the determination of molecular structure is based upon such fragmentation patterns. One of the most important pieces of information which is obtained from mass spectroscopy is the direct and exact determination of the molecular weight from the M^+ peak. As a rule, this is the peak with the highest *m/e* value in the mass spectrum. However, the ion, M^+, must live long enough to reach the detector.

Matrix effect, <*Matrixeffekt*>, the influence of the surroundings (matrix) upon a spectroscopically measured quantity. An example of the effect is the → solvent dependence of UV-VIS spectra which can make itself felt in the measured extinction coefficient, ε_λ. In analytical work this means that the values of ε_λ measured in one → solvent cannot be used for another solvent. Matrix effects are particularly problematic when the solution to be analyzed cannot be measured directly but must first be converted into a form suitable for the particular measurement. This is the case, for example, in → atomic absorption, → atomic emission and → atomic fluorescence spectroscopy. Here, the matrix effect is understood to mean the sum of all the perturbing influences, due to all the substances in the sample, the effect of which cannot be classified more exactly. To eliminate these influences, a relative measurement, the → standard addition method, is used.

MCD → magnetic circular dichroism.

MCS → multichannel spectrometer.

MDI filter → metal dielectric interference filter.

Mechanical equivalent of light, least mechanical equivalent of light, <*Lichtäquivalent, mechanisches Lichtäquivalent*>, symbol, M, inverse, K_m, the maximum value of the mechanical equivalent of light energy and the conversion factor between the units used in the physical sciences and in light technology. The mechanical equivalent of light is fixed by the statement that the energy density, B, of a black-body radiator at the temperature of solidification of platinum, T_{Pt}, (2046 K) is 60 cd/cm^2.
In general:

$$B = \frac{1}{M}\int_o^\infty B_{\delta\lambda}(\lambda T)V_\lambda \mathrm{d}\lambda.$$

where $B_{\delta\lambda}(\lambda T)d\lambda$ is the radiation density of a black-body radiator at the temperature T in the wavelength interval $d\lambda$ at the wavelength λ and, V_λ is the spectral light sensitivity of the eye. Thus, we obtain:

$$M = \frac{1}{60}\int_o^\infty B_{\delta\lambda}(\lambda, T_{Pl})V_\lambda \mathrm{d}\lambda$$
$$= 1.47 \cdot 10^{-3}\ \mathrm{W \cdot cd^{-1} \cdot sr^{-1}}$$

or $K = 680$ cd · sr/W or Lm/W (→ photometric units).

Mercury spectrum, <*Quecksilberspektrum*>, an emission spectrum based upon the outer electronic structure of the mercury atom, $6s^2$. Thus, a singlet term system ($S = 0$; $2S + 1 = 1$) and a triplet system ($S = 1$; $2S + 1 = 3$) both appear (→ multiplet structure). The two term systems are shown in the usual manner in Figure 1. The principal quantum number ($n \geq 6$) and the orbital angular momentum quantum number (s, p, d and f) are given on each energy level. The lowest triplet state, 6^3P_o, lies 4.67 eV (37,656.2 cm^{-1}) above the ground state, 6^1S_o. The → ionization energy of the Hg atom, the series limit of the singlet terms, lies at 10.38 eV (83,740.7 cm^{-1}). The → selection rules

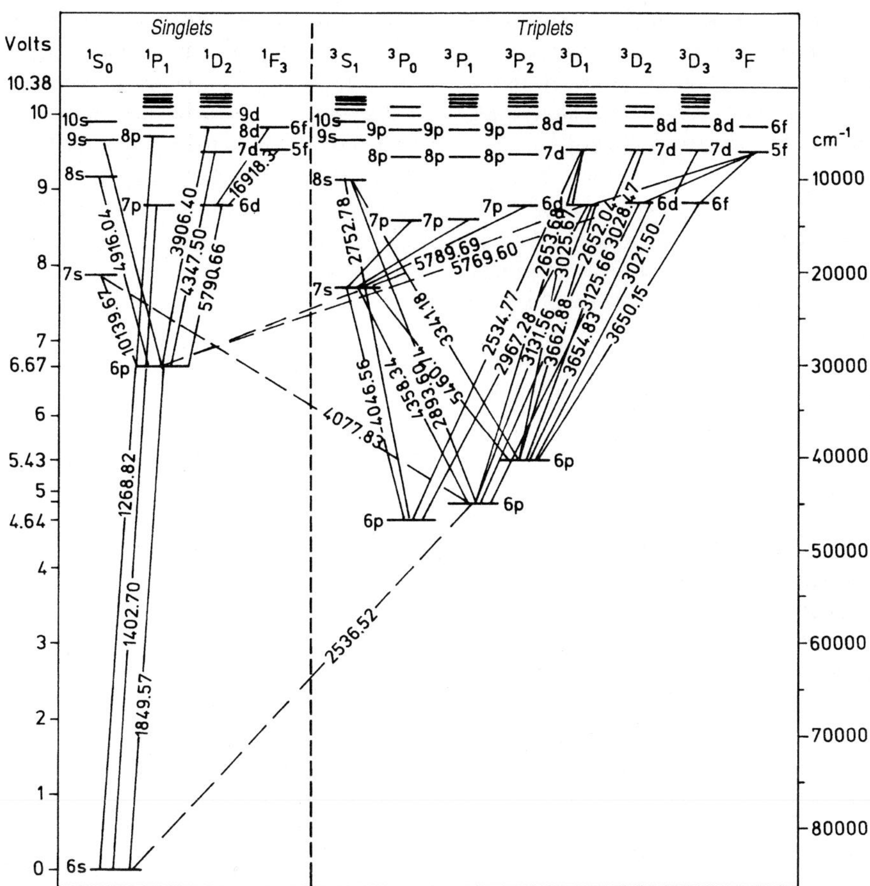

Mercury spectrum. Fig. 1. Energy-level scheme for the mercury atom; transition wavelengths in Å

for the electronic transitions are: $\Delta L = \pm 1$ and $\Delta J = 0, \pm 1$, with the addition that $J = 0$ does not combine with $J = 0$. Intercombination transitions should also be forbidden, i.e. $\Delta S = 0$. But Hg is a classical example of the failure to obey this rule on account of the → heavy atom effect and transitions between the two term systems are observed; some with high intensities. The most important of these transitions is the $6^1S_o \leftrightarrow 6^3P_1$ resonance transition at = 253.7 nm (39424.1 cm^{-1}) which, for example, is dominant in the mercury low-pressure discharge lamp. In high- and very high-pressure lamps this emission line is suppressed by → self-absorption (see also → line reversal). These intercombination lines are also seen in the → Franck and Hertz experiment. In Figure 1, those transitions are indicated which, together with the first line of the principal series, are primarily responsible for the emission in the UV-VIS region. The wavelengths of the lines are given in Å. The intercombination transitions are shown as dashed lines.

The wavelengths, λ (in nm), wavenumbers, $\tilde{\nu}$ (in cm^{-1}) and assignments of these transitions are summarized in the table. The spectral region of each line, and for the visible its color, is also given.

The mercury lines are exceptionally important for many applications in spectroscopy. The individual lines, usually isolated from the other lines by means of → interference filters or tuned filter combinations (→ combined glass filters), represent a good approximation to a monochromatic light source and are widely used as such in → photometry (see also → photometer). The lines near 365 nm, and also those in the further UV, are frequently used to excite → fluorescence and for the irradiation of photochemical reactions. The line at 253.7 nm is used in → atomic absorption spectroscopy, usually in the cold vapor technique.

Ref.: G. Herzberg, *Atomic Spectra and Atomic Structure*, Dover Publications, New York, **1944**; W. Grotrian, *Graphische Darstellung der Spektren*

Mercury spectrum. Table.

λ [nm]	$\tilde{\nu}$ [cm^{-1}]	Transition	Remarks
	Singlet system		
184.96	54066	$6^1S_O \leftrightarrow 6^1P_1$	Principal series
140.27	71291	$6^1S_O \leftrightarrow 7^1P_1$	Vacuum UV
126.88	78815	$6^1S_O \leftrightarrow 8^1P_1$	
1013.97	9862	$6^1P_I \leftrightarrow 7^1S_O$	Near IR
491.6	20342	$6^1P_I \leftarrow 8^1S_O$	Green
579.07	17269	$6^1P_I \leftarrow 6^1D_2$	Yellow
434.75	23002	$6^1P_I \leftarrow 7^1D_2$	Blue
390.64	25599	$6^1P_I \leftarrow 8^1D_2$	Near UV
	Triplet system		
404.66	24712	$6^3P_O \leftarrow 7^3S_1$	Violet
435.83	22945	$6^3P_I \leftarrow 7^3S_1$	Blue
546.07	18313	$6^3P_2 \leftarrow 7^3S_1$	Green
296.7	33704	$6^3P_O \leftarrow 6^3D_1$	UV
312.57	31993	$6^3P_I \leftarrow 6^3D_2$	UV
365.02	27396	$6^3P_2 \leftarrow 6^3D_3$	
365.48	27362	$6^3P_2 \leftarrow 6^3D_2$	Near UV
366.28	27302	$6^3P_2 \leftarrow 6^3D_1$	
313.16	31933	$6^3P_I \leftarrow 6^3D_1$	UV
	Intercombination		
253.65	39424	$6^1S_O \leftrightarrow 6^3P_1$	UV, Resonance line
407.78	24523	$6^3P_1 \leftrightarrow 7^1S_O$	Violet
578.90	17274	$6^1P_I \leftarrow 6^3D_1$	Yellow
576.9	17334	$6^1P_I \leftarrow 6^3D_2$	Yellow

von Atomen und Ionen mit ein, zwei und drei Valenzelektronen, Springer Verlag, **1928**.

Mercury-vapor discharge lamp, <*Quecksilberdampfentladungslampe, Quecksilberlampe*>, the most important of the → metal-vapor discharge lamps. Depending upon the mercury-vapor pressure in the lamp, a distinction is made between:

- Low-presure lamps, Hg vapor pressure 0.01–1 Torr,
- medium-pressure lamps, Hg vapor pressures between 100 Torr and 20 atm.,
- high-pressure lamps, Hg vapor pressures between 30 and 100 atm.

In the low-pressure lamp, essentially only the resonance lines at 184.9 nm ($6^1P_1 \rightarrow 6^1S_o$) and 253.7 nm ($6^3P_1 \rightarrow 6^1S_o$) are emitted. The line at 253.7 nm is an → intercombination line between the singlet and triplet systems. Two processes contribute to the emission. Excited Hg atoms may collide with unexcited atoms in the vapor and transfer their energy to them which they then emit. The emitted resonance line may be absorbed by atoms in the ground state and then re-emitted. In the gas discharge, both processes take place, but in low-pressure lamps emission following → self-absorption is dominant.

In the central UV region (200–300 nm) the low-pressure lamp effectively emits only the 253.7 nm intercombination line. It is therefore a suitable source for the excitation of fluorescence in molecules, which absorb in this region, and also for photochemistry. It is used as a source in the UV monitors used as HPLC detectors because, in conjunction with a filter, these detectors are easy to set up. Large low-pressure lamps are used for sterilization purposes.

The medium-pressure lamp is by far the most widely used source of UV radiation. In this discharge, the higher states of the Hg atom are excited. In contrast to the low-pressure lamp, the spectrum of the medium-pressure shows a number of strong lines throughout the UV-VIS region. The most important, and also the most intense, is the emission at 366 nm which is composed of three closely spaced lines assigned to the transitions $6^3D_1 \rightarrow 6^3P_1$ (366.23 nm), $6^3D_2 \rightarrow 6^3P_1$ (365.48 nm) and $6^3D_3 \rightarrow 6^3P_1$ (365.02 nm). Many other lines derive from transitions within the triplet system. But the line at 579.06 nm is assigned to the transition $6^1D_2 \rightarrow 6^1P_1$. Figure 1 shows the spectral energy distribution of a medium-pressure mercury source. The individual lines can be isolated by means of → interference filters or suitable filter combinations (→ combined glass filters). Medium-pressure Hg lamps are particularly suitable for the excitation of luminescence below 400 nm because, with suitable filters, the lines above 400 nm can be almost completely removed. In comparison with the low-pressure lamp, the radiation yield from the resonance line at 253.7 nm is much reduced because the atoms in the 6^3P excited state suffer collisions before they are able to undergo radiative deactivation. At the position of the resonance line, a gap is observed in the emission due to the self-absorption of the resonance line by the Hg vapor around the discharge region. Apart from the lines, there is a small continuous background which increases with Hg pressure and current density. Thus, Hg high-pressure lamps

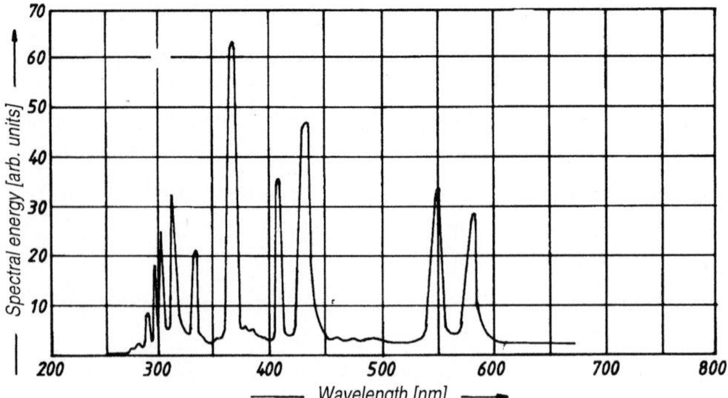

Mercury-vapor discharge lamp. Fig. 1. Spectral energy distribution of a mercury medium-pressure lamp

show a strongly rising continuous background from which the now strongly broadened lines of Figure 1 protrude. These lamps give a very high light intensity or radiation density and are therefore used in optical equipment. The short arc lamp has proved to be particularly useful for spectroscopic applications because the small discharge, 0.02 to 2 cm in length, forms a good image in optical systems.

Ref.: W. Elenbass, Ed.: *High Pressure Mercury Vapour Lamps and Their Applications*, Philips Technical Library, **1965**.

Metal-dielectric interference filter, *<Metall-Dielektrik-Interferenzfilter, MDI-Filter>*, a filter constructed of thin, partially transparent metal foils separated by transparent dielectric distance pieces. The thickness of the distance pieces largely determines the position, λ, of the transmission band of longest wavelength. They are mostly used as → band-pass filters (→ interference filters).

Metal-fluorescence indicator, *<Metallfluoreszenzindikator>* → indicator.

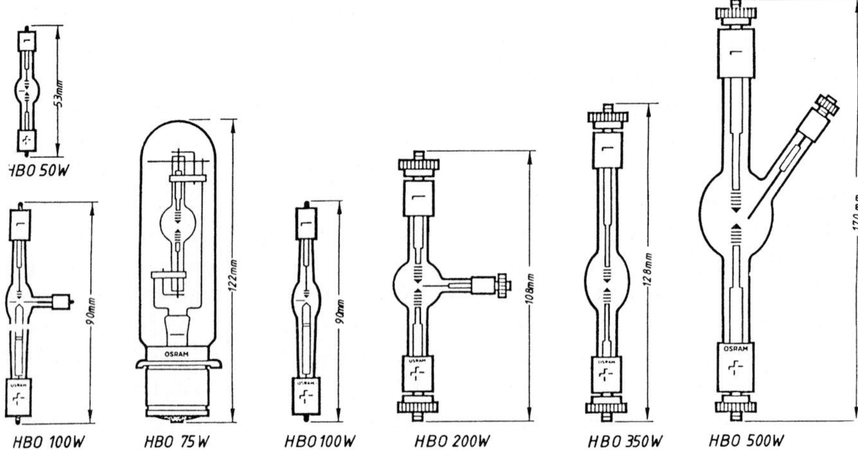

Mercury-vapor discharge lamp. Fig. 2. Examples of mercury short-arc lamps

Metallochromic indicator, *<Metallindikator>* → indicator.

Metal-vapor discharge lamp, *<Metalldampfentladungslampe>*, a gas-discharge lamp in which the metal must first be vaporized which is usually achieved by means of a subsidiary heater. Electrons traveling from cathode to anode are accelerated by an electric field and transfer their energy to atoms in inelastic collisions causing the → ionization of the atoms. The spontaneous return of the excited atoms to their ground states then takes place with the emission of the characteristic spectral lines of the particular element. For this reason, these lamps are also called spectral lamps. They are best used to obtain isolated monochromatic radiation since, in general, they emit only a few intense lines which can be easily separated out using → monochromators or → filters. The most useful spectral lamps are filled with Na, K, Rb, Cs, Zn, Cd, Hg or Tl and are generally used for special measurements. → Mercury-vapor discharge lamps are the most important and most widely used spectral lamps. They emit a variety of lines with very high intensity and are therefore very important for the excitation of luminescence and photochemical reactions in the UV-VIS region.

Michelson interferometer, *<Michelson-Interferometer>*, an → interferometer first described by A. Michelson in 1882 and used by him to measure lengths (the length of the standard meter). Figure 1 shows the optical layout from which the instrument can be seen to be a double-beam interferometer. Light from a source, *L*, is incident upon a semitransparent, mirrored glass plate, *P*, which is orientated at 45° to the light beam. The light is divided into beam 1 which passes through the plate and beam 2 which is reflected at 90°. Both beams are reflected into themselves by perpendicular plane mirrors, M_1 and M_2, and return to the plate, *P*, where they are again divided into two. The two light waves are superimposed in the plane of observation, *O*. Since beam 1 has traversed the plate *P* three times, but beam 2 only once, an equally thick, but unsilvered plate, *P'*, is placed in the path of beam 2 parallel to *P*. In this way, the asymmetry in the paths of the two beams is cancelled.

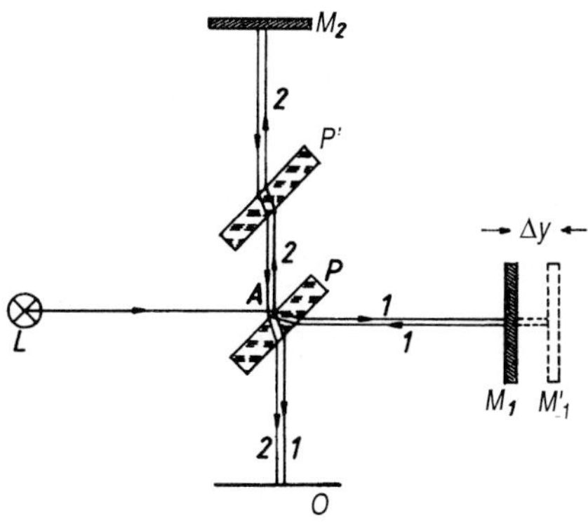

Michelson interferometer. Fig. 1. The construction of a Michelson interferometer

The phase shift or phase difference, δ, between the two beams 1 and 2 is given by $\delta = (2\pi/\lambda)\cdot 2(PM_1 - PM_2) + \Delta\phi$. $\Delta\phi$ represents the extra phase shift produced by the reflection. A detector in the plane of observation, *O*, measures a time-averaged intensity $I = \frac{1}{2}I_o(1 + \cos\delta)$. When the mirror M_1, which is movable, is moved a distance, Δy, the optical path difference changes by $\Delta s = 2n\Delta y$ (*n* is the refractive index between *P* and M_1) and the phase difference by $\delta = (2\pi/$

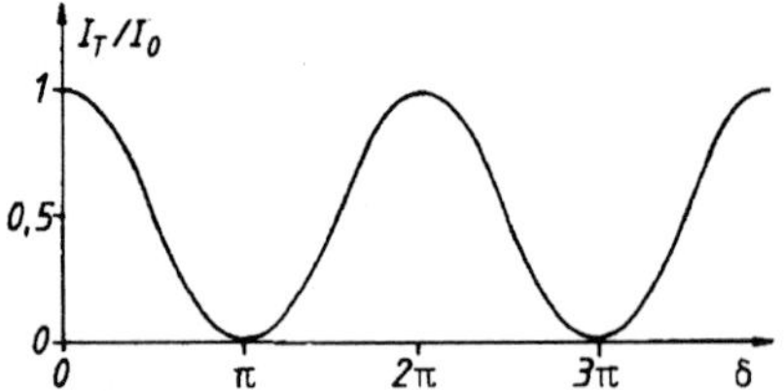

Michelson interferometer. Fig. 2. The intensity transmitted by an interferometer

$\lambda)\Delta s$. For $\delta = 2m\pi$ with $m = 0, 1, 2$ .... we always have $I_r = I_o$ ($\cos\delta = 1$); for $\delta = (2m + 1)$, on the other hand, $I_r = 0$ ($\cos\delta = -1$). In Figure 2, the transmitted intensity from the interferometer, I_T, is plotted as a function of δ for an incident plane, monochromatic wave. At the minima where $\delta = (2m + 1)\pi$, $I_T = 0$; the incident plane wave is reflected back to the source. From this it is also clear that a Michelson interferometer can be considered to be either a wavelength-dependent filter in transmission or a wavelength-selective reflector. Another description of the Michelson interferometer is of particular interest in applications. If the mirror, M_1, in Figure 1 moves with a constant velocity, $v = \Delta y/\Delta t$, then a monochromatic wave of frequency $\omega = 2\pi\nu$ and wave vector $k = 2\pi/\lambda$ experiences a Doppler shift of $\Delta\omega = \omega - \omega' = 2kv = (4\pi/\lambda)v$ upon reflection at M_1. The path difference is $\Delta s = \Delta s_o + 2vt$ which corresponds to a phase difference of $\delta = (2\pi/\lambda)\Delta s$; i.e. $\delta = (2\pi/\lambda)\cdot 2vt$ when $\Delta s_0 = \phi$. Substitution of $2v$ from the Doppler shift expression gives $\delta = \Delta\omega t$. The above equation for I_T can now be written:

$$I_T = \frac{I_o}{2}(1 + \cos\Delta\omega t)$$

or

$$I_T = (I_o/2)(1 + \cos 2\pi\Delta\nu t).$$

From the Doppler shift we obtain $\omega = (c/v)\cdot(\Delta\omega/2)$ or $\nu = (c/v)\cdot(\Delta\nu/2)$; i.e. the frequency, $\omega(\nu)$, of the wave arriving in the plane of observation can be measured via the frequency, $\Delta\omega(\Delta\nu)$, provided that the velocity of the moving mirror, M_1 is known. Thus, a Michelson interferometer can be regarded as a mechanical trick by means of which radiation of high frequency, $\nu = \omega/2\pi$ (10^{13}–10^{15} s^{-1}) can be converted into radiation of frequency $\Delta\nu = 2(c/v)\cdot\nu$ in the acoustic frequency range. If the velocity of the mirror is $v = 3$ cm s^{-1}, $v/c = 10^{-10}$ and a frequency of $\nu = 10^{14}$ s^{-1} ($\tilde{\nu} = 3300$ cm^{-1}, $\lambda = 3$ μm) will be transformed into $\Delta\nu = 2\cdot 10^{-10}\cdot 10^{14} = 2\cdot 10^4$ s^{-1}, i.e. 20 kHz. This property is exploited in → Fourier spectroscopy in the IR region, (see also → Fourier-transform IR spectrometer). Apart from this application, the Michelson interferometer is an important component of a wavemeter which is used for very accurate determination of the wavelength of laser radiation.

Ref.: W. Demtröder, *Laser Spectroscopy*, Springer Series in Chemical Physics, Vol. 5, Springer Verlag, Berlin, Heidelberg, New York, **1981**, sec. 4.2.2, pp. 141 ff.

Microscope spectrophotometer, <*Mikroskopspektralphotometer*>, an instrument for microdensitometry and microspectroscopic analysis. A microscope spectrophotometer is a combination of photometric measurement techniques with a microscope. The range of applications lies from the → UV-VIS, through the NIR to the → IR spectrum, depending upon the construction and equipment. Microscope spectrophotometers for the IR region differ in their optics from instruments

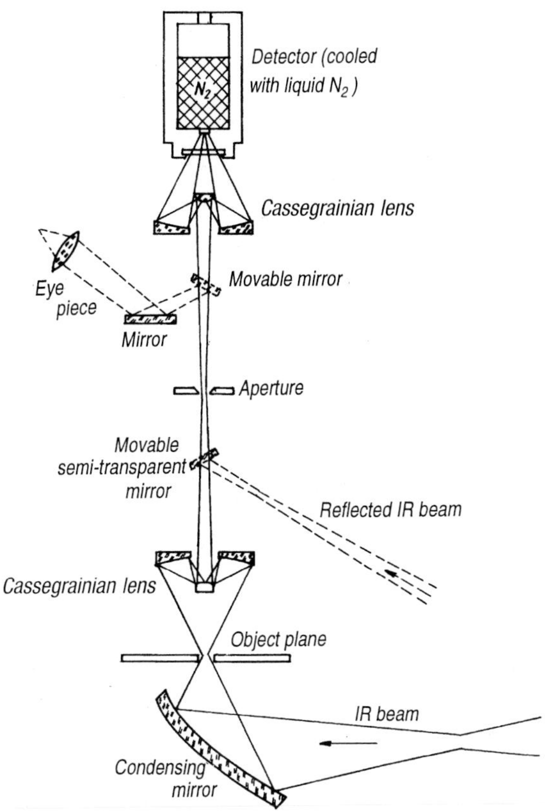

Microscope spectrophotometer. Fig. 1. Beam path in a microscope attachment for a Bruker FTIR spectrometer

for the UV-VIS region. In place of the lenses used in the latter, the imaging and magnification in the IR region requires the use of concave mirrors, e.g. the → Cassegrainian objective, because of the absorption of lens materials in the IR. Furthermore, microscopes for the IR region are constructed such that they can be built into the light path of an IR spectrometer; in recent years usually an → FTIR spectrometer. Figure 1 shows the light path of an IR microscope. In contrast, microspectrophotometers for the UV-VIS-NIR region are complete, independent instruments. The optical path and construction of the Zeiss UMSP 80 is shown schematically in Figure 2. It has two → grating monochromators, driven by stepper motors. The first monochromator (1) is used on the illumination side of the instrument to select the required wavelength from a → xenon lamp (XBO 75 W) while the second monochromator (2) is used on the detection side. The monochromators can be moved away so that the light path is open for the direct passage of the light from the source. This makes it possible to use the two monochromators in sequence and absorption measurements in the

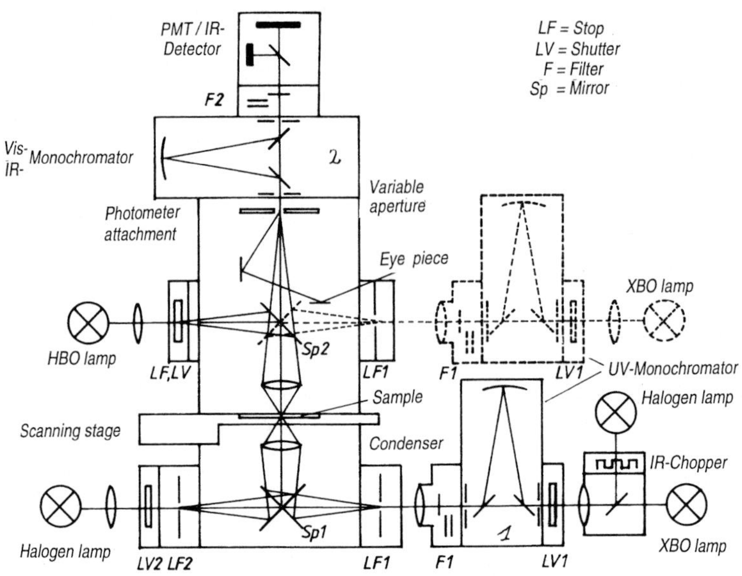

Microscope spectrophotometer. Fig. 2. Beam path and construction of the Zeiss UV-VIS microscope spectrometer MSP 80

range from 240 nm (UV) to 2100 nm (NIR) can be made. Furthermore, if monochromator (1) is set to a fixed wavelength, a fluorescence → excitation spectrum can be measured with monochromator (2). Fluorescence spectra can be measured on monochromator (2), either using the XBO lamp and a fixed wavelength on monochromator (1) or with excitation using the HBO lamp. In the second case, the monochromator (1) with the XBO lamp is raised to the height of the HBO lamp (dashed line in Figure 2). The halogen lamp is linked into the instrument by means of the diverting mirror *Sp*1. It is primarily used, in combination with monochromator (2), to make densitometric measurements on the sample. The fast scanning stage is an important part of the instrument. It can be driven manually or with a motor and the sample can be positioned very exactly. At a fixed wavelength, samples > 0.5 μm can be measured with a spatial resolution of 0.25 μm. A → photomultiplier is the detector for the UV-VIS region and a PbS cell for the NIR. The whole instrument is computer-controlled and extensive software allows a variety of applications of the UMSP.

Ref.: C. Zeiss, 73446 Oberkochen, Germany. Bruker Instruments Inc., 15, Fortune Drive, Billerica (MA), U.S.A.

Microwave spectrometer, <*Mikrowellenspektrometer, Mikrowellenspektralphotometer, MW-Spektrometer*>, a spectrometer containing the usual components (see Figure 1, → source, → sample container, → detector and amplifier, → recorder) but, unlike an IR or UV-VIS instrument, lacking a monochromator because the microwave generator itself is a tunable source of monochromatic radiation. However, the frequency and intensity of the microwave radiation must be determined with suitable measuring devices. The microwave radiation is fed to the absorption cell in waveguides, normally with a rectangular cross section. In order to prevent the development of undesirable field types, waveguides of different dimensions which match the particular frequency region (band) must be used. The data are summarized in the table.

Microwave spectrometer. Table.

Band	Frequency range [GHz]	Inner cross section [mm]
C	3.95– 5.85	47.55×22.15
XN	5.85– 8.2	34.85×15.80
X	8.2 –12.4	22.86×10.16
KU	12.4 –18.0	15.80× 7.90
K	18.0 –26.5	10.67× 4.32
R or V	26.5 –40.0	7.11× 3.56

The absorption cells for measurements in the 7 to 40 GHz range consist of a waveguide with rectangular cross section, frequently with X-band dimensions, see Table. For reasons of maximum sensitivity, the absorption cell is made as long as possible; the usual length lies between 1 and 4 m. The cell must meet the requirements of ultrahigh vacuum technology. To reduce errors in the measurements, the surfaces of the cell must be corrosion-resistant and show the smallest possible tendency to absorb gases. The absorption cell is joined to the waveguides on either end by means of step-free, single-unit couplers, see Figure 1, and sealed with thin mica windows which are quite transparent to microwaves. A sample inlet and high vacuum system are connected to the absorption cell.

Point-contact diodes, back diodes and low-barrier Schottky diodes are used as detectors. A significant improvement in the signal/noise ratio followed the introduction of → Stark modulation. In this technique, an insulated metallic electrode is fitted on the axis of the absorption cell (see Figure 1) and a strong electric field is applied

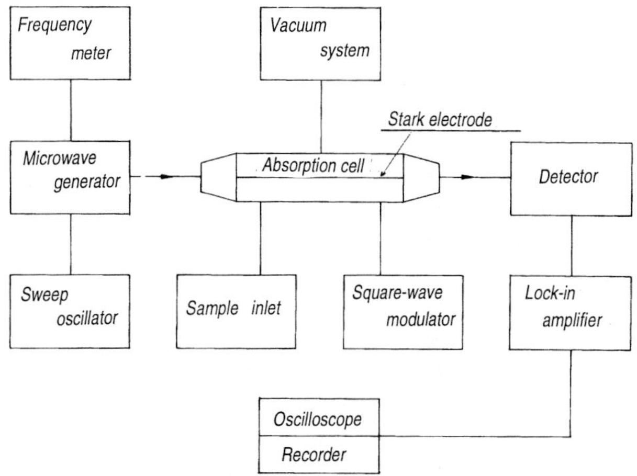

Microwave spectrometer. Fig. 1. Schematic of a microwave spectrometer

between the electrode and the cell wall. The field produces a → Stark effect in the absorbing gas which splits the observed lines. The desired increase in detection sensitivity is achieved by modulating the electric field at a frequency between 1 and 200 kHz by means of a square-wave generator while the detector system is equipped with a lock-in amplifier locked to the modulation frequency. Another form of absorption cell uses two parallel plates as the waveguide, one of which is also the Stark electrode. In this type of cell, a homogeneous Stark field is obtained and the Stark spectral lines are consequently very sharp. The whole cell must be enclosed in a vacuum chamber. In recent times, the use of microwave spectrometers, which are usually constructed out of individual components, has been made easier by automation. The frequency measurement is controlled by a computer which also sets the required frequency changes. These data, together with the output of the detectors, the vacuum system etc. are stored for subsequent analysis. Spectrometers for monitoring specific gases, such as ammonia, have been described in the literature and suggestions for the measurement of formaldehyde and water vapor in the air have been made.

Ref.: C.H. Townes and A.L. Schawlow, *Microwave Spectroscopy*, McGraw-Hill Book Co. Inc., New York, Toronto, London, **1955**.

Microwave spectroscopy, <*Mikrowellenspektroskopie, MW-Spektroskopie*>, spectroscopy which covers the frequency range from about 1 to 1000 GHz which corresponds to wavelengths in the range 30 to 0.03 cm and wavenumbers from 0.03 to 33 cm^{-1}. The rotations and inversion vibrations of gaseous, polar molecules of low molecular weight lie in this range. The most investigated region is from 8 to 40 GHz, i.e. wavelengths 3.75 to 0.75 cm and wavenumbers between 0.27 and 1.3 cm^{-1}.

Microwave spectroscopy is one of the methods of → high-frequency spectroscopy. Its theoretical foundations are those of the → pure rotation spectra of molecules. Here it is important to note that the rotational energy levels lie very close together so that for the lower rotational levels, on account of the Boltzmann distribution (→ Boltzmann statistics), the probability of finding a molecule in a particular excited rotational level is very similar for many levels because $h\nu \ll kT$ at room temperature. The excitation energies in microwave spectroscopy are some 10^3 times smaller than in the IR and about 10^4 times smaller than in the UV-VIS. The consequence of the multitude of energy levels normally occupied by molecules at room temperature and the small excitation energies is that many transitions between excited levels are observed in microwave spectroscopy. The spectra are therefore very complex. Rotation and inversion spectra can only be observed in gases and vapors; at low temperatures and pressures for the best results in intensity and resolution. Lines measured in absorption are largely determined by the geometrical structure and symmetry of the molecule. To a good approximation, the rotational properties of the molecule can be described in terms of the rotation of a → symmetric or → asymmetric top in free space. The precondition for the emission or absorpton of electromag-

netic radiation in the microwave region is the presence of a permanent dipole moment in the rotating molecule. This dipole moment provides the coupling to the electromagnetic field and determines the selection rules and the intensities of the absorption lines. Thus, microwave spectroscopy is essentially limited to polar gases. The intensity of the absorption lines is proportional to the square of the permanent electric dipole moment so that, in general, strongly polar molecules have intense microwave spectra.
Similar rotational transitions are possible in paramagnetic molecules, but they are generally much weaker than in molecules with a dipole moment. For the analysis and interpretation of microwave spectra one can make use of the rules for the → pure rotation spectra of linear molecules and top molecules. In addition, there are interactions with the vibrations and the → hyperfine structure has to be taken into account. The Stark effect and for paramagnetic molecules, the Zeeman effect, also play important roles, both in the measurement of the spectra and in their interpretation.
Microwave spectroscopy provides information about the bond lengths and bond angles of free molecules in the gaseous state and is therefore an important method for the investigation of molecular structure. In addition to the structure, the magnitude and direction of the permanent electric dipole moment of the free molecule can be determined with high accuracy by means of microwave spectroscopy. The dipole moment is linked directly to the magnitude of the line splitting by the Stark effect. From the hyperfine structure of the spectrum, the nuclear quadrupole moment can be determined from which the nuclear spin, I, can be calculated. At the present time, microwave spectroscopy is used only in a limited way for chemical analysis. The apparatus required is very expensive and the routine use common with other spectroscopic methods is therefore not possible.

Mie scattering, <*Mie-Streuung*>, a → light-scattering phenomenon by molecules or particles of dimensions which are comparable with or greater than the wavelength of the incident light; diameter greater than $\lambda/10$, say (compare → Rayleigh scattering). The electric vector of the incident monochromatic light excites the electrons of a molecule to forced vibrations of the same frequency and we first assume that no absorption takes place. The particle is now a vibrating dipole which sends out secondary electromagnetic waves of the same frequency in all directions. Because of the size of the scattering particle, the electrons in different parts of the particle are excited with different phases, so that the secondary waves emanating from these regions are coherent and therefore capable of interference. Thus, they interfere destructively with each other in certain scattering directions so that the total scattering intensity decreases in comparison with the Rayleigh scattering and the angle dependence of the scattering must depart from a symmetric distribution. Mie developed a general theory for the simple scattering of a plane wave on spherical, dielectric and absorbing particles (G. Mie, *Ann. Physik*, **1908**, *25*, 377). According to Mie, the angular distribution can be formed quite generally as the sum of the contributions of a series of electric and magnetic dipoles and multipoles arranged at the center of a sphere. The amplitudes and phases of the partial waves are functions of the scattering angle, δ_s, the relative → refractive index, $m = n/n_o$, (n is the refractive index of the scattering particle and n_o that of the surrounding medium) and the variable $x = 2\pi r/\lambda$, the ratio of the particle circumference to the wavelength of the light, λ.

If unpolarized primary light of intensity, I_o, is scattered from a dielectric, nonabsorbing particle then the intensity of the scattered light at distance, R, from the center of the sphere is given in general by:

$$\frac{I_{\delta_s}}{I_o} = \frac{\lambda^2}{8\pi^2 R^2}(i_1 + i_2) \equiv q(\delta_s).$$

Here, i_1 and i_2 are the intensities of two independent, and therefore incoherent, components of the scattered radiation whose electric vectors are respectively perpendicular and parallel to the plane defined by the directions of the incident light and of observation. They are the vertical and horizontal components of the scattered light. The scattering in all directions is partially linearly polarized and the degree of polarization is given by $(i_1 - i_2)/(i_1 + i_2)$. i_1 and i_2 can be expressed as a series of terms of which the higher members take increasing account of the influence of electric and magnetic multipoles. The functions, i_1 and i_2 (Mie functions), have been tabulated for a range of x and m values. For small particles ($x < 0.8$) the equation for pure dipole radiation is:

$$\frac{I_{\delta_s}}{I_o} = \frac{\lambda^2 x^6}{8\pi^2 R^2}\left(\frac{m^2 - 1}{m^2 + 2}\right)^2 (1 + \cos^2\delta_s).$$

Integration over a spherical surface of radius, $R = 1$, for $x = 2\pi r/\lambda$ gives the ratio of the total scattered intensity, I_{St}, to the incident intensity, I_o, for N particles/cm^3 as:

$$\frac{I_{St}}{I_o} = \frac{8\pi^4 r^6 \cdot N}{\lambda^4}\left(\frac{m^2 - 1}{m^2 + 2}\right)^2 \cdot \frac{16\pi}{3} \equiv S'.$$

Since $4\pi r^3/3 = V$, the volume of the sphere, the equation can be written:

$$\frac{I_{St}}{I_o} \equiv S' = \frac{24\pi^3}{\lambda^4} \cdot \left(\frac{m^2 - 1}{m^2 + 2}\right)^2 \cdot NV^2.$$

This is identical to the formula which Debye deduced for Rayleigh scattering by solutions where intermolecular interference between the particles could be neglected; (note that $m = n/n_o$). Thus, Rayleigh scattering is a limiting case of Mie scattering for small particles ($x < 0.8$), when only dipolar radiation is considered. The angular distribution of the scattering intensity is symmetrical. When the assumption of a pure dipole radiation no longer holds, i.e. as electric quadrupoles and magnetic dipoles gain in importance, the angular intensity distribution departs more and more from a symmetrical pattern. In particular, the forward scattering ($\delta_s = 0°$) becomes greater than the backward scattering ($\delta_s = 180°$). For the angle $\delta_s = 90°$ the scattering is largely, although not completely, linearly polarized. This also applies for the case $x > 1$. Here, terms of higher orders must be included in the series expansion of i_1 and i_2, i.e. for larger spherical particles the scattering distribution becomes very complicated. Characteristic scattering patterns are formed from which conclusions concerning the structure

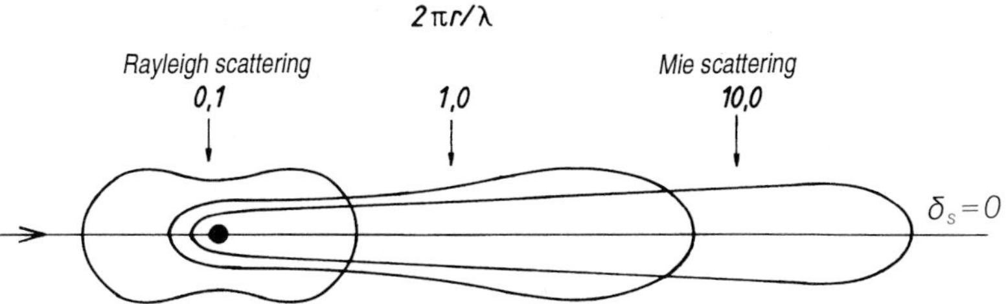

Mie scattering. Fig. 1. Scattering diagram in Mie scattering

and dimensions of the scattering particles may be drawn (see Figure 1).

In addition to the Mie functions, i_1 and i_2, the efficiency factors, $Q_{St}(x,m)$ are also frequently tabulated. $Q_{St}(x,m)$ is defined as the ratio of the optically effective particle cross section to the geometrical cross section, $N\pi r^2$. In these terms the total scattering can be expressed as:

$$(I_{St}/I_o) \equiv S' = N\pi r^2 Q_{St}(x,m).$$

The scattering coefficient or turbidity, S', is therefore equal to the scattering surface of all the particles per cm^3. If the easily accessible S' is measured as a function of λ and plotted against $1/\lambda$, and $Q_{St}(x,m)$ is calculated as a function of x for an assumed known m, then the two curves must have their maxima at the same value of x. From this result, r may be obtained and so the number of particles, N/cm^3, can also be found. In Figure 2 values of $Q_{St}(x,m)$ are plotted against x for a range of values of m as parameter. The curves are simplified and actually show more secondary maxima. (I.T. Edsall, W.D. Danliker, *Fort. der Chem. Forschg.*, **1951**, *2*, 1). For a spherical particle of radius r = 400 nm, i.e. $2\pi r \cong 2500$ nm and $\lambda = 500$ nm $x = 5$; which for $m = 1.44$ gives a scattering surface of $Q_{St} = 4$, according to Figure 2. Such a particle scatters 4 times as much light as might have been expected from its geometrical cross section of $\pi r^2 \cong 5\cdot10^5 nm^2 \approx 0.5 \mu m^2$. Q_{St} can be calculated approximately; but only for $x < 0.8$. For larger x values, higher terms must be included in the series. Here we find that the exponent, n in the λ^{-n} dependence of the scattering, which is equal to 4 in the Rayleigh regime, falls and with further increases in x finally approaches zero, i.e. the scattering is then independent of λ. This is noticeable in the scattering of white light in that the color of the scattered light gradually turns from blue to white. The complex refractive index must be introduced to describe Mie scattering by absorbing particles. $m = n/n_o$ then becomes $m' = (n/n_o) - ni\kappa/n_o = m -$

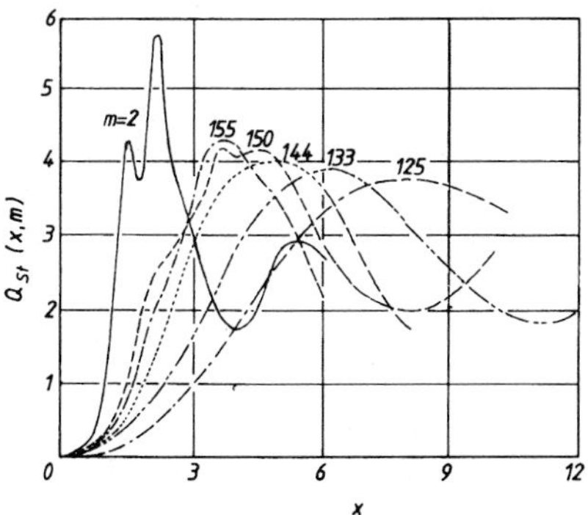

Mie scattering. Fig. 2. Efficiency factors in Mie scattering

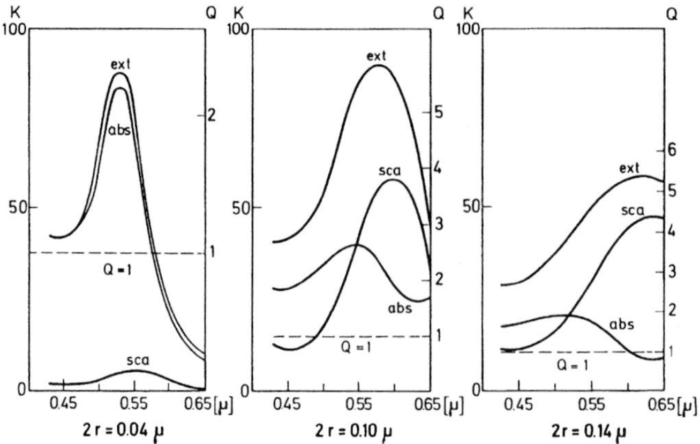

Mie scattering. Fig. 3. Efficiency factors for various particle sizes

$mi\kappa$. n_o is again the refractive index of the nonabsorbing medium and $n\kappa$ the absorption coefficient. Since n and $n\kappa$, therefore also m and $m\kappa$, vary strongly with material and wavelength, a large number of approximations have been developed from the rigorous Mie theory. The total efficiency factor, Q_{Ext}, is composed of the real part, Q_{Sca}, and the imaginary part, Q_{Abs}, of the measured apparent total absorbance, $Q_{\text{Ext}} = Q_{\text{Sca}} + Q_{\text{Abs}}$. Figure 3 shows the contributions of the efficiency factors for various particle sizes, $2r$, as a function of λ in the visible region for a gold sol in water. The figure shows that, with increasing particle size, the efficiency factor, Q_{Sca} increasingly dominates over Q_{Abs} and that Q_{Ext} moves simultaneously to lower wavelengths, i.e. the color of the solution is very dependent upon particle size. While the Mie theory is based upon spherical particles, Debye has developed formulas for scattering by particles of other shapes, e.g. rods, disks, ellipsoids and crumpled threads. These formulas are very important in colloid and polymer chemistry where they are used to determine the mean molecular mass, shape and size of the dispersed particles.

Ref.: G. Kortüm, *Reflectance Spectroscopy*, Springer Verlag, Berlin, Heidelberg, New York, **1969**; H.C. Van de Hulst, *Light Scattering by Small Particles*, J.Wiley and Sons, New York, **1957**; Dover Publications Inc., New York, **1981**.

Minimum deviation, <*Minimum der Ablenkung, Minimalablenkung*>, a frequently used value of the deviation of a ray passing through a dispersing prism. A beam of light passing through a → prism with prism angle, ε, is deviated by an angle, δ. If the prism is turned about its axis so that the angle of incidence, α_1, increases from zero the angle of deviation, δ, at first decreases, passes through a minimum and then increases. Thus, there is a minimum angle of deviation, $\delta_{\min}$, which can be calculated as follows. In general, the deviation of a ray is given by:

$$\delta = \alpha_1 + \arcsin[\sin\varepsilon \cdot (\sqrt{n^2 - \sin^2\alpha_1} - \cos\varepsilon \sin\alpha_1)] - \varepsilon,$$

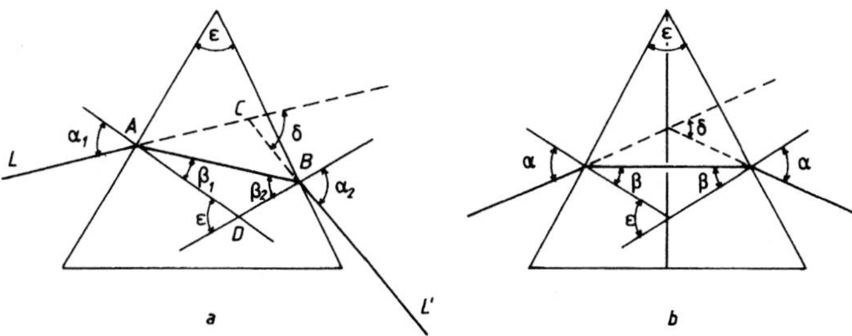

Minimum deviation. Fig. 1. a) General deviation of a ray by a prism; b) minimum deviation of a ray

where n is the refractive index of the prism material and the expression in square brackets is equal to $\sin\alpha_2$, see Figure 1a.
The minimum deviation is given by:

$$\frac{d\delta}{d\alpha_1} = 0.$$

and the differentiation gives:

$$(n^2 - 1)(\sin^2\alpha_1 - \sin^2\alpha_2) = 0.$$

Since $n > 1$, the angle of incidence is equal to the angle at which the ray leaves the prism, i.e. the ray traverses the prism symmetrically, see Figure 1b.
For the dispersion of polychromatic radiation into a spectrum, the position of minimum deviation is preferred since, although the fanning out of the spectrum is a minimum, the resolution at the position of minimum deviation is a maximum, i.e. this position gives the best possible separation of two neighboring spectral lines. Therefore, in → monochromators, prism mountings which maintain this minimum deviation for all wavelengths are preferred, e.g. the → Wadsworth mounting, the → Abbe prism and the Straubel prism (→ dispersing prisms).

Mixed indicator → indicator.

Mixing-chamber burner, expansion-chamber burner, <*Mischkammerbrenner*>, the most extensively applied device in → atomic absorption spectroscopy using the → flame technique. A mixing-chamber burner is illustrated in Figure 1. The pneumatic nebulizer sprays the sample and the oxidant into the mixing chamber. The rate of this process lies between 2 and 10 ml/min and is determined by the pressure and flow of the oxidant, usually compressed air. A change in the pressure and flow of the fuel gas (acetylene) also changes the flow of sample. In order to optimize the fuel/oxidant mixture, the burner is equipped with a separate, controllable oxidant supply which is independent of the nebulizer. The nebulizer produces an aerosol of the sample which is intimately mixed with the oxidant and the fuel gas in the mixing chamber. This aerosol, which consists of the sample solution, oxidant and fuel gas, passes through the burner slot and into the flame. Larger drops of solution are condensed onto the impact beads and surfaces, which are used as required, and the condensate flows out of a hole in the mixing chamber. This much reduces the maximum drop size, although a significant proportion of the sample (~ 90%) is lost in the process. The drop size is

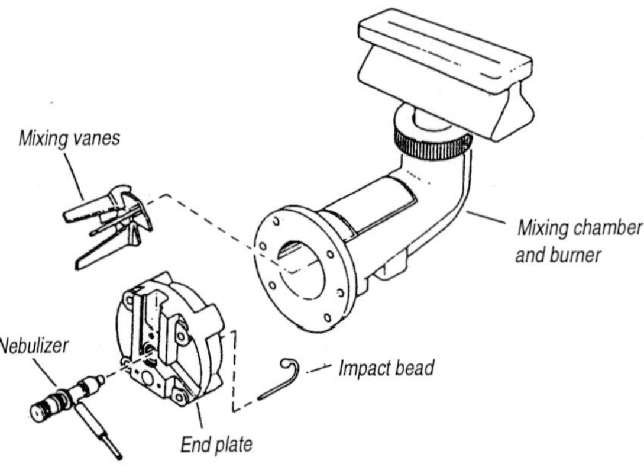

Mixing-chamber burner. Fig. 1. Construction of a mixing-chamber burner

decisive for the subsequent steps of the → atomization. The laminar flame of a mixing-chamber burner is from 5 to 10 cm long and a few mm wide, depending upon the construction of the burner slot. Mixing-chamber burners have exchangeable burner heads made of different materials. Burners made of titanium have proved valuable for high-temperature work with the nitrous oxide – acetylene flame.

Mode locking, <*Mode-Locking, Phasenkopplung*>, a technique for producing laser pulses of picosecond duration which has been developed from an amplitude modulation experiment described by Hargrove. The amplitude modulation is achieved by means of an ultrasonically excited quartz crystal which is built into the resonator of a laser. A diffraction grating inside the laser resonator produces an acoustic standing wave which increases the diffraction losses of the laser light. Twice per period of the exciting potential, which the diffraction grating induces in the crystal, the amplitude of the standing wave falls to zero and the grating disappears. In this way, the resonator losses of double the excitation frequency are reduced to the frequency, $c/2L$, the axial mode spacing (L is the length of the resonator and c the velocity of light). If the laser output is stabilized and the spectrum analyzed with an → interferometer of sufficient resolving power, then above the spectral amplification profile of the laser a number of lines, with intensity which is constant in time and a frequency separation of $c/2L = \Delta\nu$, can be seen. By modulation with a frequency of $c/2L$ the amplitude of the nth mode is coupled with the amplitudes of the $(n-1)$th and $(n+1)$th modes. This always gives rise to modulation, no matter what the degree of modulation is. Phase locking requires a degree of modulation given by:

$$M = \frac{\pi L}{\lambda Q}.$$

Q is a function of the reflectivity, R, of the laser mirror and the amplification, G, per round trip of the active medium.

$$Q \simeq \frac{2\pi L}{\lambda} \cdot \frac{(1-G)(1-R)}{1-GR}.$$

If the band width of the amplification profile is $\delta\nu$, then $m = \delta\nu/\Delta\nu$ modes can be involved in the laser oscillation which, according to the conditions above, are all locked together in phase. The superposition of these m modes gives a time-dependent total amplitude from which the resulting laser intensity, $I(t)$, can be calculated:

$$I(t) \sim \frac{\sin^2 \mid (2m+1)\Omega/2 \mid \cdot t}{\sin^2(\Omega/2)\cdot t} \cdot \cos^2 \omega_o t;$$

$$\Omega = 2\pi\Delta\nu, \; \omega_o = 2\pi\nu_o$$

The right-hand side of this equation describes an equidistant sequence of pulses separated by a time, $T = 1/\Delta\nu$ which is given by the mode separation. The pulse width, $t_p = 1/\delta\nu$, is given by the spectral width of the amplification profile.

Ref.: L.E. Hargrove, R.L. Fork, M.A. Pollach, *Appl. Phys. Letters*, **1964**, *5*, 4; D. Eastham, *Atomic Physics of Lasers*, Taylor and Francis, London, Philadelphia, **1989**, chapt. 4 and 5.

Mode locking. Table. Mode locking data for various laser types

	$\Delta\nu$ [GHz]	t_p [ps]
He-Ne-Laser continuously pumped, $\lambda = 633$ nm	1,5	600
Argon ion laser, $\lambda = 514{,}5$ nm	7	~200
Nd-YAG-Laser continuously pumped, $\lambda = 1064$ nm	12	76
Dye laser rhodamine 6G with mode-locked argon ion laser, $\lambda = 600$ nm	5000	0,4

Molecular spectrocopy, <*Molekülspektroskopie*>, methods of → spectroscopy specially used for the investigation of molecules. They are → UV-VIS, → fluorescence, → IR and → Raman, and → microwave spectroscopy from the region of → optical spectroscopy and → NMR and ESR or EPR from the → magnetic resonance spectroscopies. Although the apparatus used in molecular electronic spectroscopy is similar to that used in → atomic spectroscopy, it has nevertheless proved useful to distinguish between atomic and molecular spectroscopy. An important reason for this lies in the fact that, in the case of molecules, not only electrons, but also very molecule-specific rotations and vibrations, may be excited, which are then superimposed upon the electronic bands. The same holds for NMR and ESR spectroscopy where, upon the measurement of electron or nuclear spin, primarily a property of the atom, quite special characteristics of the molecule make themselves felt in the corresponding spectra. → Mass spectrometry can also be added to the molecular spectroscopies since special structural characteristics of the molecule are seen in the fragmentation pattern. The combination of all the molecular spectroscopies, including mass spectroscopy, constitutes the ideal set of tools for the elucidation of the structure of molecules.

Moment of inertia, <*Trägheitsmoment*>. The kinetic energy of rotation is given by $E_{rot} = \frac{1}{2}I\omega^2$. Compared with the kinetic energy, $E_{kin} = \frac{1}{2}mv^2$, we see that the mass, m, has been replaced by $I = mr^2$, the moment of inertia, and the velocity, v, by the angular velocity $\omega = 2\pi\nu$ (ν is the rotational frequency). The above definition of the moment of inertia applies to a particle of mass, m, which rotates about a fixed axis at a perpendicular distance, r, from it. In spectroscopy the excitation of the rotations of a molecule is important. Since at least two masses which rotate about a fixed axis which passes through their center of gravity are involved, the moment of inertia is calculated by means of Steiner's equation, $I = \sum m_i r_i^2$. Here, m_i is the mass of the ith particle at a perpendicular distance, r_i, from the axis of rotation. Nonlinear molecules have three, mutually perpendicular, principal axes of inertia. The moment of inertia about the → figure axis is given the symbol I_A (→ top molecules). The moment of inertia is included in the → rotational constant from which it can therefore be calculated.

Monochromatic filter, <*Monochromatfilter*>, a filter used to separate out a spectral line or a small spectral region from a discrete or continuous spectrum respectively. They are → interference filters or combinations of two or more → colored glass filters and are commercially available for the mercury lines.

Monochromatic light, <*monochromatisches Licht*>, light of a single color. Strictly speaking, light of a single wavelength, but usually light from a single spectral line (→ monochromatic radiation).

Monochromatic radiation, <*monochromatische Strahlung*>, radiation within a narrow wavelength range. The radiation emitted by natural radiation sources always consists of a mixture of various frequencies, ν, or wavelengths, λ. A small wavelength range, $\Delta\lambda$, can be selected from this radiation mixture by means of spectral dispersion or filtering. In some cases, e.g. sodium lamps, a source emits radiation which is predominantly from a small $\Delta\lambda$ range, without any special spectral filtering. Though it is not exactly true, such radiation is termed monochromatic in order to emphasize that it consists effectively of light of a single wavelength. In fact, even in monochromatic light there are different wavelengths, although the wavelength difference, $\Delta\lambda$, is very small, e.g. the sodium D lines. Strictly monochromatic light would have an energy of zero and therefore cannot exist.

The most homogeneous, naturally occurring monochromatic radiation is that of the lines selected out of a line spectrum. The selection can be made by means of filters with a narrow band pass (→ interference filter, → monochromatic filter) or with a → monochromator.

Monochromator, <*Monochromator*>, the dispersing part of a → spectrometer (→ prism monochromator, → grating monochromator). A monochromator is a device which produces → monochromatic light of a particular → wavelength, λ, from polychromatic light. The physical process, which takes place in a monochromator, is the → spectral dispersion of light or, in general, of electromagnetic radiation. For this purpose → prisms, → gratings and → filters are used and

we therefore distinguish between the monochromators using these different optical elements.
The action of a prism as a → dispersing element arises from the refraction of light at the optically dense medium. Since the → refractive index varies with wavelength, the light experiences → dispersion on passing through the prism.
In the case of a → grating, we exploit → diffraction of light and interference, which are also wavelength-dependent. For → filters, the transmission, τ, for certain large wavelength ranges in the → UV-VIS and NIR region is very small, $\tau \leq 10^{-4}$, while in other wavelength regions, broad or narrow ranges of high transmission occur. This range of high transmission is very small in → interference filters and use is made of this in filter monochromators.
The basic components of a monochromator are: entrance slit, collimator, dispersing element, collector and exit slit, as illustrated schematically in Figure 1. The light to be dispersed is imaged upon the entrance slit which then acts as a secondary source. The collimator, a concave mirror or lens, has the task of fully illuminating the dispersing element (the surface of a grating or the refracting side of a prism) with a light beam which is as

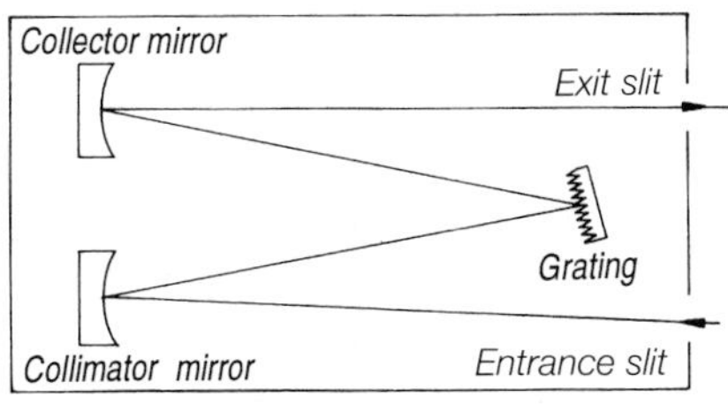

Monochromator. Fig. 1. Construction of a grating monochromator

parallel as possible. The collector, also a lens or concave mirror, focuses the dispersed light in the plane of the exit slit.
Rotation of the dispersing element brings successive, narrow wavelength ranges, $\Delta\lambda$, one after the other, to the exit slit through which they leave the monochromator. The quality of a monochromator is characterized by the following quantities:
→ resolving power; reciprocal linear dispersion;
spectral → slit width (resolution);
geometrical → slit width;
→ scattered light;
→ stray light;
reflective power of the optics.
The significant difference between a → prism monochromator and a → grating monochromator is that the spectral → slit width of the latter is independent of the wavelength, λ, and is therefore constant, which is not true of the former.

MOR, <*MOR*> → Faraday effect, → magnetooptical rotation.

MORD, magnetooptical rotatory dispersion, <*MORD*> → Faraday effect.

Morse curve, <*Morse-Kurve*>, an equation, suggested by P. M. Morse, to approximate the potential energy curve of an → anharmonic oscillator. It has the form:

$$E = D_e\{1 - \exp[a(r_e - r)]\}^2 \quad (1)$$

in which D_e is the dissociation energy (see Figure 1 → dissociation energy), r_e is the equilibrium bond length, r the actual bond length and a is a constant for any particular molecule. If the →

Schrödinger equation is solved using this function to represent the potential energy then the allowed vibrational energies of the oscillator are found to be:

$$E(v) = \omega_e(v + 1/2) - \omega_e x_e(v + 1/2)^2 \quad (2)$$

$$v = 0, 1, 2 \ldots$$

where v is the vibrational quantum number, ω_e is a frequency (expressed in wavenumbers) which will be discussed in more detail below and x_e is the anharmonicity constant which is always positive. In agreement with experiment, the energy levels crowd closer together as v increases. Since the Morse curve is an approximation to the true potential energy function, the energy levels above are also approximate. The approximation can be improved by adding cubic, quartic ... terms; → anharmonic oscillator.
If the above equation for the energy levels is rewritten in the form:

$$E(v) = \omega_e\{1 - x_e(v + 1/2)\}\{v + 1/2\} \quad (3)$$

and compared with that for the harmonic oscillator (→ vibrational spectrum):

$$E(v) = \omega(v + 1/2) \quad (4)$$

then it can be seen that:

$$\omega = \omega_e\{1 - x_e(v + 1/2)\} \quad (5)$$

Thus, the Morse oscillator behaves like a harmonic oscillator, but with an oscillation frequency which decreases with increasing v. Equation 1 shows that $E = 0$ when $r = r_e$. The energies of the simple harmonic oscillator (SHO) and the Morse oscillator are zero for the hypothetical value of $v = -1/2$ (eqtn. 3 and 4) for which $\omega = \omega_e$ (eqtn. 5). Thus, ω_e is seen to be the hypothetical frequency of the Morse oscillator at $r = r_e$. For any real vibrational state specified by a positive integer value of v, ω is less than ω_e. For $v = 0$:

$$\omega_o = \omega_e(1 - x_e/2) \quad \text{cm}^{-1}$$
$$E_o = \omega_e(1 - x_e/2)/2 \text{ cm}^{-1}$$

I.e. the zero-point energy of the Morse oscillator is $x_e\omega_e/4$ less than that of the SHO; cf. the → anharmonic oscillator.

Moseley line, *<Moseley-Gerade>*, the linear relationship between the square root of the wavenumber of comparable electronic transitions and the nuclear charge, Z, of the atom which gives Moseley's law (→ X-ray spectrum). This relation was first observed in 1913 with X-ray spectra for which all elements behave as quasi one-electron systems and the corresponding terms are therefore comparable. This idea can also be applied to → atomic spectra. Consider the spectra of the → hydrogen-like or alkali-like ions which have in each case an identical electron configuration. For the terms of the hydrogen-like ions we have $T = RZ^2/n^2$ or $\sqrt{T/R} = Z/n$. If, $\sqrt{T/R}$, is plotted against the nuclear charge, Z, a straight line passing through the origin results. The same is true for the alkali-like ions if we plot versus $Z - p$. The $\sqrt{T/R}$ values of some of the terms in the Li series are plotted in this way in Figure 1. The straight lines which result are also Moseley lines and they have a slope of $1/n$, i.e. the principal quantum number can be determined from them.

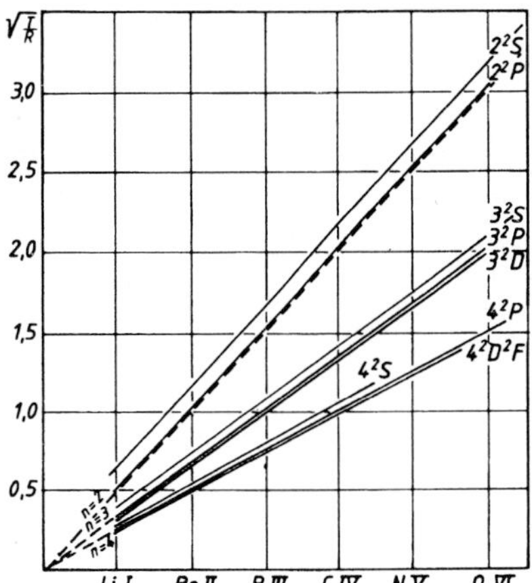

Moseley line. Fig. 1. $\sqrt{T/R}$ for the series Li I to O VI

Moseley's law, <*Moseleysches Gesetz*> → X-ray spectrum.

Mössbauer spectroscopy, <*Mößbauer-Spektroskopie*>, a spectroscopy in the gamma-radiation region based upon an effect first observed and described by R.L. Mössbauer in 1958. The γ rays used have energies in the range 10 to 200 keV, frequencies from $\sim 2.4 \cdot 10^{18}$ to $48 \cdot 10^{18}$ s^{-1}, wavelengths between 0.125 and 0.00625 nm and wavenumbers from $8 \cdot 10^7$ to $16 \cdot 10^8$ cm^{-1}; though these figures serve only to show the relationship to → optical spectroscopy.

Mössbauer spectroscopy is concerned with the resonance absorption of γ radiation by atomic nuclei. Consider an atomic nucleus, 1, (with Z protons and N neutrons) in an excited state, energy E_e. After a short time (ca. 10^{-7} s) the nucleus falls to the ground state, energy E_g, and emits a quantum of γ radiation, $E_o = E_e - E_g$. If this γ quantum should strike an identical atomic nucleus, 2, in the ground state, E_g, then the energy, E_o, can be completely absorbed whereby the nucleus 2 is excited. This process corresponds exactly to → resonance in → optical spectroscopy. The important thing is that the energy of the γ quantum is exactly the energy difference $E_o = E_e - E_g$ between the two states and is not changed, neither in emission nor in absorption. In Mössbauer spectroscopy all the emitting nuclei (γ quantum emitters) are known collectively as the Mössbauer source and all the absorbing nuclei (γ quantum absorbers) as the Mössbauer absorber, though the Mössbauer-active nuclei in source and absorber may be in different environments. For a free atom, e.g. in the gas phase, conservation of momentum requires that the emitted γ quantum with the momentum, $p = E\gamma/c$, must impart an equal but opposite recoil momentum to the atomic nucleus. In this way, the nucleus acquires a kinetic recoil energy of $E_R = p^2/2m = E\gamma^2/2mc^2 = E_o^2/2mc^2$ where m is the mass of the nucleus and c the velocity of light. Thus, the energy of the γ quantum can be written, $E\gamma = E_o - E_R$, and insertion of the corresponding physical constants shows that $E_R \ll E_o$. In an analogous way, momentum and therefore kinetic energy is also imparted to the nucleus when a γ quantum is absorbed and the energy of this γ quantum is given by $E\gamma = E_o + E_R$. Thus, in order to be absorbed a γ quantum requires an energy greater by E_R than the resonance energy E_o. Now, the natural → line width of the resonance transition, as determined by $\delta_n = h/2\pi\tau$ (where τ is the mean lifetime of the excited nucleus), is very much smaller than the recoil energy. For the γ radiation of ^{57}Fe with 14.4 keV and $\tau = 1.4\ 10^{-7}$ s we find $\delta_n = 7 \cdot 10^{-9}$ eV, while $E_R = 2 \cdot 10^{-3}$ eV. Thus,

E_R is larger by orders of magnitude than the natural line width, δ_n, and a resonance transition appears to be out of the question on energetic grounds because the emission line is shifted by E_R to lower energy values. The situation is shown schematically in Figure 1. In principle, a transfer of momentum and therefore a recoil energy, also occurs in optical spectroscopy. But because the transition energy in optical spectroscopy is of the order of a few eV, the recoil energy is some 10^8 times smaller than in γ-ray spectroscopy, i.e. about 10^{-11} eV. This recoil energy has practically no influence on the resonance absorption. At a natural line width, $\delta_n = 10^{-4}$ cm^{-1}, which corresponds to a width in energy of $\delta_n = 1.25 \cdot 10^{-8}$ eV, the recoil loss is only approximately 1/100 of the natural line width and is not observable since the line broadening, δ_{eff}, is a factor of $10^2 - 10^3$ greater.

In 1958, R.L. Mössbauer recognized that the resonance absorption of γ radiation could be observed, in spite of these difficulties, if the emitting and absorbing atoms were embedded in a solid. Then, the recoil energy is transferred to the total solid and the mass of the solid, rather than the mass of the nucleus, is entered in the equation $E_R = E_o^2/2mc^2$. E_R is then negligibly small and resonance absorption is possible. However, the consequence is that Mössbauer spectroscopy is possible only with solids, e.g. crystals, amorphous materials, polymers or frozen solutions. Mössbauer named this phenomenon recoil-free nuclear resonance absorption or nuclear resonance absorption with frozen recoil, which is what we understand in general by the Mössbauer effect.

The condition for recoil-free emission and absorption of γ quanta is $E_R = E_o^2/2mc^2 < k\Theta$; where k is the Boltzmann constant, $\Theta = h\nu_g/k$ the Debye or characteristic temperature, h is → Planck's constant and ν_g the maximum frequency of the vibrational spectrum of a crystal.

At a given temperature, the probability of recoil-free absorption and emission of a γ quantum is greater the

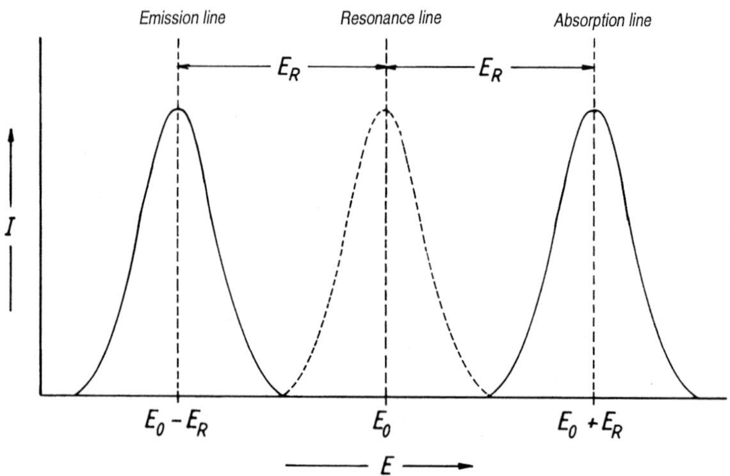

Mössbauer spectroscopy. Fig. 1. The relationship between resonance and absorption lines

greater is Θ. That is, the probability rises as the atoms are more strongly bound in the crystal lattice and as the excitation energy of the lattice vibrations rises. The probability also rises as the resonance energy, E_o, falls ($E_o \leq 200$ eV). As the temperature is lowered the lattice vibrations become increasingly frozen, the conversion of the recoil energy into internal energy becomes increasingly difficult and the probability of recoil-free absorption of γ quanta increases. For this reason, Mössbauer experiments are usually carried out at low temperatures, 77 K (liquid nitrogen), or 4.2 K (liquid helium).

An apparatus for measuring a Mössbauer spectrum is shown schematically in Figure 2. The γ source, absorber and detector are arranged in a line. The source contains a radioactive isotope which, as it decays, emits the γ radiation of interest. In the case of ^{57}Fe, this is ^{57}Co which decays to ^{57}Fe with a half-life of 220 days as a result of excited K capture. In K capture, an atomic nucleus captures an electron from the K shell thereby decreasing its proton count by one. The excited ^{57}Fe atom emits a γ ray of 14.4 eV which is registered by the detector. An absorber containing ^{57}Fe nuclei is introduced into the beam and although natural iron contains only 2% of this isotope it is quite sufficient to show resonance absorption. In order to explore the profile of the absorption line with the emission, use is made of the → Doppler effect by means of which a small movement of one line relative to the other can be achieved. Either the absorber or the source can be moved. If the source is moved, see Figure 2, the energy of the γ quantum is given by $E_\gamma(v) = E_o(1 + v/c)$ where c is the velocity of light and v is that of the source relative to the absorber. For $v > 0$, the source moves towards the absorber and for $v < 0$, away from it. Because the lines are so sharp, velocities of a few mm/s are sufficient to cover the resonance absorption completely. To obtain a Mössbauer spectrum, one simply needs to record the intensity of the γ radiation at the detector as a function of the velocity, v. A Mössbauer spectrum measured in this way is shown schematically in Figure 3a. Nowadays, the

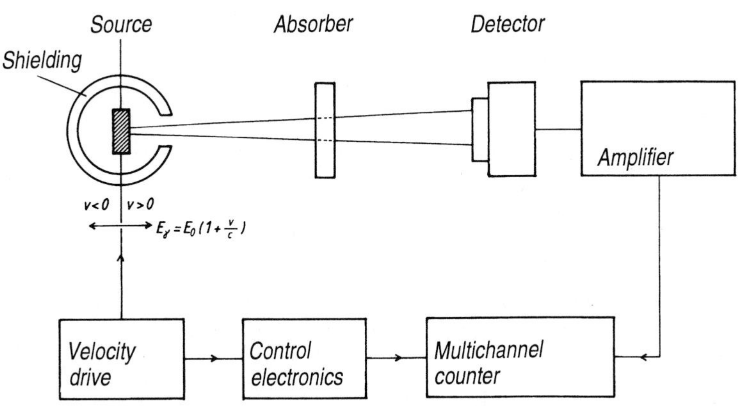

Mössbauer spectroscopy. Fig. 2. Schematic diagram of a Mössbauer spectrometer

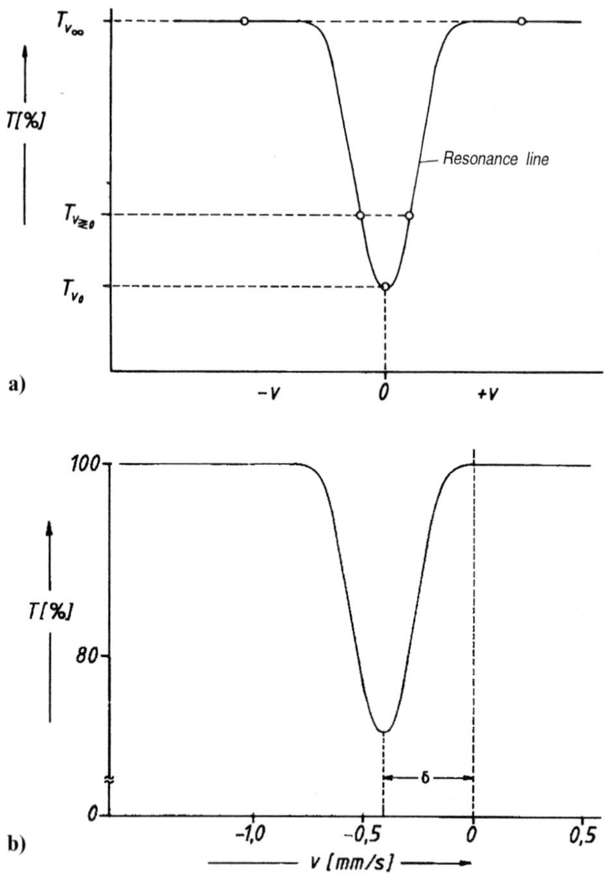

Mössbauer spectroscopy. Fig. 3. a) A Mössbauer spectrum, schematic; b) the measurement of a Mössbauer spectrum using the variable Doppler shift

measurement is made by moving the source to and fro at a frequency of a few Hertz so that it passes through all the velocities required for the region to be studied during each period of the motion; this is controlled electronically. Proportional counting tubes, → scintillation counters, or semiconductor detectors are used to detect the γ-rays. The amplifying electronics provide an electrical pulse for every γ-ray detected. These pulses are stored in a multichannel pulse store in such a way that the signal for a particular velocity, from each cycle of the source, is stored in the same memory channel. The significant fact in the application of Mössbauer spectroscopy is that the atoms of the source and the detector are frequently in different chemical and physical environments so that the absorption line is shifted with respect to the emission line. It is then possible, with the help of the variable Doppler shift, to explore the profile of the resonance line. This case is shown schematically in Figure 3b. There are two reasons for the different physical and chemical environments of source and observer. Firstly, interaction with electrons close to the nucleus (s electrons) leads to a chemical shift, usually known as a chemical isomer shift, isomer shift or center shift. Secondly, there are hyperfine interactions, two types of which may be distinguished; magnetic hyperfine and electric quadrupolar interactions. The first of these is fully analogous to the → Zeeman effect in an external magnetic field, although in Mössbauer spectroscopy the magnetic field, B, at the nucleus is not produced by a magnet but by the hyperfine field due to the electrons at the position of nucleus. The quadrupolar interaction is electronic in nature and arises from a deviation of the nuclear charge distribution from spherical symmetry (→ nuclear quadrupolar resonance). The sum total of these effects creates a multiplicity of applications for Mössbauer spectroscopy in solids.

Ref.: T.C. Gibb, *Principles of Mössbauer Spectroscopy*, Chapman and Hall, London, New York, **1976**.

MPI → multiphoton ionization.

MPS → multiphoton mass spectroscopy.

M shell, <*M-Schale*>, the third electron shell of an atom. It corresponds to the state with the principal quantum number $n = 3$. Since $l = 0, 1, 2$ and $m_s = \pm 1/2$ it can accommodate a total of 18 electrons in accordance with the → Pauli principle.

Multichannel spectrometer, MCS, <*Simultanspektrometer, MCS*>, an instrument used in → multichannel spectroscopy. The central element of the Zeiss MCS is a compact spectrometer whose functional components have been reduced to three; entrance slit, diffraction grating and detector (see Figure 1). The detector is a row of miniaturized diodes forming a → photodiode array on a silicon single-crystal chip. (Reticon photodiode array RL 512G with 512 diodes, each 25×250 μm, in a row 12.8 mm long). With the help of an integrated circuit on the chip, the signal recorded by each diode can be individually read and passed to digital electronic processing in a few microseconds. The measurement procedure of an ordinary sequential spectrometer is thus replaced by a purely electronic process which is very much quicker and, because it involves no mechanical movement, is incomparably better in wavelength accuracy. The dispersing element is a holographic → concave grating which makes it possible to form an image of the entrance slit in the plane of the detector in the ratio 1:1. In the case of a concave mirror, this implies an aperture of 1:2.3. Since

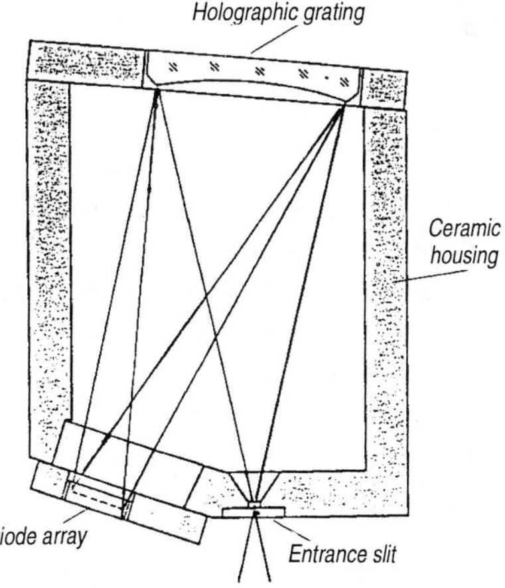

Multichannel spectrometer. Fig. 1. Construction of the Zeiss MCS spectrometer

the whole of the spectral region from 360 to 780 nm (the visible) is recorded on a row of diodes 12.8 mm long, the linear dispersion is only 33 nm/mm so that the grating-groove density of 248 lines per mm is unusually small. The large aperture requires in a large grating diameter so the total number of grooves is more than sufficient for the detector-limited band width of 0.8 nm. The theoretical resolution is 0.1 nm. The mountings of the three optical elements and the chassis of the instrument are formed from a precompressed hard ceramic which has the same, low thermal coefficient of expansion as the glass components of the optics. Consequently, the geometry of the spectrometer is exactly maintained over the very wide temperature range of –70 °C to +70 °C. The spectrometer is connected to source and sample by means of flexible light pipes (→ fiber optics) which makes it possible to separate the position of the sample and that from the spectrometer. Since the fiber optics are capable of covering distances up to 100 meters, spectroscopic studies of extended objects, in rooms where there is a danger of explosions and on running machines and production lines, are possible without great difficulty. The possible applications of the MCS are very varied. On account of the very short time required for a measurement, less than milliseconds, it is particularly suited to kinetic measurements by the → stopped-flow method. The instrument can also be used to measure color (MCS 2 × 512 VIS) or layer thickness and in → dispersion spectroscopy.

Ref.: H. H. Schlemmer, H. Mächler, *J. Phys. E: Sci. Instrum.*, **1985**, *18*, 914.

Multichannel spectroscopy, <*Simultanspektroskopie*>, all spectroscopic methods which permit a simultaneous measurement at more than one wavelength. The oldest version of this type of instrument is the spectrograph with a photographic plate as detector in the focal plane. Emission or absorption spectra over a wide spectral range (UV-VIS, NIR) could be measured simultaneously. In modern multichannel UV-VIS spectrometers, the photographic plate has been replaced by a → photodiode array (→ UV-VIS diode array spectrometer).

In analysis by emission spectroscopy simultaneous → atomic emission spectrometers with gratings in the Rowland mounting are employed. Many slits are arranged on the → Rowland circle in such a way that each corresponds to a chosen line of a particular element. Behind each slit there is a → photomultiplier so that several elements can be determined simultaneously.

Simultaneous measurement at just two wavelengths has been realized in the true → dual-wavelength spectrometers for the UV-VIS region.

In IR and NMR spectroscopy the → Fourier-transform technique has made it possible to record a complete spectrum simultaneously. The advantage of so doing is called the multiplex or Fellgett advantage in FT spectroscopy.

Multiphoton ionization, MPI, <*Multiphotonenionisation, MPI*>, a very important consequence of multiphoton excitation (→ multiphoton spectroscopy). In general it is a process of low probability, but the yield increases markedly when real states of the molecule are in resonance with the photon energy. In Figure 1 the energy-level

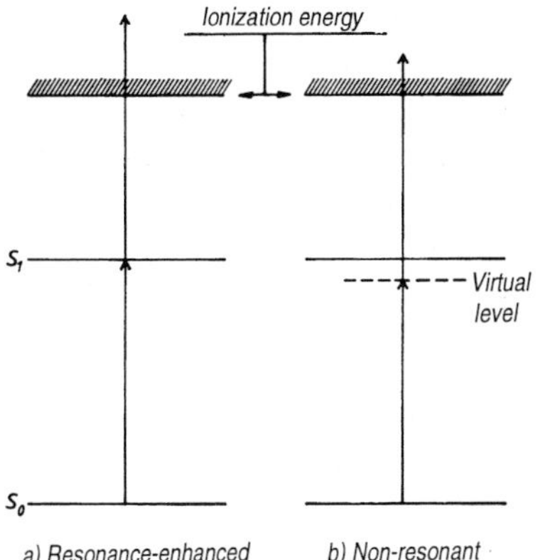

Multiphoton ionization, Fig. 1. Energy-level schemes for resonant and nonresonant multiphoton ionization

schemes for resonant and nonresonant cases are compared. Resonance-enhanced multiphoton ionization (REMPI) is also used as a new source of ions in → multiphoton mass spectroscopy. As with multiphoton spectroscopy, for multiphoton ionization powerful lasers, e.g. → nitrogen lasers or pumped → dye lasers, are used. If a beam of molecules is passed through the focal point of such a laser, then some 10^7 ions are generated in a volume of 10^{-5} cm^3 during the short laser pulse of 10 ns. These ions can be examined with a suitable → mass spectrometer. It is important that the ionization can be significantly influenced by variations in the laser power. At low laser powers ($\leq 10^6$ W cm^{-2}), high yields of the ions of the original neutral atoms are obtained; soft ionization. This type of ionization is very difficult to achieve in the usual ion sources of conventional mass spectrometers. Clearly, soft ionization is possible for all molecules which can be ionized in a two-photon ionization process via a real intermediate state which is in resonance with the photon energy. At the present state of laser technology there is an upper limit to the ionization energy of approximately 10.5 eV or 87,000 cm^{-1}. However, this limit is not a problem for analytical applications since most organic molecules have lower ionization energies.

In contrast, at higher laser powers (> 10^8 W cm^{-2}) multiphoton fragmentation begins; i.e. in addition to the molecular ions, further multiphoton excitation leads to fragmentation. The mechanism of this fragmentation is known as a ladder switching. It can be visualized as follows. A primarily formed ion is again excited and climbs one step further up the energy ladder and there immediately dissociates. The fragments thus formed are excited by further photons and fragment again.

The processes of soft ionization and hard fragmentation are compared in Figure 2. The relevant mass spectrum is shown for each process. A more exact study shows that at least 8 photons are required to produce C^+ and 11 photons for H^+.

Ref.: E.W. Schlag, N.H. Neusser, *Acc. Chem. Research*, **1983**, *16*, 355; H.J. Neusser, U. Boesch, R. Weinkauf, E.W. Schlag, *Internat. J. Mass Spectrometry*, **1984**, *60*, 147; H.J. Neusser, *ibid*, **1987**, *79*, 142.

Multiphoton mass spectroscopy, MPMS, <*Multiphotonenmassenspektroskopie, MPMS*>, a method in → mass spectroscopy which uses multiphoton ionization. It was demonstrated in 1987 that multiphoton ionization at the very small focus of an intense pulsed laser beam was a new source of ions for mass spectroscopy.

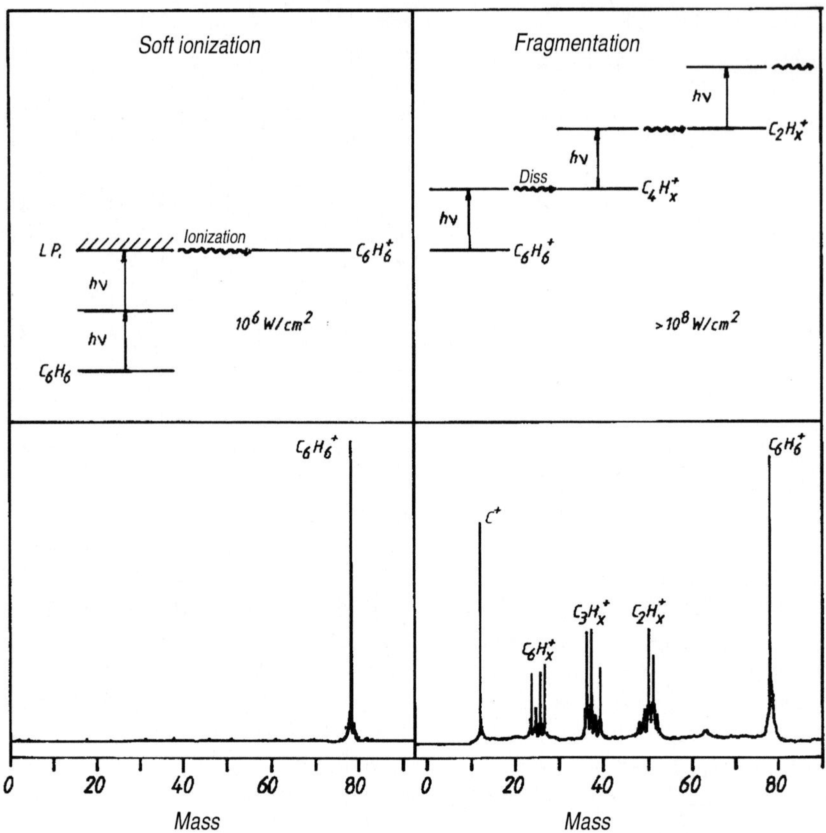

Multiphoton ionization, Fig. 2. Soft ionization and fragmentation in multiphoton ionization

At moderate light intensities of about 10^6 W cm^{-2}, obtainable, for example, with a nitrogen-pumped dye laser, neutral molecules at the focal point are ionized in good yield. About 10^7 ions are produced in a volume of 10^{-5} cm^3 during the short life of the laser pulse. The characteristics of a multiphoton ion source are a definitive time (100 μs) and small kinetic energies (0.2 eV). These properties lead to the combination of a multiphoton ionization source with a → time-of-flight mass spectrometer. A significant increase in the yield of ions occurs when real states of the molecule are excited in resonance by the multiphoton excitation (→ multiphoton spectroscopy). This is termed resonance-enhanced multiphoton ionization (REMPI) or photon-resonance ionization. This principle has been realized in the Bruker photon-resonance mass spectrometer TOF1. An important advantage of multiphoton mass spectroscopy is the ability to record a full mass spectrum in 100 μs. In contrast to a conventional → time-of-flight mass spectrometer, a reflecting electric field at the end of the flight tube turns the beam of ions through almost 360° and they return towards the source, see

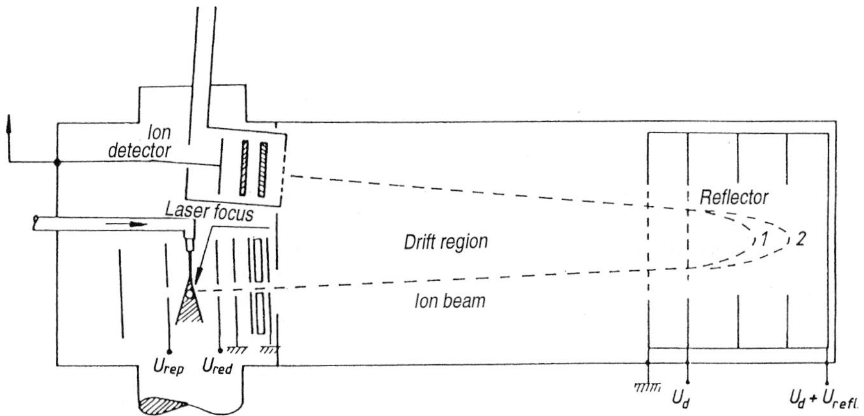

Multiphoton mass spectroscopy. Fig. 1. Construction of the photon resonance mass spectrometer TOF1

Figure 1. The Figure also shows the design of this reflectron mass spectrometer according to Neusser and Schlag, among others. The molecular ions are formed by resonance-enhanced multiphoton ionization in a molecular beam emerging from a jet. The ions formed at the focus of the laser are accelerated, in this case in a direction at right angles to the laser beam. This is effected with a variable electric field which is produced by means of a repeller electrode ($U_{\text{rep}} \sim +1100$ V) and a draw-out electrode at zero potential. The energy of the ion beam is reduced by a special reduction field ($U_{\text{red}} \sim -200$ V) which leads to a decrease in the velocity of the ions and to longer flight times. The ion beam is focused by a special electric lens system and enters the drift region (length 82 cm). After passing through the drift region, the ions are slowed by a variable field ($U_{\text{d}} \sim -300$ V) over a distance of 2 cm during which process their kinetic energy is reduced by about one half. They then enter a reflecting field ($U_{\text{refl}} \sim +800$ V) with low velocity where they remain for a time comparable with their time of flight through the drift region. The reflection towards the detector takes the ions through almost 360°, as can be seen in Figure 1. But all the ions of the same mass have a distribution of kinetic energy so that they arrive at the detector with different velocities and therefore give rise to a multiplicity of signals. However, if the retarding field, U_{d}, and the reflecting field, U_{refl}, are adjusted relative to each other in a suitable manner the differences in the kinetic energies of the ions can be compensated. This energy correction effects a considerable improvement in the resolution. The principle of the energy correction is illustrated in Figure 1. Fast ions with high kinetic energies penetrate deeper into the reflecting field (path 2) and therefore remain there longer than ions of the same mass but somewhat lower kinetic energy (path 1). In this way it is possible to reduce the flight-time differences for ions of the same mass, but differing in their kinetic energies by 10%, to below 10 ns for a total flight time of 60 μs. Figure 2 illustrates

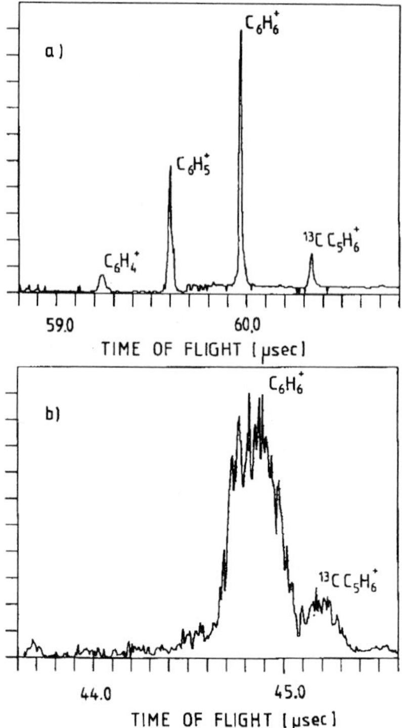

Multiphoton mass spectroscopy. Fig. 2. Mass spectrum of benzene; (a) with energy correction, (b) without correction

energy correction in a multiphoton mass spectrum using benzene as an example: (a) is measured with energy correction, (b) without. The great advantage of the reflectron mass spectrometer is clear.

A further advantage of this type of mass spectrometer lies in the fact that the ionization can be controlled through the laser power. At low powers ($\leq 10^6$ W cm^{-2}) primarily molecular ions are obtained, e.g. $C_6H_6^+$. This is called soft ionization. At higher laser powers ($\geq 10^8$ W cm^{-2}) fragmentation occurs which can go as far as C^+ and H^+; i.e. hard ionization. For this reason this special variant of time-of-flight mass spectroscopy is especially suited to the kinetic study of molecular ions.

Ref.: H.J. Neusser, *Multiphoton Mass Spectrometry and Unimolecular Ion Decay, Int. J. Mass-Spec. and Ion Processes*, **1987**, *79*, 141–181; *Bruker Report*, **1987**, *2*, 2–7.

Multiphoton spectroscopy, <*Multiphotonenspektroskopie*>, a method in → spectroscopy in which a molecule is excited by several photons. Basically, the process can take place in two different ways. One is the stepwise excitation by sequential single-photon excitations, the other is the simultaneous absorption of two or more photons which excite the transition, $E_i \rightarrow E_f$ with $E_f - E_i = h\sum_i \nu_i$ or $E_f - E_i = (h/2\pi)\sum_i \omega_i$, $(\omega = 2\pi\nu)$. In Figure 1, these two possibilities are shown in a simplified energy-level diagram:

a) Shows the stepwise excitation to the state, *f*, by two sequential single-photon processes where the intermediate state, *k*, is a definite, real state of the atom or molecule.
b) Illustrates a two-photon process where a virtual intermediate state, *v*, has been included in order to formulate the two-step transition, $E_i \rightarrow E_v \rightarrow E_f$, symbolically.

A detailed theoretical description of the two-photon process was given by M. Göppert-Mayer in 1929–1931 (*Ann. Physik*, **1931**, *9*, 273), but the phenomenon was first observed 30 years later following the development of pulsed lasers. Multiphoton spectroscopy offers many advantages for the investigation of atomic and molecular spectra:

1. Excited states may be reached with two-photon absorption which are parity forbidden for single-photon absorption.

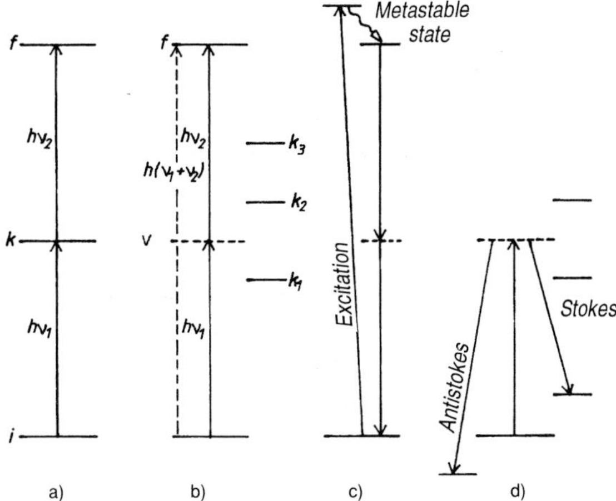

Multiphoton spectroscopy. Fig. 1. Energy-level schemes illustrating two-photon processes, see text

2. The accessible spectral range of multiphoton spectroscopy, $\nu = \sum_i \nu_i$ or $\tilde{\nu} = \sum_i \tilde{\nu}_i$, can be extended into the vacuum UV region if the photons involved, $h\nu_i$, come from a laser emitting in the visible or ultraviolet. By using a tunable laser or by combining a tunable laser with one of fixed frequency, it is possible to excite over continuous ranges in the UV and VUV.
3. If special experimental conditions are observed, multiphoton spectroscopy makes → Doppler-free spectroscopy possible.
4. Frequently, ionized states can be reached using multiphoton spectroscopy. This makes it possible to use the extremely sensitive methods of ion detection in the study of autoionizing states and opens up the new field of molecular ion spectroscopy. The practical application of this idea led to the development of a new ion source for mass spectrometry; the photon-resonance ionization source in the form of resonance-enhanced → multiphoton ionization and → multiphoton mass spectroscopy.

The probability, A_{if}, of a two-photon transition between the ground state, E_i, and an excited state, E_f, induced by the photons $h\nu_1$ and $h\nu_2$ (or $h\omega_1$ and $h\omega_2$) of two light waves with wave vectors, $\vec{k}_1$ and $\vec{k}_2$, polarization vectors, e_1 and e_2, and intensities, I_1 and I_2, can be written as the product of two factors:

$$A_{if} \propto \frac{\delta\omega_{if}}{[\omega_{if}-\omega_1-\omega_2-v(\vec{k}_1+\vec{k}_2)]^2+(\delta\omega_{if}/2)^2} \cdot \left[\sum \frac{R_{ik}\vec{e}_1 R_{kf}\vec{e}_2}{(\omega_{ki}-\omega_1-\vec{k}_1 v)} + \frac{R_{ik}\vec{e}_2 R_{kf}\vec{e}_1}{(\omega_{ki}-\omega_2-\vec{k}_2 v)}\right]^2 \cdot I_1 I_2$$

The first factor gives the line profile of the two-photon transition. It corresponds exactly to the single-photon transition of a moving molecule with a mean frequency $\omega_{if} = \omega_1 + v_2 + v(\vec{k}_1 + \vec{k}_2)$ and a homogeneous line width if $\delta\omega_{if}$. If $\vec{k}_1 = -\vec{k}_2$, the Doppler broadening vanishes, → Doppler-free spectroscopy.

The transition probability, A_{if}, is proportional to the product, $I_1 \cdot I_2$, (in the case of a single laser beam to I^2) and this means that for two-photon absorption pulsed lasers of sufficiently high power must generally be used.

The second factor describes the transition probability for the two-photon transition. It can be derived quantum mechanically using second order perturbation theory. The factor consists of the sum over all intermediate states, k, of the matrix element products, $R_{ik} \cdot R_{kf}$, for transitions between the initial state, i, and the intermediate state, k, and between k and the final state, f. However, the denominator shows that only those states which are not far removed from resonance with a Doppler-shifted laser frequency, $\omega_n' = \omega_n - \vec{k}_n \vec{v}$, contribute significantly. The second factor also describes quite generally the transition probability for all two-photon transitions, i.e. Raman scattering as well as two-photon absorption and emission. This is shown schematically in Figure 1 b, c and d. It is important to note that the same selection rules apply to all two-photon processes. For a nonzero transition probability, A_{if}, the two matrix elements, R_{ik} and R_{kf}, must be nonzero. Therefore, a two-photon transition between the states i and f is only possible when both can be connected to an intermediate state, k, by allowed single-photon transitions. Since the selection rules for single-photon transitions require that the states i and k or k and f must have opposite parities, this means that the two states, i and f, which are connected by a two-photon transition, must have the same parity. Thus, in atoms, the transitions $s \rightarrow s$ and $s \rightarrow d$ are allowed two-photon transitions though they are forbidden in single-photon spectroscopy. For molecules, e.g. homonuclear diatomics, the transition, $\sum_g \rightarrow \sum_g$, is allowed for two-photon excitation but forbidden for single-photon excitation. Therefore, it is possible to excite molecular states which cannot be reached from the ground state with single-photon excitation. Two-photon absorption spectroscopy is complementary to conventional absorption spectroscopy which is linear in intensity, I. Its results are of particular interest because they provide information about molecules which was frequently unobtainable hitherto. Furthermore, the matrix elements, $R_{ik} \cdot \vec{e}_1$ and $R_{kf} \cdot \vec{e}_2$ depend upon the polarization of the incident radiation so that it is possible to select the accessible upper state by the correct choice of the polarization. Since in single-photon spectroscopy the total transition probability is independent of the polarization of the incident light, information about the symmetry of the excited state can be obtained from the symmetry of the ground state, which is frequently known, and the polarization of the two light waves. Examples of this are two-photon excitations of benzene derivatives and polyatomic condensed aromatic hydrocarbons.

Ref.: W. Demtröder, *Laser Spectroscopy in Chemical Physics*, Springer Verlag, Vol. 5, Berlin, Heidelberg, New York, **1983**, chapt. 3, p. 438 ff.

Multiplet, <*Multiplett*>, a spectroscopic transition in absorption or emission which appears to be split into several components (lines). The reasons for this differ considerably, depending upon the type of spectroscopy. They therefore require different explanations.

The multiplet splittings in atomic spectra have been investigated in the greatest detail. The → multiplet structure of a line spectrum is due to the → multiplicity of the terms (energy levels). In atoms, because of → Russell-Saunders coupling (spin-orbit coupling), we must consider a total angular momentum, J, formed by vector addition of the total orbital angular momentum, L, and the total spin angular momentum, S. $J = L + S, L + S - 1, L + S - 2 \ldots |L - S|$ (→ electronic angular momentum). Many J values with different energies can be obtained from a particular value of the resultant orbital angular momentum, L, by various combinations with S. The result is a splitting of the terms characterized by L. The number of the terms obtained by combination with S is the → multiplicity which is given by $2S + 1$.

Thus, the value of S determines the multiplicity. For a single outer electron we always have $S = s = 1/2$, e.g. H atom, alkali-metal atoms, hydrogen-like ions. But for many-electron atoms the S values derived from all possible combinations must be considered. For two electrons and $\ell = 0$, the possible combinations are $S = 0$ and $S = 1$, giving a singlet for the former and a triplet for the latter, (e.g. He, alkaline earths, Zn, Cd, Hg). Thus, the line spectra of atoms of this type can be interpreted in terms of two independent energy-level systems. For light atoms, transitions between these two systems are strictly forbidden; the → selection rule is $\Delta S = 0$, i.e. only transitions in which the total spin, S, or the multiplicity, does not change, are allowed. In the singlet system all the lines are single lines, but in the triplet system each is split into three.

The table lists the possible S values and multiplicities, $2S + 1$, for several equivalent electrons having the same principal quantum number, n, and orbital angular momentum quantum number, ℓ_i.

Up to three electrons, the possible S values correspond to the half-filled p orbitals ($\ell = 1$), up to five electrons to the half-filled d orbitals ($\ell = 2$) and up to seven to the half-filled f orbitals ($\ell = 3$). The multiplicities which are to be expected on further filling of these orbitals can be obtained by going backwards for each additional electron, from 3 to 1 for the p orbitals, from 5 to 1 for the d orbitals and from 7 to 1 for the f orbitals. Careful thought must be given to the deter-

Multiplet. Table.

Electrons	S	$2S+1$	System
1	1/2	2	Doublet
2	1; 0	3; 1	Triplet; Singlet
3	3/2; 1/2	4; 2	Quartet, Doublet
4	2; 1; 0	5; 3; 1	Quintet; Triplet; Singlet
5	5/2; 3/2; 1/2	6; 4; 2	Sextet; Quartet, Doublet
6	3; 2; 1; 0	7; 5; 3; 1	Septet; Quintet; Triplet; Singlet
7	7/2; 5/2; 3/2; 1/2	8; 6; 4; 2	Octet; Sextet; Quartet; Doublet

mination of the → ground states of the individual electron configurations which are determined by the resultant orbital angular momentum, $L = \sum \ell_i$. The → Pauli principle and → Hund's rules, which may exclude some combinations apparently possible within the → Russell-Saunders coupling scheme, must be taken into account. In the case of two p electrons with ℓ = 1, for example, such considerations show that only the terms ^1S, ^1D and ^3P are present. The ^3P is the ground state for the electron configuration np^2.
In → X-ray spectra the splitting of the lines can be explained with reference to the multiplicities of the inner electron states.
Multiplets also occur in the → electronic spectra or → band spectra of simple molecules. They can also be explained by the coupling of the orbital angular momentum, Λ, and the spin angular momentum, S, in the form, $\Omega = |\Lambda + S|$ where, as before, $2S + 1$ gives the multiplicity. Since the vibrational and rotational structure is always superimposed upon the electronic bands in electronic band spectra, the analysis of such spectra can be very complicated. Large, especially unsaturated, molecules whose electronic (UV-VIS) spectra are usually measured in solution, often show a vibrational structure. But this should not be regarded as multiplet structure.
In → IR spectra, doublets and triplets are seen which are frequently due to inadequate resolution of the spectrum. The oldest example of this is the → Bjerrum double band. An → isotope effect or → Fermi resonance can also contribute to the doubling of a band in an IR spectrum. In → NMR spectra noticeable multiplets which are due to the electron-coupled spin-spin interaction of neighboring nuclei occur.

Multiplicity, <*Multiplizität*>, the quantity, $2S + 1$, where S is the total spin of a many-electron system obtained by forming the vectorial sum, $S = \sum s_i$. In atomic spectra singlets, doublets, triplets, quartets, quintets ... can occur, depending upon the addition of the spins (→ multiplet). The terms of atoms or ions with an even number of electrons have an odd multiplicity, those with an odd number of electrons have terms of even multiplicity. Molecules are, with very few exceptions, singlets because of the quenching of the spin in the chemical bonds. But higher lying states with $S = 1$ need to be considered and singlet and triplet energy-level systems exist side by side (→ Jablonski diagram).
Molecules with a single unpaired electron have $S = 1/2$ and have a multiplicity of 2, i.e. they have doublet energy-level schemes. Such molecules are monoradicals.

Mutual exclusion rule, <*Alternativverbot*>, in its simplest form, this rule states that vibrations that are allowed in the infrared spectrum are forbidden in the Raman spectrum, and vice versa. An exact group-theoretical treatment shows that this is a general rule only for → symmetry point groups which include an inversion center, i. The rule is especially illuminating in the case of nondegenerate point groups such as C_{2h} and D_{2h}. The $3N$-6 normal vibrations of molecules which belong to these groups can only be observed by measuring both the IR and the Raman spectra. The following limitation should be added. In the case of the $D_{2h} \equiv V_h$ point group (see →

character table, example naphthalene), all vibrations of the symmetry species A_u are both IR and Raman inactive. Thus, in the case of naphthalene, four normal vibrations cannot be measured directly.

MW spectrometer, <*MW-Spektrometer, Mikrowellenspektrometer*> → microwave spectrometer.

MW spectroscopy, <*MW-Spektroskopie, Mikrowellenspektroskopie*> → microwave spectroscopy.

N

Na D lines, sodium D lines, sodium doublet, <*Na-D-Linien, Natriumdublett*>, the first two lines in the principal series of the sodium spectrum. They correspond to the transitions $^2S_{1/2} \rightarrow {}^2P_{1/2,3/2}$ with $\lambda_{1/2}$ = 589.592 nm and $\lambda_{3/2}$ = 588.995 nm (→ alkali spectra).

Natural line width, line width, <*Linienbreite, natürliche Linienbreite*>. Spectral lines in the discrete absorption or emission spectra of gases are never exactly monochromatic. Even at the highest resolution, such as is obtained with → interferometers for example, a spectral distribution, $I(\nu)$, of the absorbed or emitted intensity about a mean frequency, $\nu_o = (E_i - E_k)/h$, corresponding to the energy difference, $\Delta E = E_i - E_k$, between upper and lower states, is always observed. The function, $I(\nu_o)$, in the neighborhood of ν_o is called the → line shape. The frequency interval, $\Delta\nu_{1/2} = |\nu_2 - \nu_1|$, between the two frequencies, ν_1 and ν_2, for which $I(\nu_1) = I(\nu_2) = I(\nu_o)/2$ is the full width at half maximum (FWHM). Figure 1 shows these relationships. The spectral region within the → half-width is known as the kernel of the line and the regions outside that as the wings.
Theoretical considerations show that the half-width, $\Delta\nu_{1/2}$, is related in a simple way to the transition probability for a spontaneous transition, i.e. the Einstein A coefficient (→ Einstein coefficient):

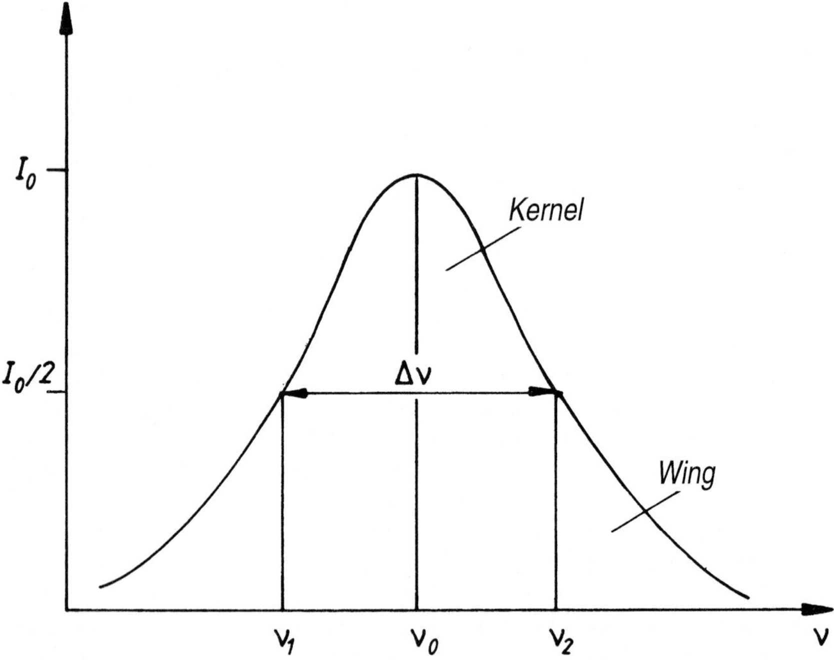

Natural line width. Fig. 1. The spectral distribution of a spectral line, see text

$$\Delta\nu_{1/2} = A_i/2\pi = 1/2\pi\tau_i$$

or in wavenumbers

$$\Delta\tilde{\nu}_{1/2} = A_i/2\pi c = 1/2\pi c\tau_i$$

Where c is the velocity of light, τ_i the natural lifetime of the emitting state and $A_i = 1/\tau_i$. This result can also be obtained directly from the → Heisenberg uncertainty relationship. The half-width, $\Delta\nu_{1/2}$, defined in this way is also known as the natural line width, $\delta\nu_n$.
For the sodium D line at $\lambda = 589.0$ nm ($3P_{3/2} \rightarrow 3S_{1/2}$) with $\tau = 16$ ns:

$$\delta\nu_n = \frac{10^9}{16 \cdot 2\pi} = 10^7\ \mathrm{s}^{-1} = 10\ \mathrm{MHz}$$

or

$$\delta\tilde{\nu}_n = \delta\nu/c = \frac{10^{-3}}{3} = 0,333 \cdot 10^{-3}\mathrm{cm}^{-1} = 3,3 \cdot 10^{-4}\ \mathrm{cm}^{-1}.$$

For molecules in the gas phase, the natural lifetime of electronic excited states also lies in the region of 10^{-8} to 10^{-9} s, so that here also the natural line width is of the order of 10^{-3} to 10^{-4} cm^{-1}.
For transitions between vibrational states in the electronic ground state, i.e. in the IR region, the transition probability is very small and the lifetime, τ, correspondingly longer than for electronic transitions. A mean value of τ of ~ 10^{-4} s corresponds to a line width, $\delta\tilde{\nu}_n = 5.3 \cdot 10^{-8}$ cm^{-1}!
Line widths less than 10^{-3} or 10^{-4} cm^{-1} are generally not observable because line broadening due to the Doppler effect (→ Doppler broadening) and → collision broadening is always present.
Ref.: W. Demtröder, *Laser Spectroscopy,* Springer Series in Chemical Physics, Vol. 5, Springer Verlag, Berlin, Heidelberg, New York, **1987**, chapt. 3, p. 78 ff.

Near infrared spectroscopy, NIR spectroscopy, *<Nahinfrarot-Spektroskopie, NIR-Spektroskopie>*, spectroscopy in the region 800–2500 nm (12,500–4000 cm^{-1}) widely used in quantitative analysis, especially of foods and agricultural products. The first NIR spectrum was recorded by W. Herschel in 1800 and in 1881 W. Abney and E.R. Festing measured the NIR spectra of several organic liquids photographically. Active development of the technique and the recognition of its potential for quantitative analysis of otherwise intractable samples took place during the 1950s and '60s. More recently NIR spectroscopy has received further stimulus from new technological developments and the application of multivariate statistical techniques for the processing of NIR analytical data. Almost all the absorption bands observed in the NIR arise from the → overtones and the hydrogenic → valence vibrations of XH_n groups or from → combinations of the valence and → bending vibrational modes of such groups. Tables of the predicted positions of the X-H overtones (→ expectation region) for a wide variety of X are available and these constitute a valuable aid to the use of NIR spectroscopy in qualitative analysis. But it is in the field of quantitative analysis that the method is most important.
NIR spectroscopy was first used for analysis because the low absorbance values in the region make it possible to perform measurements on moderately concentrated samples and with longer path lengths then in the mid IR.

Spectra can be measured in transmission through neat materials, which avoids sample preparation and saves time. NIR spectra of intact, opaque, biological samples can be obtained by → diffuse reflection and no special cells are required. These factors make NIR spectroscopy simple to perform and very well suited to both laboratory use and to on-line and automatic analysis. The low absorbance of water in the NIR region makes the method especially applicable to foods and agricultural products. Many of the early NIR analyzers were filter instruments making measurements at 2–10 wavelengths. Such devices are still valuable for on-line and automated process analysis, but in laboratory use they are being replaced by dispersive instruments.

All quantitative analyses by spectroscopic methods require a calibration. However, two factors in particular make the calibration phase of an NIR analysis more important and time-consuming than in most branches of spectroscopy. Because of the complexity of the spectra and the high degree of → band superposition it is not possible to pick out a single peak whose height or area may be correlated with analyte concentration. A large proportion of the data in each spectrum is required for the analysis which in turn dictates that the spectrum shall be available in digital form. The second problem is the strong dependence of the reflection on the scattering properties of the sample, particularly the particle size of powders. Thus, the calibration procedure prior to an NIR analysis is especially important and is one reason for the adoption of the sophisticated statistical techniques now increasingly used to evaluate NIR analyses.

The following are typical examples of quantitative NIR analyses. Determination of moisture in sand, organic solvents, fish meal, grains and milk powder and of protein nitrogen in grains, flour, malt and beans. Simultaneous determination of the bread additives ascorbic acid, L-cysteine and azodicarbonamide in starch.

Ref.: B.G. Osborne, T. Fearn, P.H. Hindle, *Practical NIR Spectroscopy with Applications in Food and Beverage Analysis*, 2nd ed., Longman Scientific and Technical, Harlow (UK) and J. Wiley and Sons, Inc., New York, **1993**.

Negative filtering, *<negative Filterung>* → nondispersive infrared equipment.

Neodymium YAG laser, *<Neodym-YAG-Laser>*, a → laser the active medium of which is either glass or garnet (yttrium aluminium garnet). The active ions are neodymium(3+) ions doped into the host crystal. Figure 1 shows the energy-level scheme of the Nd^{3+} ion; a four-level system. The primary excitation takes the system to the closely spaced levels, E_3, from which a radiationless transition to the

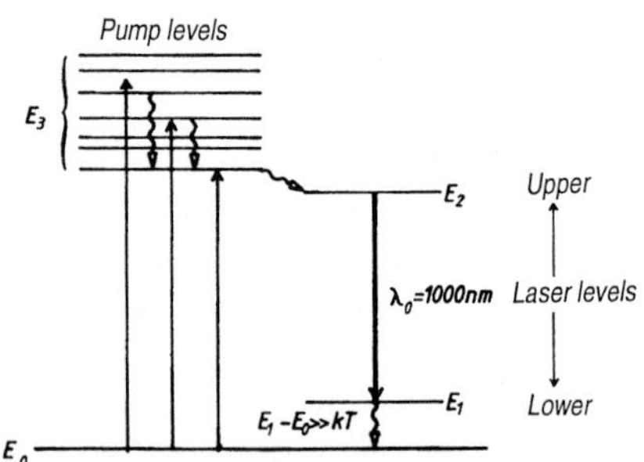

Neodymium YAG laser. Fig. 1. Energy-level diagram for the Nd^{3+} ion; four-level scheme

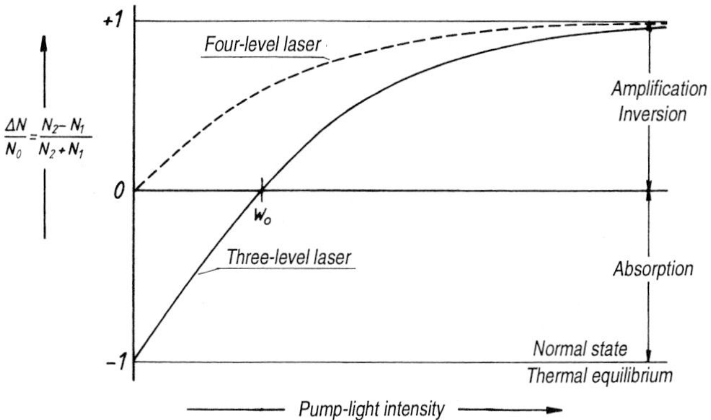

Neodymium YAG laser. Fig. 2. Relative population differences for three- and four-level lasers

upper laser level, E_2, follows. The position of the lower laser level, E_1, is decisive for the four-level scheme. It lies above the ground state, E_o, and the energy difference, $E_1 - E_o$, is sufficiently large so that E_1 is not thermally occupied; i.e. the lower laser level is always empty. Therefore a → population inversion is present between E_2 and E_1 immediately after excitation, even for the smallest exciting power. Figure 2 illustrates the dependence of the relative population difference, $\Delta N/N_o = (N_2 - N_1)/(N_2 + N_1)$, upon the pump-light intensity. For a four-level system, N_1 is always zero, i.e. $\Delta N/N_o = +1$, and we are in the regime of → population inversion, i.e. light amplification. For a three-level system (→ ruby laser), $N_1 = N_o$ and $N_2 = 0$ at zero pump-light intensity; all ions are in the ground state and $\Delta N/N_o = -1$. As the pump-light intensity is increased, at first $N_1 > N_2$, i.e. $\Delta N/N_o < 0$. The condition where $N_1 = N_2$ and $\Delta N/N_o = 0$ is first reached at an intensity, W_o. As the pump intensity is increased further, N_1 approaches zero and $\Delta N/N_o$ tends to +1. The advantage of the four-level system is immediately clear; since the lower laser level, E_1, is always empty ($N_1 = 0$), no absorption can take place from this state. The population difference, $\Delta N = N_2 - N_1$, is positive at the smallest pump intensities and the active neodymium ions can amplify. In the three-level system, as can be seen from Figure 2, an excitation power of at least, W_o, is required in order to reach the amplification region. With four-level systems, in contrast, small pump powers are sufficient and the efficiency of the four-level system is correspondingly greater.

Ref.: D. Eastham, *Atomic Physics of Lasers*, Taylor and Francis, London, Philadelphia, **1989**.

Neon ion laser, <*Neonionenlaser*> → noble-gas ion laser.

Nephelometer, <*Nephelometer, Tyndallometer, Streulichtphotometer*>, an instrument which is used in → nephelometry for measuring the Tyndall scattering (→ Tyndall effect) of turbid liquids.

Nephelometry, <*Nephelometrie*>, a method, based upon the → Tyndall effect, for measuring the scattering of light by solutions. The name derives from the Greek for cloud or mist, „nephos", and indicates a turbid and therefore light-scattering medium, e.g. a colloidal solution. The instrument used in nephelometry is called a nephelometer. In principle, the nephelometer is a special light-scattering photometer (→ light scattering) of simplified construction and the principle of the measurement is the comparison of the intensity of the scattered light, $I_{\delta s}$, with that of the primary incident light, I_o. $I_{\delta s}$ depends upon the angle, δ_s, between the direction of the incident light and the direction of observation. According to the theory of → Rayleigh scattering, there is the following relationship.

$$I_{\delta_s} = C \cdot (NV^2/R^2\lambda^4) \cdot I_o.$$

The proportionality factor, C, contains, among other things, a function of the angle, δ_s, in the form $(1 + \cos^2\delta_s)/2$. N is the number of scattering particles per cm^3, V the volume of the individual particles, I_o, the intensity of the incident light, R the distance from the irradiated volume element and λ the wavelength. The above equation holds only for the case in which the diameter of the particles is small compared to the wavelength of the light, λ, and where the distance between them is at least twice as large as their diameter. For larger, spherical particles, the equations of → Mie scattering and also empirical equations hold exactly. Measurements with a nephelometer are normally made at an angle, δ_s, of 90° to the direction of the incident radiation. The factor $(1 + \cos^2\delta_s)$ then takes the value of 1. The important fact is that, under constant external conditions, the scattering intensity, $I_{\delta s}$, is proportional to the number of particles, N, and hence to their concentration. Thus, nephelometry is primarily used to determine the concentration of turbid or colloidal solutions. Precipitation titration is an example of an analytical application of nephelometry. In principle, any photoelectric photometer can be used for nephelometric measurements. Attachments for light-scattering measurements are offered for many commercially available photometers and spectrometers; usually for an observation angle, δ_s, of 90°.

Ref.: R. Chiang in *Polymer Reviews*, vol. 6, Newer Methods of Polymer Characterization (Ed.: K.E. Bacon). J. Wiley and Sons, New York, London, Sydney, **1964**, p. 471.

Nernst glower, <*Nernst-Stift*>, a thermal radiator used as a source in → IR spectrometers. It consists of a cylindrical rod or tube of zirconia with the addition of yttrium oxide and the oxides of other rare earths. The rod is a few centimeters long and about 3 mm in diameter. The Nernst glower is not electrically conducting when cold, but it conducts at 1100 K above which it can be kept hot by resistive heating. Platinum wires attached to the end of the device serve as electrodes. The preheating is performed by a ceramic oven. Its emission is some 50–60 % of a black body at the same temperature. The average working temperature is about 1900 K. As with all thermal radiators, the emission of the Nernst glower falls off rapidly at shorter wavelengths. Its disadvantage is its low mechanical strength. Nevertheless, it was until recently one of the most widely used radiation sources in → IR spectroscopy.

Neutral density filter, <*Neutralfilter*> → colored glass filters for which the transmission, $\tau_N(\lambda)$, in a limited spectral range shows only a small dependence upon the wavelength. The Schott NG glass filters fulfill this requirement, i.e. they show an almost constant transmission in the region from 400 to 700 nm. Figure 1 shows the spectral transmission plots of 8 neutral density filters for 1 mm glass thickness. The curves of τ_i are shown in the double-logarithmic form → diabatic scale. A series of filters with uniform spectral transmission varying stepwise from filter to filter is often required for special applications in the visible spectral region. A set of filters of this type can be formed from the Schott glass filters NG 11, NG 4 and NG 9, see Figure 1. The thickness of the filters varies between 0.8 mm and 4.5 mm so that, with these three types, the spectral transmission, τ, at $\lambda = 546$ nm can be set between 0.708 and $1 \cdot 10^{-5}$. In total, two NG 11, three NG 4 and eleven NG 9 filters are required. The DIN notation for these glass filters is based upon the transmission, $\tau(546)$, at a wavelength of 546 nm and characterized by a preceding N. The glass filter NG 1 would have the DIN symbol N 10^{-4}, NG 4 would have N 0.27 and NG 11 N 0.72 for a 1 mm thickness in each case. Neutral density filters composed of wire grids or plates with evenly distributed holes drilled in them, such that they transmit only a specified percentage of the incident light, are also available commercially.

Newton's rings, <*Newtonsche Ringe*> → interference at thin films.

Nicol prism, <*Nicolsches Prisma*>, the oldest and best-known polarizer, developed in 1828 by W. Nicol. It uses the double refraction of crystalline calcite. To make a Nicol prism, the end surfaces of a rhombohedron cleaved from a calcite crystal are ground until the new surfaces make an angle of 68° with the lateral obtuse edges; instead of 71°. The crystal is then bisected diagonally such that the new surfaces make angles of 90° with the ends. When the new surfaces have been made planar and polished, the two

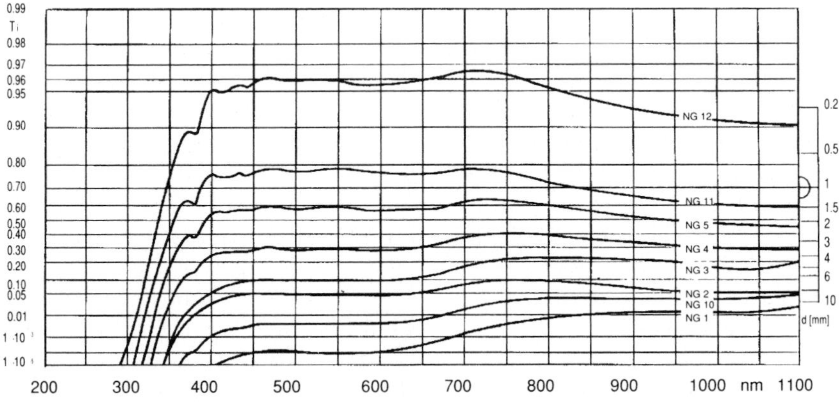

Neutral filter. Fig. 1. Transmission, $\tau_N(\lambda)$, for some NG glasses

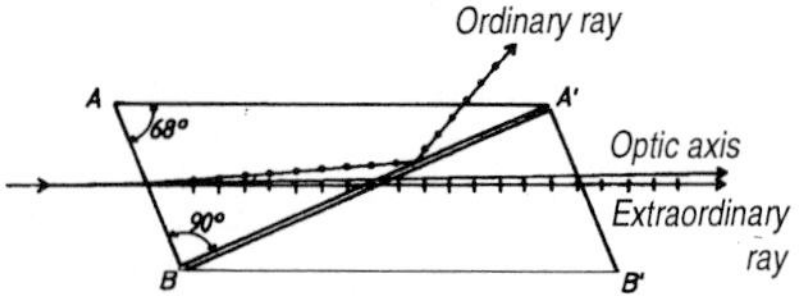

Nicol prism. Fig. 1. Section through the principal plane of a Nicol prism

parts are glued together again, in the original form, with Canada balsam. Figure 1 shows a section through a Nicol prism. A beam of natural light falling onto the prism from the left is split into an ordinary (*o*) and an extraordinary (*e*) beam. The former is polarized perpendicular to the plane of the illustration, the latter parallel to it. On entering the prism, the *o* beam is deviated more than the *e* beam as Figure 1 shows. The refractive index of Canada balsam is $n = 1.542$. For the *o* beam, the refractive index of calcite is $n_o = 1.658$ which means that for this beam the balsam is the less dense medium. Therefore, at the geometry chosen for the prism, the angle of incidence for this beam exceeds that for → total reflection and it is totally reflected at the balsam layer and is completely absorbed at the blackened side of the prism. The *e* beam emerges in the direction of the incident beam as linearly polarized light, but naturally, with one half of the intensity of the incident light. Since the surfaces, at which light enters and leaves a Nicol prism are inclined to the direction of the light beam, there is a parallel displacement of the beam when it traverses the prism and this has undesirable effects when the prism is rotated. This disadvantage is removed in the → Glan-Thompson prism.

Nitrogen discharge lamp, <*Stickstoff-entladungslampe*>, a lamp used to produce nanosecond (2.5 ns) light pulses. The emission takes place between levels which are used in the UV region for the → nitrogen laser. The technical difference between the nitrogen discharge lamp and the nitrogen laser lies in the path of the discharge. In the lamp, the discharge takes place between two pointed metal rods which prevents the establishment of laser action.

Nitrogen laser, <*Stickstoff-Laser*>, a gas laser. In 1963 H.G. Heard discovered a nitrogen emission at 337.1 nm caused by a fast, high-voltage discharge. The origin of this collision-induced emission lies in a transition in the triplet energy-level system involving the states $C^3\Pi_u$ and $B^3\Pi_g$. The 0–0 transition between the two states has a line width of 0.1 nm since many rotational transitions are involved in the emission. The optical amplification of this transition is so great, in comparison with transitions in noble-gas ions and dye molecules, that stimulated emission is observed without the use of a resonator. For this reason, the discharge tube in a nitrogen laser is generally closed by a highly reflecting end mirror and an uncoated quartz plate or output mirror.

However, the high amplification and the simple resonators produce a divergence of the laser beam which is large compared to that of noble-gas lasers.

The lifetimes of the two levels involved in the emission are very different. The upper $C^3\Pi_u$ level has a very short life of 40 ns in comparison with the 8 μs of the lower $B^3\Pi_g$. From this we see that the population inversion must be built up in a time period which is shorter than the lifetime of the upper state. Further, the long lifetime of the lower level and the great difference in the lifetimes of the two levels makes the establishment of a continuous population inversion impossible. Nitrogen lasers are divided into two groups on the basis of the method by which the electron-collisional excitation is achieved technically:

- Lasers with longitudinal excitation and low output powers.
- Lasers with transverse excitation and high output powers.

These two concepts differ primarily in the rise time of the discharge pulse in the laser gas. In transverse excitation, the discharge is ignited by electrodes separated by distances down to 1/100 times those used for longitudinal excitation. This results in a more rapid rise of the current and a higher working pressure of the nitrogen gas. The requirement for fast discharge demands correspondingly fast components. Thus, band-conductor capacitors with low impedance and hydrogen thyratrons find application here.

NMR spectrometer, nuclear magnetic resonance spectrometer, <*NMR-Spektrometer, kernmagnetisches Resonanzspektrometer, NMR-Spektralphotometer*>, a → spectrometer originally proposed by Bloch, see Figure 1. The sample tube, a, is located inside a receiver coil, d, between the poles of a magnet. The transmitter coil with transmitter, e, is positioned at right angles to the magnetic field and the receiver coil. Since the two coils are at 90° to each other, the transmitter induces effectively no voltage in the receiver coil. Only when resonance takes place is there a coupling of the

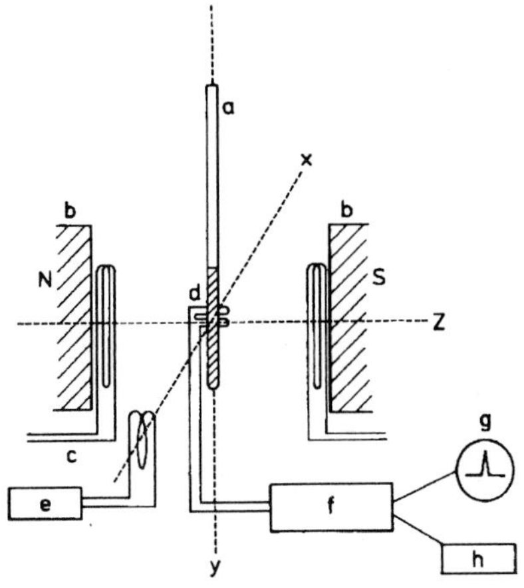

NMR spectrometer. Fig. 1. Construction of an CW-NMR spectrometer, see text

coils mediated by the nuclei. The signal induced in the receiver coil, which is of the order of μV, is amplified and displayed.

For practical reasons, earlier instruments used a fixed frequency and a variable field, (→ field-sweep method). Therefore, we have here the peculiar case of a spectroscopy at constant frequency, but which provides a complete spectrum because of the resonance condition, $\nu = (\gamma/2\pi) \cdot B$. The magnetic flux, B, of the order of 1–1.2 T is modulated with a variable magnetic field of sawtooth form applied by the sweep coils, c. The sawtooth generator also controls the x axis of the plotter, f. When an electromagnet is used a power supply for this is also required. Permanent, electro- and superconductive (cryo-)magnets are used in NMR spectrometers. Permanent magnets have the advantage that they require no electricity supply, but they have the disadvantage that they have a high temperature sensitivity. They can provide fields up to about 2 T which corresponds to a measurement frequency of 90 MHz for protons. Electromagnets consisting of coils and an iron core are made for magnetic flux densities up to about 2.2 T (100 MHz instruments). To reach this flux with the required stability, large stabilized currents (up to 50 A) must pass through the coils. This generates a great deal of heat which must be removed by a temperature-stabilized heat-exchange system. Only superconductive magnets are suitable for flux densities higher than 2.2 T. These magnets use the superconductivity of alloys such as Nb/Ti at liquid helium temperature (4.2 K). An electric current induced in a coil immersed in liquid helium continues to flow indefinitely since the coil has no resistance. This produces a magnetic field. In none of these types of magnet is the field homogeneity sufficient for an NMR measurement. Additional shim coils are required to improve the homogeneity at the position of the sample by a factor of 10^2-10^3 (max). The shim coils are wound on the pole caps of permanent and electromagnets, for superconductive magnets they are placed around the sample position but outside the liquid helium. A further improvement of the homogeneity by at least an order of magnitude is obtained by spinning the cylindrical sample tube, a in Figure 1, about its axis. In this way the effect of inhomogeneity over the sample volume is averaged out. The construction of the magnets determines that the sample tube is rotated perpendicular to the field in permanent and electromagnets and parallel to the field in cryomagnets.

Nowadays, the simplest and cheapest way of obtaining an NMR spectrum is the frequency-sweep method in which the magnetic field is held constant while a monochromatic radio frequency is scanned across the resonance range of interest. This continuous wave (CW) method is usually only used for receptive nuclei, especially protons, and both field-sweep and frequency-sweep methods have been widely superceded by the Fourier-transform technique.

In 1960, E.W. Anderson and R.R. Ernst demonstrated the possibility of obtaining an NMR spectrum by the Fourier-transform technique. Following their work, the great advantages of the FT-NMR method have resulted in the overwhelming importance of this technique in modern NMR (→ Fourier-transform NMR spectroscopy). The components of an FT-NMR instrument are arranged as described above apart from the fact that, since the magnetic field is held constant, no sweep coils are required. The first important difference from the older, frequency-sweep method is that in FT-NMR the sample is irradiated with a short pulse of radiation of a single frequency, the carrier frequency, chosen to lie approximately in the center of the range of nuclear resonances which are to be excited. In accord with Fourier's theorem, the nuclei experience the short pulse of radiation as though it were a broad band of frequencies; the shorter the pulse the broader the frequency band. The pulse of radiation disturbs the alignment of the nuclear magnetic moments in the fixed magnetic field so that, at the end of the pulse, they are no longer in thermal equilibrium. As they return to equilibrium (relax), they emit a radio-frequency signal, known as the free induction decay

(FID), which is received by the receiver coil and stored in the memory of a dedicated computer. If the signal is weak, the combination of a pulse and the recording of the subsequent FID can be repeated as many times as required and the results coadded, i.e. the FID signal, at a specific time after the pulse is added to the sum of the signals recorded at the same time for all earlier pulses. It is usual to record hundreds or even thousands of FIDs, but this may not take longer than 10–20 minutes. When a sum of FIDs, sufficient to give the required S/N ratio has been accumulated, the result is subjected to Fourier transformation which converts the FID in the time domain into an NMR spectrum in the frequency domain. The immediate result of Ernst and Anderson's work was to increase the sensitivity of the NMR experiment by a large margin. But this was only the start of an explosive development in NMR technology which was made possible by the remarkable advances in electronics and computer technology which have taken place during the last 30 years. In particular, it became clear that, by means of a sequence of pulses, the nuclei could be stimulated in many different ways. Furthermore, the FIDs recorded following these complex pulse sequences could be used to enhance the information available from the NMR experiment and to obtain quite new data about the molecule and the magnetic behavior of its nuclei. The fact that today we can obtain NMR spectra of solid samples is largely due to the use of pulse sequences and FT-NMR. Advanced, modern NMR instruments offer the spectroscopist about 50 pulse sequences. The particular pulse sequence and its variable parameters (duration and carrier frequency of each pulse type, time delays between pulses, etc.), are chosen by the operator but, having been chosen, are applied by the instrument's powerful, dedicated computer.

NMR spectroscopy, nuclear magnetic resonance spectroscopy, *<NMR-Spektroskopie, kernmagnetische Resonanzspektroskopie>*, for more than 40 years one of the most significant and important spectroscopic methods. In 1946, the experimental phenomenon of nuclear magnetic resonance was demonstrated independently by two research groups: by Bloch, Hansen and Packard and by Purcell, Torrey and Pound. NMR spectroscopy is effectively essential for the determination of molecular structure. Its theoretical basis is the → Zeeman effect (→ magnetic resonance spectroscopy) which also applies to ESR spectroscopy. The majority of atomic nuclei possess a nuclear spin angular momentum, briefly nuclear spin, given by:

$$\vec{P}_I = \frac{h}{2\pi}\sqrt{I(I+1)}$$

where I is the nuclear spin quantum number. As a result, these nuclei also possess a magnetic moment

$$\mu_I = g_N \mu_N \sqrt{I(I+1)};$$

g_N is the nuclear g factor, $\mu_N = 5.5051 \cdot 10^{-27}\ J \cdot T^{-1}$ is the nuclear magneton.

If such a nucleus, i.e. a magnetic dipole, is placed in a magnetic field of strength, H, (in Gauss) or of magnetic flux density, B, (in Tesla, T; 1 T = 10^4 Gauss) there is a → Zeeman effect, i.e. a → space quantization of the angular momentum. The energy is given by:

$$E_I = -g_N \cdot \mu_N \cdot m_I \cdot B;$$

m_I is the magnetic quantum number which gives the integral or half-integral components of the angular momentum P_I, in multiples of $h/2\pi$, in the z direction; which is the direction of the magnetic field. Every state characterized by the nuclear spin quantum number, I, splits into $(2I + 1)$ Zeeman components of different energies in the magnetic field. m_I and I are simply related by:

$$m_I = I, I-1 \ldots -(I-1), -I.$$

Thus, the proton with $I = 1/2$, has $m_I = +1/2$ and $m_I = -1/2$, and the resulting energies are:

$$E_{1/2} = -\frac{1}{2} g_N \cdot \mu_N \cdot B$$

and

$$E_{-1/2} = +\frac{1}{2} g_N \cdot \mu_N \cdot B.$$

The state, $E_{1/2}$, is the lower in energy. The situation is illustrated diagrammatically for nuclei with $I = 1/2$, 1 and 3/2 in Figure 1. The selection rule for resonance transitions is $\Delta m_I = \pm 1$, i.e. only transitions between neighboring states are allowed. Thus, the energy difference, ΔE, is independent of m_I and is given by:

$$\Delta E = h\nu_o = g_N \mu_N B$$

or

$$\nu_o = \frac{g_N \cdot \mu_N B}{h}.$$

where ν_o is the resonance frequency. The angular frequency, $\omega_o = 2\pi\nu_o$ is:

$$\omega_o = \frac{g_N \mu_N}{h/2\pi} B = \gamma_I \cdot B;$$

ω_o is known as the Larmor frequency and

$$\gamma_I = \frac{g_N \mu_N}{h/2\pi}$$

is the → magnetogyric ratio.

γ_I is a constant, specific to each nucleus, which refers to the bare nucleus. All frequency differences measured for the resonances of nuclei in real molecules are discussed with reference to a new quantity, the → chemical shift.

Not all nuclei have a nuclear spin and the associated magnetic moment; the following rules apply:

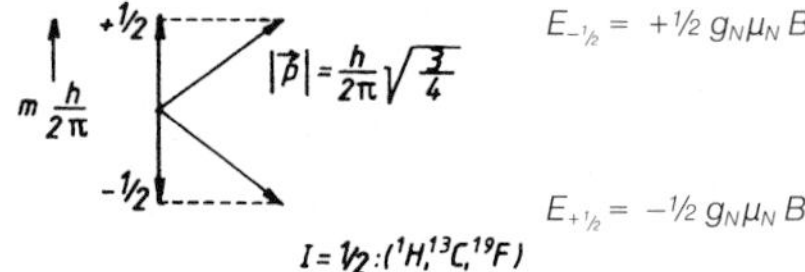

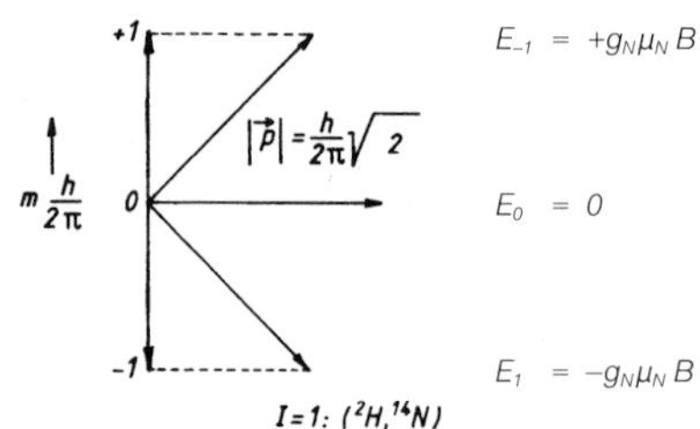

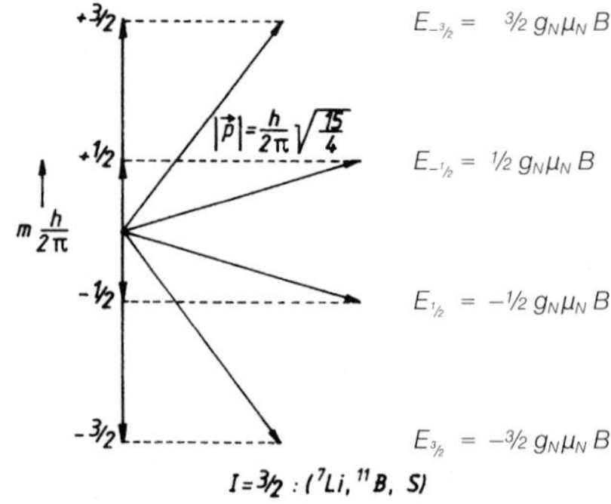

NMR spectroscopy. Fig. 1. Space quantization for the nuclear spins, $I = 1/2$, 1 and 3/2

a) For an odd mass number, the nuclear spin quantum number, I, is an odd multiple of 1/2.
b) For an even mass number and an even atomic number, $I = 0$.
c) For an even mass number and an odd atomic number, I is an integer.

The two isotopes ^{12}C and ^{16}O, which are particularly important for organic chemistry, therefore have no nuclear spin and are not visible in NMR spectroscopy. However, with modern techniques NMR measurements can be made on the isotopes ^{13}C and ^{17}O. The characteristic data for a few important nuclei are collected in the table.

A detailed list can be found in most texts on NMR spectroscopy.

Nuclei with $I > 1$ possess a quadrupole moment, Q, due to the nonspherical symmetry of the nuclear charge. Such nuclei can interact with local electric field gradients, primarily those due to the electrons in the molecule. This type of interaction is also present in the absence of an external magnetic field and leads to states of different energies between which transitions can be excited (→ nuclear quadrupole resonance spectroscopy).

As can be seen from the table, the proton has the highest sensitivity and the greatest resonance frequency. This led to an intensive study of the proton resonance spectroscopy of organic molecules from 1946 onwards.

Two phenomena are of decisive importance for the application of NMR spectroscopy in chemistry:

1. The chemical shift and
2. the spin-spin coupling.

In the → chemical shift, δ, the dependence of the resonance frequency, ν_o, and the Larmor frequency, ω_o, upon the special chemical environment (bonding) is noticeable, i.e. the same nuclei in different chemical environments have different resonance frequencies and can therefore be distinguished. From → the spin-spin coupling the structure of neighboring groups, with the same or different nuclei, can be determined from the splitting of a single resonance signal in the → NMR spectrum. From the resonance condition, $\nu_o = (g_N \mu_N / h) B$

NMR spectroscopy. Table. Data for some frequently studied nuclei

Nucleus	Nuclear spin I	Natural abundance %	Relative sensitivity for equal numbers of nuclei, relative to ^1H (constant B)	NMR frequency for $B = 1$ T MHz	Magnetic moment in multiples of the nuclear magneton μ_N
^1H	1/2	99.98	1.000	42.57	2.7927
^2H	1	0.015	$9.65 \cdot 10^{-3}$	6.53	0.8574
^7Li	3/2	92.57	0.294	16.547	3.2560
^{11}B	3/2	81.17	$1.65 \cdot 10^{-1}$	13.66	2.6880
^{13}C	1/2	1.11	$1.59 \cdot 10^{-2}$	10.70	0.7022
^{14}N	1	99.63	$1.01 \cdot 10^{-3}$	3.08	0.4036
^{15}N	1/2	0.37	$1.04 \cdot 10^{-3}$	4.31	−0.2830
^{17}O	5/2	0.04	$2.91 \cdot 10^{-2}$	5.77	−1.8930
^{19}F	1/2	100	$8.33 \cdot 10^{-1}$	40.05	2.6273
^{29}Si	1/2	4.7	$7.84 \cdot 10^{-2}$	8.46	−0.55477
^{31}P	1/2	100	$6.63 \cdot 10^{-2}$	17.23	1.1305

$= (\gamma_I/2\pi)B$, we see that the resonance frequency, ν_o, is directly proportional to the field, B. Therefore, a frequency, ν_o, can be fixed and the magnetic field, B_o, varied until the resonance frequency is reached. This can be done by adding a weak, variable field, ΔB_{var}, to a fixed field, B_o, i.e. $B_{var} = B_o + \Delta B_{var}$. When advances in technology made it possible to sweep the frequency, frequency-sweep instruments replaced field-sweep, (→ continuous wave method and → field-sweep method). But in the early 1970s the rapid growth of FT-NMR took the FT technique to its present dominant position. For NMR there is, in principle, no absolute scale since the resonance frequency and the magnetic flux density are always related by the resonance condition. Therefore, a relative condition is used. Experimentally only the frequency differences, $\Delta\nu$, between the resonance signals are determined. This difference between two resonance lines is defined as the → chemical shift. In order to compare experimental results, especially those measured on different instruments using different frequencies and fields, the chemical shifts are referred to the resonance of a standard or reference substance, → NMR spectrum. There is a second reason for measuring frequency differences with respect to a reference. It is extremely difficult to measure the exact magnetic field to which a molecule in a solvent, in a sample tube, inside a thermostatted jacket, etc. is subjected. It is experimentally far easier to measure the position of the resonance relative to that of a standard substance which is either dissolved in the same solution or contained in a concentric sample tube. The standard universally used for protons is tetramethylsilane, TMS. In principle, NMR spectroscopy of both solids and fluids is possible, but the information obtained from the two types of measurement is often very different. Of greatest significance is the high-resolution NMR spectroscopy of ^{1}H and ^{13}C nuclei which is central to the application of the NMR technique in the determination of the structure of organic molecules. The resonances of other nuclei have been the subjects of extensive studies with, as one would expect, more importance attached to the more common nuclei, such as ^{14}N, ^{15}N, ^{31}P, and ^{19}F, than to trace elements. For technical reasons, measurements on nuclei with $I = 1/2$ are particularly favorable. The proton is the optimum species for NMR, but excellent measurements on fluorine and phosphorus resonance can also be made.

As noted above, the major isotopes of oxygen and carbon (^{16}O and ^{12}C) have no nuclear spin and, therefore, no NMR spectrum. But since carbon, as the major component of organic materials is of particular interest, the ^{13}C isotope with only 1.1 % natural abundance, but $I = 1/2$, must be used. This small isotopic abundance and the low sensitivity of ^{13}C in comparison with ^{1}H (see Table), were responsible for the fact that ^{13}C NMR spectroscopy became important 10–15 years after ^{1}H NMR, from about 1970 onwards. The increase in activity grew as NMR instruments capable of obtaining ^{13}C NMR spectra from samples with natural abundance, i.e. unenriched, became commercially available.

Ref.: H. Friebolin, *Basic One- and Two-Dimensional NMR Spectroscopy*, 2nd ed., VCH, Weinheim, New York, **1993**; C.H. Yoder and C.D. Schaeffer,

Jr., *Introduction to Multinuclear NMR*, Benjamin/Cummings Publishing Co. Inc., Menlo Park, Wokingham, Sydney, Tokyo, **1987**.

NMR spectrum, <*NMR-Spektrum, kernmagnetisches Resonanzspektrum*>, a spectrum representing the intensity of the nuclear resonance signal (absorption intensity) of differently bonded similar nuclei plotted on a relative abscissa scale. Unlike optical spectra, where the abscissa is always wavelength, λ, or wavenumber, $\tilde{\nu}$, in an NMR spectrum a relative abscissa scale must be used because the resonance frequency, ν_o, and the magnetic flux density, B_o, are related by the resonance condition $\nu_o = (\gamma_I/2\pi)B_o$, → NMR spectroscopy. This relative scale is determined by reference to a simultaneously measured reference signal. For this purpose, a suitable reference substance is added to every sample before measurement (internal standard) so that the signals of the sample may be recorded with respect to that of the reference. Thus, the difference $\Delta\nu = \nu_{sample} - \nu_{ref.}$ is obtained. Tetramethylsilane (TMS) $Si(CH_3)_4$ is used as the reference substance in ^1H and ^{13}C NMR spectroscopy. It is particularly suitable from both the spectroscopic and the chemical standpoints because it gives just one sharp signal which, because of the very large shielding, is clearly separated from most of the other resonances and appears on the far right of the spectrum, i.e. at the largest magnetic field value, B. The $\Delta\nu$ values so obtained are the individual → chemical shifts referred to TMS. But these $\Delta\nu$ values are dependent upon B_o and therefore a dimensionless quantity, δ, is defined by:

$$\delta = \frac{\nu_{\text{Sample}} - \nu_{\text{Reference}}}{\nu_{\text{Reference}}} 10^6 = \frac{\Delta\nu}{\nu_{\text{Measurement frequency}}} 10^6$$

The approximation $\nu_{ref.}$ = measurement frequency is a very good one and introduces effectively no error.

For TMS as reference, $\delta_{TMS} = 0$ by definition. The δ values can usually be read directly from the spectra. They are regarded as positive when they lie to the left of TMS, i.e. in the direction of smaller shielding or correspondingly, smaller magnetic fields. The factor 10^6 is introduced in the defining equation in order to obtain numbers which are easier to handle; thus, the δ values on the abscissa of an NMR spectrum are given in parts per million, ppm. Note that ppm is not a dimension and is frequently not cited when δ values are given.

Chemical shifts measured on different instruments and at different fields and frequencies are directly comparable if they are given as δ values.

An NMR spectrum provides the following information:

1. The position of the center of gravity of the various signals, or groups of signals, gives the chemical shift, δ, of the individual nuclei directly and hence information about the bonding environment.
2. The multiplet structure gives information about the structure of neighboring groups; → spin-spin coupling.
3. From integrations of the intensities of the signals the relative numbers of nuclei contributing to each individual signal may be ascertained.

Figure 1 illustrates the information content of a representative NMR spectrum.

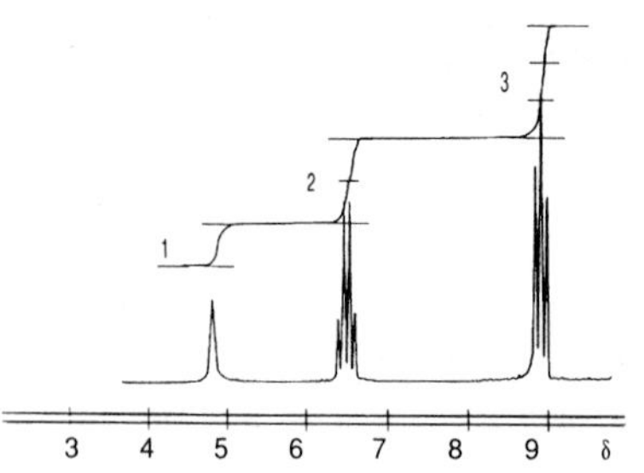

NMR spectrum. Fig. 1. The NMR spectrum of ethanol as an example of the information content, see text

Noble-gas discharge lamp, <*Edelgas-Entladungslampe*>, a → gas-discharge lamp which can be used in various pressure ranges. For this reason, noble-gas discharge lamps can have very different properties and correspondingly different areas of application.

As low-pressure discharge lamps they are used especially for the production of radiation in the → vacuum UV. They are constructed as capillary lamps. A noble gas at a pressure of 1 Torr flows through a water-cooled quartz capillary 50 mm long with an internal diameter of 5 mm. At a voltage of 13–14 KV, a current of 0.2–1 A flows. The discharge produces a line spectrum which is passed into a vacuum monochromator through an external slit. If the lamp is operated at a pressure between 100 and 500 Torr, continuum radiation is obtained the short-wavelength end of which coincides with the resonance line of the noble gas. The continua of the noble gases are summarized in the table.

Although the intensity of these continua falls off to longer wavelengths, this type of low-pressure discharge lamp is very important as a continuous light source for the vacuum UV or

Noble glas	Resonance wavelength λ [nm]	Wavelength range of the continuum [nm]
Argon	106.6	106.6–165
Helium	60.0	60.0–100
Neon	74.4	- - -
Krypton	123.6	123.6–185
Xenon	147.0	147.0–225

Schumann UV regions. As with the → hydrogen lamp, the continuum arises from a transition from a stable excited molecular state, e.g. He_2^*, to an unstable ground state, e.g. the He_2 molecule. Thus, it is an example of → decay luminescence.

At high noble-gas pressures and high temperatures, an arc discharge occurs in the gas. This gives a continuum of very high intensity in the UV-VIS region. The xenon lamp is the most important example of a high-pressure noble-gas discharge lamp.

Ref.: G. Herzberg, *Molecular Spectra and Molecular Structure I: Spectra of Diatomic Molecules*, D. Van Nostrand Co. Inc., Princeton, New York, London, **1950**, p. 403 ff.

Noble-gas ion laser, <*Edelgasionen-Laser*>, a laser which uses the primary positive ions produced by collisions with electrons as its → active medium. The most important types are the argon- and krypton-ion lasers, but neon and xenon are also used. The excitation mechanism is assumed to be a two-step process which takes place in a gas discharge, as shown for argon in a simplified manner in Figure 1.

In the first step the argon atom is ionized. In the second step a further electron impact raises the Ar^+ ion from its ground state to the upper laser level. The ground state of Ar^+ has the elec-

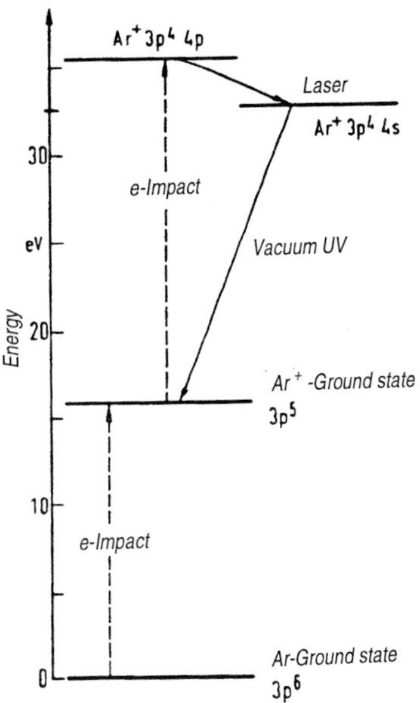

Noble-gas ion laser. Fig. 1. Two-step excitation of argon

tron configuration $3p^5$ giving rise to a $^2P_{3/2,1/2}$ state. The excitation generates an upper laser level with the configuration $3p^4 4p^1$ which gives rise to *S*, *P*, *D* and *F* states. Since the total spin, *S*, in this configuration can be either 1/2 or 3/2, ^2S, ^2P, ^2D, and ^2F states arise as well as ^4S, ^4P, ^4D, and ^4F. These *S*, *P*, *D* and *F* states can occur more than once, but with different energies. The lower laser level corresponds to excitation to the configuration $3p^4 4s^1$ which gives rise to a ^2P state. Figure 2 shows the energy-level scheme of the Ar^+ ion in which the exact positions of the states involved in the laser transition are given. The *J* values of the doublet and quartet states are given at the top of the diagram. The upper laser levels lie ~ 160,000 cm^{-1} (~ 20 eV) above the Ar^+ ground state so that a total of ~ 36 eV are required to excite the argon-ion laser; ~ 16 eV to ionize the Ar atom and ~ 20 eV to excite the Ar^+ ion. The upper laser levels are distributed over a range of 2400 cm^{-1} (0.3 eV), see Figure 2. The electron energy of the gas discharge must therefore be exactly tuned if a significant → population inversion is to be established in the upper laser levels. With the standard set of mirrors, the most important lines in the Ar^+ ion laser are the following, see Figure 2.

The lines at 488.0 and 514.5 nm are the strongest lines from which output powers of 2 to 50 W may be obtained, depending upon the specification of the laser tube. The relative intensities in the table refer to the strongest line at 514.5 nm. The amplification and power of noble-gas ion lasers are directly proportional to the densities of the electrons and ions in the discharge. A gas discharge with the highest possible degree of ionization should therefore be chosen. The other parameters of the gas discharge such as pressure and current density are fixed by the requirement for the electron energy which leads to the maximum population inversion. The current density at which laser action begins is 15–50 A/cm^2, depending upon the construction of the laser. For continuous argon-ion lasers, discharge tubes of small diameter (1–4 mm) are almost exclusively used in order to keep the discharge current low. Because of the high temperature of the gas plasma there is a danger of destroying the wall of the tube, which can be reduced by axial magnetic fields. To increase the working life, tubes of water-cooled graphite, metal segments and beryllium oxide are often used. A further effect, which should be considered when working with this type of laser, is gas pumping

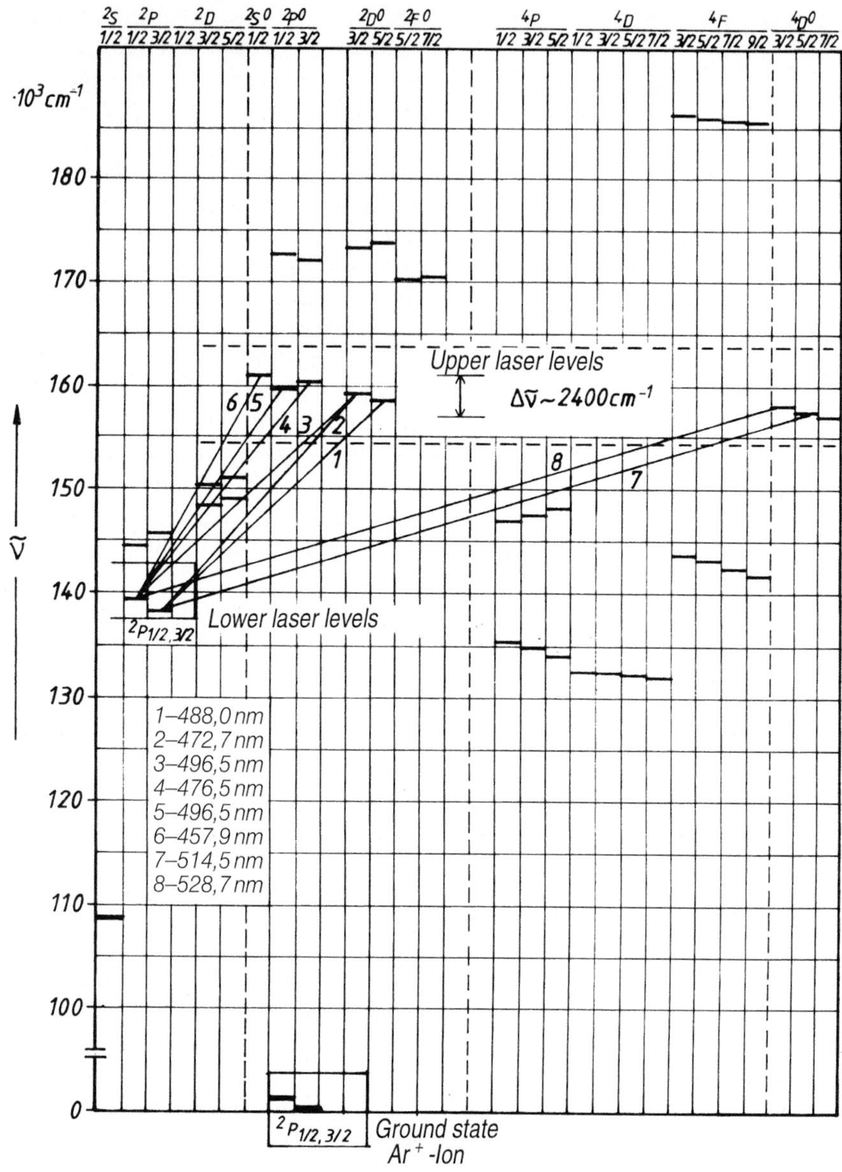

Noble-gas ion laser. Fig. 2. Energy-level diagram for the Ar^+ ion with laser transitions

whereby a decrease of pressure at the cathode leads to a pressure difference between anode and cathode. This effect leads to a drop in performance and can result in destruction of the tube. To overcome gas pumping, the neutral atom gas must be returned to the anode region and this is done by means of a connecting tube running between the anode and cathode. The

Argon ion laser lines. Table.

λ [nm]	$\tilde{\nu}$ [cm^{-1}]	Transition	Relative intensity
457.9	21839	$^2S^o_{1/2} \rightarrow {}^2P_{1/2}$	0.16
472.7	21155	$^2D^o_{3/2} \rightarrow {}^2P_{3/2}$	0.13
476.5	20986	$^2P^o_{3/2} \rightarrow {}^2P_{1/2}$	0.33
488.0	20492	$^2D^o_{5/2} \rightarrow {}^2P_{3/2}$	0.77
496.5	20141	$^2D^o_{3/2} \rightarrow {}^2P_{1/2}$	0.33
501.7	19932	$^2F^o_{5/2} \rightarrow {}^2D_{3/2}$	0.21
514.5	19436	$^4D^o_{5/2} \rightarrow {}^2P_{3/2}$	1.00
528.7	18914	$^4D^o_{3/2} \rightarrow {}^2P_{1/2}$	0.20

most important emission wavelengths of noble-gas ion lasers are given in the table.

Noble-gas ion laser. Table.

Neon$^+$	Argon$^+$	Krypton$^+$	Xenon$^+$-Laser
241.3	457.9	476.6	460.3
332.3	472.7	520.8	541.9
	476.5		
339.3	488.0	530.9	597.1
	496.5		
	501.7	647.1	627.1
371.3	514.5	676.4	714.9
	528.7		871.6

Noble-gas ion lasers are widely used in spectroscopy especially in → Raman spectroscopy and in photochemistry.
Ref.: D. Eastham, *Atomic Physics of Lasers*, Taylor and Francis, London and Philadelphia, **1989**.

NOE, <*NOE*> → nuclear Overhauser effect.

NO gamma bands, NO γ bands, <*NO-Gamma-Bande, NO-γ-Bande*>, the longest wavelength absorption bands of the NO molecule at 226.2 nm or 44,208.7 cm^{-1}. They are due to a transition from the $v'' = 0$ vibrational state of the ground electronic state to the $v' = 0$ vibrational state of the first excited electronic state. The fine structure observed is caused by a multiplicity of rotational transitions which overlie the electronic excitation. The NO molecule has an unpaired electron with orbital angular momentum, $L = \Lambda = 1$, and spin, $s = 1/2$. In the NO → energy-level diagram this corresponds to the singly occupied $\upsilon\pi$ orbital. Because of the coupling of Λ with S, the ground state is a doublet with the two components, $^2\Pi_{1/2}$ and $^2\Pi_{3/2}$. The $^2\Pi_{3/2}$ state lies 120.2 cm^{-1} above the $^2\Pi_{1/2}$ (see Figure 1). The first excited state is the state, $^2\Sigma^+$, so that the NO γ

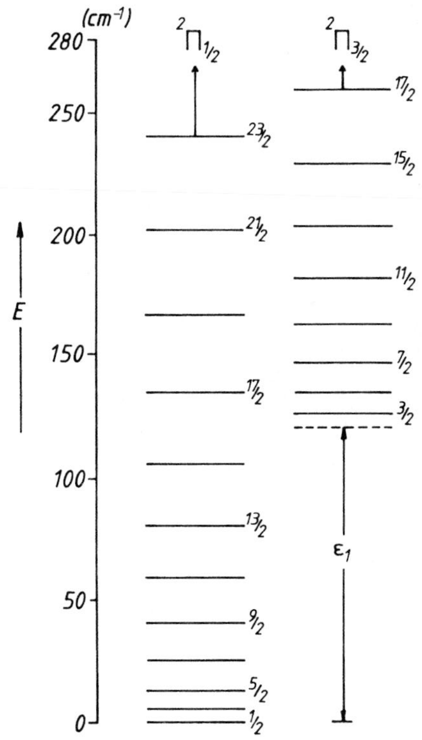

NO gamma bands. Fig. 1. The energy difference in the ground state of the NO molecule, see text

bands correspond to the transition, $^2\Pi \leftrightarrow {}^2\Sigma$ in emission or absorption. The rotational energy-level scheme for the $^2\Sigma^+$ state corresponds to → Hund's coupling case (b), while the ground state, $^2\Pi_{1/2,3/2}$, approximates to case (a), though only for small rotational quantum numbers. As these become larger, case (a) goes over into case (b). Figure 2 shows the energy-level scheme for the transitions $^2\Pi_{1/2,3/2} \rightarrow {}^2\Sigma^+$ for cases (a) and (b). The selection rules predict that 12 branches will be observed, in agreement with experiment. Figure 3 shows the exact energies of the $^2\Pi_{1/2}$ and $^2\Pi_{3/2}$ states. It is particularly important that the NO γ bands in emission can be used for the quantitative determination of NO in gases. The excited NO molecules are generated in a modified → hollow cathode lamp. In this process molecules are formed with rotational temperatures of 1540 K and 300 K; a group of hot and a group of cold excited NO molecules. If the NO gas is at room temperature (300 K) it can absorb the line spectrum emitted in a resonance process because the population of the rotational states in the upper and lower states is practically identical. But in the 1540 K group the population is such that very high rotational states of the excited NO molecules are occupied. These levels are not occupied in the gas at 300 K and no resonance absorption (→ resonance) can take place, i.e. the line spectrum emitted by the hot molecules passes through a sample of NO practically

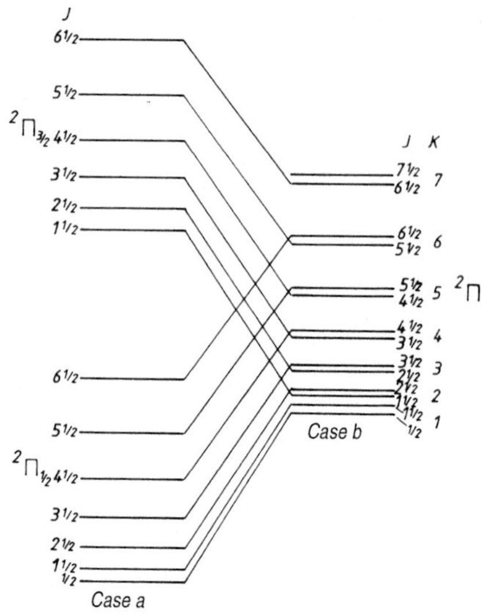

NO gamma bands. Fig. 3. Energy-level scheme for the $^2\Pi_{1/2}$ and $^2\Pi_{3/2}$ ground states of the NO molecule showing the transition from Hund's case a) to case b)

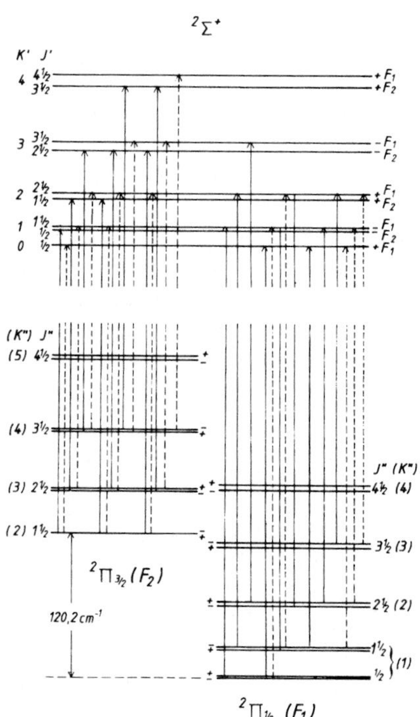

NO gamma bands. Fig. 2. Complete energy-level scheme for the NO molecule for the transitions $^2\Pi_{3/2} \rightarrow {}^2\Sigma^+$ and $^2\Pi_{1/2} \rightarrow {}^2\Sigma^+$

unabsorbed. The combination of these two possibilities and the application of the → gas filter correlation principle leads immediately to a method for the quantitative determination of NO. In effect, the cold light is completely absorbed by a reference sample of pure NO whilst the hot light passes through undiminished and provides a reference signal. If a sample containing NO is now illuminated with all the emitted light of the NO γ bands, the amount of the cold light absorbed is proportional to the concentration of NO in the sample while the intensity of the reference remains effectively unchanged. This is the principle of the on-line photometer (process photometer) → RADAS 1G.

Ref.: H. Meinel, *Z. Naturforsch.*, **1975**, *30a*, 323.

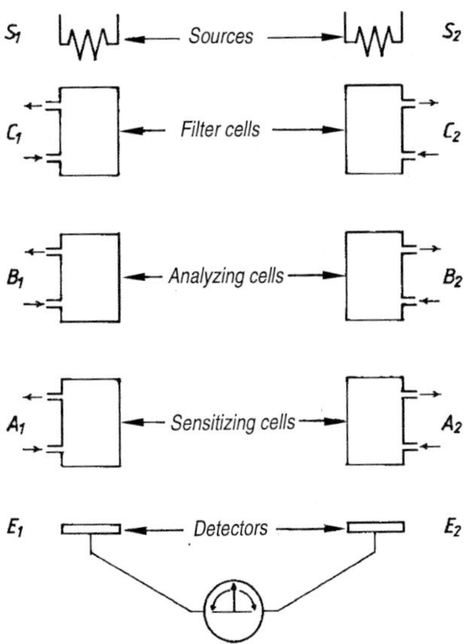

Nondispersive infrared equipment. Fig. 1. Schematic of a nondispersive IR-instrument.

Nondispersive infrared equipment, *<nichtdispersive Infrarotgeräte, nichtdispersive IR-Geräte, NDIR>.* → Nondispersive spectrometers dispense with the spectral dispersion of the measuring radiation beam. The construction of a nondispersive spectrometer is illustrated schematically in Figure 1. S_1 and S_2 are two similar IR radiation sources. A_1, A_2, B_1, B_2, C_1 and C_2 are cells. E_1 and E_2 are radiation detectors so connected that the instrument registers only the difference in their signals. As long as the intensity of the measuring light is the same in the two paths, $C_1 \rightarrow E_1$ and $C_2 \rightarrow E_2$, the instrument shows no signal. If the cell, A_1, is now filled with the gas to be detected, e.g. CO_2, while the cell, A_2, is filled with pure nitrogen, then some of the radiation in the path, $C_1 \rightarrow E_1$, will be absorbed by the CO_2 in A_1 and will be correspondingly reduced in intensity. The instrument will therefore show a signal. If now a gas mixture which contains CO_2 as the gas to be determined is allowed to flow through the cells B_1 and B_2 in both beam paths, then the intensity of radiation in both paths will be reduced because of the absorption in both cells. The measurement shown by the instrument also falls, because in the previously nonabsorbing path, $C_2 \rightarrow E_2$, there is now an absorption by CO_2 in the cell, B_2. If the CO_2 in A_1 absorbs sufficiently strongly, the situation in the beam path $C_1 \rightarrow E_1$ changes very little, or not at all. Any other absorber in the gas stream changes nothing because it is present in the same concentration in both beams. This is true only in so far as the absorption bands of this other absorber do not overlap the bands of CO_2. If there is overlap then the instrument will register the other absorber, either fully or par-

tially, as CO_2. To circumvent this problem, the other absorber is placed in sufficient quantity in both the cells C_1 and C_2. Now light of the wavelengths absorbed by the other absorber will be completely absorbed in C_1 and C_2, i.e. the cells C_1 and C_2 act as filter cells. Thus, the radiation reaching the cells B_1 and B_2 does not contain any light which is absorbed by the other absorber and in B_1 and B_2 only the absorption due to the gas to be determined is measured. This technique using nonselective detectors, which achieves selectivity by means of the sensitizing cells, A_1 and A_2, is known as the principle of negative filtering. Fundamentally, we have two equal beam paths in one of which all absorbing substances, including the one to be determined, are measured while in the other beam only the substance of interest is measured. By placing another IR absorbing gas in A_1 the instrument can be made selectively sensitive to that gas.
In the positive filtering technique, the selectivity is transferred to the detector. → Pneumatic detectors, with their measuring chambers filled with the gas to be detected as radiation absorber, are suitable for this purpose. The idea has been realized in the → IR gas analyzer URAS. For certain analytical purposes it is valuable to combine these two principles in one instrument. The principle of sensitizing in the beam path is taken over from negative filtering and the selectivity in detection from positive filtering. In 1960, Siemens introduced such an instrument under the name *Inframeter.* Apart from the double-beam, nondispersive instruments described here, single-beam instruments have been designed and are in development.

Ref.: J.C. Wright in *Laboratory Methods in Infrared Spectroscopy*, 2nd ed., (Eds.: R.G.J. Miller and B.C. Stace), Heyden and Son Ltd., London, Philadelphia, Rheine, **1979**, chapt. 6.

Nondispersive spectrometer, *<nichtdispersives Spektrometer, nichtdispersives Spektralphotometer>*, a → spectrometer in which, in contrast to dispersive spectrometers, the measuring light is not dispersed. The oldest nondispersive instrument is the → colorimeter which uses undispersed visible light and the eye as detector to compare colors. A further development is the → filter photometer in which → colored glass filters or interference filters are used to separate out quite large ranges of wavelength and photocells are used as detectors.
Nondispersive analyzers have been developed, particularly in → IR spectroscopy, for the on-line analysis (process analysis) of gases and vapors. They employ the principles of negative and/or positive filtering. The → Fourier-transform IR spectrometer is also a nondispersive instrument.

Nondispersive UV-VIS apparatus, *<nichtdispersive UV-VIS-Geräte, NDUV>*, photometers which use no dispersing element. In contrast to the NDIR instruments where the nondispersed IR radiation is filtered by the components to be determined themselves, (see, for example, positive filtering under → nondispersive infrared equipment) here the selection of spectral region is made using optical filters (→ filter photometer). Filter photometers are therefore significantly more flexible than NDIR instruments and have many applications.

Nonlinear optics, *<nichtlineare Optik>*, that part of optics which deals with high intensities and changing optical constants. Intense laser light, for example, propagates in a medium in a way different from that of the weak light from conventional light sources. In classical or → linear optics the characterizing optical constants are independent of the light intensity, but this holds only for low intensities. The following phenomena in laser spectroscopy are particularly important in nonlinear optics; → frequency doubling, optical mixing, intensity-dependent transmission and induced Raman scattering.

Nonspectral interference, *<nichtspektrale Interferenz>*, a form of interference in spectrochemical analysis (AAS, AES, AFS) which is caused by the sample itself, in contrast to → spectral interference.

Normal coordinate analysis, *<Normalkoordinatenanalyse>*, the analysis of the vibrational frequencies of a molecule. All displacements of the atoms of a molecule from their equilibrium positions can be described by → displacement coordinates. For this purpose, a 3-dimensional Cartesian coordinate system is attached to every atom. If X is a vector-displacement coordinate and X'' its second derivative with respect to time then all harmonic vibrations of the molecule can be represented by a system of coupled Lagrangian equations:

$$X'' + M^{-1}F_xX = 0. \qquad (1)$$

F_x is the matrix of the force constants and M the matrix of the atomic masses. A clearer picture is given if the atomic movements are described in terms of internal coordinates which represent the vibrations as changes in bond lengths and bond angles. The vector, R, of the internal coordinates can be expressed in terms of the displacement coordinates as:

$$R = BX. \qquad (2)$$

Using the molecular symmetry, the symmetry coordinates, R_s, may be written as orthonormal combinations of the internal coordinates:

$$R_s = UR \qquad (3)$$

The symmetry coordinates transform as irreducible representations of the symmetry point group of the molecule. Thus, the original Lagrangian equation system can be transformed into two sets of equations:

$$R'' + GFR = 0 \qquad (4)$$

$$R_s'' + G_sF_sR_s = 0 \qquad (5)$$

With this transformation, a partial blocking of the force constant matrices, F and F_s, is obtained. The matrices, G and G_s, which are derived from the mass matrix, are blocked by the same transformation. The matrices G and G_s are defined by the equations, $G = BM^{-1}B^T$ and $G_s = UGU^T$. The B matrix is derived from vectors, the S vectors, which point in the directions of the bond length and bond angle changes. The matrices, GF or G_sF_s, are diagonalized and the differential equations uncoupled by means of a transformation from the internal coordinates to the normal coordinates. The normal coordinate vector, Q, is linked to the internal coordinate vector by the L matrix, viz:

$$R = LQ. \qquad (6)$$

The product of L and its transpose gives the G matrix:

$$G = LL^T. \quad (7)$$

and equation (4) can be rearranged to give:

$$Q'' + L^T FLQ = 0 \quad (8)$$

$$L^{-1}GFL = L^T FL = \Omega. \quad (9)$$

The diagonal matrix, Ω contains the eigenvalues of the matrix product, GF, and therefore the vibrational frequencies of the molecule. The eigenvalues, σ, are determined by the condition that the secular determinant must vanish:

$$| GF - \sigma E | = 0 \quad (10)$$

(E is the unit matrix.)
The column vectors of the L matrix are then the eigenvectors of the representation of the normal coordinates in terms of internal coordinates. A direct diagonalization of GF, to obtain a solution of the secular equations, is not usually the method adopted. Normally, F is first diagonalized and a matrix, A, which has the property, $A^T A = F$, is formed. In the next step a matrix, Y, which diagonalizes the product matrix, AGA^T, is determined. Provided that G and F are not singular, there exists a matrix, S, such that $S = A^{-1}Y$ and:

$$S^{-1}GFS = \Omega. \quad (11)$$

To obtain the matrix, L, a weighting matrix, W, such that $L = SW$, must be calculated.
The determination of the force field and the eigenfrequencies and eigenfunctions for the molecule is then carried out iteratively.
A minimization technique, such as the simplex method, is used to fit the calculated and measured eigenvalues while controlling the physical parameters of the calculation such as the signs and ranges of the magnitudes of the diagonal and off-diagonal force constants. Special methods can be used for the solution of the eigenvalue problem in highly symmetric molecules.
Ref.: E.B. Wilson, J.C. Decius, P.C. Cross, *Molecular Vibrations*, McGraw-Hill Book Co., New York, **1955**.

Normal spectrum, <*Normalspektrum*>, a grating spectrum in which the deviation of the individual → spectral colors, produced by the spectral dispersion of white light with a → grating, is always proportional to their wavelengths. Dispersion by a prism gives the spectral colors in the same sequence (though in the reverse direction), but the distribution of the colors in the spectrum depends upon the → dispersion of the prism material.

Normal vibrations, <*Normalschwingungen*>, the total number of vibrations of an N atomic molecule, 3N-5 for a linear molecule and 3N-6 for a nonlinear molecule, which are determined by the number of → degrees of freedom. All the actual movements of the atoms forming a molecular structure are combinations of these normal vibrations, as with a coupled pendulum. Every normal vibration has a particular frequency and, under certain conditions, the frequencies of two different vibrations may be equal. The vibrations are then said to be degenerate.

NQR, <*KQR*> → nuclear quadrupole resonance.

N shell, <*N-Schale*>, the fourth shell of an atom which corresponds to the principal quantum number $n = 4$. Since $l = 0, 1, 2, 3$ and $m_s = \pm 1/2$, it can accommodate a total of 32 electrons in accordance with the → Pauli principle.

Nuclear magnetic resonance spectrometer, <*kernmagnetisches Resonanzspektrometer*> → NMR spectrometer.

Nuclear magneton, <*Kernmagneton*>, the magnetic moment, $\mu_{\rm I}$, of a nucleus with nuclear spin, I, is given by:

$$|\bar{\mu}_I| = g_N \frac{e \cdot \hbar}{2m_p} \sqrt{I(I+1)}.$$

Here, e is the proton nuclear charge, m_p the mass of the proton, g_N the nuclear g factor and $\hbar = h/2\pi$, → Planck's constant divided by 2π.
As the equation shows, the magnetic moment of the nucleus is given in multiples of $e \cdot \hbar/2m_p$. For this quantity the name *nuclear magneton*, μ_N, has been introduced in analogy with the → Bohr magneton:

$$\mu_N = \frac{e \cdot \hbar}{2m_p} = 5.050824 \cdot 10^{-27}\ [\mathrm{A} \cdot \mathrm{m}^2]$$

or $[J \cdot T^{-1}]$

μ_N is smaller than the Bohr magneton by a factor of 1/1836 which corresponds to the ratio of the rest masses of the proton and the electron; $m_p/m_e = 1836$.

Nuclear Overhauser effect, NOE, <*Overhauser-Effekt, dynamische Kernpolarisation*>, an effect first discovered as a result of theoretical investigations by A. Overhauser in 1953. (A. Overhauser, *Phys. Rev.*, **1953**, *91*, 476; **1953**, *92*, 411). Consider a system consisting of a nuclear spin, I, and an electron spin, S; the atomic nuclei of a metal and the conducting electrons, for example. If the electrons are irradiated with a frequency, ν_s, which saturates the electron resonance, while at the same time the nuclear resonance is measured, an increase in the intensity of the nuclear resonance signal is observed. The effect was demonstrated with alkali-metal atoms dispersed in a passive medium, e.g. ^{7}Li in glycerine and ^{23}Na in ammonia. For an interpretation of the effect, the Solomon diagram for the eigenstates of the two-spin system, IS, in a magnetic field may be used. In total there are four energy states, as can be readily seen from Figure 1. The states, 1 and 2 with $S_z = -1/2$, are the lowest in energy; they correspond to the spin combinations $\alpha\beta$ and $\beta\beta$. The selection rule, $\Delta m = \pm 1$, shows that the transitions, $1 \leftrightarrow 3$ and $2 \leftrightarrow 4$, should be assigned to the ESR and, $1 \leftrightarrow 2$ and $3 \leftrightarrow 4$, to the NMR transitions. The Solomon diagram, Figure 2, shows these four energy levels in a diamond-

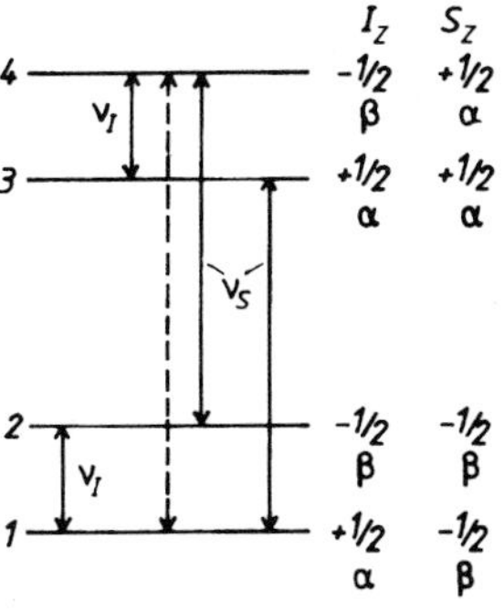

Nuclear Overhauser effect. Fig. 1. Energy-level scheme for the interaction of nuclear and electron spins

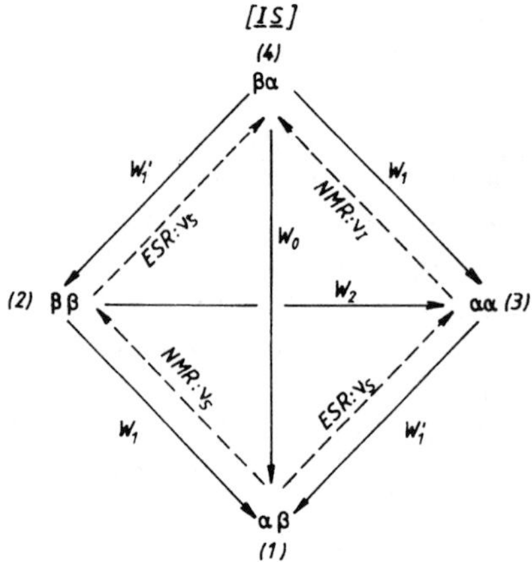

Nuclear Overhauser effect. Fig. 2. Solomon diagram for the interaction between nucleus and electron

shaped pattern and indicates, for the individual paths, the probabilities of the relaxation processes which restore the Boltzmann population after excitation. W_1 and W_1' are the longitudinal relaxations of the nuclear and electron spin respectively. The transition probabilities, W_2 and W_o, correspond to the relaxation processes in which the electron spin and the nuclear spin flip over simultaneously. They are only important when there is a spin-spin interaction between I and S.

If the ESR transitions, $1 \rightarrow 3$ and $2 \rightarrow 4$, are now saturated by irradiation with a field, B_1, of frequency, v_s, then the Boltzmann distribution between the states 1 and 3 and between 2 and 4 is disturbed, i.e. 3 and 4 have too high a population whilst for 1 and 2 it is too small. This perturbation can be alleviated by an increase in the number of relaxation transitions, i.e. by an increase in W_o. This decreases the population of state 4 and increases that of state 1. This increases the intensities of the NMR transitions, $1 \rightarrow 2$ and $3 \rightarrow 4$ because state 2 ($\beta\beta$) becomes less populated and state 3 more populated because the spin population is transported along the path, $2 \rightarrow 4 \rightarrow 1 \rightarrow 3$. The result is a polarization of the nuclear spin distribution.

It is important for this argument that $W_o \gg W_2$, i.e. that the most probable process is the simultaneous flipping, in opposite directions, of an electron spin and a nuclear spin; $\beta\alpha \rightarrow \alpha\beta$, or $\Delta m_s = -1$ coupled with $\Delta m_I = +1$. In this case the relaxation is effected by a time-dependent scalar spin-spin coupling. The relative occupations of the nuclear spin levels, N_α/N_β when the electron resonance is saturated is given by:

$$\frac{N_\alpha}{N_\beta} = -\exp\{-h(\nu_s + \nu_I)/kT\}$$

Since $h\nu_s \gg h\nu_I$ (see Figure 1), the distribution is determined by the much larger energy difference $h\nu_s$, i.e. the ground state is more populated than in a normal NMR experiment. Because the intensity of a transition is always proportional to the population difference between the two states involved, there is an increase in the intensity of the NMR signal when the ESR transitions are intensively excited.

If this discussion is now applied to a spin system consisting of two similar nuclei we have the nuclear Overhauser effect or NOE. Here also, a Solomon diagram can be used, but it differs from Figure 2 in that both spins, I_A and I_X, have the same sign and the sequence of the states is not the same, see Figure 3. If the resonance of nucleus, X, is strongly irradiated then the intensities of the lines of nucleus, A, are increased, provided that $W_o \ll$

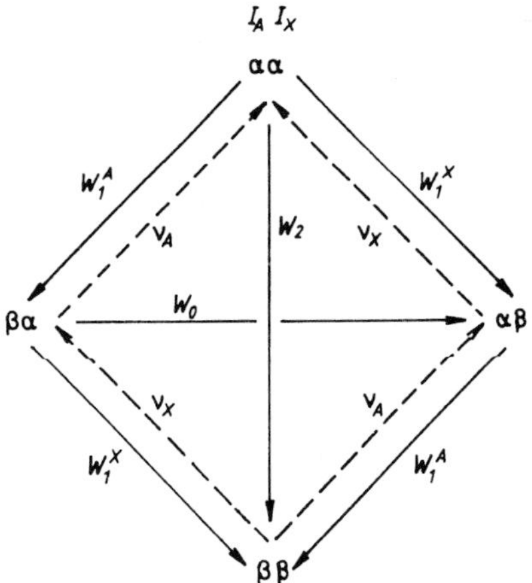

Nuclear Overhauser effect. Fig. 3. Solomon diagram for the nuclear Overhauser effect.

W_2. This condition is always fulfilled when the time-dependent dipolar coupling is reponsible for the relaxation. This corresponds to the mechanism of longitudinal relaxation. The Solomon equation gives the increase in signal intensity as:

$$\frac{M_Z^A}{M_o^A} = 1 + \frac{W_2 - W_o}{2W_1^A + W_2 + W_o} \cdot \frac{\gamma_x}{\gamma_A}.$$

It describes the z magnetization of nucleus, A, $M_Z{}^A$, produced by the extra radio-frequency field at ν_x, relative to the equilibrium magnetization, $M_o{}^A$. For pure dipole-dipole interaction between two nuclei W_2:W_1:W_o = 1:1/4:1/6. A signal enhancement of 50% is found for protons with $\gamma_A = \gamma_X$. For a spin system such as $^1H^{13}C$, on the other hand, the favorable ratio $\gamma_H/\gamma_C = 4$ leads to an enhancement of 200%. Thus, this effect is of great importance for ^{13}C NMR spectroscopy since the intensity of the ^{13}C signal can be considerably increased by saturating the proton resonance in a double resonance experiment (→ double resonance technique).

In connection with the nuclear Overhauser effect, it is also important to note that the contribution of dipole-dipole interaction to the longitudinal relaxation time of two nuclei at distance, r, apart is proportional to r^{-6}. An intramolecular NOE can therefore only be observed if the two nuclei in question are close together in space since only then can the dipole-dipole interaction make a substantial contribution to the relaxation. This connection makes the NOE a valuable tool for certain problems in molecular structure determination.

Ref.: J.H. Noggle, R.E. Schirmer, *The Nuclear Overhauser Effect*, Academic Press, New York, London, **1971**; R.K. Harris, *Nuclear Magnetic Resonance Spectroscopy*, Pitman, London, Marshfield, **1983**.

Nuclear quadrupole resonance, NQR, nuclear quadrupole resonance spectroscopy, NQR, *<Nuklearquadrupolresonanz, Kernquadrupolresonanzspektroskopie, NQR>*. Atoms with a nuclear spin $I \geq 1$ have an electric quadrupole moment. This leads to a large increase in the width of the solution state → NMR lines of such nuclei. In solids, on the other hand, nuclear quadrupole transitions can be measured directly without an external magnetic field. This is because the positions of the nuclei are fixed with respect to the crystal axes and other nuclei. This is not true in solution on account of the statistical distribution and the Brownian motion. Electrons create electric fields at the position of the nucleus which cause a space quantization of the nuclear quadrupole. The different orientations of the quadrupole have different energies so that nuclear quadrupole transitions may be induced by electromagnetic radiation of an appropriate frequency.

In general, the electric field in the solid is not very different from that in the free molecule since it is well-known that covalently bound molecules remain quite well-defined when they aggregate to form a solid and the more distant charges contribute little to the electric field. For a theoretical analysis, the symmetry axis of the inhomogeneous electric field is taken as the z axis of a space-fixed coordinate system and it is assumed that the field gradients in the x and y directions, V_{xx} and V_{yy}, are equal. In such an axially symmetric crystal, an atomic nucleus with the spin quantum number, I, and the quadrupole moment, $e \cdot Q$ has the coupling energy:

$$E_Q = \frac{eQV_{zz}}{4} \cdot \frac{3M_I^2 - I(I+1)}{I \cdot (2I-1)} \equiv \frac{B}{4} \cdot \frac{3M_I^2 - I(I+1)}{I \cdot (2I-1)} \equiv B\frac{3M_I^2 - I(I+1)}{4I \cdot (2I-1)},$$

$B \equiv e \cdot Q \cdot V_{zz}$ is the nuclear quadrupole coupling constant and V_{zz} the field gradient in the z direction. Since the components of the nuclear spin in the direction of the symmetry axis, M_I, are squared in the formula, the values $\pm M_I$ have the same energy. The selection rule is $\Delta M_I = \pm 1$, so that the nuclear spins, $I = 1$ ($M_I = \pm 1, 0$), and $I = 3/2$ ($M_I = \pm 3/2, \pm 1/2$), each have only one transition. For $I = 5/2$ ($M_I = \pm 5/2, \pm 3/2, \pm 1/2$), there are two transitions and for $I = 7/2$ ($M_I = \pm 7/2, \pm 5/2, \pm 3/2, \pm 1/2$), there are three. Figure 1 illustrates this for $I = 3/2$, 5/2 and 7/2. Appropriate examples are given

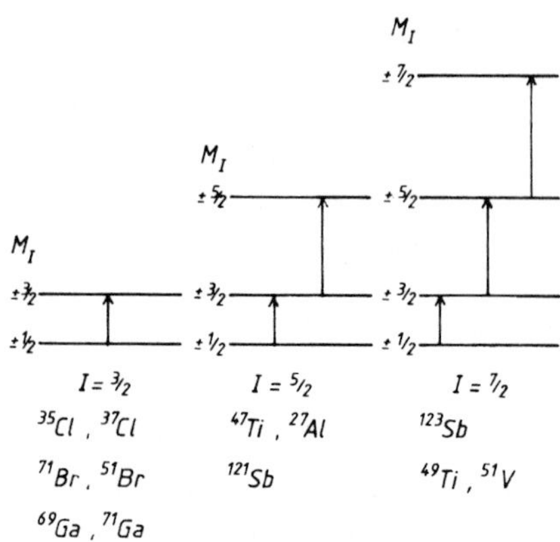

Nuclear quadrupole resonance. Fig. 1. NQR energy levels and transitions for nuclei with $I = 3/2$, 5/2 and 7/2, see text.

under each energy-level scheme. With $\Delta M_I = 1$ for $(M_{I+1} - M_I)$, the resonance frequency, $\Delta \nu_Q = 3B(2|M_I|+1)/4I(2I\text{-}1)$. The first NQR measurements on solid $^{79}Br_2$, $^{127}I_2$, $Al^{35}Cl_3$ and $Al^{37}Cl_3$ were made in the years 1950/1953 by H.G. Dehmelt *et al.* Subsequently, many measurements were made on $Al^{35}Cl_3$, $Al^{37}Cl_3$, $Al^{79}Br_3$, $^{121}Sb^{35}Cl_3$, $^{121}Sb^{37}Cl_3$, $^{123}Sb^{35}Cl_3$, $^{123}Sb^{37}Cl_3$, $^{121}Sb^{79}Br_3$, $^{121}Sb^{81}Br_3$, $^{123}Sb^{79}Br_3$, $^{123}Sb^{81}Br_3$, $Sn^{35}Cl_4$, and $Sn^{37}Cl_4$, especially in the form of their complexes with benzene derivatives and polynuclear aromatic molecules. In this context it is of particular significance that the isotopic nuclei of Cl, Br and Sb occur in almost equal natural abundances. In some cases, the measurements revealed significant deviations of the transitions from what was expected on the grounds of the assumed pure axial symmetry. A perturbation calculation for these cases produced an additional term, an asymmetry factor, η, which is defined as follows: $\eta = (V_{xx} - V_{yy})/V_{zz}$. With the help of the perturbation calculation, R.

Bersohn (*J. Chem. Phys.*, **1952**, *20*, 1505) derived equations which take account of the deviations from axial symmetry for the commonest spins (I = 1, 3/2, 5/2, 7/2 and 9/2). For the three transitions of I = 7/2 the results are:

$M_I = 1/2 \rightarrow 3/2$: $\nu_1 = (1/14)B \cdot [1 + 3.63333\eta^2 - 7.26070\eta^4]$

$M_I = 3/2 \rightarrow 5/2$: $\nu_2 = (2/14)B \cdot [1 - 0.56667\eta^2 + 1.85952\eta^4]$

$M_I = 5/2 \rightarrow 7/2$: $\nu_3 = (3/14)B \cdot [1 - 0.10000\eta^2 - 0.01804\eta^4]$

The figures 1/14, 2/14 and 3/14 are obtained when the values, I = 7/2 and M_I = 1/2, 3/2 and 5/2 are inserted into the above equation for $\Delta\nu_Q$. Values in MHz for the $M_I = 1/2 \rightarrow M_I = 3/2$ transition of the nuclei ^{121}Sb and ^{123}Sb have been measured on various Menshutkin complexes. (See H.-H. Perkampus, *Interaction of π-Electron Systems with Metal Halides*, Springer Verlag, Berlin, Heidelberg, New York, **1973**).

Instruments for measuring NQR spectra are not really available commercially. Therefore, most equipment has been developed in the laboratories where it is used. For the above examples, the NQR frequencies lie between 18 MHz and ca. 200 MHz.

Ref.: Yu. A. Buslaev, E.A. Kravčenks, L. Kolditz, *Coordination Chemistry Reviews*, **1987**, *82*, 7; J.A.S. Smith, Ed.: *Advances in Nuclear Quadrupole Resonance*, Vol. 1, Heyden and Son Ltd., London, New York, Rheine, **1974**.

Nujol technique, <*Nujol-Technik*>, a widely used method for the measurement of difficultly soluble substances in the infrared. The substance is dispersed in paraffin oil (nujol); the technique may therefore be described as a dispersion method. The principle of the method is to suspend the solid, as finely divided as possible, in an extensively nonabsorbing liquid which supports dispersions. High viscosity and good wetability by the suspending fluid aid the stability. The suspending fluid must not be a solvent for the sample since then spectra of the solid and solution would be superimposed, and this would make interpretation difficult.

Apart from nujol, hexachlorobutadiene (HCB), perfluorokerosene (Flurolube) and polymeric trifluoromonochloroethylene (KeLF10 R) are used for this purpose. The dispersions formed are usually termed *mulls*.

Ref.: W.J. Price in *Laboratory Methods in Infrared Spectroscopy*, 2nd ed., (Eds.: R.G.J. Miller and B.C. Stace), Heyden and Sons Ltd., London, Philadelphia, Rheine, **1979**.

O

Oblate symmetric top, <*Oblate-Symmetric Top*>, a → symmetric top molecule which is flattened at the poles, i.e. like a discus. Therefore, the moment of inertia with respect to the figure axis, I_A, must be larger than the other two moments of inertia, $I_B = I_C$. Thus, for the rotational constants it follows that $A < B$ (→ top molecules). The planar benzene molecule, shown in Figure 1 with the inertial axes, is an example.

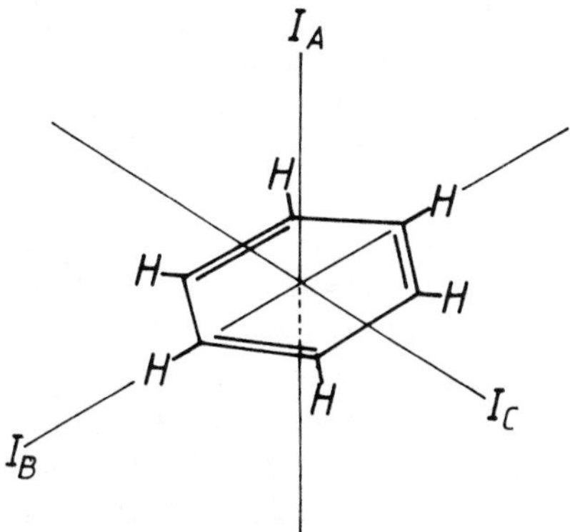

Oblate symmetric top. Fig. 1. The example of benzene

O branch, <*O-Zweig*> → rotation-vibration spectrum and → Raman effect.

One hundreth-width, <*Hundertstelwertbreite*>, $\Delta\lambda_{1/100}$, the width of the transmittance curve between the wavelengths for which the transmittance has been reduced to 1/100 of that at λ_{max}, τ_{max}. It is used, in addition to the one tenth- and one thousandth-width values to characterize → interference filters.

One tenth-width, <*Zehntelwertbreite*>, $\Delta\lambda_{1/10}$, the width of the transmittance curve between the wavelengths for which the transmittance has been reduced to 1/10 of that at λ_{max}, τ_{max}. It is used in relation to the → half-width to define the Q value which is a measure of the properties of a filter (→ band-pass filter).

One thousandth-width, <*Tausendstelwertbreite*>, $\Delta\lambda_{1/1000}$, the width of the transmittance curve between the wavelengths for which the transmittance has been reduced to 1/1000 of that at λ_{max}, τ_{max}. It is used in relation to the → half-width to define the q value which is a measure of the properties of a filter (→ band-pass filter).

On-line photometer, process photometer, <*Betriebsphotometer, Prozeßphotometer*>, an instrument used for the continuous control of chemical processes, e.g. for analyzing flue gases and the emission from coking plants and motor vehicles, or for monitoring the quality of the air and the water supply to boilers. On-line photometers are designed according to the physical principles of nondispersive absorption photometers (→ nondispersive infrared equipment) and this restricts their application to given specific control problems. In most cases they are used for the analysis of gases. The → IR gas analyzers from Hartmann & Braun (*URAS*) and Siemens (*Ultramat*) are examples of nondispersive IR instruments (NDIR instruments). The Hartmann & Braun → *RADAS 1 G* is an NDUV instrument. The Polymetron *Silkostat*, an on-line analyzer for monitoring the silicic acid in boiler-feed water, can be described as a NDVIS instrument. Depending on the components chosen, the Perkin-Elmer on-line photometer → *Spectran 647* can be used in the region from 400 to 20,000 nm. All these on-line photometers have one thing in common; the sample to be analyzed (gas or liquid) passes continuously through the sample cell or cuvette so that continuous monitoring is possible.

Optical anisotropy, <*Anisotropie, optische*>, a natural optical property of a crystal belonging to a noncubic system. This behavior can be artificially induced in optically isotropic media, e.g. glass, by deformation or temperature differences (→ deformation birefringence). The property is characterized by the fact that the velocity of light is different for different directions in the anisotropic medium. In addition, apart from special directions, the light is always split into two linearly polarized beams the vibrations (polarizations) of which are perpendicular to each other. There is a close relationship between the phase velocities, or refractive indexes, of the two beams and the wave normal which is represented by the → index ellipsoid. This phenomenon in crystal optics is known as optical → birefringence because an incoming light beam, on traversing the crystal, is split into two beams the wave normals of which, in general, experience different refraction. The absorption of the light beams traveling in different directions can also be different in optically anisotropic crystals; the two beams described above can even experience a different absorption. Because these crystals show different colors, depending upon the direction in which the light passes through them, this multicolored phenomenon is known as pleochroism.

Optical axis angle, <*Achsenwinkel, optischer*>, the angle between the two optic axes of an optically biaxial crystal. It has a specific value for every crystal type, though it can be changed by external factors such as temperature, pressure and the wavelength of light. It is normally used in the identification of crystals by optical methods.

Optical brightener, <*optischer Aufheller, optisches Bleichmittel, Weißtöner*>. It is well known that white textiles take on a yellowish tinge after long usage, i.e. the → pigmentation of the material appears yellow to the observer in daylight. The reason for this effect is the fact that the blue portion of the daylight has been lost to absorption and is not reflected. However, if the textile fibers are coated with a substance which absorbs in the UV and fluoresces in the blue, then the blue fluorescence is superimposed on the yellow reflected light and the → complementary colors, blue and yellow, give the impression of white.
In practice, the textiles are impregnated with the brightener during laundering. Stilbene derivatives are widely used for this purpose. They absorb below 360 nm and, in general, show an intense blue-violet → fluorescence.

Optical cement, <*optischer Kitt*>, a cement used to construct optical components from individual elements. The fine cements used to join optical surfaces and which actually lie in the beam

are especially important. They must have a uniform, high transparency in the spectral region in question and be free from turbidity. Their refractive indexes must be suited to the glass to which they must adhere well, while remaining sufficiently elastic so that the bond is not disrupted by rapid changes of temperature. They must solidify without shrinking so that no strains are introduced.
Canada balsam, a light yellow resin obtained from various Canadian pine trees, is the most important optical cement. Its refractive index, n = 1.53–1.55, corresponds to that of most crown glasses (→ glass, optical). The various qualities obtainable differ in their hardness and therefore in their elasticity and softening points, ca. 60–70°C. The elements to be joined are warmed and a drop of resin is placed on one surface. Then, by pressure and movement, the surfaces are brought together such that a very thin, bubble-free, cement layer is formed.
Canada balsam is not very resistant to mechanical and thermal shock. Therefore, in recent years cements composed of two-component adhesives (epoxy resins) have been introduced. The long curing period of such adhesives requires that the optical elements are supported during this time to preserve their alignment. The Merck fine epoxy resin, → Optistick®, which can be prehardened with UV radiation, has proved to be very good. The refractive index, n = 1.55, also corresponds to crown glass. The transparency of cements, especially that of Canada balsam, deteriorates rapidly below 340 nm. Therefore, special cements are required for the UV region, e.g. dimethylpolysiloxane. Glycerine, which is used as a liquid, has good transparency to below 200 nm and silicone oils have proved valuable, especially for joining large surfaces.

Optical density, <*optische Dichte*>, the → Bouguer-Lambert-Beer law, $A = \varepsilon \cdot c \cdot d$, shows that $A/d = D$ is a quantity independent of the path length d. D is the optical density. In the older literature, the term *optical density* was frequently used for A, but the correct word for A is now absorbance. In → photoacoustic spectroscopy, the quotient A_n/d, where $A_n = ln(I_o/I)$, is called the absorption coefficient, β. D and β have the units cm^{-1}.

Optical electron, <*Leuchtelektron*>, the external electron of an atom the excitation of which leads of an emission of electromagnetic radiation. The color of the emission can be recognized directly for many atoms.

Optical fiber, <*Lichtleitfaser*> → fiber optics.

Optical glass, <*optisches Glas*> → glass, optical.

Optical glass filter, <*optisches Glasfilter*> → colored glass filter.

Optically opaque, <*optisch opak, optisch undurchlässig*>, (Lat. opacus = shady), not transparent but translucent. This concept is very important in → photoacoustic spectroscopy.

Optically transparent, <*optisch durchlässig*>. The capacity of a specific substance to absorb light, varies with the wavelength across the visible spectral region. Both → optically opaque wavelength ranges and also regions which are transparent occur. → Colored glass filters exploit this fact. Whether a solid body or a solution is optically transparent or opaque depends not only upon the wavelength but also upon the path length; and in the case of solutions, on the concentration of the solute (→ Bouguer-Lambert-Beer law).

Optically uniaxial crystal, <*optisch einachsige Kristalle*>, the simplest anisotropic crystals which are very important for many applications in spectroscopy. Optically uniaxial crystals have a very high symmetry and a principal crystallographic axis which, because of its symmetry, defines an optically preferred direction in the crystal. In all such crystals the → refractive index depends upon the directions of propagation and polarization of the light. But there is one special direction in the crystal for which this is not true and the refractive index is independent of the direction of polarization. This preferred direction is the optic axis, or *c* axis, of the crystal. When a Cartesian coordinate system is used, the optic axis is frequently orientated in the *z* direction.

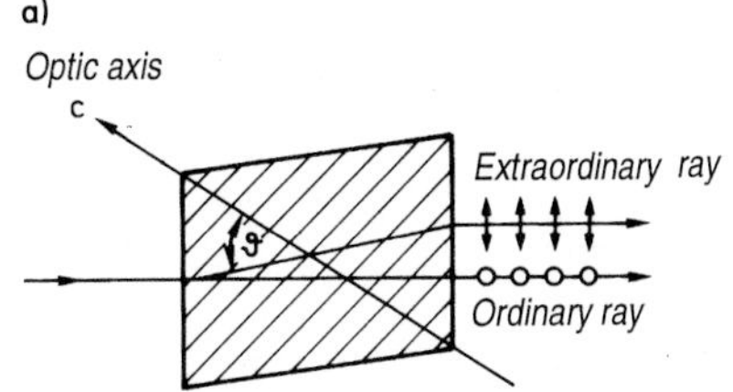

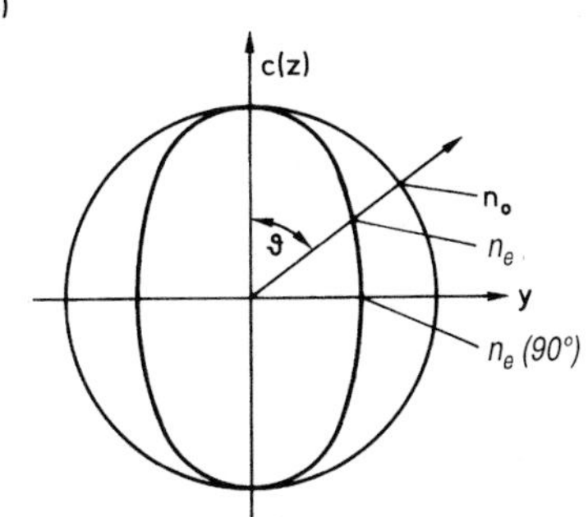

Optically uniaxial crystals. Fig. 1. a) Ordinary and extraordinary rays, refractive indexes, n_o and n_e; b) the dependence of n_o and n_e upon propagation direction, ϑ

If a light beam is incident on the crystal at an angle, ϑ, to the optic axis, the refractive index depends upon the plane of polarization. A light ray, polarized perpendicular to the optic axis, is called an ordinary ray and is characterized by the refractive index, n_o. A ray polarized perpendicular to the ordinary ray is termed an extraordinary ray and characterized by the refractive index, n_e (Figure 1a). The refractive index n_o is independent of the propagation direction, ϑ, of the light in the crystal, but n_e varies with the angle ϑ between n_o and a minimum value. This is illustrated in Figure 1b. n_o describes circles in the *zy* or *zx* planes (a sphere in space) while $n_e = f(\vartheta)$ describes an ellipsoid in space; hence the name *index ellipsoid*. For $\vartheta = 0$, $n_o = n_e$, i.e. this is the direction in which the incident light is parallel to the optic axis and for which the refractive index is independent of the direction of polarization.

If an unpolarized light beam is incident upon the crystal at an angle, ϑ, to the optic axis, as shown in Figure 1b, it is split, because of the two different refractive indexes, into two beams polarized perpendicularly to each other. In this medium, these beams propagate with different velocities, $c_e = c/n_e$ and $c_o = c/n_o$, where c is the velocity of light in a vacuum. This is the basis of the → double refraction of

Optically uniaxial crystals. Table.

Crystal	λ [nm]	n_o	n_e
Calcite	589.2	1.6584	1.4864
Corundum (Al_2O_3)	589.2	1.7682	1.6598
Quartz (SiO_2)	546.0	1.5462	1.5554
Rutile (TiO_2)	589.2	2.6158	2.9029
Ice (H_2O)	589.2	1.309	1.313
KDP (KH_2PO_4)	530.0	1.5131	1.4711
ADP ($NH_4H_2PO_4$)	530.0	1.5280	1.4819
Lithium niobate ($LiNbO_3$)	530.0	2.325	2.232
Potassium sulfate (K_2SO_4)	589.2	1.4550	1.5153

optically uniaxial crystals. Here also n_o and n_e are dependent upon the wavelength (wavenumber) of the light.
Because of their properties uniaxial crystals find application in the construction of → polarizers (→ Nichol, → Glan-Thompson prism etc.). KDP crystals are used in → Lyot filters and for → frequency doubling.
Ref.: E.E. Wahlstrom, *Optical Crystallography*, 4th ed., J. Wiley and Sons, Inc., New York, London, Sydney, Toronto, **1969**.

Optical null principle, *<optischer Nullabgleich>*, a technique in double-beam spectrometers whereby the energy difference, at the detector, between the sample and reference beams, is continually maintained at zero (optical null) by means of an adjustable aperture or comb in the reference beam. Thus, in regions where the sample absorption increases the comb is moved into the reference beam (or the aperture is closed) to reduce the light throughput until it is equal to that coming through the sample beam. When the absorption of the sample decreases, the comb is withdrawn from the beam. The comb is driven by a servomotor which is also coupled to the pen of a chart recorder. If the wavelength drive of the instrument is also coupled to the paper feed of the recorder, an IR spectrum can be recorded. This method, once common in IR spectrometers, has now been superceded by → ratio recording.

Optical penetration depth, penetration depth, *<optische Eindringtiefe, optische Tiefe>*, the inverse of the absorption coefficient, β in cm^{-1}, symbol, ℓ_β, (units cm). β is defined by $\beta = A_n/d = 2.303 \cdot \varepsilon \cdot c$ where A_n is the absorbance, d the path length in cm, ε the molar decadic → extinction coefficient and c the concentration. According to the → Bouguer-Lambert-Beer law, $A_n = \ln(I_o/I) = 2.303 \cdot \varepsilon \cdot c \cdot d = \beta \cdot d$, so that $I/I_o = \exp(-\beta d)$. Thus, when $d = \ell_\beta$, $I/I_o = \exp(-1)$, i.e. at the position $\ell_\beta = 1/\beta$, the intensity of the irradiating light has been reduced to I_o/e, one e-th of its original intensity. The optical penetration depth is especially important in → photoacoustic spectroscopy.

Optical pumping, *<optisches Pumpen>*, the excitation of the → active medium to establish the required → population inversion in a laser; e.g. the use of high-intensity light to populate a state, E_3, which lies above the upper laser level, E_2. From E_3 the system falls to E_2 by a radiationless transition; a mechanical analogy would be that the excitation energy flows from the upper to the lower level. The exciting light acts like a pump which pumps the atoms or ions of the active medium over a higher lying level and into the lower lying laser level. Hence, the use of the terms *optical pumping* and *pump light*. In the process described above, the energy-difference, $E_3 - E_2$, is given up as heat to the active medium, and therefore represents an energy loss.

Optical resonator, <*optischer Resonator*>, a laser resonator the main function of which lies in the extended lifetime of the photons in the laser medium (→ active medium). The probability of induced emission, which is decisive for the start of the laser process, increases proportionally to the lifetime.

In its simplest form, an optical resonator consists of two parallel, flat mirrors, facing each other with a separation of L, the resonator length (see Figure 1). Usually, the resonator length is very long when compared with the actual diameter $2a$. This distinguishes an optical resonator from → interference filters and → Fabry-Perot interferometers, the lengths of which are generally very short when compared with the mirror diameter ($L < 2a$). Light within an optical resonator is constantly reflected to and fro between the mirrors, if the light wave falls vertically upon the mirror surface. However, at each reflection a fraction, $\delta_R = 1 - R$, where R is the reflectivity of the mirror, is lost. The time taken by the light to travel the distance L is $t_o = L/c$. Due to the reflection at the mirrors, the light is concentrated in the resonator volume over a longer time which can be expressed as an increase in the transit time t_o.

$$t = \frac{L}{c} \cdot \frac{1}{1-R} = t_o \frac{1}{1-R} = t_o \cdot \delta_R^{-1}$$

According to this expression, it may be expected that the lifetime of the photons in the resonator can be significantly increased, provided that the reflection losses, $1-R$, can be sufficiently reduced. However, this is possible, in principle, only as far as a limit which is determined by the losses due to diffraction, δ_B. In addition, there are the alignment losses, δ_A, which arise if the two mirrors are not exactly plane-parallel to each other. The accuracy of adjustment required is $\beta < 10^{-4}$

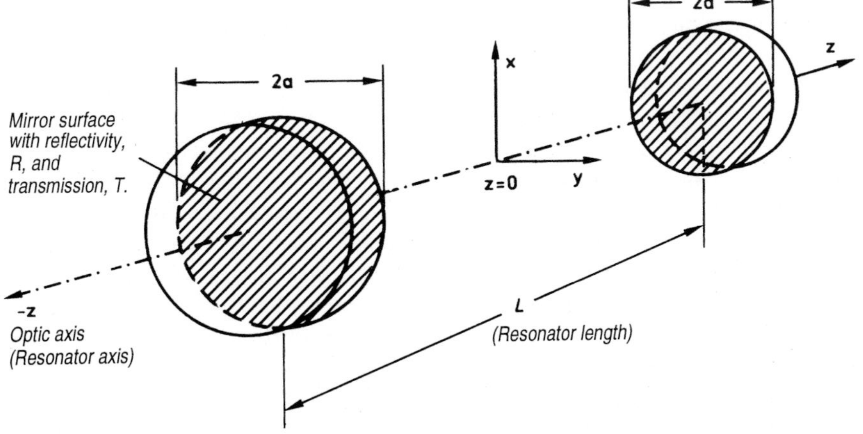

Optical resonator. Fig. 1. The simplest form of an optical resonator

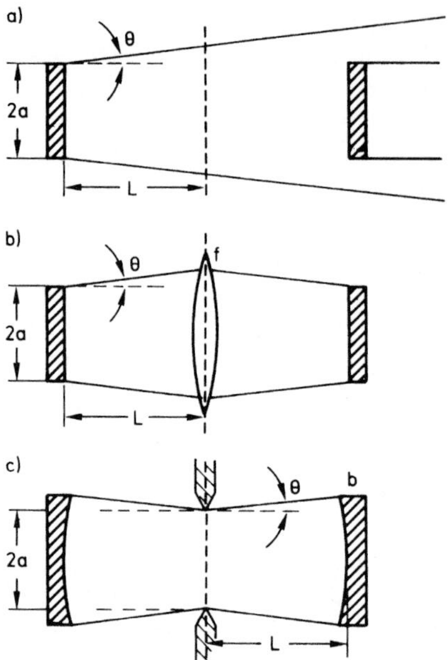

Optical resonator. Fig. 2. In b) and c) confocal resonators are compared with the simple form in Figure 1a

radians, where β is the deviation from planarity. To reduce the diffraction losses, a lens of suitable focal length, f, which focuses the light from one mirror to the other, may be placed in the resonator (see Figure 2b). The same effect may be obtained by replacing the plane mirrors and lens by two concave mirrors with the same imaging properties (see Figure 2c). This type of resonator is called a confocal resonator. Overconfocal and hemispherical or concentric resonators are also used.
It is clear that the plane or concave mirrors used in lasers to reduce light losses must be of the highest quality.

Ref.: D.L. Andrews, *Lasers in Chemistry*, 2nd ed., Springer Verlag, Berlin, Heidelberg, New York, London, **1990**.

Optical rotation, OR, *<optische Drehung, optische Rotation, OR>*, the rotation of the plane of → polarized light in an optically active medium. This effect arises because the velocities of → left- and right-circularly polarized light are not equal in such a medium; circular birefringence. Therefore, the two circular components emerging from the medium have a phase difference, γ, and their superposition gives linearly polarized light which vibrates in a plane rotated by an angle of $\gamma/2$ from that of the incident light. Since the rotation, α, caused by an optically active solid is proportional to the path length, d, of the light through the solid, the rotation per unit length can be defined as the specific rotation (specific rotatory power), $[\alpha]^\lambda$, at the wavelength λ: $[\alpha]^\lambda = \alpha/d$. By general agreement, d, is measured in mm. Values of $[\alpha]^\lambda$ for quartz at 20°C are given in the table for some wavelengths in the visible region.
According to Biot's law (1817), the rotation produced by a solution of an optically active substance in an inactive solvent is proportional to the concentration, c, and the path length, d:

$$\alpha = [\alpha]\frac{c \cdot d}{100}.$$

Since the rotation of solutions is normally considerably less than that of quartz, d is given in decimeters and c is defined as the number of grams of

Optical rotation. Table.

λ [nm]:	760.8	656.3	589.3	486.1	396.8
$[\alpha]^\lambda_{20}$ [deg./mm]:	12.704	17.324	21.724	32.766	51.119

active substance in 100 ml of solution. Thus, the specific rotation, $[\alpha]$, of a solution is the rotation produced by a 1 dm path length through a solution containing 100 g of substance in 100 ml of solution. An optically active substance is sometimes characterized by the molar or molecular rotation, $[m]$, rather than the specific rotation. $[m]$ is defined by:

$$[m] = \frac{M}{100}[\alpha]$$

M is the molecular weight (RMM) and the factor 1/100 is introduced to give more tractable values for $[m]$. The molar rotation is used in stoichiometric comparisons. The specific rotation is a quantity which is characteristic of a substance; it depends upon experimental conditions such as temperature, concentration, solvent and the wavelength of the light.

Optical rotatory dispersion, ORD, *<optische Rotationsdispersion, ORD>*, the dependence of the → optical rotation of an optically active substance, measured as the specific rotation, on the wavelength. In 1817, Biot had already published his observation that, in general, the angle of rotation becomes larger from longer to shorter wavelengths. This behavior of the absolute value of the specific rotation is the normal rotatory dispersion (see Figure 1). It is found over broad spectral ranges, if they are sufficiently distant from the absorption bands of the optically active substance. Drude expressed the rotation in the theoretical form:

$$\alpha = \Sigma_i \frac{A_i}{\lambda^2 - \lambda_i^2}$$

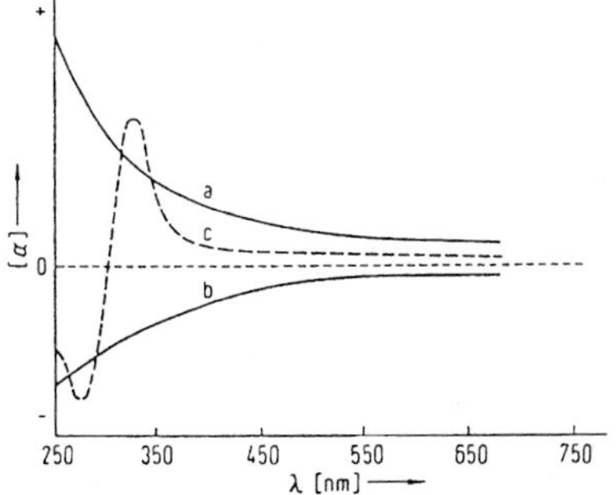

Optical rotatory dispersion. Fig. 1. Normal rotatory dispersion curves, a) and b); anomalous curve, c)

A_i and λ_i are constants, λ_i being the wavelength of the center of gravity of the absorption band. λ is the wavelength of the measurement. From Figure 1 we see that the ORD curves show a positive or negative gradient towards shorter wavelengths for positive (+) or negative (–) rotations respectively.

The wavelength dependence of the specific rotation in the region of an absorption band reveals → circular dichroism. In the absorption region, the normal ORD curve is found to be S-shaped, which is termed anomalous rotatory dispersion, (see Figure 1c). A plot of $[\alpha]$ against wavelength, λ, or wavenumber, $\tilde{\nu}$, is the ORD spectrum. If an ORD curve shows several peaks and troughs it is described as a complex ORD spectrum.

Optical spectroscopy, *<optische Spektroskopie, optische Spektrometrie>*, → spectroscopy in the optical region of the spectrum which is limited to the ultraviolet, visible and infrared → spectral regions. The name arises from the fact that, in these regions the instruments and methods are based upon the classical phenomena of opt-

ics, e.g. → dispersion, → diffraction, the focusing and reflection of light, etc., and also upon the use of optical radiators as light sources.
Instruments for the micro- and radio-wave regions (FIR, ESR and NMR) are based on totally different concepts so that the separation is reasonable.

Optics, <*Optik*> → linear optics, → nonlinear optics.

Optistick®, a fine optical cement, developed by Merck, and based upon a two-component epoxy adhesive. (→ optical cement). Similar products are available from other suppliers, e.g. HE-80 from Eastman.

Optoelectronic components, <*opto-elektronische Bauelemente*>, components in which electromagnetic radiation in the optical spectral region is converted into electrically measurable quantities such as current, voltage, changes of resistance; or vice versa. These devices may be classified according to various criteria, e.g. solid-state and vacuum tube types. The most sensible division would appear to be that of application, since the significance is then immediately clear to the user. The schematic diagram, Figure 1, gives a summary of optoelectronic components. For the sake of completeness, conventional lamps and natural light sources are

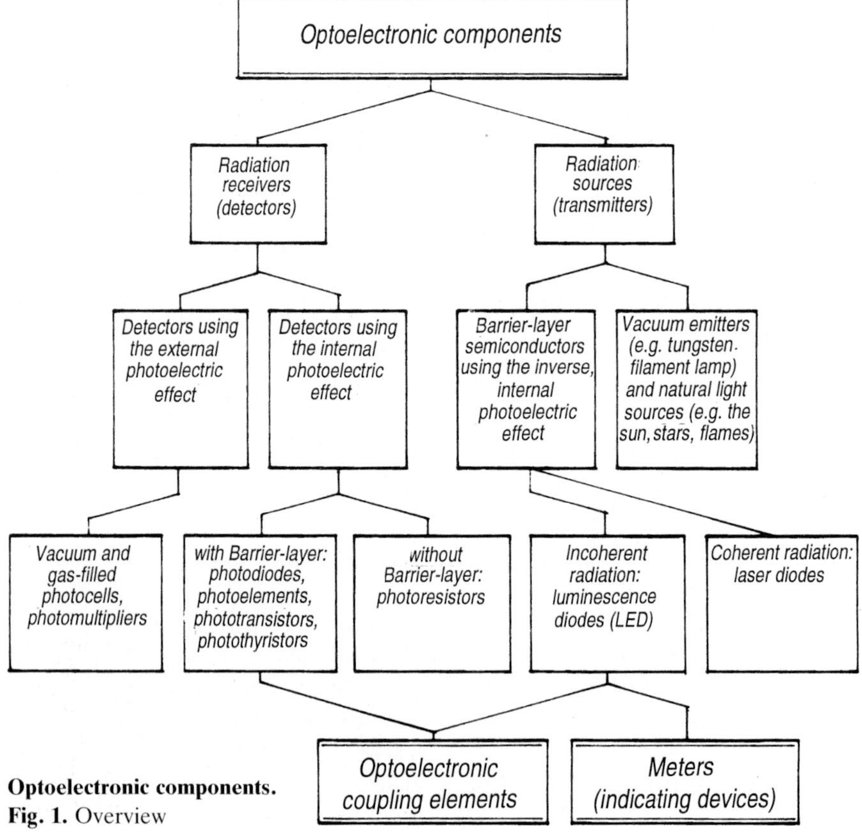

Optoelectronic components. Fig. 1. Overview

included with the radiation emitters, although they do not really belong with the electronic transmitters. We see from this overview that, for example, all radiation receivers (detectors) should be classified as such.
Ref.: J. Wilson, J.F.B. Hawkes, *Optoelectronics: An Introduction*, Prentice Hall International, Englewood Cliffs (NJ), London, Sydney, Toronto, **1983**.

Optoelectronics, *<Optoelektronik>*, a technology in which optical and electronic components are combined together for special tasks. Optoelectronics embraces all technologically valuable interactions between optical radiation and electronic processes, including the ways in which they can be integrated in suitable circuits and made accessible for technical applications.

OR → optical rotation.

Orbital angular momentum quantum number, *<Bahndrehimpulsquantenzahl>* → electronic angular momentum.

ORD → optical rotatory dispersion.

Ordinal number, *<Ordnungszahl>*.
1. For a diffraction grating; the order of the diffracted radiation. (→ grating and → grating monochromator).
2. The number of an element in the periodic system (atomic number), also the number of protons in that element (nuclear charge).

Ordinary ray, *<ordentlicher Strahl, ordinärer Strahl, ordentliche Welle>*. When light is refracted by an → optically uniaxial crystal, e.g. → calcite, it is split into two linearly polarized rays (→ birefringence). The ordinary ray is that one of the pair which, in contrast to the → extraordinary ray, obeys Snell's law of refraction. The ordinary ray propagates with the same velocity in all directions in the doubly refracting crystal and the wave front is spherical.

ORD spectrometer, *<ORD-Geräte>* → spectropolarimeter.

Orientation birefringence, *<Orientierungsdoppelbrechung>*, optical double refraction which occurs in solids, especially high polymers, as a result of the orientation of molecular chains, or the prescence of whole crystalline regions. The orientation can be brought about by the effect of external forces, e.g. stretching. If the crystalline regions are initially birefringent, i.e. they show an intrinsic double refraction which depends upon the structure of the region, and if they are also birefringent on account of their form, i.e. nonspherical, then mechanical stress and the consequent orientation of these regions can give rise to a relatively large birefringence. In contrast to → deformation birefringence, orientation birefringence is time-dependent because it takes a finite time for the molecular chains to orientate themselves. This relaxation phenomenon can give a certain amount of information about the visco-elastic behavior of high polymers. By measurement of the relaxation, an insight into the reordering processes of the molecular chains can be obtained, for example. Since these materials, when mechanically stressed, show both deformation and orientation birefringence, it has become the practice to call the sum of the two stress birefringence.

Oscillator, harmonic oscillator, <*Oszillator, harmonischer Oszillator*> → vibrational spectrum.

Oscillator strength, <*Oszillatorenstärke*>, according to classical dispersion theory, the number of virtual oscillators equivalent to a transition from the *l*th to the *k*th electronic state in quantum theory. It can be viewed as a correct measure of the intensity of an absorption band and determined from the → integral absorption, A_g, of the band. For solutions:

$$f_{l,k} = \frac{10^3 mc^2}{\pi e^2 N_L n} 2.303 \int_{\text{Band}} \varepsilon(\tilde{\nu}) d\tilde{\nu} = \frac{4.32 \cdot 10^{-9}}{n} \int_{\text{Band}} \varepsilon(\tilde{\nu}) d\tilde{\nu} \qquad (a)$$

The constants may be collected together to give a value of $4.32 \cdot 10^{-9}$. The refractive index, *n*, is usually set equal to one. The experimentally determined oscillator strength represents the connection to theory since it is linked to the Einstein *B* coefficient (→ Einstein coefficients) and to the electronic → transition dipole moment. The relevant relations are:

$$f_{l,k} = \frac{mh\nu_{lk}}{\pi e^2} B_{l,k} \qquad (b)$$

$$f_{l,k} = \frac{8\pi^2 mc\tilde{\nu}_{lk}}{3he^2} G \mid \bar{M}_{l,k} \mid^2 \qquad (c)$$

$$f_{l,k} = 4.70 \cdot 10^{29} \tilde{\nu}_{l,k} \cdot G \mid \bar{M}_{l,k} \mid^2 \qquad (d)$$

$\nu_{l,k}$ and $\tilde{\nu}_{l,k}$ are the frequency and wavenumber respectively, of the electronic transition. *G*, the statistical weight, is one for a singlet-singlet transition, $S_o \rightarrow S_p$. The constants in equation (c) are collected together to give $4.7 \cdot 10^{29}$ in equation (d). $|\overrightarrow{M}_{l,k}|$ is the mean transition dipole moment which is defined without reference to the vibrational transitions. Where vibrational excitations are superimposed upon an electronic transition, each vibrational component band may be characterized by its oscillator strength, $f_{l,o\rightarrow k,n}$ as:

$$f_{l,o\rightarrow k,n} = \frac{f_{l,k} \cdot \tilde{\nu}_{l,o\rightarrow k,n}}{\tilde{\nu}_{l,k}} \cdot \mid S_{l,o\rightarrow k,n} \mid^2$$

with $n = 0, 1, 2, 3 \ldots$
$S_{l,o\rightarrow k,n}$ is the vibrational overlap integral; its magnitude determines the intensity of the component bands in the vibrational structure of a UV-VIS absorption band (→ Franck-Condon principle). If the → integral absorption in equation (a) is replaced by the approximate form:

$$\frac{\pi}{2} \varepsilon_{\tilde{\nu}\max} \Delta\tilde{\nu}_{1/2},$$

an estimate of the magnitudes of oscillator strengths can easily be obtained. For $\varepsilon_{\max} = 5 \cdot 10^4 \text{ l} \cdot \text{mol}^{-1}$ and $\Delta\tilde{\nu}_{1/2} = 5000 \text{ cm}^{-1}$, equation (a) gives a value of $f \cong 1.72$. In the literature the factor $\pi/2$ in this estimate is often neglected and the oscillator strengths given are correspondingly smaller. Thus, intense bands, which according to the theory of electronic excitation are generally to be regarded as allowed, have oscillator strengths, $f \geq 1$. Weak bands with $f \ll 1$ (i.e. 10^{-2}–10^{-4}) are forbidden bands.

O shell, <*O-Schale*>, the fifth electron shell of an atom. It corresponds to the principal quantum number, $n = 5$. Since $l = 0, 1, 2, 3, 4$ and $m_s = 1/2$, it can accommodate a total of 50 electrons in accordance with the → Pauli principle.

Overhauser effect, *<Overhauser-Effekt, dynamische Kernpolarisation>* → nuclear Overhauser effect.

Overtone, *<Oberschwingung>*, vibrations which, to a first approximation, are integer multiples of the fundamental vibration. For an → anharmonic oscillator, the selection rule for vibrational transitions is $\Delta\upsilon = 1, 2, 3$ ... which means that overtones, as well as the fundamental, $\Delta\upsilon = 1$ ($\upsilon'' = 0 \rightarrow \upsilon' = 1$) can be excited. But as the anharmonic oscillator model shows, the higher vibrational levels crowd closer and closer together so that the overtones are not integer multiples of the fundamental. $\Delta\upsilon = 2$ gives the first overtone, $\Delta\upsilon = 3$ the second, and so on through the higher overtones. The intensity of the corresponding IR transitions decreases strongly from the first to the higher overtones because of the rapid fall in the transition probability (→ oscillator strength). The → group frequency of the C-Cl bond lies in the range 830–560 cm^{-1} in aliphatic compounds and shows very intense overtones. In carbon tetrachloride, the first overtone is seen at ~ 1550 cm^{-1} and can cause problems when CCl_4 is used as a solvent.

Oxygen, influence of, *<Sauerstoff-einfluß>*. Molecular oxygen can cause problems in → molecular spectroscopy in a number of different ways.

1. Because it absorbs below 200 nm, oxygen sets the lower limit for the use of conventional UV-VIS spectrophotometers. Instruments which measure down to 180 or 160 nm must be flushed with dry nitrogen.
2. The O_2 molecule is paramagnetic. Therefore it facilitates → intersystem crossing and works as a fluorescence → quencher. Thus, solutions for fluorescence measurements must be purged of oxygen.
3. Many molecules in the excited state, especially organic compounds, react with molecular oxygen (photooxidation), and irreversible changes can take place in the solutions to be measured. Here also, the solutions must be purged of oxygen. These side reactions can be particularly troublesome in photochemical work.

P

PAE, *<PAE>* → photoacoustic effect.

Parallel bands, *<Parallelbanden>*, bands in the rotation-vibration spectrum of a → symmetric top molecule for which the change in dipole moment during the vibration is parallel to the → top axis, (figure axis). Vibrations of this type are illustrated in Figure 1, using CH_3-Cl as the example. In this case the selection rules are:

$$\Delta K = 0;\ \Delta J = 0,\ \pm 1,\ \text{when } K \neq 0$$

and

$$\Delta K = 0;\ \Delta J = \pm 1,\ \text{when } K = 0.$$

The corresponding energy-level scheme showing the rotation-vibration transitions of a → prolate symmetric top is given in Figure 2. For $K = 0$, $\Delta J = \pm 1$, and there is only an R branch ($\Delta J = +1$) and a P branch ($\Delta J = -1$). For $K \geq 1$, the Q branch ($\Delta J = 0$) is added so that the → rotation-vibration spectrum of a parallel band is the superposition of all the R, Q and P branches, which are known as subbands. The number of bands is determined by the number of states, K, which are excited in the vibrational state $v'' = 0$. The branches are characterized more exactly by a left superscript, Q, which stands for $K = 0$: ${}^QR$, ${}^QQ$ and ${}^QP$ branches, see Figure 3. The K value of the lower state can be given as a right subscript. Figure 4 shows a typical parallel band for a prolate top, using CH_3Br as the example. In the case of linear polyatomic molecules, bands for which the change of dipole moment during the vibration is along the bond direction are termed parallel bands.

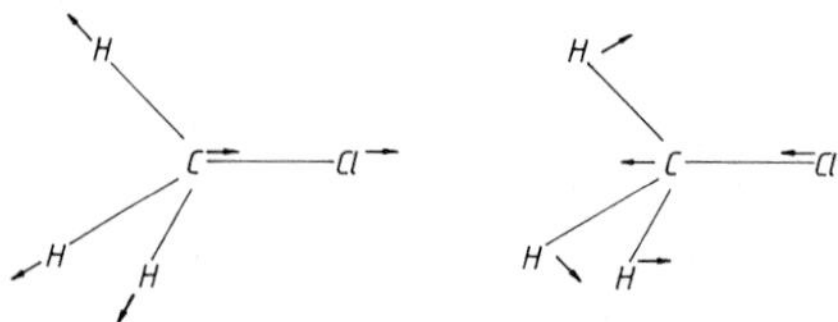

Parallel bands. Fig. 1. Examples of molecular vibrations parallel to the figure and top axis.

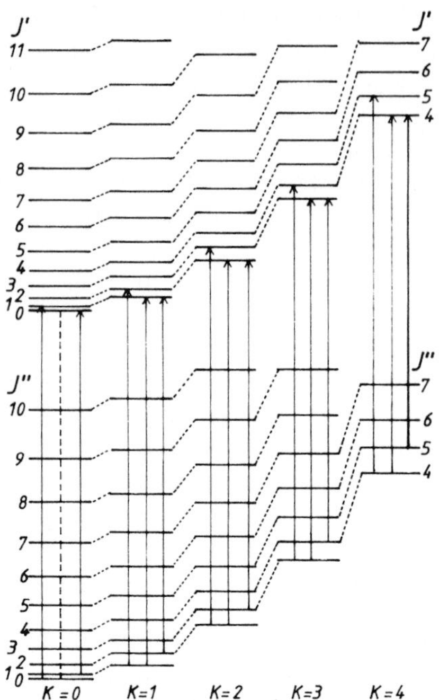

Parallel bands. Fig. 2. Energy-level scheme for a prolate symmetric top.

Partition function, *Z*, *<Zustandssumme, Z>*. The canonical partition function, Z, is defined by the equation:

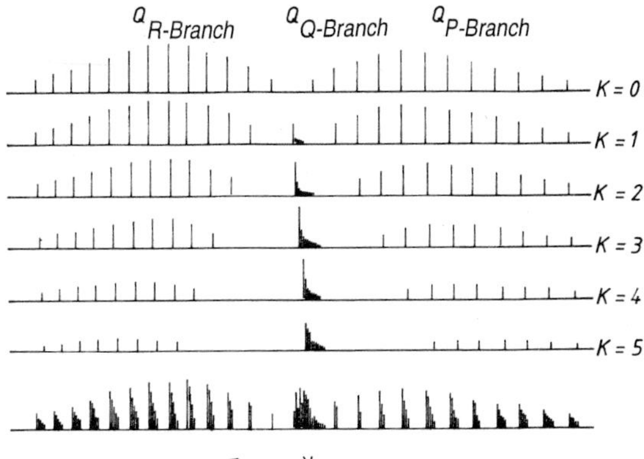

Parallel bands. Fig. 3. Details of the subbands in a parallel band.

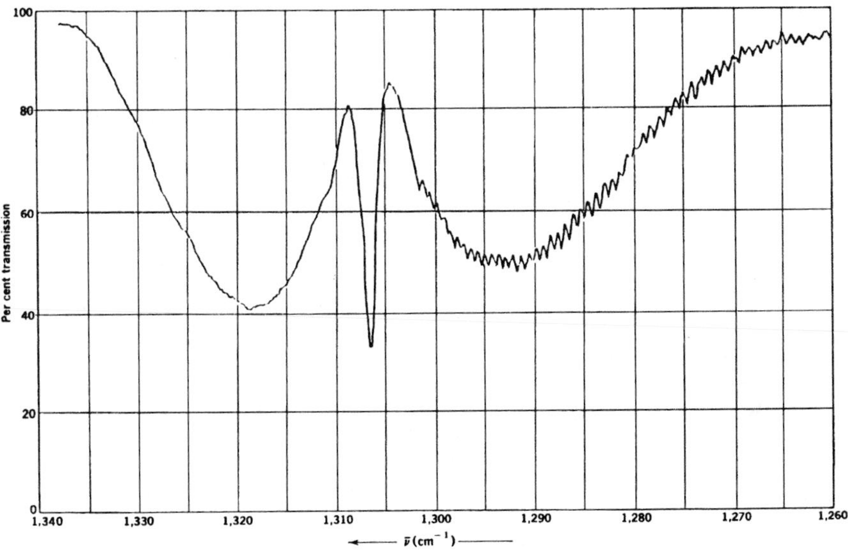

Parallel bands. Fig. 4. A parallel band of CH_3Br.

$$Z = \frac{Q^N}{N!} = \left(\frac{\sum_i g_i e^{-\varepsilon_i/kT}}{N!}\right).$$

$$Q = \sum_i g_i e^{-\varepsilon_i/kT}$$

Q is the molecular partition function in which the sum is to be extended over all the energy states, ε_i, of the atoms or molecules of an ideal gas. N is the total number of particles, generally the Loschmidt number, N_L. The partition function appears as denominator in the Boltzmann distribution (→ Boltzmann statistics). Its significance lies in the fact that it provides the connection between molecular spectroscopy and statistical thermodynamics through the relation, $F = -kT\ln Z$, in which F is the Helmholtz function, $k = R/N_L$ the Boltzmann constant and T the absolute temperature. The total energy, ε_n, of a molecular system at room temperature is a sum of the translational, ε_T, rotational, ε_R, and vibrational, ε_V, i.e. $\varepsilon_n = \varepsilon_T + \varepsilon_R + \varepsilon_V$. The molecular partition function, Q, is then given by:

$$Q = \sum g_n e^{-\varepsilon_n/kT} = \sum g_n e^{-(\varepsilon_T + \varepsilon_R + \varepsilon_V)/kT}$$

in which g_n is the degeneracy of the state of energy ε_n. In terms of the individual contributions:

$$Q = \sum g_{tr} e^{-\varepsilon_T/kT} \cdot \sum g_R e^{-\varepsilon_R/kT} \cdot \sum g_V e^{-\varepsilon_V/kT}$$

or

$$Q = Q_T \cdot Q_R \cdot Q_V.$$

For $F = -kT\ln Z$ this gives

$$F = -kT \frac{\ln Q_T^N}{N_L!} - kT\ln Q_R^N - kT\ln Q_V^N$$

or

$$F = -RT \frac{\ln Q_T}{N_L!} - RT\ln Q_R - RT\ln Q_V.$$

Here the factor $1/N_L!$ is placed with the translational energy sum, Z_T. Thus, the partition function can be written as a product of three factors and the free energy as a sum of three terms.

Using the Schrödinger equation, the partition function for translation can be calculated for a particle of mass, M, the molecular mass, in a cube of side, $V_{mol}^{1/3}$. Expressions for the quantized vibrational and rotational energies must be entered in the vibrational and rotational partition functions. From this it is immediately clear that molecular spectroscopy is of considerable importance to statistical thermodynamics. Because the function, F, can be related to the other thermodynamic functions such as the free energy, G, the internal energy, U, the molecular specific heats, C_p and C_v and the entropy, S, all these quantities can be separated into translational, rotational and vibrational contributions. For diatomic molecules, the rotational partition function, Q_R, can be readily calculated for the rigid rotator (→ dumbbell model) or the nonrigid rotator. This is also the case for the vibrational partition function, Q_V, with the assumption of either a → harmonic oscillator model or an → anharmonic oscillator. For polyatomic molecules, $N \geq 3$, the partition function for each normal mode, i, of frequency ν is

$$Q_{V,i} = \sum_v \exp[-(v+½)h\nu/kT] = \exp(-h\nu/2kT)/(1-\exp[-h\nu/kT]),$$

the total vibration partition function is the product of $3N$-6 ($3N$-5) such terms, i.e. $Q_V = \Pi_i Q_{V,i}$, and the free energy is the sum:

$$F_V = -RT \sum_i \ln Q_{V,i}$$

In calculating the rotational partition function, Q_R, allowance must be made for the type of molecule, i.e. linear, → symmetric, → asymmetric or spherical top.

Ref.: J.H. Knox, *Molecular Thermodynamics*, J. Wiley and Sons Ltd., London, New York, Sydney, Toronto, **1971**.

PAS, <*PAS*> → photoacoustic spectroscopy.

Paschen-Back effect, <*Paschen-Back-Effekt*>, an effect first observed by F. Paschen and E. Back in 1921. On going from a weak to a strong magnetic field, the anomalous → Zeeman effect changes to a normal Zeeman effect, i.e. the normal Zeeman (Lorentz) triplet appears. An exact theoretical treatment of this effect shows that the description of the magnetic field as weak or strong is a relative one. The field is said to be weak when the Zeeman splitting of the individual multiplet lines, $\Delta\tilde{\nu}$, is small compared with the normal multiplet splitting in the absence of a field, $\Delta\tilde{\nu}_o$. $\Delta\tilde{\nu}_o = 17$ cm^{-1} for the transitions, $^2P_{1/2} \rightarrow {}^2S_{1/2}$ and $^2P_{3/2} \rightarrow {}^2S_{1/2}$ of the sodium atom. At 1 T (10^4 Gauss), the Zeeman splitting of the $^2S_{1/2}$ is $\Delta\tilde{\nu}_{\text{norm}} = g \cdot 4.668 \cdot 10^{-1}$ cm^{-1}, so with $g = 2$, $\Delta\tilde{\nu}_{\text{norm}} = 0.934$ cm^{-1}. The field strength for which the Zeeman splitting is of the same order as the normal multiplet splitting, is calculated to be 18 T. For sodium therefore, this field strength would be required to achieve a full Paschen-Back effect. For potassium, with a term separation of 92 cm^{-1}, a field of ~ 98 T would be needed. This simple illustration shows that for multiplets with, $\Delta\tilde{\nu}_o \leq 1$ cm^{-1} a weak magnetic field is capable of producing a full Paschen-Back effect. The relationship may be represented by the ratio, $\nu = \Delta\tilde{\nu}_o/\Delta\tilde{\nu}_{\text{norm}}$. $\Delta\tilde{\nu}_{\text{norm}} = g \cdot 0.4668\ H$ (H in Tesla), so, ν is inversely proportional to H. $\nu \gg 1$ characterizes a weak field, $\nu \ll 1$ a strong field.

Sommerfeld drew a diagram (Figure 1) to illustrate the transition from an anomalous Zeeman effect to the Paschen-Back effect with increasing magnetic field, using the sodium D lines as an example. The diagram also gives the change in the polarization. With increasing magnetic field there is a complete change in the combination of, π and σ components (π = parallel, σ = circular or perpendicular polarization) which arise from the lines of the multiplet in the anomalous Zeeman effect. Some components come together while others loose their intensity. Of the original six plus four

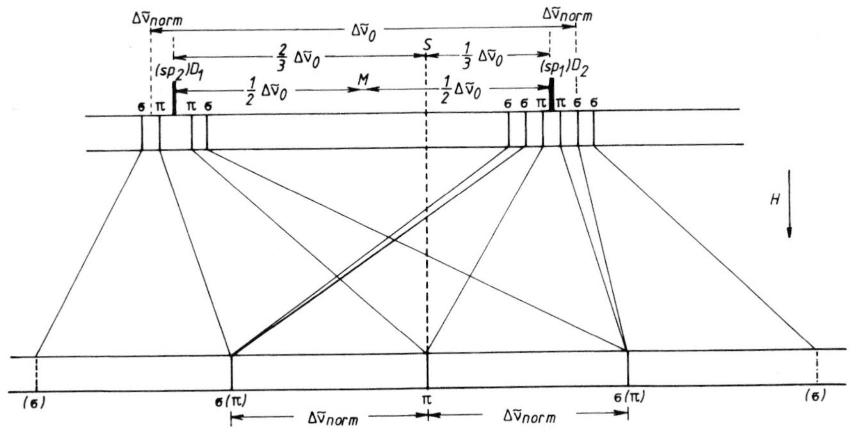

Paschen-Back effect. Fig. 1. Sommerfeld diagram for the transition from the anomalous Zeeman effect to the Paschen-Back effect; Na D lines

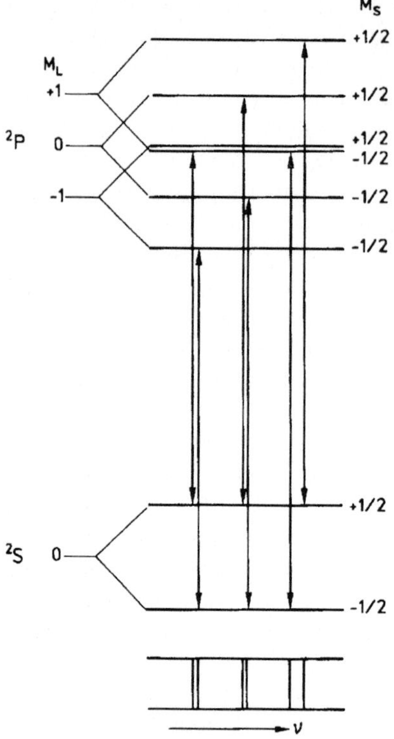

Paschen-Back effect. Fig. 2. Energy-level scheme for the decoupling of L and S by the Paschen-Back effect; Na D lines, $^2S_{1/2} \rightarrow {}^2P_{1/2}$, see text

components at low field, there finally remain just three at high field. These form a normal Zeeman triplet with two outer components, σ, and a central component, π. The π and σ components in brackets then have zero intensity.

The → Zeeman effect can be interpreted in terms of the space quantization of the total angular momentum, J, while for the Paschen-Back effect, the uncoupling of L and S by the field must be considered. The result of the uncoupling is that the resultant orbital angular momentum, L, and the total spin angular momentum, S, are, to a good approximation, independently space quantized according to their components, M_L and M_S. The magnitude of the term splitting is then given by:

$$\Delta\tilde{\nu} = \mu_B M_L H + 2\mu_B M_S H$$

($g = 1$ for L and $g = 2$ for S!)

Application of the selection rules, $\Delta M_L = 0, \pm 1$ and $\Delta M_S = 0$, gives the normal Zeeman triplet for which the energy-level scheme is shown in Figure 2.

Ref.: E.U. Condon and G.H. Shortley, *The Theory of Atomic Spectra*, Cambridge University Press, Cambridge, **1935**.

Paschen-Runge mounting, <*Paschen-Runge-Aufstellung*>, a mounting of a → concave grating in which the grating and radiation detector are fixed on the → Rowland circle (→ Rowland mounting) as in Figure 1. In the direction of the dispersion, every point on

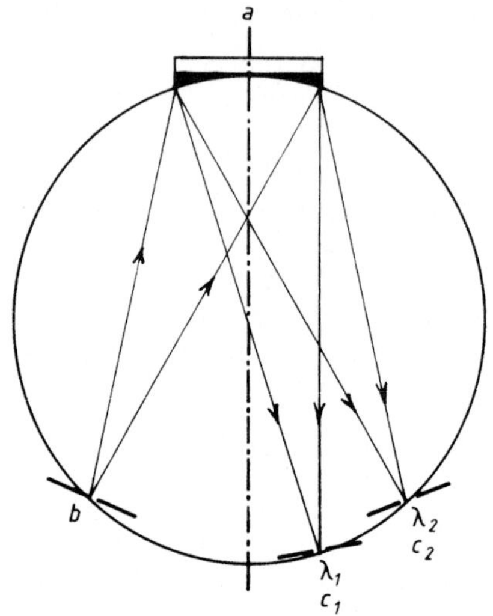

Paschen-Runge mounting. Fig. 1. Paschen-Runge grating mounting

the circle is imaged again at a point on the circle. The spectrum, which is a series of images of the entrance slit, therefore lies on the periphery of the circle, Figure 1. This mounting is used almost exclusively for → polychromators with many exit slits for the simultaneous photoelectric measurement of radiation.

Paschen terms, *<Paschen-Terme>*. The order of the deep energy levels of most known spectra can be interpreted in terms of → Russell-Saunders (spin-orbit) coupling. The values of the total angular momentum, J, so obtained explain the multiplet structure of most spectra. The J splittings of the individual terms are generally small compared to the distance to the next higher term. This type of coupling is also called LS coupling. Deviations from LS coupling occur when the J splitting is comparable with those of L and S. This is the case in heavy atoms and in the later columns of the periodic table (noble gases, Ni, Pd, Pt). In these spectra, the J splitting increases enormously (up to a few thousand cm^{-1}) and the terms appear to move with respect to each other in a random manner. To explain this phenomenon another coupling mechanism must be adopted, jj coupling. The total angular momentum, J, retains its mechanical significance however. For several electrons, jj coupling can be expressed as follows:

$$(l_1s_1)\ (l_2s_2)\ \ldots = (j_1j_2\ \ldots) = J$$

In contrast to Russell-Saunders (RS) coupling, jj coupling is characterized by the fact that the electrons are, to a first approximation, independent of each other. In 1920, Paschen analyzed the neon spectrum in which the coupling is not of the RS type. The ground state of neon has the configuration, $2p^6$, and the ground term has $J = 0$. The higher terms of neon are formed from the configurations, $2p^53s^1$, $2p^54s^1$ ... and $2p^52p^1$, $2p^53p^1$ ... For $2p^53s^1$, since $l_1 = 1$, $l_2 = 0$ and $s_1 = s_2 = ½$, the J values found for jj coupling are J = 2, 1, 1, 0. They are expected to be the first group of levels above the ground state, $2p^6$. In the event, Paschen found four very deep terms which he called s terms (s_2, s_3, s_4, s_5). The next group of neon terms derive from the configuration, $2p^53p^1$. Since $l_1 = l_2 = 1$ and $s_1 = s_2 = ½$, the J values found are J = 3, 2, 2, 2, 1, 1, 1, 1, 0, 0. These give ten levels which Paschen called P_1, P_2 ... P_{10}. In the literature, these terms are called Paschen terms, and the corresponding transitions, Paschen transitions. The Paschen terms play an important role in the interpretation of the laser lines of the → helium-neon laser.

Ref.: A. Sommerfeld, *Atomic Structure and Spectral Lines*, Translator H.L. Brose, E.P. Dutton, New York, **1934**; E.U. Condon and G.H. Shortley, *The Theory of Atomic Spectra*, Cambridge University Press, Cambridge, **1935**.

Path length of a cuvette (cell), *<Schichtdicke einer Kuvette>*, the internal distance between the plane-parallel windows of a cuvette or cell, in cm. (see determination of the path length of a cuvette under → interference spectroscopy).

Pauli [exclusion] principle, exclusion principle, *<Pauli-Prinzip, Pauli-Verbot, Ausschließungsprinzip>*, a principle formulated by W. Pauli in his work on the connection between the completion of electron groups in an atom and the complex structure of the spectra, (*Zeit. f. Phys.*, **1925**, *31*, 765). The exact statement of the principle is: "In an atom, the same quantum state can be occupied by only one electron". The quantum state is more exactly defined by the values of the four quantum numbers, n, l, m_l and m_s. In these terms, the Pauli principle can be stated in another way: "In one and the same atom no two electrons can have the same four values of the four quantum numbers, n, l, m_l and m_s". Since n, l and m_l are uniquely related to each other and there are only two possible values for m_s, $\pm 1/2$, the maximum number of electrons in a state having the principle quantum number, n, is determined exactly. This leads to the *aufbau (building up) principle* and the structure of the periodic system of the elements. It is the quantum-mechanical and therefore also the energetic foundation of that system. The universal character of the Pauli principle is seen in the fact that it applies to all the electrons in any molecule and also to the conduction electrons of an arbitrarily extended metal. Furthermore, it is important in → quantum statistics.

Ref.: P.W. Atkins, *Physical Chemistry*, 5th ed., Oxford University Press, Oxford, **1994**.

P branch, *<P-Zweig>* → rotation-vibration spectrum.

Peak, *<Pik>*, the individual mass lines in → mass spectroscopy. In other branches of spectroscopy, and in gas chromatography, sharp lines or bands are frequently described as peaks.

Pellicle beam splitter, NPC Pellicle, *(NPC = National Photocolor Corporation)*, *<Pellicel>*, a membrane beam splitter with a thickness of 8 μm or 2 μm which has the advantage, in comparison with glass beam splitters, of being largely free of image and refraction errors. The membranes are made of cellulose nitrate supported on a rectangular or circular frame. The ratio of reflection to transmission, R/T, can be set by the treatment of the surface. In the visible region, the R/T ratio depends upon the wavelength, but in the NIR, the value is relatively constant at about 10/90 up to 2400 nm. The curves for R and T cross between 350 nm and 700 nm so that in this region there is a wavelength for which $R/T = 1$. The wavelength at which $R/T = 1$ can be fixed for every pellicle by the surface treatment. Because of the absorption of cellulose nitrate itself, pellicles cannot be used in the UV and IR regions.

Permanent white, (barium oxide), *<Barytweiß>*, powdered barium oxide which is used as a white pigment in paints. It is distinguished by its high reflecting power (~ 90 %), (diffuse → reflection).

Perpendicular band, *<Senkrecht-bande>*, bands in the rotation-vibration spectrum of a symmetric top molecule for which the change in dipole moment during the vibration is perpendicular to the → top axis. Figure 1 shows a vibration of this type in CH_3Cl. In this case the selection rules are: $\Delta K = \pm 1$; $\Delta J = 0, \pm 1$. Figure 2 shows the corresponding energy-level scheme and the rotation-vibration transitions. Only transitions with $\Delta K = +1$ are possible from $K = 0$, so that since $\Delta J = 0, \pm 1$ the well-known pic-

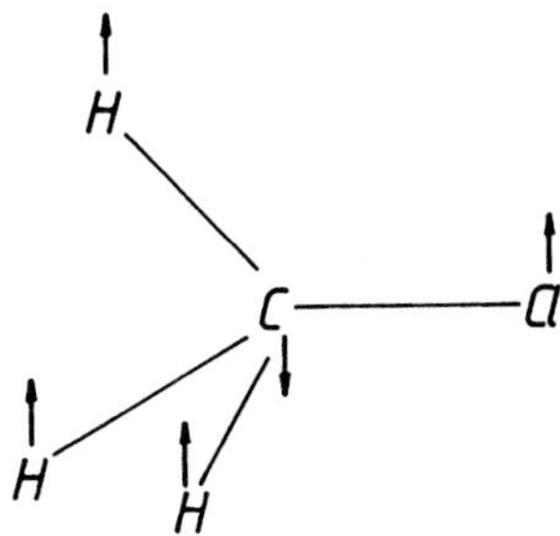

Perpendicular bands. Fig. 1. A vibrational mode of CH_3Cl perpendicular to the top axis

ture with *P*, *Q* and *R* branches is obtained. Note that the first line of the *Q* branch derives from the lower level $J'' = 1$, since all the states with $J < K$ are missing in the *K* states. Since $\Delta K = \pm 1$, when $K \geq 1$ two sets of *P*, *Q* and *R* branches, quite widely separated in energy, are possible. The $\Delta K = +1$ structure lies at higher wavenumbers than ν_o and the $\Delta K = -1$ structure at lower wavenumbers. Thus, for every value of *K* there are two series of bands; known as → subbands. The resulting IR spectrum is a

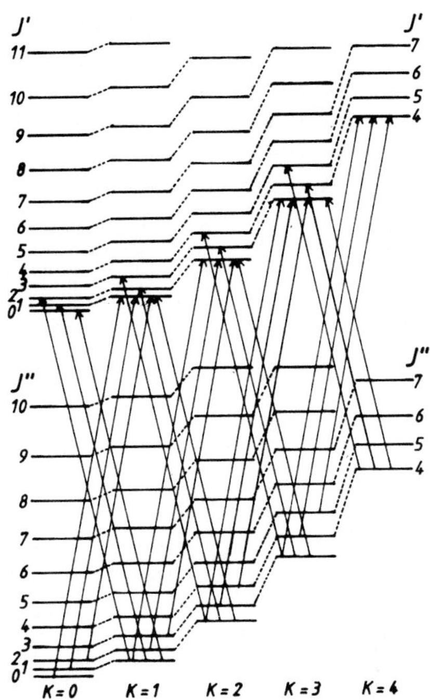

Perpendicular bands. Fig. 2. Energy-level scheme for the rotational transitions of a prolate symmetric top

superposition of all the subbands as shown schematically in Figure 3. The

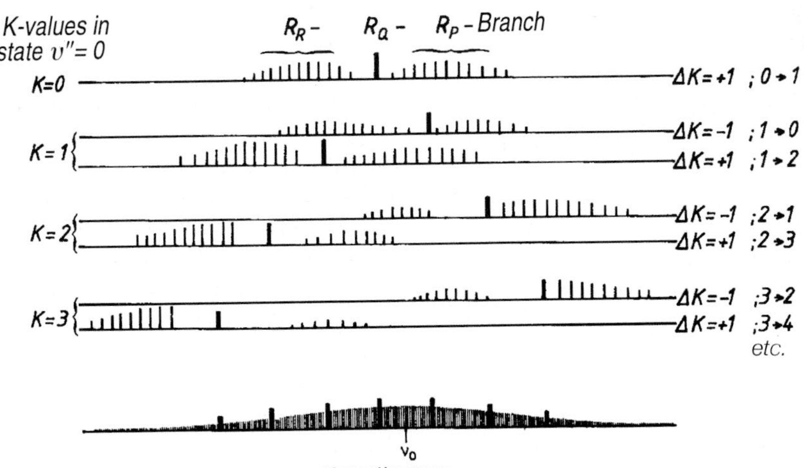

Perpendicular bands. Fig. 3. The subbands of a perpendicular band

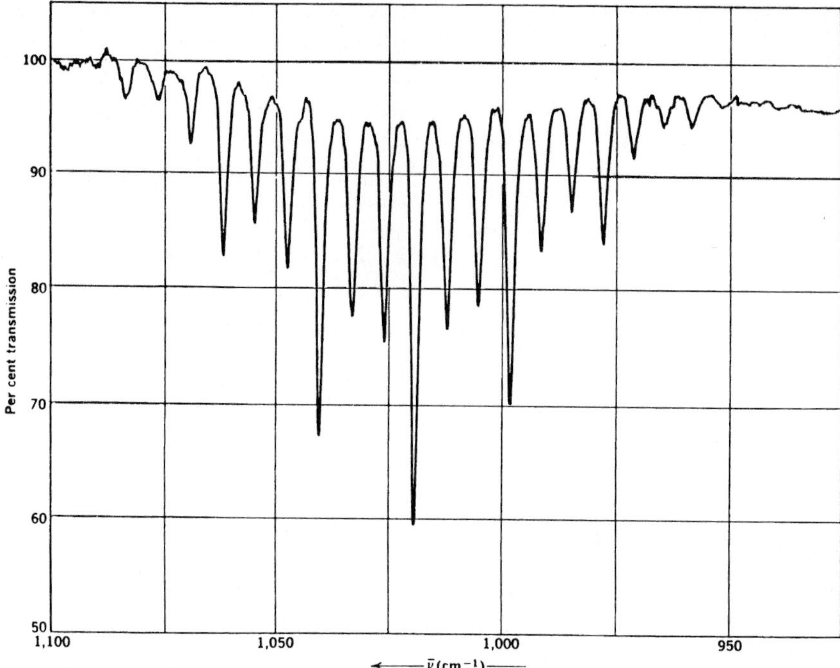

Perpendicular bands. Fig. 4. A perpendicular band of CH_3Cl

individual subbands are distinguished by a left superscript, i.e.:

$$\Delta K = +1: {}^{R}R,\ {}^{R}Q,\ {}^{R}P$$
$$\Delta K = -1: {}^{P}R,\ {}^{P}Q,\ {}^{P}P$$

The K value of the lower state can be added as a right subscript. In practice, with a perpendicular band, only the dominant Q branch can be picked out above the background of strongly overlapping P and R branches; even with good resolution in the gas phase. Figure 4 shows CH_3Cl as an example.

PES, <*PES*> → photoelectron spectroscopy.

Pfund series, <*Pfund-Serie*>, a series of lines in the hydrogen-atom emission spectrum which lies in the IR. It corresponds to transitions from states with principal quantum number, $n > 5$, to the state with $n = 5$. The first line occurs at $\lambda = 7.404$ μm or $\tilde{\nu} = 1351$ cm^{-1}.

Phase, <*Phase*>.

1. State: The solid, liquid and gaseous states of a material are each phases. Three phases are distinguished in liquid crystals; smectic, nematic and cholesteric. The solid state of a pure compound, e.g. a crystal, can be present in different phases characterized by different crystal structures, e.g. γ Al_2O_3 and α Al_2O_3.
2. In vibrations and waves, phase means the displacement of a vibration from a defined zero position. In physical optics, the words *vibrational phase* are used.

Phase angle, <*Phasenwinkel*>, the argument of the sine function, $\sin(2\pi\nu t \pm \delta_o)$, in oscillations and waves; ν = frequency, t = time, δ = phase constant (→ phase of an oscillation).

Phase difference, <*Phasendifferenz*>, in general, the time and position-dependent difference between the → phases of two oscillations or waves having the same frequency (→ phase of an oscillation). It is only sensible to speak of the phase difference of waves or vibrations of the same frequency at the same place, or at two different but specified places, at a specified time; or of the phase difference of a single wave at different places at a specified time. In crystal optics, the phase difference is proportional to the thickness of the crystal in the direction of the light and the difference in the relevant refractive indexes, $n''-n'$. For monochromatic light of wavelength, λ, we distinguish between:

a) the optical path difference $\Delta s = d(n'' - n')$ [cm];
b) the specific path difference $m = \Delta s/\lambda = d(n'' - n')/\lambda$;
c) the optical phase difference, $\delta = 2\pi d(n'' - n')/\lambda$ [radians];
d) the optical phase-angle difference $\delta = 360°\, d(n'' - n')/\lambda$ [degrees].

Phase jump, phase loss, phase shift, <*Phasensprung, Phasenverlust, Phasenverschiebung, Phasenverzögerung*>, in general, the change in the vibrational phase of a light wave upon reflection at an optically more dense medium.

Phase of an oscillation, <*Schwingungsphase*>, the displacement of an oscillation from its defined zero point (origin). To illustrate their relationships, Figure 1 shows two wave packets which can be represented by undamped sine waves:

$$a = A\cdot\sin(2\pi\nu t - \delta_o) = A\sin 2\pi\nu(t - t_o).$$

a is the momentary displacement at time t, A is the maximum displacement (amplitude), ν the frequency, $\nu^{-1} = \tau$ the period of the oscillation and $\delta_o = 2\pi\nu t_o$ is a quantity which says that a does not equal zero at times $t = 0, \tau, 2\tau \ldots$ but rather at the times $t = t_o$, $\tau + t_o$, $2\tau + t_o$ etc. The argument of the sine function $(2\pi\nu t - \delta_o)$ minus the integer multiple of 2π, is then an angle between 0 and 2π and is known as the phase angle of the oscillation. δ_o is called the phase constant. E.g. in Figure 1 the points B and B' and P and P' have the same phase (see the continuous curve).

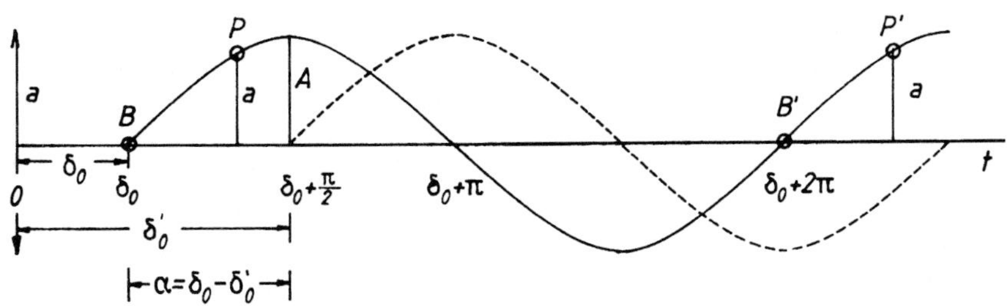

Phase of an oscillation. Fig. 1. The phase relationship of two wave trains, see text.

The two curves, continuous and dashed, in Figure 1 represent two oscillations with the same frequency, ν, and phase constants δ_o and δ_o'. The phase difference between them is $\alpha = \delta_o - \delta_o'$. If $\alpha = 0, 2\pi, 4\pi$ ... then the oscillations are said to be in phase. Phase difference and amplitude are important in the explanation of the → interference of coherent waves (→ coherent light) of the same frequency. If a light wave falls upon the surface of an optically more dense medium there is a phase change of $\pi = 180°$ in the reflected wave. This is known as a phase loss, phase jump or phase inversion because the displacement, a, changes sign. The wave which enters the optically more dense medium (→ reflection, → refraction of light) suffers a phase retardation, in comparison with a light wave in air having the same origin, because the wave velocity (velocity of light) is less in the denser medium.

Phase velocity, <*Phasengeschwindigkeit, Wellengeschwindigkeit*>, the velocity with which the maximum amplitude of a sine-form wave propagates. For electromagnetic radiation, this is the → velocity of light. Apart from its phase velocity, a wave also has a group velocity which is the speed at which the beginning or the end of a wave train moves. The phase velocity, c, and the group velocity, υ, are related by the equation:

$$\upsilon = c\left(1 + \frac{\lambda}{n}\frac{dn}{d\lambda}\right).$$

Thus, when there is → dispersion, the two velocities are different, but in vacuum, they are the same; $c = c_o$, the velocity of light in a vacuum, and $dn/d\lambda = 0$. From the phase velocity, c, and the wavelength, λ, the → frequency of a wave is given by: $\nu = c/\lambda$ or $c = \nu \cdot \lambda$.

Thus, the velocity of light depends upon the refractive index; $c = c_o/n$ (→ refraction of light). From this it follows that the wavelength also shows this dependence, i.e. $\lambda = \lambda_o/n$. λ_o is then the wavelength of the radiation in a vacuum which is frequently required for the calculation and calibration of wavelength scales.

Phosphorescence, <*Phosphoreszenz*>, like → fluorescence, one of the phenomena of → photoluminescence. The name derives from the element phosphorus on account of the blue glow of white phosphorus in the dark. (Greek: phosphorus = light bearing). The luminescence of white phosphorus, however, is not photoluminescence but → chemiluminescence.

Phosphorescence following excitation by absorption of radiation is observed in inorganic solids and in organic substances in solid solutions or as adsorbates. It is characterized by the fact that it persists for several seconds after the excitation ceases. In the typical inorganic solids known as phosphors, it can persist for minutes or even hours.

Phosphorescence and → fluorescence occur together in organic compounds, under the above conditions, and they were formerly generally distinguished by their lifetimes: fluorescence $< 10^{-4}$ s; phosphorescence $> 10^{-4}$ s and up to seconds.

In 1935 Jablonski explained the phosphorescence of organic compounds with the assumption of a metastable state lying below the first excited singlet state. Later (1944/45) Lewis and Kasha identified this state as a triplet state. Thus, its radiative decay to the ground state is a spin-forbidden → intercombination transition, which explains the long life of the triplet state and the long lifetime of the phosphorescence as compared with the fluorescence. In this way, the difference between fluorescence and phosphorescence is seen to lie in the → multiplicity of the electronic states involved. Fluorescence is a spin-allowed transition ($S_1 \rightarrow S_o$) which takes place without change of multiplicity while phosphorescence is a spin-forbidden transition ($T_1 \rightarrow S_o$) with a change of multiplicity. The transition, $S_o \rightarrow T_1$, in absorption is also an intercombination transition and therefore forbidden, i.e. the → transition probability is very small. The decadic molar → extinction coefficients for singlet → triplet transitions are of the order of 10^{-4} to 10^{-5} l mol^{-1} cm^{-1}. Direct excitation of the phosphorescent triplet state is therefore extremely difficult. For this reason most studies of phosphorescence begin with the excitation of a singlet state from which

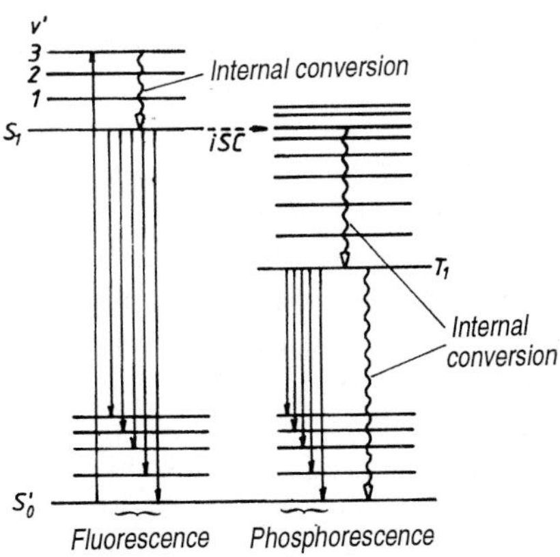

Phosphorescence. Fig. 1. A Jablonski energy-level scheme illustrating fluorescence and phosphorescence

the triplet state is populated by → intersystem crossing from S_1 to T_1. This process is very efficient, especially in solid solutions. The transitions are illustrated in the → Jablonski energy-level scheme, Figure 1. The intersystem crossing is an isoenergetic transition from S_1 into higher excited vibrational levels of the T_1 state. The fluorescence transitions are also shown for comparison. When the triplet state, T_1, lies significantly below the first excited singlet state, S_1, it is clear that the → phosphorescence spectrum will show a strong → bathochromic shift with respect to the → absorption and → fluorescence spectra. This is particularly true of the aromatic hydrocarbons. The solid solutions are frequently produced in glass-forming solvents or solvent mixtures at low temperatures (rigid solvents) and the phosphorescence observed is also called → low-temperature phosphorescence. As well as the rigid solvents, solid polymers, inorganic glasses and adsorbates are also used. The low-temperature phosphorescence, the spectrum of which differs very clearly from the fluorescence, is sometimes accompanied by a → high-temperature phosphorescence which has the same spectrum as the fluorescence and can be recognized only by the time delay in the emission. It is known as → delayed fluorescence of the *E* type. In principle, the same kind of energy-level scheme can be drawn for inorganic solids, although more intercombination transitions need to be considered for inorganic complexes on account of the diversity of their states. The luminescence of the typical inorganic phosphors is the result of a different process. It is also true, in general, that intercombination transitions are facilitated by heavy atoms (→ heavy atom effect).

Phosphorescence excitation spectrum, *<Phosphoreszenzanregungsspektrum, Ph-A-Spektrum>*. The spectrum which is obtained when the wavelength, λ_e, for the excitation of the phosphorescence, is scanned across the absorption spectrum while the observation wavelength, λ_{ph}, is held constant. Like the → fluorescence excitation spectrum, the phosphorescence excitation spectrum is very similar to the absorption spectrum, provided that dilute solutions are used. The spectrum must be corrected for the spectral energy distribution of the exciting light, $I_o(\lambda)$, and any fluorescence must be excluded from the measurement by means of suitable → filters, or better, a → monochromator.
Ref.: H.-H. Perkampus, *UV-VIS Spectroscopy and its Applications*, Springer Verlag, Berlin, Heidelberg, New York, **1992**, section 5.5.

Phosphorescence polarization spectrum, *<Phosphoreszenzpolarisationsspektrum, Ph-P-Spektrum>* → photoselection.

Phosphorescence spectrum, *<Phosphoreszenzspektrum>*. The spectrum which is obtained when the phosphorescence of a sample is spectrally dispersed and the intensity plotted as a function of the wavelength or wavenumber. The spectrum so obtained is not the true phosphorescence spectrum of the sample. Like the → fluorescence spectrum, it must be corrected for the spectral sensitivity of the instrument with which it was measured.

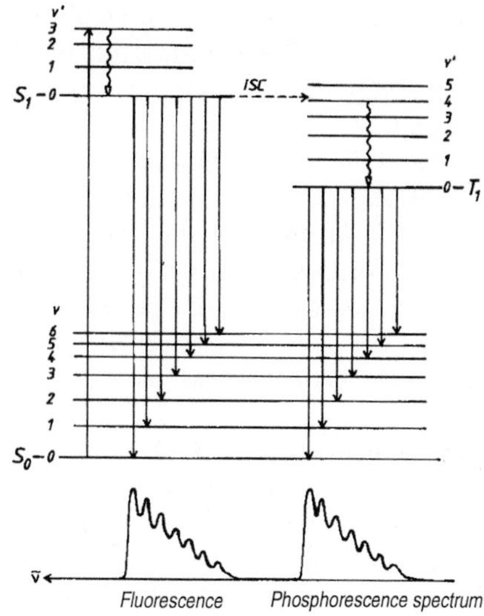

Phosphorescence spectrum. Fig. 1. Energy-level scheme comparing phosphorescence and fluorescence spectra

The → low-temperature phosphorescence spectrum frequently shows structures due to contributions from vibrations in the electronic ground state, S_o. The simplified energy-level diagram in Figure 1 indicates the origin of this structure and compares it with the fluorescence. There is a large → bathochromic shift of the phosphorescence relative to the fluorescence, which arises because the triplet state, T_1, lies much lower in energy than the first excited singlet, S_1. If the two states S_1 and T_1 lie close together in energy the two spectra may overlap. However, because of the different lifetimes of the fluorescence ($\sim 10^{-8} - 10^{-9}$ s) and phosphorescence ($> 10^{-4}$ s) it is possible to separate the two spectra by using modulated exciting light and a lock-in amplifier locked to this modulation. At a sufficiently high modulation frequency, the phosphorescence gives a time-independent signal which is not seen by the amplifier which then amplifies only the fluorescence.

Photo-, <*Photo-*>, a prefix for many important concepts, methods and effects in spectroscopy and technology. It is derived from the Greek, phos = light. Photography = writing by light; photometry = measurement of light; photoeffect = effect of light; etc.

Photoacoustic effect, PAE, <*photoakustischer Effekt, PAE*>, originally known as the optoacoustic effect, this phenomenon was discovered and studied in solids by A.G. Bell in 1880/81. At about the same time, Tyndall and Röntgen observed the same effect in gases. The photoacoustic effect manifests itself in the following manner. If modulated light falls on a solid body in contact with a gas in a closed vessel, then a receiver (in the early experiments, a stethoscope) also enclosed in the gas, registers an acoustic signal at the frequency of the light modulation. As Bell also observed, the strength of the acoustic signal depends upon the absorbing power of the solid and therefore also upon the wavelength of the light. This stimulated Bell to construct his *spectrophone*. Despite Bell's very detailed description of the effect, it was regarded simply as a curiosity and forgotten for many years. But in the early 70's there was a renaissance, partially due to the development of modern amplification techniques, improvements in intense light sources and advances the construction of sensitive microphones. This led to the development of → photoacoustic spectroscopy.

Photoacoustic spectrometer, <*Photoakustikspektralphotometer, Photoakustikspektrometer*>, an instrument used in → photoacoustic spectroscopy. Figure 1 is a schematic of a modern, computer-controlled, double-beam photoacoustic spectrometer. The individual components are labeled on the diagram. The optics, with a xenon high-pressure lamp (Osram, XBO 450 W/1) and monochromator, are ordered as in conventional → atomic absorption spectroscopy. The beam splitter and measurement cells are also analogously arranged. But two fixed microphones (Bruel and Kjaer capacitance microphone 4166 + preampli-

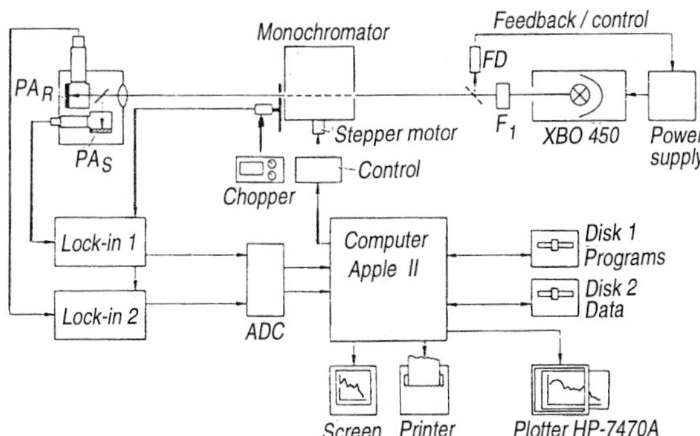

Photoacoustic spectrometer. Fig. 1. A double-beam photoacoustic spectrometer showing light path and components, see text

fier), for the reference (PA_R) and for the sample (PA_S), take the place of the photomultiplier. The signal from each microphone passes to the computer via a lock-in amplifier and an ADC. The lock-in amplifiers are locked to the chopper frequency which can be varied mechanically between 10 Hz and 4000 Hz. The lamp intensity is controlled to ±1% over long operating periods by a feedback mechanism in the power supply, and this makes single-beam operation feasible. In this case, the reference spectrum of the carbon standard is stored in the computer so that the recorded spectrum can always be referred to the standard. In this standard, also called a black standard, use is made of photoacoustic signal → saturation, i.e. the reference signal is dependent only upon the spectral energy distribution of the light source, and not upon the optical properties of the sample.

The beam splitter is a partially transparent mirror which transmits 5–10% of the light to the reference and reflects 90–95% to the sample. All results can be viewed on the computer screen during measurement, stored on a diskette and finally sent to a printer or plotter.

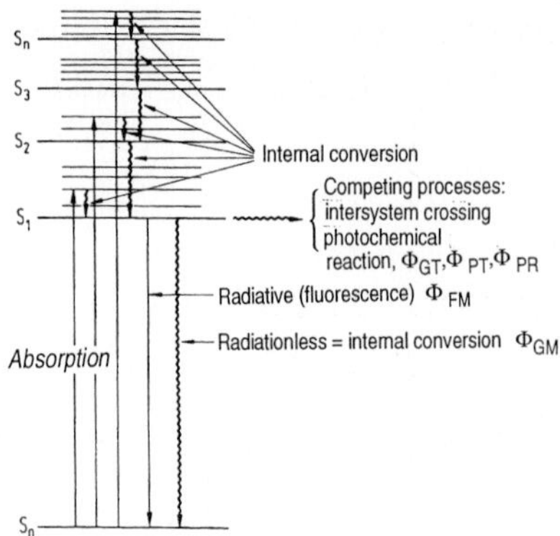

Photoacoustic spectroscopy. Fig. 1. Energy-level scheme with radiative and radiationless transitions

Photoacoustic spectroscopy, PAS, <*Photoakustikspektroskopie*>, a spectroscopic method based upon the photoacoustic effect, discovered by Bell in 1880/ 81. According to Bell, the primary step in producing the → photoacoustic effect is an absorption process. An acoustic signal or sound wave is observed. In the first step, a molecule absorbs electromagnetic radiation and within $10^{-15}-10^{-14}$ s is raised to the excited singlet states, S_1, S_2 ... S_n, together with superimposed vibrational states, see Figure 1. Within $10^{-13}-10^{-12}$ s, the molecule relaxes from the higher states, S_2 ... S_n, by radiationless deactivation whereby the stored excitation energy is given up as heat in vibrational relaxation processes, (→ internal conversion). After $10^{-9}-10^{-8}$ s the state, S_1, is also deactivated. This can occur in two ways:

1. By fluorescence and
2. nonradiatively, i.e. by conversion of the excitation energy, $S_o \rightarrow S_1$, into heat (internal conversion).

If a closed system, capable of absorbing light, is illuminated with light modulated at a specific frequency, which is low compared with the speed of the deactivation processes, then the heat production due to the radiation will follow the period of the modulation.

Thus, the pressure of a gas in a closed system will change periodically. This alternating pressure wave is a sound wave, and if a microphone is attached to the cell wall this acoustic signal can be directly recorded. In the case of solids, liquids and solutions, the situation is more complicated. Figure 2 shows,

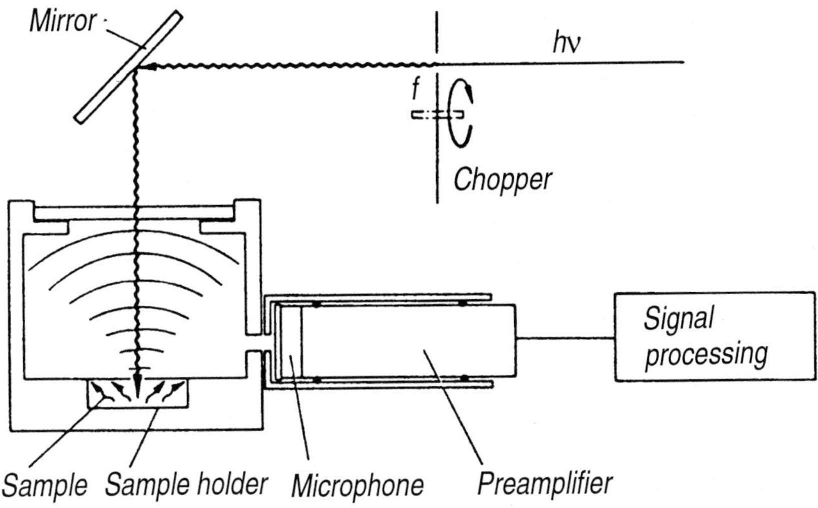

Photoacoustic spectroscopy. Fig. 2. A photoacoustic cell for solid samples (gas-coupled PAS)

schematically, a cell for measuring such samples. The microphone is coupled to the cell through a hole in the wall. There must be an atmosphere of gas above the sample. An angled mirror directs the modulated light through a quartz window and onto the sample where it is absorbed. The deactivation processes described above rapidly release heat and the local warming of the inside of the sample is followed by diffusion of the heat to the sample surface. At the interface between sample and gas, the heat wave passes into the gas and becomes a pressure (sound) wave which is registered by the microphone. This arrangement is known as a gas-coupled photoacoustic cell. Theoretical considerations show that the heat wave reaches only a thin gas layer at the sample/gas interface which therefore changes its thickness with the modulation frequency, $\omega = 2\pi f$. This is communicated to the rest of the gas adiabatically.

The relationship to the absorption of light is expressed in photoacoustic spectroscopy through the → absorpton coefficient, β:

$$A_{\tilde{\nu}} = \ln(\frac{I_o}{I})_{\tilde{\nu}}$$
$$= 2.303 \cdot \varepsilon \cdot c \cdot d;$$
$$\beta = \frac{A_{\tilde{\nu}}}{d}$$
$$= 2.303 \cdot \varepsilon \cdot c.$$

The units of the absorption coefficient are cm^{-1}. If the measured PA signal (S^{PA}) is directly proportional to β, then a graph of S^{PA} against wavelength or wavenumber would be expected to give a photoacoustic spectrum in agreement with the known absorption spectrum. In fact, considerable deviations from this behavior occur, as can be seen from Figure 1. The deactivation of the higher excited states, S_2, S_3 ... S_n, to S_1 by internal conversion generally produces heat. Compared with

S_2, S_3 ... S_n, S_1 has a relatively long life. From there several deactivation processes can begin; those relevant to the quantum yield are:

1. Radiative deactivation = fluorescence Φ_{FM};
2. intersystem crossing to the triplet state with Φ_{ISC} followed by
   a) radiative deactivation = phosphorescence Φ_{PT};
   b) radiationless deactivation Φ_{GT};
3. photochemical reaction Φ_{PR};
4. radiationless deactivation, $S_1 \rightarrow S_o$, Φ_{GM}.

The desirable processes, as far as PAS is concerned, are 2b and 4. All the other deactivation mechanisms are competing processes and do not therefore contribute to the PA signal. It can generally be assumed that, at room temperature, the processes under 2 do not occur, i.e. Φ_{ICS}, Φ_{PT} and $\Phi_{GT} = 0$. If, in addition, no photochemical reactions occur under the given conditions, then:

$$\Phi_{GM} = 1 - \Phi_{FM}.$$

Frequently, fluorescence cannot be neglected. Therefore, it represents the most important competing process for the photoacoustic effect. From another viewpoint, this relationship offers the possibility of an absolute determination of the fluorescence quantum yield. The → Rosencwaig-Gersho theory deals with the production of PA signals in condensed phases.

Photocell, <*Photozelle*>, a device depending upon the application of the external → photoelectric effect. Following the discovery of this effect by H. Hertz and W. Hallwachs in 1887, J. Elster and H. Geitel built the first photocell for light measurement in 1889/91. They constructed the first photoelectric → photometer with the photocells which they developed.

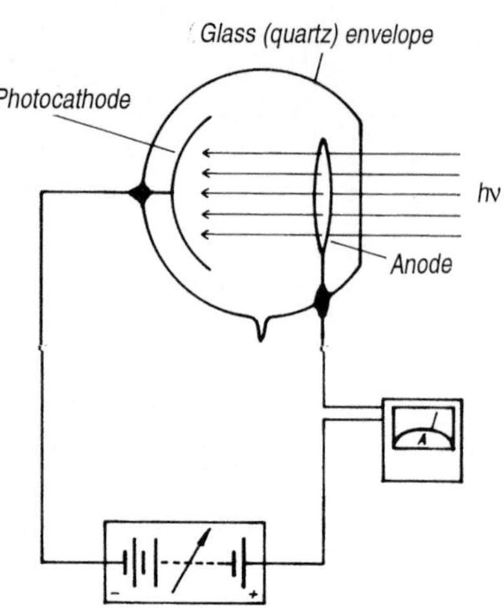

Photocell. Fig. 1. The principles of a photocell

Figure 1 shows the principles of the construction of a photocell. When light falls on the light-sensitive cathode, the photocathode, electrons (primary electrons) are released. The electrons are drawn to the anode by an applied voltage and a photocurrent flows which can be measured. This current is directly proportional to the light intensity so that the photocell can be used for intensity measurements. Metals with small electron work functions, $e_o\varphi$ in eV, are used as photocathodes. In general, these are the alkali metals, K, Rb, Cs, either alone or as bialkali or multialkali photocathodes. (For more details about photocathode materials and their spectral sensitivities, → photomultiplier). The current-voltage characteristic of a photocell may be determined by varying the applied voltage. Figure 2 shows an idealized current-voltage curve for a vacuum photocell. With increasing voltage, a saturation current is produced because, from a certain voltage onwards, all emitted electrons should reach the anode. However, exact measurements show that even in this region a slow growth of the photocurrent is observed. Nevertheless, this region, which is reached with a voltage of about 30 V, is of particular importance for practical intensity measurement because a change in the voltage of a few volts produces only an extremely small change in the photocurrent. Consequently, photocells are used in the regime of quasi-constant saturation current, i.e. with a voltage $\geq$ 30 V. The photocurrents of vacuum photocells are of the order of 10^{-9} to 10^{-14} A, depending upon construction through the insulation resistance of the cell. In a gas-filled photocell (argon is usually the added gas), the current-voltage curve is quite different. It is shown schematically in Figure 2, curve *b*. From a particular voltage onwards, the photocurrent is increased by the onset of collision ionization by energetic electrons. The current climbs steeply until discharge begins at the glow-discharge voltage. Photocells of this type are preferred for switching applications because of their significantly higher photocurrents. Figure 3 shows the first photocell constructed by Elster and Geitel in 1891.

Ref.: B.H. Vine in *Applied Optics and Optical Engineering*, (Ed.: R. Kingslake), Academic Press, New York, London, **1965**, vol. II, chapt. 6; A.H. Sommer, *Photoemissive Materials*, J. Wiley and Sons, Inc., New York, London, Sydney, Toronto, **1968**.

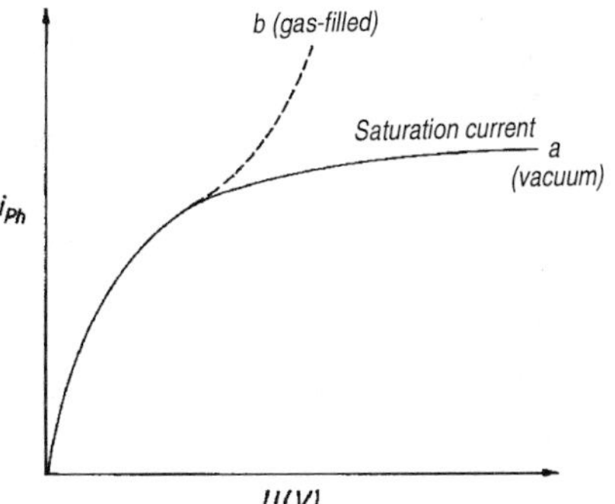

Photocell. Fig. 2. Current-voltage characteristcs of photocells

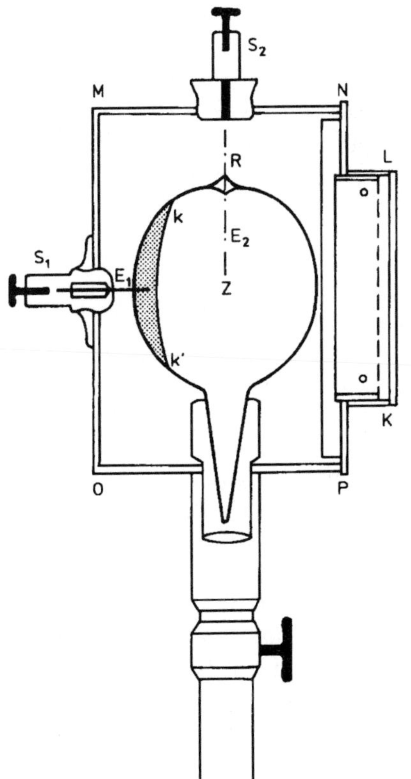

Photocell. Fig. 3. Elster and Geitel's first photocell

Photochemical hole burning, PHB, <*photochemisches Lochbrennen, PHB*> → hole burning.

Photochemical reaction, <*Photoreaktion, photochemische Reaktion*>, a chemical reaction which is initiated when a molecule, M, absorbs a light quantum of appropriate energy, $h\nu$; $M + h\nu \rightarrow M^*$. This produces the excited species, M^*, from which the photochemical reaction begins. Ideally, the reaction continues as long as the radiation persists. When the light is interrupted the reaction stops, provided that there are no sequential or back reactions which take place in the dark. In this, the most simple case, the course of the photochemical reaction can be followed spectrophotometrically by measurements taken during dark periods. It is at once clear that the course of a photochemical reaction depends quite generally upon the intensity of the irradiation and its duration.

Photoreaction is a further possibility for the deactivation of an excited state and it competes with the other → radiative and → radiationless processes. The → quantum yield is of decisive significance in the evaluation of photochemical reactions. In this context a distinction is made between simple and complex photochemical reactions. Simple reactions are those in which only one substance absorbs the light and initiates the reaction, and in which all dark reactions are so fast that the Bodenstein (steady-state) hypothesis can be applied to unstable intermediates. The other reaction partners can absorb light, but they must not initiate a photochemical reaction. Reactions which do not fulfill one of these conditions are termed complex photochemical reactions. Typical photochemical reactions are, photoisomerization, cyclization, dimerization, reduction, elimination, addition etc.

Ref.: R.P. Wayne, *Principles and Application of Photochemistry*, Oxford University Press, Oxford, New York, **1988**.

Photochemistry, <*Photochemie*>, a collective name for the study of chemical reactions which, in contrast to dark reactions, are induced by the absorption of light. → Photokinetics.

Photochromism, <*Photochromie*>, a process whereby substance A is converted into substance B by absorption of electromagnetic radiation in the UV-VIS region. It is important that this process is reversible, i.e. that the species B can be converted to A by irradiation at another wavelength or thermally. The general reaction scheme can be written as follows:

$$A \underset{h\nu' or\ \Delta T}{\overset{h\nu}{\rightleftharpoons}} B.$$

Since the absorption spectra of A and B differ considerably, the conversion, $A \leftrightarrow B$, is seen in the visible region as a color change. All the reversible *trans-cis* photoisomerizations are examples of photochromism, if sequential reactions are excluded. The bianthrones and spiropyrans are particularly typical photochromic compounds. In the photochromic form, B, they show an intense color which disappears on irradiation at the wavelength of the absorption maximum of B, or on raising the temperature, because A is reformed. These compounds also show → thermochromism.

Ref.: H.H. Jaffe , M. Orchin, *Theory and Application of Ultraviolet Spectroscopy*, John Wiley and Sons, Inc., New York, London, **1962**, p. 553.

Photodiode, <*Photodiode*>, an → optoelectronic component which depends for its operation on the → barrier-layer photovoltaic effect. In principle, photodiodes and photoelements are constructed in an identical manner. The presence of a *pn* junction always gives rise to a barrier layer which acts like a rectifier, i.e. current can only flow in one direction. If a photodiode is operated with reverse bias (*n* type positive, *p* type negative), only a very small dark current flows. This current is of the order of 2 nA–1 μA for silicon and 5–15 μA for germanium photodiodes. The voltage applied is generally in the region of 10–100 V. When the photodiode is irradiated, the dark current is augmented by a photocurrent due to charge carried by the electrons and holes produced by the impinging photons. The photocurrent is directly proportional to the intensity of the light falling on the diode up to values of 10^5 lx (lumen/m^2), i.e. the whole of the illumination-intensity scale which is of technical interest. Photodiodes are very well suited to the quantitative measurement of light. In applications, the spectral sensitivity distribution of photodiodes is of particular interest.

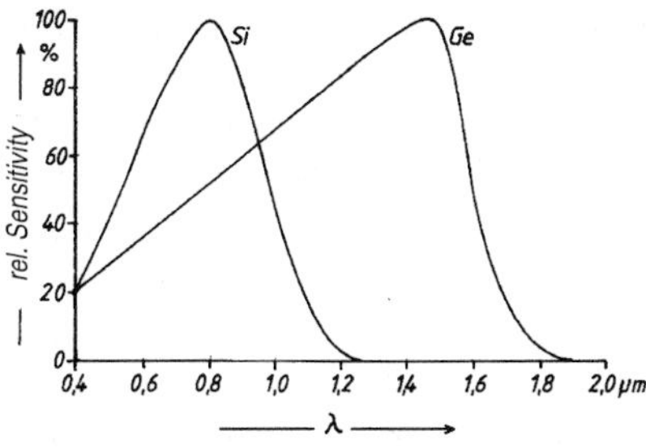

Photodiode. Fig. 1. Spectral sensitivity curves for a silicon and a germanium photodiode

Typical curves for a silicon and a germanium photodiode are shown in Figure 1 from which it can be seen that Si photodiodes are more suitable for the visible and Ge photodiodes for the near IR. The rise time of photodiodes lies in the range 1 ns to 30 μs, depending upon material, technology, surface, conditions of the measurement etc. The PIN, avalanche, Schottky, and differential photodiodes are special types.

Ref.: A.H. Sommer, *Photoemissive Materials*, J. Wiley and Sons, Inc., New York, London, Toronto, Sydney, **1965**.

Photodiode array, <*Photodiodenarray*>, a modern detector for the UV-VIS-NIR spectral region used in combination with a → polychromator. The simultaneous measurement of a complete spectrum which is thus possible is reminiscent of the earlier use of photographic plates as detectors. A photodiode array is a row of miniature → photodiodes on a single-crystal silicon chip giving a position-dependent detector. The individual photodiodes on the semiconductor chip are connected in parallel with correspondingly microscopic capacitors which are charged to a starting voltage of 5–10 V before the measurement begins. If spectrally dispersed light falls on the diodes during a measurement, then a photocurrent, which is proportional to the intensity and duration of the light, flows which discharges the capacitors. After the measurement, the controlling electronics determine the state of charge of each capacitor which is recorded against its position in the row, i.e. as a function of wavelength. A photodiode row consists of 256, 512 or 1024 individual photodiodes (pix-

els) from which the resolving power of the spatial detector with respect to the spectral range covered can be calculated. The length of a photodiode row lies between 1 and 3 cm. As an example: the Reticon photodiode array RL 512 G has 512 diodes, each 25×250 μm, in a length of 1.28 cm.
Examples of the application of these detectors are the Zeiss → multichannel spectrometer MCS, the Perkin-Elmer Lambda 3840 and the Hewlett-Packard HP 8450 A and WP 8452 A.
Photodiode arrays are also used in vidicon tubes where several rows are arranged, one above the other, to give a two-dimensional detector.

Photoelastic modulator, PEM *<photoelastischer Modulator>*, an optical device due to Billardon and Badoz which induces a phase shift in linearly polarized light. It is especially valuable for converting linearly polarized light into right- and left-circularly polarized light (RCP and LCP). A typical PEM for the UV-VIS region (Figure 1) consists of two blocks of quartz, one crystalline and the other fused, bonded together. By means of electrodes deposited on two surfaces of the crystalline quartz, an alternating voltage can be applied to the crystal at a typical frequency of 50 kHz. The piezoelectric effect causes vibrations in the crystal along the z direction and these vibrations induce resonant vibration in the fused quartz block. The size of the two quartz blocks is chosen so as to favor this sympathetic vibration of the fused material. The vibration causes periodic strain in the fused quartz which, in turn, affects the polarization of light passing through it (→ stress birefringence). At 400 nm voltages of the order of 10 V, peak-to-peak, are required to produce a → quarter-wave plate, i.e. to convert linearly polarized light to LCP or RCP. Moreover, by changing the applied voltage pure CPL can be produced over the entire UV-VIS range. For measuring circular dichroism (CD), the PEM is driven by an AC voltage in such a way as to produce LCP and RCP at the extremes of the voltage range. The resulting AC signal is detected and fed to a suitable lock-in amplifier. For measuring linear dichroism (LD) the voltage range is chosen so as to produce perpendicularly polarized beams of linearly polarized light at the extremes of the voltage cycle. Clacium fluoride is used as PEM material in the → vacuum UV region and was originally used in IR CD instruments; PEMs for the IR region are now normally made from zinc selenide.

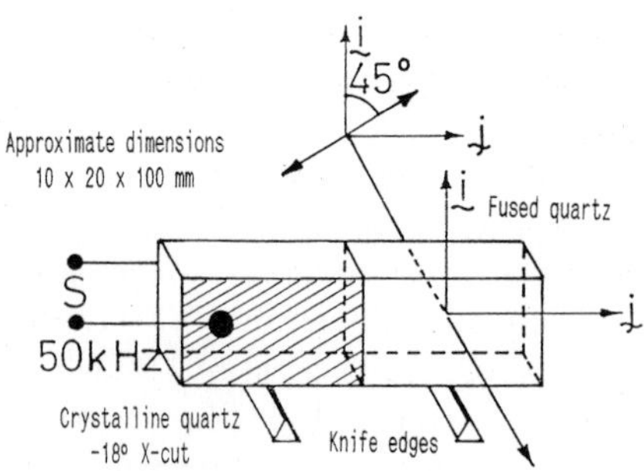

Photoelastic modulator. Fig. 1. Schematic view of a quartz PEM

Photoelectric amplification, *<lichtelektrische Verstärkung>* → photocell.

Photoelectric cell, *<lichtelektrische Zelle, Photozelle>*, a photoelectric detector. The word is frequently used as a general term for → photocell, → photomultiplier, → photoresistive cell, → photodiode and → phototransistor.

Photoelectric effect, <*lichtelektrischer Effekt*>, the appearance of a radiation-induced photocurrent. Three effects can be distinguished, depending upon the primary event responsible for the generation of the photocurrent:

a) An external photoelectric effect, (→ photoelectric effect, external);
b) an internal photoelectric effect, (→ photoelectric effect, internal);
c) a → barrier-layer photovoltaic effect which is observed when the phase boundary between a semiconductor and a metal electrode is irradiated.

The barrier layer, which can be partially chemical and partially physical in its nature, has a unidirectional conductivity for electrons which causes a voltage (photovoltage) to develop when the layer is illuminated. For this reason, the device is called a photoelement. Although the effect was first observed by H. Becquerel in 1839, and described for selenium especially in 1887, it was first used in the construction of selenium and silicon photoelements by B. Lange in the years 1926–1930. Today, → photodiodes, → phototransistors and photothyristors are common in → optoelectronic components. The barrier-layer effect is also an internal photoelectric effect so that b) and c) can also be distinguished as internal photoelectric effects *without* and *with* the → barrier-layer photoeffect.

Photoelectric effect, external, <*äußerer, lichtelektrischer Effekt*>, the emission of electrons, into the adjacent vacuum, by metals when irradiated with light which was discovered by Hertz in 1887 and by Hallwachs in 1888. If we fit an anode opposite the emitting cathode (photocathode) and apply a voltage between them a photocurrent is observed. This principle forms the basis for the construction of the → photocell and → photomultiplier. A classic experiment by Lenard in 1899 and 1900 proved that the emission was actually electrons. However, the explanation of the effect was first given by Einstein in 1905. He proposed the following energy-balancing equation for the external photoelectric effect:

$$h\nu = e_o\varphi + E_{kin}.$$

A photon of energy, $h\nu$, (light corpuscle, Planck's quantum of energy) shares its energy momentarily with the electron concerned by changing partially into the kinetic energy of the electron, E_{kin}, and also doing the work required to eject the electron from the metal (the work function, $e_o\varphi$). The kinetic energy of the electrons can be measured experimentally by determining the countervoltage, U, at which the photocurrent disappears. The energy, e_oU, is then equal to the kinetic energy. Since the work function, $e_o\varphi$, is a substance-specific quantity, photons with a different energy, $h\nu_i$, i.e. light of varying frequency, require different values of the countervoltage, U_i. Therefore, Einstein's equation can also be written as:

$$h\nu_i = e_o\varphi + e_oU_i \quad \text{or} \quad U_i = \frac{h}{e_o}\nu_i - \varphi.$$

Planck's constant, h, and the work function of the metal can be determined using these equations. When $U = 0$, $h\nu_o = e_o\varphi$ and the photon energy at this frequency, $h\nu_o$, corresponds to the work function. However, for frequencies smaller than ν_o, the photon

Photoelectric effect, external. Table.

Element	$e_{0\varphi}$ [eV]	λ_o [nm]	Element	$e_{o\varphi}$ [eV]	λ_o [nm]
Li	2.46	504	Be	3.92	316
Na	2.28	543	Mg	3.70	335
K	2.25	551	Ca	3.20	387
Rb	2.13	582	Sr	2.74	452
Cs	1.94	639	Ba	2.52	492

energy is insufficient to eject electrons from the metal. This is the threshold frequency or the long-wavelength limit of the photoelectric effect. The work function and limiting wavelengths for alkali metals and alkaline earth metals are compiled in the table.

All other metals have work functions which correspond to limiting wavelengths, $\lambda_o < 300$ nm. Thus, alkali metals, especially K, Rb and Cs, are suitable for the construction of → photocells and → photomultipliers, as Elster and Geitel showed in 1889/90.

Ref.: P.W. Atkins, *Quanta*, 2nd ed., Oxford University Press, Oxford, **1991**.

Photoelectric effect, internal <*innerer lichtelektrischer Effekt, innerer Photoeffekt*>, an effect discovered in selenium by W. Smith in 1873. It is observed pre-eminently in semiconductors and may be explained using the band model. Electrons are raised

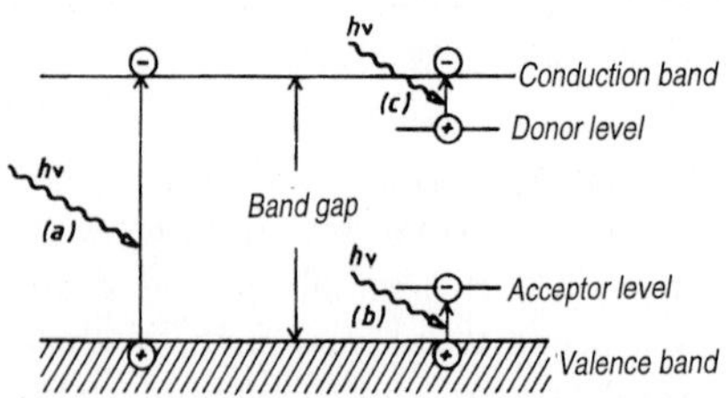

Photoelectric effect, internal. Fig. 1. The internal photoelectric effect in semiconductors

from the valence band into the empty conduction band by absorption of → photons which increases the conductivity. The electrons in the conduction band and the holes in the valence band both contribute to the photoconductivity. The basic band-gap absorption (Figure 1a) determines the internal photoelectric effect in a pure semiconductor. However, in the case of a doped semiconductor, which is generally obtained by doping a pure semiconductor, an impurity absorption can contribute to the internal photoelectric effect (Figures 1b and 1c). Figure 1 shows these three possibilities schematically. In case *a* we obtain freely moving electron/hole pairs and in cases *b* and *c* moving charge carriers with only one sign, i.e. electrons (*n* type) when doping with donors or holes (*p* type) when doping with acceptors. In the case of band-gap absorption, the long-wavelength limit of the photoelectric effect coincides with the absorption edge of the optical absorption of the semiconductor. The long-wavelength limit can reach very far into the near infrared spectral region in doped absorption.

The increase of conductivity which results from the internal photoelectric effect led to the construction of → photoresistive cells as detectors, especially for the infrared spectral region. The → barrier-layer photoelectric effect, which is also used for the construction of detectors, occurs in semiconductors with *pn* junctions (→ photoelement, → photodiode).

Photoelectric fatigue, *<lichtelektrische Ermüdung>*, the loss of sensitivity, especially spectral sensitivity, of photodetectors with time. Photoelectric fatigue can be reversible or irreversible and usually increases with increasing intensity and decreasing wavelength of the light. In the internal → photoelectric effect, the fatigue is also known as photoelectric excitation and is caused by a lack of available photoelectrons.

Photoelectric measurement methods, *<lichtelektrische Meßmethoden>*, a general expression for light-measuring methods using photoelectric radiation detectors such as → photocells, → photomultipliers, → photoelements or → photoresistive cells. In contrast to many other methods, an almost inertia-free and wattless transfer of the measured value from the test object to the measuring instrument can be achieved. These special properties have led to a widespread adoption of photoelectric methods of measurement in science and technology. The intensity of light sources, absorption, scattering, reflection and emission can be measured. Also, nonoptical properties such as pressure, velocity and the number of objects can be measured with photoelectric methods, provided that the original quantity to be measured can be converted into an optical property. More details can be found under → photometer, → spectrometer for the UV-VIS and NIR spectrum (→ spectral regions), → photoelectric modulated light methods.

Photoelectric modulated light methods, *<lichtelektrische Wechsellichtmethoden>*, special → photoelectric measurement methods which are used, for example, in → ratio recording and in applications of → intensity-modulated light.

Photoelectric quantum yield, *<lichtelektrische Quantenausbeute>*, the ratio of the number of measured photoelectrons to the number of absorbed photons. The quantum yield of photocells with combined cathodes is approximately 0.01 to 0.3 at the maximum of the spectral sensitivity curve. (→ photoelectric effect, external; → photocell; → photomultiplier).

Photoelectric semiconductor elements, *<lichtelektrische Halbleiterzellen>*, a general term for a group of photoelectric → radiation detectors with semiconductor properties. Particular members of the group are → photoresistive cells, → photoelements and → phototransistors.

Photoelectric sensitization, *<lichtelektrische Sensibilisierung>*, a reduction of the electron work function of metallic photocathodes and a raising of the → light sensitivity by deposition of intermediate layers, e.g. antimony/cesium oxide/cesium cathodes. K/KH cathodes. (→ photocells).

Photoelectric surface effect, *<lichtelektrischer Oberflächeneffekt>*, a component of the → photoelectric effect. Only a small part of the light energy falling onto the surface of a solid metal cathode is absorbed in the surface layer ($<10^{-7}$ cm) of the metal. Nevertheless, the electrons thus ejected from the surface of the metal contribute the major part of the photocurrent.

Photoelectric volume effect, <*lichtelektrischer Volumeneffekt*>, a part of the external → photoelectric effect. Of the great number of electrons released by the action of light in the deeper layers (> 10^{-7} cm) of a metal, only a few leave the metal surface. The contribution of these electrons to the photocurrent is therefore small.

Photoelectron, <*Photoelektron*>, an electron which is ejected from a solid or a molecule by the action of light. (external → photoelectric effect, → photoelectron spectroscopy).

Photoelectron spectroscopy, PES, <*Photoelektronenspektroskopie, PES*>, a modern spectroscopic method for which the energy balance and hence the fundamental equation is analogous to that of the external → photoelectric effect; $h\nu = I + E_{\text{kin}}$. A photon of energy, $h\nu$, ejects an electron (→ photoelectron) from a molecule with a kinetic energy, E_{kin}, which is the difference, $h\nu - I$, where I is the ionization energy of the molecule. Thus, if the energy, $h\nu$, is known, e.g. the HeI line at 21.21 eV, and E_{kin} is measured, the ionization energy of the molecule can be found. Compare the external photoelectric effect, in which the electron work function is determined. Figure 1 shows the construction of a UV photoelectron spectrometer schematically. Photons with an energy of $h\nu = 21.21$ eV, produced in the helium-gas discharge tube, collide with the sample molecules. The ejected electrons enter an electrostatic potential maintained between two curved electrodes. By varying the voltage between the plates, electrons having a specific kinetic

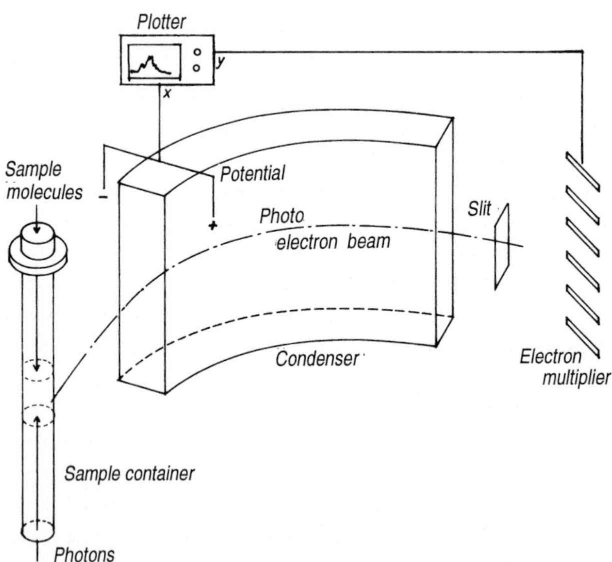

Photoelectron spectroscopy. Fig. 1. The principle (schematic) of a photoelectron spectrometer

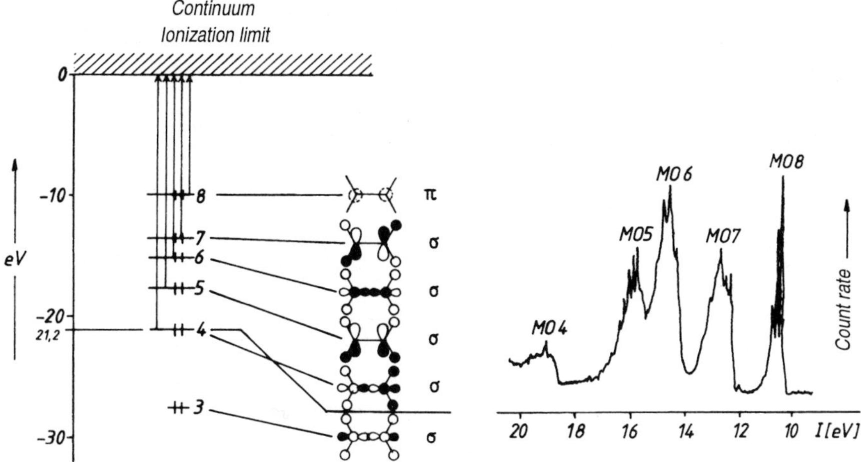

Photoelectron spectroscopy. Fig. 2. The photoelectron spectrum of ethylene and an MO energy-level diagram

energy can be selected to pass through the exit slit and be registered by the detector. The output of the detector is amplified and recorded and the number of electrons arriving at the detector is plotted as a function of the kinetic energy or ionization energy. The result is the photoelectron spectrum of the molecule. In their ground states, molecules possess a specific number of occupied molecular orbitals (MO). Those orbitals having energies above –21.21 eV can be ionized, i.e. the photoelectron spectrum gives the number of these orbitals and their absolute energies with respect to the continuum, 0.0 eV. Vibrational structure is frequently seen in the spectra because the vibrational states are superimposed upon the electronic states. Figure 2 shows the example of the PES of ethene (ethylene). The connection to the MO energy-level scheme is indicated. It is important to note that MO energy levels are frequently calculated quantum mechanically so that, in this case, PES is the experimental method which permits a confirmation and assignment of the orbital energies.

Ref.: T.A. Carlson, *Photoelectron and Auger Spectroscopy*, Plenum Press, New York, London, **1975**.

Photoelement, <*Photoelement*>, a radiation detector which makes use of the internal → photoelectric effect. Like the → photodiode, the photoelement is an → optoelectronic component the function of which is based upon the → barrier-layer photoelectric effect. Its exact description should therefore be → barrier-layer photoelement. Selenium, in which Smith discovered the effect in 1873, was formerly used as the semiconductor with the *p–n* junction.

The construction of a selenium barrier-layer cell is shown schematically in Figure 1. It is important that all the radiation falls on the *p–n* interface. This is achieved by making the *p* layer so thin that the incident radiation is transmitted as completely as

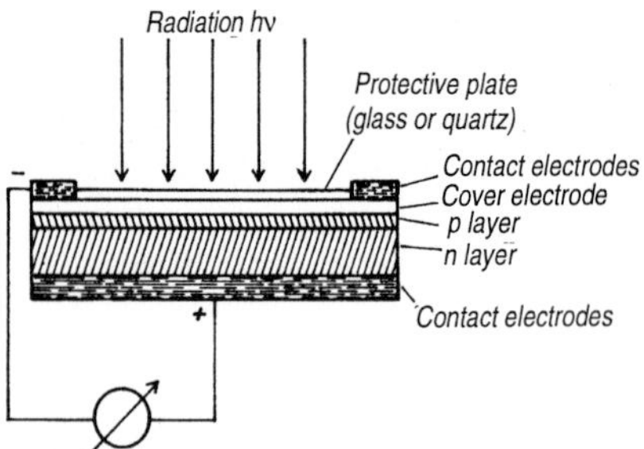

Photoelement. Fig. 1. The construction of a selenium barrier-layer cell

possible by it. The semiconductor is covered with a thin, transparent electrode with a good electrical conductivity, e.g. CdO. The electrical connections are made via evaporated metal films. The vulnerable front electrode is usually protected with a glass or quartz plate or a thin coating of lacquer. Nowadays, silicon or germanium are predominantly used as the semiconductor and the *p–n* junction is made by the well-known techniques of the semiconductor industry. Silicon photoelements have an energy output which is about 10 times as large as that of selenium devices, and they are also superior to selenium with respect to constancy (fatigue), temperature dependence (–60 to 170°C) and → spectral sensitivity.

Photoionization, <*Photoionisation, Photoionisierung*>, the positive charging of atomic or molecular ions when light acts on a gas or a vapor of atoms or molecules and sets → photoelectrons free. The applications lie in → photoelectron spectroscopy and in → ion sources.

Photokinetics, <*Photokinetik*>, the course or velocity of processes which are induced by light. In particular, we distinguish:

a) The kinetics of excited states in which the molecule, excited in 10^{-14} s by the primary process of light absorption, returns unchanged to its ground state. Since the deactivation processes can be very fast (10^{-6} – 10^{-12} s), we are dealing here with events which fall within the time scale of very fast reactions.
b) The alternative is that the primarily excited molecule is converted into another species, e.g. *trans-cis* isomerization, or it may react with a second species to form a new compound. The kinetics of such processes is the → kinetics of photochemical reactions. The velocities of these processes are comparable with those of normal chemical reactions in the ground electronic state, so this part of photokinetics is really a special area in the study of chemical reaction kinetics. Very fast photochemical reactions can compete with the deactivation of excited states and complicate the kinetics of that process.

In the → kinetics of chemical reactions, the change with time of the concentration of a substance involved in the reaction is followed. With photochemical reactions we need, in addition, to know the quantity of light absorbed during the reaction. The quantity of light absorbed, I_{a_i}, in unit time and in unit volume, by a substance, a_i, depends in a complicated manner upon where, within the reaction mixture, the reaction takes place and the duration of the illumination. It is possible to determine the quantity of light absorbed if the following conditions pertain:

1. The photoreaction takes place in a cuvette with plane-parallel windows. The area of the cross section is $F(\mathrm{cm}^2)$.
2. The exciting light is monochromatic and,
3. falls perpendicularly onto the entrance window of the cuvette.
4. The intensity of the incident light, I_o, (mol sec^{-1} cm^{-2}) is constant both in time and over the whole entry surface.
5. The reaction product is not turbid. It may absorb light, but it must not scatter light.
6. The exit window does not reflect any light.

For the derivation of the explicit differential equations of photokinetics it is further assumed that the irradiated solution is stirred so vigorously that the individual reactive species are homogeneously distributed throughout the cuvette at all times. Further, in accord with the Bodenstein (steady-state) hypothesis, it is assumed that the short-lived intermediate products do not themselves absorb light appreciably.
Starting with the definition of the true differential → quantum yield of a photochemical reaction of the type $a \rightarrow b$ $\Phi_b^{\ a} = c_a/I_a$, the photochemical rate law is:

$$\frac{\mathrm{d}c_a}{\mathrm{d}t} = 1000 \cdot \Phi_b^a \varepsilon'_{a,\lambda_\mathrm{e}} I_o c_a(t) \cdot \frac{1 - \mathrm{e}^{-A'_{\lambda_\mathrm{e}}}}{A'_{\lambda_\mathrm{e}}}.$$

The factor $(1 - \exp[-A'_{\lambda_e}])/A'_{\lambda_e}$, which appears in this equation, is known as the photokinetic factor which must be considered in all photokinetic equations. In the derivation of the equation it is also assumed that the quantum yield is independent of the quantity of light absorbed. For the determination of the quantum yield see → kinetics of photochemical reactions.
For the derivation of the explicit differential equations of photokinetics see: H. Mauser, *Z. Naturforsch.*, **1967**, *22b*, 367.

Photoluminescence, <*Photolumineszenz*>, the → luminescence following excitation of a material by absorption of → electromagnetic radiation, i.e. of photons. In atoms and molecules a distinction is made between → fluorescence and → phosphorescence. At first, the distinction was based upon the decay of the luminescence when the excitation was suddenly stopped. If the fading of the luminescence could be seen with the eye, or with simple optical-mechanical aids, it was termed phosphorescence. If not, it was called fluorescence. On the basis of this criterion even today, photoluminescence with a decay time of $< 10^{-4}$ s is generally described as fluorescence and if the decay is longer, as phosphorescence.
A different criterion was suggested by Lewis and Kasha in 1944. They proposed that phosphorescence was the result of an → intercombination transition between states of different → multiplicities while fluorescence was a radiative transition between states of the same multiplicity. In quantum-mechanical terms, phosphorescence is a spin-forbidden and fluorescence a spin-allowed transition. For the majority of organic compounds, which do not contain heavy atoms, these last two criteria can generally be regarded as valid.

Photometer, <*Photometer*>, an instrument which, like a → spectrometer or a → spectrophotometer, is used to measure relative light intensities caused by the absorption of light of a defined wavelength. The measurement is based upon the → Bouguer-Lambert-Beer law. The quantity measured is the → transmittance, $T = I/I_o$, and the absorbance which can be calculated from it.

$$A = \log\frac{1}{T} = \log\frac{I_o}{I}.$$

If the wavelength selection is made by means of a filter, the instrument is called a → filter photometer. If a → dispersing element is available for the spectral dispersion of the light (→ grating or → prism) the instrument is called a → spectrometer or → spectrophotometer. The term *photometer* is usually reserved for simple instruments for use in the visible or near UV. Spectrophotometers are more sophisticated instruments for use in all the regions of → optical spectroscopy where the terms *spectrometer* and *spectrophotometer* are often used interchangeably.

Photometric units, <*photometrische Einheiten*>, the units commonly used (DIN 5031) are summarized in the table (Figure 1). The following remarks are made by way of explanation:

1. The supplementary SI unit of solid angle is the steradian, sr;
2. emitter/transmitter quantities carry the subscript 1; detector/receiver quantities the subscript 2;
3. to avoid confusion, quantities from radiation physics sometimes carry the index, e; those from light technology the index v;
4. in optoelectronics, the radiant energy, Q_e, is given by:

$$Qe = \int_{100\text{nm}}^{1\text{mm}} Q_\lambda d\lambda,$$

Physical quantity	Equation	Radiation-physics designation	SI	Light-technology designation	SI
Energy	Q	Radiant energy	Ws	Luminous energy	lms
Power	$\varnothing = \frac{dQ}{dt}$	Radiant flux	W	Luminous flux	lm
Transmitter quantities					
Radiated power per unit surface	$M = \frac{d\varnothing}{dA_1}$	Radiant exitance	$\frac{W}{m^2}$	Luminous exitance	$\frac{lm}{m^2}$
Radiated power per unit solid angle.	$I = \frac{d\varnothing}{d\omega_1}$	Radiant intensity	$\frac{W}{sr}$	Luminous intensity	$\frac{lm}{sr} = cd$
Radiated power per solid angle and projected unit surface	$L = \frac{dI}{dA_1 \cos \varepsilon_1}$	Radiance	$\frac{W}{m^2\, sr}$	Luminance	$\frac{cd}{m^2}$
Receiver quantities					
Incident power per unit surface	$E = \frac{d\varnothing}{dA_2}$	Irradiance	$\frac{W}{m^2}$	Illuminance	$\frac{lm}{m^2} = lx$
Time integral of the incident power unit surface	$H = \int E\, dt$	Radiant exposure	$\frac{Ws}{m^2}$	Light exposure	$\frac{lms}{m^2} = lxs$

Photometric units. Fig. 1. Summary table.

where Q_λ is the spectral energy density of the radiation:

$$Q_\lambda = \frac{dQ}{d\lambda}$$

The range of the optical spectrum as defined by DIN 5031 is 100 nm – 1 mm.

5. cd → candela;
6. photometric radiation equivalent: the ratio of the light flux of a beam of monochromatic radiation to the radiation flux of this beam. At the maximum sensitivity of the eye (555 nm) the photometric radiation equivalent, according to DIN 5031, is 673 lm/W.

Photometry, <*Photometrie*>, the quantitative analytical application of → absorption spectroscopy in solutions. The basic principle underlying a photometric analysis is the → Bouguer-Lambert-Beer law:

$$\log(\frac{I_o}{I})_\lambda = A_\lambda = \varepsilon_\lambda \cdot c \cdot d.$$

A_λ is the → absorbance at the wavelength λ, c the concentration of the dissolved substance, ε_λ the molar decadic → extinction coefficient and d the path length of the cuvette (cell). The concentration is given by the above law as:

$$c = \frac{A_\lambda}{\varepsilon_\lambda d};$$

If $\varepsilon_\lambda = 10^4$ l mol^{-1} cm^{-1}, $d = 1$ cm and $A = 0.1$ then $c = 10^{-5}$ mol l^{-1} which, for a molecular weight of $M = 200$, is exactly 2 ppm. The smallest measurable concentration or the detection limit depends very much upon the value of the extinction coefficient, ε_λ, and upon the accuracy with which small absorbance values, $A < 0.1$, can be measured. An increase of the path length can be helpful. Since the extinction coefficient varies over a range $20 \leq \varepsilon_\lambda \leq 10^5$, the detection limits for different substances are similarly very variable. This is true for both organic and inorganic compounds. For metallic cations, which normally have only very weak colors themselves, complexation can be used. The complexes obtained with organic and inorganic ligands have intense colors and high extinction coefficients. With the help of special methods such as → extraction photometry and → floatation photometry, very small quantities of metal ions can be detected. In the case of organic compounds, derivatives which absorb more strongly than the original compound are often prepared. Indirect methods are sometimes used for the analysis of anions.

Ref.: H.-H. Perkampus, *UV-VIS Spectroscopy and its Applications*, Springer Verlag, Berlin, Heidelberg, New York, **1992**.

Photomultiplier, PM, <*Photovervielfacher*>, the most commonly used detector in → UV-VIS spectroscopy. The PM is an evacuated → photocell the photocurrent of which is multiplicatively increased by → secondary electron emission. The secondary electrons are ejected from special emitting cathodes (dynodes) which release several secondary electrons for every primary electron. The emission coefficient, δ, which is defined as the ratio of the number of secondary electrons, *S*, to the number of primary electrons, *P*, depends upon the accelerating voltage. In general, it has a flat maximum between 100 and 500 eV. At 200 eV,

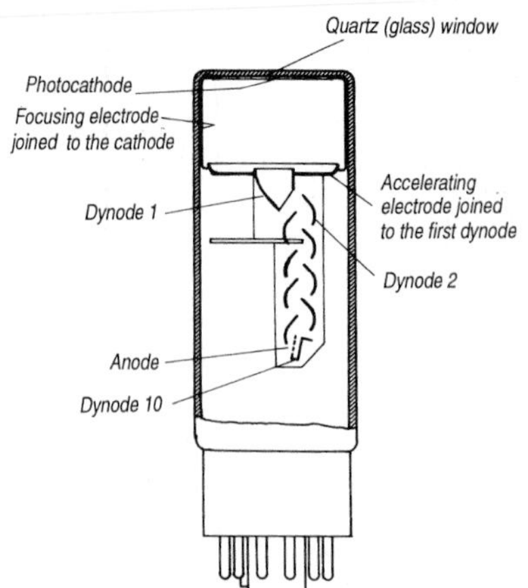

Photomultiplier. Fig. 1. The construction of a photomultiplier

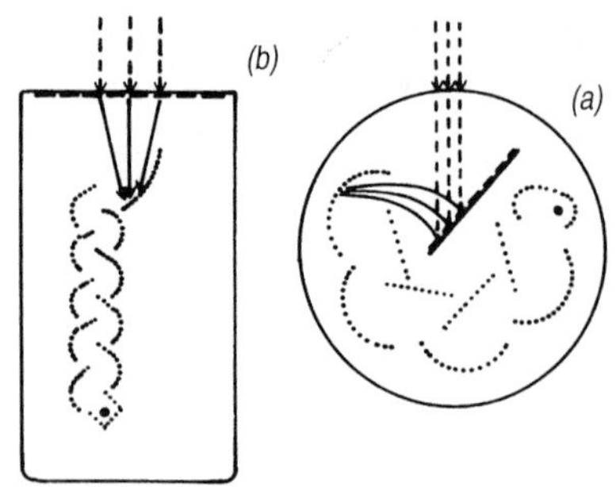

Photomultiplier. Fig. 2. Comparison of a side-on (a) and a head-on (b) photomultiplier

the emission coefficient lies between 3 and 8, depending upon the particular form of the dynode. The construction of a photomultiplier is illustrated in Figure 1. 10 to 12 dynodes follow a photocathode, at successively more positive potentials with respect to it. The final unit in the series is an anode. The amplified photocurrent is given for n steps by: $i = i_o \cdot \delta^n$, where i_o is the primary photocurrent leaving the photocathode and δ is the emission coefficient defined above. If $n = 10$ and $\delta = 4$, $i \cong i_o \cdot 10^6$. In this it is assumed that all the electrons leaving a dynode reach the next, i.e. an efficiency of 1. To operate a PM a direct voltage of 2000–3000 V is required, and this must be available in (from 9 to a maximum of 14) individual steps of 200–250 eV.
Head-on and side-on types of PM are produced. The majority of side-on types has a fully opaque photocathode with the dynodes arranged around it in a circle. The head-on types have a semitransparent photocathode immediately behind the front window with the dynodes stacked one behind the other, see Figure 1. The difference in the two forms of PM is shown schematically in Figure 2. Because of the different features in their construction and the way in which the primary electrons are generated, these two types of PM are sometimes called reflection mode and transmission mode PMs. No one material is known which would permit the construction of a PM to cover the whole spectral range from 180 to 900 nm. Therefore, there are a large number of PMs, differing in the composition of their photocathodes and each optimally suited to a particular wavelength region. The alkali metals plus As, Sb, Ga and Te are used as photocathode materials, either alone or in combination as bialkali or multialkali cathodes; e.g. Sb-Cs, Ga-As and Cs-Te. Figure 3 shows some spectral sensitivity curves in mA/W for named PMs.
Photomultipliers are not only used in optical spectrocopy (vacuum UV to VIS); they have also found applications in many other techniques, e.g. → scintillation measurements, X-ray and γ-ray spectroscopy.
Ref.: *BURLE Photomultiplier Handbook*, Burle Industries, Inc., Tube

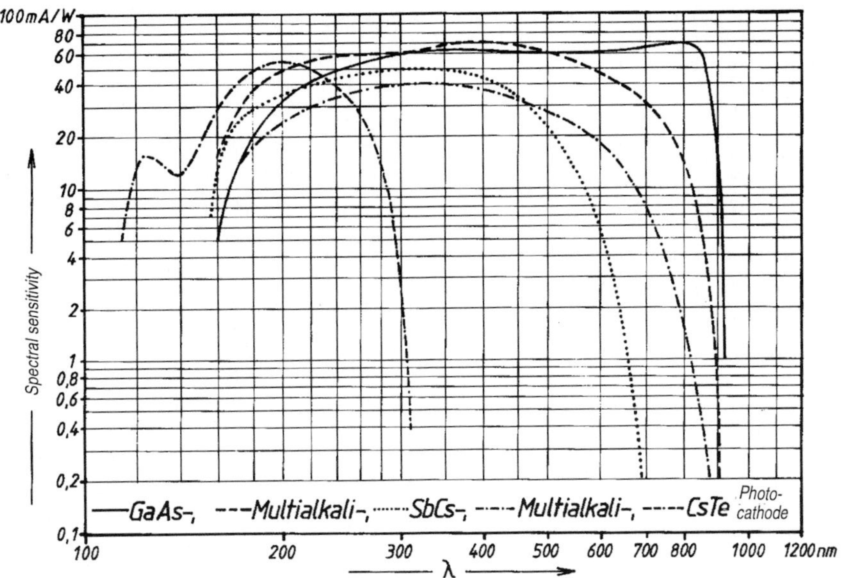

Photomultiplier. Fig. 3. Spectral sensitivity curves for photocathodes of different compositions

Products Division, Tube Marketing Dept., Lancaster, PA 17601–5688, USA.

Photon, <*Photon*>. The → Planck radiation law for the → black-body radiator introduced the light quantum, *hv*, which Einstein (1905) identified as a light particle or photon. According to Einstein, monochromatic light of frequency, *v*, consists of a stream of light quanta or photons which travel in straight lines with the velocity of light, *c*. As a particle, each photon has the energy, *hv*, the → photon energy.

Photon counting, <*Photonenzähltechnik, Impulszähltechnik, Lichtquantenzähler*>, a method of measuring very small light intensities or charges. It is used, for example, in → mass spectroscopy, → Raman spectroscopy, → UV-VIS spectroscopy and in the measurement of → luminescence.

The basis of the technique lies in the fact that a → photomultiplier is capable of measuring one photoelectron which produces an anode pulse 5 to 20 ns (i.e. the time required to pass through a chain of 8–12 dynodes) after its emission. The number of anode-current pulses per second, N_a, is given by:

$$N_a = Q(\lambda) \cdot F \cdot P(h\nu).$$

Here, $Q(\lambda)$ is the quantum yield of the photocathode (number of electrons emitted per incident photon; in the VIS region from 400–500 nm, 20–25 % is a typical value), F is the product of the electron-collecting efficiency of the individual dynodes with typical values of 0.7 to 0.8, and $P(hv)$ the incident light flux in number of photons per second. For $Q(\lambda) = 0.2$, $F = 0.8$ and $P = 10^5$ photons per second, $N_a = 1.6 \cdot 10^4$ anode pulses per second

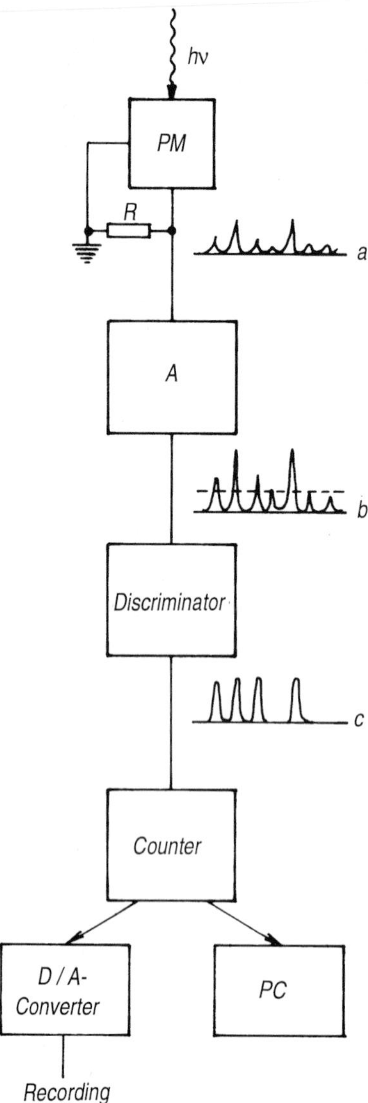

Photon counting. Fig. 1. The principle of photon counting

which corresponds to an anode current of ca. 2.6 nA. This current will produce a voltage of 2.6 mV on a load resistance of 1 MΩ This is the principle of direct-current amplification which is used in UV-VIS spectroscopy, and also to some extent in luminescence spectroscopy, when sufficiently large numbers of photons are produced in the measurements, i.e. at relatively high light intensities.

Only at very low light intensities can individual pulses be detected and the photon counting technique applied. Below saturation (the transition to DC amplification), the number of pulses is directly proportional to the light intensity. The principle of photon counting is illustrated in Figure 1. The anode-current pulses from the photomultiplier are converted to voltage pulses (*a*) by the load resistor, *R*, amplified by the amplifier, *A*, and, if necessary, inverted (*b*). The amplified voltage pulses pass through a discriminator which eliminates weak pulses due to noise, as indicated by the dashed line in (*b*), and selects those above the threshold. Normally, the discriminator also incorporates a shaper which shapes the selected signals so as to obtain a sequence of equal strength (*c*) which can then be sent directly to the counter. The counter is usually a linear counter with an adjustable time constant between 0.1 and 50 s which can be linked directly to a computer (PC) or, via a D/A converter, to a recorder.

The principle described above is exactly that which is used in nuclear physics to count radioactive particles. The concept of impulse counting is therefore quite general, although the expression *photon counting* is preferred in optical spectroscopy. The advantages of this technique over other signal amplification methods, especially DC amplification, are:

- A significantly better signal-to-noise ratio which can be increased to any desired level;
- very well adapted to use with a computer;

– high sensitivity;
– no baseline drift;
– variations of the high voltage on the PM cause no problems.

The signal-to-noise ratio, *S/N*, can be expressed approximately by:

$$S/N = [N_s T_i/(1+2N_n/N_s)]^{1/2}$$

Where

N_s = signal count rate in pulses s^{-1};
N_n = noise count rate, i.e. with closed monochromator slit. In most cases $N_n = N_d$ where N_d is the count rate for the dark noise;
T_i = duration of count or signal integration time in seconds.

S/N is proportional to $(T_i)^{1/2}$, which means that *S/N* can be readily improved by raising T_i. Thus, an increase of the counting time by a factor of 10 gives an *S/N* which is 3.16 times greater.

Ref.: D.V. O'Connor and D. Phillips, *Time-correlated Single Photon Counting*, Academic Press, London, **1979**.

Photon energy, <*Photonenergie*>, the energy of a → photon given, according to Einstein, by $E = h\nu$. Since the photon is a particle we can speak of 1 mole of photons and this is defined as 1 Einstein $= N_L \cdot h\nu$. The energy of 1 mole of photons in the UV-VIS region is given in the table for $h = 6.63 \cdot 10^{-34}$ J·s[1)] $= 1.584 \cdot 10^{-34}$ cal·s $= 0.687 \cdot 10^{-38}$ eV s.

Photon momentum, <*Photonenimpuls*>. According to the theory of special relativity, mass and energy are equivalent and related by the equation, $E = mc^2$, where *c* is the velocity of light. According to Planck and Einstein, the photon has an energy of $E = h\nu$, where *h* is → Planck's constant and ν the frequency of the light. From these two equations the photon mass, m_{Ph}, is found to be $m_{Ph} = h\nu/c^2$. A moving particle of mass, *m*, also has a momentum, *P*, given by the equation, $\vec{P} = m \cdot \vec{v}$. With $|\vec{v}| = c$, the equation for the momentum of the photon is:

$$|\vec{P}_{Ph}| = (h\nu/c^2)c = h\nu/c = h/\lambda,$$

since $c/\nu = \lambda$. In vector form the equation is:

$$\vec{P}_{Ph} = (h/2\pi)\vec{k},$$

where $|\vec{k}| = 2\pi/\lambda$, is the wave vector of the light.

Photon energy. Table.

			$N_L \cdot h\nu$ = Einstein mol^{-1}		
l [nm]	$\tilde{\nu}$ [cm^{-1}]	ν [s^{-1}]	[kJoule]	[kcal]	[eV]
200	50000	$1.5 \cdot 10^{15}$	599	143.0	6.29
250	40000	$1.2 \cdot 10^{15}$	479	114.5	4.96
333	30000	$9.0 \cdot 10^{14}$	359	85.6	3.72
400	25000	$7.5 \cdot 10^{14}$	299	71.4	3.09
500	20000	$6.0 \cdot 10^{14}$	239	57.1	2.48
800	12500	$3.8 \cdot 10^{14}$	152	36.2	1.57
1000	10000	$1.0 \cdot 10^{14}$	40	9.5	0.41

[1)] The best value of → Planck's constant currently available is $h = 6.6262 \times 10^{-34}$ J·s or W·s^2

Experimental and theoretical investigations have shown that the photon also has an angular momentum or spin, $\vec{s}$. The spin vector is always parallel or antiparallel to the direction of motion of the photon (propagation direction of the light), and has the values $\pm(h/2\pi)$; $|\vec{s}| = h/2\pi$. Thus, an electromagnetic radiation field carries an angular momentum which is related to the polarization of the light.
Right-circularly polarized light

$$j = -q(h/2\pi),$$

linearly polarized light

$$j = 0,$$

left-circularly polarized light

$$j = +q(h/2\pi);$$

j is the angular momentum density and q the photon density. For the wavelength of the Na D line $\lambda_D = 5.893 \cdot 10^{-5}$ cm, or $\nu_D = 5.091 \cdot 10^{14}$ s^{-1}, the following values of the above properties of the photon are found: ($h = 6.625 \cdot 10^{-27}$ erg·s)
Photon energy,

$$E_{Ph} = 3.37 \cdot 10^{-12} \text{ erg}$$

Photon mass,

$$m_{Ph} = 3.75 \cdot 10^{-33} \text{ g}$$

Photon momentum,

$$|\vec{P}|_{PH} = 1.12 \cdot 10^{-22} \text{ g} \cdot \text{cm} \cdot \text{s}^{-1}$$

Photon angular momentum,

$$|\vec{s}| = 1.05 \cdot 10^{-27} \text{ erg} \cdot \text{s}.$$

A distinction must be made between the rest mass, m_o at $v = 0$, and the mass, m, when the photon is in motion. When the velocity, $\vec{v}$, is comparable with the velocity of light, c, we have:

$$m_o = m[1 - (\frac{\vec{v}}{\vec{c}})^2]^{1/2}.$$

Since the photon moves with the velocity of light, $\vec{v} = \vec{c}$, its rest mass is therefore zero, i.e. $m_{oPh} = 0$.
Ref.: E. Goldin, *Waves and Photons*, J. Wiley and Sons, Chichester, New York, Brisbane, Toronto, Singapore, **1982**.

Photophysical hole burning, NPHB, nonphotochemical hole burning, *<photophysikalisches Lochbrennen>* → hole burning.

Photoresistive cell, *<Photowiderstandszelle>*, an → optoelectronic component made from a semiconductor such as PbS, PbSe, PbTe, CdS or TlS. The last two are used especially for the near IR (as far as 1.4 μm). Their mode of action depends upon the internal → photoelectric effect in which electrons from the valence band are promoted into the conduction band, thereby increasing the electrical conductivity. The excitation of impurity centers also contributes to the conductivity; or decrease of resistance. The region of application of these devices is the near IR where lead tellu-

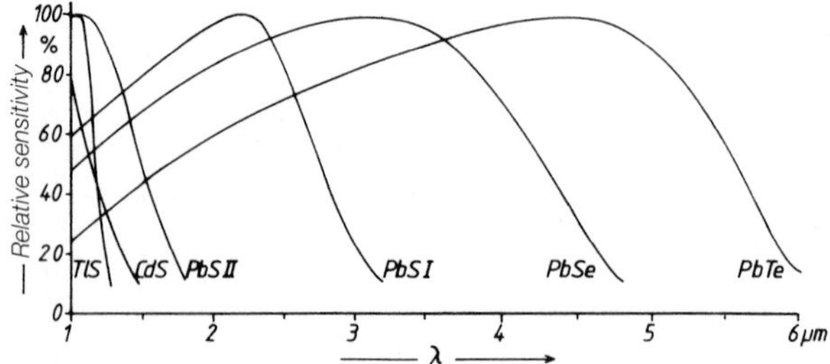

Photoresistive cell. Fig. 1. The spectral sensitivity of some photoresistive cells

ride (PbTe), which is used from 1 μm (10,000 cm^{-1}) to about 6 μm (1600 cm^{-1}), has the widest sensitivity range; Figure 1.

Photosedimentometer, <*Photosedimentometer*>, a → photometer which utilizes the principle of gravimetric sedimentation and is predominantly used to determine the size and distribution of particles in the dispersed state. It can also be used to study sedimentation processes. An instrument of this type, known as the *Lumosed*, has been developed by Messrs F.K. Retsch. It is illustrated schematically in Figure 1. The light of a halogen lamp, undispersed, is divided into three beams by the plane mirrors, P_1, P_2 and P_3. The beams pass horizontally through a cuvette and onto three detectors (sensors), S_1, S_2 and S_3. The three beams are arranged in geometric sequence, one above the other, and relatively rapid measurements can be made with each. The change of → turbidity with time, i.e. the apparent absorption (→ absorbance, apparent) which gives the concentration of the solid matter, is measured during a sedimentation process. The primary result, naturally, is an intensity-time curve. The data measured by the three detectors are passed via an A/D converter to a computer which takes over the complete evaluation process and reports the particle-size distribution in the form of a linear or logarithmic granulation equation. The software contains all the important physical and mathematical laws, i.e. the relationships derived from them, suitably programmed. The instrument is calibrated with a test sample for which the particle-size distribution is known. For powders the range of measurable particle sizes is 1–250 μm.
Source: F.K. Retsch, D-42781 Haan 1, Germany.

Photoselection, <*Photoselektion*>, a use of the fact that fluorescence always derives from the lowest excited state of the molecule, no matter which electronic state was excited; → Kasha's rule. However, there is an → anisotropy of light absorption in many unsaturated molecules, especially the planar species. The effects are marked in the case of the condensed aromatic hydrocarbons and their heterocyclic analogues. As an example, consider a fixed anthracene molecule (Figure 1a) which has D_{2h} symmetry. The long-

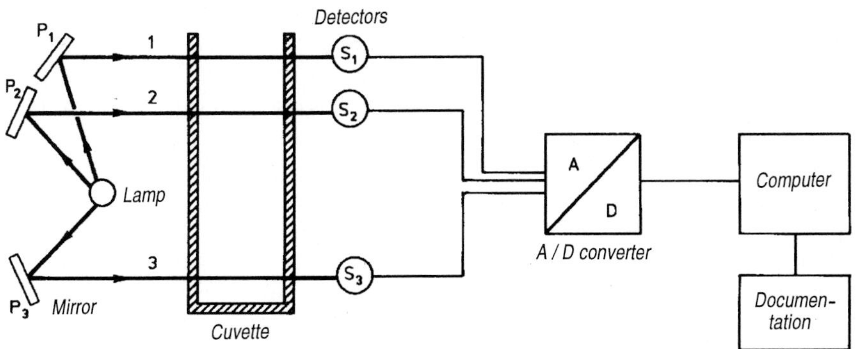

Photosedimentometer. Fig. 1. Schematic diagram of the *Lumosed*

a) Anthracene

b) 2,7 Diazaphenanthrene

Photoselection. Fig. 1. Axes in anthracene (a) and 2,7-diazaphenanthrene (b)

wavelength transition, ^1L_a, (→ Platt classification) is polarized along the short molecular axis (y). The second transition, ^1B_b, is polarized in the perpendicular direction, along the long axis of the molecule (x). If the molecule is irradiated with light polarized parallel to the short axis then the ^1L_a transition will be excited. To excite the ^1B_b transition, the plane of polarization must be turned through 90° at the appropriate wavelength. If the wavelength of the polarized light is varied over the spectrum, only those molecules are excited whose axes lie parallel to the plane of polarization of the light. Since fluorescence always takes place from the lowest singlet excited state, the plane of polarization of the fluorescence rotates through 90° when the second excited state is irradiated because the transition dipole moments of these two states are perpendicular to each other. If this effect is to be observed, the molecule must not change its orientation in space during the lifetime of the lowest singlet excited state; 10^{-9} – 10^{-7} s. Thus, the experiments are carried out in frozen glasses at liquid nitrogen temperature or in polymer films. Under these conditions, the polarized light selects from the statistically distributed population those molecules for which the transition dipole moment at the particular wavelength is parallel to the plane of polarization; hence the name of the method.

The experiment may be performed in a number of different ways each of which gives rise to a different spectrum:

a) The wavelength of the exciting light, λ_e, is varied and the degree of polarization of the fluorescence at a fixed wavelength, λ_f, is measured. The resulting spectrum is the absorption-polarization-fluorescence (APF) spectrum.
b) The exciting wavelength is held constant while the degree of polarization throughout the fluorescence spectrum is measured. This gives the fluorescence- polarization (FP) spectrum.
c) The same procedure as in a), but the phosphorescence is measured giving the absorption-polarization-phosphorescence (APPh) spectrum.
d) Measurement of the phosphorescence as in b) gives the phosphorescence-polarization (PhP) spectrum.

In the APF spectrum, the transition from one electronic state to the next highest is usually accompanied by a change in the degree of polarization from positive to negative values. The example of 2,7-diazaphenanthrene (Figure 1b) is shown in Figure 2. The change in the degree of polarization at ca. 34,000 cm^{-1} is clearly seen and makes a definitive assignment of the electronic transitions possible.

The degree of polarization, which is plotted as a function of wavelength or wavenumber, is given by:

$$P = \frac{I_{\parallel} - I_{\perp}}{I_{\parallel} + I_{\perp}}$$

or

$$P = \frac{3\cos^2\alpha - 1}{\cos^2\alpha + 3}$$

where, α is the angle between the transition moments of the absorption and emission processes. For $\alpha = 0$ (parallel) $P = 0.5$ and for $\alpha = 90°$ (perpendicular) $P = -0.33$. These figures hold for all molecules with twofold axes of rotation. Figure 2 shows that the APF spectrum is often structured which is due to the coupling of electronic and vibrational excitation. For further details and experimental methods see: F. Dörr, *Angew. Chem.*, **1966**, *78*, 457; and F. Dörr, *Polarized Light in Spectroscopy and Photochemistry in Creation and Detection of the Excited State*, vol. 1, pt. 1, (Ed.: A.A. Lamola), Dekker, New York, **1971**, p. 53 ff.

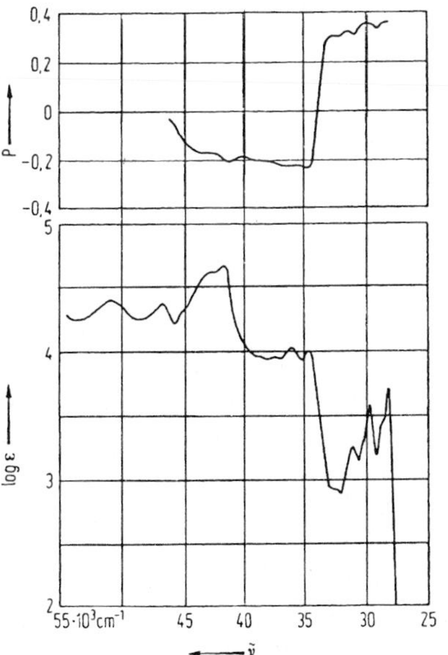

Photoselection. Fig. 2. The APF spectrum of 2,7-diazaphenanthrene and the corresponding absorption spectrum

Photosensitivity, *<Lichtempfindlichkeit>* → light sensitivity.

Photostationary state, *<photostationärer Zustand>*, a state in a dynamic process where the velocities of the forward and backward reactions are equal. This means that, macroscopically, no change in an observable quantity can be detected and there is effectively an equilibrium. In the case of excitation by light, we speak of a photostationary state when the number of molecules excited by the light in unit time is equal to the number deactivated in unit time. For → fluorescence this means that the intensity remains constant in time, which is an important precondition for the evaluation of fluorescence measurements. If a back reaction, e.g. a dark reaction, is possible in a system which is deactivated by a → photochemical reaction, then a photochemical equilibrium will be established.

Phototransistor, *<Phototransistor>*, an → optoelectronic component which, in its normal form, consists of three alternating *p* and *n* semiconductor (Si or Ge) layers. There are two types, the *pnp* or the *npn* transistor, according to the sequence of the layers. A *pnp* transistor and its associated electrodes and power supply is shown in Figure 1. The middle layer, which in reality is very thin, is called the base; the layer to the left is the emitter and to the right the collector. In the circuit shown here, the right-hand unit (base/collector) acts as a barrier layer while the left-hand unit is biased in the conducting direction. This gives a voltage amplification, i.e. a small change in the current of positive holes coming from the emitter causes a correspond-

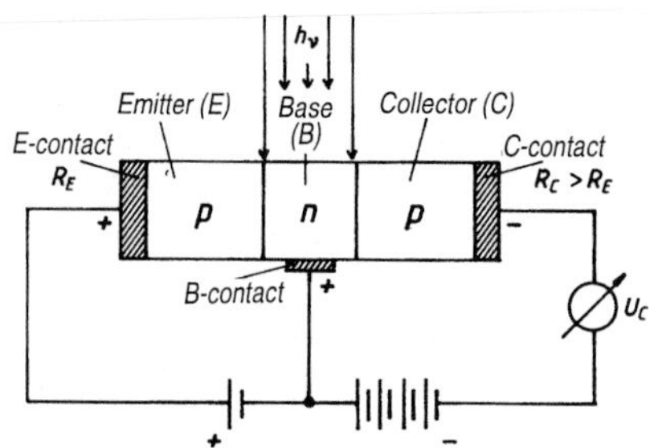

Phototransistor. Fig. 1. Schematic diagram of a *pnp* transistor

ing change in the electron current flowing from the collector to the base. But the high resistance, R_C, of the collector contact converts this small change of current into a large change of voltage and the ratio of the gate voltage to the output voltage, U_E/U_C, is equal to the ratio of the resistances of the emitter and collector contacts, R_E/R_C. The functional mode as a phototransistor depends upon the fact that an emitter lies at the entrance of the reverse biased collector base diode. Because of the internal → photoelectric effect, irradiation of the *np* junction increases the radiation-controlled reverse junction current in the transistor (→ barrier-layer photoeffect). Typical amplification values are 100 to 700 times. The amplification depends upon the current. Thus, phototransistors do not show the linear relation between light intensity and photocurrent that there is in → photodiodes. Therefore, their main applications lie in switching functions which are triggered by changes in illumination, e.g. dark/light changes. The photodarlington transistor is a special development.

Ref.: J. Wilson and J.F.B. Hawkes, *Optoelectronics: An Introduction*, Prentice Hall International, Englewood Cliffs (NJ), London, Sydney, Toronto, **1983**.

PhP spectrum, <*PhP-Spektrum, Phosphoreszenzspektrum*> → photoselection.

Pigment, <*Pigmentfarbstoff, Pigment*>, a → colorant which is insoluble in water and soluble only with difficulty in organic solvents. Its properties are very dependent upon its crystal structure and particle size. For this reason, the same pigment can have different colors and, accordingly, great care must be taken in the technical preparation of these materials. Pigments are very important in lacquers, paints, textiles, plastics, etc. → Reflection spectroscopy has been successfully used to measure pigments. In recent years, → photoacoustic spectroscopy has also proved very suitable for this purpose. Like → dyes, pigments are classified by the Color Index.

Ref.: K. McLaren, *The Colour Science of Dyes and Pigments*, 2nd ed., Adam Hilger Ltd., Bristol, Boston, **1986**.

Pigmentation, pigment, <*Körperfarbe*>, the color which a body shows when illuminated with white light. It is caused by the fact that not all the → spectral colors are reflected to the same extent, since some of the incident light is absorbed by the body. The color of the body thus appears to be that of the → complementary color to that which is absorbed. Since we understand by color the color of a body in white light, a body viewed in artificial light, which has a much reduced blue component in the short-wavelength region, often appears to

be changed in color. The difficulty of choosing a colored material in artificial light is well enough known.
The term *pigmentation* appears to be applied exclusively to the color of living organisms (plants, animals and man) and *pigment* in all other cases.
Ref.: K. McLaren, *The Colour Science of Dyes and Pigments*, 2nd ed., Adam Hilger Ltd., Bristol, Boston, **1986**.

Pixel, <*Pixel*>, a single photodiode in a → diode array detector.

Planck-Einstein relation, <*Planck-Einstein-Beziehung*>. In his 1905 interpretation of the external → photoelectric effect, Einstein started with Planck's formulation (1900) of the quantum hypothesis (→ Planck's constant) and went on to postulate that light propagates as individual light quanta which travel with the velocity of light. These light quanta, also known as → photons, are tiny energy packets of energy $E = h\nu$, i.e. they may be compared to corpuscles or particles. The equivalence of electromagnetic radiation and energy in the above equation is also known as the Planck-Einstein relation and leads directly to the → Bohr-Einstein relation, the fundamental equation of spectroscopy.

Planck's constant, <*Plancksches Wirkungsquantum, Planck-Konstante*>, a natural constant, symbol h, which has the dimensions energy·time and is measured in $\mathrm{W} \cdot s^2$ or J·s. These are the units of action. The constant was first proposed by Planck in his general radiation (distribution) law of 1900. At the same time, Planck introduced the quantum hypothesis that energy is not continuous but quantized in units given by $\varepsilon = nh\nu$ ($n = 1, 2, 3 \ldots$ or $n = 1/2, 3/2, 5/2 \ldots$), where $h\nu$ is the energy of a → photon of frequency ν. The numerical value of Planck's constant is $h = 6.6252 \cdot 10^{-34}\ \mathrm{W \cdot s^2}$.

Planck's radiation (distribution) law, <*Plancksche Strahlungsformel*>. This formula was derived by M. Planck in 1900. It describes the dependence of the energy of a → thermal radiator, in the form of the ideal → black-body radiator, upon the parameters of temperature, T, and wavelength, λ.

$$S_\lambda\ \mathrm{d}\lambda = \frac{c^2 h}{\lambda^5 (\mathrm{e}^{hc/\lambda kT} - 1)}\ \mathrm{d}\lambda.$$

$S_\lambda \mathrm{d}\lambda$ is the energy in the wavelength range, λ to $\lambda + \mathrm{d}\lambda$, radiated in unit time from unit-surface area into the surrounding hemisphere. S_λ ($\mathrm{Wm^{-3}}$) is known as the spectral radiation density, h is → Planck's constant and c is the velocity of light. In wavenumbers, Planck's formula reads:

$$S_{\tilde{\nu}} \mathrm{d}\tilde{\nu} = \frac{c^2 h}{\mathrm{e}^{\ hc\tilde{\nu}/kT} - 1}\ \mathrm{d}\tilde{\nu}.$$

In Figure 1, the spectral energy density, $S_\lambda \mathrm{d}\lambda$, is shown as a function of wavelength, λ, for four temperatures (isotherms). The general characteristics of a thermal radiator can be recognized in these spectral energy-distribution curves for an ideal black body. The maximum of the distribution lies in the NIR region. The energy decreases rapidly towards the visible and ultraviolet, i.e. to shorter wavelengths. The energy also falls towards the mid IR, but not so steeply. The form of the curves shows that the normal thermal radiator is not a suitable source in the UV, but can be used in the VIS, NIR and IR regions. The curves also show that the position of the maximum moves towards shorter wavelengths with increasing temperature. This shift of the maximum agrees with Wien's displacement law (→ Wien's radiation law): $\lambda_{max} \cdot T$ = constant = 0.2898 cm K. Further, the total radiated energy, i.e. the area under the isotherms in Figure 1, increases extremely rapidly with increasing temperature. The increase follows the Stefan-Boltzmann law and is proportional to the fourth power of the absolute temperature:

$S = \sigma T^4$ where $\sigma = 5.6697 \cdot 10^{-12}$ W cm^{-2} K^{-4}.

Plane grating, <*Plangitter*>, a → grating obtained by ruling on a plane mirror. A concave mirror is required in order to form an image of the spectrum produced by a plane grating (→ grating monochromator).

Plasma, <*Plasma*>, a gas in which a certain percentage of the atoms or molecules is dissociated into positive (ions) and negative (electrons) charge carriers so that freely mobile, electrically charged particles exist as well as neutral species. An important prerequisite for the maintenance of the plasma is the presence of sufficient numbers of electrons which move in the electric field and transfer their energy to other particles (molecules, atoms) in the plasma in collisions. In this way, for example, the required high temperatures for → atomization and → ionization in → atomic emission spectroscopy are reached. Argon is usually used as plasma gas, especially in → inductively coupled plasmas, because it is relatively easy to ionize. However, nitrogen can also be used.

Plasma burner system, <*Plasmabrennersystem*>, consists, in → atomic

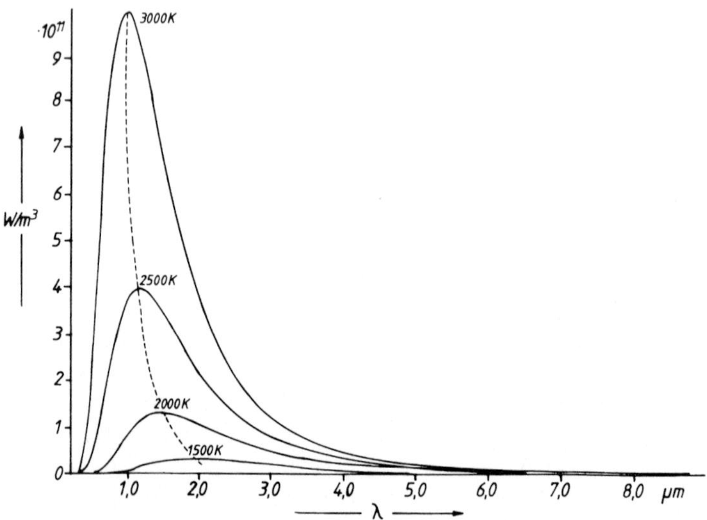

Planck's radiation (distribution) law. Fig. 1. Spectral radiation density as a function of wavelength for 4 temperatures

emission spectroscopy with → inductively coupled plasma, of the plasma torch, expansion chamber and end cap, Figure 1. The plasma torch consists of three concentric quartz tubes. In the innermost tube, the nebulized sample is transported by a stream of argon (aerosol-carrier gas) into the plasma. Through the next tube, an auxiliary argon supply enters and through the outer tube, argon as the plasma gas. Argon therefore fulfills the functions of both plasma-forming gas and coolant. The tulip-shaped middle tube at first restricts the flow and then causes a high acceleration along the inside of the outer tube. The result is that essentially those argon atoms are ionized in the induction field which are in the turbulent region in the middle of the tube, while the fast gas stream along the outer tube performs the cooling function. However, if powers higher than 2 kW are used the fast flowing gas is also ionized and the temperature becomes so high that the quartz tube begins to melt.

The auxiliary argon supply is not required for the analysis of aqueous samples. It is only required for the nebulizing of organic solutions. The argon usage is 0.5–1.5 l/min for the aerosol-carrier gas, 0–1 l/min for the auxiliary gas and 10–20 l/min for the plasma gas.

The plasma torch is connected to the expansion chamber with a ball joint. The function of the chamber is to remove large droplets of liquid which would reduce the stability of the plasma. Some 90% of the sample drawn up is deposited again in the expansion chamber. It flows to waste through a tube which must dip under the surface of the liquid so that no unwanted air is drawn in. The expansion chamber is normally made of glass, but versions in teflon or other plastics are available. It is closed by means of the PTFE end cap through

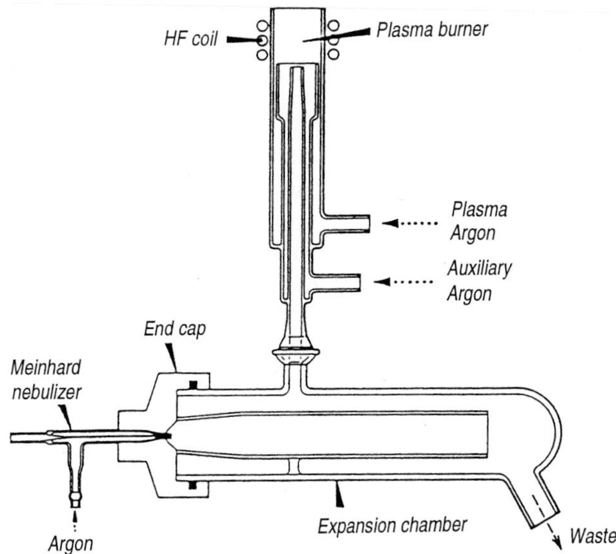

Plasma burner system, Fig. 1. The construction of a plasma torch for ICP

which the nebulizer jet enters. The nebulizer should produce as fine an aerosol as possible, but the rate of suction is limited to ca. 1–2 ml/min by the power of the plasma generator and the construction of the torch. This limit is necessary in order to maintain a stable plasma. The passage from sample to plasma can be divided schematically into three phases; liquid transport, nebulization and aerosol transport. The nebulization can be achieved pneumatically or ultrasonically. The commonest instrument is the concentric (Meinhard) nebulizer. In principle it is comparable with the nebulizers used in atomic absorption spectroscopy, but it has an exit capillary with a smaller cross section and is also not adjustable.

Ref.: A. Montaser and D. Golightly, *Inductively Coupled Plasmas in Analytical Atomic Spectrometry*, VCH Publishers, Deerfield Beach, Florida, **1986**.

Platinum temperature, platinum point, <*Platinpunkt*>, the melting temperature of platinum (2046 K) used in the definition of the → candela, the unit of light intensity.

Platt classification, <*Platt-Klassifizierung*>. In 1949 Platt published a free-electron theory of the electronic spectra of the aromatic hydrocarbons based upon the recognition that in many cases all the carbon atoms in the molecule lie on its periphery. Platt suggested that the π electron spectrum of a molecule such as phenanthrene ($C_{14}H_{10}$) could be described in terms of the wave functions of an electron confined to move on the circumference of a circle. The solutions of → Schrödinger's wave equation for this problem are the energies:

$$E_n = n^2h^2/2ml^2 \qquad n = 0, \pm 1, \pm 2 \ldots.$$

with the corresponding wave functions (free-electron orbitals):

$$\Psi_{+n} = \sqrt{(1/2\pi)}\exp[+in\phi] \quad \text{and}$$
$$\Psi_{-n} = \sqrt{(1/2\pi)}\exp[-in\phi]$$

where h is Planck's constant, m the mass of the electron, l the circumference of the ring, ϕ the angle which determines the position of the electron around the ring and n a → quantum number. We see immediately that there is a single solution for $n = 0$ while the solutions for which $n > 0$ are doubly degenerate so that the number of π electrons required for a stable closed shell is 2 plus an integer multiple of 4, i.e. exactly the number given by the Hückel $4N + 2$ rule. The orbital angular momentum of the electrons occupying the above orbitals is $nh/2\pi$, and we have an → energy-level diagram in which there are pairs of degenerate energy levels. For each pair of levels, one corresponds to electrons rotating in a clockwise direction and the other to electrons circulating in a counterclockwise manner around the ring. For a hydrocarbon having $P = 4r + 2$ carbon atoms, and therefore P π electrons, the levels $n = 0 \pm 1 \ldots \pm r$ are fully occupied by electrons. The ground state is a closed shell system with zero angular momentum. The lowest excited states arise from the configuration $r^3(r+1)$, and to reach these configurations we require to change the angular momentum either by one unit of $h/2\pi$, $r \to r+1$ or $-r \to -(r+1)$ or by $2r+1$ units, i.e. $r \to -(r+1)$ or $-r \to (r+1)$. Promotion of electrons to higher levels therefore gives rise to states of higher angular momentum and Platt denoted states having angular momenta of 0, ±1, ±2 ... $h/2\pi$ by A, B, C ... respectively and those having $\pm(2r+1)$, $\pm(2r+2)$, $\pm(2r+3)$ ... by L, M, N ...; independent of the value of r. Thus, the ground state of an aromatic hydrocarbon is an A state and the four lowest excited states are L and B states. Since the → selection rule for electronic transitions requires that there should be a change of angular momentum of $\pm h/2\pi$ only, transitions to the B states are allowed while those to the L states are forbidden. Thus, in the most simple view, the low-energy region of the spectrum of an aromatic hydrocarbon will consist of two strong and two very weak bands. The periodic potential of the nuclei in a real hydrocarbon removes the degeneracy of the L states, but not that of the B states. The degeneracy of the B states is, however, removed by the distortions and addition of bonds required to convert the circle of carbon atoms into a real hydrocarbon and this process also mixes the four states. To accommodate this loss of degeneracy and mixing, Platt extended his notation using the symbols, L_a, L_b, B_a and B_b which are widely used to describe the spectra of the aromatic hydrocarbons. Left superscripts are frequently added to the Platt symbols to indicate singlet and triplet states. The relationship to the → Clar classification is ${}^1L_b = \alpha$, ${}^1L_a = p$, ${}^1B_a$ and ${}^1B_b = \beta$ and β'.

Ref.: H.H. Jaffe and M. Orchin, *Theory and Applications of Ultraviolet Spectroscopy*, J. Wiley and Sons Inc., New York, London, **1962**; J.N. Murrell, *The Theory of the Electronic Spectra of Organic Molecules*, Methuen and Co. Ltd., London, J. Wiley and Sons Inc., New York, **1963**.

Pleochroism, <*Pleochroismus*>, a word derived from the Greek to describe the property of the many colors which can be observed with anisotropic, transparent crystals. Such crystals absorb different wavelength regions of light incident from different directions, so that different colors emerge depending upon the side from which the crystal is viewed. If the pleochroism is limited to two directions it is called → dichroism. See also → anisotropy of light absorption.

Pneumatic detector, <*pneumatischer Empfänger*>, a thermal radiation detector used preferably in the IR. In pneumatic detectors, the expansion of a gas in a small, closed chamber, which results from absorbing radiation, is measured. The technique used to measure this expansion depends upon the construction of the detector. The oldest method is that which makes use of a membrane capacitor. In this technique, the gas together with a radiation absorber is enclosed in a metal container which has a window on one side, see Figure 1. The opposite wall of the container is made of a thin, gastight aluminium foil which forms a capacitor in conjunction with a fixed metal plate. When no radiation is falling on the detector, the capacitance is determined by the geometry of the device, primarily the separation between the plates. If radiation now falls on the absorber it is warmed and gives up heat to the surrounding gas which expands and pushes out the flexible chamber wall. This reduces the plate separation of the capacitor and increases its capacitance. This

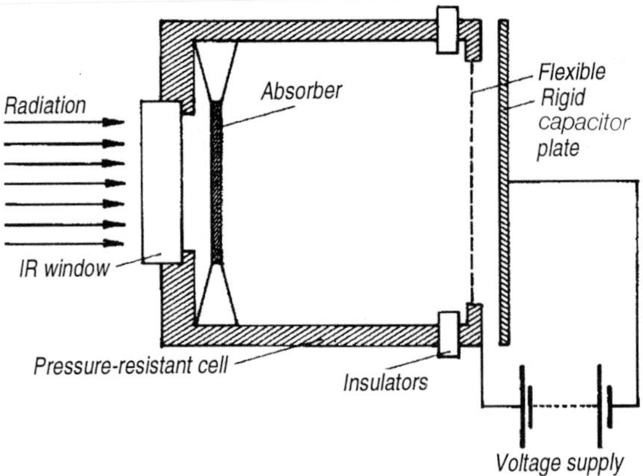

Pneumatic detector. Fig. 1. Section through a pneumatic detector

increase in capacitance is the measure of the incident radiation.

The static capacitance measurement can be replaced by a dynamic procedure if the incident light is modulated and the → time constant of the whole detector made sufficiently small so that it can follow the modulation. The output of the detector can then be connected directly to an electronic amplifier to amplify the small alternating voltage produced.

Another model uses the → photoacoustic effect in which the pressure wave produced in the detector chamber by the absorption of modulated light is detected as a sound wave by a microphone. In this case, the gas in the chamber is frequently itself capable of absorbing the radiation. When working with a black standard as radiation absorber, i.e. in the → signal saturation region of → photoacoustic spectroscopy, then a nonabsorbing gas can be coupled in. The → Golay detector is a further very efficient variant of the pneumatic unit.

Nonselective pneumatic detectors, i.e. equally effective at all wavelengths, can be constructed and so also can expressly selective versions. If a black absorber is used in the chamber the detector responds to radiation of every wavelength from the UV into the IR region. If a selective absorber is used then the detector responds to those wavelengths which it absorbs. The gas itself can be the selective absorber, e.g. in effect, CO_2 absorbs only at the wavelengths 2.72, 4.25 and 14.97 μm, while CO absorbs at 2.37 and 4.66 m. This selective absorption is used in the → IR gas analyzer URAS which was used for on-line (process) analysis at an early date.

Ref.: G.K.T. Conn and D.G. Avery, *Infrared Methods*, Academic Press, New York, London, **1960**.

Pockels effect, <*Pockels-Effekt*>, the change in the optical properties of uniaxial or cubic crystals under the influence of an electric field and named after its discoverer. Examples are crystals of potassium dihydrogen phosphate (KDP, see Figure 1) and ammonium dihydrogen phosphate (ADP). Both crystals are naturally birefringent (doubly refracting). But there is a direction in which the → birefringence is not noticeable, namely when the light passes through the crystal along the optic axis. However, if an electric field is applied parallel to the optic axis then this direction also shows birefringence. The two crystallographic axes, a and b, are no longer equivalent and the refractive index for light polarized along axis, a, is different from that for light polarized along axis, b. The difference between the refractive indexes and the applied voltage are related by the equation:

$$\Delta n = n_1 - n_2 = k \cdot U.$$

The difference, Δn, increases linearly with the applied field and the Pockels effect is also called the linear electrooptical effect for that reason. Figure 2 is a schematic illustration of a Pockels cell. The shaded section in Figure 1 represents the Z cut plate which is required to construct such a cell. Like the Kerr cell (→ Kerr effect) it can also be used as a light shutter, but it has the great advantage that, if suitably constructed, it can be used in the UV region as far as ca. 200 nm.

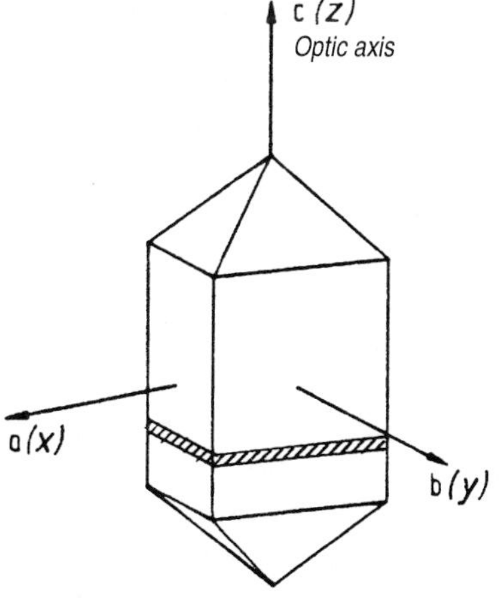

Pockels effect. Fig. 1. A KDP crystal with the crystallographic axes a(x), b(y) and c(z). c(z) is the optic axis

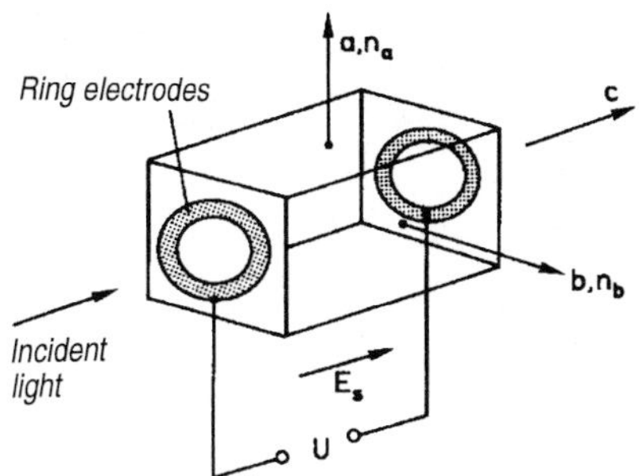

Pockels effect. Fig. 2. Schematic illustration of a Pockels cell

Polarimeter, <*Polarimeter*>, a group of instruments to which both visual and photoelectric polarimeters belong. The → spectropolarimeters form an important subgroup of the latter. The basic requirements for a simple visual polarimeter are a light source, a → polarizer, an → analyzer for which the angle of rotation can be read off from a scale, and optics for illumination and observation. The substance to be measured is placed in a liquid or gas cuvette between the polarizer and the analyzer. The sample might also be an optically active transparent solid.

Figure 1 shows the optics. A monochromatic beam from the source, *a*, falls upon the lens, *b*, which focuses *a* at the aperture, *f*, in such a way that the diameter of the image of *a* is slightly larger than the free diameter of *f*. The beam emerging from *b* passes through the polarizer, *c*, leaving it as linearly polarized light which then passes through the aperture, *d*, and into

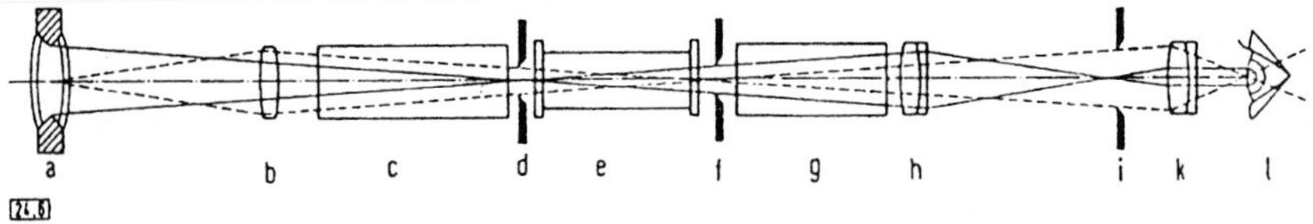

Polarimeter. Fig. 1. The basic components and optical path of a simple visual polarimeter, see text

the sample, *e*, (here in a liquid cuvette). After passing through *f* the beam enters the analyzer, *g*, which is fitted with a sector scale. *g* is identical to *c* in construction. Following *g* there is a telescope which consists of an objective, *h*, an ocular aperture, *i*, and an eyepiece, *k*. *k* is equipped with a screw with which the telescope can be focused sharply on *d*. The diameter of *f* and the brightness of the source together determine the brightness of the measuring field. A measurement is made by turning *g* to obtain the darkest field, first without and then with the sample. The scale reading is taken each time and the difference is the angle of rotation of the sample. A clockwise rotation of the analyzer can mean either a right-handed rotation by the sample of $\alpha°$ or a left-handed rotation by $(180-\alpha)°$. Therefore, for substances whose sense of rotation is not known, this must be determined separately by means of measurements on solutions of different path lengths or concentrations. Setting the instrument to the minimum brightness is always less accurate than setting the two halves of a split field to equal brightness. To improve setting accuracy, half-shadow polarimeters have been constructed. In these instruments the measuring field is divided into two or three adjacent regions which are set to equal brightness.

Half-shadow polarimeters are normally constructed using a Lippich half-prism and a Laurent half-shadow plate. Figure 2 shows a self-equalizing photoelectric polarimeter which uses a → Faraday modulator. Polarizer, *a*, and analyzer, *e*, are crossed. The Faraday modulator, *d*, lies behind the → Wollaston prism, *c*. The light emerging from the polarizer is split by the Wollaston prism into two weakly diverging beams polarized perpendicularly to each other. The prism, *c*, is positioned such that, in the absence of a sample, the planes of polarization of the two beams of light leaving the prism both make angles of 45° with the plane of polarization of the light entering the prism. The two beams then have equal intensities. On passing through the

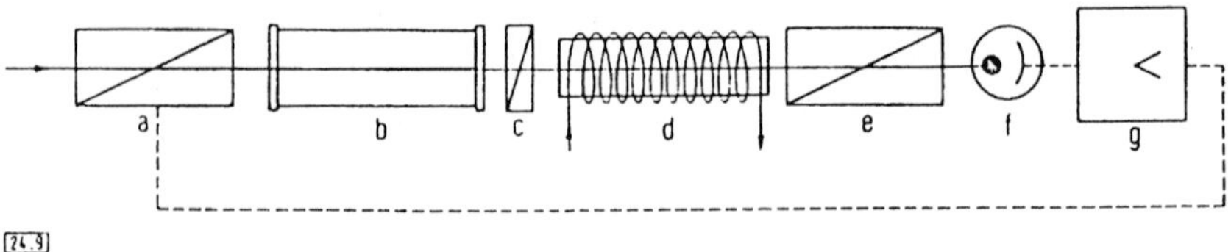

Polarimeter. Fig. 2. The principle of a self-balancing polarimeter

Faraday modulator, *d*, the planes of polarization of the two beams experience an alternating rotation and the beams emerging from *d* add together to give a constant light intensity. However, when an optically active substance is introduced, there is also an alternating intensity superimposed upon the constant signal. Both constant and alternating intensities fall upon the → photomultiplier, *f*, which delivers a photocurrent proportional to the total light flux. The amplifier, *g*, picks out the alternating component of the current and the amplified signal drives a phase-dependent servomotor which rotates the polarizer, *a*, in the sense opposing the rotation, *α*, until the signal disappears. The dashed line in Figure 2 represents this electromechanical connection. At the moment of balance, i.e. at the disappearance of the signal, the angle of rotation of *a* is equal to the rotation of the sample and can be read off on a screen. Precision polarimetry is possible with this technique.

A photoelectric polarimeter which does not use the Wollaston prism, *c*, in Figure 2 and works only with the Faraday modulator, *d*, has been developed by Zeiss. It is the digital polarimeter model OLD, see Figure 3. In other instruments, e.g. the Perkin-Elmer models 241 and 241 MC, instead of using a Faraday modulator, modulation is achieved by rocking the polarizer, and therefore the plane of polarization of the light passing through it, through an angle of ±0.7 degrees about the long optic axis with a frequency of 50 Hz.

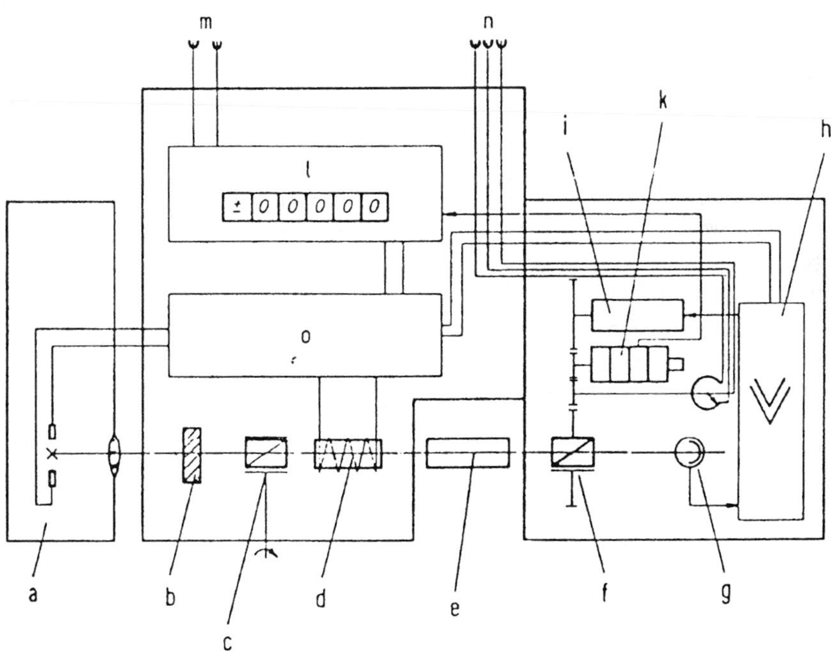

Polarimeter. Fig. 3. Digital polarimeter, Zeiss model OLD

Ref.: H.G. Krüger, *Polarimetry*, in: Ullmann's Encyclopedia of Technical Chemistry, 5th ed., VCH Publishers, Weinheim, New York, **1988**.

Polarizability, <*Polarisierbarkeit*>, a shift of the positive and negative charge with respect to their mutual center of gravity under the influence of an electric field. The shift produces an induced electric dipole moment, $\vec{\mu}_{\text{ind}}$, which is directly proportional to the applied field, $\vec{E}$.

$$|\vec{\mu}_{\text{ind}}| = \alpha\,|\vec{E}|.$$

In this equation, α is the polarizability which, for gases at low pressure, can be determined from the relative permittivity, ε_r.

$$\varepsilon_r = 1 + \frac{N_A \rho}{M\varepsilon_o}\alpha.$$

Here, M is the molar mass, ϱ the density, ε_o the vacuum permittivity and N_A the Loschmidt or Avogadro number. For liquids and solutions the Clausius-Mosotti equation is used giving the molar polarizability, P_{Mol}, as:

$$P_{\text{Mol}} = \frac{\varepsilon_r - 1}{\varepsilon_r + 2}\frac{M}{\rho} = \frac{1}{3}\frac{N_A}{\varepsilon_o}\alpha.$$

With an alternating electromagnetic field in the form of light in the visible region, α can be obtained from a measurement of the → refractive index, using the Lorentz-Lorenz relation and the molar refractivity, R_M. This can be derived directly from the Clausius-Mosotti equation since Maxwell has shown that $\varepsilon_r = n^2$.

$$R_M = \frac{n^2 - 1}{n^2 + 2}\frac{M}{\rho} = \frac{1}{3\varepsilon_o}N_A\alpha_e$$

where, n is the refractive index. However, one obtains only the electron polarizability, α_e, in this way. For molecules, the above expression for the induced dipole moment applies generally for all directions of the applied electric field vector, $\vec{E}$, the Cartesian components of which are E_x, E_y and E_z. This is true in an analogous manner for the vector, $\vec{\mu}_{\text{ind}} \equiv \vec{P}$. If we label these components, P_x, P_y and P_z, we have:

$$P_x = \alpha_{xx}E_x + \alpha_{xy}E_y + \alpha_{xz}E_z$$

$$P_y = \alpha_{yx}E_x + \alpha_{yy}E_y + \alpha_{yz}E_z$$

$$P_z = \alpha_{zx}E_x + \alpha_{zy}E_y + \alpha_{zz}E_z.$$

These three equations show that the x component of P, P_x, depends not only on the x component, E_x, of the vector, E, but also on the y and z components. The same is true of P_y and P_z. x, y and z are the axes of a molecule-fixed coordinate system. α_{xx}, α_{xy} ... are constants which are independent of the directions of $\vec{E}$ and $\vec{P}$. They are the components of the polarizability tensor and the three equations above are generally written in the form:

$$\begin{bmatrix} P_x \\ P_y \\ P_z \end{bmatrix} = \begin{bmatrix} \alpha_{xx} & \alpha_{xy} & \alpha_{xz} \\ \alpha_{yx} & \alpha_{yy} & \alpha_{yz} \\ \alpha_{zx} & \alpha_{zy} & \alpha_{zz} \end{bmatrix} \begin{bmatrix} E_x \\ E_y \\ E_z \end{bmatrix}.$$

The polarizability tensor can be represented in a visual way by the polarizability ellipsoid.

The above relationships provide an important foundation for the theoretical study of the → Raman effect.

Polarization foil, sheet polarizer, <*Polarisationsfolie*>, an artificial → polarizer of sheet-like form the size of which is effectively unlimited. The most useful are microcrystals of herapathite (quinine sulfate periodide). They possess a marked → dichroism and are orientated and embedded in a cellulose film. Another possiblility is to induce → strain birefringence in a film of cellulose hydrate or polyvinyl alcohol by first stretching the film and then coloring it with special dyes. These polarization foils transmit about 25% of the light in the parallel orientation and appear light gray in color. In the crossed orientation the transmission is less than 0.01%. Polarizers of this type are easy to use and have many applications. Polarization filters made from iodine-colored polyvinyl alcohol films are sold commercially under the trade name *Polaroid*. Manufacturer: Polaroid Corporation, Boston (MA), 021394687

Polarized light, <*polarisiertes Licht*>. In 1808, the French physicist, E.L. Malus, made an observation which led to the following conclusion. If → light is reflected from a transparent medium (e.g. glass or water), the symmetry about the direction of propagation is lost, whereas natural light is completely axially symmetric about the direction of propagation. This observation was the origin of the study of polarized light.
A more exact study of the phenomenon showed that light waves incident upon a reflecting surface at a particular angle, α_p, the polarization angle or Brewster angle (→ Brewster's law), are only reflected when the plane of vibration is perpendicular to the plane of incidence. The reflected light then vibrates only in that plane and is said to be linearly polarized. Light vibrating parallel to the plane of incidence is not reflected but rather transmitted. In general, the plane of vibration of polarized light produced by reflection is perpendicular to the plane of incidence (reflection).
Brewster showed that the polarization angle, α_p, which gives the maximum polarization of the light depends upon the refractive index, n, of the glass used. According to → Brewster's law, the polarization angle, α_p, is the angle of incidence for which the tangent is equal to the refractive index: $\tan(\alpha_p) = n$. The values of n and α_p for a few materials are given in the table.
Polarizers which make use of reflection at the Brewster angle are rarely used in the UV-VIS since superior → polarizers are available for this region. However in the IR, mirrors are still used for polarization, especially selenium mirrors; though AgCl polarizers are also found. Selenium has a rather large polarization angle (see Table) which hardly changes above 5 μm (2000 cm^{-1}) because the refractive index in this region shows only a small → dispersion.
Linearly polarized light represents only one limiting form for the pro-

Polarized light. Table.

Material	n	α_p
	$\lambda = 589$ nm	
Water	1.333	53°7′
Quartz glass	1.4589	55°35′
Boron crown glass	1.5076	56°28′
Heavy flint glass	1.7473	60°33′
	$\lambda > 5$ μm	
Selenium	2.420	67°
Sulfur	2.008	63°32′

pagation of light. Another form is circularly polarized light, and the most general form is elliptically polarized.

A reflection experiment can again be used to explain elliptically polarized light. If linearly polarized light falls on the boundary surface between an optically dense and an optically rare medium at an angle of incidence, α, which is greater than the critical angle, α_c, for → total reflection it is totally reflected. It is found experimentally that the reflected light is no longer linearly polarized. Investigation of the state of polarization shows that no orientation of the → analyzer can be found for which the light is completely extinguished. Positions differing by 90° show alternating maxima and minima of brightness, but complete darkness is never achieved. The physical interpretation of this result is as follows. The electric field vector, $\vec{E}(t)$, of the light can be resolved into two components, $E_{||}$ which lies in the plane of incidence (x direction) and $E_{\perp}$ which is perpendicular to the plane of incidence (y direction). Both components are rigidly coupled in phase, though their → phase difference, Δ, is not zero (or π). This situation corresponds to the most general state of polarization which is described as elliptical polarization. The time dependence of the electric field vector is $\vec{E}(t) = \vec{E} \cdot \exp[i2\pi \nu t]$ where ν is the frequency and $\vec{E} = \{E_{||}, E_{\perp}\}$. The real parts of the components of $E(t)$ in the x and y directions are important and they are given by:

$$x(t) = |\vec{E}_{||}| \cdot \cos(2\pi\nu t + \Delta);$$

$$y(t) = |\vec{E}_{\perp}| \cos 2\pi\nu t.$$

Here it is assumed that $E_{||}$ leads by the phase angle Δ. The two components combine to form an elliptical motion, i.e. the head of the vector, $E(t)$, describes a complete ellipse in the xy plane during one period of the vibration, ν^{-1}. As the ellipse is drawn around the direction of propagation, z, the light wave advances one wavelength, $\lambda = c/n\nu$, along z. The equation of the ellipse is:

$$\frac{x^2}{|E_{||}|^2} - 2\frac{x}{|E_{||}|} \cdot \frac{y}{|E_{\perp}|} \cdot \cos\Delta + \frac{y^2}{|E_{\perp}|^2} = \sin^2\Delta.$$

The form and orientation of the major axis, a, depends upon the phase difference, $\Delta = \delta_{||} - \delta_{\perp}$, as can be seen from Figure 1. The equation includes the two cases of linearly polarized light for which $\Delta = 0$ or $\Delta = \pi$, i.e. $\cos\Delta = \pm 1$ and $\sin\Delta = 0$. Hence:

$$y = \frac{|E_{\perp}|}{|E_{||}|} \cdot x.$$

If the contributions of $E_{||}$ and $E_{\perp}$ are equal, then for $\Delta = \pm\pi/2$ the equation of the ellipse becomes the equation of a circle because, with $\cos\Delta = 0$ and $\sin\Delta = 1$, $x^2 + y^2 = E^2$. This equation describes circularly polarized light. Thus, we have here the connection between linearly, circularly and elliptically polarized light. Elliptically polarized light is characterized by its ellipticity, Ψ, which is defined as $\Psi = \arctan(b/a)$ where, a is the major and b the minor axis of the ellipse. Figure 1 also shows that there are right and left ellipses and we speak of right- and left-elliptically polarized light. If, when looking at the light source, i.e. in the negative z direction, the E vector rotates clockwise, then we have right-elliptically polarized light; counter-clockwise rotation corresponds to left-

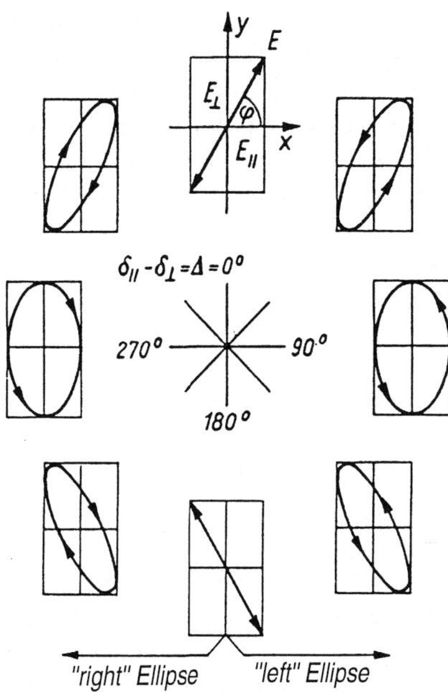

Polarized light. Fig. 1. Elliptically polarized light, see text

elliptically polarized light. The same applies to circularly polarized light.

The connection between linearly and circularly polarized light is important in applications. Every linearly polarized light beam can be formed as the sum of a right- and a left-circularly polarized beam of the same frequency, velocity and intensity. This is illustrated in Figure 2. In circularly polarized light, the head of the electric field vector, E, describes a helix on the surface of a cylinder which lies axially symmetrically about the propagation direction (z) of the light, Figure 2. A perpendicular cross section of the cylinder at any time gives the position of the E vector on a circular surface. We see, at once, that the two E vectors for left- and right-circularly polarized light can be added to give a vector sum. If the sections across the cylinder are made at specific distances, e.g. every $\lambda/8$, then the time dependence

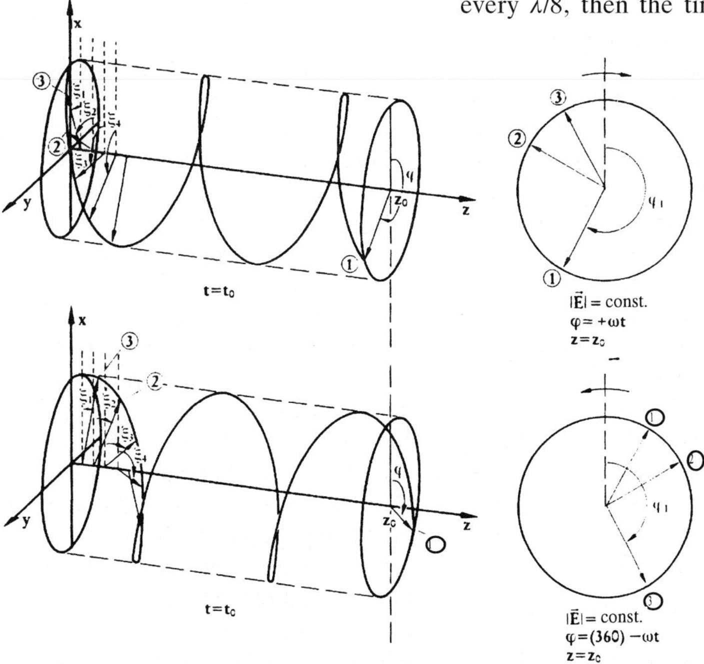

Polarized light. Fig. 2. Circularly polarized light, see text

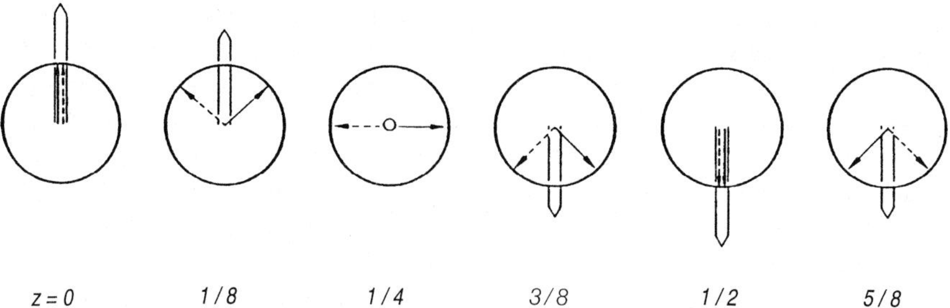

Polarized light. Fig. 3. Representation of linearly polarized light as a sum of right- and left-circularly polarized light

of the vector sum can be seen, Figure 3. It is exactly that of linearly polarized light of the same frequency and the same velocity, but twice the amplitude. One can also immediately recognize from the connection between circularly and elliptically polarized light, that when a component of circularly polarized light is weakened, e.g. by absorption where $\varepsilon_R \neq \varepsilon_L$, then the two E vectors are no longer equal. This process gives elliptically polarized light (→ circular dichroism).

Polarized light is produced by means of → polarizers. The following table gives a summary.

Ref.: D.S. Kliger, J.W. Lewis, C.E. Randall, *Polarized Light in Optics and Spectroscopy*, Academic Press, Inc., New York, London, Sydney, Toronto, Tokyo, **1990**.

Polarizer, *<Polarisator>*, an optical device which converts natural light into linearly or circularly polarized light. The generally available polari-

Polarized light. Table.

Polarized light	Polarizer	Effect used
Linear	→ Nicol prism	
	→ Glan-Thompson prism	
	→ Glan prism	Birefringence
	→ Wollaston prism	
	→ Rochon prism	
	→ Tourmaline	Dichroism
	→ Foils	Dichroism
	→ Pockels cell	→ Pockels effect
Circular	→ Fresnel rhomb	Total reflection of linearly polarized light
	$\lambda/4$ Plate	Birefringence
Elliptical	Mirror	Total reflection of linearly polarized light
	Absorption by chiral substances	Circular dichroism
	$\lambda/4$ Plate	Birefringence

zers make direct use of the → birefringence of optically uniaxial crystals. Quartz and → calcite are crystals of this type and they are used, for example, in the → Nicol, → Glan-Thompson, → Glan, → Rochon and → Wollaston prisms. In conjunction with an electric field, optically uniaxial crystals such as ammonium and potassium dihydrogen phosphate can be used for the construction of polarizers. Other methods utilize the → total internal reflection of quartz prisms, e.g. the → Fresnel rhomb, or phase difference, e.g. → lambda plates and → photoelastic modulators. The → dichroism of various materials, e.g. tourmaline, can also be exploited in the production of polarizers and there are also the → polarization foils.

Ref.: W.A. Shurcliff, *Polarized Light: Production and Use*, Harvard University Press, Cambridge (MA) and Oxford University Press, Oxford (UK), **1962**.

Polarizing angle, Brewster angle, <*Polarisationswinkel, Brewster-Winkel*> → reflection.

Polarizing beam splitter, <*Polarisationsstrahlenteiler*>, an optical component developed by Foster and based upon the → calcite → Glan-Thompson prism. The device produces beams of light polarized perpendicularly to each other with an angle of either 45° or 90° between them. The → Wollaston and → Rochon prisms also produce light beams polarized perpendicularly to each other, but their use as beam splitters is made problematic by the small divergence between the beams. The two models of beam splitter, with 45° and 90° angles, are shown in Figure 1.

Polychromator, <*Polychromator*>, a → monochromator from which the exit slit has been removed so that the complete spectrum is focused in the image (exit) plane. Instead of monochromatic radiation, spectrally dispersed polychromatic radiation is obtained and it is then possible to make measurements over a wide spectral region.

The oldest instrument of this type is the → spectrograph where a photographic plate is mounted in the exit plane and with which the simultaneous measurement of a complete spectrum is possible. In modern polychromators fitted with → plane gratings, → diode array detectors, which allow a rapid collection and processing of the spectrum, have replaced the photographic plates. In recent years, this type of → diode array detector spectrometer for the UV-VIS region has been developed for kinetic studies and for combination with HPLC by several manufacturers. The classical polychromator system, now in use for many years, uses a → concave grating as dispersing element. The properties

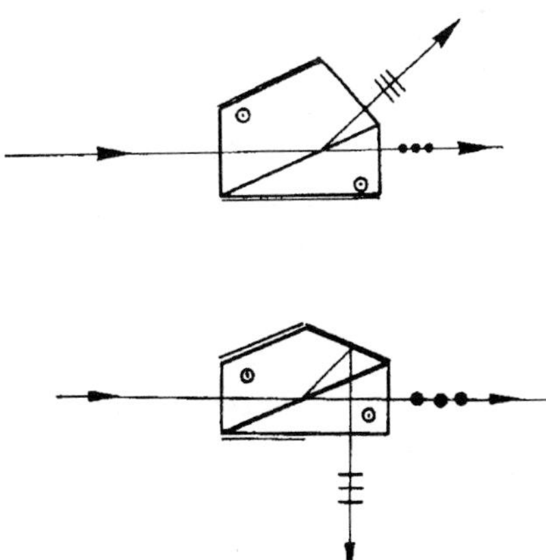

Polarizing beam splitter. Fig. 1. Examples of the 45° and 90° beam separation.

of such gratings make it possible to produce a series of adjacent images of the entrance slit, each for a different wavelength. Today, essentially, a concave grating is used in a → Rowland mounting of the Paschen-Runge type (→ Paschen-Runge mounting). In this arrangement, entrance slit, grating and exit slit (up to 60) are all fixed on the → Rowland circle. The concave grating serves simultaneously as both dispersing and focusing element. This design is realized in simultaneous (multielement) → atomic emission spectrometers. The exit slits in these instruments are always fixed in alignment with a specific spectral line. Behind each slit there is a → photomultiplier with its own high-voltage supply, data processing etc. With this type of polychromator 40 or more elements can be determined in 1–3 minutes. A disadvantage is the fact that it must be decided at the time of construction which elements are to be measured and at what wavelengths. A subsequent change is very expensive and can only be carried out by the manufacturers of the instrument.

Polymethine dyes, <*Polymethinfarbstoffe*>, a large group of colored compounds which may be formulated in the following simplified mesomeric forms:

$$[X = CH - (CH = CH)_n - Y]^q \leftrightarrow$$
$$[X(CH = CH)_n - CH = Y]^q$$

The substituents X and Y are atomic groupings of which one is an electron acceptor and the other an electron donor. They are connected together by a conjugated chain containing an odd number of methyne (CH) groups. The polymethine dyes can be divided into groups on the basis of the charge, q; cationic with q positive; anionic with q negative; and neutral, $q = 0$. In the most important group, the cationic polymethines, groups X and Y are nitrogen-containing substituents. These compounds are named according to whether or not the nitrogen atom is part of a ring. Thus, we have streptocyanines (Figure 1a), cyanines (Figure 1b) and hemicyanines (Figure 1c). All these dyes show a marked charge resonance as the resonance structures of Figure 1d clearly indicate. H. Kuhn (*J. Chem. Phys.*, **1948**, *16*, 840; **1949**, *17*, 1198) described the free movement of the electrons along the conjugated chain with a simple electron-gas model and was able to calculate the light absorption of the compounds in good agreement with experiment. From the spectroscopic point of view, it is important that the

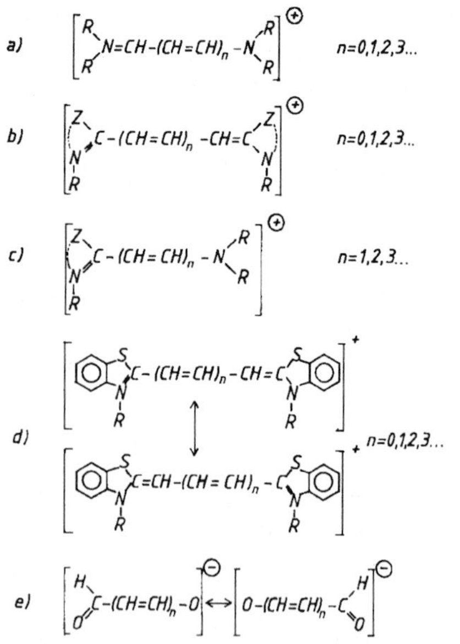

Polymethine dyes. Fig. 1. Structural formulas for various polymethine dyes, see text

absorption maximum of the polymethine dyes shows a strong bathochromic shift (→ bathochromism) with increasing n, i.e. as the number of methyne groups increases. Figure 2 shows the absorption spectra of the streptocyanines, which are vinyl homologues of the amidinium ion ($n = 0$), for $n = 0, 1, 2, 3$ and 4. The bathochromic shift is accompanied by an increase in intensity as Figure 2 clearly shows. In the case of the cyanines, a further distinction is made between symmetric and asymmetric compounds, depending upon whether the nitrogen atoms are located in identical or different heterocyclic rings. Figure 1d shows a symmetrical cyanine.

In 1873, H.W. Vogel discovered that the polymethines known at that time had a photosensitizing effect. They are therefore used as → sensitizers for the → sensitizing of photographic emulsions. The oxonols (Figure 1e), which are vinyl homologues of the carboxylate ion, are examples of anionic polymethines. The merocyanines, or neutrocyanines, are examples of the neutral species.

Ref.: H. Zollinger, *Syntheses, Properties and Applications of Organic Dyes and Pigments,* 2nd ed., VCH, Weinheim, New York, **1991**.

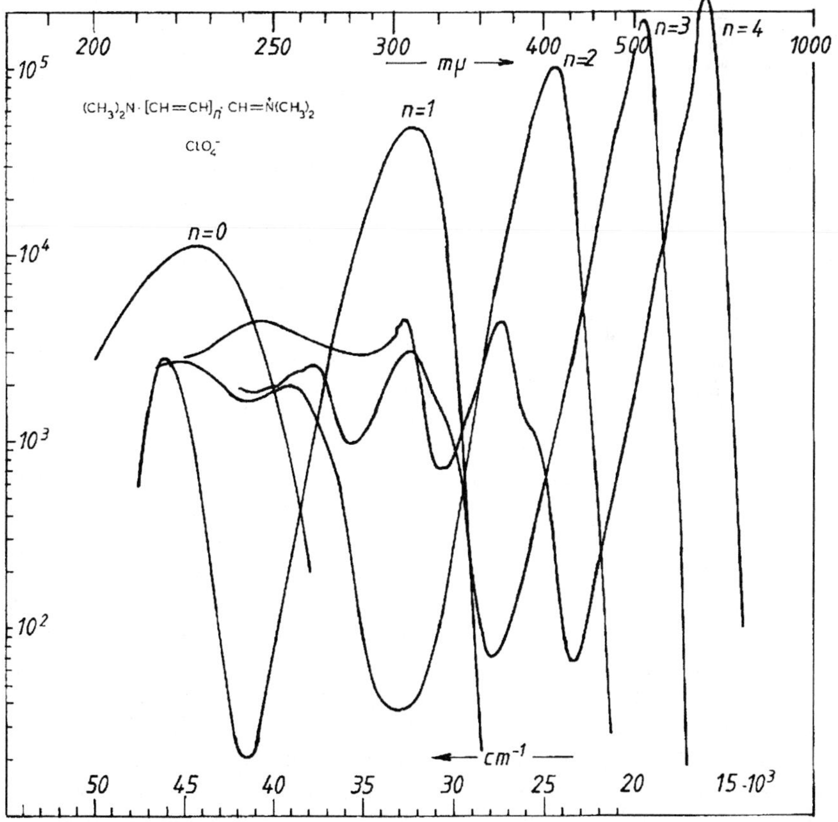

Polymethine dyes. Fig. 2. Absorpton spectra of streptocyanines for $n = 0, 1, 2, 3$, and 4

Population inversion, <*Besetzungsinversion*>, a situation in which the population in the excited state is greater than in the ground state. It is particularly important for → lasers. According to → Boltzmann statistics, the distribution of population between two states in thermal equilibrium is given by:

$$\frac{N_1}{N_o} = e^{-\Delta\varepsilon_1/kT}$$

N_1 is the population of the excited state, N_o the population of the ground state and $\Delta\varepsilon_1$ the excitation energy. kT in the exponential function is the thermal energy per molecule which corresponds to $N_L \cdot kT = RT$, the thermal energy per mole. The limiting value of the exponential function at $T \to \infty$ is:

$$\lim_{T\to\infty} e^{\Delta\varepsilon_1/RT} = 1.$$

At this limit, both states are equally populated at thermal equilibrium, i.e. $N_1 = N_o$.
A population inversion cannot be produced thermally; it must be achieved by adding energy by other means, e.g. by making use of other conveniently positioned energy levels (→ laser). Figure 1 shows the ratio N_1/N_o as a function of temperature. $RT/\Delta\varepsilon_1$ is plotted as the quantity proportional to temperature on the abscissa. This representation has the advantage of being independent of the actual value of the excitation energy.

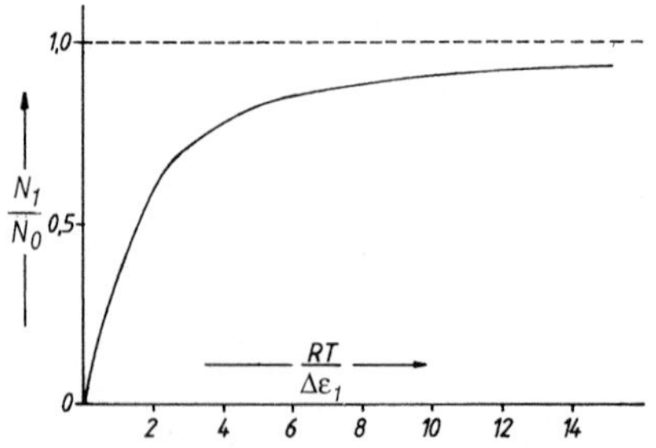

Population inversion. Fig. 1. N_1/N_o as a function of $RT/\Delta\varepsilon_1$

Positive filtering, <*positive Filterung*> → IR gas analyzer.

Positron, <*Positron*> → electron.

Potassium-bromide technique, KBr technique, KBr disk technique, <*Kaliumbromid-Technik, KBr-Technik*>, a method for the measurement of IR spectra (→ IR spectrum) in which use is made of the plasticity of alkali-metal halides at high pressures (7000–10,000 bar). Among the available halides, potassium bromide has proved to be the most useful. To obtain a transparent pellet, all the air must be pumped out of the press. The KBr used must be very pure and it is especially important that it is dry so as to exclude OH-bands from the spectrum. A number of procedures have been proposed to obtain a finely ground and intimate mixture of sample and KBr; grinding in an agate mortar, grinding and mixing in a vibratory mill or lyophilization. Evacuable hydraulic presses for pressing the KBr disks are available commercially. Apart from the widely used KBr, other substances such as KCl, CsI, TlBr and AgCl, as well as crystalline polyethylene (useful down to 200 cm^{-1}), are used for special purposes.
Ref.: W.J. Price in *Laboratory Methods in Infrared Spectroscopy,* 2nd ed., (Eds.: R.G.J. Miller and B.C. Stace), Heyden and Son Ltd., London, Philadelphia, Rheine, **1979**, chapt. 8.

Pressure broadening, <*Druckverbrei-*

terung von Spektrallinien> → collision broadening.

Primary colors, <*Urfarben*>, the colors red, yellow, green and blue which, in comparison with other colors, appear to be simple and not composed of mixtures of other colors, i.e. primary red, yellow, green and blue. The colors lying in between two primary colors, e.g. blue-green, orange or purple, appear to be mixtures of the primary colors. They are called intermediate colors. The wavelength of the radiation which is perceived as a pure primary color is 469 ±1.2 nm for blue, 504.5 ±0.5 nm for green and 568 ±1 nm for yellow.
No corresponding wavelength can be given for red because even the long-wavelength end of the spectrum does not show a pure red but has a yellow tinge. Red is the complementary color of 510 nm. The primary colors are frequently described as a set of three; magenta, yellow and cyan or red, yellow and blue. They are quite different from the primary stimuli red, green and blue. → Color theory.

Principal series, <*Hauptserie*>, the series of lines in an atomic spectrum which derives from the ground state and is observable in both emission and absorption. See, for example, → alkali spectra.

Prism, <*Prisma*>, a body composed of transparent material which is bordered by at least two intersecting planes, Figure 1. The edge where the two planes intersect is called the refracting edge of the prism at right angles to which is the principal section. The angle which lies in the principal section at the refracting edge of the prism

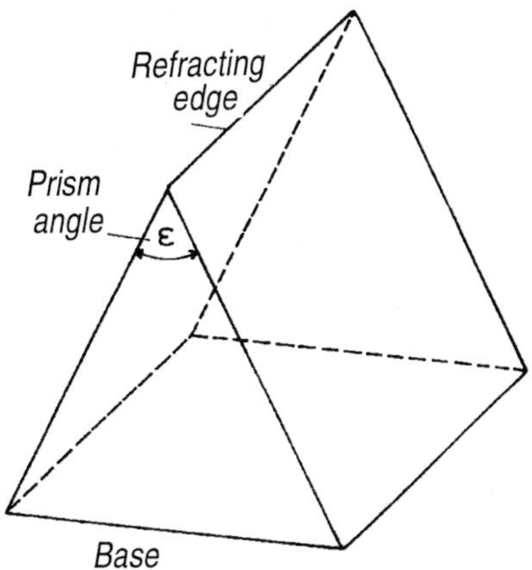

Prism. Fig. 1. Nomenclature in a prism

is the prism angle, ε Figure 1. A light ray passing through the prism is deflected from its original path by the angle of deviation. The prism as a → dispersing element is of fundamental importance to → optical spectroscopy (→ prism monochromator). For this purpose, various materials (→ prism materials) are used, depending upon the → spectral region, UV-VIS, NIR or IR. The prisms are normally used under conditions of → minimum deviation. In addition to their use as dispersing elements, prisms also find application as deviating and reflecting prisms and in → polarizers.

Prism angle, <*brechender Winkel*>, the angle between two surfaces of a → dispersing prism where, upon passage of light, → dispersion occurs. See also → prism.

Prism materials, <*Prismenmaterialien*>. The materials used to make → dispersing prisms. They must be transparent in the required spectral region and have a good dispersion. The region from ca. 160 nm (UV) to $3 \cdot 10^5$ nm (FIR) (63,000 cm^{-1} to 30 cm^{-1}) can-

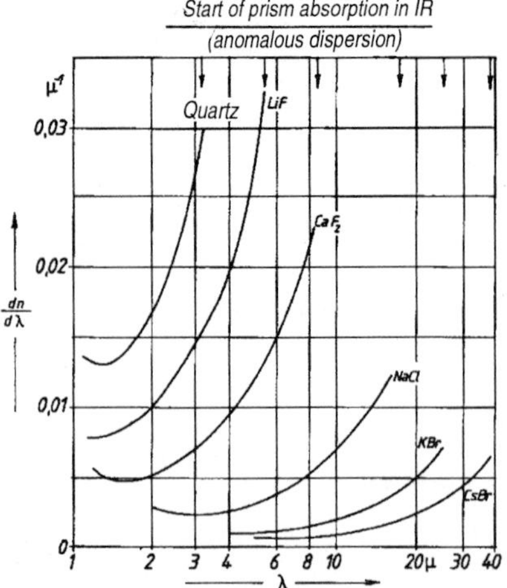

Prism materials. Fig. 1. Dispersion curves of prism materials commonly used in the IR

not be covered by a single prism because the above requirements are to some extent mutually exclusive. Thus, the dispersion, $dn/d\lambda$, always increases near an absorption band of the prism material while the transparency, naturally, decreases. This is true of both the UV and the IR region. Therefore, a variety of prisms must be used through the accessible spectral region. The quartz prism has proved valuable for the UV-VIS region and it can still be used into the NIR, ca. 3300 cm^{-1}. In the visible region glass prisms are frequently used since glass has a better → dispersion than quartz in that range. Figure 1 shows the dispersion curves for a number of materials used for IR prisms. Note the steep rise in the curves in the neighborhood of the absorption band of the material. If the best materials for each wavelength range are selected, then the optimium series of materials for the IR region is given in the table.

In IR instruments with prisms, NaCl was preferred for the range 4000 cm^{-1} to 625 cm^{-1} and KBr for the continuation to 400 cm^{-1}. The spectrometers were constructed so as to make exchanging of prisms rapid and easy. But one can see from Figure 1 that the dispersion is significantly smaller than with quartz, LiF and CaF_2. Larger prisms were used to counteract the consequent loss of → resolution which is proportional to the length of the base of the prism. Today, → prism monochromators have been largely replaced by → grating monochromators which, in principle, can be used in all spectral regions. Only in the UV below 300 nm is a quartz prism superior to a grating, because the dispersion rises rapidly on approaching an absoption band of the quartz (anomalous dispersion). This is also true for a CaF_2 prism below 200 nm.

Prism method, <*Prismenmethode*>, a method of determining the → dispersion of a material using a → prism spectrograph.

Prism materials. Table.

Quartz	700– 2700 nm;	14000–3700 cm^{-1}
LiF	2700– 6000 nm;	3700–1667 cm^{-1}
CaF_2	5000– 8000 nm;	2000–1250 cm^{-1}
NaCl	8000–16000 nm;	1250– 625 cm^{-1}
KBr	15000–27000 nm;	670– 370 cm^{-1}
CsBr	15000–39000 nm;	670– 250 cm^{-1}

Prism monochromator, *<Prismenmonochromator>*, a → monochromator used for many years for the UV, VIS and IR spectral regions for which the prism was the preferred dispersing element. The use of prisms is limited by the requirement that the → prism material must be transparent in the region of application. The efficiency of a prism as a → dispersing element depends upon the → refraction of the light on passing from an optically rare to an optically dense medium (air/prism), or the reverse. Since the → refractive index, n, depends upon the wavelength, λ, which is the property known as → dispersion, $\mathrm{d}n/\mathrm{d}\lambda$, the passage of a beam of light through a prism results in a spectral decomposition of the beam. Figure 1 shows the path of a parallel, monochromatic light beam passing symmetrically through a 60° prism. φ is the prism angle, ϑ is the angle of deviation, b is the length of the base and a the width of the light beam (aperture). The quality of the dispersion is defined by the → angular dispersion of the prism. The greater the angular dispersion of a prism, the greater will be the distance between the spectral lines to be separated. The value of a prism in use is determined by its resolving power or → resolution, R:

$$R = \lambda/\Delta\lambda$$

$\Delta\lambda$ is the difference in the wavelengths of two spectral lines in the region of interest which are just separated (resolved) by the prism. R is given by:

$$R = \lambda/\Delta\lambda = b\mathrm{d}n/\mathrm{d}\lambda$$

where b is the length of the base of the prism, see Figure 1. Thus, the resolving power of a prism depends upon the dispersion of the prism material and upon the length of the base. The above equation shows that, in addition to the required transparency of the prism material, the dispersion, $\mathrm{d}n/\mathrm{d}\lambda$, is also of decisive importance and should be as large as possible. However, the further into the IR or UV spectrum the material is transparent, the smaller is its dispersion. It is for this reason that various → prism materials are required to cover the whole of the accessible spectral region.

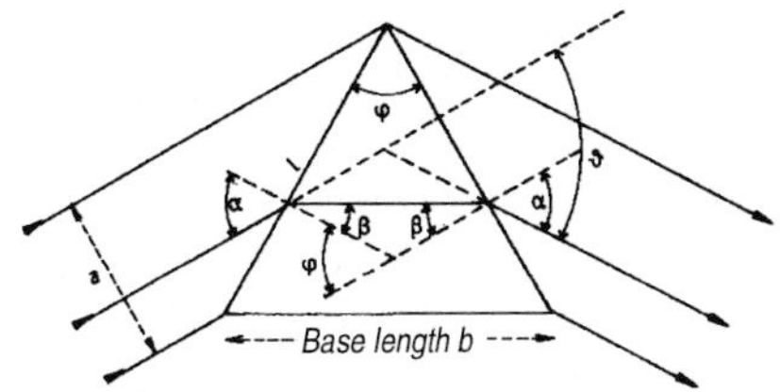

Prism monochromator. Fig. 1. A parallel light beam passing through a 60° prism

Figure 2 shows, schematically, the construction of the Zeiss M4Q II prism monochromator. It employs a 30° quartz prism of 2 cm base length, silvered on the back so that the light pas-

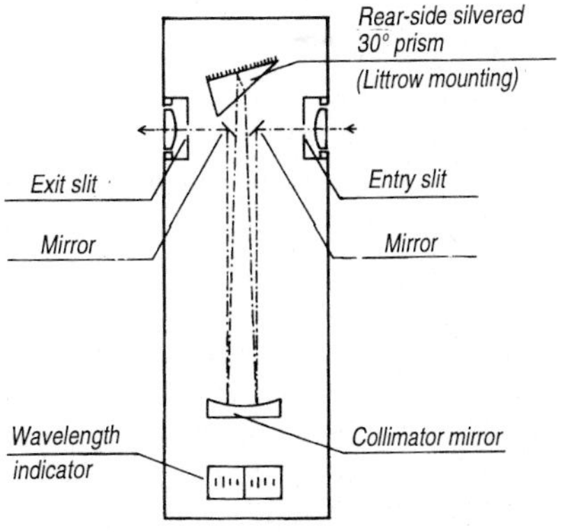

Prism monochromator. Fig. 2. The Zeiss prism monochromator M4Q II

ses through it twice. This is a → Littrow mounting so the effective base length is 4 cm. The dispersed light is moved across the exit slit by rotating the prism. The slit is variable (0–2 mm) so that the geometrical → slit width can be freely chosen. Because of the connection between geometric slit width and spectral slit width, the latter can also be set as required. The non-linear dispersion of the prism material is a disadvantage for prism monochromators and it leads to varying slit widths within a single spectrum. In practice, this means that a prism monochromator must work with a variable slit, and in order to maintain the same spectral slit width, the geometric slit width must be increased as the instrument scans towards shorter wavelengths. At the same time, the → light conductance of the monochromator decreases continuously towards shorter wavelengths. Below 200 nm (→ Schumann UV), a quartz prism monochromator must be flushed with dry nitrogen to remove the absorption of atmospheric oxygen. Then, if suitable → solvents are chosen, such a monochromator can be used down to 185 nm. Today, prism monochromators have been largely replaced by → grating monochromators.

Prism spectrograph, *<Prismenspektrograph>*, a spectroscopic instrument in which the → dispersing element is a prism and the spectrum is recorded on a photographic plate or film. Prism spectrographs can be used in the range from 110 nm (vacuum UV) to ca. 1300 nm (NIR). The lower wavelength limit is determined by the fact that there are no → prism materials which are transparent below this wavelength. The upper limit is a result of the lack of sufficiently sensitive photographic emulsions for longer wavelengths.

The essential components of a prism spectrograph are: slit, collimating lens, dispersing prism, camera lens and film cassette or photographic plate holder. The prism and the camera and collimating lenses are normally made of the same material. An important characteristic of a prism spectrograph is that slit plus collimating lens and camera lens plus cassette are mounted at a fixed angle to each other. The angle is determined by the → minimum deviation condition for a wavelength in the middle of the appropriate range, e.g. the → sodium *D* lines. For every wavelength, the prism spectrograph produces an image of the slit at a specific position on the focal curve, i.e. in the exit plane where the photographic plate is mounted. All wavelengths are measured simultaneously. As an example of a prism spectrograph, the famous apparatus of Bunsen and Kirchhoff is illustrated schematically in Figure 1. The light from the source to be studied enters

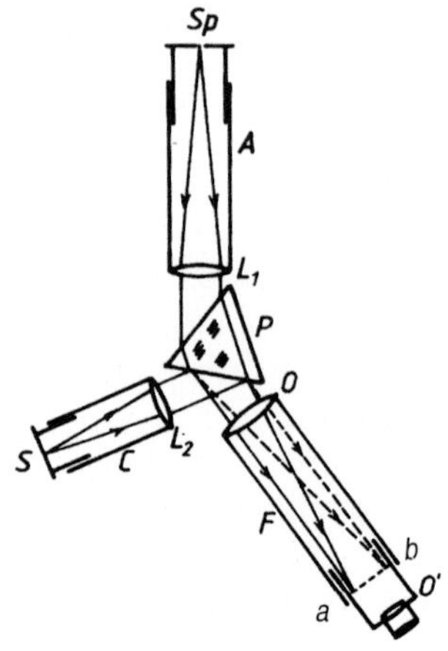

Prism spectrograph. Fig. 1. Bunsen and Kirchhoff's spectrograph

the collimator, A, through the slit, Sp, where it is made parallel by the lens, L_1, which is located at its focal length away from the slit. The parallel beam passes through the prism, P, which is set for minimum deviation of a wavelength in the middle of the range, and enters the telescope objective, F. The camera objective, O, in F focuses the spectrum along the line, a-b. If there is a photographic plate at this position then it records the spectrum. Alternatively, as shown in the diagram, an eyepiece, O', can be used to view the spectrum directly. This visual method of working was the only possibility before the use of photographic plates became routine. In order to measure the spectrum, a third tube, the scale tube, C, is required. This contains a small scale with transparent markings, S, located at the focus of the lens, L_2. If the scale is illuminated, the light passing through it is reflected from the front surface of the prism and into the telescope, F. An observer can then see the spectrum plus a sharp image of the scale in the plane, a-b. The unknown wavelengths can then be calculated from known wavelengths with the help of the scale.

If the length of the spectrum is greater than the field of view of the telescope, the tube, F, must be mounted on a rotatable arm so that it can be moved to the required position in the spectrum. Cross wires in the eyepiece can be used to position the moving telescope on a particular spectral line. In order to measure the relative positions of the individual lines, the rotation of the telescope can be read off from a circular scale or a micrometer. It has become customary to call an instrument of this type with visual observation, a → prism spectrometer. If the eyepiece is replaced by a camera to record the spectrum photographically the instrument is known as a prism spectrograph. The fundamental optical and mechanical construction is identical in both cases.

Glass and quartz spectrographs are distinguished by the → prism material used and the wavelength range for which this is appropriate. Glass-prism instruments are preferred for the visible spectral range and the bordering regions of near UV and near IR. Their dispersion is very large in the visible and exceeds that of quartz in this region. Quartz-prism spectrographs are fitted exclusively with quartz optics. They are used particularly in the UV spectral region down to 200 nm. The → Schumann UV and → vacuum UV regions can be covered using calcium fluoride prisms in evacuated spectrographs.

Many prism spectrographs and spectrometers have been developed in the last 100 years. The primary objectives of these efforts were to increase light intensity and improve dispersion, i.e. resolution. This was achieved by using specially constructed prisms such as the → Abbe and the → Rutherford and also prism combinations such as the Försterling three-prism arrangement. → Autocollimation prisms in the → Littrow mounting were also widely used.

If the Littrow prism is rotatable, the wavelength range appearing on the photographic plate can be continuously changed, i.e. the spectral lines can be directed, one after the other, through a slit in the focal plane of the instrument. Such an autocollimation spectrograph can be viewed as a → prism monochromator. This is also true of the Abbe spectrometer which works with a constant deviation and a collimator and telescope which are fixed at 90° to each other. This configuration is imposed by the use of an → Abbe prism. The prism system in this instrument always operates at → minimum deviation. The prisms are mounted on a rotatable table so that, by means of a wavelength drum, any desired wavelength can be brought to the exit slit. The spectrum can also be viewed through a telescope.

Prism spectrometer, <*Prismenspektrometer*>, an instrument which, in principle, differs from a → prism spectrograph only in the way in which the spectrum is observed. In a spectrometer, observation is visual or photoelectric and is made from behind the exit slit. *Spectrograph* is the term used for an instrument with photographic recording.

Process photometer, <*Betriebsphotometer*> → on-line photometer.

Prolate symmetric top, <*Prolate-Symmetric Top*>, a symmetric top molecule which is extended in the direction of the poles, i.e. like a (rugby) football; but not a soccer ball which is a spherical top. In this case, the moment of inertia about the → figure axis, I_A, is smaller than the two other, equal, moments of inertia, i.e. $I_A < I_B = I_C$. Therefore, for the rotational constants, $A > B$ (→ top molecules). CH_3Cl is an example of a prolate top.

Pump source, pumping light, <*Pumplicht*> → optical pumping.

Pure rotation spectrum, <*Rotationsspektrum*>.

1. Linear molecules, <*lineare Moleküle*>. The energy states of a linear rotating molecule with closed electron shells can be described in terms of the simple rigid rotator (→ dumbbell model). The quantum-mechanical expression for the energy is:

$$E_{\text{rot}} = \frac{h^2}{8\pi^2 I_B} J(J+1);$$

and in wavenumbers:

$$F(J) = \frac{h}{8\pi^2 I_B c} J(J+1) = BJ(J+1)$$

where I_B = the moment of inertia about the axis which is perpendicular to the chemical bonds, h = Planck's constant, B = the rotational constant and c = velocity of light. For diatomic molecules, $I_B = \mu r^2$, where μ = the → reduced mass, r = the distance between the two atoms and J is the rotational quantum number which can take the values $J = 0, 1, 2, 3 \ldots$

As shown in Figure 1, the rotational energy levels increase as the square of J. The selection rule for changes of J is $\Delta J = \pm 1$ resulting in energy differences (for absorption: $\Delta J = +1$) given in wavenumbers by:

$$\tilde{\nu}_{\text{rot}} = F(J+1) - F(J) = 2B(J+1).$$

The rotational spectrum consists of a series of equally spaced lines; the separation being $2B$, see Figure 1. In reality, the molecule cannot be considered to be rigid because the centrifugal force makes itself felt as the rotational quantum number, J, rises, which leads to an increase of the bond length. This effect is allowed for by the introduction of the elongation term, D. The rotational energy levels of a nonrigid

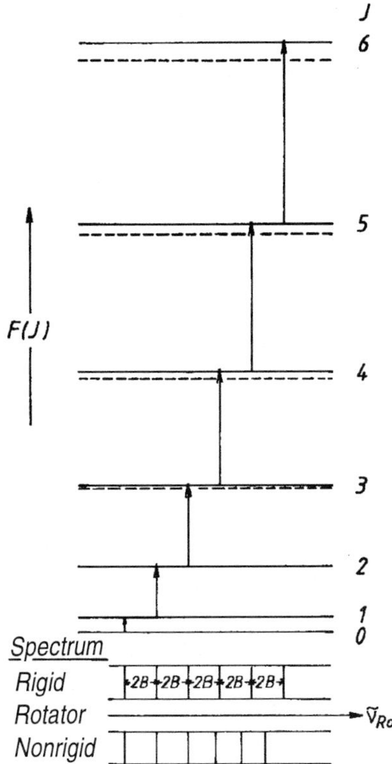

Pure rotation spectrum (linear molecules). Fig. 1. Rotational energy levels of a linear molecule

rotator are then given, to a first approximation by:

$$F(J) = BJ(J+1) - DJ^2(J+1)^2.$$

With the selection rule, $\Delta J = +1$, the wavenumbers of the rotational spectrum are:

$$\begin{aligned}\tilde{\nu}_{\mathrm{rot}} &= F(J+1) - F(J) \\ &= 2B(J+1) - 4D(J+1)^3.\end{aligned}$$

In this equation, $D \ll B$, so that the effect of the elongation term is noticeable only for high rotational states. These states move closer together and the equal separation of the spectral lines is lost, as indicated by the dashed lines in Figure 1. In the diagram, this effect is exaggerated. In reality the corrections for low J values are so small that they cannot be shown on the energy-level scheme.

The elongation constant, D, is simply related to the rotational constant, B, and the vibrational frequency of the molecule, ω:

$$D = \frac{4B^3}{\omega^2}.$$

For the HCl molecule, $B \cong 10.4\ \mathrm{cm}^{-1}$, $\omega \cong 2885\ \mathrm{cm}^{-1}$, giving $D \cong 5 \cdot 10^{-4}\ \mathrm{cm}^{-1}$, which, compared with B, is indeed a very small number.

Although the model of the nonrigid rotator describes the real behavior of a diatomic dipolar molecule on rotational excitation, the rigid rotator model can be used in good approximation for excitation to the lower levels. For excitation by electromagnetic radiation in the appropriate spectral region (FIR or microwave), the molecule must have a permanent dipole moment. Thus, homonuclear diatomic molecules show no pure rotation spectrum. But these species do show a → rotational Raman spectrum. The discussion above concerning rigid and nonrigid molecules also applies in an analogous manner to linear polyatomic molecules with $N \geq 2$.

The → moment of inertia, I_B, is defined by the equation, $I_B = \sum m_i \cdot r_i^2$, where r_i is the perpendicular distance of the mass m_i from the axis of rotation. For pure rotational excitation it is again the case that the molecule must possess a permanent dipole moment. Thus, molecules belonging to the → point group $D_{\infty h}$ show no pure rotation spectrum, while molecules belonging to $C_{\infty v}$ do.

2. Symmetric top molecules, <*symmetrische Kreiselmoleküle*>. For a symmetric top molecule, it is conventional to denote the moment of inertia about the figure axis with I_A, while the other two equal moments of inertia about the two axes perpendicular to the figure axis are called I_B and I_C. $I_A \neq I_B = I_C$. The → rotational constants are then defined by:

$$A = \frac{h}{8\pi^2 I_A c}; \; B = C = \frac{h}{8\pi^2 I_B c}.$$

The rotational energy levels of a rigid symmetric top are given in wavenumbers by:

$$F(J,K) = BJ(J+1) + (A-B)K^2;$$

where J, the rotational quantum number, can take the values 0, 1, 2, 3 ... and K is a further quantum number which can take the values 0, 1, 2, 3 ... J. K gives the component of the angular momentum vector, P_J, in the direction of the figure axis in units of $h/2\pi$. J and K are not independent since:

$$J = K, K+1, K+2, K+3 \ldots$$

Thus, J can never take smaller values than K.

Two cases of the above energy-level scheme may be distinguished.

1. $I_A < I_B$, i.e. $A > B$; a → prolate symmetric top; $A - B > 0$.
2. $I_A > I_B$, i.e. $A < B$; an → oblate symmetric top; $A - B < 0$.

In the first case, the energy for a given value of J increases as K increases. In the second, the energy falls as K increases. The energy-level schemes corresponding to these two situations are illustrated in Figures 1a and 1b.

In a symmetric top, the figure axis is the symmetry axis and the dipole moment must also lie in this direction. In this case, the selection rules for the absorption and emission of electromagnetic radiation are:

$$\Delta K = 0,\ \Delta J = 0, \pm 1;\ K \neq 0$$
$$\Delta K = 0\ \Delta J = \pm 1;\ K = 0$$

For absorption, $\Delta J = +1$ and $\Delta K = 0$.

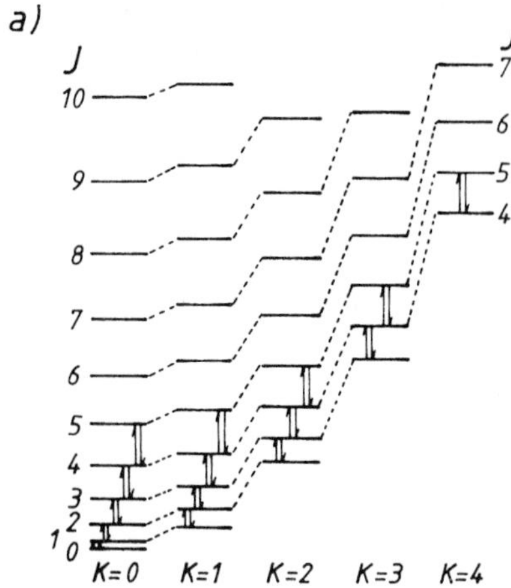

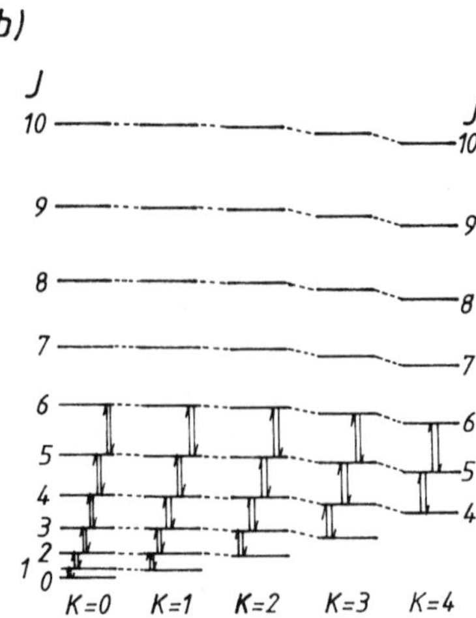

Pure rotation spectrum (symmetric top molecules). Fig. 1. Rotational energy levels of a symmetric top; a) prolate top, b) oblate top

The result, that electromagnetic radiation can induce no transition between levels having different values of K is explained by the fact that a rotation about the figure axis causes no change in the dipole moment and cannot therefore interact with the radiation, i.e. K remains constant. Thus, for a rigid molecule, the rotational IR spectrum consists of a series of equally spaced lines, like the → rotational spectrum of a linear molecule. If allowance is made for nonrigidity, the energy levels are found to be:

$$F(J,K) = BJ(J+1) + (A-B)K^2 - D_J J^2(J+1)^2 - D_{JK}J(J+1)K^2 - D_K K^4$$

and, applying the above selection rules, the wavenumbers of the absorption bands are; $\tilde{\nu}_{\text{rot}} = 2B(J+1) - 2D_{JK}J(J+1)K^2 - 4D_J(J+1)^3$. Because of the term, $D_{JK}J(J+1)K^2$, the same transition $J \to J+1$ for different K values splits into individual lines which also makes it possible to distinguish between a linear molecule and a symmetric top. But this requires a high resolving power in the measurement of the pure rotation spectrum and the influence of nonrigidity can usually be neglected. For the Raman spectrum a consideration of the change of polarizability during rotation leads to the selection rules:

$$\Delta J = 0, \pm 1, \pm 2 \text{ and } \Delta K = 0$$

For a true symmetric top, one axis of the polarizability ellipsoid (→ polarizability) always coincides with the figure axis so that rotation about this axis causes no change in the polarizability. Therefore, no change of K can take place in Raman scattering. For the pure rotational Raman effect the selection rules $\Delta J = -1$ and $\Delta J = -2$ are not satisfied. Two branches are possible for $\Delta J = +1$, the R branch, and $\Delta J = +2$, the S branch. The corresponding wavenumbers are given by:

$$\Delta J = 2 : \tilde{\nu}_S = \tilde{\nu}_o \pm 4B(J + \frac{3}{2})$$

or

$$|\Delta\tilde{\nu}_S| = 4B(J + \frac{3}{2})$$

$$\Delta J = 1 : \tilde{\nu}_R = \tilde{\nu}_o \pm 2B(J+1)$$

or

$$|\Delta\tilde{\nu}_R| = 2B(J+1)$$

The Stokes shifts are distinguished by a left super- script: the Stokes R and S branches are labeled ${}^PR$ and ${}^OS$ and the anti-Stokes R and S branches ${}^RR$ and ${}^SS$.

Figure 2 shows the energy-level scheme for a → prolate symmetric top and the corresponding Raman spectrum. Allowance for nonrigidity leads to the following formulas for the band series:

S branch:

$$|\Delta\tilde{\nu}_S| = (4B - 6D_J)(J + \frac{3}{2}) - 4D_{JK}(J+2)K^2 - 8D_J(J + \frac{3}{2})^3$$

R branch:

$$|\Delta\tilde{\nu}_R| = 4B(J+1) - 2D_{JK}(J+1)K^2 - 4D_J(J+1)^3$$

The terms, $D_{JK}(J+2)K^2$ and $D_{JK}(J+1)K^2$, lead to a splitting of the transition, $J \to J+1$, for different values of K.

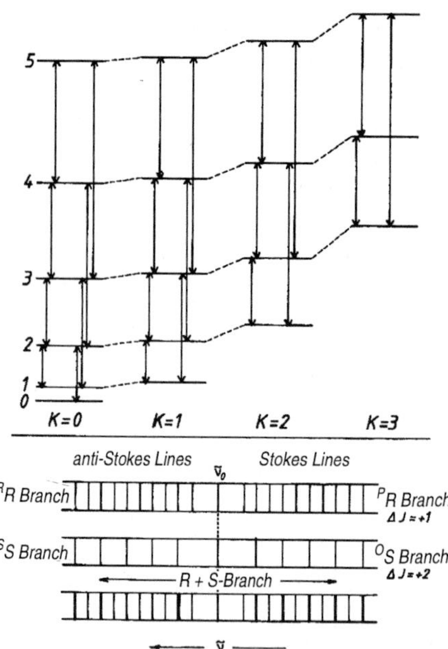

Pure rotation spectrum (symmetric top molecules). Fig. 2. Rotational energy levels for a prolate top and the corresponding Raman transitions

Ref.: G. Herzberg, *Molecular Spectra and Molecular Structure, II. Infrared and Raman Spectra of Polyatomic Molecules*, D. Van Nostrand Co. Inc., New Jersey, **1945**.

Pyroelectric detector, <*pyroelektrischer Detektor*>, a detector increasingly used in dispersive IR spectrometers since the beginning of the 80's. It is based upon a triglycine sulfate crystal doped with L-alanine; an L-ATGS detector. The mechanism of the detector depends upon the pyroelectric effect. Below a specific temperature, known as the Curie temperature or Curie point, ferroelectric materials such as L-ATGS show a strong, momentary electrical polarization when subjected to a change of temperature. This polarization, which is a result of changes in the distances between atoms in the crystal caused by the temperature change, can be used to detect IR radiation. The change in polarization can be measured as an electrical signal if the two surfaces of a thin L-ATGS crystal are furnished with electrodes so as to form a capacitor. This effect is analogous to the piezoelectric effect where the application of a mechanical pressure on a crystal produces an electrical voltage. A high impedance FET amplifier is connected to the L-ATGS crystal to form the IR radiation detector. The voltage measured at the detector electrodes is proportional to the temperature and gives a signal which is linearly dependent upon the incident radiation. The magnitude of the initial voltage is inversely proportional to the frequency of the chopper mirror which is maintained at at least 10 Hz. With this arrangement, plus some simple electronics, the detector can be used for → ratio recording in a manner comparable with the procedure in UV-VIS spectrophotometers. An L-ATGS detector can be exposed to a considerable overload of IR radiation without suffering adverse effects on its performance or service life.

Q

Q branch, <*Q-Zweig*> → rotation-vibration spectrum, → electronic band spectrum.

Quadrupole mass spectrometer, <*Quadrupolmassenspektrometer*>, a → spectrometer in which mass separation is brought about by oscillations of the ions in a high-frequency electric quadrupolar field. The field is produced by four rod electrodes or hyperbolic cylinders which are arranged in a circle of radius, r_o, see Figure 1. The potential on two opposite electrodes is the sum of a constant voltage, U, and an alternating voltage, $V \cdot \cos\omega t$. The ions to be separated are injected as a fine beam into the field in the direction of the long axis. As they pass through the system, the highfrequency field causes the ions to execute oscillations at right angles to the long axis. For set values of the field parameters, U, V, ω and r_o, the amplitude of the oscillation is dependent upon the mass of the ion. Only for ions of a specific mass, or mass range, is the amplitude of these oscillations limited, so that they can pass through the quadrupolar field; provided that the amplitude is less than r_o. The amplitudes of the oscillations of the other ions, however, increase very rapidly with time. Thus, if they spend sufficient time in the field, the ions strike the electrodes or the housing and are discharged, i.e. removed. This is the action of a mass filter.

By means of a proportional and simultaneous change of U and V, a mass spectrum can be scanned. The assignment of the observed peaks to their masses is very simple because a linear relationship exists between these volt-

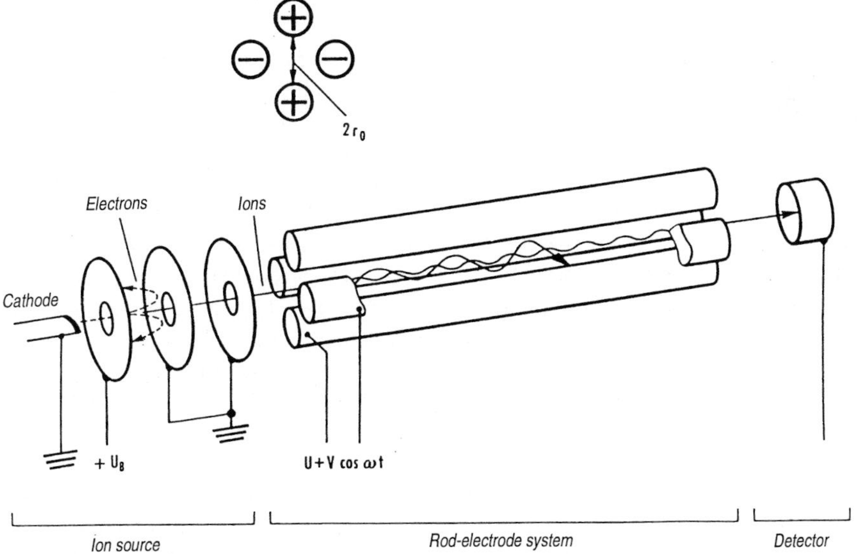

Quadrupole mass spectrometer. Fig. 1. The construction of a quadrupole mass spectrometer

ages and the masses of the ions. Mass separation can also be achieved by changing the frequency, ω. A → Faraday cup or a → secondary electron multiplier can be used as a detector for the ions.

Quadrupole mass spectrometers are especially used for gas analysis where they permit measurements at very low partial pressures, $\leq 10^{-12}$ Torr.

Ref.: J. Roboz, *Mass Spectrometry*, J. Wiley and Sons, New York, London, **1968**.

Quanta, <*Quanten*>, a name introduced by M. Planck to describe small quantities which can take only discrete values or multiples of them. The multiples are given by → quantum numbers which are also allowed to take only discrete values. The energy quanta such as rotational and vibrational quanta are examples.

Quantum counter, <*Quantenzähler*>, usually solutions of strongly fluorescing dyes which are used in the measurement of → fluorescence excitation spectra so that allowance can be made for the → spectral energy distribution of the excitation source. These solutions must fulfill the following conditions:

a) The concentration of the solution must be such that all the light in its absorption range is absorbed.
b) The fluorescence spectrum must be independent of the wavelength of excitation.
c) The → fluorescence quantum yield must also be independent of the wavelength of excitation.

When these requirements are fufilled, the fluorescence intensity is directly proportional to the number of incident photons (light quanta) from the excitation source in the relevant wavelength region, provided that allowance is made for the transparency of the windows of the cuvette containing the quantum counter. In a → fluorescence excitation spectrum, the fluorescence intensity, I_f, at a fixed wavelength, λ_f, is recorded as a function of the excitation wavelength, λ_e. Thus, with the aid of a quantum counter, I_f can always be related to the number of quanta absorbed at any particular wavelength and a corrected fluorescence spectrum evaluated. This should correspond to the UV-VIS spectrum.

The most useful quantum counters are rhodamine-B or rhodamine-101 solutions in ethanol, ethylene glycol or similar solvents. The concentrations required lie between 3 and 8 g l^{-1}. These quantum counters have been carefully tested by several authors. The fluorescence output is constant to within 2% in the range 350–600 nm (the fluorescence maximum lies at 610–620 nm) and to within 5% from 350 nm to ca. 250 nm. Apart from rhodamine-B or rhodamine-101, dimethylaminonaphthalene-5-sulphonate has been suggested for the range 210–400 nm. A 0.01–0.02 M solution in 0.1 M NaOH or Na_2CO_3 is the most useful. Drexhage suggested the carbocyanine dye HIDC, which is also used as a → laser dye. The useful range extends to 700 nm. A quantum counter can also be used for fluorescence measurements. In this application, the fluorescence light of the sample is converted by the quantum counter into its own secondary fluorescence. This combination has been described as a light transformer, see Figure 1. If the fluorescence is spectrally dispersed, then a fluorescence spectrum proportional to the number of quanta is mea-

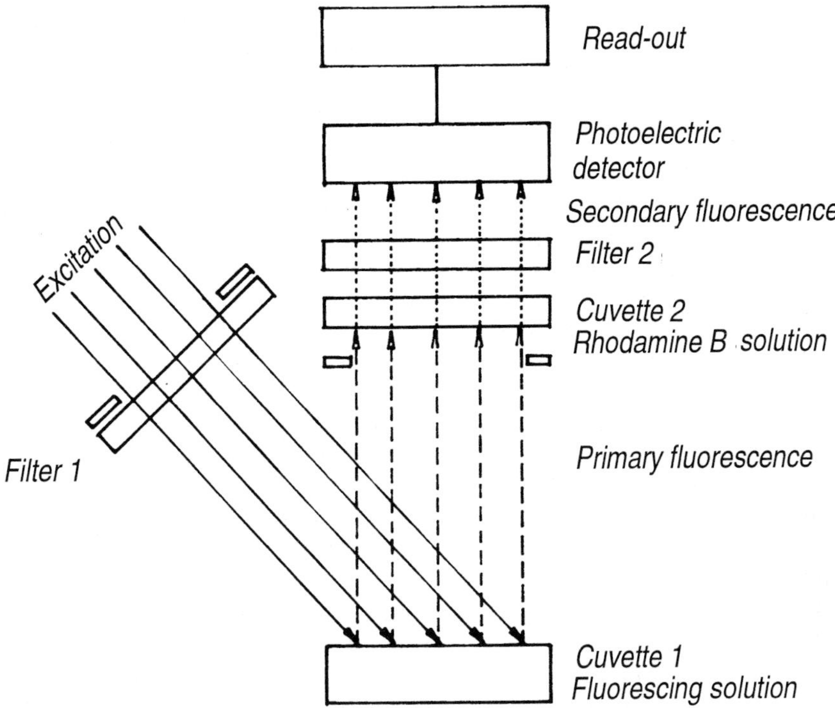

Quantum counter. Fig. 1. Conversion of primary into secondary fluorescence – light transformer

sured via the quantum counter. Clearly, a → photomultiplier must be placed behind the quantum counter to detect the secondary fluorescence and deliver the corresponding photocurrent.

Ref.: J.N. Miller, Ed.:, *Standards in Fluorescence Spectrometry*, Chapman and Hall, London, New York, **1981**.

Quantum number, <*Quantenzahl*>, a number which characterizes an eigenfunction and eigenvalue of a particular quantum-mechanical problem. Normally, several quantum numbers (*q.n.s*) will be required to characterize any one eigenfunction. The *q.n.s* obtained for the solutions (eigenfunctions) of → Schrödinger's equation for the hydrogen atom, and the corresponding energies (eigenvalues) provide a good example. They are:

1. The principal *q.n.*, n, with values 1, 2, 3, 4 ...
2. The orbital angular momentum *q.n.*, ℓ, with values 0, 1, 2, 3 ... $n-1$.
3. The magnetic *q.n.*, m_ℓ, with values $\ell, \ell-1 \ldots 0 \ldots -\ell+1, -\ell$.
4. To complete the description of the electron in a state defined by these three *q.n.s*, a fourth *q.n.*, the spin angular momentum *q.n.*, s, with a value of 1/2 is required. Its components are characterized by the m_s values of +1/2 and –1/2.

The principal *q.n.*, n, determines the energy of the state of the H atom, the energy eigenvalue, through the equation:

$$E(n) = -\frac{m_e e^4}{8\varepsilon_o^2 h^2} \cdot \frac{1}{n^2}.$$

For $n = 1$ this gives a value of $E_{n=1} = -13.595$ eV. If this figure is converted into cm^{-1}, the → Rydberg constant, $R_H = 109{,}677.578\ cm^{-1}$ is obtained. For any given value of the principal *q.n.*, there are a number of states having different values of the orbital angular momentum, *q.n.*, ℓ, ($\ell = 0 \rightarrow \ell = n - 1$), but the same energy. There is degeneracy in both the n and the ℓ states. The states having the same *q.n.*, ℓ, but different values of the magnetic *q.n.*, m_ℓ, are also degenerate. Here the degeneracy is $(2\ell + 1)$-fold. The same is true of the components, m_s, of the spin *q.n.*, s.

The degeneracy of the states with the same value of n is lost (the technical expression is *lifted*) for atoms having more than one electron due to interelectronic repulsion. For such atoms, states having the same value of n ($n > 1$), but different values of ℓ, have different energies. The states having the same ℓ but different m_ℓ remain degenerate for the higher atoms. This degeneracy can only be removed by a magnetic field, → Zeeman effect. This is also true for the spin states with $m_s = \pm½$. The systematic application of the connections between the *q.n.s*, n, ℓ, m_ℓ, s and m_s, while adhering to the → Pauli exclusion principle, leads to the building up (aufbau) principle for the electron shells of the elements.

Quantum statistics, <*Quantenstatistiken*>, a statistical theory which, in contrast to the classical → Boltzmann statistics, assumes that the particles forming a given ensemble are indistinguishable. There are two branches of the subject; Fermi-Dirac (F.D.) and Bose-Einstein (B.E.) statistics.

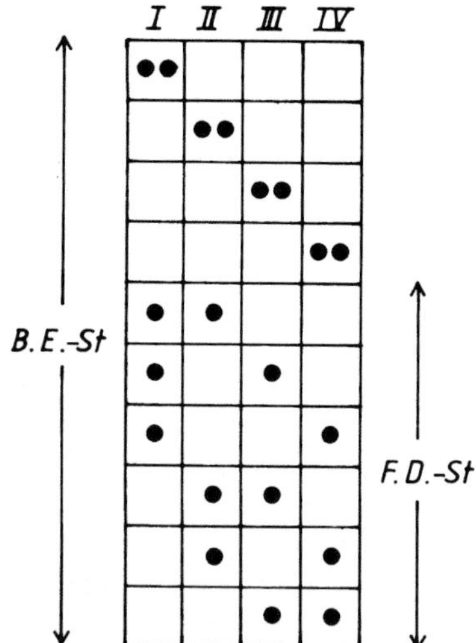

Quantum statistics. Fig. 1. An example of the occupation of states according to B.E. and F.D. statistics

It is common to all statistical methods that they derive a function which describes the most probable distribution of N_i particles over A_i energy levels. B.E. statistics starts from the assumptions that the particles are indistinguishable and that any number of them can occupy the quantum state, A_i. F.E. statistics requires, however, that the particles must obey the → Pauli exclusion principle, i.e. a quantum state may be occupied by only one particle. Figure 1 shows the occupation of four states by two particles for these two cases. The top four possibilities are not allowed in F.D. statistics. The distribution functions are compared with Boltzman statistics below.

Bose-Einstein:

$$\frac{N_i}{A_i} = \frac{1}{B \cdot e^{\varepsilon_i/kT} - 1}$$

Fermi-Dirac:

$$\frac{N_i}{A_i} = \frac{1}{B \cdot e^{\varepsilon_i/kT} + 1}$$

Boltzmann:

$$\frac{N_i}{A_i} = \frac{1}{B \cdot e^{\varepsilon_i/kT}}$$

These three equations differ only in the presence of a 1, with a sign, in the denominator on the right-hand side. If $B \cdot \exp[\varepsilon_i/kT] \gg 1$, then both quantum statistical equations reduce to Boltzmann statistics. $B = \exp[-\alpha]$, where α is introduced in the derivation of the distribution function by the method of Lagrangian multipliers. B is very closely connected with the → partition function $Z = \sum \exp[-\varepsilon_i/kT]$. Detailed theoretical studies show that, for gases, quantum statistics come into effect only at very low temperatures, < 2 K. This deviation of gases from classical behavior at low temperatures is known as gas degeneracy. But the electron gas of a metal, for which the temperature of degeneracy lies between 10^4 and 10^5 K, behaves quite differently and is fully degenerate at room temperature. The conduction electrons of a metal therefore obey F.D. statistics.

Boltzmann statistics can normally be used in spectroscopy at room temperature. For certain problems, e.g. the → intensity alternation in rotation-vibration and electronic band spectra, the statistics of the nuclei have an influence and, depending on the nuclear spin, B.E. or F.D. statistics apply.

Ref.: B.J. McClelland, *Statistical Thermodynamics*, Chapman and Hall, London, **1973**.

Quantum yield, <*Quantenausbeute*>, Q is the number of secondary steps which result from the absorption of light divided by the total number of absorbed light quanta (photons). The general definition is therefore:

$$Q = \frac{\text{Number of resulting steps}}{\text{Total number of light quanta (photons) absorbed}}$$

Excluding the case of light-induced chain reactions, the quantum yield, Q, must have a maximum value of 1 when every absorbed photon results in one consequent step. The various quantum yields can be distinguished by the nature of the consequent steps:

a) If the consequent step is a photochemical reaction of a primary excited species formed by light absorption, then this is the quantum yield of a photochemical reaction.
b) If the consequent step is the emission of light from the primary excited species, then this is the quantum yield of fluorescence or phosphorescence.
c) If the excitation energy is lost by radiationless decay, we have the quantum yield of radiationless decay (transitions).
d) If the consequent step produces electrical conductivity, we speak of the quantum yield of photoconductivity.
e) If the absorbed photons eject electrons from the surface of a metal, it is the quantum yield of photoelectrons which is of interest.

The quantum yield is of decisive importance in the assessment of a → photochemical reaction. One can find four definitions of quantum yield in the literature, and they have not been clearly distinguished.

1. The apparent integral quantum yield, γ_c, for any reaction partner, C, is defined by:

$$\gamma_c = \pm\frac{\Delta N_c}{N_{h\nu}} = \pm\frac{c(t) - c(o)}{\int_o^t I_T \mathrm{d}t}.$$

2. The definition of the true integral quantum yield, γ_c^A, is:

$$\gamma_c^A = \pm\frac{\Delta N_c}{N_{h\nu_A}} = \pm\frac{c(t) - c(o)}{\int_o^t I_A \mathrm{d}t}$$

In both equations, ΔN_c is the number of moles of the reaction partner, C, converted in time, t, and N_{hv} is the number of moles of light quanta, in Einsteins, absorbed in the same time. $c(t)$ and $c(o)$ are the volume concentrations of C at time, t, and at the start of the irradiation, respectively. In case 1, I_T is the quantity of light absorbed by the system per unit volume per second, i.e. in Einsteins l^{-1} s^{-1}. In case 2, the conversion is related to the quantity of light absorbed by the substance, I_A, it being assumed that A initiates the photoreaction. Since the general definition of quantum yield is positive, the positive sign should be used when C is formed during the reaction and the negative sign when it is lost. γ_c is relatively easy to measure, but it is not clear how it depends upon the reaction time.

3. The apparent differential quantum yield, Φ_c:

$$\Phi_c = \pm\frac{\mathrm{d}c/\mathrm{d}t}{I_T} = \frac{\dot{c}}{I_T}.$$

This is calculated for an infinitesimal conversion of substance. $\mathrm{d}c/\mathrm{d}t = \dot{c}$ is the change of the concentration of C with time. It is referred to the total quantity of light absorbed per unit volume and per unit time, I_T. Like γ_c and γ_c^A, in general Φ_c depends upon the duration of the irradiation in a way which is not clear.

4. The true differential quantum yield, Φ_c^A:

$$\Phi_c^A = \frac{\mathrm{d}c/\mathrm{d}t}{I_A} = \frac{\dot{c}}{I_A}.$$

This is the most important definition for the mathematical treatment of simple → photoreactions. Again, the yield is referred to the quantity of light, I_A, absorbed by the substance, A, and the assumption is made that A alone initiates the reaction. In the case of complex photoreactions, it is expedient to divide the process into partial photoreactions which are defined by partial quantum yields which are also constant in time:

$$\Phi_k^{a_i} = \pm\frac{\mathrm{d}x_k/\mathrm{d}t}{I_{a_i}} = \pm\frac{\dot{x}_k}{I_{a_i}}.$$

The definition of the true differential quantum yield must be the starting point for the interpretation of the → kinetics of photochemical reactions.

Ref.: H.-H. Perkampus, *UV-VIS Spectroscopy and Its Applications*, sec. 7.6, Springer Verlag, Berlin, Heidelberg, New York, **1992**.

Quarter-wave plate, Quarter-lambda plate, λ/4 plate, <*Lambda-Viertel-Plättchen, Viertelwellenlängenplättchen, λ/4-Plättchen*>, a doubly refracting (birefringent) phase plate, usually made of mica or gypsum (→ half-wave plate). For optical studies of polarization with a large field, doubly refracting dichroic sheet polarizers are used. The methods of manufacture have been developed to ensure that:

1. When placed between crossed → polarizers, the sheet appears completely dark in specific positions.
2. The two beams which arise because of the double refraction have an optical path difference of exactly one quarter of the desired wavelength.

The crystal plates are reduced to the thickness required to achieve an optical path difference of $\lambda/4$ by repeated cleaving or grinding. Quarter-wave plates are used to produce circularly polarized light and in the measurement of optical phase differences.
If a quarter-wave plate follows a polarizer then circularly polarized light is produced for one specific wavelength if the direction of vibration in the polarizer is orientated at 45° to the two in the quarter-wave plate.

Quartz, <*Quarz*>, crystalline silicon dioxide (SiO_2), two modifications of which exist between room temperature and 1143 K (870°C). The form stable at room temperature, the α or low-temperature form, is stable to 846 K (573°C) at which temperature it is in equilibrium with the β or high-temperature form. At 1143 K (870°C) there is a further transition point where β quartz changes into β tridymite which changes into β christobalite at 1743 K (1470°C). α quartz crystals have a rhombohedral structure and are → optically uniaxial crystals. They exhibit → birefringence, → optical rotatory dispersion and circular refraction. The crystals also occur naturally in two enantiomorphic forms, one of which rotates the plane of polarized light to the left and the other to the right. The optical properties of quartz are utilized in work with → polarized light (quartz wedge), and to produce polarized light; → Rochon prism. If a prism made of crystalline quartz is used as a → dispersing element, the circular birefringence presents a noticeable problem; → Cornu prism.

Quartz glass, <*Quarzglas*>, the only single-component glass which has found many applications in research and technology. In contrast to the optical → glasses, which have a SiO_2 content of 40–90%, quartz glass has a SiO_2 content of 99.9% with natural impurities in the ppm region. Normal quartz glass is made from powdered crystalline quartz by fusion in an electric vacuum furnace or in an oxyhydrogen flame. In two other processes, $SiCl_4$ is either hydrolyzed in a oxyhydrogen flame or oxidized with oxygen in a plasma flame. The quartz glass thus obtained is also called synthetic quartz glass. It is purer than most *analytically pure* materials and its purity approaches that of semiconductor materials. The requirements of → fiber optics technology have been a great stimulus to the production of very pure quartz. Heraeus produces a range of quartz glasses by the above methods. FLUOSIL® is the quartz glass made especially for the manufacture of optical fibers.
For spectroscopic applications, the transparency of the quartz glass in the UV-VIS or IR regions is of interest and examples of these, for 10 mm path length, are given in Figure 1. The data show that the SUPRASIL® types, which are synthetic quartz glasses, have high transmission in the UV region even below 200 nm. SUPRASIL® 300 also shows very good transmission in the IR and is comparable with INFRASIL® which is obtained by electric vacuum fusion. Both SUPRA-

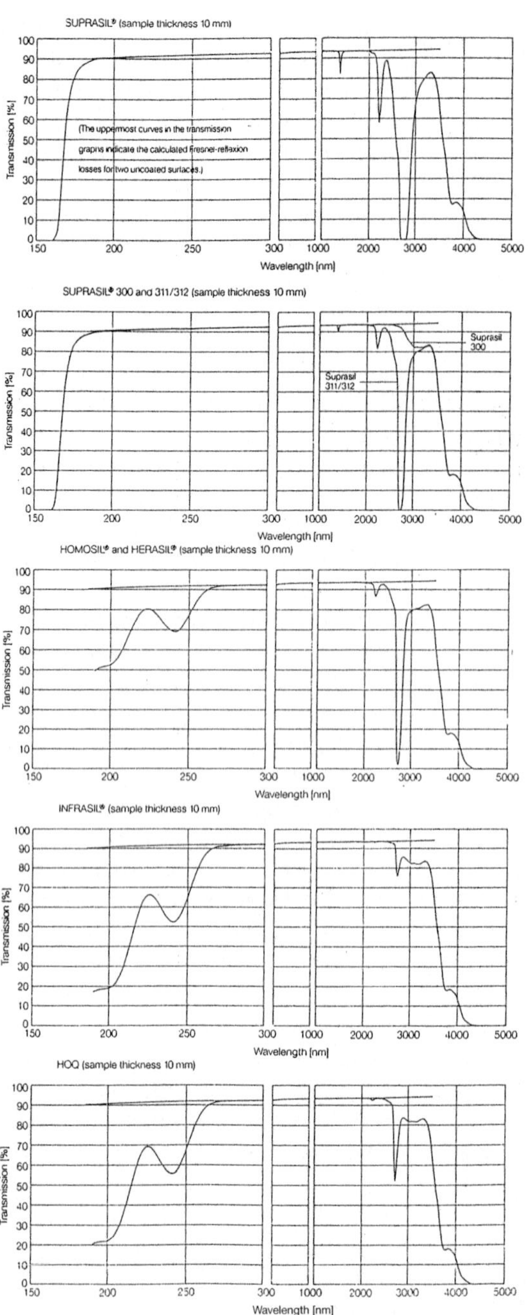

Quartz glass. Fig. 1. The transmittance of various quartz glasses

SIL® 300 and INFRASIL® are prepared under exclusion of water so that the Si-OH groups, which give rise to absorption in the IR, are effectively absent. SUPRASIL® 300 has the widest spectral transmission range of all known grades of quartz glass. All the quartz glasses made by the above methods are optically isotropic and, in contrast to crystalline → quartz, show no double refraction (→ birefringence) or rotatory dispersion. Information on the physical, electrical, optical and chemical properties of the above quartz glasses can be found in the Heraeus information literature: *Quartz Glass for Optics, Data and Properties* POL-O/102E; *Quartz Glass for Optics, Products, Properties, Applications* POL-O/202E; *FLUOSIL® Fibre Optic Preforms* PLW-B1. Lit.: Hereaus Quarzglas GmbH, D63405 Hanau, Germany.

Quartz-lens method, *<Quarzlinsenmethode>*, a method, developed by H. Rubens and R.W. Wood in 1910, for selecting very long-wavelength infrared radiation. The principle of the method is shown in Figure 1; it is based upon the fact that for long wavelengths, > 50 μm, → quartz has a → refractive index, $n > 2$, while for the UV and VIS regions $n \sim 1.5$. According to the simple lens formula, the focal length of a lens depends upon the refractive index, n, of the material. To be exact, it depends upon, $1/(n - 1)$. Thus, a quartz lens, which for visible light in the middle of the range, has a focal length of f = 26 cm, has a focal length of f = 13 cm in the long-wavelength infrared, i.e. one half of the visible value. Suppose that the two quartz lenses, Q_1 and Q_2, in Figure 1 have these focal lengths and that the light source, L, is placed more than 13 cm but less than 26 cm away from Q_1. Radiation leaving L with wavelength ⩽ 50 μm leaves Q_1 divergently (dashed lines) while long-wavelength radiation ≥ 50 μm is focused at P_1. A small circular aperture allows only these longwavelength rays to pass through the screen, B_1, while the short-wavelength rays close to the axis are blocked by the small disk, S_1. To make quite certain that there is no possible contamination with short-wavelength radiation, the same process is repeated a second time. The light from P_1 passes through the lens, Q_2, and produces at P_2 a very long-wavelength radiation which is dependent upon the emission of the source. In this way, infrared radiation with wavelengths reaching into the range of millimeter waves (microwaves) can be selected.

Quencher, *<Quencher>*, substance or foreign molecule able to quench photophysical or photochemical processes, e.g. → fluorescence. (See also → fluorescence quenching).

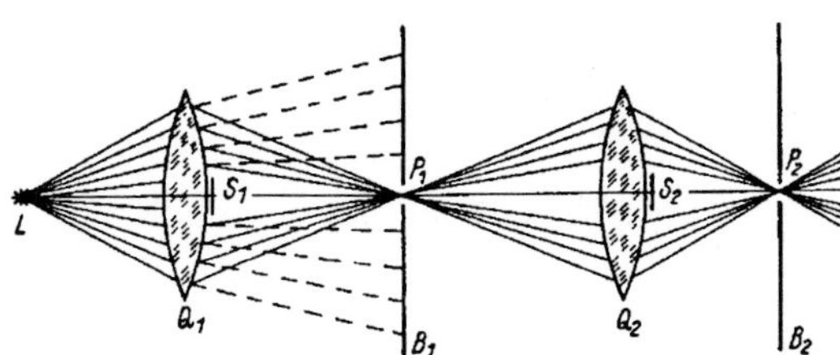

Quartz-lens method. Fig. 1. The principle of the quartz-lens method

R

RADAS, <*RADAS*>, the model type of an → on-line (process) photometer manufactured by Hartmann & Braun which works according to the principle of → nondispersive UV-VIS spectroscopy. The instrument was developed for the analysis of gases, especially for NO, NO_2, NO_x (NO + NO_2), Cl_2, SO_2, H_2S, $COCl_2$ etc. The analytical part of the instrument is shown in Figure 1. The UV radiation is produced by a modified → hollow cathode lamp, *L*. The modification is such that, rather than the elemental lines of the cathode material, this lamp emits a specific NO emission. The source contains a mixture of nitrogen and oxygen at a pressure of 5 mbar. The lamp runs with a few milliamperes and electronically excited NO molecules in the $^2\Sigma^+$ state are formed in the negative glow discharge. As they decay to the $^2\Pi_{1/2,3/2}$ ground state, these molecules emit a specific NO emission between 220 and 290 nm in the form of a line spectrum with rotational structure, the → NO γ bands. There is also a line spectrum emission between 290 and 430 nm which is due to coexcited N atoms. The lamp is followed by a wheel, *B*, by means of which either a gas filter cell and a quartz plate or two interference filters can be placed in the beam. The wheel divides the radiation into two parts, the measurement and reference beams, separated in time. The beam splitter, *S*, divides the light into two beams separated in space. The measurement and reference beams reach the detector, *E*, after passing through the sample cell, *MK*. Sample and reference beams, uninfluenced by the sample, are also received by a second detector, the correction detector, *KE*. Thus, during one rotation of the wheel each detector produces two sequential signals (four-channel system) by means of which the influence of nonselective absorption, e.g. contamination of the cell, or lack of stability in the source or detector, can be eliminated. A collimator lens, *K*, and an optical interference filter, *F*, are placed in front of the beam splitter, *S*. The instrument can be calibrated by means of calibration

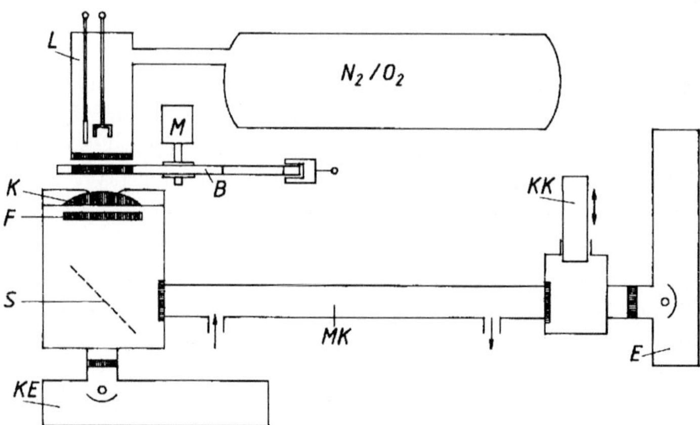

RADAS. Fig. 1. The analytical component of the RADAS instrument, see text

cells filled with the various gases to be determined and inserted at position, *KK*. The sample cell, *MK*, is designed for measurements on flowing gases. For the measurement of NO, the instrument operates according to the → gas filter correlation principle. The gas filter cell in the wheel, *B*, is charged with NO so that that part of the source NO emission which can be resonantly absorbed is absorbed. The superimposed NO emission which is not resonantly absorbed serves as the reference beam. After half a rotation period, all the light passes through a quartz plate so that now both sampling and reference light pass through the sample cell where only the sampling light capable of resonant absorption will be absorbed. An interference filter, *F*, is placed in front of the beam splitter to ensure that only a small wavelength region around the center of gravity of the emission is used for the measurement. All other applications of the RADAS depend upon the interference filter correlation method. When used in this way, the wheel, *B*, is equipped with two appropriate interference filters which are transparent only to the sampling and reference light necessary for the particular application. In this case, the interference filter, *F*, in the beam splitter unit is removed.

Radiating (radiative) transition, *<strahlender Übergang>*, a transition from an excited state to the ground state which takes place with the emission of radiation, normally → fluorescence or → phosphorescence.

Radiation, electromagnetic radiation, *<Strahlung, elektromagnetische Strahlung>* → light.

Radiation conversion, *<Strahlungsumwandlung>*, the conversion of radiation (light) of a specific frequency, ν, and therefore a specific energy, $h\nu$, into light of another frequency, ν', and energy, $h\nu'$. Here, → Stokes rule generally applies, i.e. $\nu' < \nu$. The best-known process is the appearance of → luminescence (→ fluorescence, → phosphorescence); UV light absorbed by a substance in the primary process is emitted again as light shifted to longer wavelengths (red). This form of light conversion is used in lighting technology where the inner walls of fluorescent lighting tubes are coated with suitable luminescent materials.
The conversion of X-rays into visible light should also be included. Energy-rich radiation such as → corpuscular radiation can also be converted into visible radiation, → scintillation counter, cathode ray tube, etc.

Radiation detector, *<Strahlungsempfänger>*, a device used most frequently to measure radiation flows, radiation flux densities or, in the UV-VIS and IR regions, light intensities. The measurement is usually a quantitative one, but relative measurements are also often made. Suitable radiation receivers, → detectors, have been developed for the many spectroscopic methods. The most important detectors are the → thermal radiation detector and the photoelectric detectors such as the → photocell, → photomultiplier, → photoelement, → photoresistive cell and the → photodiode. They all have the ability to convert incoming radiation intensity into a proportional electrical signal. Depending upon type, some have a pronounced → spectral sensitivity.

Radiationless transition, nonradiative transition, <*strahlungsloser Übergang*>, a transition in which an excited molecule or atom moves to a lower energy state without emitting radiation. A transition of this type between states of the same multiplicity is called → internal conversion; a transition in which there is a change of multiplicity is called → intersystem crossing.

Radiation standard, <*Strahlungsnormal*>, a light source for the quantitative measurement of radiation and for the calibration of radiation detectors (receivers). The → black-body radiator is an important radiation standard as is also the → light intensity standard or scientific lamp (e.g. Osram Wi 15 or Wi 40 → tungsten lamp), which provide a defined emission, when operated with specified values of current and voltage. These lamps are delivered by the manufacturers with a calibration certificate and are used in the VIS and NIR spectral regions. → UV standard lamps have been developed for the UV region.
Ref.: J. Kiefer, Ed.: *Ultraviolette Strahlen (Ultraviolet Radiation)*, de Gruyter, Berlin, New York, **1977**, p. 75 ff and p. 118 ff.

Raman active, <*Raman-aktiv*>, a property of molecular vibrations in the → Raman effect. In the absence of simultaneous electronic excitation, the Raman effect can only occur when the vibration is accompanied by a change in the → polarizabilty. Vibrations for which this condition is not fulfilled are said to be Raman inactive. It is well known that for homonuclear diatomic molecules (e.g. H_2, N_2, O_2, Cl_2), the only possible normal vibration is IR inactive. However, it is Raman active because there is a change of polarizability with the vibration.
For larger molecules, the symmetry species to which the Raman active vibrations belong can be determined from the → symmetry point group of the molecule and the → character table. The character table also contains all the data necessary to calculate the number of vibrations spanned by (belonging to) a particular symmetry species.

Raman effect, <*Raman-Effekt, Smekal-RamanEffekt*>. If a substance is irradiated with a light source which emits one or more intense lines, e.g. a → mercury high-pressure discharge lamp or a → laser, then, when spectrally dispersed, the light scattered by the substance shows not only the lines in the spectrum of the light source, the Rayleigh lines, but also further lines which show a wavenumber shift with respect to the exciting lines. These shifted lines are called Raman lines after the Indian physicist, C.V. Raman, who published his experimental discovery of them in 1928. The appearance of more lines in the scattered light than in the incident light is called the Raman effect. Smekal had predicted the effect theoretically in 1923. Practical and theoretical studies of the effect led to → Raman spectroscopy and the recording of the lines provides the → Raman spectrum. If the exciting light has the wavenumber $\tilde{\nu}_o$, and the Raman scattered light the wavenumber $\tilde{\nu}_s$, then a Raman shift, $\Delta\tilde{\nu}_s = \tilde{\nu}_o - \tilde{\nu}_s$, can be defined. This shift is the energy difference between the exciting and scattered photons, $\Delta\tilde{\nu}_s = \Delta E_s/hc = (E_o - E_s)/hc$. The following three cases can be distinguished.

1. $\Delta\tilde{\nu}_s > 0$, i.e. $E_o > E_s$
   The wavenumber of the scattered light, $\tilde{\nu}_s$, is smaller than that of the exciting light, $\tilde{\nu}_o$, i.e. the corresponding energy, $E_s = hc\tilde{\nu}_s$, of the scattered photon is also smaller than $E_o = hc\tilde{\nu}_o$. Some energy, $\Delta E_s = E_o - E_s$, must therefore have remained in the molecule which scattered the radiation, i.e. that molecule is in an excited state.
2. $\Delta\tilde{\nu}_s = 0$, i.e. $E_o = E_s$ and $\tilde{\nu}_o = \tilde{\nu}_s$. The scattered light has the same frequency as the exciting light. This is → Rayleigh scattering.
3. $\Delta\tilde{\nu}_s < 0$, i.e. $E_o < E_s$
   The scattered light has a larger wavenumber than the exciting light, and therefore a greater photon energy. In this case, the molecule which scattered the light has transferred energy to the scattered photon, which is only possible if that molecule was in an excited state.

The first case, $\Delta\tilde{\nu}_s > 0$, is described as a Stokes, and the third, $\Delta\tilde{\nu}_s < 0$, as an anti-Stokes Raman shift, → Stokes rule. We speak of a small or large Raman shift depending upon whether $\tilde{\nu}_s$ lies close to, or distant from, $\tilde{\nu}_o$. It is an important precondition for the Raman effect that the exciting light, which usually lies in the visible region, is not absorbed by the scattering molecules, i.e. there must be no electronic excitation. (But see also the → resonance Raman effect.) This means that the energy differences, ΔE_s, must be assigned to the quantized excitation of vibrational and rotational states. Therefore, small Raman shifts are identified with the rotational Raman spectrum and the large Raman shifts with the vibrational Raman spectrum. Thus, the Raman spectrum represents a rotation and rotation-vibration spectrum transposed to the visible spectral region.

The classical wave theory of scattering shows that a change of the → polarizability during the vibration or rotation of the molecule is essential for the appearance of a Raman spectrum. For the IR spectrum, on the other hand, the requirement is a change of dipole moment. Consequently, molecules which have no changing dipole moment and cannot be excited by the absorption of radiation in the IR region can, nevertheless, show a Raman spectrum. The homonuclear diatomic molecules are good examples of this. The Raman spectrum of a molecule provides important information complementary to that obtained from its IR spectrum.

Raman inactive, <*Raman-inaktiv*> → Raman active.

Raman spectrometer, <*Raman-Spektrometer, Raman-Spektralphotometer*>, a → spectrometer used in → Raman spectroscopy. A dispersive instrument built by B. Schrader and colleagues is shown schematically in Figure 1. The individual components are: light source (→ laser), sample holder, *P*, → monochromator and → detector (→ photomultiplier). A hard coated → interference filter, *I*, with threefold reflection is placed between the sample and the monochromator which reduces the unshifted (Rayleigh) scattering by a factor of $5 \cdot 10^{-5}$. Lasers as light sources have now completely replaced the → mercury-vapor discharge lamps which were always used before the former became available. Their radiation density is some 10 powers of ten higher. The → noble-gas

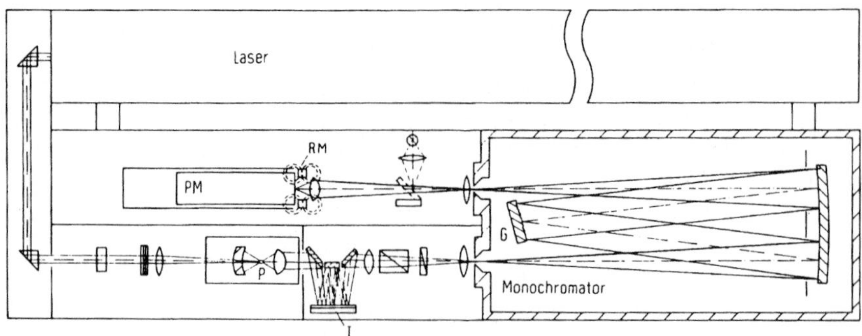

Raman spectrometer. Fig. 1. Schematic of a modern dispersive Raman spectrometer

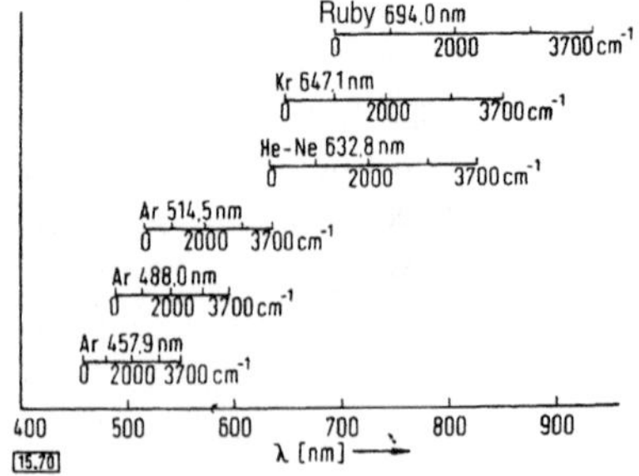

Raman spectrometer. Fig. 2. Laser lines and associated Raman spectral ranges

ion lasers, in particular argon- and krypton-ion lasers, have proved to be very good for continuous application in chemistry. Figure 2 shows the range of the Raman spectra for different laser lines. Special cells have been developed for measurements on gases, liquids and crystalline powders. The practical aspects of sampling for Raman spectroscopy have been discussed by Louden (→ literature, chapt. 22). In the spectrometer (Figure 1), the various samples are arranged on a sample table and can be readily changed without realigning the instrument. A multichannel analyzer can be used as a detector in place of the → photomultiplier. Recently, Raman modules for → Fourier-transform IR spectrometers have been developed. A → neodymium laser, which emits at 1064 nm in the NIR, is used as source in these instruments, → FT-Raman spectrometer. The great advantage of this form of Raman spectroscopy is the reduction in the very troublesome fluorescence which excitation with visible radiation so often produces. With NIR excitation, this problem is effectively nonexistant.

Ref.: H.A. Willis, J.H. van der Mass and R.G.J. Miller, Eds.: *Laboratory Methods in Vibrational Spectroscopy*, J. Wiley and Sons, Chichester, New York, Brisbane, Toronto, Singapore, **1987**.

Raman spectroscopy, <*Raman-Spektroskopie*>, that part of molecular spectroscopy which is concerned with the utilization of the → Raman effect. The molecular physics of Raman spectroscopy is essentially that of → IR spectroscopy since both are concerned with the observation of the excitation of molecular vibrational and rotational energy states. The important difference is that whereas for the observation of a Raman spectrum with rotation and vibration there must be a change in the polarizability of the molecule, for the IR spectrum there must be a permanent dipole moment for the excitation of rotation and an associated change in dipole moment for the excitation of a vibration. Raman and IR spectroscopy are therefore complementary, and both are required for the full interpretation of the vibrational spectrum of a molecule.

Raman spectrum, *<Raman-Spektrum>*, the spectrum obtained when Raman scattered light is spectrally dispersed and the intensity of the individual lines or bands is plotted against their wavenumber, $\tilde{\nu}$. Starting from the exciting line, $\tilde{\nu}_o$, the wavenumber axis represents the Raman shift; $\tilde{\nu}_o - \tilde{\nu}_s > 0$ in the case of a Stokes shift (→ Stokes rule). Since the rotational and vibrational spectra of molecules are measured by the Raman effect, the Raman spectrum embraces the range, 0 – ca. 4000 cm^{-1}. The intensity of the scattering is measured with a → photomultiplier or by → photon counting. Very recently, increasing use has been made of the charge-coupled device (CCD). When the rotational structure is resolved, the spectra are described as rotational or rotation-vibration Raman spectra. In the case of large molecules, where the rotational structure is not resolved, the term *vibrational Raman spectrum* is used. The measurement of a Raman spectrum is not very dependent upon the state of aggregation of the sample.

Rate constant, *<Geschwindigkeitskonstante>*, a constant which characterizes the rate (velocity) of a chemical process. In spectroscopy it is generally a photochemical process; e.g.:

→ fluorescence
rate constant k_{FM}
→ phosphorescence
rate constant k_{PM}
→ intersystem crossing
rate constant k_{TM}
→ radiationless transition
rate constant k_{GM}.

The dimension of k is s^{-1}. The inverse of k is usually given the symbol, τ, and indicates the time duration of the process; e.g. → fluorescence decay time, → phosphorescence decay time, etc. Quenching processes and the formation of excimers and exciplexes are generally second order reactions and the appropriate dimensions for the rate constant are $mol^{-1} \cdot s^{-1}$.

Ratio recording, *<Ratiorecording>*, a procedure, used in recording spectrometers, where the intensities in the sample and reference beams are compared with each other directly. The ratio of the signals generated in the detector by a wide range of intensities is formed electronically or digitally and the result presented as % transmission (→ transmittance) or (→ absorbance).
The advantages of this approach were recognized more than 25 years ago when recording → spectrophotometers for the UV-VIS and NIR were being designed. The significant point was that, in photocells and → photomultipliers, detectors with a very rapid response were available. Thus, an important prerequisite for the simple implementation of ratio recording was fulfilled. With the advent of fast → pyroelectric detectors, the technique could also be applied, without too great an expenditure on electronics, to IR spectrophotometers which, up to that time, had generally used the →

optical null principle. (Philips, Series SP-3; Perkin-Elmer, Series 881–883).

Rayleigh scattering, <*Rayleigh-Streuung*>, molecular → light scattering by gases observed by Tyndall and analyzed theoretically by Rayleigh in the years 1881–1899. A quantitative description can be based upon the concept that the electric vector of a beam of incident, monochromatic light excites the electrons in a molecule to forced vibrations of the same frequency as that of the light. It is assumed that the frequency of the radiation, ν, is much less than that of the electrons, ν_o, i.e. there is no absorption. If the dimensions of the scattering molecule are small compared with the wavelength of the light, λ, it can be assumed that all the electrons move in phase. Then the molecule is an oscillating dipole and behaves like a transmitter of molecular dimensions. The dipole sends out secondary electromagnetic waves of the same frequency, ν, in all directions, i.e. the primary wave is scattered. The scattered light is incoherent because of the statistical distribution of the secondary emitters. The amplitude of the scattered wave depends upon the magnitude of the electric dipole moment, $\vec{\mu}_i$, induced in the molecule which, in turn, depends upon the movement of charge and is proportional to the field strength, $\vec{E}$: $\vec{\mu}_i = \alpha \cdot \vec{E}$. The factor of proportionality, α, is known as the → polarizability of the molecule. In an isotropic molecule, α is the same in all directions and $\vec{E}$ and $\vec{\mu}_i$ are parallel. If the molecule is anisotropic, α is an average value formed from the three principal polarizabilities in the three, mutually perpendicular, spatial directions, (polarizability ellipsoid).

Consider a scattering molecule placed at the origin of a Cartesian coordinate system, Figure 1. A beam of light, incident in the x direction with its electric vector, $\vec{E}$, vibrating in the z direction (→ polarized light), induces a dipole vibrating in the z direction. The primary wave is scattered equally in all directions in the xy plane, independent of the value of φ, the azimuthal angle in the xy plane. In contrast, the amplitude, A, of light scattered in other directions falls off as $\sin\delta$, where δ is the angle between the radius vector and the z axis. The vibrating dipole scatters no light in the direction of its own axis. Since the intensity of the scattering, I_s, is proportional to the square of the amplitude, a polar diagram of I_s is shaped like a doughnut a cross section of which is shown in Figure 2a. This figure is called a radiation characteristic. If the incident light

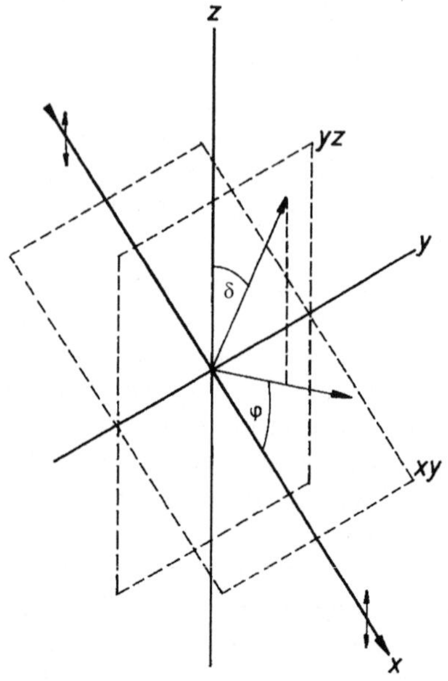

Rayleigh scattering. Fig. 1. Axis system for Rayleigh scattering

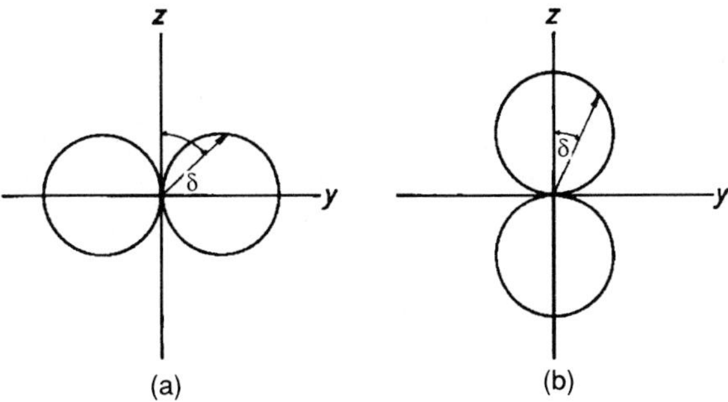

Rayleigh scattering. Fig. 2. Intensity distribution of scattered radiation, see text

comes from the x direction and is polarized along y, then the induced dipole also vibrates in the y direction. Using the coordinate system of Figure 1 again, the intensity distribution in the polar diagram is now proportional to $\cos^2\varphi$, i.e. the radiation characteristic is rotated by 90°, in the plane of the paper, from that shown in Figure 2a; Figure 2b.

However, for natural light, there is no preferred direction perpendicular to the direction of propagation (x direction). Therefore, the intensity of scattered light arising from natural, incident light must be rotationally symmetrical in the plane perpendicular to this direction. Because of this symmetry, the angle dependence of scattering can be represented by only one angle, the scattering angle, δ_s. It is the angle between the exit direction of the primary light (x direction) and the direction of observation, which can be freely chosen. In the xy plane it is the angle equivalent to φ.

If N is the number of molecules per cm^3, and under the assumption that the scattering of the individual molecules is incoherent, the relative intensity of the scattered light, I_δ, per cm^3 to the incident natural light, I_o, is:

$$I_\delta/I_o = (2\pi/\lambda_o)^4(N\alpha^2/R^2)\cdot [(1+\cos^2\delta_s)/2] = q(\delta_s).$$

The polarizability, α, of a dilute gas is given by the well-known equation $\alpha = (\varepsilon-1)/4\pi N$, which, using Maxwell's result $\varepsilon = n^2$, gives $\alpha = (n^2-1)/4\pi N$. When this is substituted into the expression above we obtain:

$$I_\delta/I_o = (\pi^2/2\lambda_o^4 \cdot R^2N)(n^2-1)^2(1+\cos^2\delta_s) \quad \text{(a)}$$

The relationship between the total intensity, I_{St}, scattered by N molecules per cm^3, to the incident intensity, I_o, can be obtained by integrating over the surface of a sphere of radius $R = 1$. The double integral, $\int_o^\pi \int_o^{2\pi}(1+\cos^2\delta_s)\sin\delta_s d\delta_s d\varphi$, has the value $16\pi/3$, and it therefore follows that:

$$\frac{I_{St}}{I_o} = \frac{8\pi^3}{\lambda_o^4\cdot 3N}\cdot(n^2-1)^2 = S' \quad \text{(b)}$$

S' is identical with the apparent extinction coefficient (→ absorbance, apparent, → light scattering), observed in transmission and also called → turbidity. Frequently, the expression $(n^2 - 1)$ is approximated by $2(n - 1)$ which, when substituted into equation a above gives:

$$I_\delta/I_o = (2\pi^2/N\lambda_o^4R^2)(n-1)^2\cdot(1+\cos^2\delta_s) \quad \text{(a')}$$

and with equation b:

$$\frac{I_{St}}{I_o} = \frac{32\pi^3}{3\lambda_o^4 N}\cdot(n-1)^2 \equiv S' \quad \text{(b')}$$

Equations (a,b) or (a', b') are the Rayleigh scattering formulas for dilute gases. The scattered intensity is inversely proportional to the fourth power of the wavelength and depends, according to a, on the angle, δ_s, between the incident beam and the direction of observation. The angle dependence of I_δ/I_o is illustrated as a polar diagram in Figure 3. It is rotationally symmetric in the *yz* plane perpendicular to the propagation direction, *x*, of the primary beam. According to equations a and a', the intensity is a maximum for $\delta_s = 0$ and $\delta_s = 180°$ because then $1 + \cos^2\delta_s = 2$. These are the cases of forward and backward scattering. Also, the scattered light is unpolarized. The intensity is only half as large at $\delta_s = 90°$ and $270°$, since $1 + \cos^2\delta_s = 1$, and the scattered light is polarized. Elliptically polarized light is observed in all other directions. Attention is explicitly drawn to the fact that equations a, a', b and b' apply only to isotropic molecules.

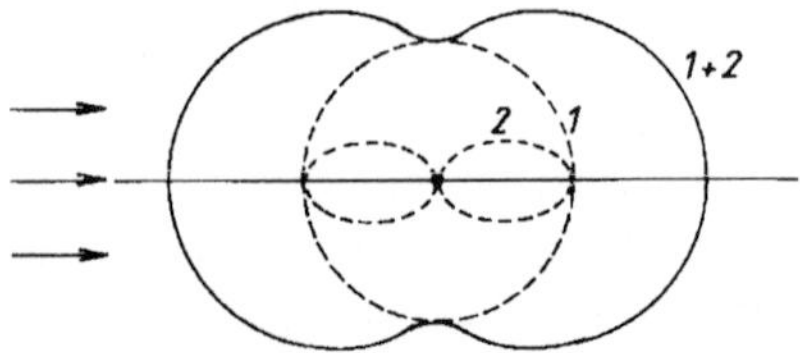

Rayleigh scattering. Fig. 3. Angle dependence of I_δ/I_o, see text

The λ_o^{-4} dependence means that white light, e.g. sun light, will be colored after a long passage through a gas. Since blue light is scattered more than red light, the latter is dominant when the sun is just above the horizon which explains the red of the evening and morning skies. The superposition of the sensitivity curve of the eye, which has its maximum in the green, and the wavelength dependence of scattering, is responsible for the blue color of the sky.

The number of particles per cm^3 can be expressed as: $N = 3G/(4\pi R^3\varrho)$, where G is the mass of all particles in one cm^3 and ϱ is their density. If N in equation b or b' is replaced by this expression we see that the scattering coefficient, S', is proportional to the third power of the radius of the particles, and therefore to their volume. Also, as above, $n^2 - 1 = 4\pi N\alpha$ or $n - 1 \sim 2N$. Thus, S' is proportional to N and this offers the possibility of determining the Loschmidt number (N) with the help of Rayleigh scattering.

In 1947, Debye extended the theory of Rayleigh scattering to solutions. The equation analogous to b' is:

$$\frac{I_{St}}{I_o} = 24G^3\frac{n^2 - n_o^2}{n^2 + 2n_o^2}\cdot\frac{NV}{\lambda^4}.$$

in which n and n_o are the refractive indexes of the solute and solvent respectively. λ is the wavelength for the solution replacing λ_o for the gas.

Ref.: G. Kortüm, *Reflectance Spectroscopy*, Springer Verlag, Berlin, Heidelberg, New York, **1969**, chapt. III.

R branch, <*R-Zweig*> → rotation-vibration spectrum.

Reaction spectrum, <*Reaktionsspektrum*>, the result obtained when the → absorption spectrum (e.g. UV-VIS, IR) of a chemical reaction mixture is measured at a sequence of time intervals. A reaction spectrum represents the course of a chemical reaction in the form of spectra which change with time and conclusions concerning the nature of the reaction can be derived from the spectra. In evaluating the results, it must be remembered that the changes in the → absorbance values do not correspond to the same point in time because the reaction continues as the spectrum is recorded. However, the absorbance differences, $\Delta A_\lambda(t)$, correspond to the same time difference, Δt, at all wavelengths and the uncertainty concerning the time of the start of the reaction, t_o, is thereby eliminated.

Receiver, <*Empfänger*> → detector.

Recombination continuum, <*Rekombinationskontinuum*>. Consider two atoms which approach each other in such a way that their potential energy for every internuclear distance is given by the upper potential energy curve in Figure 1. The horizontal line, *A-B*, represents the total energy of the two atoms. At every position, the relative kinetic energy is given by the distance between *A-B* and the potential curve. At *A*, the kinetic energy has been completely converted into potential energy and the classical turning point of the motion has been reached. The molecule now returns along the potential energy curve in the opposite direction and the two atoms separate again. A continuing union of the two atoms can only take place if energy is removed during their short (about 10^{-13} s) meeting (collision time). This can happen in two ways; either by collision with a third particle and without emission of radiation, or by emission of the excess energy as radiation. The first process requires a triple collision, the second a binary collision. If the second process occurs, then there is a transition, mainly from the classical turning point, *A* in Figure 1, to a vibrational level of the ground state for which the turning point lies approximately vertically below *A*. Since the kinetic energies of the colliding particles can take a wide range of possible values, which corresponds to a shift of the point *A* along the upper potential energy curve, a continuous emission, known as the recombination continuum, results. It corresponds to the dissociation continuum in absorption. But, combination as a result of a two-body collision is a very rare event

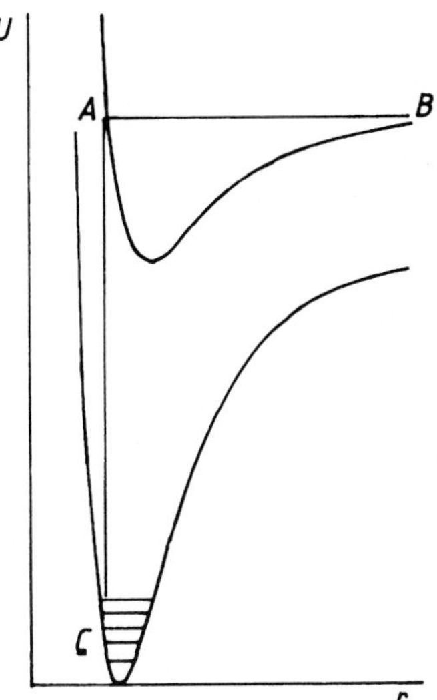

Recombination continuum. Fig. 1. Potential energy curves for the explanation of a recombination continuum

because the lifetime of an excited molecule at 10^{-8} s is very large compared with the duration of the collision, 10^{-13} s, during which the electronic transition must take place. Consequently, two-body recombinations occur for only an extremely small proportion of the collisions; about one in 10^{-5}. In spite of their very small probability, two-body recombinations have been substantiated spectroscopically by the observation of recombination continua, e.g. in halogens and Te_2.

Redox indicator, *<Redoxindikator>* → indicator.

Red shift, *<Rotverschiebung>* → bathochromic shift.

Reduced mass, *<reduzierte Masse>*, the reduced mass, μ, of the two masses, m_1 and m_2, is given by $\mu = m_1 \cdot m_2/(m_1 + m_2)$. This result can be readily derived for the rotation of two masses, m_1 and m_2, separated by a rigid distance, r, about the axis through their center of gravity, using the → dumbbell model. The same can be done for vibration in the case of an elastic force between the two masses. Even for the movement of an electron, mass m, around a nucleus, mass M, the reduced mass, $\mu = m \cdot M/(M + m)$, should be used because the rotation takes place around the center of gravity and not around the center of the nucleus. Thus, in the → Rydberg constant, $R = 2\pi^2\mu e^4/ch^3$, the mass of the nucleus enters in the reduced mass, although only as a correction term.

Reflectance attachment *<Reflexionszusatz>* → integrating sphere.

Reflection of light, *<Reflexion>*. If a monochromatic, parallel beam of light falls, at an angle of incidence, α, upon the boundary surface between two nonabsorbing media with refractive indexes, n_o and n_1 ($n_1 > n_o$), then on entry into the medium with refractive index, n_1, the beam is refracted with an angle of refraction, β, (→ refraction of light). But a part of the light beam is reflected at the surface and, in accordance with the law of reflection, the angle of incidence = the angle of reflection. This type of reflection is known as normal or specular reflection. The incident, reflected and refracted light beams are related by energy conservation in that the intensity (radiant power per cm^2) of the incident light, I_i, must be equal to the sum of the intensities of the reflected, I_l, and refracted, I_r, light. But the intensity of an electromagnetic wave is proportional to the square of the electric vector, $\vec{E}$, and the velocity, $v = c/n$, of the light in the particular medium, so that the energy conservation requirement can be written:

$$n_o E_i^2 = n_o E_\ell^2 + n_1 E_r^2(\cos\beta/\cos\alpha)$$

The factor, $\cos\beta/\cos\alpha$, allows for the fact that the cross section of the beam has increased as a result of the refraction. If $\vec{E}$ is now decomposed into two components, $\vec{E}_\perp$ and $\vec{E}_{||}$, perpendicular and parallel to the plane of incidence respectively, then, using Snell's law of refraction, we obtain Fresnel's equations for the relationship between $E_{l\perp}$ and $E_{i\perp}$, $E_{r\perp}$ and $E_{i\perp}$, $E_{l||}$ and $E_{i||}$, $E_{r||}$ and $E_{i||}$ and their dependence upon the angles α and β. If the ratio of the intensities $I_{l\perp}/I_{i\perp}$ and $I_{l||}/I_{i||}$ is measured as a function of the angle of incidence, α, then the square roots of

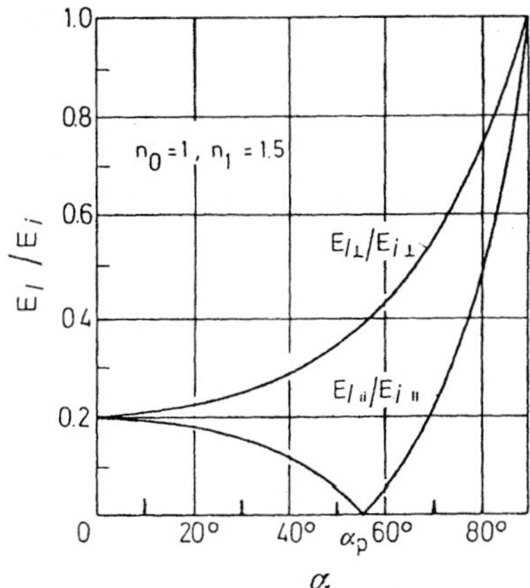

Reflection. Fig. 1. The reflection coefficients, E_l/E_i, as a function of the angle of incidence

these ratios give the reflection coefficients, $E_{l\perp}/E_{i\perp}$ and $E_{l||}/E_{i||}$. If these ratios are plotted as functions of α, curves such as the two shown in Figure 1 are obtained. For perpendicular ($\alpha = \beta = 0$) and grazing ($\alpha = 90^o$) incidence, the coefficients are equal. $E_{l||}/E_{i||}$, however, is equal to zero for a quite definite angle, α_P. This is the case when $\alpha_P + \beta = \pi/2$, as is clear from the relevant Fresnel equation. In this way, using the law of refraction, we can obtain → Brewster's law; $\tan(\alpha_P) = n_1$. For crown glass with $n_1 = 1.5$, $\alpha_P = 56.3^o$. For this angle only the perpendicularly polarized component of the light beam will be reflected from a crown glass surface. For this reason, the Brewster angle, α_P, is also called the polarization angle. This law is important for the generation of polarized light in spectroscopy.

When the reflection takes place on strongly absorbing media the complex → refractive index, $n' = n(1-i\kappa)$, must be used. Then, for specular (or regular) reflection, R_{reg}, at perpendicular incidence, the Fresnel equation gives:

$$R_{reg} = \frac{(n_1 - n_o)^2 + (n_1\kappa_1)^2}{(n_1 + n_o)^2 + (n_1\kappa_1)^2}.$$

For comparison; for specular reflection in the case of nonabsorbing media, i.e. $\kappa = \beta = 0$, we have:

$$R\mathrm{reg} = \frac{(n_1 - n_o)^2}{(n_1 + n_o)^2}$$

κ is the absorption index of the absorbing medium and $n\kappa$ the → absorption coefficient. If $n_1\kappa_1 \gg n_1$, i.e. if the absorbance of the material 1 is very large, then the reflectance must also be very high, nearing a value of unity, which is known in metals and in the → residual ray method in the IR region. For crown glass at perpendicular incidence, only ca. 4% of the light is reflected while 96% is transmitted, but silver, an absorbing medium, reflects 99.5% of the light at a wavelength of 4.04 μm, so that the proportion of the light passing through is very small. The reflectance of metals is very important in selecting suitable mirrors for the various spectral regions.

As well as specular reflection, diffuse reflection must also be considered. This occurs whenever there is reflection at an uneven phase boundary, e.g. at a rough or matt surface. Let us consider the surface to be an arrangement of crystallites lying close together; a crystalline powder in which the dimensions of the crystallites are large compared with the wavelength of the incident light. Under these conditions, the phenomena of reflection, refraction and diffraction are well defined for each individual crystallite, see above. But because the crystal surfaces are

statistically orientated in all possible directions, the incident light will now be reflected in every direction into the hemisphere from which the light is incident, i.e. the reflected radiation is independent of the angle of incidence. An ideal diffuse reflection can therefore be defined as one in which the angular distribution is independent of the angle of incidence. We owe the fact that we are able to see nonluminous bodies, to this diffuse reflection. As with specular reflection, we must also allow for absorption. This led to the development of diffuse → reflection spectroscopy.
Ref.: G. Kortüm, *Reflectance Spectroscopy*, Springer Verlag, Berlin, Heidelberg, New York, **1969**, chapt. II.

Reflection grating, *<Reflexionsgitter>* → grating.

Reflection-measurement attachments, *<Reflexionszusätze, Remissionszusätze>* → integrating sphere.

Reflection spectroscopy, *<Reflexionsspektroskopie>*, a spectroscopic method which is based upon reflection. Incident light is diffusely reflected by strongly scattering or nontransparent bodies. The reflecting power of such bodies depends upon their ability to absorb light. Basically, it is this fact which allows us to detect a colored body. Since diffuse reflection is due to simple and multiple scattering on the surface and in the interior of a solid, the remissive power of a body can, as a first approximation, be represented by its → absorption coefficient (β in cm^{-1}) and its scattering coefficient (s in cm^{-1}). This two-constant theory led to the theoretical treatment of the problem by Kubelka and Munk. In reflection spectroscopy the Bouguer-Lambert-Beer law is replaced by the Kubelka-Munk function $F(R)$, which establishes a relationship between the diffuse reflectance, R, the absorption coefficient, $K = 2\beta(\text{cm}^{-1})$ and the scattering coefficient, $S = 2s(\text{cm}^{-1})$:

$$F(R_\infty) = \frac{(1 - R_\infty)^2}{2R_\infty} = \frac{K}{S} \qquad \text{(a)}$$

In the terms of this theory, R_∞ represents the situation where the sample thickness $d \to \infty$ and, simultaneously, the reflection of the background $R_\infty \to 0$. The factor of two which occurs in the absorption and scattering coefficients, K and S, as they are defined in the Kubelka-Munk theory, arises because the radiation flux of the incident and scattered light in both directions of the sample surface normal must be considered. Since the absolute value of R_∞, which is important in the applications of reflection spectroscopy, cannot be measured with the usual apparatus, diffuse reflection, R_∞, is always measured with respect to a white standard, i.e. it is obtained as a relative quantity, R':

$$R_\infty' = R_{\text{sample}}/R_{\text{standard}} \qquad \text{(b)}$$

If the absolute reflectance, ϱ, of the white standard were equal to one, i.e. $R_{\text{st.}} = \varrho = 1$, then the absolute and relative reflectances of the sample would be equal. But there is no known white standard which has this property over the whole spectral region of interest; UV-VIS-NIR.
To date, MgO has proved to be the most useful white standard in practice because it is easy to prepare under

defined conditions. Its ϱ values in the visible region are: 0.983 at $\lambda = 420$ nm, 0.986 at $\lambda = 680$ nm, and a maximum of 0.988 at $\lambda = 620$ nm. Apart from MgO, the following white standards are frequently used: Li_2CO_3, NaF, NaCl, $MgSO_4$, $BaSO_4$, Aerosil, Al_2O_3, SiO_2 and glucose.

The Kubelka-Munk function applies only to diffuse reflection. As soon as a component of specularly reflected light is present, significant deviations can occur. The contribution of specular reflection can be removed by use of the → dilution method. The powdered sample is diluted with an inert, nonabsorbing, solid standard (e.g. MgO, NaCl, $BaSO_4$, SiO_2, TiO_2) in such an excess that, in a relative measurement against the same pure standard, the specular component of the reflection is less than the accuracy of the measurement. These cases are of particular practical interest and it can be shown that the absorption coefficient, K, in equation (a) above is proportional to the concentration of the absorbed substance. But this means that the scattering coefficient, S, is constant for a series of dilutions by the same standard diluent, and that the Kubelka-Munk function depends only upon the absorption coefficient. The absorption coefficient can then be written as the product of the molar decadic → extinction coefficient, ε, and the concentration, c. Equation (a) then becomes:

$$F(R_\infty) = \frac{\varepsilon \cdot c}{S}$$

or

$$\log F(R_\infty) = \log \varepsilon + C. \qquad \text{(c)}$$

$$C = \log \frac{c}{S}$$

This relationship has been confirmed for many systems and provides the justification for the wide variety of applications of reflection spectroscopy. Equation (c) corresponds to the Lambert-Beer law in transmission measurements. Both laws are limiting laws for high dilutions. Since the Kubelka-Munk function depends upon wavelength, a graph of $\log F(R_\infty)$ as a function of wavelength gives the absorption spectrum in the form of a typical → color curve which can be brought into coincidence with the true absorption spectrum, measured in transmission, by a parallel shift along the ordinate. This simple relationship between the typical color curve in a reflectance spectrum and the true absorption spectrum is only found when the scattering coefficient is independent of wavelength and the standard shows no absorption of its own. Reflection spectroscopy may be applied to the spectroscopy of insoluble substances or those which change upon dissolution, spectroscopy of absorbed substances, kinetic measurements, spectroscopy of crystal powders, dynamic reflection spectroscopy, analytical photometric measurements such as color measurement and color matching. The quantitative photometry of thin layer chromatograms is a special analytical application for which instruments for routine analysis have been developed in recent years.

Ref.: G. Kortüm, *Reflectance Spectroscopy,* Springer Verlag, Berlin, Heidelberg, New York, **1969**.

Refraction of light, <*Lichtbrechung*>, the deviation of a light beam passing from one medium to another. According to Snell's law, when a monochromatic light beam is incident at angle α on the surface between two media (air/medium II) the following equation holds:

$$\frac{\sin\alpha}{\sin\beta} = \text{const} = n.$$

n is the refractive index which is defined with reference to air as medium I, α is the angle of incidence and β the angle of refraction. When two materials are compared the one with the higher refractive index is said to be optically more dense. The origin of the refraction of light lies in the fact that the velocity of light, c, is always smaller in a medium than in a vacuum where the value is $3 \cdot 10^8$ ms^{-1}. The exact value is $[2.99792458 \pm 1.21 \cdot 10^{-8}]$ 10^8 ms^{-1}. Thus, the refractive index, n, for monochromatic light may be written; $n = c_{air}/c_{material}$ = constant. Since the velocity of light, c, (more accurately the → phase velocity) is simply related to the frequency, v, and wavelength, λ, $c = v\lambda$, the refractive index of every material depends upon the wavelength, λ. Therefore, a measurement of n must be accompanied by a statement of the wavelength used. The dependence of the refractive index upon the wavelength, $dn/d\lambda$, is the → dispersion which is of fundamental importance in the use of prisms as dispersing elements in spectroscopy.

Refractive index, <*Brechungsindex, Brechungszahl, Brechzahl*>, for absorbing materials, the refraction of light (→ refraction of light) is defined as:

$$n' = n(1 - i\kappa) = n - in\kappa.$$

For transparent materials, the real refractive index, n, is defined. The absorption index, κ, comes from dispersion theory and is defined by the Lambert law, viz:

$$\Phi_\lambda = \Phi_0 \exp\{-(4\pi n\kappa/\lambda_0) \cdot s\}$$

This equation states that when light of wavelength (in vacuum) λ_0 passes through a medium of thickness, $s = \lambda_0$, then the radiant power of the light is reduced by the fraction, $\exp[-4\pi n\kappa]$, where $n\kappa$ is the → absorption coefficient. The use of complex numbers permits the two quantities, the refractive index, n, and the absorption coefficient, $n\kappa$, which are linked by dispersion theory, to be written in one symbol, n'.

Refractometer, <*Ablenkungsrefraktometer*> → refractometry.

Refractometry, <*Refraktometrie*>, the measurement of the → refractive index, in the region of normal → dispersion, for analytical purposes in chemistry, physics and technology. The instrument used for this purpose is called a refractometer and the basis of the measurement is the → refraction of light. Solid, liquid and gaseous samples can be studied. The applications of refractometry may be divided into three:

1. Identification of substances.
2. Purity checks and controls.
3. Quantitative analysis of mixtures containing two or more components.

Since the refractive index, n, depends upon the wavelength of the light used to measure it (→ dispersion), a fixed wavelength must be chosen for analytical work, e.g. the sodium D line characterized by the subscript, D, as in n_D. With reference to the laws of →

refraction of light, the following principles used in refractometry may be distinguished:

a) The deviation of light by a → prism;
b) the critical angle for total reflection.
c) the measurement of reflection intensity;
d) interference.

The best-known refractometer is the Abbe refractometer which depends upon a measurement of the critical angle. The range of n, which can be measured, is $1.32 \le n_D \le 1.70$ with an uncertainty of ca. $1 \cdot 10^{-4}$. The immersion refractometer, in which the measuring prism is immersed in the solution, uses the same principle and is more accurate. The instrument has a total of 10 exchangeable prisms and the range of measurement is $1.325 \le n_D \le 1.642$ with an uncertainty of $1 \cdot 10^{-5}$ at $n_D = 1.33$.

The Pulfrich refractometer is also a critical angle instrument which has a moving observation telescope and a line source. The uncertainty is ca. $1 \cdot 10^{-5}$.

In deviation refractometers, the principle of the deviation of light by refraction is used. The classical instrument for exact measurements is the → spectrometer or spectrograph. The deviation method has found extensive application in differential refractometers such as are used, for example, as detectors in liquid chromatography.

In the reflection method, the ratio of the intensity of the light incident upon a reflecting surface, I_o, to that of the reflected light, I_r, is measured, i.e. I_r/I_o. Since I_r changes rapidly near the critical angle, the method is particularly sensitive in that case. Refractometers making use of interference are usually applied to gases. Refractive index differences of the order of 10^{-7} are measurable. The → Jamin and → Mach-Zehnder interferometers have proved to be particularly valuable.

Ref.: N. Bauer. K. Fajans and S.Z. Lewin in *Technique of Organic Chemistry* (Ed.: A. Weissberger), 3rd ed., Interscience Inc., New York, London, **1960**, vol. I., part II, chapt. XVIII.

Regular reflection → specular reflection.

Relaxation spectroscopy, <*Relaxationsspektroskopie*>, an area of spectroscopy in which the relaxation method is used. In conventional kinetic measurements, the transition from the initial state at $t = 0$ to the final state at $t = \infty$ is followed with time. In the relaxation method, the final state, in general an equilibrium state, is itself used as the initial state. If the conditions of this equilibrium are suddenly disturbed, the system moves to a new position of equilibrium appropriate to the change of conditions. The system is said to relax to the new equilibrium conditions. Since a chemical equilibrium depends upon the variables, temperature and pressure, the temperature-jump and pressure-jump methods have been widely applied. For work with solutions, the temperature-jump method is the most usual.

The general application of the temperature-jump technique is based upon the fact that reaction enthalpy, ΔH_o, is finite during chemical changes, i.e. $\Delta H_o \neq 0$. Thus, an increase in temperature results in a displacement of the equilibrium and hence a change in the chemical composition of the system. According to the van't Hoff equation, the relative change of the equilibrium constant, $\Delta K/K$, is given by:

$$\frac{\Delta K}{K} = \frac{\Delta H_o}{RT} \cdot \frac{\Delta T}{T}.$$

Thus, the relative change of the equilibrium constant is proportional to the relative temperature change. This causes a continuous change of the chemical composition until the system has established itself in a new condition of chemical equilibrium at $T + \Delta T$. Therefore, the effect is a function of the stoichiometry and the original concentrations of reactants and products. If the profile of the temperature jump is assumed to be rectangular, then equilibrium is re-established according to the simple exponential function:

$$\frac{\mathrm{d}c_t}{\mathrm{d}t} = \frac{c_o - c_t}{\tau}, \; c_t = c_\infty(1 - e^{-t/\tau}).$$

where c_t is the deviation of the concentration from its equilibrium value, c_o is the sudden displacement from equilibrium at $t = 0$ and τ is the relaxation time. But the relaxation time is related to the rate constants, k_{ij}, of the reaction steps which proceed in the direction of attainment of the new equilibrium. For the simple system:

$$\mathrm{A} + \mathrm{B} \underset{k_{21}}{\overset{k_{12}}{\rightleftharpoons}} \mathrm{AB}$$

the result is:

$$\tau = \frac{1}{k_{21} + k_{12}(c_A + c_B)}$$

or

$$\frac{1}{\tau} = k_{21} + k_{12}(c_A + c_B).$$

The following also applies to the above equilibrium:

$$K = \frac{c_{AB}}{c_A \cdot c_B} = \frac{k_{12}}{k_{21}}.$$

A relaxation time of $\tau = 3 \cdot 10^{-6}$ s = 3 μs results from the above equation if $K = 10^5$ l mol^{-1}, $c_A = c_B = 10^{-1}$ mol l^{-1} and the diffusion-controlled rate constant of the observed association equilibrium, k_{12}, equals 10^{10} l mol$^{-1}\cdot$s^{-1}. This shows that the heating period must be of the order of microseconds if we are to be able to follow this fast reaction. Therefore, the temperature jump must occur extremely rapidly. This is achieved by discharging a condenser in the sample cuvette itself. Depending upon the details of the equipment (voltage, capacitance of the discharged condenser, separation between the electrodes, resistance of the solution and volume of the cuvette) temperature jumps between 4 and 10 K can be realized in a few microseconds with this technique. A time constant less than 175 ns has been achieved with a specially constructed temperature-jump cell. In another version, the electrical energy was fed in via a coaxial cable which made heating pulses as short as 80 ns possible. In addition to the conventional means of supplying energy as Joule heating by an electric current, other possibilities have been proposed. The use of microwaves and of lasers should be mentioned.

For the detection of the time-dependent concentration changes following the temperature jump we use the Bouguer-Lambert-Beer law in the form:

$$A = -2.303 \cdot \ln\frac{I}{I_o} = \varepsilon_i d \cdot c_i. \quad \text{(a)}$$

For a differential concentration change as a result of a temperature jump this becomes:

$$\Delta A = -2.303 \cdot \Delta \ln I = \varepsilon_i d \cdot \Delta c_i,$$

From which we obtain:

$$\Delta c_i = \frac{\Delta A}{\varepsilon_i d} = -\frac{2.303}{\varepsilon_i d} \cdot \frac{\Delta I}{I}. \qquad \text{(b)}$$

The change of light intensity, ΔI, is measured by a → photomultiplier and followed on an oscilloscope or stored in a data aquisition system as a function of time from the beginning of the temperature jump onwards. Since the signal is obtained as a voltage, U, and because $I \sim U$, equation (b) may be written:

$$\Delta c_i = -\frac{2.303}{\varepsilon_i d} \cdot \frac{\Delta U}{U}. \qquad \text{(c)}$$

Thus, the relative signal change is proportional to the change in concentration of the absorbing species, if $\Delta c_i/c_i$ is small. Figure 1 shows a temperature-jump apparatus, suitable for teaching purposes, schematically. The sample is placed between two electrodes in a cell. A 0.5 μF condenser is charged with a voltage of 4 KV. The volume of the cell is 0.5 ml. The discharge supplies 4 J of energy which causes a temperature jump of ca. 4 K in 100 μs. Instruments developed for research purposes operate at higher voltages (ca. 20 KV) and smaller capacitances (0.04–0.01 μF) and provide temperature jumps of 6 to 10 K within a few microseconds. The time scale obtainable with these instruments covers $5 \cdot 10^{-6} - 5 \cdot 10^{-1}$ s.

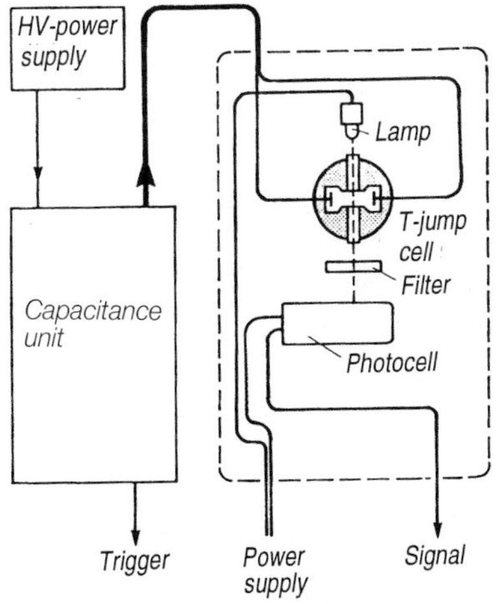

Relaxation spectroscopy. Fig. 1. Apparatus for the temperature-jump method

Proton-transfer reactions, metal-ligand reactions and enzyme-substrate reactions are examples of processes which can be studied with the temperature-jump technique. In the case of protolytic reactions, which cannot be followed directly by spectrophotometric means, or which lie in a spectral region unfavorable for measurement, the dissociation equilibrium under investigation can be coupled with an indicator equilibrium. The temperature-jump method has been found to be suitable for following chemical relaxation processes on the 10^{-6} to 1 s time scale by spectroscopic means. However, in the great majority of experiments using the pressure-jump method, the reaction has been followed by the change in conductivity as a result of the sudden pressure change. But a pressure-jump relaxation method with spectroscopic detection has been reported. (W. Knoche, G. Wiese, *Rev. Sci. Instrum.*, **1976**, *47*, 220; H.-H. Buschmann, E. Dutkiewicz, W. Knoche, *Ber. Bunsenges. Phys. Chem.*, **1982**, *86*, 129). The reversible hydration of carbonyl compounds was studied with this technique.

In addition to the above methods, the field-jump technique, which utilizes the dissociation-field effect, can also be mentioned. Here again, the electri-

cal conductivity of an electrolyte solution is used as the detection method.
→ Flash photolysis is another spectroscopic method for following fast reactions.
A brief but clear description of all the possible techniques for investigating chemical relaxation has been given by L. de Maeyer, (*Z. Elektrochem. Ber. Bunsenges. Phys. Chem.*, **1960**, *64*, 65).
Ref.: M. Eigen, L. de Maeyer in *Technique of Organic Chemistry*, vol. VIII (Ed.: A. Weissberger), Interscience, New York, **1963**; J.F. Holzwarth in *Techniques and Applications of Fast Reactions in Solutions* (Eds.: W.J. Gettins, E. Wyn-Jones), Reidel Publishing Co., Dordrecht, **1979**, pp. 47–59; *ibid*, **1979**, pp. 61–70.

Residual rays method, <*Reststrahlenmethode*>. The → reflectance or reflecting power, R, of a lightabsorbing body (→ reflection of light) is:

$$R = \frac{(n^2 - 1) + n^2\kappa^2}{(n^2 + 1) + n^2\kappa^2}$$

where n is the refractive index, and $n\kappa$ the absorption coefficient. κ is also called the absorption coefficient. Outside the region of absorption $n\kappa = 0$ and the equation goes over into the one which describes the reflectance of transparent materials. In the region of an absorption band, $n\kappa \gg (n-1)^2$, and the reflectance increases strongly to reach values which are 80–90% of those for reflection by metals. Outside the absorbing regions, the reflectance is only a few percent of this. This observation led H. Rubens to propose his residual rays method in 1897. Let the incident radiation intensity at wavelength, λ, be $S(\lambda)$, and the reflectance, $R(\lambda)$. The light reflected is then $S(\lambda) \cdot R(\lambda)$, which is very small because $R(\lambda) \ll 1$. $S(\lambda_{max}) \cdot R(\lambda_{max})$ is almost as large as $S(\lambda_{max})$ itself only in the region of an absorption band, λ_{max}, because $R(\lambda_{max})$ is of the order of 1. Rubens now allowed the undispersed radiation, covering a wide spectral range in the infrared, to be reflected several times by a suitable substance. The contribution of the reflected radiation following n reflections is $S(\lambda)R(\lambda)^n$ and, at a wavelength other than λ_{max}, it will be very small while for λ_{max} the product will still be significant. Thus, n-fold (n = 4–5) reflection suppresses all wavelengths with the exception of an almost homogeneous radiation of wavelength, λ_{max}. The reflectance of → calcite, for example, at $\lambda_{max} \sim 7$ μm is $R \cong 0.8$. After four reflections, $0.8^4 = 41\%$ of the radiation at this wavelength remains. The refractive index, n, is 1.6, outside this wavelength region giving $R = 0.06$ and $R^4 = 0.00001$ or 0.001%. At the wavelength corresponding to the → halfwidth, $R = 0.4$ by definition and $R^4 = 0.026$ or 2.6%. Thus, the residual rays method works like a selective filter.

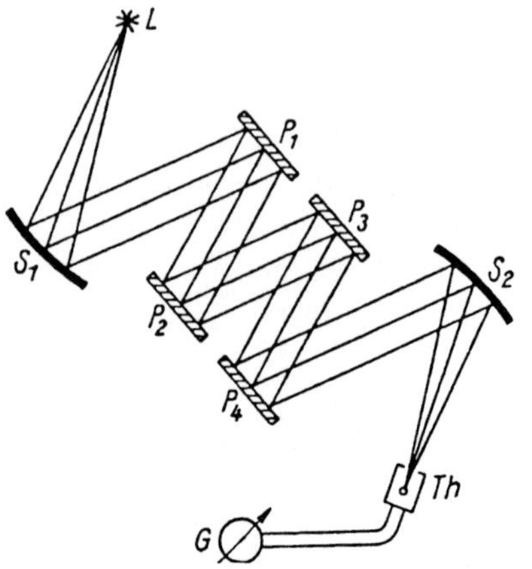

Residual rays method. Fig. 1. The residual rays method of Rubens

Residual rays method. Table.

Material	λ [μm]
$CaCO_2$	6.56
SiC	12.0
ZnS	30.9
NaCl	52.0
KCl	63.0
AgCl	81.5
KBr	82.6
TlCl	91.6
KI	94.7
AgBr	112.7
TlBr	117.0
TlI	152.8

The advantage is that one can isolate one or more discrete wavelengths without spectral dispersion of the light. In the years following 1900 long-wavelength infrared rays, which could not be detected using prismatic dispersion, were found by this method. Figure 1 illustrates the principle of the residual rays method and the table gives some residual ray wavelengths.

Resolution, resolving power, <*Auflösungsvermögen*>, is defined in spectroscopy as:

$$R = \lambda/\Delta\lambda$$

$\Delta\lambda$ is the wavelength difference which can just be separated by a → dispersing element set for wavelength, λ. The following expressions can be derived from this definition.
For a prism:

$$R = b \cdot dn/d\lambda$$

the resolving power of the prism depends upon the dispersion of the → prism material and upon b, the length of the base of the prism.

For a grating:

$$R = N \cdot Z$$

the resolving power is the product of the total number, N, of the illuminated lines of the grating and the order used, Z.
The dependence of resolving power on the base length, b, of a prism is the reason for the fact that, in → IR spectroscopy, NaCl and KBr prisms with base lengths up to 15 cm have been used. With gratings, a higher resolution may be obtained by increasing the number of lines or by working in a higher order.
A general expression for the resolution of any method of measurement may be defined as follows:

$$R = x/\Delta x$$

where Δx is the difference in two measured values, x, which the method or apparatus is just capable of separating. For example, in → mass spectroscopy, $R = m/\Delta m$, where m = mass and Δm = the mass difference.

Resonance, <*Resonanz*>, a process which takes place when an alternating electromagnetic field of frequency, ν, interacts with matter under conditions where, according to the → Bohr-Einstein relation, the quantity, $h\nu$, is exactly equal to the energy difference $\Delta E = E^* - E_o$ (E^* and E_o are the energies of excited and ground state respectively). Then, discrete atomic or molecular energy differences are in resonance with the energy of the radiation, $h\nu$. The corresponding absorption/emission frequencies are often known as resonance frequencies.
The expression is particularly applied

to situations in which the energy of the absorption process is exactly the same as that of the emission, e.g. in atomic spectroscopy, the lines of the principal series in emission and absorption are known as → resonance lines. This equality is also found in → resonance fluorescence and → nuclear magnetic resonance.

Resonance fluorescence, *<Resonanzfluoreszenz>*, the situation in which the frequency (wavelength or wavenumber) of the fluorescence of an atom or molecule is exactly the same as that of its absorption. Thus, the → fluorescence takes place directly from the state which has been reached by the absorption process and the absorbed energy, $h\nu_a$, is equal to the emitted energy, $h\nu_e$.
In the case of molecules, a resonance fluorescence can only take place from higher excited states when the → internal conversion, which converts the molecule to the lowest excited state, is blocked. For molecules in solution, the radiationless deactivation brought about by collisions with the surrounding molecules is dominant so that, in general, the fluorescence is from the lowest excited state; → Kasha's rule. In the gaseous state at low pressure, collisions with neighboring molecules are rare because of the large mean free path, and radiationless deactivation is severely hindered. Resonance fluorescence from higher excited singlet states can be observed under these conditions.

Resonance ionization mass spectroscopy, *<Resonanzionisationsmassenspektroskopie>* → resonance ionization spectroscopy.

Resonance ionization spectroscopy, RIS, *<Resonanzionisationsspektroskopie, RIS>*, a method in atomic spectroscopy suggested in 1971 by Letokhov and coworkers and, since 1983, developed further into a very promising analytical method, especially for trace-element analysis. In this technique, an atom is promoted from the ground state via a number of intermediate states by a series of resonant photons, i.e. photons whose energy matches the energy gap between the atomic states exactly. Finally, the atom is ionized and the ion or electron produced detected. Figure 1 shows three typical possibilities for the resonance ionization of an atom in three excitation steps. The photoions are usually detected with an attached mass spectrometer. This technique is called resonance ionization mass spectroscopy; RIMS. Since pulsed lasers are usually used in resonance ionization spectroscopy, the → time-of-flight (TOF) mass spectrometer is the instrument of choice. The construction of a RIMS apparatus is shown schematically in Figure 2. An instrument for plutonium RIMS

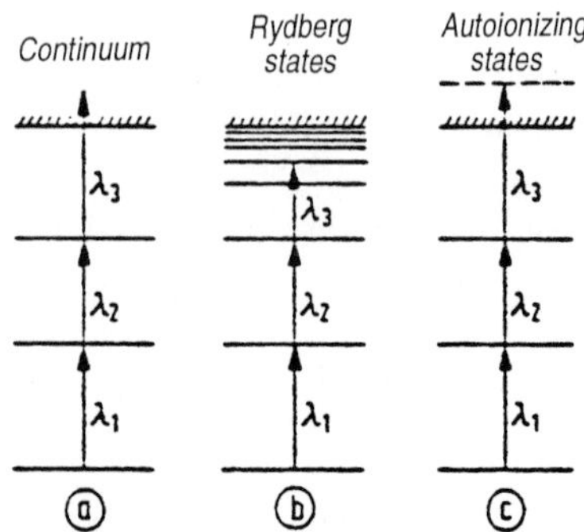

Resonance ionization spectroscopy. Fig. 1. Typical possibilities for three-step resonance ionization

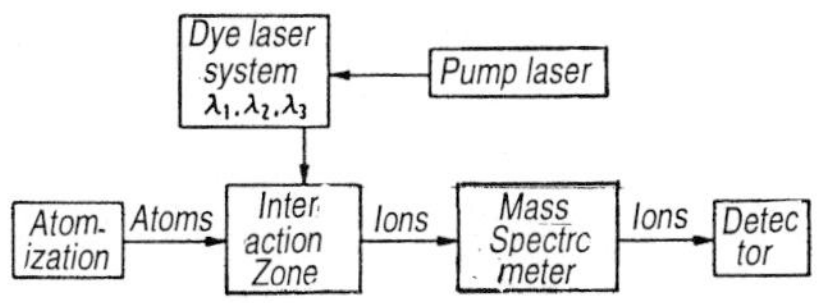

Resonance ionization spectroscopy. Fig. 2. Block diagram of a resonance ionization mass spectrometer

developed by Kluge and Trautmann is shown in Figure 3. A 30 W copper-vapor laser with a pulse repetition rate of 6.5 KHz pumps three dye lasers the light from which is conducted to the interaction zone by means of light pipes to ensure a good spatial overlap. The plutonium sample is deposited electrolytically on a rhenium filament and covered with a thin rhenium layer. If the filament is heated to some 1700°C, the plutonium evaporates, essentially as neutral atoms. The ions formed in the interaction zone are accelerated and registered by the detector, a TOF mass spectrometer. The authors report a detection limit for plutonium by the RIMS method of $2 \cdot 10^6$ atoms of plutonium, or $8 \cdot 10^{-16}$ g. This is two orders of magnitude bet-

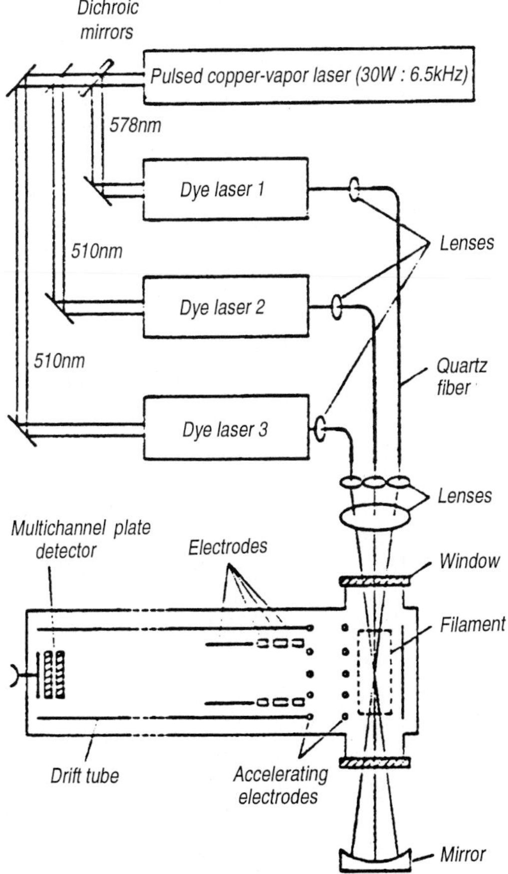

Resonance ionization spectroscopy. Fig. 3. Schematic view of a plutonium RIMS apparatus

ter than by alpha spectroscopy, the method generally used to date. It is important for the future wide-scale use of RIS and RIMS to note that these methods are suitable for almost all elements and show a high sensitivity and excellent selectivity including an isotopic sensitivity.
Ref.: D.M. Lubman, Ed.:, *Lasers and Mass Spectrometry*, Oxford University Press, Oxford, New York, **1990**.

Resonance line, <*Resonanzlinie, letzte Linie*>, the first line, and thus the line with the smallest wavenumber in the principal series of an atom or ion; → atomic spectra, → Balmer formula, → alkali spectra. The first line of each principal series corresponds to the transition of lowest excitation energy. Viewed from the series limit, i.e. from higher energies, the first line in the series is the last line.
For molecules, the resonance line corresponds to the first transition in the → Jablonski energy-level scheme.

Resonance Raman spectroscopy, <*Resonanz-Raman-Spektroskopie*>.
In normal → Raman spectroscopy, the Raman scattering is excited using a laser frequency outside the absorption range of the sample and the probability that the light will be absorbed is low. If the excitation frequency is moved into a wavelength region where the sample has an electronic absorption band, the sample absorbs more light and the intensities of some, but not all, of the Raman bands are much enhanced. This is the resonance Raman (RR) effect. It can be observed with a normal → Raman spectrometer and provides important information for the spectroscopist, if the problems of sample heating can be overcome.
The vibrations which are enhanced are almost always the totally symmetric vibrations of the sample (→ vibrational species) and they are frequently closely related to the electronic transition within which the exciting radiation is falling. Thus, use of an excitation frequency corresponding to the $\delta \rightarrow \delta^*$ transition in the metal-metal bond of $[Mo_2X_8]^{4-}$ results in a highly enhanced Raman spectrum of the symmetric Mo–Mo stretching vibration and members of the progression can be seen up to the eleventh overtone, i.e. vibrational transitions involving up to 11 quanta of the vibration are measurable in the RR spectrum. The ability to explore the vibrational potential energy function over such a wide range is one of the major advantages which RR spectroscopy has over conventional Raman spectroscopy.
But for the wide-ranging applications it is the intensity enhancement, which may be several orders of magnitude, which has proved most valuable. Thus, RR spectra of the metallic cores of important biological molecules, e.g. heme proteins such as hemoglobin and cytochrome c, have been measured. In these molecules, the concentration of the metal is so low that it is difficult to obtain vibrational spectra in any other way.
Compounds at concentrations of 10^{-6} $mol \cdot l^{-1}$ in water can be detected enabling RR spectroscopy to be used as an analytical method.
Ref.: R.J.H. Clark and T.J. Dines, *Angew. Chem. Int. Ed. Engl.*, **1986**, *25*, 131.

Resonator, <*Resonator*>, a laser component tuned to a particular frequency. The quality of a resonator is given by the resonance frequency, ω_o, divided by the → half-width, $\Delta\omega$, of the resonance curve. Low quality means high losses and a low time constant for the resonator. The time constant, T, of the resonator is related to the quality or Q value, Q, by:

$$\frac{Q}{\omega_o} = T$$

It can be thought of as the mean lifetime of the photons in the resonator. The half-width of the resonance curve is identical with the inverse of the time constant, T. The quality of a resonator is a measure of the coherence length of the light in the resonator because the light is coherent within the time T.

RGB laser, <*RGB-Laser*>, a designation of a → He-Cd$^+$ laser specific to a manufacturer. The initials derive from the red, green and blue lines emitted.

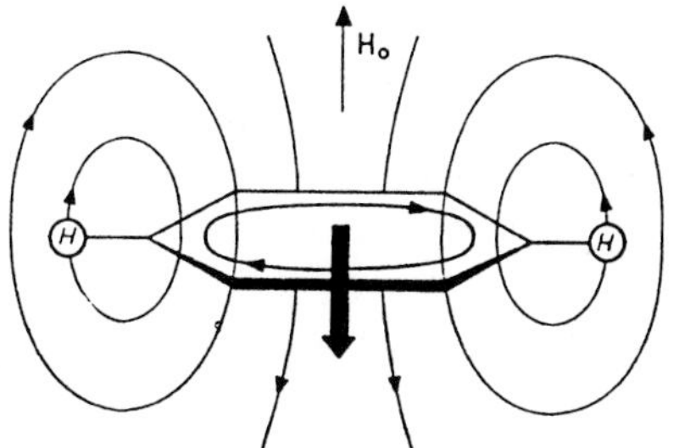

Ring current. Fig. 1. The ring current effect in benzene

[18] - Annulene

Ring current. Fig. 2. [18]-annulene

Ring current, <*Ringstromeffekt*>. The protons of aromatic compounds are collectively less shielded than those of olefins. This effect has its origin in the ring currents which are induced when a molecule with delocalized electrons is placed in a magnetic field. This constitutes a fourth component, σ_{ind}, of the → shielding constant, σ. The ring currents produce an extra magnetic field with lines of force having the directions shown in Figure 1. This gives rise to regions of higher and lower shielding. Protons bonded directly to the aromatic ring are in regions of reduced shielding. Ring current effects have been studied in large planar conjugated ring systems. A particularly clear example of the effect is seen in [18]-annulene, Figure 2. The molecule is planar and has 12 outer and 6 inner protons. The value of σ for the outer protons is 8.9 and for the inner protons it is 1.8. The outer protons lie in the region where the shielding is reduced by the ring current, while the inner protons lie in a region where it is increased.

Ring laser, <*Ring-Laser*>, a group of mirrors and an active laser medium arranged such that they form a closed path around a limited area. Typical arrangements are triangles or light paths formed from two triangles which meet at a corner. Ring lasers are of interest because, by rotating the whole of the resonator, two frequencies can be produced. The total light path length corresponds to an integer multiple of the wavelength of each and one propagates in the direction of the rotation while the other propagates in the opposite sense. The frequency difference of the two waves, $\Delta\nu$, is given by:

$$\Delta\nu = \frac{4\omega_R A}{L \cdot \lambda},$$

where ω_R is the rotational frequency, A the enclosed area and L the path length. The frequency difference arises from a linear → Doppler effect. If the light of the two waves is uncoupled and the power measured with a detector which rotates with the resonator, then the frequency difference can be determined from the beating of the two waves. Ring lasers are used in this way to determine absolute rotational frequency and in navigation systems.

RIS, <*RIS*> → resonance ionization spectroscopy.

Rochon prism, <*Rochon-Prisma*>, a → prism described by A.M. Rochon in 1801 which divides a light beam into two, perpendicularly polarized, beams. A Rochon prism consists of two right-angled calcite prisms cemented together at their bases, or two right-angled quartz prisms in optical contact at their bases; see Figure 1. Prism, I, has its optic axis parallel to the direction of the incident light, i.e. parallel to AB. In prism, II, the optic axis is parallel to the prism edge, i.e. perpendicular to the plane of Figure 1. Incident light normal to the face of entry, AC, is split by prism I into an ordinary (o) and an extraordinary (e) ray. These two beams travel in the same direction as the incident beam, but with different velocities, until they

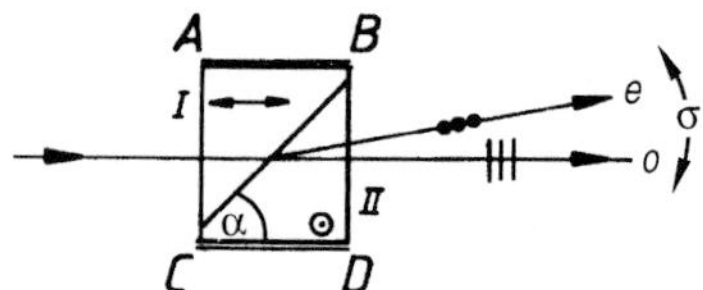

Rochon prism. Fig. 1. A section through a Rochon prism, see text

reach the plane BC. For the orientation of the optic axes given above, the o beam is effectively undeviated at BC and passes out of prism II in the direction of the incident light. The e beam, however, departs from this direction by the divergence angle, σ, which depends upon the difference in the refractive indexes for the o and e beams and also upon the section angle, α. The smaller angle, α, the greater is the divergence angle; but the length of the prism must be increased if the → aperture is to be maintained. For → calcite Rochon prisms, the divergence angle lies between 2.5° and 10° for $\alpha = 45°$; but for quartz prisms it is only 0.25° to 0.5°. For $\alpha = 30°$ it is about 1°. MgF_2 Rochon prisms are made especially for the UV region. For a ratio of length (C-D) to height (C-A) of 3.4, a divergence angle of 2.6° is found for $\lambda =$ 175 nm.

Rosencwaig-Gersho theory, <*Rosencwaig-Gersho-Theorie*>, the theory formulated by Rosencwaig and Gersho in 1975/76 which has been responsible for the rapid development of the → photoacoustic spectroscopy of condensed phases in gas-coupled photoacoustic cells. Rosencwaig and Gersho based their theory upon a one-dimensional model which is shown schematically in Figure 1. An optically and thermally homogeneous sample with

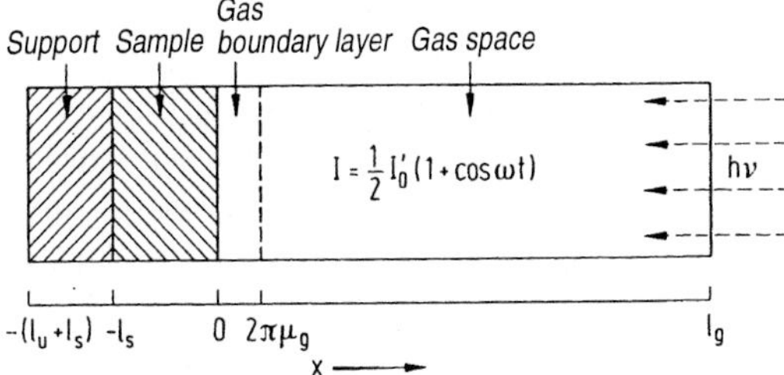

Rosencwaig-Gersho theory. Fig. 1. The model for the RG theory, see text

path length, l_s, is placed on a support of thickness, l_u, in a cylindrical photoacoustic cell. The sample is irradiated with sinusoidally modulated light. The energy absorbed by the sample is completely deactivated, nonradiatively, i.e. converted into heat. The incident light and the heat flow in the sample both take place only at right angles to the sample surface. The Bouguer-Beer-Lambert law applies to the absorption of light by the sample. From considerations of the resultant heat distribution, three thermal diffusion equations, one each for the sample, sample support and gas phase, can be obtained. These equations can be solved for the stationary state by reference to the boundary conditions which arise from the requirements for continuity of the heat flow and temperature at the interfaces between sample and support and between sample and gas phase. It is further assumed that the temperature at the cell walls is the ambient temperature and that convection in the gas phase can be neglected. The quantities listed in the table occur in the equations and their solutions.

There is a relationship between the thermal diffusivity, α, and the thermal conductivity, κ, $\alpha = \kappa/(\varrho \cdot C)$. Further, the thermal diffusion length, $\mu_{s,g,u}$, is introduced. It is given by $\mu = (2\alpha/\omega)^{1/2}$ in centimeters. The solutions for $\Theta(0,t)$ and $\Theta(x,t)$ are of interest. $\Theta(0,t)$ gives the amplitude of the periodic temperature change at the surface of the sample ($x = 0$). $\Theta(x,t)$

Rosencwaig-Gersho theory. Table.

θ	temperature	K	
x	position variable	cm	
t	time	s	
$l_{s,g,u}$	path length	cm	s = sample
$\alpha_{s,g,u}$	thermal diffusivity	cm^2s^{-1}	g = gas
$\varkappa_{s,g,u}$	thermal conductivity	$Wcm^{-1}\,K^{-1}$	u = support
$C_{s,g,u}$	thermal capacity	$J\,g^{-1}\,K^{-1}$	
$\varrho_{s,g,u}$	density	$g\,cm^{-3}$	
I_o	radiation intensity	Wcm^{-2}	
β	absorption coefficient	cm^{-1}	
$\omega = 2\pi f$	circular frequency	$rad\,s^{-1}$	

describes a heat wave in the coupled gas ($x > 0$) which is fully damped within a layer $2\pi\mu_g$ thick. This $2\pi\mu_g$ thick gas layer works like an acoustic piston on the remaining gas and, accordingly, this model is known as the piston model.

When all the optical and thermal data and also the intensity of the radiation are known, the photoacoustic signal, Q^{PA}, in pressure units can be calculated from the solutions of the equations. For practical applications using the gas microphone technique, Q^{PA} must be converted to volts, i.e. the microphone sensitivity, S^M, must be known in mV/Pa. The solutions of the equations are complex expressions, i.e. they consist of a real, Q_{real}, and an imaginary, Q_{im}, part. The amplitude of the photoacoustic signal is given by the modulus:

$$Q^{PA} = (Q^2_{\text{real}} + Q^2_{\text{im}})^{1/2}.$$

The phase shift of the signal is also of interest and the corresponding phase angle, ψ, is given by:

$$\psi = \arctan \frac{Q_{\text{im}}}{Q_{\text{real}}}.$$

These two equations are very important in actual photoacoustic measurements for which the lock-in amplifier technique is always used and either the amplitude or the phase angle may be chosen as the primary signal.

Since the explicit solutions of the Rosencwaig-Gersho theory are very complicated, limiting cases of them are used in practical applications. In this connection, the optical penetration depth, l_β , and the thermal diffusion length, μ_s, are important. l_β is the inverse of the absorption coefficient: $l_\beta = \beta^{-1}$ in cm. Since $I = I_o\exp[-\beta l_\beta]$, the intensity, I_o, has dropped to the value of $I = 0.37\, I_o$ at $l_s = l_\beta$. The thermal diffusion length, μ_s, is a function of the thermal diffusivity, α_s, and of the modulation frequency, ω The latter is very important. μ_s gives information about the depth in the sample from which the photoacoustic signal can still be detected.

Taking the path length of the sample, l_s, into account we can differentiate between:

1. Optically transparent samples with $l_\beta > l_s$,
2. optically opaque samples with $l_\beta < l_s$ and
   a) thermally thin samples with $\mu_s > l_s$ and
   b) thermally thick samples with $\mu_s < l_s$.

Here, a more exact distinction between $\mu_s > l_s$ and $\mu_s < l_\beta$ should be made. The following combinations are important for applications:

1. $l_\beta > l_s$ and $\mu_s < l_s$, but $\mu_s < l_\beta$.
   The amplitude is:

$$S^{PA} = \frac{\mu_g}{\sqrt{2}} \cdot \frac{\mu_s}{\kappa_s} \cdot (\beta\mu_s) \cdot I_o B;$$

$$[B = \frac{\gamma P_o}{2\sqrt{2}T \cdot V} \cdot S^M]$$

   $\gamma = C_p/C_v$, P_o is the external pressure, T_o the ambient temperature, V, the volume of the gas phase, i.e. the cell volume, and S^M is the sensitivity of the microphone.
   The signal, S^{PA}, depends upon β and the thermal properties of the sample and is proportional to $\omega^{-3/2}$.
2. $l_\beta < l_s$ and $\mu_s < l_s$, but $\mu_s > l_\beta$.
   The amplitude is:

$$S^{PA} = \frac{\mu_s}{\sqrt{2}} \cdot \frac{\mu_s}{\kappa_s} \cdot I_o \cdot B$$

   where the expression for B is given above.

The signal, S^{PA}, does not depend upon β, but it is proportional to ω^{-1}. There is signal → saturation. This limit is very important for the production of a reference signal using a → black standard, because S^{PA} depends only on the → spectral energy distribution, $I_o(\lambda)$.

3. $l_\beta < l_s$ and $\mu_s < l_s$, but $\mu_s < l_\beta$.
The amplitude is given by the same equation as in case 1.
If all the thermal quantities are gathered together and the modulation frequency is held constant at ω, then the following simple expression is obtained:

$$S_1^{PA}(mV) = S_3^{PA}(mV) = K_1 \cdot I_o \cdot \beta;$$

or

$$S_2^{PA}(mV) = S_{\text{sat}}^{PA}\,(mV) = K_{\text{sat}}\, I_o;$$

$$K_{\text{sat}} = B \cdot \frac{\mu_g \cdot \mu_s}{\sqrt{2}\,\kappa_s}.$$

If the measurement, S_1^{PA}, is made relative to the saturation signal, S_{sat}^{PA}, as reference then at constant I_o a normalized photoacoustic is obtained:

$$S_{\text{norm}}^{PA} = \frac{S_1^{PA}}{S_{\text{sat}}^{PA}} \cdot \beta = K' \cdot \beta.$$

A graph of S_{norm}^{PA} against the wavelength, λ, is the photoacoustic spectrum under the assumptions made above.
Ref.: A. Rosencwaig, *Photoacoustics and Photoacoustic Spectroscopy*, J. Wiley and Sons, New York, Chichester, Brisbane, Toronto, **1980**.

Rotating sector, <*Sektor, rotierender Sektor*> a device for → light attenuation which usually consists of two disks

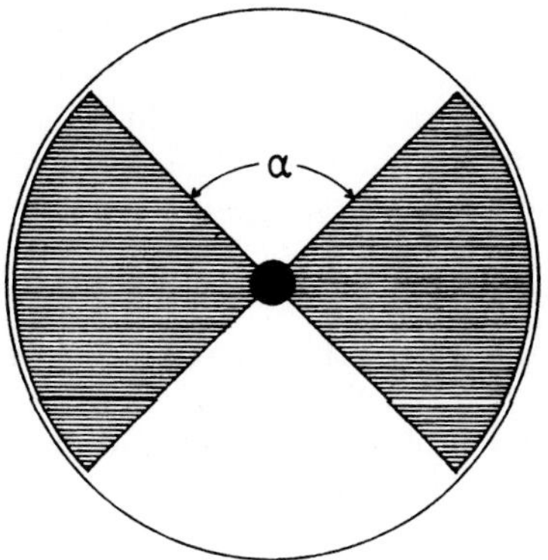

Rotating sector. Fig. 1. A rotating sector, see text.

which can be rotated with respect to each other. Each disk has two 90° sectors removed from it so that the transmission can be varied from 0 to 50%. Figure 1 shows the situation where the disks are positioned one exactly above the other so that a 90° sector transmits the light while the next blocks it off. If the unattenuated light intensity in front of the sector is denoted by, Φ_o, and the attenuated intensity after the sector by, Φ, and if α is the angle of the opening then the transmission is:

$$T = \frac{\Phi}{\Phi_o} = \frac{\%\text{ open}}{100} = \frac{2\alpha}{360} = \frac{\alpha}{180}$$

or the absorbance:

$$A = \log\frac{\Phi_o}{\Phi} = \log\frac{180}{\alpha}.$$

At the maximum value of α, 90°, the attenuation is 50% or A = log2 = 0.3010. Thus, → absorbance values from 0.3 upwards can be set with such a sector. For a discussion of the accuracy of measurements with a rotating sector and other types of sector see G. Kortüm, *Kolorimetrie, Photometrie und Spektrometrie*, Springer Verlag,

Berlin, Heidelberg, **1962**, 4th. ed., chapt. II.3, pp. 87 ff. In addition to its use in the attenuation of light, the rotating sector also finds application in the production of intensity-modulated light. For this purpose, disks with a fixed sector division are used. They are known as → choppers and are usually supplied with a number of readily interchangeable sector disks of differing sector division. The speed of rotation is controlled electronically and modulation frequencies can be varied from a few Hz to ca. 4000 Hz. Choppers are particularly important in → photoacoustic spectroscopy.

Rotational branch, <*Rotationszweig*>, individual series or branches which arise in the interpretation of rotational Raman or rotation-vibration spectra. They are distinguished by different values of the change of the rotational quantum number, J. The following nomenclature has been agreed internationally:

$\Delta J = +2$: S branch
$\Delta J = +1$: R branch
$\Delta J = 0$: Q branch
$\Delta J = -1$: P branch
$\Delta J = -2$: O branch

In the rotational Raman spectrum of a symmetric top, lines with an anti-Stokes shift are known as the ^{S}S branch and those with the Stokes shift as the ^{O}S branch for $\Delta J = +2$. Similarly, ^{R}R and ^{P}R are used for $\Delta J = +1$. Further, in rotation-vibration spectra, the → subbands of a → perpendicular band are designated ^{R}P, ^{R}Q, ^{R}R for $\Delta K = +1$ and ^{P}P, ^{P}Q, ^{P}R for $\Delta K = -1$. For parallel bands, $\Delta K = 0$, and the subbands are called ^{Q}P, ^{Q}Q and ^{Q}R. In the corresponding Raman spectrum, the branches ^{O}Q and ^{S}Q with $\Delta K = \pm 2$ must be included.

Rotational constant, <*Rotationskonstante*>, for a linear molecule, which has two identical inertial axes perpendicular to the internuclear axis, this is defined as:

$$B = \frac{h}{8\pi^2 I_B c} = \frac{27.986}{I_B} 10^{-47} \ |\, \mathrm{cm}^{-1} \,| \, .$$

h is → Planck's constant, c the velocity of light and $I_B = \sum m_i r_i^2$, where m_i is the mass of the ith atom and r_i its distance from the axis of rotation. In the case of a diatomic molecule, $I_B = \mu r^2$, where μ is the → reduced mass and r the bond length. Since the moment of inertia depends upon the masses of the atoms and the distance between them, the rotational constant, B, is a molecule-specific quantity and its measurement is an important task for → IR spectroscopy.
For nonlinear molecules, the rotation is referred to three principal axes (→ top molecules). The three moments of inertia, I_A, I_B and I_C, then define the rotational constants:

$$A = \frac{h}{8\pi^2 I_A c};$$

$$B = \frac{h}{8\pi^2 I_B c};$$

$$C = \frac{h}{8\pi^2 I_C c}.$$

The corresponding → pure rotation spectra are observed in the far IR or microwave regions. If the resolution of the spectra is good, evaluation of the rotational constants of linear and symmetric top molecules is possible. Since the pure rotation spectrum in the FIR is, in general, difficult to measure, the → rotation-vibration spectrum in the mid IR, which is readily accessible, is often used for the determination of rotational constants. But the change of the interatomic distances due to the vibration must be taken into account, i.e. the rotational constant is no longer a constant. Allowance is made for this by the introduction of an average distance, $\bar{r} > r_e$ (r_e = equilibrium distance). The interaction between vibration and rotation is represented by a relationship which describes the dependence of the rotational constant upon the vibrational quantum number, v, of the vibration involved:

$$B(v_1, v_2 ...) = B_{|v|} =$$
$$= B_e - a_1(v_1 + \frac{1}{2}) - a_2(v_2 + \frac{1}{2}) + ...$$

B_e is the rotational constant for the equilibrium interatomic distance, r_e, $|v|$ represents all the vibrational quantum numbers, v_1, v_2 ... and a_i is a quantity which is small compared with B_e. The above equation applies to a polyatomic molecule ($N > 3$) with $3N$-6 or $3N$-5 → normal vibrations. For a diatomic molecule:

$$B_v = B_e - a_e(v + \frac{1}{2})$$

where $a_e \ll B_e$.

It follows from these equations that the rotational constant in the vibrational ground state, $v = 0$, is greater than that in the first excited vibrational state, $v = 1$, and that the interaction is greater in the upper state. These interactions must be considered for all three rotational constants of a nonlinear molecule:

$$A_{|v|} = A_e - \sum a_i^A (v_i + \frac{d_i}{2});$$
$$B_{|v|} = B_e - \sum a_i^B (v_i + \frac{d_i}{2});$$
$$C_{|v|} = C_e - \sum a_i^C (v_i + \frac{d_i}{2});$$

Here, d_i is the degree of vibrational degeneracy equal to 1, 2 ... for nondegenerate, twofold degenerate ... etc.

The simplest rules for the determination of the rotational constant from the → pure rotation spectrum in the FIR or microwave region are those for diatomic molecules with a permanent dipole moment, linear molecules of the point group, $C_{\infty v}$, and symmetric top molecules. If the → centrifugal term is neglected, the energy difference between the rotational levels is given by:

$$\tilde{\nu} = F(J') - F(J'') = 2B(J + 1)$$

for linear molecules, and by:

$$\tilde{\nu} = F(J', K') - F(J'', K'') = 2B(J + 1)$$

for symmetric top molecules.

Thus, there is a series of equidistant rotational lines separated by $2B$. For the → Raman spectrum, the rotational Raman lines are found at:

$$| \Delta\tilde{\nu}_R | = 4B(J + \frac{3}{2})$$

for linear molecules, and at:

$$| \Delta\tilde{\nu}_R | = 2B(J + 1)$$

for the symmetric top. The result is two series of lines corresponding to the selection rules, $\Delta J = 0, \pm 1, \pm 2$ and $\Delta K = 0$; the S branch ($\Delta J = +2$) and the R branch ($\Delta J = +1$). $|\Delta\tilde{\nu}|$ is the absolute Raman shift with respect to the wavenumber of the exciting line, $\tilde{\nu}_o$; $|\Delta\tilde{\nu}| = |\tilde{\nu} - \tilde{\nu}_o|$.

It is not possible to give the rotational energy levels of an asymmetric top in a simple manner and the analysis of the rotational spectrum is complicated and tedious.

The series formulas for the R and P branches of the → rotation-vibration spectra of linear molecules are usually given together in one equation, viz:

$$\tilde{\nu} = \omega_o + (B_{\upsilon'} + B_{\upsilon''})m + (B_{\upsilon'} - B_{\upsilon''})m^2$$

where $B_{\upsilon'}$ and $B_{\upsilon''}$ are the rotational constants for the upper and lower vibrational states respectively and $m = 1, 2, 3 \ldots$ for the R branch or $m = -1, -2, -3 \ldots$ for the P branch. The difference between neighboring rotational lines, $\Delta\tilde{\nu}(m) = \tilde{\nu}_{m+1} - \tilde{\nu}_m$, is:

$$\Delta\tilde{\nu}(m) = 2B_{\upsilon'} + 2(B_{\upsilon'} - B_{\upsilon''})m$$

The second difference is:

$$\begin{aligned}\Delta^2\tilde{\nu}(m) &= \Delta\tilde{\nu}(m+1) - \Delta\tilde{\nu}(m) \\ &= 2(B_{\upsilon'} - B_{\upsilon''})\end{aligned}$$

The second difference gives a good mean value for the difference between the two rotational constants, $B_{\upsilon'}$ and $B_{\upsilon''}$, and should therefore be constant. A graphical analysis of the first difference for $m = 0$ gives $2B_{\upsilon'}$ directly as the intercept on the ordinate.

$B_{\upsilon'}$ and $B_{\upsilon''}$ may also be found by combining data from lines in the P and R branches which belong to the same rotational level in the upper state. The result is:

$$\begin{aligned}&R(J-1) - P(J+1) \\ &\quad = F'(J+1) - F'(J-1) \\ &\quad = \Delta^2 F'(J).\end{aligned}$$

Inserting explicit expressions for the R and P branches and neglecting the centrifugal term gives:

$$\Delta^2 F'(J) = 4B_{\upsilon'}(J + \frac{1}{2}).$$

A graph of $\Delta^2 F'(J)$ against J gives a straight line of slope, $4B_{\upsilon'}$. If the corresponding lower rotational levels of the R and P branches are similarly combined $4B_{\upsilon''}$ may be obtained from $\Delta^2 F''(J)$. A Q branch is often found for linear molecules with more than two atoms which, if the resolution is sufficient, permits a direct measurement of the difference, $B_{\upsilon'} - B_{\upsilon''}$.

The rotational energy levels of a → symmetric top are given by:

$$T = G(\upsilon_1, \upsilon_2, \upsilon_3 \ldots) + F_{|\upsilon|}(J, K)$$

where

$$\begin{aligned}F_{|\upsilon|}(J, K) = {} & B_{|\upsilon|}J(J+1) \\ & + (A_{|\upsilon|} - B_{|\upsilon|})K^2.\end{aligned}$$

The subscript, $|\upsilon|$, indicates that account is being taken of the interaction of vibration and rotation, i.e. $X_{|\upsilon|}$ is the average value of X during a vibration. The quantum number, K, gives the component of the total angular momentum in the direction of the top axis (figure axis) in multiples of $h/2\pi$. In symmetric top molecules, different selection rules apply to → parallel and → perpendicular bands and we must distinguish between them. For parallel bands, $\Delta J = 0, \pm 1$ when $K \neq 0$ and $\Delta J = \pm 1$ when $K = 0$, i.e. only transitions within the same K level are allowed. For every K value, except $K = 0$, there are P, Q and R branches; → subbands which obey laws corresponding to those of diatomic and linear polyatomic molecules. To a good approximation, the separation between the lines is $\sim 2B$; more exactly, $3B_{\upsilon'} - B_{\upsilon''}$ in the R branch and $B_{\upsilon'} + B_{\upsilon''}$ in the P branch. The explicit expressions for the individual subbands of a parallel band are:

$$\tilde{\nu}_R^{\text{sub}} = \omega_o + 2B_{\upsilon'} + (3B_{\upsilon'} - B_{\upsilon''})J + (B_{\upsilon'} - B_{\upsilon''})J^2 + [(A_{\upsilon'} - A_{\upsilon''}) - (B_{\upsilon'} - B_{\upsilon''})]K^2$$

$$\tilde{\nu}_Q^{\text{sub}} = \omega_o + (B_{\upsilon'} - B_{\upsilon''})J + (B_{\upsilon'} - B_{\upsilon''})J^2 + [(A_{\upsilon'} - A_{\upsilon''}) - (B_{\upsilon'} - B_{\upsilon''})]K^2$$

$$\tilde{\nu}_P^{\text{sub}} = \omega_o - (B_{\upsilon'} - B_{\upsilon''})J + (B_{\upsilon'} - B_{\upsilon''})J^2 + [(A_{\upsilon'} - A_{\upsilon''}) - (B_{\upsilon'} - B_{\upsilon''})]K^2.$$

For $J = 0$, there is an equation for the band origin of every subband, i.e.:

$$\tilde{\nu}_o^{\text{sub}} = \omega_o + [(A_{\upsilon'} - A_{\upsilon''}) - (B_{\upsilon'} - B_{\upsilon''})]K^2.$$

If the K fine structure of a parallel band is not resolved, the structure of the perpendicular band of a linear molecule is obtained. The rotational constants, $B_{\upsilon'}$ and $B_{\upsilon''}$, can be determined from the differences, $\Delta^2F'(J) = R(J) - P(J)$ and $\Delta^2F''(J) = R(J-1) - P(J+1)$, as described above. The rotational constant, A, cannot be determined from a parallel band. If a parallel band is resolved, the analysis of every subband can proceed in an analogous manner. A criterion for the reliability of the results is that the values of $\Delta^2F(J)$ for every subband must agree. If the formula for the band origin (see above) is used for a particular value of J, then, $\tilde{\nu}_o^{\text{sub}}$ plotted against K^2, gives a straight line with a gradient of $[(A_{\upsilon'} - A_{\upsilon''}) - (B_{\upsilon'} - B_{\upsilon''})]$. This slope should be the same for all values of J, provided that the influence of the centrifugal term can be neglected. $B_{\upsilon'}$ and $B_{\upsilon''}$ can be separately determined (see above), but for A only the difference, $A_{\upsilon'} - A_{\upsilon''}$, can be obtained. However, the rotational constant, A, can be found from the perpendicular bands. The combination differences, $\Delta^{2K}F''(J,K)$ and $\Delta^{2K}F'(J,K)$, are given by the equations:

$$\Delta^{2K}F''(J,K) = F''(J,K+1) - F''(J,K-1) = 4(A_{\upsilon''} - B_{\upsilon''})K$$

$$\Delta^{2K}F'(J,K) = F'(J,K+1) - F'(J,K-1) = 4(A_{\upsilon'} - B_{\upsilon'})K.$$

Since these relations hold for every J value, they can also be applied to unresolved Q branches.
The evaluation of the spectra of asymmetric tops is significantly more complicated.
Ref.: G. Herzberg, *Molecular Spectra and Molecular Structure, II. Infrared and Raman Spectra of Polyatomic Molecules*, D. Van Nostrand and Co. Inc., Princeton, New Jersey, **1945**, chapt. IV.

Rotational Raman spectra of linear molecules, <*Rotations-Raman-Spektrum linearer Moleküle*>. If a → Raman spectrum is to be observed there must be a change of → polarizability in a particular direction during the vibration or rotation of the molecule. In linear molecules, the polarizabilities along the internuclear axis and perpendicular to it are always different. Therefore, the polarizability in a particular direction changes as the molecule rotates about an axis perpendicular to the internuclear axis. Homo- and heteronuclear diatomic molecules and also linear polyatomic molecules belonging to the → point groups $C_{\infty\upsilon}$ and $D_{\infty h}$, therefore show a rotational Raman spectrum. However, to give a → pure rotation spectrum, a diatomic or polyatomic molecule must have a dipole moment and therefore only those linear molecules of $C_{\infty v}$ symmetry show such a spectrum. The rotational Raman spectrum of a linear molecule can be interpreted in terms of the energy levels of the rigid or non-rigid rotator (→ dumbbell model). In contrast to the excitation in the → IR region, the → selection rule is $\Delta J = 0$, ± 2. Although the change, $\Delta J = +1$ (absorption), is the only one of importance in the IR spectrum, the change, $\Delta J = \pm 2$, must be considered in the → Raman effect, which gives rise to the Stokes ($\Delta J = +2$) and anti-Stokes ($\Delta J = -2$) lines in the → Raman spectrum, (→ Stokes rule). $\Delta J = 0$ corresponds to the unshifted exciting wavenumber, $\tilde{\nu}_o$. Using the rigid rotator model, two series of lines spreading symmetrically from the exciting wavenumber, $\tilde{\nu}_o$, to higher and lower wavenumbers are found; the Stokes lines:

$$\Delta J = 2; \tilde{\nu}_{\text{rot}} = \tilde{\nu}_o - 4B(J + \frac{3}{2})$$

and the anti-Stokes lines:

$$\Delta J = -2; \tilde{\nu}_{\text{rot}} = \tilde{\nu}_o + 4B(J + \frac{3}{2})$$

These two series formulas are usually combined into one by introducing the absolute value of the Raman shift $|\Delta\tilde{\nu}| = |\tilde{\nu}_{\text{rot}} - \tilde{\nu}_o|$, giving:

$$|\Delta\tilde{\nu}| = |\tilde{\nu}_{\text{rot}} - \tilde{\nu}_o| = 4B(J + \frac{3}{2}).$$

If the rotator is not assumed to be rigid, the result is:

$$|\Delta\tilde{\nu}| = (4B - 6D)(J + \frac{3}{2}) - 8D(J + \frac{3}{2})^3.$$

Here, B is the → rotational constant and D the → centrifugal term. If $D \ll B$, the expression reduces to the equation above. Figure 1 shows the corresponding energy-level scheme with the transitions, $\Delta J = \pm 2$. The observation of a series of lines is seen to be due to the fact that there are large numbers of molecules in excited rotational states, even at room temperature (Boltzmann energy distribution → Boltzmann statistics). Unlike the → pure rotation spectrum, the separation of the rotational Raman lines is $4B$, and the first line appears at $6B$. The two series are also called branches. That to higher wavenumbers, i.e. $\Delta\tilde{\nu} = \tilde{\nu}_{\text{rot}} - \tilde{\nu}_o > 0$ and $\Delta J = -2$ is the O

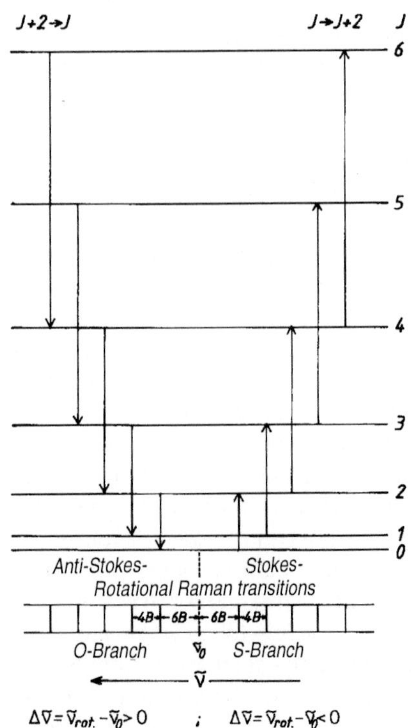

Rotational Raman spectra of linear molecules. Fig. 1. Energy-level scheme for the rotational Raman transitions of a molecule

branch while that to lower wavenumbers, i.e. $\Delta\tilde{\nu} = \tilde{\nu}_{rot} - \tilde{\nu}_o < 0$ and $\Delta J = +2$, is the S branch. The transitions with $\Delta J = -2$ correspond to a transfer of energy to the scattered photon and those with $\Delta J = +2$ to a transfer of excitation energy to the molecule, → Raman effect. The designation of the two branches as O and S branches is not quite correct. According to the agreed nomenclature for the designation of → rotational branches, the correct form would be ${}^{O}S$ branch and ${}^{S}S$ branch. The left superscript, O, refers to the anti-Stokes ($\Delta J = -2$) and left superscript, S, to the Stokes lines of the S branch ($\Delta J = +2$). For some homonuclear diatomic molecules and polyatomic linear molecules of symmetry point group, $D_{\infty h}$, the nuclear spin statistics must be invoked to interpret the finer points of the rotational Raman spectrum.

Rotation spectrum, *<Rotationsspektrum>* → pure rotation spectrum.

Rotation-vibration spectrum, *<Rotationsschwingungsspektrum>*. Vibrational energy is always greater than rotational energy. Therefore, when vibrations are excited the rotations of a molecule are also excited. The situation is illustrated with an energy-level diagram in Figure 1. A series of rotational levels, $F(J)$ or $F(J,K)$, is built upon every vibrational level, $G(v)$. Only the rotational levels for linear molecules, $F(J)$, are shown in the Figure. The levels for nonlinear molecules are complicated (→ pure rotation spectrum of a symmetric top). For the interpretation of rotation-vibration spectra, the total energy of a level may be given as the sum of the individual energies, $G(v)$ and $F(J)$ or $F(J,K)$:

$$T = G(v) + F(J)$$

in the case of linear molecules, and:

$$T = G(v) + F(J,K)$$

for nonlinear molecules.

In the case of polyatomic molecules, the rotational energies are superimposed upon $3N$-6 or $3N$-5 normal vibrations with energies, $G(v_1, v_2, v_3 \ldots)$. In analyzing the transitions between the levels the corresponding selection rules must be considered.

When the molecules are excited by the absorption of electromagnetic radiation, a typical → IR spectrum, which always shows rotational structure superimposed on the vibrational structure, is obtained. If the interaction with the radiation is a scattering process, then it is the → Raman spectrum

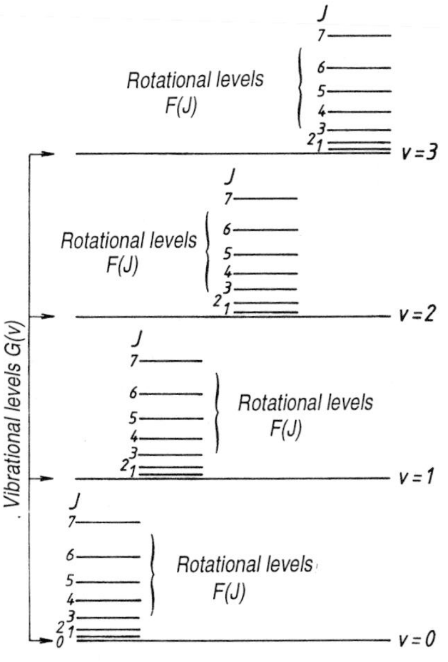

Rotation-vibration spectrum. Fig. 1. An energy-level scheme

which results. The question of in how far the individual superimposed rotational transitions in each spectrum are resolved, and therefore susceptible to exact analysis, depends upon various factors:

1. The resolution of the IR spectrometer;
2. the state of aggregation of the sample;
3. the magnitude of the rotational constant, B which, to a first approximation, gives the separation between two neighboring rotational lines. The smaller the rotational constant, the smaller the distance between the lines.

In solution, the rotation of the molecules is strongly hindered. Consequently, the vibrational transitions are broadened giving the typical IR or Raman spectrum of a solution or liquid. In solids, the rotations are quasi-frozen and only vibrational transitions are seen and the individual IR bands are sharper. The same is true of the Raman spectrum.

In a gas or vapor, the rotation-vibration spectra are readily measurable, provided that the resolution of the IR instrument is sufficient.

The rotation-vibration spectrum can be observed either as an → IR spectrum or as a → Raman spectrum, depending upon the method of excitation. In analyzing the spectra, the cases of diatomic, linear polyatomic and nonlinear molecules must be distinguished because of the different formulas for the rotational levels.

Linear molecules

The → harmonic oscillator and rigid rotator (→ dumbbell model) models give the rotation-vibration energy levels of a diatomic molecule as:

$$\begin{aligned} T &= G(v) + F_v(J) \\ &= \omega_o(v + \frac{1}{2}) + B_v J(J+1) \end{aligned}$$

where ω_o is the wavenumber of the → fundamental vibration, B_v the rotational constant of the vibrating rotator, v the vibrational quantum number 0, 1, 2, 3 … and J the rotational quantum number 0, 1, 2, 3 … The → rotational constant, B_v, takes account of the interaction between vibration and rotation and can be written as a function of the vibrational quantum number, v.

$$B_v = B_e - a_e(v + \frac{1}{2}) \qquad \text{(a)}$$

a_e is the correction factor which is small compared with B_e, the rotational constant for the equilibrium internuclear distance. The selection rules for the transitions are: $\Delta v = \pm 1$ and $\Delta J = \pm 1$; $\Delta J = 0$ is forbidden. For the transition $v'' = 0 \rightarrow v' = 1$ we have:

$$\begin{aligned} \tilde{\nu} &= \mathrm{G}(v') - G(v'') + F'_{v'}(J') - F''_{v''}(J'') \\ &= \omega_o + F'_{v'}(J) - F''_{v''}(J). \end{aligned}$$

The energy difference, $G(v') - G(v'')$, corresponds to the fundamental vibration, ω_o. For the energy difference, $F'_{v'}(J) - F''_{v''}(J)$, it must be noted that $\Delta J = \pm 1$, i.e. $\Delta J = +1$: $J' = J'' + 1$ and $\Delta J = -1$: $J' = J'' - 1$. Two → rotational branches, the R and P branches, result:

$\Delta J = +1$, R branch: $J' = J'' + 1$

$$\begin{aligned} \tilde{\nu}_R &= \omega_o + 2B_{v'} + (3B_{v'} - B_{v''}) \\ &\quad + (B_{v'} - B_{v''})J^2; \qquad \text{(b)} \end{aligned}$$

$J = 0, 1, 2 \ldots$

$\Delta J = -1$, P branch: $J' = J'' - 1$

$$\tilde{\nu}_P = \omega_o - (B_{v'} + B_{v''})J + (B_{v'} - B_{v''})J^2; \quad \text{(c)}$$
$$J = 1, 2, 3 \ldots$$

If the rotation-vibration interaction is neglected, $B_{v'} = B_{v''}$, and we have:

$$\tilde{\nu}_R = \omega_o + 2B(J+1); \; \tilde{\nu}_P = \omega_o - 2BJ.$$

In this formulation, J is always the rotational quantum number of the lower state. Thus, for every branch there is a series of lines with a separation of $\Delta\tilde{\nu} = 2B$. The pure vibrational transition is not observed because transitions for which there is no change in J, i.e. $\Delta J = 0$, are forbidden. A gap, the → zero gap, therefore appears at this position. The series formulas can be written as one equation in which the running index, m, is introduced:

$$\tilde{\nu} = \omega_o + (B_{v'} + B_{v''})m + (B_{v'} - B_{v''})m^2 \quad \text{(d)}$$

$m = 1, 2, 3 \ldots$ for the R branch and $m = -1, -2, -3 \ldots$ for the P branch. Figure 1 shows the corresponding energy-level scheme with the transitions $\Delta J = +1$ and $\Delta J = -1$. The resulting spectrum is shown for the case, $B_{v'} = B_{v''}$, in Figure 1a. The actual behavior in a rotation-vibration excitation is determined by the rotational constants, $B_{v'}$ and $B_{v''}$. Since $B_{v'} < B_{v''}$, the difference, $B_{v'} - B_{v''} < 0$, and the quadratic term, $(B_{v'} - B_{v''})J^2$ gains in importance with increasing J. The result is that the separation of the individual lines decreases in the R branch (compression) and increases in the P branch (expansion), as shown schematically in Figure 1b. Figure 2 shows the experimental rotation-vibration spectrum of the

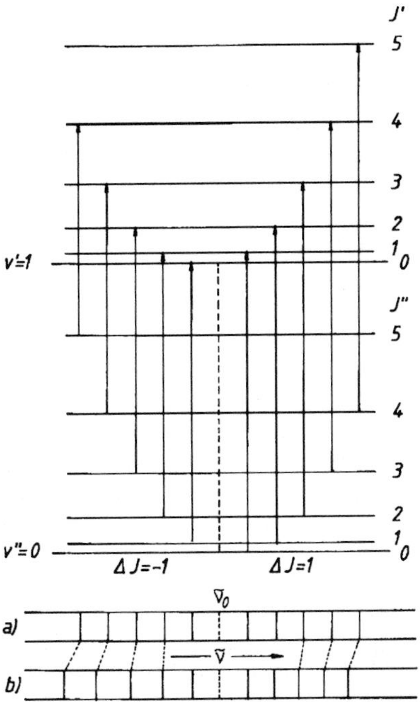

Rotation-vibration spectrum of linear molecules. Fig. 1. Energy-level scheme for the rotation-vibration spectrum of a diatomic molecule

HCl molecule in which the compression and expansion are clearly seen. The characteristic decrease in the intensities of the rotational lines can be explained with the help of the Boltzmann distribution (→ Boltzmann statistics). The doubling of the lines is due to the → isotope effect. For the Raman transitions, the selection rules are: $\Delta v = \pm 1$ and $\Delta J = 0, \pm 2$. In contrast to the IR spectrum, transitions without change of J, i.e. $\Delta J = 0$, are also allowed, giving a Q branch. The other two rotational branches are an S branch ($\Delta J = +2$) and an O branch ($\Delta J = -2$). The series formulas for the Stokes shifted lines (→ Stokes rule) are:

$\Delta J = +2$:

$$\Delta\tilde{\nu}_S = \tilde{\nu}_e - | \omega_o + 6B_{v'} (5B_{v'} - B_{v''})J + (B_{v'} - B_{v''})J^2 |; \quad \text{(e)}$$

$J = 0, 1, 2, 3...$

$\Delta J = 0$:

$$\Delta\tilde{\nu}_Q = \tilde{\nu}_e - | \omega_o + (B_{v'} - B_{v''})J + (B_{v'} - B_{v''})J^2 |; \quad \text{(f)}$$

$J = 0, 1, 2, 3...$

$\Delta J = -2$:

$$\Delta\tilde{\nu}_O = \tilde{\nu}_e - | \omega_o - (3B_{v'} + B_{v''})J + (B_{v'} - B_{v''})J^2 |; \quad \text{(g)}$$

$J = 2, 3, 4...$

Since the difference between $B_{v'}$ and $B_{v''}$ is very small for the $v'' = 0 \rightarrow v' = 1$ vibrational transition, the lines of the Q branch fall very close together and are not generally resolved. Thus, in the → Raman spectrum an intense band, which may be broadened towards higher wavenumbers, is observed at the position, $\Delta\tilde{\nu} = \omega_o - \tilde{\nu}_e$. The lines of the S and O branches are very much weaker and are generally not observed. For linear molecules with three or more atoms, all 3N-5 vibrations must be entered in equation (a), i.e.

$$B_{[v]} = B_e - a_1(v_1 + \frac{1}{2}) - a_2(v_2 + \frac{1}{2}) + ...$$

where $a_1 \ll B_e$. $[v]$ is the collective symbol for all the vibrational quantum numbers v_1, v_2, v_3 ... The general formula, including degenerate vibrations, is:

$$B_{[v]} = B_e - \sum_i a_i(v_i + \frac{d_i}{2});$$

d_i = 1, 2 ... is the degree of degeneracy.

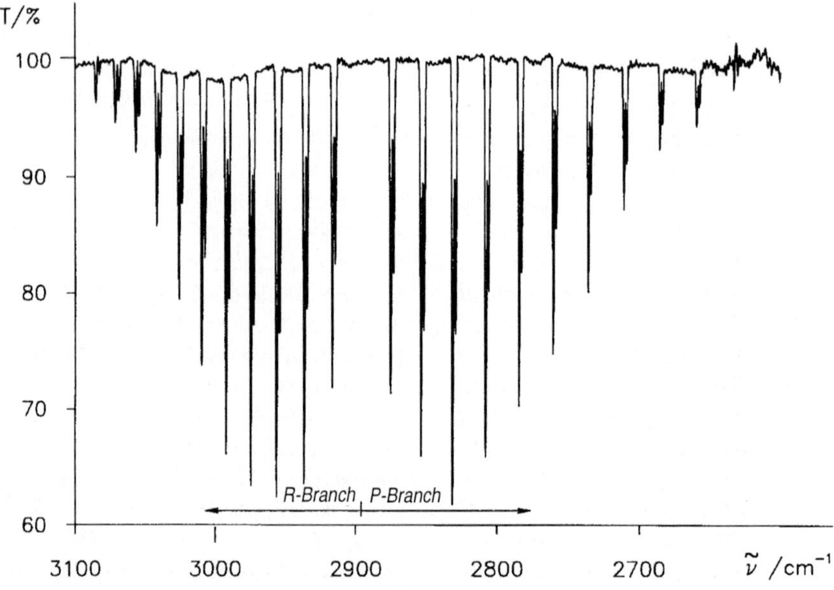

Rotation-vibration spectrum of linear molecules. Fig. 2. Rotation-vibration spectrum of HCl

If the anharmonicity of the vibration and the centrifugal term in the rotation are neglected, the rotation-vibration energylevels are given by:

$$T = G(v_1, v_2...) + F_{[v]}(J)$$
$$= \sum \omega_i(v_i + \frac{d_i}{2}) + B_{[v]}[J(J+1) - \ell^2]$$

ℓ is the quantum number for the angular momentum ($\ell h/2\pi$) about the bond axis for degenerate vibrations (perpendicular vibrations) and takes the values 0, 1, 2, 3 ... J and ℓ are related; $J = \ell, \ell + 1, \ell + 2$ ... Therefore, rotational levels with $J = 0, 1 ... \ell - 1$, do not occur. Taking account of the anharmonicity, the selection rules are: $\Delta v = +1, +2$ ... with the addition of $\Delta\ell = 0, \pm 1$ and $\Delta J = 0, \pm 1$, with the restriction that $\Delta J = 0$ for the case where $\ell = 0$ in both upper and lower states.

Because of these selection rules, quite specific types of rotation-vibration spectrum are seen in the IR region.

a) $\ell = 0$ in both upper and lower states; parallel bands. $\Delta J = \pm 1$ and gives an R and a P branch, see series formulas (b) and (c). For molecules having the symmetry $D_{\infty h}$ the lines of the branches show → intensity alternation.
b) $\Delta\ell = \pm 1$; perpendicular bands. $\Delta J = 0, \pm 1$ and all three branches are seen, though the Q branch ($\Delta J = 0$) is stronger than the P and R branches.
c) $\Delta\ell = 0$, but $\ell \neq 0$ in the upper and lower states; parallel bands. $\Delta J = 0, \pm 1$ giving R, Q and P branches, though the Q branch is weak.

   Naturally, all three band types appear in the resulting IR spectrum of a linear molecule. The selection rules for the Raman spectrum of polyatomic linear molecules ($N \geq 3$) are:

$$\Delta v = \pm 1, \pm 2 ...$$
$$\ell = 0: \quad \Delta J = 0, \pm 2$$
$$\ell = 1, 2 ...: \quad \Delta J = 0, \pm 1, \pm 2$$

   Thus, various band types are found for the excitation of the fundamental vibration, $v'' = 0 \rightarrow v' = 1$.

d) $\ell = 0$, $\Delta J = 0$ and ± 2 gives three branches, S, Q and O. Analogously to (a), we are dealing here with parallel bands of totally symmetric vibrations which show intensity alternation in molecules of $D_{\infty h}$ symmetry. The rotational lines in the Q branch fall very close together ($B_{v'} - B_{v''}$ very small) so that it is the dominant line in the Raman spectrum.
e) $\ell = 1, 2...$ gives perpendicular bands due to vibrations perpendicular to the bond axis. Since $\Delta J = 0, \pm 1, \pm 2$ there are five branches in total; S, R, Q, P and O. There is again an intensity alternation for molecules of $D_{\infty h}$ symmetry.

Nonlinear molecules are described as → top molecules, or simply tops, with respect to their rotation. A rotation-vibration energy level is the sum of the vibrational energy, $G(v_1, v_2 ...)$, and the appropriate rotational energy of the top. Therefore, a distinction is made between the rotation-vibration spectrum of a → spherical top, a → symmetric top and an → asymmetric top. The important thing is that the symmetry of the molecule, i.e. the possible vibrational modes, enters into the considerations. For a spherical top of point group T_d, the rotation-vibration energy for a nondegenerate vibration is:

$$T = G(v_1, v_2 ...) + B_{[v]}J(J+1)$$

with

$$B_{[v]} = B_e - \sum a_i(v_i + \frac{d_i}{2})$$

which allows for the interaction between vibration and rotation. For degenerate vibrations the → Coriolis coupling causes a splitting of the rotational energies. For threefold degenerate levels of symmetry species, F_2, every rotational energy level, $F(J)$, splits into three levels which are designated, $F^+(J)$, $F^o(J)$ and $F^-(J)$:

$$F^+(J) = B_{[v]}J(J+1) + 2B_{[v]}\zeta_i(J+1)$$
$$F^o(J) = B_{[v]}J(J+1)$$
$$F^-(J) = B_{[v]}J(J+1) - 2B_{[v]}\zeta_i J.$$

The quantity ς_i is the Coriolis factor which gives the magnitude of the angular momentum associated with a particular degenerate vibration in units of $h/2\pi$. For a T_d molecule, only vibrations of symmetry species, F_2, can be excited in the IR region. The selection rules are: $\Delta J = 0, \pm 1$; and with $\Delta J = +1, 0, -1$ only the states, $F^-(J)$, $F^o(J)$, and $F^+(J)$ respectively can be excited. If the Coriolis coupling is neglected, then a simple picture with *P*, *Q* and *R* branches is obtained.

For the Raman spectrum the selection rule is $\Delta J = 0, \pm 1, \pm 2$. In this case vibrations of symmetry species, A_1 and E, as well as those of species, F_2, can be excited.

For $A_1 \rightarrow A_1$ transitions, $\Delta J = 0$ and a strong *Q* branch is seen. For the $A_1 \rightarrow E$ transitions, the above selection rules apply without any limitation and five branches, *O, P, Q, R* and *S* branches are seen. In contrast to the IR spectrum, in the Raman there is no restriction upon transitions involving the F^+, F^o and F^- levels of the $A_1 \rightarrow F_2$, so that a total of 15 branches result from this transition.

For a → symmetric top the rotation-vibration energy levels are given by:

$$T = G(v_1, v_2 ...) + B_{[v]}J(J+1) + (A_{[v]} - B_{[v]})K^2$$

with $B_{[v]}$ as above and

$$A_{[v]} = A_e - \sum a_i^{(A)}(v_1 + \frac{d_i}{2})$$

The selection rules for ΔJ depend upon the direction of the dipole moment change during the vibration.

1. If the change is parallel to the top axis (→ parallel bands) then:

$$\Delta K = 0; \Delta J = 0, \pm 1 \text{ if } K \neq 0$$

and

$$\Delta K = 0; \Delta J = \pm 1 \text{ if } K = 0.$$

2. If the dipole moment change is perpendicular to the top axis (→ perpendicular bands) then:

$$\Delta K = \pm 1; \Delta J = 0, \pm 1$$

3. If the dipole moment change is both parallel and perpendicular to the top axis i.e. a → hybrid band, then the selection rules above apply simultaneously.

For nonplanar, symmetric top molecules → inversion doubling must also be considered. Furthermore, symmetric tops must be divided into → prolate ($A > B$) and → oblate ($A < B$) tops since the rotational energy levels of the two are different.

Again, Coriolis coupling must be considered in the case of degenerate

vibrations which adds a further term, $\pm 2A_{[v]}\varsigma_i K$, to the energy expression. If the top axis is not simultaneously a symmetry axis, then for the Raman spectrum the selection rules are: $\Delta K = 0, \pm 1, \pm 2$ and $\Delta J = 0, \pm 1, \pm 2$, with the limitation that $J' + J'' \geq 2$. The symmetry and degeneracy of the vibrational transitions must be considered if the selection rules are to be applied explicitly to rotation-vibration transitions in a Raman spectrum.
The rotation-vibration energy levels of an → asymmetric top are given by:

$$T = G(v_1 v_2 ...) + \frac{1}{2}(B_{[v]} + C_{[v]}J(J+1) + A_{[v]} - \frac{1}{2}(B_{[v]} + C_{[v]})W_{\tau\,[v]}$$

The quantity, $W_{\tau[v]}$, depends in a complicated manner upon the three rotational constants, $A_{[v]}$, $B_{[v]}$ and $C_{[v]}$. There are $2J + 1$ different values of $W_{\tau[v]}$ for every J value. In addition to the selection rule $\Delta v = \pm 1$, which applies generally, there is also $\Delta J = 0, \pm 1$ with the limitation that $J'' = 0 \rightarrow J' = 0$ is forbidden. Since, in addition, the selection rules differ depending upon the different orientations of the changing dipole moment, the situation is rather complicated. For molecules of the → symmetry point groups, C_{2v}, V and V_h, there are in general three types of band. If the change of dipole moment is in the direction of the axis of the smallest moment of inertia, then it is called a type *A* band. Similarly, bands arising from changes of dipole moment in the direction of the axes of the intermediate and largest moments of inertia are known as type *B* and type *C* bands respectively (→ type *A, B* and *C* bands). Again, the selection rule for the Raman spectrum is $\Delta J = 0, \pm 1, \pm 2$ with the limitation $J' + J'' \geq 2$. But it is usually very difficult to obtain a well-resolved Raman spectrum of an asymmetric top molecule and an exact analysis is frequently not possible.
Ref.: G. Herzberg, *Molecular Spectra and Molecular Structure, II. Infrared and Raman Spectra of Polyatomic Molecules*, D. Van Nostrand Co. Inc., Princeton, New Jersey, **1945**, chapt. IV.

Rotatory dispersion, *<Rotationsdispersion>* → optical rotatory dispersion.

Rowland circle, *<Rowland-Kreis>*, the circle which has the radius of curvature of a particular → concave grating. It is named after Rowland who introduced the concave grating which is the basis of the → Rowland mounting.

Rowland grating, *<Rowland-Gitter>* → concave grating.

Rowland mounting, *<Rowland-Aufstellung>*, the mounting in which the arc of a → concave grating (introduced by Rowland) describes a circle, the Rowland circle, of the same radius as the radius of curvature of the grating. With this mounting, the image of every point on the circle is another point on the circle, in the direction of the dispersion. Thus, the spectrum is formed as a series of images of the entrance slit along the periphery of the circle. The mounting has been implemented in a variety of ways among which the → Paschen-Runge, → Eagle and → Seya-Namioka methods have proved valuable.

Ruby laser, <*Rubinlaser*>, a → laser whose active medium is a crystal of ruby. Ruby is essentially aluminium oxide (Al_2O_3) containing a small quantity of chromium(3+) ions which are responsible for the red color. The ruby laser, built by Maimann in 1960, was the first solid-state laser.

The → population inversion in the active medium can be explained with the aid of an → energy-level scheme for the Cr^{3+} ion in the ruby crystal, Figure 1. It is a three-level system. Excitation with light of a suitable wavelength, λ_3 or λ_4, raises the Cr^{3+} ion to the excited state, E_3 or E_4. There is a very fast (~ 10^{-12} s) radiationless transition from E_4 to E_3. The state, E_3, has a lifetime of ca. 10^{-8} s and deactivates radiationlessly to the upper laser level, E_2, which has a mean lifetime of $3 \cdot 10^{-3}$ s. During this time there are some 10^5 transitions from E_3 to E_2 so that, if the intensity of the pump light is sufficient (→ optical pumping), a population inversion with respect to the lower laser level, the ground state, E_1, can be produced. For a lower exciting light intensity, only the red luminescence of the ruby is seen. A xenon flashlamp is normally used to excite the ruby laser. Since this is a very intense source, the Cr^{3+} ions arrive in the upper laser level faster than they return from it to the ground state. If the intensity of the lamp is sufficiently high, then finally there are more ions in the upper laser level than in the ground state. An inversion of population, $\Delta N = N_2 - N_1 > 0$, has been created and the crystal can now amplify red light of wavelength 694.3 nm. The addition of resonator mirrors then gives a laser which produces laser light of the above wavelength. The disadvantage of a three-level laser is that it requires a minimum energy input before laser operation can begin. The four-level → neodymium laser is much more convenient.

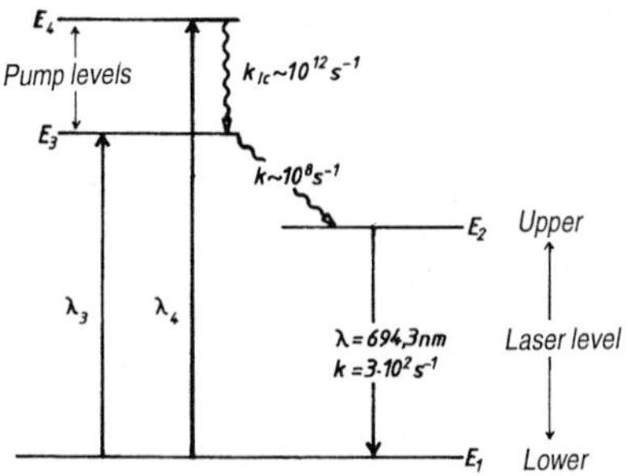

Ruby laser. Fig. 1. Energy-level diagram for the Cr^{3+} ion; a three-level scheme

Russell-Saunders coupling, RS coupling, <*Russell-Saunders-Kopplung*>, a description of spin-orbit coupling which applies particularly to light atoms. In the case of many electrons, each with a specific value of ℓ_i, it is assumed that the ℓ_i are strongly coupled together and combine to form a resultant orbital angular momentum vector, L (→ electronic angular momentum). For example, two electrons with ℓ_1 and ℓ_2 give the L values (→ multiplicity):

$$L = \ell_1 + \ell_2, \ell_1 + \ell_2 - 1, \ell_1 + \ell_2 - 2, \ldots \\ \ldots |\ell_1 - \ell_2|$$

States having different L values have very different energies. The individual electron spins, s_i, are also strongly coupled together to give a total spin vector, S (→ electronic angular momentum). If the total number of electrons is N, the greatest possible value of $S = N/2$, because $s_i = 1/2$. Other values of S are possible, decreasing in unit steps from $N/2$, i.e.:

$$S = \frac{N}{2} - 1;\ \frac{N}{2} - 2....1/2 \text{ or } 0.$$

The smallest value of S is 1/2 or 0, depending upon whether N is odd or even.
The individual L and S values couple together to give a total angular momentum. J (→ electronic angular momentum). J can take the values: $J = L + S, L + S - 1, L + S - 2 \ldots |L - S|$. Each J value corresponds to a different energy so that a term having particular values of L and S splits into a number of levels of different energy equal to the number of values of J. The addition of the vectors, L and S, is such that their vectorial sums have integer differences. Symbolically, Russell-Saunders coupling can be written:

$$(s_1, s_2 \ldots)\ (\ell_1, \ell_2 \ldots) \Rightarrow (S, L) \Rightarrow J$$

The number of possible J values is equal to the number of components into which a state having a particular value of L is split.
This type of coupling provides the explanation of the → multiplet structure of atomic spectra.

Rutherford prism, <*Rutherford-Prisma*>, a → prism named after its constructor. The total length of a spectrum produced by a prism increases with the prism angle, ε. It therefore appears possible to produce more extended spectra by constructing prisms with larger prism angles. However, if the angle of a flint-glass prism is increased to ca. 100°, then rays entering the prism can suffer total internal reflection and do not re-emerge. But if two crown-glass prisms with prism angles of ca. 25° are cemented to the flint-glass prism in the

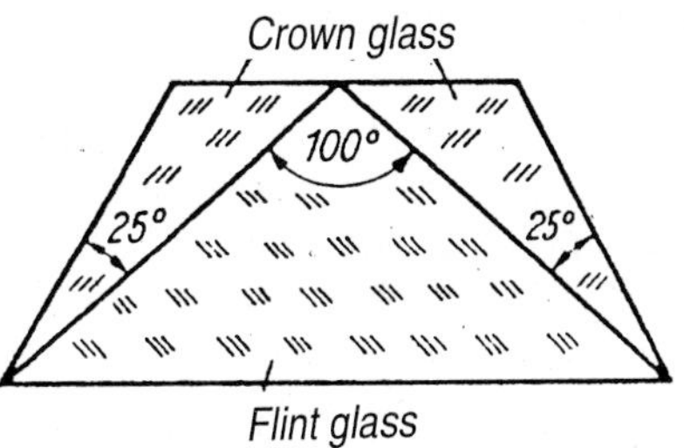

Rutherford prism. Fig. 1. The construction of a Rutherford prism

manner shown in Figure 1, this total reflection can be prevented and we have a Rutherford prism. The dispersion of this prism is about the same as that of a train of two 60° flint-glass prisms.

Rydberg constant, <*Rydberg-Konstante*>, a numerical factor which occurs in the → Balmer formula the value of which has been found to be $R = 109{,}677.578\ \text{cm}^{-1}$ for the hydrogen atom. In 1885, Balmer suggested a series formula for the line sequence of the hydrogen atom in the visible spectral region. The formula was written in the form which we now know by Rydberg in 1890. The factor also applies to other series formulas, in the → alkali spectra for example. According to the quantum-mechanical theory of the hydrogen atom, and the older Bohr theory, the Rydberg constant is given by:

$$R = \frac{\mu e^4}{8\varepsilon_o^2 h^3 c}$$

in which μ is the reduced mass.

$$\mu = \frac{m \cdot M}{M + m}$$

m is the mass of the electron and M the mass of the nucleus. Since $m = 9.109534 \cdot 10^{-31}$ kg and $M = M_H =$

$1.6726485 \cdot 10^{-27}$ kg, $M_H \gg m$ and μ can be set equal to m in the above equation. But the Rydberg constant has a dependence, albeit small, upon the mass of the nucleus:

$$R' = R \cdot \frac{M}{M+m}$$

The Balmer formula also applies to hydrogen-like ions which have only a 1*s* electron as → optical electron. But the very small change of the Rydberg constant must be considered together with the change of nuclear charge, Z. In wavenumbers:

$$\tilde{\nu} = R \cdot Z^2 (\frac{1}{n_1^2} - \frac{1}{n_2^2});$$

where n_1 = constant = 1, 2, 3 ... and n_2 = variable $\geq n_1 + 1$.
The values of the Rydberg constant for the hydrogen-like ions, He^+, Li^{++}, Be^{+++}, are given in the table.
The deviation from the value for hydrogen is of the order of 40–50 cm^{-1}, i.e. ~ 0.5%.

Rydberg constant. Table.

Ion	Z	R [cm^{-1}]
He^+	2	109722.263
Li^{++}	3	109728.723
Be^{+++}	4	109730.624

Rydberg correction, <*Rydberg-Korrektur*>, the corrections, *s, p, d, f,* introduced, for example, into the fixed and running terms of the series formulas for the → alkali spectra. These corrections are pure decimal fractions, < 1, the magnitudes of which represent the deviations from the hydrogen atom states and decrease in the order, *S, P, D, F.* Thus, the higher states become increasingly hydrogen-like.

Rydberg-Ritz combination principle, <*Rydberg-Ritzsches Kombinationsprinzip*>, the inverse of the fact that every spectral line can be represented by the difference of two → terms. It states that, with certain limitations (→ selection rules), the difference between any two terms of an atom gives the wavenumber of a spectral line. This combination principle was discovered in 1908 by Ritz who formulated it as follows. "By additive or subtractive combination of either the series formula itself or of the constants, which are entered into it, new series formulas may be formed". This "makes it possible to calculate the positions of newly discovered lines from ones which were already known".

S

Saturation, <*Sättigungseffekt*>, an effect in → photoacoustic spectroscopy (PAS) which leads to saturation of the signal, i.e. the photoacoustic signal is independent of the → absorption coefficient, β, of the sample. In the limiting cases of the → Rosencwaig-Gersho theory, a signal saturation always occurs when the thermal diffusion length, μ_s, of the sample is greater than the optical penetration depth, l_β; $\mu_s > l_\beta$. In this case, the PA signal depends only upon the intensity of the incident light, I_o. Since the intensity of the light source has a spectral distribution, the PA signal as a function of wavelength, then corresponds to the spectrum of the source. A body which completely absorbs all the incident light over a wide range in the UV-VIS region can be described to a good approximation as a → black standard. The PA spectrum of this standard is therefore the lamp spectrum or, more exactly, the combination of lamp and monochromator. This saturation effect with a black standard can be used to provide a reference signal for PA measurements which allows the signal from a sample to be related to the incident light intensity, I_o, at the same wavelength. The saturation effect is frequently present in highly absorbing samples and often presents an undesired problem in the PAS of these samples since the spectra become flat in the regions of the absorption maxima. But PAS itself offers a way of eliminating the saturation effect. The thermal diffusion length is given by:

$$\mu_s = \sqrt{\frac{2\alpha}{\omega}}; \; \omega = 2\pi f.$$

where α is the thermal diffusivity. If the modulation frequency, ω, is increased, the thermal diffusion length, μ_s, decreases and by variation of ω can be made less than l_β. When this is true we are approaching the limiting case in the Rosencwaig-Gersho theory where the PA signal again depends linearly upon the absorption coefficient, β. The excursions into the saturation effect show that this effect also contributes to the fact that a PA spectrum does not always correspond to the known absorption spectrum. A similar effect is known in → reflection spectroscopy and occurs whenever intensely colored solids, where the specular reflection outweighs the diffuse reflection, are measured. In such cases the → dilution method, in which the sample powder is diluted with an inert, nonabsorbing standard (MgO, NaCl, $BaSO_4$, SiO_2, TiO_2, etc.), can be used. This technique can be transferred to PAS.

In addition to the examples above, saturation effects are also observed in microwave and magnetic resonance spectroscopies. They occur when the thermal equilibrium is disturbed by high intensity irradiation. This easily happens in → NMR spectroscopy since the differences in the populations of the energy levels involved are very small and the deactivation rate of the excited states is determined by relaxation processes which are relatively slow in comparison with the excitation.

Ref.: A. Rosencwaig, *Photoacoustics and Photoacoustic Spectroscopy*, J. Wiley and Sons, New York, London, **1980**.

Saturation spectroscopy, <*Sättigungs-spektroskopie*>, a method of studying optical transitions with high spectral resolution. A molecular transition is irradiated with monochromatic light so as to produce a change in the populations of the molecular energy levels involved.

For small irradiating powers, the total absorption of a molecular transition between the states, i and k, is proportional to the population, N_i, of the absorbing molecules since the population, N_k, of the energetically higher state can be neglected. If N_k is no longer small with respect to N_i, then the intensity decrease, dI, of a light wave traveling in the z direction is given by a modified → Bouguer-Lambert-Beer law.

$$dI = -\sigma_{ik} \cdot I(N_i - g_i/g_k N_k)dz \quad (1)$$

Here, σ_{ik} is the absorption cross section, g_i and g_k are the statistical weights of the two levels and I is the intensity of the incident radiation. Nonlinear absorption begins when the populations, N_i and N_k, of the two states become functions of I so that dI is no longer linearly dependent upon I.

In the simplest case of a two-level system with the populations, N_1 and N_2, the time dependence of the populations is given by:

$$\frac{dN_1}{dt} = -\frac{dN_2}{dt} = -B_{12}\rho(\nu_{12}) \cdot N_1 + B_{21}\rho(\nu_{21})N_2 + A_{21}N_2 \quad (2)$$

B_{12} and B_{21} are the → Einstein B coefficients for absorption and induced emission, A_{21} is the → Einstein A coefficient for spontaneous emission and ϱ is the spectral radiation density (→ photometric units). If the two statistical weights are equal, the total population is $N = N_1 + N_2$ and $B_{12} = B_{21}$. In the photostationary state, the two derivatives in equation 2 are zero and, using the equation $I(\nu) = c\varrho(\nu)$, the population of the ground state is:

$$N_1 = N\frac{1+S}{1+2S} \quad (3)$$

in which the saturation parameter, S, describes the ratio of induced to spontaneous emission.

$$S = \frac{I(\nu)}{c}\frac{B_{21}}{A_{21}}$$

If I tends to ∞ then N_1 tends to $N/2$. In this case the two levels are equally populated and the irradiated substance is fully transparent. These changes of population brought about by high light intensities can be used for the selective saturation of inhomogeneously broadened lines. In this way, a degree of resolution can be realized which is not limited by the line width of the inhomogeneous line.

Suppose that the molecules of a gas with a thermal velocity distribution are irradiated with monochromatic laser light with an electric field strength of $E = E_o \cos(\omega \cdot t - kz)$ traveling in the positive z direction.

The probability that a molecule having velocity, v, absorbs light and is excited from state, 1 to state 2, is proportional to the absorption cross section:

$$\sigma_{12} = \frac{\hbar\omega}{c} \cdot B_{12}g(\omega_o - \omega + k \cdot v) \quad (4)$$

The function g in equation 4 is the absorption line profile in the coordinate system of the moving molecule:

$$g(\omega_o - \omega + k \cdot v) = \frac{\gamma}{(\omega_o - \omega + kv)^2 + (\gamma/2)^2} \quad (5)$$

γ is the homogeneous line broadening of the transition, $1 \rightarrow 2$, with the mean frequency ω_o.
But the homogeneous line width depends upon the intensity of the laser light because γ is given by the equation:

$$\gamma = \gamma_o \cdot \sqrt{1+S}$$

The line width, γ_o, which can be reached at low laser intensity, is determined by the natural line width, $1/\tau$ (τ = lifetime). The absorption of the monochromatic wave of frequency, ω, *burns a hole* in the Doppler distribution around the velocity:

$$v_z = \frac{(\omega - \omega_o)}{k}$$

The frequency, $k \cdot v$, is the magnitude of the Doppler shift relative to ω_o.

S branch, <*S-Zweig*>, → rotation-vibration spectrum and → Raman effect.

Scattered light, <*Streulicht*>, the light deflected from a directed light beam by small particles (→ Tyndall effect) and which can be measured, for example, with a → nephelometer. In spectroscopy, the light scattered in an optical instrument (→ spectrometer) and which, therefore, does not pass through the instrument along the intended path, is known as → stray light. This stray light is superimposed upon the → useful light at any particular wavelength setting and leads to false results.

Scheibe cuvette, <*Scheibe-Küvette*>, a cuvette, named after its constructor G. Scheibe, which consists of two plane-parallel quartz plates and a glass ring of specified thickness, d. The glass ring has optically ground end surfaces and when the three components of the cell are held together by a spring fastening the cuvette is ether-tight. The thickness, d, of the glass ring varies between 10 and 0.1 cm. Cuvettes of this type have also proved valuable for pathlengths down to 0.01 cm.
A set of Scheibe cuvettes consists of 21 cuvettes of the above type with path lengths between 0.1 and 10 cm in logarithmic steps of $\Delta \log d = 0.100$.

Scheme of band heads → band-head scheme.

Schott glass filter, tempered glass filter, <*Anlauffilter*>, a filter based upon a glass <*Anlaufglass*> having the composition K_2O (10–20%), ZnO (10–20%) and SiO_2 (50–60%) with additional CaO, Na_2O and B_2O_3. The color is produced by addition of 1–3% of CdS, CdSe and CdTe. The glass is formed under reducing conditions and is almost colorless when it solidifies. The color is generated by heat treatment (tempering) at 500–700 °C in which process an absorption edge which lies in the UV or blue region is shifted, with increasing temperature, towards the visible and near infrared. The position of this edge determines the color of the glass. Since glass filters of this type absorb the short-wavelength radiation effectively completely and transmit the long wavelengths they are known as → edge filters or → long-wave pass filters. The appearance of the color, i.e. the shift of the absorption edge to the red upon tempering, has a variety of causes. Mixed crystals of CdS and CdSe may be formed, changes in the crystal form

(ZnS ↔ wurtzite) are possible and precipitation of submicroscopic crystals of cadmium chalcogenides can occur. In the last case, the position of the edge depends upon the size of the precipitated particles and moves to longer wavelengths as the particle size increases. In general, the maximum wavelength is determined by the band gap (→ band model) of the Cd chalcogenide semiconductor; e.g. for CdSe the band gap of 1.73 eV corresponds to a wavelength of 715 nm (red glass RG 715). But the glass base has a greater significance in the tempering process.

1. The glass base must form submicroscopic precipitates under heat treatment.
2. The separated regions must be distributed homogeneously in large numbers throughout the glass (they are formed preferentially in phase-separated glasses, such as the K_2O-ZnO-SiO_2 system), because:
   a) The 3d-elements are enriched in a phase where there is the possibility of primary coordination.
   b) The highly charged, network-changing cations, when enriched in a phase, exert a positive influence upon the formation of seeds and crystals (seed formation without diffusion).
   c) Seed and crystal formation occurs simultaneously at many locations, independent of external influences.
3. The phases must be good solvents for the 3d-elements and must crystallize with the impurities.
   a) The phases (drops) are enriched with K and Zn.
   b) The drops crystallize (willemite Zn_2SiO_4).
   c) The impurities crystallize (CdS and ZnS).
4. The potassium content must ensure that:
   a) The tendency in the ZnO-SiO_2 system to separate is reduced.
   b) The drops formed are small.
   c) A large number of drops are formed per volume element.
   d) There is no marked growth of the drops during tempering, which would lead to turbidity.

Apart from the tempered glasses with crystalline precipitates, there are also systems in which the coloring components are distributed as a colloidal dispersion in the glass base. Examples of glasses of this type are, Sb_2S_3 (antimony ruby), FeS (iron ruby), PbS, CuS, NiS, Mo_2S_3, W_2S_3 and Ag_2S which are collectively known as the heavy-metal rubies. The families of colored glasses; the yellow (from GG 400), orange (OG) and red (RG) → colored glass filters from Schott are tempered glasses. The internal (pure) → transmittance of these filter glasses is shown in Figure 1 in log(log) form (→ diabatic scale). The three- or four-digit number following the letters indicating the colored glass family give the position of the absorption edge in nm.

Ref.: Schott, *Optical Glass Filters*; W. Vogel, *Glaschemie*, VEB Verlag, Leipzig, **1979**.

Schrödinger's equation, <*Schrödinger-Gleichung*>. In 1924, L. de Broglie suggested a union of the wave and particle views of atomic and subatomic particles (→ de Broglie relationship). In 1926, E. Schrödinger, following de Broglie's lead, sought a fundamental fusion of these two concepts which would lead to a wave mechanics in complete correspondence with wave optics. For this purpose, he began with the wave equation of wave optics

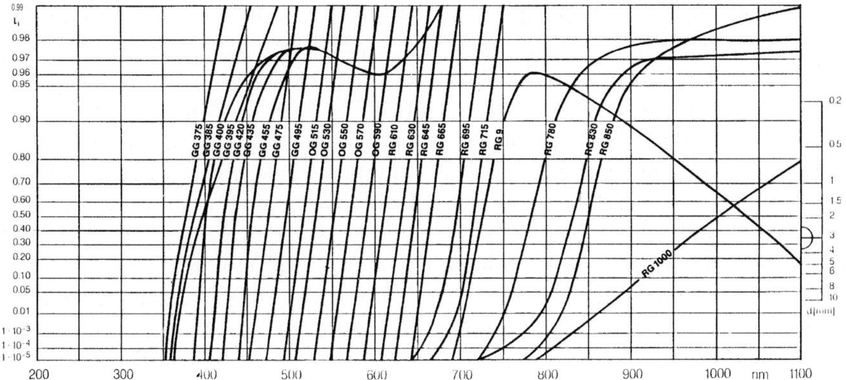

Schott glass filters, Fig. 1. Transmittance curves for 3mm thick GG, OG and RG filters

and the function for a harmonic wave train. Starting with a one-dimensional optical wave in the e.g. x direction, and basing his argument upon the experimentally verified relationships of Bohr, $E = h\nu$, and de Broglie, $p = h/\lambda$, Schrödinger obtained a *material* wave in the form:

$$\Psi(x,t) = a \cdot \sin \frac{2\pi}{h}(E \cdot t - px)$$

or

$$\Psi(x,t) = a \cdot \cos \frac{2\pi}{h}(E \cdot t - px)$$

In these equations, E is the total energy and $p = mv$ is the momentum of a material particle of mass, m, and velocity, v. Using Euler's formula and after separating the space- and time-dependent parts of the equation, it can be written:

$$\Psi(x,t) = \Psi(x)\exp \pm \left\{ \frac{2\pi i}{h} \cdot E \cdot t \right\}$$

By differentiating twice with respect to x and t, substituting into the one-dimensional wave equation and using $p = E/c$, the following differential equation is obtained:

$$\frac{d^2\Psi}{dx^2} + \frac{4\pi^2}{h^2} \cdot p^2 \Psi = 0$$

This equation is known as the amplitude equation since it determines the position-dependent amplitude of the material wave. If the velocity, v, in $p = mv$ is replaced with the help of:

$$E = E_{\text{kin}} + E_{\text{pot}} = \frac{m}{2}v^2 + E_{\text{pot}}$$

one obtains

$$\frac{d^2\Psi}{dx^2} + \frac{8\pi^2 m}{h^2}(E - E_{\text{pot}})\Psi = 0 \qquad (1)$$

or in three dimensions:

$$\nabla^2\Psi + \frac{8\pi^2 m}{h^2}(E - E_{\text{pot}})\Psi = 0 \qquad (2)$$

where

$$\nabla^2 = \frac{\partial^2}{\partial x^2} + \frac{\partial^2}{\partial y^2} + \frac{\partial^2}{\partial z^2}$$

is the Laplacian operator.
This is the time-independent Schrödinger equation. As the brief sketch above shows, the equation is obtained by *deduction* rather than by *derivation*. It should therefore be regarded in the

same way as the laws of thermodynamics which cannot be derived but are confirmed by experience. The significance of the Schrödinger equation lies in the fact that a solution, which is everywhere finite and continuous, can only be found for discrete values of the energy parameter, E, which leads directly to the quantization of energy. These values of E are known as eigenvalues and the corresponding wave functions as eigenfunctions and, in fact, the title which Schrödinger chose for his first paper on wave mechanics was, *Quantization as an Eigenvalue Problem*. Today, we are used to describing the behavior of atoms and molecules, particularly the interaction of matter and light, in terms of this equation. And the results of these theoretical calculations are in excellent agreement with experiment, as is well known in the case of atomic and molecular spectra, for example. For practical purposes, the Hamiltonian form of the equation, which can be readily obtained from equation (2), is used:

$$\left(-\frac{\mathrm{h}^2}{8\pi^2\mathrm{m}}\nabla^2 + \mathrm{E}_{\mathrm{pot}}\right)\psi = E\psi$$

The first term in the brackets is the kinetic energy operator, T, and the second term, $E_{pot} \equiv V$ is the potential energy; $T + V \equiv H$, the Hamiltonian operator. Equation (2) can now be written in the simple form:

$$H\Psi = E\Psi \qquad (3)$$

The Schrödinger equation is frequently applied in this form, but many generally valid results can be obtained, even when the exact form of the Hamiltonian operator for the problem in question is not known, as e.g. in the linear combination of atomic orbitals – molecular orbitals (LCAO-MO) method.

On account of the great theoretical significance of this equation, particularly because of the time dependence of the equation in special cases, the reader is referred at this point to the many text books in physics, physical chemistry and especially quantum theory and quantum chemistry.

Schumann UV region, <*Schumann-UV-Bereich, Schumann-Ultraviolettbereich*>, the far UV region below 200–180 nm. In the early investigations of the emission spectra of gases it was found that such spectra were not observable below 180 nm. By excluding the interfering influence of the surrounding atmosphere from his experiments, V. Schumann was the first to show that emission spectra could be found at much shorter wavelengths. In the range 200–160 nm, it is in general sufficient to flush the apparatus with nitrogen in order to remove the absorption of molecular oxygen. For work at even shorter wavelengths the apparatus must be evacuated. (→ vacuum UV spectrum).

Scintillation, <*Szintillation*>, the phenomenon that highly energetic electromagnetic radiation, e.g. gamma or X-rays, produce a flash of visible light when they strike some crystals. This effect is used in → scintillation counters in X-ray spectroscopy. The most energetic particle radiation, e.g. alpha rays, can also be detected by scintillation. In optics, scintillation is the word used to describe the phenomenon whereby the fixed stars are observed to twinkle and change in intensity or color. The origins of this effect lie in changes and irregularities in the → refractive index of the atmosphere.

Scintillation counter, *<Szintillationszähler, Szintillationsdetektor>*, a device used as a → detector in wavelength-dispersive → X-ray spectroscopy. Scintillation counters are proportional counters which produce voltage pulses which are proportional to the energy of the incident X-ray photon. They therefore give a spectrally resolved measurement of the radiation. They consist of a scintillation crystal and a → secondary electron multiplier (SEM). An incident X-ray photon produces a flash of light in the crystal which generates a voltage pulse in the SEM. The amplitude of the pulse serves as a measure of the energy of the X-ray photon. In general, the crystal used is sodium iodide doped with thallium ($\varepsilon \approx 300$ eV) and protected by a window from light and moisture. Scintillation counters can be used in the range 0.02 to 0.15 nm and their quantum yields are almost 100%. → Gas flow counters are sensitive in the region from 0.15 to ca. 5 nm. Combinations of scintillation counters and gas flow counters (tandem arrangement), which cover the X-ray range from 0.02 to 5 nm, are frequently used.

Screened indicator, *<Screened Indicator>* → indicator.

Secondary electron emission, *<Sekundärelektronenemission>*. If an electron (primary electron) strikes the surface of a material then it can give up some of its energy to the electrons of the material. If the primary energy is very much greater than the work function, one or more electrons may leave the surface. E.g. primary electrons, generated in a photocell by the external → photoelectric effect and raised in kinetic energy to 100–200 eV by an electric field difference of 100–200 V, have an energy much greater than the work function which lies between 3 and 5 eV. The ratio of the number of secondary electrons to the number of primary electrons is called the emission coefficient, δ, of the secondary electrons. Figure 1 shows the dependence of this coefficient upon the energy of the primary electrons for an Ag-Mg alloy. The emission coefficient depends upon a number of factors, apart from the work function of the material used. Of substantial importance are the energy of the primary electrons (see Figure 1), the angle of incidence of the electrons and the form of the surface. The curves of δ against energy generally show broad maxima and the emis-

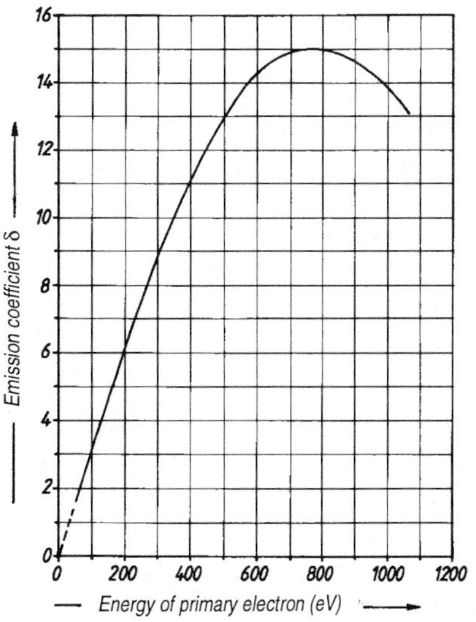

Secondary electron emission. Fig. 1. The emission coefficient, δ, as a function of the energy of the primary electrons

sion coefficient value at ca. 200 eV lies between 3 and 8. Secondary electron emission is the basis for the construction of the → photomultiplier which is a vacuum → photocell with a secondary electron multiplier attached to it.

Secondary electron multiplier, SEM, *<Sekundärelektronenvervielfacher, SEV>*, an important component of a → photomultiplier which consists of a vacuum → photocell combined with an arrangement of from 9 to 14 → dynodes. The amplification of the SEM is produced by the → secondary electron emission.

Secondary ion mass spectroscopy, SIMS, *<Sekundärionenmassenspektroskopie, Sekundärionenmassenspektrometrie, SIMS>*, a method of analyzing surfaces or substances absorbed on surfaces. The technique is primarily used for studying thermally labile substances, e.g. in biochemistry.

The construction of a SIMS apparatus can be described as follows. A beam of high-energy ions is produced by an ion source in a high-vacuum chamber. Because of their high kinetic energies, these ions eject numerous neutral particles and a smaller number of charged particles from a limited area of the sample surface. The ions produced are separated from the neutral particles by an inhomogeneous electric field. Some of the ions pass through an energy filter, into a → quadrupole mass spectrometer and are recorded by an electron multiplier. Of all the processes involving the interaction of the primary ions with the sample surface, which have to be considered, desorption is the one which needs the least energy. The → ionization of particles from the sample requires the most energy.

Ref.: S.M. Scheifers, R.C. Hollar, K.L. Busch, R.G. Cooks, *International Laboratory*, **1982**, *14*, 12.

Secondary ion source, *<Sekundärionenquelle>* → ion-bombardment ion source.

Secondary series, *<Nebenserie>*, a series of spectral lines in an → atomic spectrum which, in contrast to the → principal series which is visible in absorption and in emission, terminates in a higher electronic state and, consequently, can in general be seen only in emission. In the alkali spectra the state in which the series terminates is the $2^2P_{1/2,3/2}$.

Selection rule, *<Auswahlregel>*, a rule in quantum mechanics according to which not all transitions between the energy states of an atom or a molecule are allowed. There is a selection rule for every quantum number which characterizes the states involved. The prohibitions of certain transitions which arise from these selection rules are not absolutely strict; they can be partially mitigated by external influences such as strong electric fields. For example, the transition $1^1S \rightarrow 2^3P$ is strictly forbidden in the He atom. But in the series Zn, Cd, Hg, which have the same electron configuration (ns^2) in the ground state as the He atom, the intensity of this same transition, $^1S \rightarrow {}^3P$, increases strongly and for Hg it is the most intense in the spectrum, → mercury-vapor discharge lamp. This observation is attributed to the → heavy atom effect. Under certain circumstances, transitions which are forbidden by the selection rules can also appear with much diminished intensity. These transitions are called forbidden transitions and the lines in the spectrum, forbidden lines.

Selective Excitation Probe Ion Luminescence, SEPIL, *<SEPIL>*, a variant of fluorescence excitation spectroscopy (→ excitation spectrum) using excitation with tunable → dye lasers. This method has proved valuable in the investigation of inorganic solids doped with lanthanide ions. The selective excitation of luminescence with tunable lasers makes it possible to obtain fluorescence and excitation spectra from ions which are located in a single crystallographic environment. The rare-earth ions serve as selectively excitable probes. The SEPIL method is suitable for trace-element analysis at extremely low levels. The detection limit for the Er^{3+} ion is 25 fg/ml, i.e. $25 \cdot 10^{-15}$ g/ml. Naturally, the method assumes that excitable fluorescing levels of the ions to be analyzed lie within the tunable range of the laser.

Selective fluorescence excitation, *<selektive Fluoreszenzanregung>*, a technique used in the analytical application of → fluorescence spectroscopy. In this type of spectroscopy there is frequently the problem of proving the presence of several fluorescing components in a sample where the → fluorescence spectra of the species more or less overlap. If the various components have different absorption spectra it is possible to select excitation wavelengths such that just one or more components fluoresce so that a simpler, and perhaps even a known, fluorescence spectrum results.

Ref.: C.A. Parker, *Photoluminescence of Solutions with Applications to Photochemistry and Analytical Chemistry*, Elsevier, New York, **1968**, chapt. 5E, p. 438 ff.

Selective wavelength modulation, *<selektive Wellenlängenmodulation>*, like → synchronous fluorescence excitation spectroscopy, a method for the analysis of overlapping → fluorescence spectra. In many cases it can be superior to the synchronous method. The principle of the technique will be briefly explained with reference to Figure 1. The fluorescence excitation spectra of two components, A and B, overlap as shown. We select a wavelength interval, $\Delta\lambda$, which is symmetrically placed with respect to the maximum of component A, but only includes the rising flank of component, B. If the wavelength over the region $\Delta\lambda$ is now changed with a repetition frequency of v, the fluorescence of com-

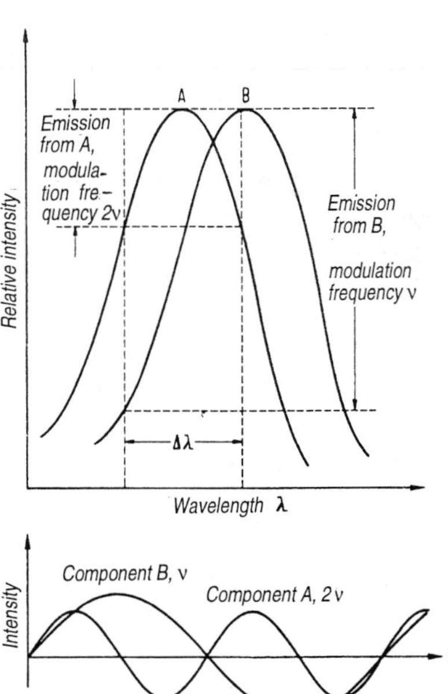

Selective wavelength modulation. Fig. 1. The principle of the technique, see text.

ponent, A, will be modulated with frequency 2ν, but the fluorescence of component, B, only with ν. In the case of component A we pass through the absorption maximum in the wavelength interval, $\Delta\lambda$, twice per cycle, but with component, B, we observe only the rise and fall of the absorption with frequency, ν, as shown in Figure 1. It is therefore possible to measure the fluorescence of the two components independently with the help of a lock-in amplifier. Since the amplitude of the modulated fluorescence intensity is a constant fraction of the total fluorescence intensity, the complete fluorescence spectrum of each component can be recorded free of interference from the other component.

Ref.: Th.C. O'Haver, W.M. Parks, *Anal. Chem.*, **1974**, *46*, 1886.

Self-absorption, reabsorption, *<Selbstabsorption, Reabsorption, Eigenabsorption>*, the phenomenon whereby the light emitted by a system is absorbed again within the system. A well-known example is the Hg resonance line at 253.7 nm which, in the → mercury-vapor discharge lamp, is increasingly reabsorbed by unexcited Hg atoms as the pressure of the mercury vapor is increased. The reabsorption of emitted spectral lines is closely related to → line reversal. Self-absorption is particularly important in fluorescence measurements in solution and can lead to errors, because of the considerable overlap of the fluorescence and absorption bands. The absorption of the fluorescence takes place as the light is on its way out of the cuvette and, in accordance with the → Bouguer-Lambert-Beer law, it is proportional to path length and concentration. Therefore, these two parameters are available for variation in order to optimize the measurement. In the construction of → fluorescence spectrometers, the geometry of the cuvette position with respect to the exciting light and the direction in which the fluorescence is observed are chosen so as to minimize the influence of self-absorption.

Self-quenching of fluorescence, concentration quenching of fluorescence, *<Selbstlöschung der Fluoreszenz, Konzentrationslöschung der Fluoreszenz>*, a decrease in the → fluorescence quantum yield, q_{FM}, which is generally brought about by an increase in the molar concentration of a fluorescing molecule in solution. In the case of aromatic molecules in solution, the self-quenching of the fluorescence of the excited molecules $^1M^*$ is due to the formation of associates or complexes between an excited molecule, $^1M^*$, and a molecule in the ground state, ^1M. The → photostationary state can be described by means of the rate constant for complex formation, k_{DM}, and the rate constants for radiative and radiationless deactivation, k_{FM} and k_{IM} respectively, in a manner analogous to that used for → fluorescence quenching by impurities:

$$\frac{d|^1M^*|}{dt} = I_o - (k_{FM} + k_{IM} + k_{DM}[^1M])[^1M^*] = 0.$$

The quantum yield of the fluorescence is then given by:

$$q_{FM} = \frac{k_{FM}}{k_{FM} + k_{IM} + k_{DM}[^1M]}.$$

Aromatic molecules form a complex, $(^1M^* \cdot {}^1M)$, that is also called an →

excimer. The monomer fluorescence is quenched by this process and there is frequently a structureless fluorescence, shifted to longer wavelengths, which is assigned to the → excimer fluorescence. Concentration quenching is also common with dyes. In that case, in contrast to the aromatics, the cause is an association in the ground state which can be detected, and in part quantitatively analyzed, by the changes in the UV-VIS absorption spectrum. The dimers formed in this process are frequently not capable of fluorescence which is the reason for the reduction of the quantum yield.

Self-reversal of spectral lines, <*Selbstumkehr der Spektrallinien*> → line reversal.

Sensitizer, <*Sensibilisator*>, organic dyes or suitable organic compounds which transfer their primarily absorbed excitation energy to a photochemical system capable of reaction but which, because of its absorption spectrum, is not itself capable of taking up this excitation energy. Such a process is called → sensitizing or sensitization. → Polymethine dyes such as pseudoisocyanines, pinacyanoles, cryptocyanines and polymethinecyanines are usually used for photographic sensitizing. Figure 1 shows the general structural formula of a polymethine dye. Such systems show a pronounced charge resonance as shown by the two resonance structures in Figure 1. The position of the long-wavelength absorption maximum, which has a → molar decadic extinction coefficient, $\varepsilon \cong 10^5\ l\ mol^{-1}\ cm^{-1}$, is strongly shifted to longer wavelengths as the length of the conjugated system is increased, $n = 1, 2, 3 \ldots$ (→ bathochromic shift).

a)

n = 1,2,3... X = O,S,Se R,R' = Alkyl groups y⁻ = Anion

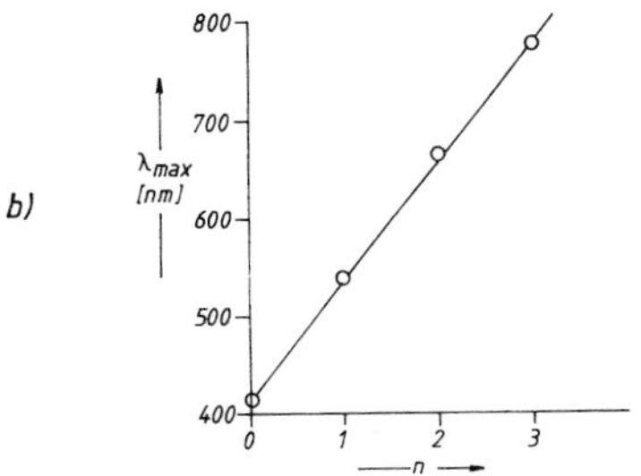

Sensitizer. Fig. 1. a) General structural formula of a polymethine dye; b) dependence of the position of the long-wavelength absorption maximum upon the number of double bonds

When $n = 3$, the absorption is already in the red, as can be seen from Figure 1b. The activity of these dyes as sensitizers derives from the fact that they transfer the excitation energy which they take up to the AgBr, which is itself only light sensitive to 480 nm. Electron transfer and energy transfer have been discussed as possible mechanisms for this photographic sensitizing. Aldehydes and ketones, and also some special dyes, frequently serve as sensitizers in photochemistry where the sensitizing usually proceeds via a triplet state of the sensitizer to a triplet state of the reactive molecule. Chlorophyll is the sensitizer in photosynthesis.

Sensitizing, <*Sensibilisierung*>, the appearance of a higher sensitivity as a result of adding a → sensitizer to the original system. Sensitizing is important in photochemistry, photography, fluorescence and photosynthesis.

In photochemical sensitizing, the addition of a sensitizer to a chemical system makes a photochemical reaction possible. The sensitizer is the primary absorber of the light energy which it then transfers from its excited state to the system, without undergoing any change itself. It acts rather like a catalyst. Acetophenone, benzophenone and other ketones are among the substances used as sensitizers for photochemical reactions. They have a long wavelength, $n \rightarrow \pi^*$, excitation which the substances of the reaction mixture do not. The excitation energy is transferred from the $^1(n \rightarrow \pi^*)$ state or, after a transition from the $^3(n \rightarrow \pi^*)$, to an energetically equal or lower lying triplet state of a component of the potentially reactive system. The *trans-cis* isomerization of olefins, especially stilbene, has been exhaustively studied. If sensitizers of various energies such as, acetophenone, benzophenone, Michler's ketone, α- and β-naphthaldehyde, diacetyl, benzil, fluorenone, dibenzalacetone, pyrene and eosin are studied, then a comparison of the triplet levels of the sensitizers with those of *cis*- and *trans*-stilbene can give information about the mechanism of triplet energy transfer.

Photographic sensitizing is the process of increasing the range of sensitivity of silver-halide emulsions. A pure, unsensitized AgBr emulsion has an absorption edge at ~ 480 nm, i.e. light of longer wavelengths is not absorbed by AgBr. But by adding suitable → sensitizers to the emulsion, photographic emulsions can be prepared which are sensitive throughout the whole of the visible region and into the NIR as far as ca. 1200 nm. Emulsions which are sensitized to 600 nm for blue and green light are called orthochromatic; those sensitized to 700 nm are called panchromatic. Because the process is aimed at particular spectral regions, photographic sensitizing is also called spectral sensitizing.

The sensitizing dyes are absorbed on the surface of the AgBr crystallites. They absorb the light corresponding to their absorption spectra and transfer this energy, probably in the form of an electron, to the silver halide. Investigations of the fluorescence and phosphorescence of the absorbed dyes have shown that it is the first excited singlet state, S_1, which is responsible for the sensitizing effect. A dye can be effective as a sensitizer if its excitation energy is smaller than that of the silver halide, provided that the S_1 level lies within the energy range of the silver-halide conduction band, as in Figure 1a. The excited electron can then pass from the S_1 level into the conduction band. The positive holes, which result from the electron transfer, are probably filled by electrons from bromide ions in energy-rich defect centers, D, of the AgBr crystal. If the S_1 level lies below the conduction band of the AgBr, sensitizing is not possible, Figure 1b. There is then the opposite effect of desensitizing. In addition to the electron transfer, a mechanism for sensitizing which involves the direct transfer of energy has also been suggested. A chemical sensitizing process can also increase the sensitivity of silver-halide emulsions. The process depends upon the introduction of defects in the silver-halide crystal lat-

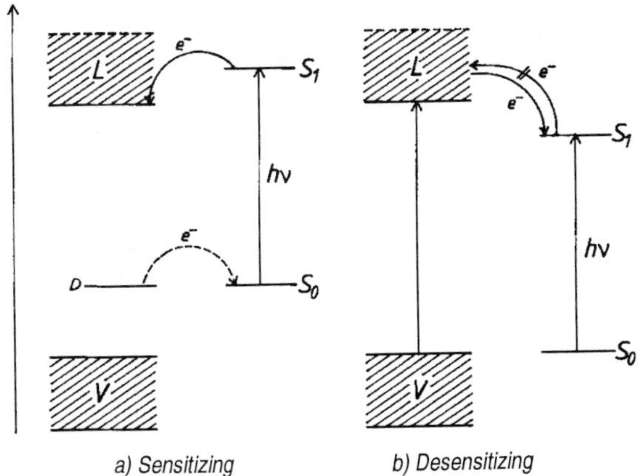

Sensitizing. Fig. 1. The mechanisms of a) sensitizing and b) desensitizing

tice, e.g. by incorporating heavy metal ions such as gold or mercury, sulfide ions, colloids or reducing substances. A chemical sensitizing also occurs during the ripening process in the manufacture of photographic emulsions. Interaction with the ripening substances (compounds containing weakly bound sulfur) deposits small quantities of silver and silver sulfide on the silver halide. The positive effect of mercury is well known in the technology of photographic plate production. The plates are stored in the dark in the presence of mercury vapor, which enhances their sensitivity.

Ref.: J. Okabe, *Photochemistry of Small Molecules*, J. Wiley and Sons, Chichester, New York, Brisbane, Toronto, **1978**, pp. 144–149.

Sequential spectroscopy, <*sequentielle Spektroskopie*>, a form of spectroscopic measurement in which the measured values for different wavelengths are recorded sequentially, i.e. one after the other, with e.g. a dispersive → spectrometer. → Fluorescence and → atomic emission spectrometers use this principle. Fourier-transform methods represent an alternative method of measuring spectra, → Fourier-transform NMR spectroscopy, → Fourier-transform infrared spectrometer.

Seya-Namioka mounting, <*Seya-Namioka-Aufstellung*>, a variant of the → Rowland mounting of a → concave grating in a → monochromator. The entrance and exit slit are fixed. The difference between the angle of incidence and the angle of diffraction is constant at 70°30', see Figure 1. To

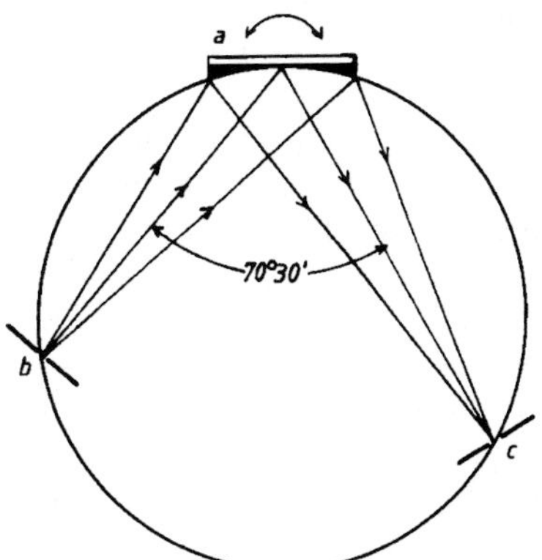

Seya-Namioka mounting. Fig. 1. A grating on the Rowland circle in a Seya-Namioka mounting

change the wavelength only the grating is moved. It is rotated about an axis in the center of the concave grating and parallel to the rulings. Thus, the mechanics of this mounting are simple. However, it is not a true Rowland mounting because rotation of the grating moves the → Rowland circle away from the fixed slits. In many cases, the resulting loss of focus can be tolerated since it is very small for the angles chosen. The Seya-Namioka mounting is used mainly in monochromators for measuring → vacuum UV spectra because it is easy to build the whole unit into an evacuable housing.

Sharp secondary series, *<scharfe Nebenserie, zweite Nebenserie>* → alkali spectra.

Shielding constant, screening constant, *<Abschirmungskonstante>*, σ describes the local magnetic field, H, interacting with an atomic nucleus when an external field, H_0, is applied. Since, in general, the nucleus is shielded by the surrounding electrons, σ is defined as follows:

$$H = H_0(1 - \sigma)$$

The shielding constant, σ, of an atomic nucleus involved in a chemical bond (→ chemical shift) can be written as the sum of four contributions:

1. σ_{dia}, the diamagnetic contribution for that particular atom,
2. σ_{para}, the paramagnetic contribution for that particular atom,
3. σ_A, the contribution from neighboring atoms,
4. σ_{ind}, the contribution due to induced currents in the molecule.

σ_{dia} gives the chemical shift which would occur if the electron distribution around the atom was spherically symmetrical and the electrons could precess without restraint. The magnetic field produced by this circular motion is proportional to the magnetic flux density, B_0, and in the opposite direction. Thus, the magnetic field at the nucleus is reduced and the resonance signal appears only when B_0 is raised. There is a shielding of the nucleus and a diamagnetic shift of the resonance signal to higher B_0 values, i.e. smaller δ values. σ_{para} takes into account the fact that the electron distribution around the nucleus is not spherically symmetrical. This is principally the case in covalently bonded molecules. The paramagnetic contribution describes the difference in the chemical shifts of a linear, covalently bonded molecule and a similar species, which is ionically bonded and therefore has an approximately spherically symmetrical electron distribution. The paramagnetic contribution increases as the energy difference between the ground and excited electronic states decreases. Since hydrogen has no low-lying excited states, this effect is of little significance in proton NMR spectroscopy. However, the effect is the decisive one in the case of heavier nuclei.

σ_A is the contribution to the shielding from currents on neighboring atoms or groups of atoms. It takes account of the → magnetic anisotropy of neighboring groups. This contribution is particularly important, also in proton resonance, for the interpretation of the chemical shift.

σ_{ind} is the result of intramolecular currents. This type of effect is especially marked in aromatic compounds, → ring current.

Short-wave pass filter, <*Kurzpaßfilter*>, an → edge filter which has a range of high transmission in the short-wavelength region which is adjacent to a range of low transmission, the blocked range, in the long-wavelength region. The Schott → colored glass filters, BG 12, BG 38, BG 39, BG 40, are examples of short-wave pass filters. So also are the heat filter glasses. → ADI filters can be made into either long- or short-wave pass filters.

Shutter, wedge-shaped → wedge-shaped shutter.

Signal amplification, <*Signalverstärkung*>, a variant of → dual-wavelength spectroscopy which is used when the assumptions of the → equiextinction method are not fulfilled, i.e. the extinction coefficients, ε_{1b} and ε_{2b}, are different at every pair of wavelengths, λ_1 and λ_2. With the help of a simple modification of the signal the absorbance difference can be made independent of the concentration of component, b. Figure 1 gives an example of this. The equality of the absorbances, $A_{\lambda 1}$ and $A_{\lambda 2}$, required by the equiextinction method is achieved by instrumental means. Most UV-VIS spectrophotometers fitted with microprocessors or microcomputers permit such conversions without difficulty.

At wavelength λ_1:

$$\varepsilon_{1b} \cdot c_b = A_{1b},$$

but at wavelength λ_2:

$$\varepsilon_{2b} \cdot c_b = A_{2b}.$$

It can be seen immediately from Figure 1 that:

$$A_{1b} \neq A_{2b}.$$

If A_{2b} is multiplied by the factor, k, so that:

$$\varepsilon_{1b} c_b = k \cdot \varepsilon_{2b} c_b, \qquad \text{(a)}$$

then, in principle, the requirement of the equiextinction method has again

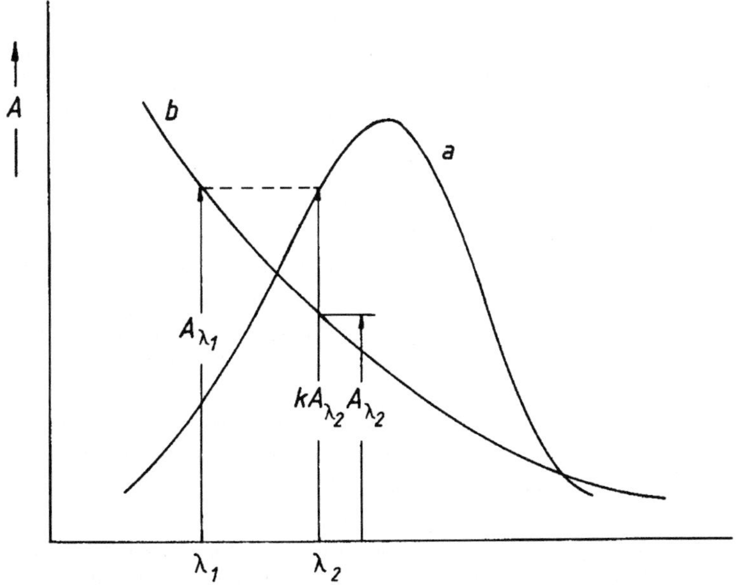

Signal amplification. Fig. 1. Illustration of the signal amplification method, see text

been met. The following now applies for the measured absorption difference:

$$A_{\lambda_2} - A_{\lambda_1} = k(\varepsilon_{2b}c_b + \varepsilon_{2a}c_a) - (\varepsilon_{1b}c_b + \varepsilon_{1a}c_a),$$

$$A_{\lambda_2} - A_{\lambda_1} = (k\varepsilon_{2b} - \varepsilon_{1b})c_b + (k\varepsilon_{2a} - \varepsilon_{1a})c_a.$$

And using equation (a):

$$A_{\lambda_2} - A_{\lambda_1} = \Delta A = (k\varepsilon_{2a} - \varepsilon_{1a})c_a. \quad \text{(b)}$$

The absorption difference, ΔA, in equation (b) is again independent of the concentration of component, b. Therefore, component, a, in the mixture can be determined quantitatively. In effect, the factor, k, in equation (a) is equivalent to an amplification of the signal at wavelength, λ_2, which is the reason for the name given to this procedure.

SIMS, <*SIMS*> → secondary ion mass spectroscopy.

Slit, <*Spalt*>, a narrow and, in general, rectangular aperture the width and height of which can frequently be varied. It is used in spectroscopic apparatus to limit the extent of a beam of light in a defined way. Every spectroscopic instrument possesses at least one slit, the entrance slit, which is also called the spectrograph slit in spectrographs. It is placed at the focus of the collimating lens or the collimating concave mirror, or on the → Rowland circle. A slit usually consists of two parallel plates (jaws), ground to a wedge-like form at their meeting edges. The distance between the edges, the → slit width, can be adjusted with a micrometer screw. A slit in which, when the micrometer is turned, the two jaws move either towards each other or apart, in such a way that the position of the center line of the slit remains constant, is known as a symmetrical or bilateral slit. Therefore, when a symmetrical slit is used, the center of a spectral line remains in its original position when the slit width is changed. This is especially true of precision, bilateral slits. The optimal slit width, S_{opt}, is that where the width of the geometrical image of the slit is equal to half the distance between the first two diffraction maxima which would be obtained if the slit were infinitely narrow, i.e.:

$$S_{opt} = \lambda \frac{f}{d}$$

or for a concave grating:

$$S_{opt} = \lambda \frac{r}{b}$$

where f is the focal length of the collimator lens, d is the width of the beam, r is the radius of curvature of the grating and b is its width.

If the width of the slit is smaller than the optimum, intensity is lost without a significant gain in resolution. If the slit width is greater than the optimum, resolution is lost without a significant gain in intensity. In recording → double-beam spectrometers, especially → IR spectrophotometers, the slit width is adjusted automatically by a slit program during the recording of a spectrum. This is done to compensate for the steep drop in the intensity of the radiation source at longer wavelengths. In single-beam instruments this is done by hand or with a mechanical device.

A normal slit is one with width such that two rays traveling from one edge to the nearest and farthest points in the collimator lens differ at most by $\lambda/4$. There is usually a → slit diaphragm in front of the entrance slit of a spectroscopic instrument. In addition to the entrance slit, spectroscopic devices, e.g. monochromators, usually possess one further slit, the exit slit. This is either fixed or capable of movement along the spectrum. Its function is to take a fairly narrow wavelength interval out of the spectrum and it is therefore generally adjustable in width, and often in height. The change of height may then be used to determine the proportion of → stray light. Some special instruments, e.g. quantometers, are equipped with several exit slits. → Double monochromators have an additional central slit which serves as both the exit slit for the first monochromator and the entrance slit for the second. If very high light intensities are to be used, the slits are made from heat-resistant materials such as tantalum, and they can also be cooled.

Slit diaphragm, <*Spaltblende*>, an aperture placed before the entrance slit of a → spectrograph which by sliding or turning serves to limit the effective length of the slit and therefore of the spectrum. Slit diaphragms make it possible to take several spectra, which are to be compared with each other, one above the other, without changing the position of the photographic plate or film. This frequently simplifies the evaluation of the resulting spectra.

Slit function, <*Spaltfunktion*> → slit width.

Slit width, <*Spaltbreite*>. In → monochromators, a distinction is made between the geometrical, Δs (mm), and the spectral, $\Delta\lambda_s$ (nm), slit width. The geometrical slit width is the effective mechanical width, in mm, of the entrance and exit slits of the monochromator. The spectral slit width is the section of the spectrum, in nm, which for a given geometrical slit width fills the exit slit with light.
The relationship between the two quantities is given by the reciprocal linear dispersion, $1/DL$:

$$\frac{1}{DL} = \frac{\text{spectral slit width } (\Delta\lambda_s)\ [\text{nm}]}{\text{geometrical slit width } (\Delta s)\ [\text{mm}]}$$

it gives the number of wavelength units in nm which appear in 1 mm of the image plane. It therefore depends upon the optical properties of the imaging system, which are different for → prism and → grating monochromators. The ratio defined above is also called the dispersion characteristic of the monochromator system, D.
In a more exact view, the quantity, $\Delta\lambda_s$, should be understood as the effective slit width. This arises from the fact that the finite width, Δs, of the beam of light emerging from the slit of a monochromator is not monochromatic. Rather, it contains a range of wavelengths, $\Delta\lambda$, or of wavenumbers, $\Delta\tilde{\nu}$, which spreads to either side of the set wavelength and depends upon the geometrical slit width, Δs. The intensity distribution of the radiation across the width of the slit therefore takes the form of a triangle, as shown in Figure 1. This distribution is known as the slit function. The effective band width, $\Delta\lambda_e$, is determined from the geometrical slit width and the reciprocal linear dispersion (or dispersion characteristic, D) as explained above. Explicitly, for a → prism monochromator we have the expression:

$$\Delta\lambda_e = \frac{d\lambda}{ds} \cdot \Delta s = \frac{1}{d\vartheta/d\lambda} \cdot \frac{\Delta s}{f}$$
$$= D_P \cdot \Delta s$$

where $d\vartheta/d\lambda$ is the angular dispersion, f the focal length and D_p the dispersion characteristic of the prism monochromator:

$$(D_P = \frac{1}{f \cdot (d\vartheta/d\lambda)});$$

For grating monochromators the expression is:

$$\Delta\lambda_e = \frac{d\cos\beta}{z} \cdot \frac{\Delta s}{f} = D_G \cdot \Delta s$$

where d is the grating constant, z the order of interference, f the focal length of the monochromator, β the angle of diffraction and D_G the dispersion characteristic of the grating monochromator:

$$(D_G = \frac{d \cdot \cos\beta}{f \cdot z}).$$

The spectral band width, or effective spectral band width, is defined as the → half-width of the triangular slit function, as in Figure 1. The effective slit width is approximately equal to the spectral slit width. The spectral slit width in nm of → grating monochromators is almost constant over the whole of the spectral region. Given in wavenumbers as $\Delta\tilde{\nu}$, this is naturally no longer true. For prism monochromators, $\Delta\tilde{\nu}_e$ is a function of the wavelength.

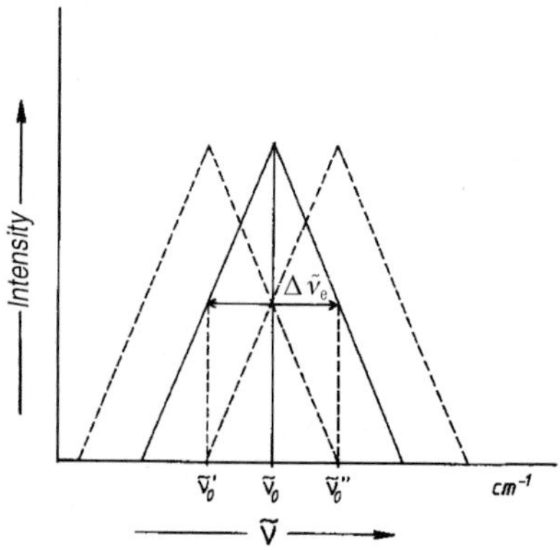

Slit width. Fig. 1. The slit function

The width of the slit of the spectrometer used has a decisive influence upon the quality of an absorption or a fluorescence spectrum. The spectral slit width, $\Delta\lambda_e$, represents that section of the wavelengths from the continuous radiation which the exit slit of the monochromator passes for a given geometrical slit width. To a good approximation, this is inversely proportional to the → resolving power of the spectrometer. The greater $\Delta\lambda_e$, the greater is the wavelength interval picked out for the measurement.

In the case of structured fluorescence or absorption spectra, this has the effect that the absorption or emission signals, within the wavelength interval, arrive at the detector as integrated values. Thus, if the spectral slit width is large, information concerning the structure of the spectrum is lost. If the spectrum is plotted on a wavenumber scale, the selectable spectral slit width on the instrument (in nm) should be converted into $\Delta\tilde{\nu}_e$ and recorded at the maxima of the absorption bands. The reporting of the spectral slit width in cm^{-1} has the additional advantage that the figures can be compared directly with the wavenumbers of the vibrational structure. It is also important that the → resolution, $\Delta\tilde{\nu}_e$, of a grating monochromator increases with increasing wavelength, i.e. $\Delta\tilde{\nu}_e$ becomes smaller.

Therefore, in practical spectroscopic measurements the spectral slit width used must always be as small as possible if none of the spectral details are to be lost. → Band superposition, Figure 1.
Ref.: A.P. Thorne, *Spectrophysics*, Chapman and Hall, London, **1974**.

Smekal-Raman effect, *<Smekal-Raman-Effekt>* → Raman effect.

Sodium D lines, sodium doublet, *<Natrium-D-Linien, Natriumdublett>*, the first two resonance lines in the principal series of the sodium atom spectrum. They correspond to the transitions, $^2S_{1/2} \rightarrow {}^2P_{1/2,3/2}$, with $\lambda_{1/2}$ = 589.592 nm and $\lambda_{3/2}$ = 588.995 nm. (→ alkali spectra).

Sodium flame, *<Natriumflamme>*, the emission of a gas flame made luminous by the addition of a sodium salt. The emission consists primarily of the → sodium *D* lines.

Sodium spectrum, *<Natriumspektrum>*, → alkali spectra.

Sodium-vapor lamp, *<Natriumdampflampe>*, a → metal-vapor discharge lamp which emits mainly the → sodium *D* lines. (→ alkali spectra).

Solar constant, *<Solarkonstante>* → global radiation.

Solar simulator, *<Solarsimulator>*, a light source which is similar to the sun in its spectral distribution of the radiation from the near UV to the near IR. A → xenon lamp offers the best prerequisites for service as an artificial source for a solar simulator. 300 W or 1000 W lamps are used. A complete solar simulator consists of the light source in a housing with specially positioned optics and filters, plus the lamp power supply. Figure 1 shows the beam paths in a solar simulator from the company Oriel, schematically. The purpose of the integrator is to produce

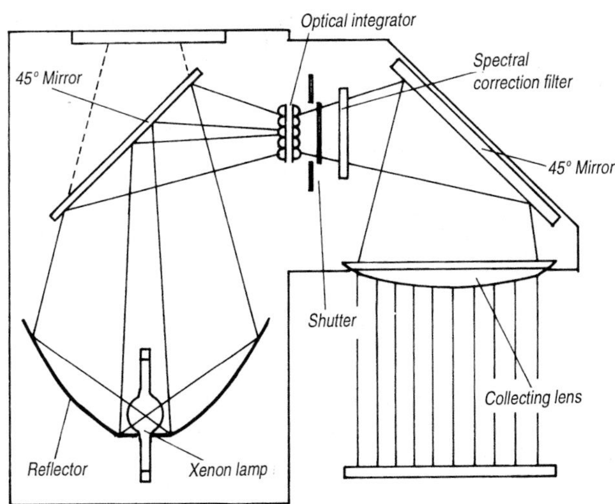

Solar simulator. Fig. 1. The optics of the Oriel solar simulator

a uniform diverging beam. The spectral correction filter ensures a better match between the spectra of the xenon lamp and the sun in the 850–1050 nm region. For special purposes in the UV spectrum, the 45° mirror above the lamp is replaced by a dichroic mirror which reflects the UV radiation and allows the visible and NIR radiation to pass through. The UVC and UVC+UVB regions can be blocked by further filters (→ long-wave pass filters). For particular applications, the xenon highpressure lamp can be replaced by a mercury high-pressure lamp (→ mercury-vapor discharge lamp) or by a xenon-mercury lamp.

Solid-state laser, <*Festkörperlaser*>, a laser in which the active atoms are embedded in a host crystal and give up two or three electrons to the host, depending upon the exact nature of the crystal. They are therefore present as metal ions. Examples of systems and laser wavelengths are given in the table. The most remarkable thing about solid-state lasers is their extremely high output powers which can reach 10^9 W in pulsed operation. Therefore they have many technical applications. All laser systems operate according to the same principle. Because the active medium is a solid crystal, it can be machined like an optical component and it normally also forms the optical resonator. The first practical solid-state laser was the → ruby laser; the → neodymium laser is also important.

Solvatochromism, <*Solvatochromie*>, a general term, introduced by Hantzsch in 1922, for the → solvent dependence of UV-VIS absorption spectra, i.e. the phenomenon of the change of color of a dissolved substance with change of solvent. If increasing solvent polarity induces a red shift of the absorption bands (→ bathochromic shift) the solvatochromism is said to be positive; in the case of a blue shift (→ hypsochromic shift) it is negative. Substances which show solvatochromism are found to have fairly large, permanent dipole moments. Substances such as benzene, naphthalene, anthracene, and also symmetrical → polymethine dyes, which have no dipole moment, show only very small bathochromic shifts of

Solid-state laser. Table.

Atom/Ion	Z	Host crystal	λ [nm]
Holmium	3+	CaF_2	551.2
Praseodymium	3+	LaF_3	598.5
Europium	3+	Y_2O_3	611.3
Chromium	3+	Al_2O_3 (Saphire-ruby)	694.3
Samarium	2+	SrF_2, CaF_2	696.9
Neodymium	3+	Glass, $Y_3Al_5O_{12}$ (YAG)	1060.0
Neodymium	3+	CaF_2, $CaWO_4$, LaF_3	900–1060
Ytterbium	3+	$Y_3Al_5O_{12}$	1020
Thulium	2+	CaF_2	1100
Nickel	2+	MgF_2	1600
Cobalt	2+	MgF_2, ZnF_2	1700–2600

their long-wavelength absorption bands on going from a nonpolar to a polar solvent, e.g. n-hexane → water. A systematic discussion and classification of solvents has been given by Reichardt.
Ref.: C. Reichardt, *Solvents and Solvent Effects in Organic Chemistry,* VCH, Weinheim, New York, **1990**.

Solvent, <*Lösemittel, Lösungsmittel*>, a medium used in molecular spectroscopy to make possible the measurement of solutions. Since the majority of organic and inorganic compounds are solids or liquids at room temperature, liquids are required which will dissolve them. The solvents must fulfill a number of conditions, one of the most important being that they have no absorption in the → spectral region to be studied. For UV-VIS spectroscopy above 250 nm, this condition is met by a number of solvents, as the schematic summary in Figure 1 shows. Some chemical suppliers offer highly purified solvents, e.g. Uvasole (Merck), Spectrosol (BDH), Spectrophotometric Grade (Aldrich) and Spectroscopic Grade (Baker). The requirements are the same for → IR spectroscopy (see Figure 2), but the number of solvents which can be generally used is much smaller. CCl_4 and CS_2 are suitable and they complement each other well with respect to their own absorption bands.

For ESR spectroscopy, it is further required that a solvent forms a glass at 77 K. For proton NMR spectroscopy, the solvent should contain no protons. D_2O, perdeuteriated or perfluorinated hydrocarbons or CCl_4 can be used. In addition, the solvent should have good dissolving properties, which is particularly important for preparing concen-

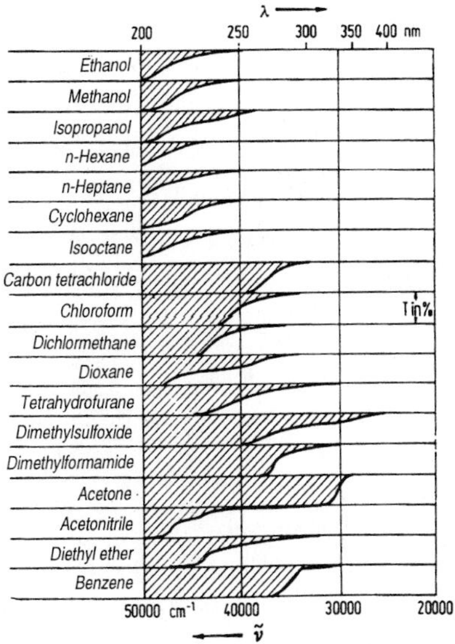

Solvent. Fig. 1. Transmittance of solvents in the UV- VIS spectral region

trated solutions. The required concentration of the sample solution increases from the UV-VIS through the IR to the NMR region. The solvent must not react with the materials from which the sample containers are made. This requirement is usually easy to meet in the UV and VIS regions where quartz and glass cuvettes are used. In IR spectroscopy it means that water can never be used as a solvent when alkali halides are used as cuvette windows, which is very common in practice. The final requirement is that the solvent shall not react chemically with the dissolved material, e.g. solvolysis or hydrolysis. This could lead to irreversible changes in the system to be measured and thence to departure from the → Bouguer-Lambert-Beer law.

The → potassium bromide technique and the → nujol technique are special methods used in the IR region for difficultly soluble substances.

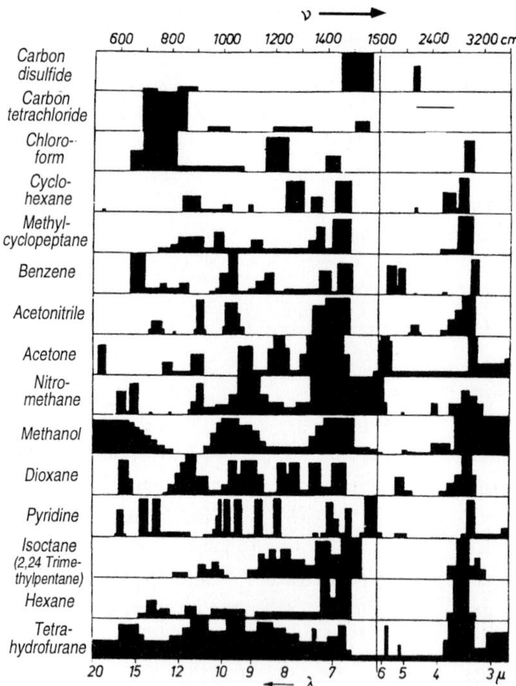

Solvent. Fig. 2. Transmittance of solvents in the IR spectral region

Solvent dependence of UV-VIS spectra, <*Lösemittelabhängigkeit der UV-VIS-Spektren*>. Absorption and fluorescence spectra in solution are normally red shifted (→ bathochromic shift) with respect to the position of their 0–0 transition in the gas phase. Let $\tilde{\nu}_f$ be the wavenumber of the Franck-Condon (→ Franck-Condon principle) 0–0 transition in the fluorescence and $\tilde{\nu}_a$ be the corresponding quantity in the absorption spectrum: $\Delta\tilde{\nu} = \tilde{\nu}_a - \tilde{\nu}_f > 0$. In the gas phase ($g$), however, $\Delta\tilde{\nu}_g = 0$ since the 0–0 transition in absorption and fluorescence fall at the same wavenumber, $\tilde{\nu}_{a,g} = \tilde{\nu}_{f,g}$. Thus, for each separately, we can write:

$$\Delta\tilde{\nu}_a = \tilde{\nu}_{a,g} - \tilde{\nu}_a > 0$$

and

$$\Delta\tilde{\nu}_f = \tilde{\nu}_{f,g} - \tilde{\nu}_f > 0,$$

in which, because of different influences upon the ground and excited states, $\Delta\tilde{\nu}_a \neq \Delta\tilde{\nu}_f$. It would be correct to refer the shifts of $\tilde{\nu}_a$ and $\tilde{\nu}_f$ to the corresponding values in the gas phase. But, because these values are not experimentally obtainable in many cases, the shifts are referred to a standard solution such as n-hexane or n-heptane.

Since absorption and fluorescence are linked, the relation $\Delta\tilde{\nu} = \tilde{\nu}_a - \tilde{\nu}_f$ provides an adequate formulation of the solvent dependence. Thus, a general theoretical description of solvents must also link the emission and absorption processes with each other. In such a theory two types of intermolecular interaction between the solvent and the dissolved molecules should be distinguished. First there is a universal interaction due to the collective influence of the solvent, the dielectric medium, which depends upon the static dielectric constant and the refractive index (→ refraction of light) of the solvent. In addition, in certain solvents there are specific interactions, such as hydrogen bonding, formation of complexes and exciplexes, which are determined by the particular molecular properties of the solvent and solute. These additional interactions play an important role in the characteristic phenomena of → solvatochromism, but as specific interactions they are at first excluded in the theoretical treatment of solvent dependence. The bases of the general interaction theory are Onsager's reaction field and the Franck-Condon principle. Figure 1 shows the energy levels involved in the 0–0 transition for the gas and solution states. A dissolved molecule with a dipole moment, μ_o, in its ground state, S_o, and located in a spherical cavity of radius, a, in a medium with a static dielectric constant, ε, polarizes the medium and generates a reaction field:

$$R_o = \frac{2\mu_o}{a^3} f(\varepsilon)$$

where $f(\varepsilon) = \varepsilon - 1/2\varepsilon + 1$.
The energy of the ground state, S_o, of an unexcited molecule with a dipole moment, μ_o, in a solvent of dielectric constant, ε, is reduced, relative to the gas phase, by this reaction field. The 0–0 transition takes place into an excited state, S_1' with dipole moment, μ_1. According to the Franck-Condon principle, the electronic transition is very much faster (~ 10^{-15} s) than the dielectric relaxation time, τ_d, of the solvent. Therefore, the transition

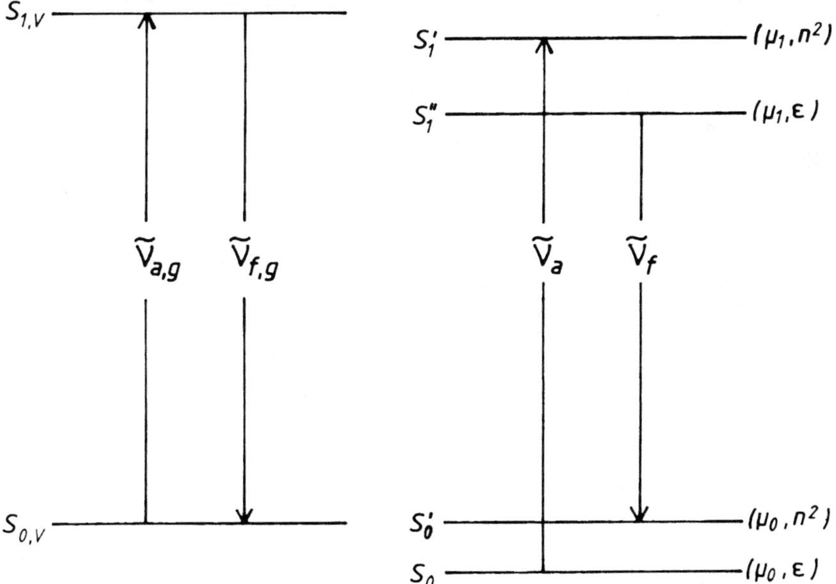

Solvent dependence of UV-VIS spectra. Fig. 1. States participating in the 0–0 transition. Comparison of the free and dissolved molecules

takes place without any re-orientation between the solvent and solute molecules and the reaction field in the state, S_1', is given by:

$$R_1 = \frac{2\mu_o}{a^3} f(n^2),$$

with:

$$f(n^2) = \frac{n^2 - 1}{2n^2 + 1}.$$

n^2 is the high-frequency dielectric constant (refractive index) at the wavenumber, $\tilde{\nu}_a$. The excitation energy $\tilde{\nu}_a$ in wavenumbers is then given by:

$$\tilde{\nu}_a = \tilde{\nu}_{a,g} + \frac{2\mu_o}{a^3}\left[\frac{n^2 - 1}{2n^2 + 1} - \frac{\varepsilon - 1}{2\varepsilon + 1}\right] \cdot \frac{\mu_1 - \mu_o}{hc}$$

where $\tilde{\nu}_{a,g}$ is the gas-phase wavenumber.

If the lifetime, τ_M, of the excited state, S_1', is much greater than the dielectric relaxation time, τ_d, the molecular dipole μ_1 will polarize the solvent statically, producing the reaction field:

$$R_1' = \frac{2\mu_1}{a^3} f(\varepsilon)$$

The result of this is that the state, S_1', is lowered in energy to S_1. The fluorescence takes place from S_1 into the Franck-Condon state, S_o', for which the reaction field is given by:

$$R_o' = \frac{2\mu_1}{a^3} \cdot f(n^2).$$

from which, the wavenumber of the fluorescence is found to be:

$$\tilde{\nu}_f = \tilde{\nu}_{f,g} + \frac{2\mu_1}{a^3}\left[\frac{n^2 - 1}{2n^2 + 1} - \frac{\varepsilon - 1}{2\varepsilon + 1}\right] \cdot \frac{\mu_1 - \mu_o}{hc}.$$

The following expression is obtained for $\Delta\tilde{\nu}$:

$$\Delta\tilde{\nu} = \tilde{\nu}_a - \tilde{\nu}_f = \frac{2(\mu_1 - \mu_o)^2}{hca^3}\left[\frac{\varepsilon - 1}{2\varepsilon + 1} - \frac{n^2 - 1}{2n^2 + 1}\right] = \frac{2(\mu_1 - \mu_o)^2}{hca^3} \cdot F(\varepsilon, n^2)$$

(Note: $\mu_o - \mu_1 = -(\mu_1 - \mu_o)$!)

In addition to the expression for $\Delta\tilde{\nu}$, the sum, $\tilde{\nu}_a + \tilde{\nu}_f = \sum\tilde{\nu}_{a,f}$ is used. It is obtained from the above formulas as:

$$\sum\tilde{\nu}_{a,f} = -\frac{2(\mu_1^2 - \mu_o^1)}{hca^3}\left[\frac{\varepsilon - 1}{2\varepsilon + 1} - \frac{n^2 - 1}{2n^2 + 1}\right] = -\frac{2(\mu_1^2 - \mu_o^2)}{hca^3} \cdot F(\varepsilon, n^2).$$

For a nonpolar solvent, $\varepsilon \cong n^2$, so that $\Delta\tilde{\nu}$ is approximately zero. Because of the mirror-image symmetry between the fluorescence and absorption spectra it is generally true that:

$$\tilde{\nu}_o = \frac{1}{2}(\tilde{\nu}_a + \tilde{\nu}_f);$$

from which it follows that:

$$\tilde{\nu}_o = \frac{1}{2}\sum\tilde{\nu}_{a,f} = -\frac{\mu_1^2 - \mu_o^2}{hca^3} \cdot F(\varepsilon, n^2).$$

The equations for $\Delta\tilde{\nu}$ and $\sum\tilde{\nu}$ contain the quantities μ_o, μ_1 and a. μ_o is in general known and μ_1 can be determined from an analysis of the experimental data with the aid of the above equations. A plot of $\Delta\tilde{\nu}$ against $F(\varepsilon,n^2)$ gives a straight line of gradient, $2(\mu_1 - \mu_o)^2/hca^3$. If μ_o is known, then the assumption of a reasonable value for a, allows the dipole moment μ_1 to be determined. The uncertainty in this analysis lies in the quantity, a, which is assumed to be the diameter of the spherical cavity, which is only a crude approximation. Furthermore, in the derivation of the above equations, approximations have been made and other effects neglected. In particular, the polarizability, α_μ, has been neglected relative to the permanent dipole moment, μ. Nevertheless, these equations are frequently used to evaluate the solvent dependence of absorption and fluorescence spectra; the determination of the dipole moment in the excited state being of particular interest. In the case, where a molecule possesses no dipole moment, any shift of the absorption or emission is attributed to dispersion interaction. This gives the simple relation:

$$\tilde{\nu}_a = -\frac{n^2 - 1}{2n^2 + 1} D',$$

in which D' is a molecule-specific constant which is composed of several terms. This simple expression has been verified in measurements on lycopene, naphthalene and phenanthrene. See the literature for the theory. Ref.: *International Color Symposium*, Verlag Chemie, Weinheim, **1966**. See particularly papers by W. Liptay, pp. 263–341; H. Labhart, pp. 342–355 and G. Briegleb, pp. 391–452.

Soret band, <*Soret-Bande*>, the very intense band around 400 nm (25,000 cm^{-1}) in the UV-VIS absorption spectrum of porphyrins. Its → molar decadic extinction coefficient, $\varepsilon_{max} > 10^5$ l mol^{-1} cm^{-1}, and its form depend upon the symmetry of the porphyrin. For example, the band broadens on going from D_{4h} to D_{2h} symmetry. The Soret band is a ^1B band in the → Platt classification. The bands in the visible from 500–650 nm (20,000–15,500 cm^{-1}) are significantly weaker ($5 \cdot 10^3 \leq \varepsilon_{max} \leq 2 \cdot 10^4$ l mol^{-1} cm^{-1}) and are classified as ^1Q bands.

Space quantization, <*Richtungsquantelung*>. The orbital motion of electrons in the incomplete outer shell of an atom always gives rise to a magnetic moment, $\vec{\mu}_J$, which is given by:

$$\vec{\mu}_J = -\frac{\mathrm{e}}{2m_\mathrm{e}} \cdot \vec{J}.$$

Here, e is the elementary charge, m_e the mass of the electron and $\vec{J}$ the total angular momentum vector which, for → Russell-Saunders coupling, is given by $\vec{J} = \vec{L} + \vec{S}$. $\vec{J}$ is quantized and using the total angular momentum quantum number J, we have:

$$|\vec{\mu}_J| = \frac{\mathrm{e}}{2m_\mathrm{e}} \frac{h}{2\pi} \sqrt{J(J+1)} = \mu_B \sqrt{J(J+1)}$$

where μ_B is the → Bohr magneton. According to quantum mechanics, an angular momentum vector can orientate itself in a magnetic field only in specific, discrete directions. Thus, the angular momentum vector, $\vec{J}$, and therefore also its magnetic moment, $\vec{\mu}_J$, are space quantized in a magnetic field. For every discrete orientation of $\vec{J}$ there is a component of $\vec{J}$, M_J, in the direction of the magnetic field. This component is an integer multiple of $\hbar = h/2\pi$ if J is integer, and a half-integer multiple if J is half-integer. Thus, for M_J we have:

$$M_J = J, J-1, J-2 \ldots -J+1, -J$$

These are $2J + 1$ different values. Figure 1 shows the possible orientations of J with respect to the direction of the magnetic field, H, (usually, the z direction) for the example of $J = 3/2$. The potential energy of μ_J in the magnetic field is given by:

$$E_{\mathrm{mag}} = E_o + g \cdot M_J \cdot \mu_B \cdot \mu_o \cdot H.$$

where μ_o is the vacuum permeability and E_o the energy of a state characterized by J in the absence of the magnetic field. Since M_J can take a total of $2J + 1$ different values there are $2J + 1$ different energies and the space quantization causes a splitting of the initial $(2J + 1)$-fold degeneracy of the state. These states are shown in Figure 1 in which the state with $M_J = -J$ is the lowest in energy. In the equation for E_{mag} g is the → Landé g factor which takes the value $g = 1$ for orbital angular momentum, l or L, and the value $g = 2$ for spin angular momentum s or S. For the total angular momentum, J, the Landé factor, g, must be calculated from the quantum numbers J, L and S. It was found necessary to introduce g into the expression for the magnetic moment since the theory was not in agreement with the experimental observations (anomalous → Zeeman effect). The difficulty was resolved by assigning the two different g values to orbital and spin angular momentum.

Space quantization and the consequent splitting of degenerate states in a magnetic field leads directly to the → Zeeman effect and finally to → electron spin resonance and nuclear magnetic resonance; → NMR spectroscopy.

Spark discharge, <*Funkenentladung*>, a rapid AC discharge across an electrode gap of 2–4 mm. Temperatures reached in the spark are of the order of 10,000 K which leads to spectra of atoms and ions which are extremely rich in lines. → Spark spectra have a relatively strong continuous background which tends to obscure weak lines. Together with → arc discharges and flames, sparks are the oldest form of excitation in → atomic emission spectroscopy.

Spark ionization, <*Funkenionisation, Funkenionisierung*>, the striking of a

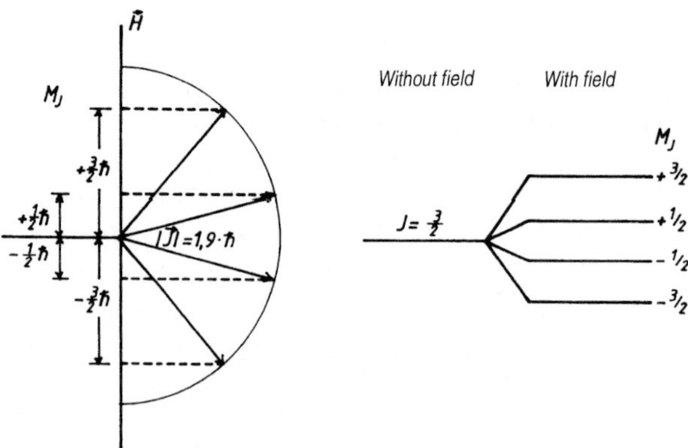

Space quantization. Fig. 1. Example of space quantization for a total angular momentum, $J = 3/2$:

discharge between two electrodes, about 0.1 mm apart and made of the material to be studied, with the requirement that the applied voltage is of the order of 10^4 to 10^5 V.
Spark ionization is used in vacuum-discharge ion sources. Under fields of 10^6 to 10^7 V · cm^{-1}, electrons emitted from the cathode by field emission are accelerated towards the anode by the field. There they heat up the anode and vaporize sample material from its surface. The vaporized sample is ionized by further electrons. In the practical implementation of the procedure, a distinction is made between high-frequency spark-discharge ion sources and low-voltage discharge ion sources. Because of their form, these ion sources are only suitable for work with solids. The application of the spark discharge, which was already well known in the context of emission spectroscopy (→ spark spectrum), was suggested by Dempster in 1936.
Ref.: J. Roboz, *Mass Spectrometry*, Interscience, New York, London, Sydney, Toronto, **1968**.

Spark spectrum, *<Funkenspektrum>*, the spectra of ions which occur predominantly in → spark discharges because the excitation voltages for ions are much greater than they are for neutral atoms, especially on account of the necessity first to ionize the atom several times. The neutral atom spectra are seen in arc discharges, → arc spectrum. The alkali-like spark spectra are designated with Roman numerals, after the symbol for the element, and starting at II. Thus, II is the first, III the second ... spark spectrum. The series with the electron configuration $3s^1$, for example, is:

Na I, Mg II, Al III, Si IV, P V, S VI, Cl VII

Thus, the sixth spark spectrum is assigned to Cl VII, a sixfold ionized chlorine atom; with corresponding assignments for the earlier elements in the series.

Specific rotation, *<spezifische Drehung>* → optical rotation.

Spectral accumulation, *<Spektrenakkumulation>*, a method frequently used in spectroscopy when the signal to be measured cannot be clearly seen above the background noise. Repeated recording of the spectrum and the addition of each successive spectrum, with the help of a microcomputer, gives a much improved signal-to-noise (S:N) ratio in such cases. This improvement arises because the noise is of a purely statistical nature and is therefore averaged by the addition process. At those places in the spectrum, where a real signal is present, there is always a positive contribution to be added. The improvement in the S:N ratio obtained by spectral accumulation is proportional to the square root of the number of scans, $\sqrt{n}$; i.e. for 100 accumulated spectra the S:N ratio is improved by a factor of 10.
Spectral accumulation is very frequently used in → NMR spectroscopy, in both the CW method, where the spectra are directly accumulated (co-added), and in → Fourier-transform NMR spectroscopy where many free induction decays are co-added before final Fourier transformation to give the NMR spectrum. Spectral accumulation is also widely used in other branches of spectroscopy, e.g. → UV-VIS, → fluorescence and → Raman

spectroscopy. However, it must be borne in mind that spectral accumulation can last many hours and during this time conditions must not change.

Spectral dispersion, *<spektrale Zerlegung>*, the dispersion of white light into the colors of the spectrum (→ spectrum, colors of). The observation of this physical phenomenon in nature (rainbows) and in cut-glass objects is ancient, but the first to investigate it systematically was I. Newton who introduced the glass prism as a dispersing instrument. He also founded → color theory which he based upon the dispersion of white light into its spectral colors. It was recognized very early that gratings as well as prisms could be used to disperse white light. Prisms and gratings are used in monochromators as dispersing elements.

Spectral energy distribution, energy spectrum *<spektrale Energieverteilung, Energiespektrum>*. In addition to the total radiated energy, which is an integral measure of the emission of a radiator (→ light source), the distribution of this energy within the spectral region concerned is also of interest. The energy, E, may be shown as a function of frequency, ν, wavenumber, $\tilde{\nu}$, or wavelength, λ. This distribution is given by → Planck's radiation law for a black-body radiator (→ thermal radiator).
Such a closed, theoretically well-founded representation is not possible for line sources. For these, the measured line intensities are plotted as vertical lines against the frequency, wavenumber or wavelength (→ mercury-vapor discharge lamp). The spectral energy distribution is measured using the light source (radiator), a → monochromator and a detector. Since the sensitivity of most common detectors depends upon the wavelength, the → spectral sensitivity of the detector must be known. Furthermore, allowance must be made for the → spectral transmittance of the monochromator.

Spectral interference, *<spektrale Interferenz>*, difficulties experienced in spectrochemical analysis (AAS, AES) due to incomplete separation of lines in a light source in the case of AAS, or in an emission spectrum in AES; → nonspectral interference.

Spectral region, *<Spektralbereich>*. The complete range of the interaction of electromagnetic radiation (→ light) with matter is split up into a number of adjoining regions with which particular spectroscopic methods can be associated. These spectral regions are summarized in Figure 1. The physical quantity used to characterize the electromagnetic radiation is the → wavelength, λ, → wavenumber, $\tilde{\nu}$, or → frequency, ν.
From → X-ray spectra ($\lambda \cong 10^{-2}$ nm; $\tilde{\nu} \cong 10^9$ cm^{-1}) to → NMR spectra ($\lambda \cong$ 10m; $\tilde{\nu} \cong 10^{-3}$ cm^{-1}), the range of the electromagnetic spectrum which is used in spectroscopy covers 12 powers of ten. In the Figure, the wavelength is also given in → Ångström units (1 Å = 10^{-10} m) which were widely used in earlier work. The definition of the various spectral regions from vacuum UV to microwave is laid down in *Report No. 6 of the Joint Committee on Nomenclature in Applied Spectroscopy, Anal. Chem.*, **1952**, *24*, 1349.

Spectral sensitivity, spectral distribution of photoelectron yield, *<spektrale*

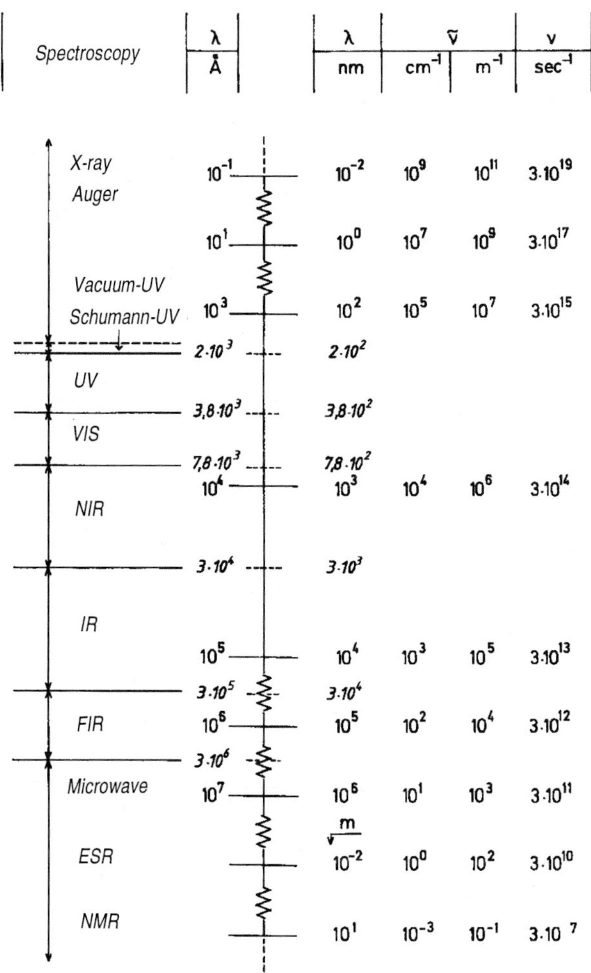

Spectral regions. Fig. 1. Overview of the various spectral regions

Empfindlichkeit, spektrale Verteilung der Ausbeute an Photoelektronen>, the dependence of the yield of photoelectrons upon the wavelength of the radiation which is shown by all radiation detectors which utilize the external → photoelectric effect. The various photocathodes which are used in → photocells and → photomultipliers show a differing spectral sensitivity curve which is determined by their construction. (See Figure 1, → photomultiplier) The → quantum yield is usually plotted against wavelength, λ, to characterize the spectral sensitivity. The value of the quantum yield is obtained by relating the measured yield, E, in μA/lumen at the various wavelengths to the maximum yield, E_{max}, which is given by the → Stark-Einstein equation for $\lambda = 555$ nm as 656 μA/lumen. If the measurements are normalized to the maximum value, the relative quantum yield is obtained which is often used to illustrate the spectral sensitivity. Naturally, in this type of measurement the → spectral energy distribution of the source in the UV-VIS region, and perhaps also in the NIR, must be known, together with the → spectral transmittance of the monochromator used. The manufacturers of photocells and photomultipliers supply spectral sensitivity curves for their products. Spectral sensitivity curves are also required in the case of detectors based upon the inner photoelectric effect such as → photoresistive cells, barrier-layer photocells (→ barrier-layer photovoltaic effect), → photoelements, and → photodiodes. They can be determined in an analogous way.

Spectral transmittance, <*spektrale Durchlässigkeit*>, the measured → transmittance; the primary measured quantity as a function of wavelength or wavenumber, which leads directly to the → absorption spectrum in the UV, VIS, NIR and IR regions. In practical measurements with a → spectrometer, the spectral transmission of the → monochromator itself must be taken into account. This is a part of the calibration of a spectrometer. The transmittance is a function of the optical components of the monochromator such as mirrors, lenses, prisms and gratings. The transmission of prisms and lenses falls steeply as the absorption of the material increases in the UV and NIR. Figure 1 shows some transmittance curves for prism monochromators provided by Zeiss. Note that the transmission of the quartz prism decreases rapidly towards 200 nm and is only ca. 40% (or 20%) in the usable UV-VIS region.

Spectran, <*Spectran*>, a nondispersive on-line (process) photometer produced by Perkin-Elmer which operates by means of the single-beam, two-frequency or gas-filter correlation principle. In the two-frequency method, light from the source, which has passed through the sample cell, is chopped by a rotating filter wheel fitted with two → interference filters.

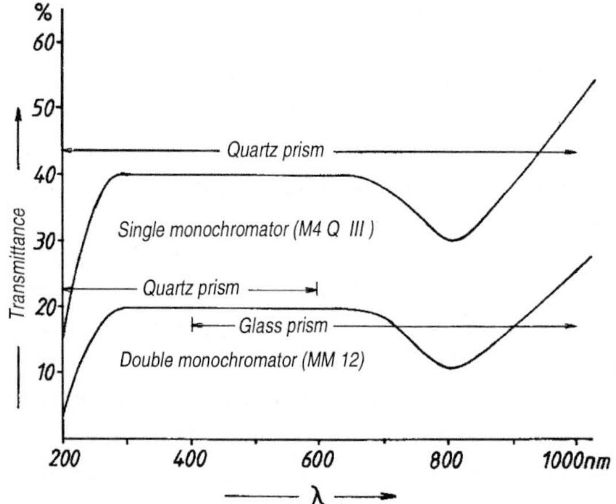

Spectral transmittance. Fig. 1. The transmittance of two prism monochromators

One of these is the reference filter (RF) and the second is the measurement filter (MF). The latter is chosen to transmit the absorption region of the substance to be measured, while the former is chosen to transmit a spectral region where neither the substance to be measured nor any impurity has an absorption. Thus, depending upon the position of the filter wheel, a reference signal or a measurement signal is obtained and from these the absorbance can be calculated by forming their ratio and taking logarithms. (The absorbance of the empty cell must be subtracted from the measurement signal). When using the gas-filter correlation principle, the reference filter is replaced by a cell filled with the gas to be measured. This filter absorbs the spectrum of the gas under investigation out of the light of the source. The corresponding signal is therefore independent of the concentration of the gas in the sample and is used as a reference. The second opening in the filter wheel is free and a stationary filter is placed in front of the detector. The absorbance is again obtained by forming a quotient and taking the logarithm. The model 647 with the long path length cell, for which path lengths of 2–20 m are possible depending on the particular cell, is useful for trace analysis, e.g. monitoring of workplace atmosphere concentrations. The instrument can be used in the VIS, NIR, IR and FIR regions, depending on the detector fitted, and can therefore be adapted to various problems; e.g. gases, vapors, solutions; stationary or flowing.

Spectrograph, <*Spektrograph*>, a → spectroscope in which visual observation in the exit plane is replaced by a photographic plate to photograph the spectrum.

Spectrometer, <*Spektrometer*>, an instrument for making relative measurements in the optical spectral region, especially for analytical applications. The concept has become established outside optical spectroscopy and it is now necessary to distinguish between a multiplicity of spectrometers according to the type of spectrum which they produce; e.g. atomic absorption, atomic emission, UV-VIS, IR, fluorescence, Raman, microwave, ESR, NMR, X-ray, photoelectron, mass, Mössbauer, photoacoustic, reflectance.
Until recently, these instruments were described as spectrophotometers in → optical spectroscopy. This made sense in that, in contrast to the photometers, these instruments use light which is spectrally dispersed by means of a → dispersing element. Thus the syllables *spectro* drew immediate attention to the principal difference between the photometer and the spectrophotometer.
In spite of the number and variety of the spectrometers listed above, a general constructional scheme can be discerned even though the individual elements differ because of the technology applied. Figure 1 shows a series of simplified block diagrams for various spectrometers.

Spectrophotometer, <*Spektralphotometer*>, a photometer which is coupled with a spectrometer. A spectrophotometer is normally constructed according to the single-beam or double-beam principle. It is the instrument of choice for absorption measurements on gases and solutions in

Spectrometer	Source	Ist Dispersion Element	Sample Space	2nd Dispersion Element	Detector + amplifier	Read-out
UV-VIS	D_2-Lamp Tungsten lamp	Monochromator	Cuvette /cell	None	Photo-multiplier SEM	Measuring instrument Plotter, PC
IR	Globar, Nernst glower, glowing filament	" Order changed	"	None	Thermopile Golay detector, etc.	"
Raman	Laser	None	"	Monochromator	Photomultiplier	"
Fluorescence	Xenon Lamp Hg high-pressure	Monochromator Filter	"	Monochromator	"	"
PAS	Xenon-Lamp	Monochromator	"	None	Microphone	"
AAS	HCL	None	Flame	Monochromator	Electron multiplier	"
AES	Flame or ICP with Sample	Monochromator	None	None	"	"
AFS	HCL	None	Flame or ICP and Sample	Monochromator	"	"
NMR	HF Coil	"	Tube	Magnetic field	Receiver coil	"
ESR	MW Source	"	Resonating cavity and Sample	"	Diode, Bolometer	"
Mass	Ion beam	Magnetic field	None	None	Faraday Cup, SEM	"

Spectrometer. Fig. 1. Schematic construction of various types of spectrometer

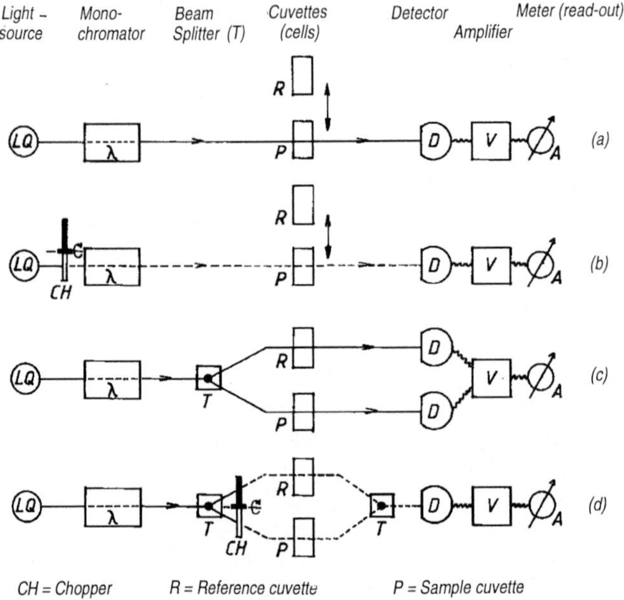

Spectrophotometer. Fig. 1. Schematic representations of the various modes of construction of single- and double-beam spectrophotometers

the UV-VIS, NIR, and IR regions. With the addition of an → integrating sphere, spectrophotometers can also be used for → reflectance spectroscopy. The principles of the construction of spectrophotometers are shown schematically in Figure 1. The single-beam instruments, a) and b), use the substitution principle. First, the reference cell/cuvette, filled with the pure solvent, is placed in the instrument and a 100% reading set on the meter, *A*. The reference is then replaced by the sample and the transmittance of the sample read directly from *A*. The instrument b) differs from a) only in the fact that the light can be modulated by means of the → chopper, *CH*. This makes selective alternating voltage amplification possible giving a significant improvement in the signal/noise ratio. The Zeiss PMQ II and PMQ III are examples of instruments of this type and they are extremely flexible as components in a modular construction. The light is modulated by an oscillating aperture, the disadvantage is the point-by-point measurement by hand. Thus, instruments c) and d) represent a technical improvement since they are double-beam instruments using the → ratio-recording technique, i.e. they are designed as double-beam recording spectrophotometers. Type c) differs from d) in that it uses two detectors. The equalizing of their sensitivities requires additional electronics. Most modern instruments are of type d) and they are increasingly equipped with computers to control the functioning of the apparatus and to process the data acquired for analytical purposes. Spectrophotometers of this type are available from numerous manufacturers for the UV-VIS, VIS and into the NIR spectral regions. They differ in their optical components and detectors. (→ UV-VIS-NIR spectrometer). Spectrophotometers for the IR region have a modified construction. The radiation beam is divided immediately after it leaves the source and the cells are placed before the monochromator, (→ IR spectrophotometer).

Spectropolarimeter, ORD instrument, *<Spektralpolarimeter, ORD-Geräte>*, recording spectropolarimeters were constructed some 30 years ago because of the importance of → optical rotatory dispersion (ORD) in stereochemical structural analysis. The measuring unit of a spectropolarimeter corresponds to the measurement optics in a photoelectric → polarimeter.

The important difference lies in the monochromator placed in front of the polarizer (Figure 1). This must be a quartz prism or a grating monochromator since measurements are required far into the UV region, at least to 200 nm. The working range of some instruments reaches to 185 nm. The use of a double monochromator is advantageous since it has a much reduced → stray light level. But it is expensive.

The polarizer and analyzer must be transparent in the short-wavelength UV spectrum which means that prism combinations cemented together with Canada balsm cannot be used. In general, calcite is not useful since its absorption limit lies in the 220–250 nm region. In modern instruments quartz → Rochon prisms are usually used as polarizers or, in order to push the absorption limit to below 185 nm, ammoniumdihydrogen phosphate immersed in cyclohexane. The source

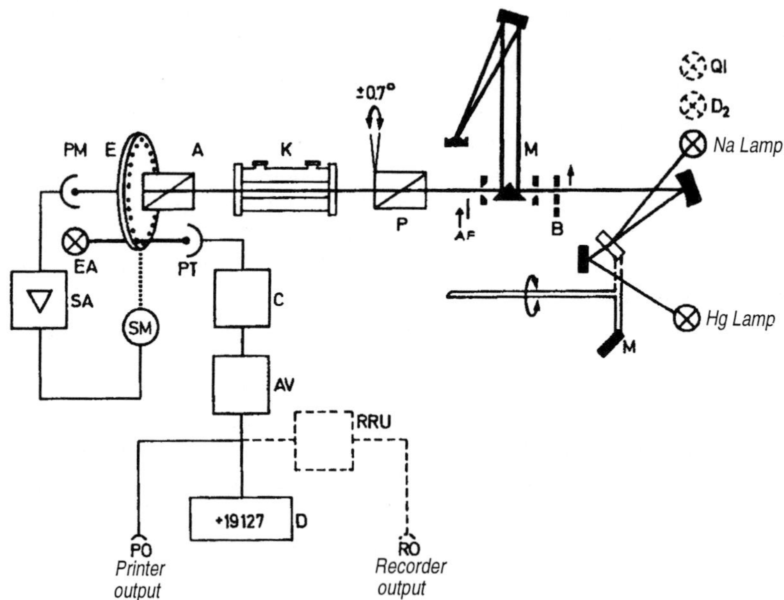

Spectropolarimeter. Fig. 1. The beam path in the Perkin-Elmer spectropolarimeter model 241 MC

is frequently a xenon lamp of 450 W or 500 W. The spectral emission characteristcs of these lamps make them suitable for use throughout the range 185 to 800 nm. The Perkin- Elmer spectropolarimeter 241 MC (Figure 1) uses a deuterium lamp for the region between 250 and 420 nm and a quartz-Hg-iodine lamp for the region from 350 to 650 nm.

The instrument has a rocking polarizer which forms a polarizer and modulator in one component, i.e. it performs both the Faraday modulation (→ Faraday modulator) and also the angular modulation by the rocking of the polarizer (or analyzer) through a fixed angle, e.g. ±0.7°.

A stationary polarizer controls the automatic counter-rotation through a servomotor electrically coupled to the amplifier output. The wavelength drive of the monochromator and the recorder chart drive are linked. Since the dependence of optical rotation upon either wavelength or wavenumber can be of interest, depending upon the particular problem, a choice of scale linear with respect to wavelength or wavenumber is desirable.

Spectroscope, *<Spektroskop>*, a spectroscopic instrument for the visual observation of spectra in the visible region. Most spectroscopes are fitted with a telescope focused on infinity. A distinction is made between prism, grating or diffraction and interference spectroscopes. A prism spectroscope differs from a → prism spectrograph or prism spectrometer in principle only in the way in which the spectrum is observed. This is also true of the grating spectroscope. The simplest and most frequently used instrument is the hand or pocket spectroscope which is

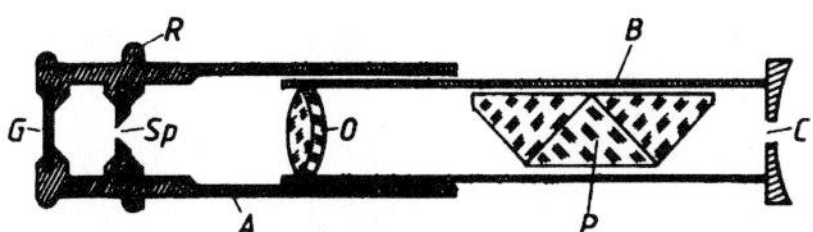

Spectroscope. Fig. 1. A section through a pocket spectroscope

usually fitted with a direct vision prism (→ Amici prism). Figure 1 shows a cross section of such an instrument. It consists of two telescopic metal tubes, *A* and *B*. The slit, *Sp*, is mounted in *A* and can be adjusted with the ring, *R*. The entrance to *A* is closed by a glass plate, *G*, to keep the slit clean. In the metal tube *B* there is an object lens, *O*, which is a magnifying lens with which to observe the slit through the opening *C*, and a three-part → direct vision prism, *P*.
If a transmission grating is used to disperse the light, the instrument is known as a grating hand spectroscope.
The best known, and at the same time the oldest, spectroscope is that of Bunsen and Kirchhoff; it should be called a prism spectrometer, → prism spectrograph.
The dispersing element in an interference spectroscope is either a Fabry-Perot interferometer or a Lummer-Gehrke plate. In contrast to prism and grating spectroscopes, an interference spectroscope requires no entrance slit. However, because of the high resolution which is possible with interference spectroscopes, a preliminary dispersion of the light by means of a prism or grating is required, i.e. the range of the visible spectrum is partitioned into smaller wavelength intervals. Interference spectroscopes are normally equipped with a photographic camera and they should therefore really be called interference spectrographs.

Spectroscopic analysis, <*Spektralanalyse*>, an analytical method which uses the spectroscopic properties of atoms and molecules in the free, the solution and, to some extent, the solid state for analytical purposes. Following the classical spectroscopic analyses of Bunsen and Kirchhoff from the years 1859/60, and the advances in spectroscopic research, many analytical methods have been developed which are predominantly based upon the absorption or emission of electromagnetic radiation. As examples we may cite: → atomic emission, → atomic absorption, → atomic fluorescence, → UV-VIS, → IR, → fluorescence, → Raman and → X-ray spectroscopy.
If structural analysis is included, then we may add → ESR, → NMR and → photoelectron spectroscopy.

Spectroscopic displacement law, displacement law, <*Verschiebungssatz, spektroskopischer Verschiebungssatz, Sommerfeld-Kosselscher Verschiebungssatz*>. A law which states that the first → spark spectrum of an element is similar in all details to the → arc spectrum of the preceding element in the periodic table. For the same reasons, the second spark spectrum is similar to the first spark spectrum of the preceding element and to the arc spectrum of the element with atomic number two less, and so on. Therefore:

The spectrum H I is similar to He II, Li III, Be IV ...

and

the spectrum Li I is similar to Be II, B III, C IV, N V, O VI.

(I is the arc spectrum, II, III, IV ... the 1st, 2nd, 3rd ... spark spectra) See also the spectra of the → hydrogen-like ions.

Spectroscopic instrument, <*Spektralapparat*>, a general term for apparatus which makes a spectral dispersion of light possible, and hence spectroscopic measurements. We distinguish between prism, grating and interference equipment, depending upon the method of dispersing the radiation. We differentiate between spectroscopes, spectrographs and spectrometers according to the way in which the spectrum is detected. In a spectroscope the spectrum is observed visually, with a spectrograph the spectrum is recorded on a photographic plate, and with a → spectrometer the spectrum is usually observed at an exit slit using an eyepiece and with the help of a device by means of which the wavelengths can be determined. A spectrometer with an exit slit can be viewed as the forerunner of the → monochromator. Today, a spectrometer is defined as a spectroscopic instrument in which, in contrast to a spectrograph, the intensity of the radiation leaving the exit slit at different wavelengths is measured photoelectrically.
A distinction is made between spectroscopic instruments according to the → dispersion element which they employ, i.e. between prism and grating spectrographs or between prism and grating spectrometers.

Spectroscopic moment, <*spektroskopisches Moment*>, the transition moment which a substituent induces in the long wavelength, ^1L_b transition of benzene (260 nm or 38,500 cm^{-1}) → Platt classification. The concept was introduced by Platt in 1951. For monosubstitution we have:

$$\varepsilon_s - \varepsilon_o = Km^2$$

or

$$m = \left[\frac{\varepsilon_s - \varepsilon_o}{K}\right]^{1/2}$$

m is the spectroscopic moment which can be obtained directly from the experimental data as the square root of the difference between the → extinction coefficients of substituted benzene, ε_s, and unsubstituted, ε_o. The proportionality constant, K, is usually set equal to 1. The units of m are (l mol^{-1} cm^{-1})$^{1/2}$, or, as they are usually given in the literature (mol cm l^{-1})$^{-1/2}$. The sign of m can be determined from studies of heterosubstituted benzene molecules. In 1961, Petruska showed that there is a linear relationship between the inductive parameter, δ, and the spectroscopic moment, m.
In the case of disubstituted benzenes the equation:

$$\varepsilon_s - \varepsilon_o = K(m_a^2 + m_b^2 + 2m_a m_b)$$

holds for *para* substitution and:

$$\varepsilon_s - \varepsilon_o = K(m_a^2 + m_b^2 - m_a m_b)$$

for *ortho* and *meta* substitution.
These relations are obtained by vector addition of the spectroscopic moments taking account of the angles of 180°, 120° and 60° for p, m and o substitution respectively. The sign of the moment must also be considered in this calculation. m_a and m_b are the moments of the individual substituents. The idea of the spectroscopic moment is related to that of the → transition dipole moment, which is proportional to the square root of the → oscillator strength of an electronic transition. The last named is propor-

tional to the → extinction coefficient, ε, or, more exactly, to the integral extinction, $\int \varepsilon d\tilde{\nu}$. Thus, $(\varepsilon_s - \varepsilon_o)^{1/2}$ is the contribution of the substituent to the transition dipole moment of the monosubstituted benzene.
Ref.: J.R. Platt, *J. Chem. Phys.*, **1951**, *19*, 263; J. Petruska, *J. Chem. Phys.*, **1961**, *34*, 1120.

Spectroscopy, <*Spektroskopie*>, all the methods in the field of electromagnetic radiation which are important for research and applications. The word is derived from Latin, spectrum = ghost, and Greek, skopos = watcher. In this sense, a spectroscopist is a ghost-watcher. This etymology arose in the following way. The colors produced by the → spectral dispersion of light by I. Newton in 1666 appeared to contemporary observers as ghosts and Newton coined the word *spectrum* to describe the colors of dispersed white light. Nowadays, spectroscopy is not confined to the phenomenon of the spectral dispersion of sunlight alone. The extension of the subject, in both research and applications, proceeded with the development of the physical methods of studying the electromagnetic spectrum and with the development of the quantum theory, which provides an explanation of the interaction of electromagnetic radiation with matter. At the present time, the region of the electromagnetic spectrum accessible to physicists and chemists covers a range of 12 to 14 powers of ten in energy units and embraces numerous methods (→ spectral regions).

Spectrum, colors of, <*Spektralfarben*>, the sequence of colors produced by dispersing white or visible light with a → grating or → prism: red, orange, yellow, green, blue, indigo and violet. In this order, the colors run from longer to shorter wavelengths. In a grating spectrum, the red light is the most strongly deviated, in a prism spectrum it is the violet light. The sequence of the colors is independent of the dispersing element though it runs in opposite directions. If all the colors are mixed together white light is obtained (→ complementary colors).

Spectrum shifter, <*Spektrum-Shifter*> a quartz plate, which can be rotated about an axis, mounted behind the entrance slit of a spectrometer. The spectrum shifter is used in multichannel spectrometers to make possible the measurement of another wavelength range, independent of the fixed range, e.g. for background measurements or to include other analytical lines.

Specular reflection, regular reflection, <*reguläre Reflexion, gerichtete Reflexion*>, → reflection of light.

Spherical aberration, <*sphärische Aberration*>, the inability of a simple lens to unite at one point rays which have passed through it at different distances from the axis. For convex lenses, rays passing through the periphery of the lens meet before those closer to the axis; the opposite is the case for concave lenses. The problem can be alleviated by forming a combination of suitably formed convex and concave lenses so that both the spherical and the chromatic aberration are reduced; the former by aspherical grinding of the surfaces of the lenses. This procedure is frequently followed for the lenses of quartz spectrographs. Concentric lenses, the two surfaces of which form part of two concentric spherical surfaces, can also be used.

Spherical top molecule, <*Kugelkreiselmolekül, sphärisches Kreiselmolekül*>, a → top molecule characterized by three equivalent, mutually perpendicular, inertial axes corresponding to three equal moments of inertia; $I_A = I_B = I_C$. Molecules with tetrahedral or octahedral → symmetry point groups, T_d and O_h, are typical spherical tops, e.g. CH_4, CCl_4, SF_6, $[PtCl_6]^{2-}$.

Spin decoupling, <*Spinentkopplung*>, → double resonance technique.

Spin-lattice relaxation, longitudinal relaxation, <*Spin-Gitter-Relaxation*>. In → magnetic resonance spectroscopy the excitation energies are very small. Therefore, according to the Boltzmann distribution law (→ Boltzmann statistics), there is only a very small difference in the populations of the ground and excited states at thermal equilibrium. The thermal equilibrium is sensitively disturbed by excitation or resonance. At the start, there are usually more transitions from the lower to the higher level than in the reverse direction. This would lead, in a short, time to saturation, i.e. $N_1 = N_2$, were it not for the fact that the relaxation processes work in the other direction. These effects make radiationless transitions from the higher state to the lower possible, and the energy is given up as thermal energy to the surroundings or *lattice*. By lattice we understand neighboring molecules in the solution or in the walls of the containing vessel capable of taking up the energy. The use of the word arises from the fact that these relaxation phenomena were first observed in solid samples where there is indeed a lattice. The speed of these processes is characterized by the spin-lattice relaxation time, T_1. In the terminology of chemical kinetics, $1/T_1$ is the rate constant of the first order relaxation process.

The interaction of the elementary magnetic dipoles with their surroundings takes place through the agency of fluctuating magnetic fields. These are produced, for example, by movements of magnetic dipoles and their frequencies cover a wide range of values determined by the mobility of the dipoles. If the band of frequencies produced by the fluctuation has components in the direction of the radio frequency field, B_1, with the correct frequency, ν_1, then transitions can be induced. This is equivalent to a reduction in the lifetime of the nuclei (or electrons) in the higher energy state. In a classical description, the peturbation, B_1, rotates the macroscopic magnetization, M, away from the z direction (field direction) so that only the component, M_z, remains in this direction; Figure 1. The magnetization vector, M, which is the sum of many individual vectors, precesses, like the individual vectors, about the field direction, z, at the Larmor frequency, ν_L. When the perturbation ceases the equilibrium value of M in the field direction is again established; Figure 1, bottom. The spin-lattice relaxation time, T_1, describes the speed of this process. T_1 is also called the longitudinal relaxation time because, in this process, the magnetization along the field direction changes. Spin-lattice relaxation is always associated with a change in the

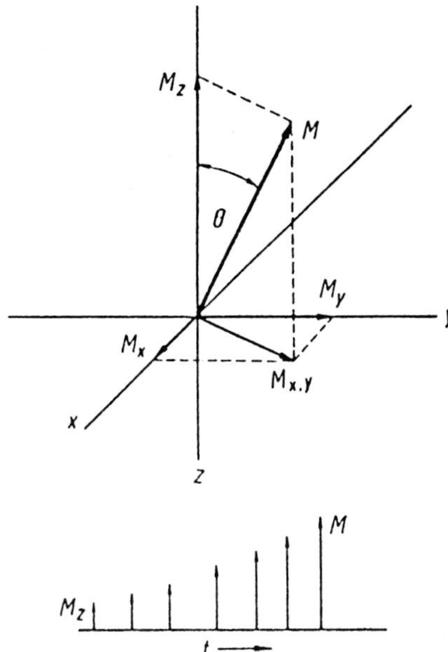

Spin-lattice relaxation. Fig. 1. Illustration of spin-lattice relaxation, see text

energy of the spin system; compare → spin-spin relaxation.

Ref.: N. Bloembergen, *Nuclear Magnetic Relaxation*, W.A. Benjamin, Inc., New York, **1961**.

Spin-spin coupling, *<Spin-Spin-Kopplung>*, together with the → chemical shift, the most important effect in the application of → NMR spectroscopy to structure determination. The chemical shift alone would give a spectrum in which there are as many signals as there are groups of nuclei with the same magnetic shielding. However, the spectra are much more complicated and the signals mostly show a multiplet structure. The origin of this splitting lies in the interaction between neighboring magnetic nuclei in the same molecule. In the ethyl group of ethanol, for example, the two methylene protons interact with the three methyl protons, and vice versa. This interaction is known as spin-spin coupling. The effect produces an extra perturbation of the applied magnetic field at the site of the observed nucleus whereby the magnetic flux density, B, may be enhanced or reduced. This changes the value of the resonance frequency and thus the value of δ, the chemical shift.

The nuclei can couple directly through space. This dipole-dipole coupling plays a decisive role in solid-state NMR spectroscopy, but in high-resolution NMR in solvents of low viscosity it averages to zero. The origin of the → multiplet structure observed in NMR spectra is the indirect spin-spin coupling; indirect because it takes place via the chemical bonds. Such indirect spin-spin coupling occurs between all magnetic nuclei, even unlike nuclei, e.g. protons and fluorine or phosphorus. The multiplet splitting of an NMR resonance signal depends in a characteristic way upon the number of magnetic nuclei in the neighboring group and their orientation with respect to the magnetic field, B, which is usually represented by arrows, ↑ and ↓. Each spin orientation produces an additional field at the site of the observed nucleus which may increase the field there, decrease it, or have no effect. If the observed nucleus has two equivalent neighboring nuclei, e.g. in a molecular fragment such as $CH_{(A)}-CH_{2(X)}$, then the spectrum of proton, A, shows three signals corresponding to the three different possibilities for the orientation of the two X protons: parallel and in the field direction (↑ ↑), parallel and in the opposite direction (↓ ↓) or antiparallel (↑ ↓ or ↓ ↑), Figure 1. In the last two

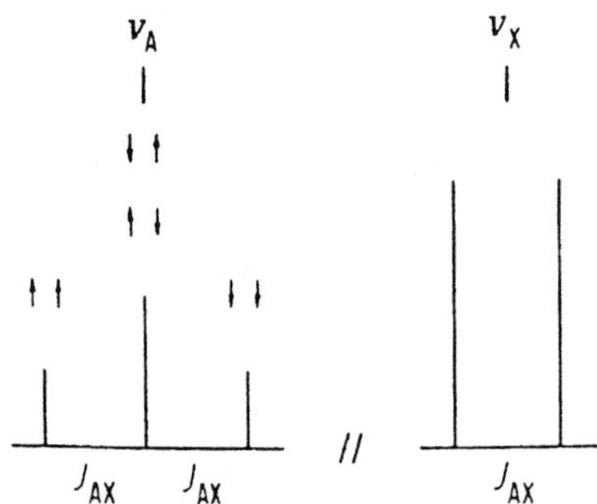

Spin-spin coupling. Fig. 1. Spin-spin coupling in the spin system $=CH\text{-}CH_2$-

cases, the additional fields due to the two spins cancel each other so that the resonance of A lies where it would be in the absence of spin-spin coupling. The separation of the three signals of the triplet is known as the spin-spin coupling constant and is given the symbol J; or J_{AX} to specify the nuclei involved. For protons, coupling constants lie in the range 0–20 Hz. In contrast to the chemical shift, coupling constants do not depend upon the magnetic flux density, B, since they are produced by the nuclear magnetic moments. Therefore, J can also be given in Hz. The relative intensities of the lines in the above example are 1:2:1 because in a macroscopic sample of the molecule there would be twice as many molecules with the antiparallel nuclear spin orientation as with either of the other two possible arrangements with parallel moments. For the two X protons of the CH_2 group a doublet, separated by the same coupling, J_{AX}, is seen because they couple with only one neighboring nucleus, H_A. The ratio of the total intensity of doublet to triplet is 1:2. The splitting pattern for coupling with more than two equivalent neighbors can be constructed in the same way. For coupling with three neighbors, the protons of a CH_3 group for example, a quartet with the intensity ratio 1:3:3:1 is expected. The number of signals in a multiplet, the → multiplicity, M, is easy to calculate. The general result for the coupling with n spins of nuclear spin quantum number, I, is that: $M = 2nI + 1$ whence, for $I = 1/2$, $M = n + 1$.

The intensity ratios within a multiplet correspond to the coefficients of the binomial series and can be taken from Pascal's triangle:

$n=0$						1					
$n=1$					1		1				
$n=2$				1		2		1			
$n=3$			1		3		3		1		
$n=4$		1		4		6		4		1	
$n=5$	1		5		10		10		5		1

If a nucleus, A, is coupled with two nonequivalent nuclei, M and X, which are themselves not coupled ($J_{MX} = 0$) then, in general, $J_{AM} \neq J_{AX}$. The resonances of the nuclei M and X each show a doublet as a result of their coupling to A. The splitting pattern for A can be readily determined as illustrated in Figure 2, which, *mutatis*

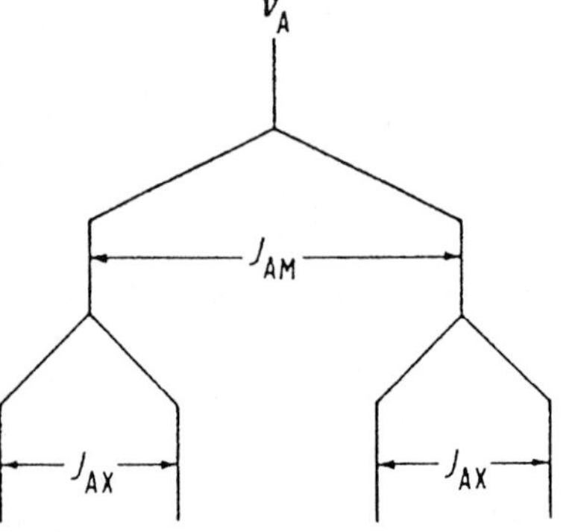

Spin-spin coupling. Fig. 2. Spin-spin coupling in an AMX spin system

mutandis, also shows how a spectrum can be analyzed. The multiplicity for n_M and n_X neighboring nuclei with I = 1/2 is:

$$M = (n_M + 1) \cdot (n_X + 1)$$

Only one line is seen in the ^{1}H NMR spectrum of an isolated methyl group or for benzene, even though several protons are present. However, all the protons are equivalent and, although they couple, the result cannot be observed in the spectrum.

If an exerimental NMR spectrum can be analyzed in the way described above then it is a first order spectrum. The requirement is that the difference between the chemical shifts of the coupled nuclei, $\Delta \nu_{A,X} = \nu_A - \nu_X \gg J_{AX}$. Noticeable departures from first order behavior, with changes in intensities and number of lines, take place even for $\Delta \nu/J$ values < 10. The analysis of such spectra is complicated and is often only possible with a computer. Spin-spin coupling constants can have both positive and negative signs, but the sign of the coupling has no influence on first order spectra.

Indirect spin-spin coupling of protons in saturated compounds does not normally extend over more than three bonds. The protons can be bound to heteroatoms such as sulfur or oxygen. But for an H-H coupling along a path such as H-C-O-H, the proton must remain bound to the oxygen atom for about J^{-1} s which, because of the ready exchange, is often not the case.

In unsaturated compounds, coupling over more bonds is frequently observed.

Spin-spin relaxation, transverse relaxation, <*Spin-Spin-Relaxation*>, a second relaxation process which occurs in → NMR spectroscopy, in addition to → spin-lattice relaxation. It is characterized by the spin-spin or transverse relaxation time, T_2. In this process, a nucleus (or an electron in ESR) on falling from a higher to a lower energy level, gives up its energy to another similar nucleus in the lower level whereby the latter is raised to the higher level. The energy given out is exactly equal to that taken up. In contrast to → spin-lattice relaxation, in this case the total energy of the spin system is unchanged. There is a change only in the phase relationship of the precessing spins to each other. The process is best illustrated with a classical description (see → spin-lattice relaxation, Figure 1). The macroscopic magnetization vector, M, is tipped out of the z direction by the radio frequency field, B_1. This produces components of M in the xy plane, the transverse magnetization, $M_{x,y}$. When the perturbation is removed this transverse magnetization decays to zero again. The time constant for this relaxation process is the spin-spin or transverse relaxation time, T_2. It is a measure of the speed with which the equilibrium state, $M_{x,y} = 0$, is reached. The energy of the system is not changed in this process since energy depends only upon the component of M in the z direction, and that is not changed.

Ref.: F. Bloch, *Phys. Rev.*, **1946**, *70*, 460; N. Bloembergen, *Nuclear Magnetic Relaxation*, W.A. Benjamin, Inc., New York, **1961**.

Splitter cube, <*Teilungswürfel*> → beam splitter.

Splitter plate, <*Teilungsplatte*> → beam splitter, → pellicle.

Standard addition method, <*Additionsverfahren, Aufstockverfahren, interne Eichung, Standardadditionsverfahren*>, a calibration method used where the measurement is subject to sample → matrix effects. In these circumstances the → standard calibration method fails. We have the general linear equation, $y = a + bx$. In using it the constant, a, already contains the unknown quantity, x_u, which we wish to determine. The → absorbance is given, for example, by:

$$\begin{aligned} A_\lambda &= \varepsilon_\lambda \cdot d(c_u + c_v) \\ &= \varepsilon_\lambda \cdot d \cdot c_u + \varepsilon_\lambda \cdot d \cdot c_v \\ &= A_{\lambda,u} + bc_v \end{aligned}$$

c_v is the concentration of a solution prepared by adding a known quantity of a solution of known concentration to the solution of unknown concentration, c_u. Now, if A_λ is plotted against c_v a calibration line is obtained which does not go through the coordinate origin, but cuts the ordinate for $c_v = 0$ at the value, $A_{\lambda,u}$. The unknown concentration, c_u, can then be found by extrapolating the line to cut the negative abscissa. It may sometimes be necessary to correct the result with a blank value (reagent value) in which the identical procedure is carried out on a sample which does not contain the component to be determined.

This technique is especially important in → atomic absorption spectroscopy, but it has also been applied in conjunction with many other analytical methods because its use makes possible the elimination of → matrix effects. Figure 1 compares the standard addition method of calibration and the → standard calibration method.

Ref.: D.C. Harris, *Quantitative Chemical Analysis*, 3rd ed., W.H. Freeman and Co., New York, **1991**.

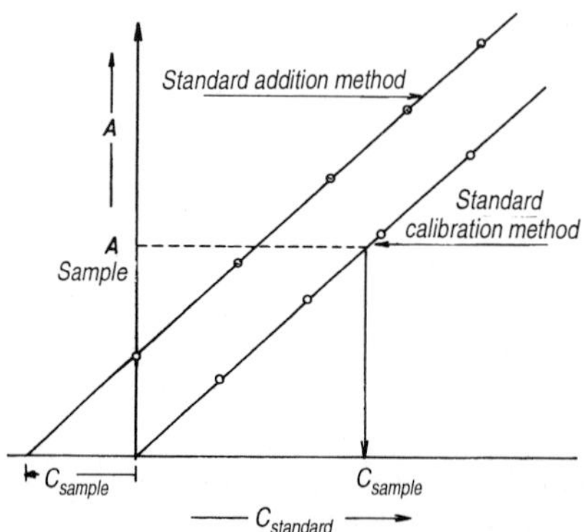

Standard addition method. Fig. 1. Comparison of the standard addition method and the standard calibration method

Standard calibration method, <*Standardkalibrierverfahren*>, the graphical representation of a function, $y = F(x)$, where y is the result of a measurement which depends linearly upon the variable quantity, x. In → atomic absorption spectroscopy and → photometry, this relationship is given by the → Bouguer-Lambert-Beer law:

$$A_\lambda = \varepsilon_\lambda \cdot d \cdot c.$$

If A_λ is plotted as a function of concentration, a linear calibration is obtained (→ standard addition method, Figure 1) by means of which unknown concentrations can be determined without a knowledge of ε_λ. Alternatively, $\varepsilon_\lambda \cdot d$ can be determined from the gradient.

Stark effect, <*Stark-Effekt*>, the splitting of atomic spectral lines by an electric field, first described by J. Stark in 1913. The splitting of the energy levels (terms) arises because application of the field reveals the → space quantization of the angular momentum, J, which requires that the components of J can take only the values, $M = +J, J-1, J-2 \dots -J$, with respect to some preferred direction. However, the electric field does not interact with the magnetic moment due to J, which distinguishes the Stark effect from the → Zeeman effect. The action of the electric field is due to the polarization of the atom, i.e. an electric dipole moment, $\vec{\mu}_{ind}$, is induced in the atom. This dipole is proportional to the field and it also depends upon the orientation of the angular momentum, J, with respect to the field. The energy of the induced electric dipole in the electric field, $\vec{E}$, is given by the product, $\vec{E} \cdot \vec{\mu}_{ind}$. And since $\vec{\mu}_{ind}$ is itself proportional to the field strength, $\vec{E}$, the shift of the energy levels is proportional to the square of the field. This lies at the heart of the fact that energy levels which differ only in the sign of M have the same energy. A change in the sense of rotation, i.e. going from $+M$ to $-M$, does not change the field-induced dipole moment so that the energy shift for $+M$ and $-M$ is the same. Thus, the number of components separated by the splitting is $J + 1/2$ for half-integral values of J and $J + 1$ for integral, J. There are always $2J + 1$ components for the → Zeeman effect. There is therefore a qualitative difference between the Stark and Zeeman effects. The magnitude of the shift due to the electric field is quite complicated to describe, but it is in general true, that the components with the smallest M values, i.e. $M = 0$ or $M = 1/2$ are the lowest in energy. Figure 1 shows the splitting of the Na D lines by the Stark effect. In contrast to the → Zeeman effect, the splitting of the lines or energy levels is not symmetrical with respect to their positions in the absence of the field. In this case, the lines are shifted to longer wavelength.

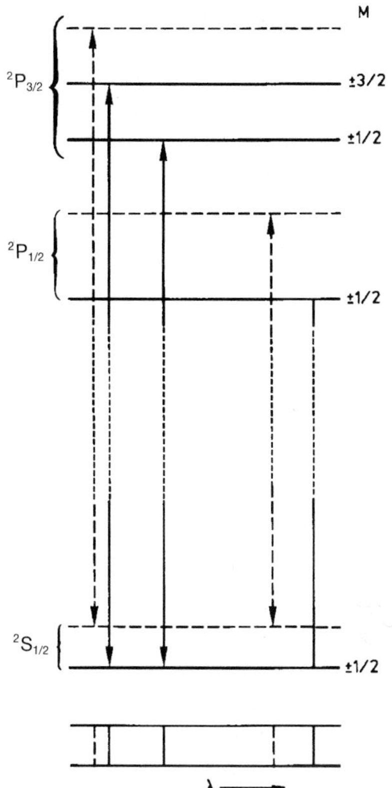

Stark effect. Fig. 1. The Stark effect on the Na D lines

At high electric fields, there is an uncoupling of L and S, as with the Zeeman effect. This is analogous to the → Paschen-Back effect.

As an aid to the interpretation of atomic spectra, the Stark effect is much less important than the Zeeman effect. But it plays an important role in the theory of the chemical bond, the broadening of spectral lines, the dielectric constant and in → microwave spectroscopy.

The Stark effect in microwave spectroscopy.
The rotational energy of a linear molecule with a permanent dipole moment along the internuclear axis is changed when the molecule is placed in an electric field, *F.* The quantum-mechanical analysis of such a linear rotator shows that the rotational energy in an electric field is determined by two quantum numbers, J and m. The rotational quantum number, J, can take integer values 0, 1, 2 ... and the rotational quantum number, m, the integer values $-J$ ... 0 ... $+J$. In the absence of an external electric field the rotational energy depends only upon J and is given by:

$$E_J = \frac{h^2}{8\pi^2 I} \cdot J(J+1).$$

where I is the moment of inertia about an axis perpendicular to the internuclear axis and h is Planck's constant.
When an external electric field is applied, the original degeneracy of all the levels having the same J but different m values is lifted and the levels split. In this case, the rotational energies can be expressed by the addition of further terms, in powers of the field strength, *F,* to the equation above, i.e.:

$$E_{J,m} = \frac{h^2}{8\pi^2 I} J(J+1) + a_{J,m}\vec{F} + \frac{1}{2} b_{J,m}\vec{F}^2 + \ldots$$

$a_{J,m} \cdot F$ is the linear or first order Stark effect term and $b_{J,m}F^2$ is the quadratic or second order term. $a_{J,m}$ is zero for linear molecules, i.e. there is only a second order term. $b_{J,m}$ is given by the equation:

$$b_{J,m} = \frac{8\pi^2 I\mu^2}{h^2} \cdot \frac{3m^2 - J(J+1)}{J(J+1)\cdot(2J-1)(2J+3)}$$

where m depends upon J.
For the $J = 0 \rightarrow J = 1$ transition, $m = 0$ and does not change during the transition. Thus we have:

$$E_{1,0} - E_{0,0} = h\nu = \frac{2h^2}{8\pi^2 I} - [b_{1,0} + b_{0,0}]F^2$$

$$\nu = \frac{2h}{8\pi^2 I} + \frac{4\pi^2 I\mu^2}{h^3} \cdot \frac{8}{15} F^2$$

Since $2h/8\pi^2 I \equiv \nu_o$, the frequency of the rotational line $0 \rightarrow 1$ in the → pure rotation spectrum of the rigid rotator in the absence of an external field, the shift of the $0 \rightarrow 1$ line in the Stark effect is given by:

$$\Delta\nu = \nu - \nu_o = \frac{8}{15} \cdot \frac{\mu^2 F^2}{h^2 \nu_o}.$$

For this transition only there is a shift, but no splitting. For the transition $J = 1 \rightarrow J = 2$ the following differences are found for $m = \pm 1$ and $m = 0$ with $\Delta m = 0$:

$$E_{2,1} - E_{1,1};\ E_{2,0} - E_{1,0} \text{ and } E_{2,-1} - E_{1,-1}$$

Because the energies depend upon m^2 rather than m, the first and third of the above three expressions are equal. But they differ from the second. The resulting splittings are:

$$\Delta\nu = \frac{13}{105} \cdot \frac{\mu^2 F^2}{h^2 \nu_o}$$

for $m = \pm 1$ and:

$$\Delta\nu = (-16/105)(\mu^2 F^2 / h^2 \nu_o)$$

for $m = 0$.

Thus, the line v_o splits into two with separations from the zero field position in the ratio 13:16. Since all m values are equally likely and the transitions with $m = +1$ and $m = -1$ have the same shift, the intensity of the $m = \pm 1$ line is twice that of the $m = 0$ line, Figure 1. The selection rule, $\Delta m = 0$, brings out a specific property of microwave absorption in an electric field. Although the selection rule is $\Delta m = 0, \pm 1$, in normal microwave spectroscopy only the $\Delta m = 0$ transitions are observed. This is also the case when the applied, constant electric field is orientated in the same direction as the electric vector of the radiation field. This arrangement is easy to set up experimentally and the $\Delta m = \pm 1$ components of the lines are not seen when using it.

For nonlinear molecules, the starting point is the → pure rotation spectra of the symmetric or asymmetric top. In the case of a → symmetric top, the quantum number, K, is introduced. K is related to J and can take all values from $+J$ to $-J$, including 0. When $K = 0$, $\Delta m = 0$ and the expression for the Stark shift of $J'' = 0 \rightarrow J = 1$ is the same as for a linear molecule, i.e. a second order (quadratic) effect. For all other cases, $K \neq 0$, $J'' \geq 1$, $\Delta J = +1$ and $\Delta m = 0$, the shift, Δv, with respect to the frequency, v_o, in the absence of a field is given by:

$$\Delta \nu = \frac{2Km}{J(J+1)(J+2)} \cdot \frac{\mu F}{h}.$$

In this case, there is a first order (linear) Stark effect and the second order effect can, in general, be neglected. Since K can take $2J + 1$ values, $2J + 1$ Stark components, symmetrically distributed about v_o, are obtained. The

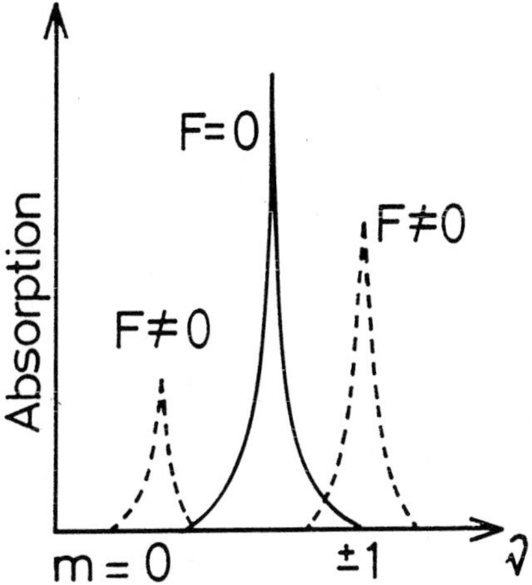

Stark effect in microwave spectroscopy. Fig. 1. The Stark effect in molecules, see text

Stark effect is also very important in microwave spectroscopy because an exact measurement of the dependence of the shifts, Δv, upon the strength of the electric field is a way of determining the dipole moments of gases. The dipole moment is given by the gradient of the graph of Δv against F or F^2, depending upon the order of the Stark effect.

Ref.: W. Gordy, *Microwave and Radio Frequency Spectroscopy, in Technique of Organic Chemistry*, Ed.: W. West, Wiley Intersc. Publ. Inc., New York, London, **1956**, p. 93 ff.; J.W. Smith, *Electric Dipole Moments*, Butterworths Scientific Publications, London, **1955**, Chapt. 1D p. 11 ff; chapt. 3B, p. 66 ff.

Stark-Einstein equation, Stark-Einstein law, <*Einsteinsches Äquivalentgesetz, Stark-Einsteinsches Äquivalentgesetz, Quantenäquivalentgesetz*>, the law, discovered by A. Einstein, that the absorption of light must always occur in whole light quanta or photons because, according to the photon theory of light, light consists of a stream of photons of energy, $E = h\nu$ (h = Planck's constant and ν = the frequency of the light). The absorption of every individual photon corresponds to an elementary process of equivalent energy in the absorbing atom or molecule. This equivalence holds for the absorption of photons by atoms or molecules, which can only take place when, at the same time, an electron is promoted to an energy level of the atom or molecule which is higher by exactly the amount of the energy of the absorbed photon. It is also true when the absorbed energy is sufficient to remove the electron completely from the atom or molecule. In that case, the sum of the ionization energy and the kinetic energy, $mv^2/2$, is equal to the energy of the incident photon (→ photoelectron spectroscopy). Apart from absorption, ionization and emission, the Stark-Einstein law also applies to photochemical processes and the external → photoelectric effect.

Stark modulation, <*Stark-Modulation*>, modulation of the → Stark effect in microwave spectroscopy. The Stark effect is important in microwave spectroscopy, not only for purely scientific reasons, but also because it can be used to improve the signal/noise ratio and hence increase sensitivity. An alternating electric field of a frequency between 1 and 200 kHz is applied to the Stark electrode. The field is supplied by a square-wave generator which produces up to 2 kV at frequencies up to 200 kHz. The Stark electrode is a metal wire, insulated throughout its length, and mounted along the long axis of the microwave cell; → microwave spectrometer, Figure 1. The wall of the cell forms the other electrode. The alternating voltage modulates the Stark effect and produces a modulated shifting and splitting of the rotational lines to be measured. The signal from the detector is processed with the aid of a lock-in amplifier which is tuned to the modulation frequency.

Stern-Volmer relation, <*Stern-Volmer-Beziehung*>, the empirical relationship for the dependence of the fluorescence quantum yield, Φ_{FM}, on the concentration of impurity molecules, $[Q]$, found in 1919 by O. Stern and M. Volmer as a result of their work on the quenching of the fluorescence of gases by impurities. The relation is:

$$\frac{\Phi_{FM}}{(\Phi_{FM})_o} = \frac{1}{1 + K[Q]}$$

$(\Phi_{FM})_o = \Phi_{FM}$ at $[Q] = 0$.
In 1929, the relationship was extended to solutions by S. Vavilov. Since that time, the Stern-Volmer relation has been the simplest way of describing → self-quenching and → fluorescence quenching by impurities. The Stern-Volmer coefficient, K, is defined as the concentration of impurity molecules at which the fluorescence yield, $(\Phi_{FM})_o$, has been reduced to half its original value. It can be interpreted more exactly through considerations of kinetics (see quantum yield of → excimer fluorescence).
Ref.: O. Stern, M. Volmer, *Physik. Zeitschr.*, **1919**, *20*, 183; S.I. Vavilov, *Z. Physik*, **1929**, *53*, 665.

Stokes rule, <*Stokessche Regel, Stokessches Fluoreszenzgesetz*>. If atoms or molecules are excited by the absorption of radiation, they can return to a lower state, in general the ground state, by emitting radiation. Stokes rule then states that the wavelength (wavenumber), $\lambda_e(\tilde{\nu}_e)$, of the emitted radiation will be equal to or greater (smaller) than the absorbed wavelength (wavenumber), $\lambda_a(\tilde{\nu}_a)$. Lines or bands for which λ_e is greater than λ_a are therefore called Stokes lines. Correspondingly, lines for which the emission wavelength is less than the absorption wavelength are called → anti-Stokes lines.

Stopped-flow method, *<Stopped-Flow-Methode>*, a → flow method for the spectroscopic study of the kinetics of chemical reactions which has become increasingly important in recent years. In this method, separate solutions of the reactants are mixed very rapidly (within a few ms) in a mixing chamber and the flow is suddenly stopped. Figure 1 shows schematically a simple thermostatted stopped-flow unit which can be easily fitted as an option to practically any spectrophotometer. The mixing chamber is a cell with a very small volume which permits both absorption and fluorescence measurements. The advantage of the stopped-flow method over the flowing stream is that the reaction can be followed further in time. Thus, a complete concentration-time curve or absorbance-time curve can be obtained with one injection. An oscilloscope or a fast recorder, depending upon the speed of the reaction, can be used to present the results. The use of stopped-flow techniques is particularly important when the overall reaction has to be divided into several main steps. This occurs frequently in enzyme-catalyzed reactions.

The stopped-flow technique, and flow methods in general, suffer from the disadvantage that measurements can only be made at one wavelength and that frequently single-beam instruments must be used. For that reason, double-beam, rapid scanning spectrometers, which enable the complete spectrum of the reaction mixture to be measured, have been described in the literature. These instruments have fast moving optical elements which make it possible to measure a reaction spectrum in 1–10 ms. A significant advance was made in this technique with the introduction of the → UV-VIS photodiode-array spectrometer. The stopped-flow mixing cuvette is com-

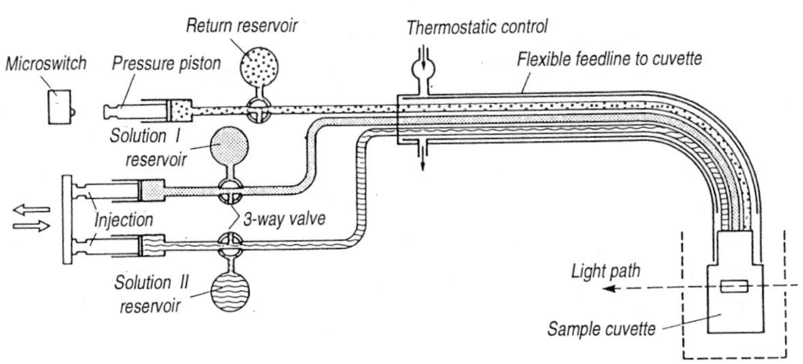

Stopped-flow method. Fig. 1. Schematic of a stopped- flow system

bined with a polychromator and a diode-array detector. The advantage of the instrument is that the complete spectrum of the reaction mixture can be measured simultaneously at a specific time, *t*. The disadvantage is the fact, that the cuvette preceeds the polychromator entrance slit and the sample is therefore exposed to the full intensity of the undispersed light. This can cause significant problems in the case of photochemically unstable systems.

Ref.: E.F. Caldin, *Fast Reactions in Solution*, Blackwell, Oxford, **1964**, chapt. 3.

Stray light, <*Falschlicht*>, light of other wavelengths generated in a monochromator by scattering at the optical surfaces and superimposed upon the → useful light. It is sometimes mistakenly called scattered light. The proportion of the photoelectric current produced by the detector and reaching the display which is due to stray light is the critical measure of the stray light effect when making spectrophotometric measurements. Thus, the term *proportion of stray light* means the ratio of the photoelectric current arising from stray light to the total photoelectric current. Although the proportion of the photoelectric current arising from stray light is generally low (< 0.1%), it can reach significant proportions if the useful light photocurrent becomes small. In practice, this occurs in the following situations:

1. The useful light can be weakened by absorption occurring in the light path whilst the stray light is hardly diminished. This occurs particularly below 230 nm and is due to the optical elements in the light path, which absorb increasingly with decreasing wavelength, and also to solvents which absorb in the short-wavelength UV region.
2. In some spectral ranges, the intensity of the radiation from a light source is relatively small in relation to the intensity in the region generating the stray light.
3. In certain spectral ranges, the detector sensitivity in the useful light region is relatively small, while it is large in the stray light region. This is the case at the long-wavelength limit of the detector sensitivity; i.e. above approximately 620 nm for → photomultipliers and for → photocells above 1.1 μm.

Case 1 is particularly important in practice and requires a check on the proportion of stray light during the measurement. This is especially true if the solvent has a significant absorption in the useful light region. A correction of the measured transmission, τ, can easily be made if the proportion of stray light, p, is known.

Let the useful light leaving the reference cuvette induce a photocurrent, I_o, in the detector and let the photocurrent due to the useful light leaving the sample cuvette be I. Then, the true transmission of the sample = $\tau = I/I_o$. The stray light produces an additional photocurrent, I_f, so that the measurement gives a false transmission $\tau' = (I + I_f)/(I_o + I_f)$. This assumes that the stray light is reduced to the same extent by the sample and the reference. Introducing the proportion of stray light, $p = I_f/(I_o + I_f)$, gives $\tau' = \tau(1 - p) + p$. If p is known, the true transmission can be calculated from the value which is incorrect due to stray light as:

$$\tau = \frac{\tau' - p}{1 - p}.$$

Using the equations, $A = -\log\tau$ and $A' = -\log\tau'$, the absorbance error due to the stray light is found to be:

$$\Delta A = A' - A = \log\tau - \log[\tau(1-p) + p].$$

To measure the proportion of stray light, the useful light must be removed from the beam. Then the photocurrent, I_f, due solely to the stray light can be measured. The useful light can be removed either by absorption or by shielding.

For the determination of p by absorbing out the useful light we proceed as follows. Cuvettes containing solvent (L), sample solution (P) and a very strongly absorbing solution (K) are used. The cuvette, K, has the same path length and contains the substance to be measured at such a concentration that the actual transmission lies below 0.1% in the useful light region. We then measure P against L (result τ') and K against L (result p) and calculate the corrected transmission, τ, using the above equation. It follows from the definitions of τ, τ', and p that we must use the values as fractions of 1.0, and not as percentage figures, in the formula.

Ref.: H.-H. Perkampus, *UV-VIS Spectroscopy and Its Applications*, Springer Verlag, Berlin, Heidelberg, New York, **1992**.

Stress birefringence, <*Spannungsdoppelbrechung*>, the optical birefringence induced in optically isotropic bodies by mechanical stress, e.g. pressure and tension. The phenomenon was discovered in glasses by Brewster in 1815. The degree of mechanical stress can be determined by measurements of the optical birefringence. Rapidly cooled glasses also become optically anisotropic under the influence of internal forces. The strains can be made visible with the help of linearly polarized light. This is very important in the manufacture of → optical glasses in order that these states can be detected and, if necessary, removed by annealing. Stress birefringence can also be seen in optically anisotropic crystals when they are subjected to external forces. In this way, for example, → optically uniaxial crystals can be made optically biaxial by mechanical stress. In general, high polymeric substances also become optically birefringent if they are cooled too rapidly or deformed. In these materials, the stress birefringence is regarded as a sum of → deformation birefringence and → orientation birefringence.

Subband, <*Subbanden*>, the P, Q and R branches, which arise according to the various selection rules, and superimpose to give the total → rotation-vibration spectrum of → symmetric top molecules. The quantum number K in the expression for the rotational energy of a symmetric top molecule is responsible for the appearance of the subbands:

$$F(J,K) = B_{[v]}J(J+1) + (A_{[v]} - B_{[v]})K^2.$$

The number of subbands which overlap to form the total spectrum depends upon the number of K levels which are excited at the temperature of measurement. The position of the subbands relative to the band origin, ω_o, (the pure vibrational transition, $v'' = 0 \to v' = 1$) varies depending upon whether a → parallel band or a → perpendicular band, which have different selection rules for K, is involved. For parallel bands, $\Delta K = 0$; for perpendicular bands, $\Delta K = \pm 1$.

Surface layer, <*Oberflächenschicht*>, a thin layer which is deposited on a substrate or is produced in a natural way. Surface layers of metals are produced to make → surface mirrors for → optical spectroscopy. Protective surface layers are deposited on other surface layers in order to protect them; e.g. an aluminium layer may be protected from oxidation by a layer of quartz deposited from the vapor phase. Aluminium reacts with oxygen to form a surface oxide layer which protects the metal from further corrosion.

Surface mirror, <*Oberflächenspiegel*>, a mirror in which the radiation-reflecting layer is deposited on the surface facing the light.

Symmetric top molecule, <*symmetrisches Kreiselmolekül*>, a → top molecule which is characterized by two equivalent, mutually perpendicular inertial axes which are perpendicular to the third, usually the figure or top axis. The moments of inertia are related by $I_A \neq I_B = I_C$. The total angular momentum vector, $\vec{P}_J$, of a symmetric top is not perpendicular to the figure axis and has a constant component, $\vec{P}_Z$, in the direction of the figure axis, as can be seen from the vector diagram in Figure 1. This component is characterized by the quantum number K, in multiples of $h/2\pi$, while the total angular momentum vector, $\vec{P}_J$, is assigned the rotational quantum number, J. The rotational motion in this case may be described as nutation of the figure axis, P_Z, about the direction of P_J which is constant in space. It is important to note that the superposition of the two motions, the nutation of the figure axis, P_Z, about the direction of P_J and the rotation of the molecule about P_Z, is not simply a rotation of the molecule about the axis, P_J, since P_J, unlike P_Z, is not fixed in the molecule. The molecule rotates about an instantaneous axis whose position in the molecule changes continuously in the following way. Imagine a cone fixed in space with P_J as its axis, the center of mass of the molecule as vertex, and with an angle of $2(\Theta - \Psi)$, where Θ is the angle between P_J and P_Z. A second cone with the figure axis as axis and angle 2Ψ may be fixed to the molecule, see Figure 2. If now the second cone rolls without slipping on the first, at a uniform speed, then it represents the motion of the molecule. The line of contact of the two cones is the instantaneous axis of rotation. This axis rotates about the axis of P_J as does the figure axis also, and with the same angular velocity. As can be seen from Figure 2, both the instantaneous axis of rotation and the axis of P_J, which is fixed in space, continuously change their positions with respect to the molecule.

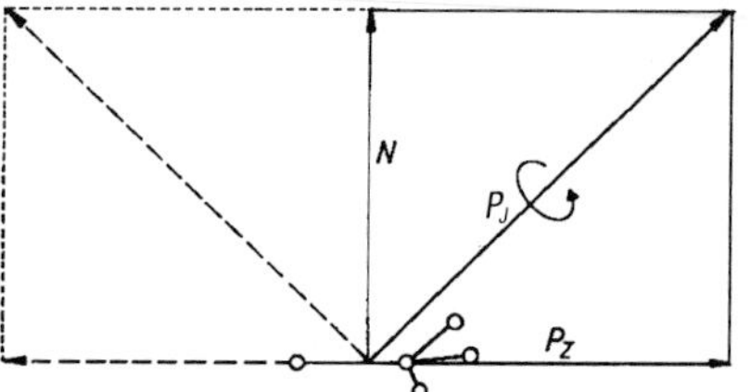

Symmetric top molecule. Fig. 1. Vector diagram for a symmetric top

The rotational energy of a symmetric top is given by:

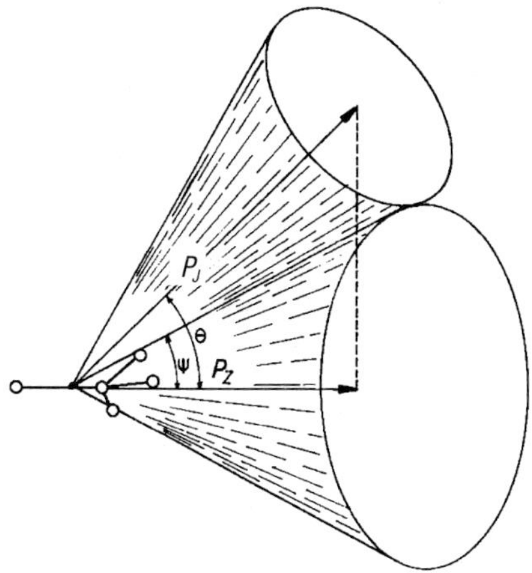

Symmetric top molecule. Fig. 2. Rotational motion of a symmetric top

$$F(J,K) = BJ(J+1) + (A-B)K^2.$$

K is the quantum number mentioned above (→ pure rotation spectrum of symmetric top molecules).
Ref.: G. Herzberg, *Molecular Spectra and Molecular Structure, II. Infrared and Raman Spectra of Polyatomic Molecules*, Van Nostrand Co., Princeton, New Jersey, New York, **1945**, chapt. I,2.

Symmetry, <*Symmetrie*>. A spatial form is generally described as symmetric when linear orthogonal coordinate transformations exist which convert the original conformation into another conformation which is indistinguishable from the original. Every such coordinate transformation which brings a spatial form into coincidence with itself is a → symmetry operation.

Symmetry element, <*Symmetrieelement*>, geometrical locus of the points which go into themselves under a → symmetry operation. Thus, the plane is the symmetry element of reflection, the axis that of rotation and the point (also called a center of inversion) that of inversion. In crystal lattices, the operation of translation must be added, but this has no corresponding symmetry element. The mirror plane, axis of rotation and inversion center are the elements of simple symmetry. In addition, there are the compound symmetry operations, i.e. rotation-reflection and rotation-inversion, corresponding to rotation-reflection and rotation-inversion axes. The geometrical representation of these symmetry elements should be a point. But since this is already the symbol for the inversion, it is customary to represent these elements of combined symmetry by means of the simple symmetry elements from which they are built up, i.e. axis of rotation and mirror plane or axis of rotation and inversion center. In crystal lattices, these combined symmetry elements must be augmented with the screw axis and the glide plane. The usual symbols for the symmetry elements and operations are summarized in the table. The symmetry operations are denoted by the operator, S.

Symmetry group, <*Symmetriegruppe*> → symmetry point group.

Symmetry operation, <*Symmetrieoperation, Deckoperation*>, a coordinate transformation which brings a spatial form into coincidence with itself. Examples of such operations (S) are, rotation by $2\pi/p = S(C_p)$, reflection in a plane $\sigma = S(\sigma)$, inversion in a center of symmetry $i = S(i)$ and rotation reflection $= S(S_p)$. For molecules, only those symmetry operations need be considered which leave at least one

Symmetry element. Table. Space and point group symmetry elements and symmetry operations

Spatial element	Symmetry element		Symmetry operation			
	–		Translation			
Plane	Mirror plane	σ	Reflection in the plane	$S(\sigma)$	*Simple* symmetry elements or operations	*Point* symmetry
Straight line	Axis of rotation	C_p	Rotation by $2\pi/p$	$S(C_p)$		
Point	Center of symmetry or inversion	i	Inversion in the center of inversion	$S(i)$		
	Rotation-reflection axis	S_p	Rotation-reflection	$S(S_p)$	Elements or operations of *combined* symmetry	
	Inversion axis		Rotation-inversion			
	Screw axis		Translation-rotation			
	Glide plane		Translation-reflection			

point in space fixed. They are called point symmetry operations. It is usual to characterize symmetry operations by means of → symmetry elements.

Symmetry point group, point group, symmetry group, <*Punktgruppe, Symmetriegruppe*>, a complete set of the → symmetry operations of a molecule. In general, a molecule can have several symmetry operations. The more symmetry operations it has the higher is the → symmetry of the molecule. For every molecule, there is at least one point which does not change its position in space, no matter how many symmetry operations are performed. Hence the term *symmetry point group*. In total, 45 complete sets of rotations, reflections and rotation reflections can be distinguished in molecules. Of these, 32 play a role in crystallography as the symmetry groups of the unit cells of the 32 crystal classes. Since parallel displacements are possible symmetry operations in a crystal (though no point remains fixed in space), the symmetry of crystals is described in terms of operations which are composed of combinations of parallel displacements and symmetry point group operations. These form the symmetry space groups of crystals and in this way the 32 crystal classes form the total of 230 possible space groups. The 45 possible point groups for molecules are given in the table together with the complete set of symmetry operations, Ng, in each case.

Symmetry species, <*Symmetrierassen*>, the irreducible representations of a → symmetry point group shown in the → character table; sometimes called → vibrational symmetry species in vibrational spectroscopy. The nomenclature goes back to Placzek (1931/34). The letters A, B, C, D, E and F(T) have the following meanings:

Symmetry point group. Table. The point groups.

Point group	Symmetry operations	N_g
C_s	$E,$	1
C_i	E, i	2
C_1	E	1
C_2	E, C_2	2
C_3	E, C_3, C_3^2	3
C_4	E, C_4, C_2, C_4^3	4
C_5	$E, C_5, C_5^2, C_5^3, C_5^4$	5
C_6	$E, C_6, C_3, C_7, C_3^2, C_6^5$	6
C_7	$E, C_7, C_7^2, C_7^3, C_7^4, C_7^5, C_7^6$	7
C_8	$E, C_8, C_4, C_8^3, C_2, C_8^5, C_4^3, C_8^7$	8
C_{2v}	$E, C_2, \sigma_v\,(xz), \sigma_v'\,(yz)$	4
C_{3v}	$E, 2C_3, 3\sigma_v$	6
C_{4v}	$E, 2C_4, C_2, 2\sigma_v, 2\sigma_\alpha$	8
C_{5v}	$E, 2C_5, 2C_5^2, 5\sigma_v$	10
C_{6v}	$E, 2C_6, 2C_3, C_2, 3\sigma_\alpha$	12
$C_{\infty v}$	$E, 2C_\infty, \infty\sigma_v$	∞
D_2	$E, C_2\,(z), C_2\,(y), C_2\,(x)$	
D_3	$E, 2C_3, 3C_2$	4
D_4	$E, 2C_4, C_2, 2C_2', 2C_2''$	6
D_5	$E, 2C_5, 2C_5^2, 5C_2$	8
D_6	$E, 2C_6, 2C_3, C_2, 3C_2', 3C_2''$	10
C_{2h}	E, C_2, i, σ_h	4
C_{3h}	$E, C_3, C_3^2, \sigma_h, S_3, S_3^5$	6
C_{4h}	$E, C_4, C_2, C_4^3, i, S_4^3, \sigma_h, S_4$	8
C_{5h}	$E, C_5, C_5^2, C_5^3, C_5^4, \sigma_h, S_5, S_5^7, S_5^3, S_5^9$	10
C_{6h}	$E, C_6, C_3, C_2, C_3^2, C_6^5, i, S_3^5, S_6^5, \sigma_h, S_6, S_3$	12
D_{2d}	$E, 2S_4, C_2, 2C_2', 2\sigma_d$	8
D_{3d}	$E, 2C_3, 3C_2, i, 2S_6, 3\sigma_d$	12
D_{4d}	$E, 2S_8, 2C_4, 2S_8^3, C_2, 4C_2', 4\sigma_d$	16
D_{5d}	$E, 2C_5, 2C_5^2, 5C_2, i, 2S_{10}^3, 2S_{10}, 5\sigma_d$	20
D_{6d}	$E, 2S_{12}, 2C_6, 2S_4, 2C_3, 2S_{12}^5, C_2, 6C_2', 6\sigma_d$	24
D_{2h}	$E, C_2\,(z), C_2\,(y), C_2\,(x), i, \sigma\,(xy), \sigma\,(xz), \sigma\,(yz)$	8
D_{3h}	$E, 2C_3, 3C_2, \sigma_h, 2S_3, 3\sigma_n$	12

Symmetry point group. Table. Continued.

Point group	Symmetry operations	N_g
D_{4h}	$E, 2C_4, C_2, 2C_2', 2C_2'', i, 2S_4, \sigma_h, 2\sigma_v, 2\sigma_d$	16
D_{5h}	$E, 2C_5, C_5^2, 5C_2, \sigma_h, 2S_5, C_5^3, 5\sigma_v$	20
D_{6h}	$E, 2C_6, 2C_3, C_2, 3C_2', i, 2S_3, 2S_6, \sigma_h, 3\sigma_d, 3\sigma_v$	24
$D_{\infty h}$	$E, 2C_\infty, \infty\sigma_v, i, 2S_\infty, \infty C_2$	∞
S_4	E, S_4, C_2, S_4^3	
S_6	$E, C_3, C_3^2, i, S_6^5, S_6$	6
S_8	$E, S_8, C_4, S_8^3, C_2, S_8^5, C_4^3, S_8^7$	8
T_d	$E, 8C_3, 3C_2, 6S_4, 6\sigma_d$	24
T	$E, 4C_3, 4C_3^2, 3C_2$	12
O_h	$E, 8C_3, 6C_2, 6C_4, 3C'_2, i, 6S_4, 8S_6, 3\sigma_h, 6\sigma_d$	48
O	$E, 8C_3, 6C_2, 6C_4, 3C_2'$	24
K_h	$E, \infty C_\infty, \infty S_\infty$	∞

A indicates symmetry with respect to rotation about a designated axis, usually that of highest order. In point groups with no unique axis, e.g. D_2, $D_{2h} = V_h$, and in the cubic groups T, T_d, O, T_h, and O_h, A means symmetry with respect the three mutually perpendicular twofold axes. For the point group C_s, A is merely a basis for the indices ′ and ″ (A′ and A″).

B indicates antisymmetry with respect to rotation about a designated axis; in the case of D_2 and D_{2h}, antisymmetry with respect to two of the twofold axes.

E indicates twofold degeneracy.

F(T) indicates threefold degeneracy, both symbols are in common use.

G indicates fourfold degeneracy. In the character tables, the number in the column for the identity operation (E) is equal to the degree of degeneracy.

g, u are additional right subscripts indicating symmetry with respect to inversion (i) in the center of symmetry. (g = gerade = even; u = ungerade = odd).

′ and ″ serve to indicate the symmetry with respect to reflection in a plane of symmetry, σ. If several planes of symmetry are present, then it is the plane perpendicular to the designated axis.

+ and − refer to $C_2(z)$. This distinction is only required in the case of degeneracy.

1, 2 ... are used as right subscripts to characterize further symmetry properties. The subscript 1 always indicates symmetry with respect to the remaining → symmetry operations.

Example: Naphthalene
Point group: $D_{2h} = V_h$ → character table.
Symmetry species:
A_g, sym. w.r.t. $C_2(z)$, $C_2(y)$ and $C_2(x)$, also w.r.t. i;

B_{1g}, sym. w.r.t. $C_2(z)$ and i, but antisym. w.r.t. $C_2(y)$ and $C_2(x)$;
B_{2g}, sym. w.r.t. $C_2(y)$ and i, but antisym. w.r.t. $C_2(z)$ and $C_2(x)$;
B_{3g}, sym. w.r.t. $C_2(x)$ and i, but antisym. w.r.t. $C_2(z)$ and $C_2(y)$;
B_{1u}, B_{2u}, B_{3u}, sym. w.r.t. the three mutually perpendicular C_2 axes, as above, but antisym. w.r.t. inversion in the center of symmetry, i.

Synchronous fluorescence excitation spectroscopy, <*synchrone Fluoreszenzanregungsspektroskopie*>. A further development of → selective fluorescence excitation spectroscopy in which both the excitation wavelength, λ_e, and the observation wavelength, λ_o, are synchronously changed while keeping a constant difference, $\Delta\lambda$, between the two. The basic idea is the following. The fluorescence intensity of a mixture of different components is a function of both the excitation and the observation wavelengths. In normal fluorescence spectroscopy for one component, λ_e is kept constant and the fluorescence spectrum is recorded as a function of λ_o. However, if the solution contains several components λ_e cannot correspond to the absorbance maximum of the individual components and this has a strong influence upon their contribution to the total fluorescence spectrum. Synchronous fluorescence excitation spectroscopy links the variation of the excitation and observation wavelengths which has the advantage that the absorption maximum of every component is automatically chosen as excitation wavelength. When we remember that the fluorescence maximum is shifted some 10–20 nm to the red of the absorption maximum (→ bathochromic shift), then with $\Delta\lambda = 20$ nm, the absorption maximum of a particular compound can be correlated with its fluorescence maximum. This method can be quite effective when used with aromatic hydrocarbons which show structured fluorescence spectra at room temperature and for which, in dilute solutions, the 0–0 transition is the most intense in absorption and emission.

Ref.: J.B.F. Lloyd, *J. Forensic Sci. Soc.*, **1971**, *11*, 83, 153, 235; P. John, I. Soutar, *Anal. Chem.*, **1976**, *48*, 520.

T

Tandem cuvette, <*Tandemküvette*>, a special version of a → cuvette in which a normal rectangular cuvette is divided by a partition into two cuvettes lying one directly behind the other as shown in Figure 1. Depending upon the thickness of the partition, each of the two cuvettes has a path length of ca. 4.25 mm in a cuvette of conventional external dimensions. Tandem cuvettes are of particular importance in kinetic and enzymatic studies since the reactants can be placed in the two cuvette compartments before reaction, which permits the measurement of a spectrum at the time $t = 0$. The reaction is started by inverting the cuvette several times to mix the solutions. The deadtime is 10–15 s.

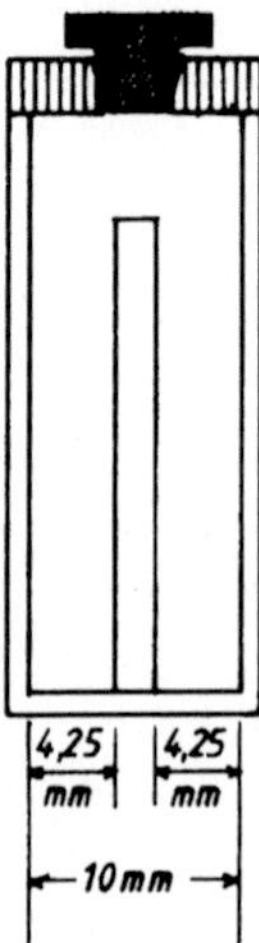

Tandem cuvette. Fig. 1. Cross section of a tandem cuvette

Target, <*Target*>, a solid sample in solid-state spectroscopy and solid-state physics, e.g. the surface of a solid.

Tenth-width → one tenth-width.

Term, <*Term*>, defined, using the → Bohr-Einstein relationship, as $T_i = E_i/hc$ (E = energy, h = Planck's constant and c = velocity of light). The units of term value are therefore cm^{-1} or m^{-1}. By definition, a term is always a particular, discrete energy state of an atom or molecule and, according to the type of energy involved, we speak of atomic, electronic, vibrational, and rotational terms. Although, in principle, the word might also be applied to energy states in other branches of spectroscopy, e.g. NMR and ESR, this is not the case in practice. In fact, the expression *energy level* appears largely to have displaced the word *term* in the English-language scientific literature. This may have something to do with the other uses of the word in English. However, it retains an important use in the expression *term symbol*; see below.

In the empirical representation of → atomic spectra, before the advent of quantum mechanics, a term was defined with reference to the → Balmer formula for one-electron atoms, as RZ^2/n^2. R is the → Rydberg constant, Z the nuclear charge (atomic number) and n is a running index which can take the integer values, n = 1, 2, 3 … ∞. Since the units of the Rydberg constant are cm^{-1} and Z and n are dimensionless, the term so defined has the units cm^{-1}. For multi-electron atoms, n is not an integer; → atomic spectrum.

The use of the word *term* in this context arose from the empirical observation that the wavenumber of any atomic spectral line could be expressed as the difference between two such terms; using the word in its algebraic sense.

An exact notation has been defined for the characterization of terms by means of appropriate symbols. For atoms the term symbol is $^{2S+1}L_J$. $2S + 1$ is the → multiplicity where S is the total electron spin angular momentum. L is the total orbital angular momentum and J the total angular momentum of the electrons; in units of $h/2\pi$ in all cases. (→ Multiplet structure, → electronic angular momentum). In writing the term symbol for any specific case, $(2S + 1)$ and J are written as numbers, but L is indicated by an upper case letter according to the rule; $L = 0, 1, 2, 3 \ldots \rightarrow S, P, D, F, G, H, I \ldots$, e.g. $^2P_{1/2}$, $^1F_3$.

For diatomic molecules, the only component of the electronic orbital angular momentum which is a constant of the electronic motion is that which lies along the internuclear axis. The magnitude of this component, Λ, in units of $h/2\pi$, is denoted by the Greek letters, Σ, Π, Δ … for values of 0, 1, 2 … in analogy with the Roman S, P, D used to denote electronic orbital angular momentum in atomic term symbols. The multiplicity is given, as before, as a left superscript. Other characterizing symbols are given as right superscripts and subscripts.

For larger molecules, especially unsaturated organic molecules, terms (energy levels) in the singlet system are denoted with S_o, S_1, S_2 … S_p and in the triplet system with T_1, T_2, T_3 … T_q. The subscripts give only the sequence of the terms, in order of increasing energy. The terms of such molecules can be characterized more precisely by their → symmetry species derived by means of group theory. Transition metal complexes can be handled in a similar way, → ligand field theory.

Ref.: G. Herzberg, *Molecular Structure and Molecular Spectra, I. Spectra of Diatomic Molecules*, D. Van Nostrand Co. Inc., Princeton, New Jersey, New York, **1950**. (Term symbols for both atoms and diatomic molecules are described.)

Term diversity, <*Termmannigfaltigkeit*>. A one-electron atom shows several → term series in its spectrum. Each is distinguished by a particular value of the electronic orbital angular momentum, ℓ_1: $S, P, D, F \ldots$ terms for $\ell_1 = 0, 1, 2, 3 \ldots$ respectively. For atoms with several outer electrons, it is the total resultant orbital angular momentum, L, rather than the individual ℓ_i values, which determine the types of terms involved. L is the vector sum of the individual ℓ_i values (→ electronic angular momentum). For two electrons, ℓ_1 and ℓ_2, the L values are:

$$L = \ell_1 + \ell_2,\ \ell_1 + \ell_2 - 1,\ \ldots, |\ell_1 - \ell_2| + 1, |\ell_1 - \ell_2|\,.$$

If the coupling is sufficiently strong, which is almost always the case, the energies of the states having different values of L are not the same, i.e. they also correspond to a variety of terms. Thus, the diversity of the terms is determined by the many ways in which the orbital angular momenta of individual electrons can be coupled. In the table, the individual ℓ_i values, the

Term diversity. Table.

Electron configuration	ℓ_1 ℓ_2 ℓ_3	L	Term symbol
sp	0 1	1	P
p^2	1 1	0 1 2	S P D
pd	1 2	1 2 3	P D F
d^2	2 2	0 1 2 3 4	S P D F G
p^3	1 1 1	0 1 1 1 2 2 3	S P P P D D F

resultant L values and the corresponding term symbols (but without the multiplicity) of a number of two- and three-electron configurations are given. If all ℓ_i values apart from one are equal to zero, then, naturally, the L value is equal to the nonzero ℓ_i. Then, this single ℓ_i retains its physical meaning as an angular momentum. This is the case for most of the (normal) terms of helium and the alkaline earths, for example. The form of the terms, S, P, D, F ... then depends only upon the ℓ value of this single → optical electron, as in the alkali metal atoms.

Term scheme → energy-level diagram.

Term series, <*Termfolge*>, in atomic spectra, the totality of all the terms of differing energies which are assigned to a definite orbital angular momentum, ℓ_i for one electron or for many electrons, the resultant, L (→ electronic angular momentum); S series, P series, D series etc. A diagram with the term series next to each other is known as a term or energy-level scheme. In the case of larger molecules, in particular organic molecules, the singlet states themselves form a term series, and so also do the triplet states. Each → normal vibration, v_i of the $3N$-6 normal vibrations of a molecule forms a term series the sequence of which is numbered by the vibrational quantum number, v_i.

This applies in an analogous way for the rotational spectra of linear molecules. The rotational quantum number, J, is then decisive. The term series are more complicated in the case of → top molecules. For → symmetric tops there is a series of terms with increasing rotational quantum number, J, for every value of the quantum number K.

Term symbol, <*Term-Symbol*> → term.

Ternary combination, <*ternäre Kombination*> → combination vibrations.

Thallium bromide iodide crystal, <*Thalliumbromidiodidkristall*> → KRS-5.

Thermal blooming, <*Thermal-Blooming-Verfahren*>, a special application of → thermal lensing spectroscopy which is used to determine → fluorescence quantum yields. A laser beam is focused onto the sample. If the light is absorbed by the sample then, shortly after the opening of a photoshutter, a temperature gradient develops between the point of absorption and the remainder of the sample because of the radiationless deactivation. This temperature gradient causes a change in the refractive index which broadens the laser beam rather like a diverging lens (thermal lensing). This blooming of the laser beam reduces its intensity at its center which is proportional to the heat produced and can be detected with a photodiode. The fluorescence quantum yield can be determined by comparing the intensity loss for the fluorescing sample with that of an equally strongly absorbing, but not fluorescing, standard sample. The technique has been used to determine the fluorescence quantum yield of various xanthene, oxazine and carbazine dyes in solution. For strongly fluorescent samples ($q_{FM} > 0.80$) an error of $< 0.5\%$ has been quoted, which rises to 5 to 15% for more weakly fluorescent samples (q_{FM}: 0.30 to 0.10).

The advantage of this method, which depends upon the measurement of radiationless deactivation processes, lies in the fact that the determination of the fluorescence quantum yield can be made relative to a nonfluorescent comparison sample, as in → photoacoustic spectroscopy. Errors arising from an inexact knowledge of the fluorescence quantum yield of a standard substance, as in conventional methods, are therefore avoided.

Ref.: J.H. Brannon and D. Madge, *J. Phys. Chem.*, **1978**, *82*, 705; R. Sens, *Dissertation*, Universität-GH Siegen, **1984**.

Thermal ionization, <*Thermoionisation, Thermoionisierung, thermische Ionisation, thermische Ionisierung*>, a technique used in → mass spectrometry, predominantly in the analysis of inorganic solids and especially metals. It is a thermal surface ionization. A thermal ion source consists of a heatable metallic strip from which the sample is evaporated; the sample filament. In the immediate neighborhood, the atoms or molecules strike one or two more very high-temperature ionization filaments where they are ionized or dissociated and ionized.

The degree of ionization, α, is given the Saha-Langmuir equation:

$$\alpha = \frac{n^+}{n_o} = \frac{g^+}{g_o} \exp\left\{\frac{e(W-I)}{k \cdot T}\right\},$$

where

- n^+, n_o are the number of ions and neutral atoms respectively, emitted from the surface per unit time and area,
- eW is the work function of the surface,
- eI is the ionization energy,
- k is the Boltzmann constant,
- T is the absolute temperature,
- g^+, g_o are the statistical weights of the electronic ground states of the ion and atom respectively.

Thermal ionization sources are suitable for elements with low ionization energies such as the alkali metals, the alkaline-earth metals, elements of groups, IIIA and IIIB, the lanthanides and the actinides.

Thermal lens, <*thermische Linse*> → thermal lensing spectroscopy.

Thermal lensing spectroscopy, <*Thermal-Lensing-Spektroskopie*>, a spectroscopic method which is based upon the thermal lensing effect first described by J. Gordon and coworkers in 1965. The thermal lens is formed when a cw laser beam is passed through a cuvette filled with a weakly absorbing liquid. The reason for the development of the thermal lens is the change in the refractive index, n_o, of the liquid which is caused by a change of temperature. This, in turn, is due to the conversion of the absorbed light intensity, I_{abs}, into heat by radiationless deactivation of the primarily excited solvent molecule, i.e. a process analogous to → photoacoustic spectroscopy. Since most solvents have a positive temperature coefficient of expansion, the temperature coefficient for the refractive index, dn/dT, is negative and the thermal lens is diverging. Thus, the laser beam is broadened, and most of the experimental equipment designed to measure this effect is based upon this broadening.

Account must also be taken of the fact that the thermal lens takes a finite time to develop and reach a stationary state. According to Gordon among others, $n(r,t)$ is given by:

$$n(r,t) = n_o + (\frac{dn}{dT})\Delta T(r,t).$$

$n(r,t)$ is the refractive index at a distance, r, from the center of the laser beam at time, t, after the start of the laser heating. Similarly, $T(r,t)$ describes the spatial and temporal development of the temperature profile. The detailed theoretical treatment gives the following expression for the time dependence of the focal length, F, of the thermal lens:

$$F(t) = F_\infty(1 + \frac{t_c}{2t})$$

where:

$$F_\infty = \frac{\pi n_o \kappa \omega_o^2}{0{,}24 I_o \cdot \beta \cdot l(dn/dT)}.$$

Here, n_o is the refractive index of the sample, κ the thermal diffusivity of the sample in W cm^{-1} K^{-1} and ω_o the radius of the laser beam in cm. For small values of $\beta \cdot l$, i.e. $\beta \cdot l \ll 1$, $I_{abs} = I_o \cdot \beta \cdot l$ is the light intensity, in W, absorbed by the sample, β is the absorption coefficient in cm^{-1} and l the length of the cuvette in cm. dn/dT is the temperature coefficient of the refractive index. t_c represents a characteristic time defined by $t_c = \omega_o^2 \varrho c_p / 4\kappa$, where ϱ is the density in g cm^{-3} and c_p the specific heat in J g^{-1} K^{-1}.

Figure 1 shows the change of $F(T)$ with time. We note that the time required to reach the stationary state is some 6–8 times the characteristic time, t_c. F_∞ is the focal length in the stationary state, and if this is known then the absorption coefficient, β, can be determined. For example, with F_∞ = 100 cm, I_o = 0.8 W, ω_o = 0.05 cm, dn/dT = -10^{-3}, n_o = 1.5, l = 1 cm, and $\kappa = 4 \cdot 10^{-4}$ a value for β of $2.5 \cdot 10^{-4}$ cm^{-4} is found. Very small absorption coefficients (10^{-5} – 10^{-4} cm^{-1}) can be measured with this technique. But the focal lengths of the thermal lenses are very long which has consequences for the construction of the apparatus.

The original design of Gordon *et al.* was refined by C. Hu and J. Whimery and today provides the basis for the construction of thermooptical spectrometers. Figure 2 shows, schematically, the light path in a single-beam instrument used by R. Swofford, M. Long and A. Albrecht. A rhodamine-6G → dye laser is pumped with an 8 W argon ion laser (→ noble-gas ion laser). The

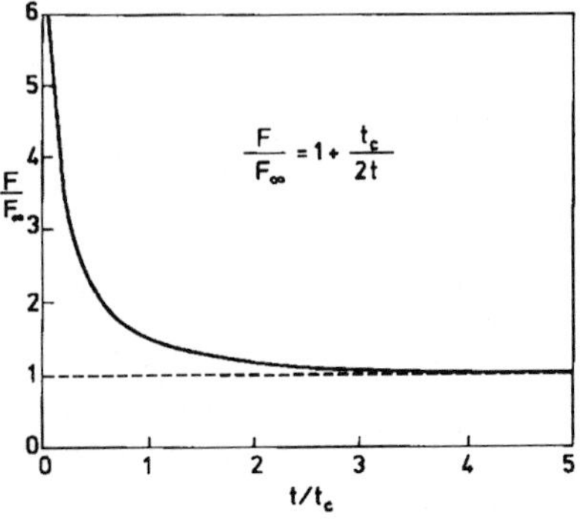

Thermal lensing spectroscopy. Fig. 1. The time dependence of $F(T)$, see text

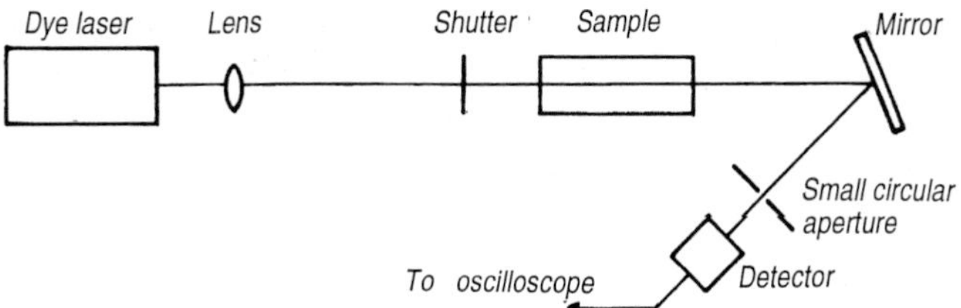

Thermal lensing spectroscopy. Fig. 2. Simplified beam path of a thermooptical spectrometer

sample (cuvette: l = 6.8 cm) is placed confocally in the laser beam behind the focal point of a lens with a long focal length. The broadening of the laser beam is observed on a target which has a small circular aperture ($\phi \sim 0.5$ mm) in the center. Immediately behind this there is a → photomultiplier as detector. The temporal development of the thermal lens is followed via the intensity of the laser beam with the help of the aperture and detector. If I_o is the intensity of the beam immediately after opening the shutter and I_s the intensity when the stationary state has been reached then, as Hu and Whimery have shown, the absorption coefficient, β, in cm^{-1}, is given by the following equation:

$$\beta = \frac{I_o - I_s}{I_s} \cdot \frac{4,18\lambda\kappa}{I_o \cdot l(\mathrm{d}n/\mathrm{d}T)}$$

Figure 3 shows a spectrum of benzene in the range 17,400–15,800 cm^{-1} (574.7–632.9 nm) measured with this equipment. The circles are the data for benzene-D_6 which clearly shows no absorption in this region. This is interpreted as confirmation that it is a high

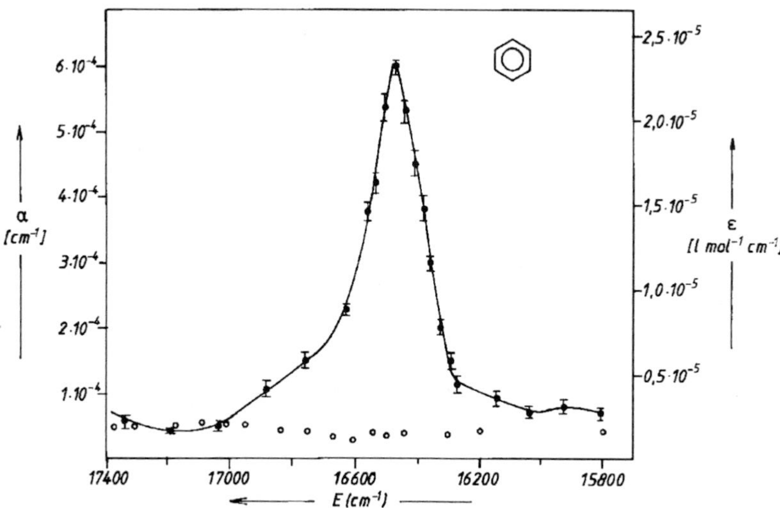

Thermal lensing spectroscopy. Fig. 3. Thermal lensing spectrum of benzene and benzene-D_6, see text

→ overtone of a benzene fundamental vibration in the IR region ($\Delta v = 6$ in the case of benzene) which is being measured. Swofford *et al.* have also described a double-beam instrument in addition to the single-beam instrument shown here.
A further interesting application is the use of thermal lensing to determine fluorescence quantum yields (→ thermal blooming). Laser-beam initiated exothermic photochemical reactions, which enhance the thermal lens effect, can also be followed using thermal lensing spectroscopy.
Ref.: J.P. Gordon, *et al.*, *J. Appl. Phys.*, **1965**, *3*, 36; C. Hu, J.R. Whimery, *J. Appl. Phys.*, **1965**, *12*, 72; R.L. Swofford, M.E. Long, A.C. Albrecht, *J. Chem. Phys.*, **1976**, *65*, 179; H.L. Fang, R.L. Swofford in *Advances in Laser Spectroscopy*, Eds.: B.A. Garetz, J.R. Lombardi, vol. I, Heyden and Sons Inc., **1982**, chapt. 1.

Thermal radiation detector, <*thermischer Empfänger*>, a detector which, unlike the photoelectric detectors such as → photocells, → photomultipliers and → photodiodes, is not selective because it converts the incident radiation, independent of its wavelength, into heat. The asborbed radiation, which is converted as completely as possible into heat, produces a temperature difference, ΔT, between the detector element and the surroundings which then induces the actual effect which is measured. Thermal detectors are classified according to the nature of this measurement, e.g. → thermocouple, → bolometer, → pneumatic detector (→ Golay detector), → pyroelectric detectors. These devices are used almost exclusively in the mid infrared where no other sensitive detectors are available. Photoelectric detectors, such as photodiodes, lead sulfide detectors etc., are available only for the NIR region, as far as 5 or 6 μm.

Thermal radiator, <*Temperaturstrahler*>, the most commonly used continuous → light source. For spectroscopic applications, the spectral emissive power or radiation density is of great importance. Here, one may first consider the properties of a → black-body radiator which can be described by → Planck's radiation law which gives the spectral radiation density, S_λ, as:

$$S_\lambda = \frac{c^2 h}{\lambda^5 (e^{hc/\lambda k T_s} - 1)} \mid Wm^{-3} \mid;$$

where λ is the wavelength in m, h → Planck's constant, c the velocity of light and T_s the temperature of the black body. Note that there are no properties specific to a material in this equation; S_λ depends only on the temperature of the body and the wavelength. As is customary, a plot of the function S_λ at constant temperature is called an isotherm. Figure 1 shows an isotherm for T_s = 3000 K. As far as spectroscopy is concerned, the most important fact about the isotherm is that the radiation density below 400 nm, i.e. in the UV region, is so small that the use of a black-body radiator as a light source for the UV is, in general, out of the question. The maximum of the isotherm lies in the NIR and the thermal radiator is therefore the radiation source of choice for the VIS, NIR and IR spectral regions.
In practical applications of Planck's law, the fact that thermal radiators, usually glowing solids, are not ideal → black-body radiators must be taken

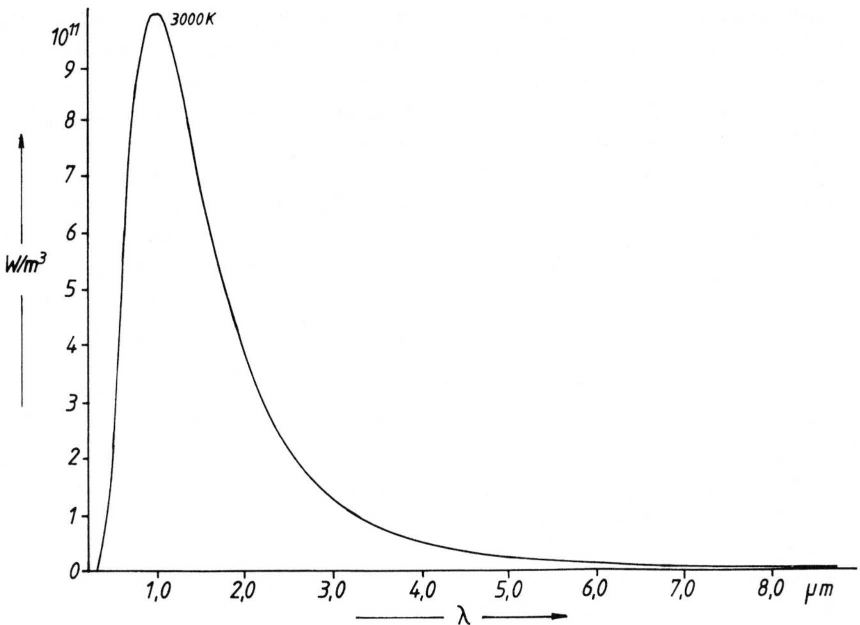

Thermal radiator. Fig. 1. Radiation isotherm for T_s = 3000 K

into consideration; at all wavelengths they emit less energy than a black body at the same temperature. Thus, the emissive power is less than that of a black body and is now dependent upon the nature of the material, as well as upon temperature and wavelength. The radiation isotherm of a nonblack body can be given in terms of Planck's formula if the latter is multiplied by the function $\varepsilon(\lambda,T)$ which takes account of the emissive power of the material. In addition, the concepts of → black-body temperature, T_b, and color temperature, T_c, are used to characterize the radiative properties of a thermal radiator. These two temperatures should be distinguished from the true temperature, T_t. If, for a particular wavelength, a radiator has the same radiation density as a black body at temperture, T_b, then the radiator has a black-body temperature of T_b. For tungsten (→ tungsten lamp) the black-body temperature, T_b, is some 200–400° lower than the true temperature, T_t, because the emissive power of tungsten is significantly less than unity. The color temperature, T_c, is the temperature which a black-body radiator should have to give the same impression of color to the eye as the nonblack radiator. In general, color temperature says nothing about the energy distribution of the emission but is related to the integral visual process in the VIS region.

Ref.: M.A. Bramson, *Infrared Radiation*, Plenum Press, New York, **1968**.

Thermochromism, <*Thermochromie*>, the phenomenon whereby solutions of specific compounds become intensely colored as the temperature is raised. In compounds which have been extensively studied, e.g. bianthrones, bixanthylenes and spiropyrans, the intensity of a band in the visible region grows strongly with rising temperature, while bands in the UV region remain unchanged. The phenomenon is reversible. It cannot be due to a dissociation of the molecules because, at a fixed temperature, the intensity of the bands depends upon concentration in accordance with the → Bouguer-Lambert-Beer law. The cause is actually a true equilibrium between a normal form, *A*, and a thermochromic form, *B*, which, in the case of bianthrone, differ in that in the form *B* the molecule is essentially planar with a small angle of twist between the two halves of the molecule, while in the form *A* the two halves are twisted out of the plane. This is also the case for the bixanthylenes. A reversible ring opening is responsible for the thermochromism of the spiropyrans.

Ref.: W.T. Grubb, G.B. Kistiakowsky, *J. Amer. Chem. Soc.*, **1950**, *72*, 419; G. Kortüm, W. Theilacker, V. Braun, *Z. Physikal. Chem.*, **1954**, *2*, 179; W. Theilacker, G. Kortüm, G. Friedhelm, *Chem. Ber.*, **1950**, *83*, 508.

Thermocouple, <*Thermoelement*>, a → thermal radiation detector in which the radiation to be measured is first absorbed and converted quantitatively into heat. This heat is then used to generate an effect which can be measured; in the case of a thermocouple, the effect is a thermoelectric EMF.

For radiation measurements, the decisive temperature difference between the two junctions is produced by irradiating the *hot* junction which is warmed by the conversion of the absorbed radiation into heat, while the *cold* junction is not irradiated.

The simplest form of thermocouple consists of two wires of different metals, M_A and M_B, soldered or welded together at *L*, Figure 1a. *L* is covered with the detector surface, *F*. The free ends of the wires are connected to a sensitive galvanometer. The temperature difference, ΔT, between the hot junction, *L*, and the cold junctions, *l*, is usually very small, and so also is the EMF. To increase the effect, several thermocouples are wired together to form a thermopile, as shown in Figure 1b. If all the junctions of the thermopile can be brought to the same temperature difference as in the case of the single thermocouple, then a multiplication of the EMF by the number of single elements is obtained. Commercially available thermopiles contain up to 50 thermocouples, though they are not always arranged in the simple manner illus-

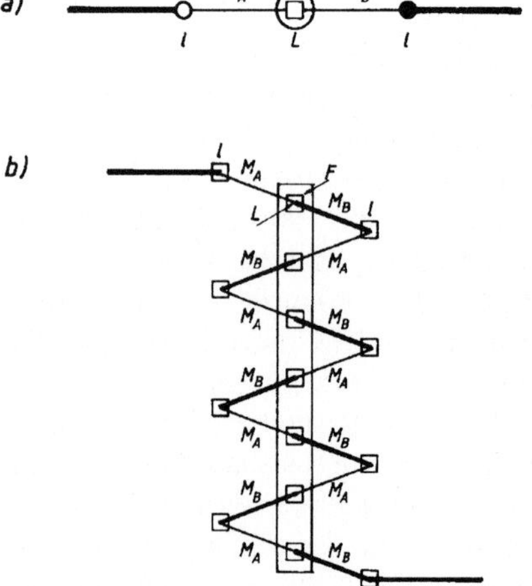

Thermocouple. Fig. 1. A thermocouple (1a) and a thermopile (1b)

trated in Figure 1b. Accordingly, thermopiles are characterized by their mode of construction as wire, band, plug and layer devices.
The following data are important in judging the quality of a thermocouple. The sensitivity, S_o, the smallest detectable radiation power, in Watt, and the → time constant, τ, in seconds. The sensitivity, S_o, is usually defined as the thermoelectric EMF, E_{th}, in volts divided by the incident radiation in Watt:

$$S_o\ [\text{Volt/Watt}] = E_{th}/\Phi A$$

where Φ in W cm^{-1} is the radiation density and A is the area of the detector in cm^2. In ordinary thermocouples, A is a few square millimeters. S_o values lie in the range from tenths to ca. 50 V/W according to the materials used. The smallest detectable radiation power, i.e. that which can just be seen above the noise level, is ca. $5 \cdot 10^{-11}$ W for a thermocouple with a surface area of 1 mm^2. This value naturally depends upon the properties of the metals, alloys and semiconductors used in the construction. The alloys, 97% Bi + 3% Sb and 95% Bi + 5% Sn have proved to be particularly suitable. The time constant is given by $\tau = C\ (dT/d\Phi)$, where $dT/d\Phi$ is the temperature change per Watt of incident radiation and C is the heat capacity. To reduce C, the detector surface and the wires must be made as thin as possible. But note that a high sensitivity, i.e. large $dT/d\Phi$, implies a long time constant and rapid response can therefore only be obtained at the cost of sensitivity. Most thermocouples have time constants $\tau < 0.13$ s.
Thermocouples and thermopiles are the preferred radiation detectors for → IR spectrometers, but they are now being challenged by → pyroelectric detectors.
Ref.: *The Infrared Handbook* (Eds.: W.L. Wolfe and G.J. Zissis), Office of Naval Research, Department of the Navy, Washington, DC, **1978**, sections 11–24 and 11–63.

Thin film interference colors, <*Farben dünner Blättchen*> → interference at thin films.

Thousandth-width → one thousandth-width.

Three-level system, <*Dreiniveausystem*> → ruby laser.

Time constant, <*Zeitkonstante*>, the time which passes, after the end of the process causing the signal, before the signal falls to exp[-1] times its orignal value, i.e. to 36.8% of it. The response time, on the other hand, is the time required after the stimulus until a signal near the theoretically possible result is obtained. A figure of 99% of the final value is usually chosen. There is also the half-life, which is the time for a 50% change in the signal. Its reciprocal is the half-frequency. All these concepts are interrelated so that one of them is sufficient to characterize a measuring device.

Time-of-flight mass spectrometer, TOF-MS, <*Flugzeitmassenspektrometer*>, a spectrometer in which the ions are drawn out of the electron-impact ion source and accelerated by a briefly applied electric field. The ions then traverse the field-free zone between A and B (see Figure 1) as an ion packet. The conditions are con-

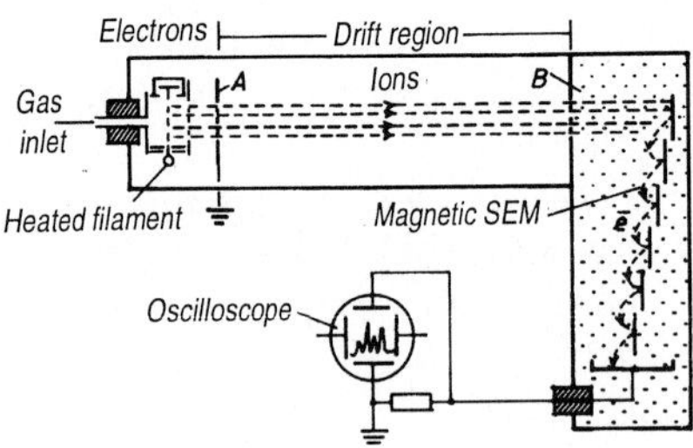

Time-of-flight mass spectrometer. Fig. 1. A section through a TOF mass spectrometer

trolled such that the ion packet is as precisely limited in space and as monoenergetic as possible. When they have the same energy, ions of different *m*/*e* values have different velocities in accord with the equation, (→ mass spectrometer):

$$v = \sqrt{\frac{2U}{m/\mathrm{e}}}$$

Thus, the light ions arrive first and the heavier ions later at the end of the separation region, $A - B$. Figure 1 illustrates the principle of a time-of-flight (TOF) mass spectrometer. The ions are detected by means of a special type of → secondary electron multiplier and, after broad-band amplification, the signal can be followed on an oscilloscope. The time interval within which the different ions arrive at the detector is only 10 to 100 μs. If the horizontal axis of the oscilloscope is synchronized with the ion pulses a stationary picture of the mass spectrum is obtained on the screen. In this manner, 10,000 mass spectra per second can be taken so that TOF mass spectrometry is particularly useful for the study of fast reactions in the gas phase.

Ref.: *Time-of-Flight Mass Spectrometry and its Applications*, Ed.: E.W. Schlag, Elsevier, Amsterdam, New York, **1994**.

TOF, TOF-MS, <*TOF*> → time-of-flight mass spectrometer.

Top molecule, top, <*Kreiselmolekül*>, a nonlinear polyatomic molecule. It is known from classical mechanics that the inertial behavior of a body can be described by defining the → moments of inertia about three specific, mutually perpendicular axes which pass through the center of gravity. These moments of inertia are called the principal moments of inertia. The rotational behavior of such a body is that of a top. The three moments of inertia of the top molecule with respect to the three principal axes are denoted with I_A, I_B and I_C. The subscript, A, always indicates the figure axis or the symmetry axis if it exists. The rotational constants are defined by:

$$A = \frac{h}{8\pi^2 I_A c};$$

$$B = \frac{h}{8\pi^2 I_B c};$$

$$C = \frac{h}{8\pi^2 I_C c}.$$

The moments of inertia are conventionally placed in the order, $I_A < I_B < I_C$, so that the corresponding → rotational constants are in the order $A > B > C$. Three classes of top molecules can be distinguished:

1. All moments of inertia are different: $I_A \neq I_B \neq I_C$. This is an asymmetric top.

2. Two moments of inertia are equal to each other but they differ from the third: $I_A \neq I_B = I_C$. This is a → symmetric top and we distinguish between the prolate symmetric top, $I_A < I_B = I_C$ ($A > B = C$) and the oblate symmetric top, $I_A > I_B = I_C$ ($A < B = C$).
3. All three moments of inertia are equal: $I_A = I_B = I_C$. This is a spherical top.

Total internal reflection, <*TIR*> → attenuated total reflection.

Total reflection, <*Totalreflexion*>. Consider the passage of a light beam from an optically dense medium (refractive index $= n_1$) into a less dense one (refractive index $= n_o$); from quartz or glass into air for example. From the law of refraction (→ refraction of light) we have:

$$\sin\beta/\sin\alpha = n_1/n_o$$

Since in this case we always have $\beta > \alpha$; for $\beta = 90°$ and $\sin\beta = 1$:

$$\sin\alpha_c = n_o/n_1$$

Therefore, for an angle of incidence, α_c, the light emerges parallel to the boundary between the two media. This angle is known as the critical angle. For angles of incidence, $\alpha > \alpha_c$, the light is totally reflected as illustrated in Figure 1. For $\alpha < \alpha_c$, the laws of specular → reflection hold. Thus, there is a polarization angle, α_p, in this case too, but it is always smaller than the critical angle, α_c.

As a result, in spite of the total reflection, there is radiation present in the optically less dense medium which arises because of refraction phenomena

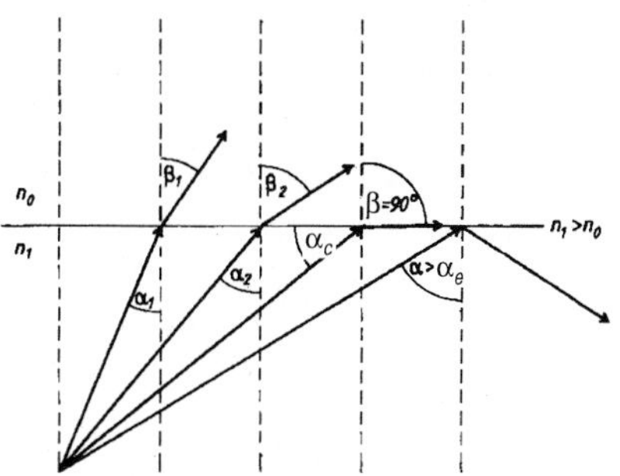

Total reflection. Fig. 1. Dependence of reflection upon the angle of incidence

at the edges of the incident light beam. Radiation energy enters the less dense medium as a transversely damped surface wave. This phenomenon is the basis of internal reflection spectroscopy and the → attenuated total reflection method.

Ref.: G. Kortüm, *Reflection Spectroscopy*, Springer Verlag, Berlin, Heidelberg, New York, **1969**, chapt. II.

Tourmaline plate, <*Turmalinplatte*>, a plate of tourmaline, a mineral which has a very pronounced dichroism. Tourmaline is a complex borate/silicate mixed crystal which may contain, for example, the elements aluminium, magnesium, iron, titanium or chromium in adddition to the required sodium and aluminium. Because of this multiplicity of compositions, the colors of tourmaline crystals vary very widely from colorless through blue, green, red, pink and brown to black. The pink and green crystals are suitable for optical applications. The green crystals are of particular interest because they absorb the ordinary ray, which is polarized perpendicularly to the optic axis, so strongly that a thickness of 1 mm transmits effectively only the extraordinary ray, which is polarized parallel to the optic axis. A 1 mm plate of tourmaline cut parallel to the principal crystallographic axis is therefore a usable polarizer. Although the light passing through it is green, this presents no problem in many applications. Two tourmaline plates which can be rotated relative to each other form a complete polarization apparatus which was very much used in earlier times.

Transition dipole moment, *<Dipolübergangsmoment, elektrisches Dipolübergangsmoment>*, gives the probability of a transition between two states l and k as a result of a photon resonance absorption obeying the → Bohr-Einstein condition; $E_k - E_l = hc\tilde{\nu}$. The transition dipole moment is defined by:

$$\vec{M}_{lk} = e \int \Psi_k \left[\sum_i z_i \vec{r}_i \right] \Psi_l \mathrm{d}r_i.$$

Ψ_k and Ψ_l are the eigenfunctions of the states involved, k and l. r_i is the position vector of the ith particle in the molecule (electron or nucleus) with charge $z_i e$. The transition dipole moment is a vector, i.e. it has both magnitude and direction. It can therefore be written in terms of its three spatial components:

$$|\vec{M}_{lk}| = (|\vec{M}_x|^2 + |\vec{M}_y|^2 + |\vec{M}_z|^2)^{1/2}.$$

If, for example, $\overrightarrow{M_z} = 0$, then M_{lk} lies in the xy plane. This is frequently the case with planar molecules such as the aromatic hydrocarbons. With these molecules it can also occur that for one transition $M_z = M_y = 0$ and for another, $M_z = M_x = 0$; i.e. the electronic transitions are polarized in either the x or the y direction. In such cases there is → anisotropy in the light absorption. The Einstein B coefficient (→ Einstein coefficient) is directly related to the transition dipole moment:

$$B_{lk} = \frac{8\pi^3 G}{3h^2} |\vec{M}_{lk}|^2.$$

$\overrightarrow{M_{lk}}$ is also related to the → oscillator strength, f_{lk}, which is quite easy to obtain experimentally:

$$f_{kl} = \frac{8\pi^2 mc\tilde{\nu}_{kl} G}{3he^2} |\vec{M}_{lk}|^2$$
$$= 4.70 \cdot 10^{29} \tilde{\nu}_{kl} G |\vec{M}_{kl}|^2.$$

The constants have been gathered together in one numerical factor. The statistical weight, G, is one for a pure singlet electronic transition.
If the molecular eigenfunctions, Ψ_k and Ψ_l, are known, $\overrightarrow{M_{lk}}$ can be calculated using the above equation. In order to write down the complete eigenfunctions with their superimposed vibrational states, use is made of the → Born-Oppenheimer approximation, i.e. the electronic and nuclear motions are separated. An average transition dipole moment, $\overline{M}_{lk}$, which represents the pure electronic transition, can then be defined and the transition moment for each vibronic transition, $lo \rightarrow kn$, can be decomposed into a product of two factors:

$$M_{lo \rightarrow kn} = \overline{M}_{lk} \cdot S_{lo,kn} \quad (n = 0, 1, 2, 3, ...).$$

$S_{lo,kn}$ is the overlap integral for the vibrational states, o and n, involved in the electronic transition. This analysis is important in connection with the → Franck-Condon principle.

Ref.: J.N. Murrell, *The Electronic Spectra of Organic Molecules*, J. Wiley and Sons, Inc., New York, Methuen, London **1963**, chapt. 1 and 2.

Transmittance, <*Durchlässigkeit, Transmissionsgrad*>, the fraction, $\Phi_T/\Phi_0 = \tau$, when light of intensity Φ_0 (radiation flux) enters a homogeneous, isotropic medium and, on leaving it, is attenuated to intensity Φ_T as a result of absorption processes. Considering only that light flux which has actually entered the medium, Φ_{in}, and the flux actually leaving, Φ_{ex}, then the pure, internal or true transmittance, τ_i, is the ratio Φ_{ex}/Φ_{in}. It is clear from Figure 1 that the true transmittance can only be obtained if the reflection losses on entry and exit of the light are eliminated. In the spectroscopy of solutions, this is usually done by making measurements on a cuvette of the same material which contains only the solvent. In this it is assumed that the dilute solution in the sample cuvette has the same refractive index as the solvent. The relationship between τ and τ_i is: $\tau = R \cdot \tau_i$, where R is the reflection factor which for perpendicularly incident light is:

$$R = 2n/(n^2 + 1)$$

where n is the refractive index of the medium at a wavelength $\lambda = 587.6$ nm. This relationship is of particular importance in measurements on solid samples, e.g. glass filters.

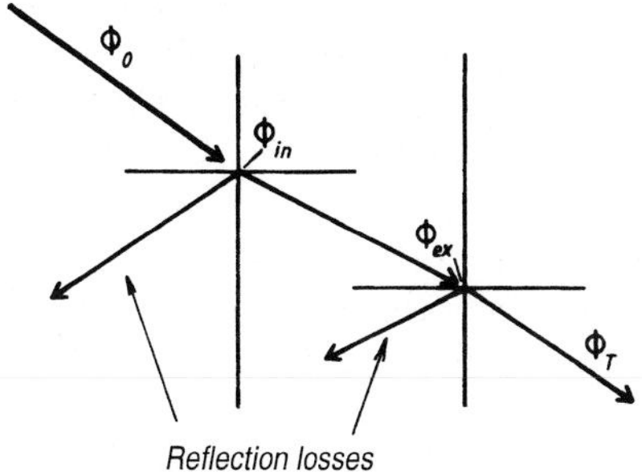

Transmittance. Fig. 1. For the definition of internal, pure or true transmittance

Tunable diode laser absorption spectrometer, TDLAS, <*TDLAS, abstimmbares Diodenlaserabsorptionsspektrometer*>, a modern variant of the IR absorption spectrometer. However, in contrast to conventional IR spectrometers, the TDLAS is

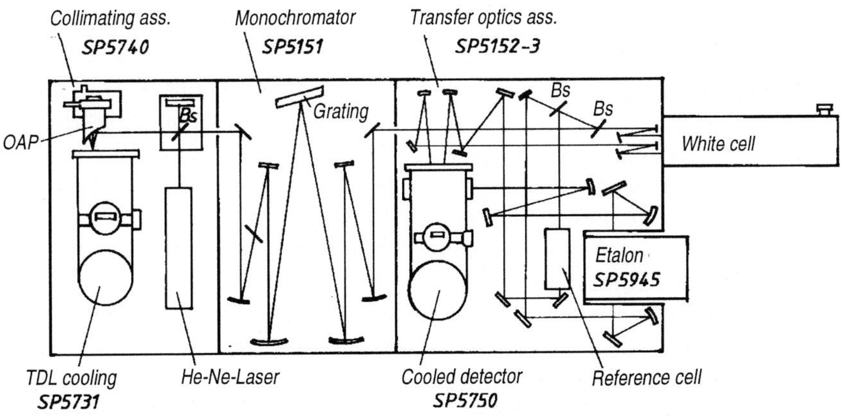

TDLAS. Fig. 1. Optical layout of the Spectra Physics TDLAS

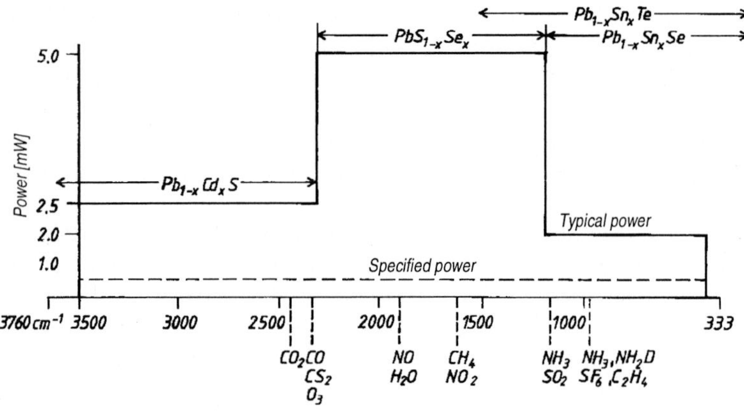

TDLAS. Fig. 2. Tunable regions for various laser diodes

designed for high-resolution measurements on gases and for analytical purposes. Figure 1 shows the optical layout of the instrument according to information provided by Spectra Physics. It consists essentially of three or four modules: the laser source with collimator and alignment aids (He-Ne laser), a long path length absorption cell and the detector unit with reference cell and tunable → etalon as frequency standard. There is usually also a grating monochromator to select the modes of the diode laser radiation. The monochromator is graduated so as to give an approximate value of the wavelength. The active media of the diode laser are lead chalcogenides and their exact chemical composition allows a crude choice of the wavelength region. Figure 2 gives an overview of the compositions of tunable laser diodes for the region 3500 to 333 cm^{-1}. Analytically relevant gases whose absorption maxima fall in this region are also included.

Tungsten lamp, tungsten filament lamp, <*Wolframlampe, Wolframglühlampe*>. Tungsten is a suitable material for a → thermal radiator because it has a very high melting point; 3655 K. The common tungsten lamp consists of a thin tungsten spiral mounted in an evacuated or gas-filled glass envelope. Many forms and models are available, depending upon the intended application. They can be operated at the supply voltage or as low-voltage lamps. For most applications a knowledge of the → spectral energy distribution of the lamp is not necessary.

To obtain the maximum yield of radiation the filament must be operated at as high a temperature as possible. In this, the limiting factor is not only the melting point of tungsten, but also its vapor pressure because the evaporated metal condenses on the cooler glass envelope and reduces its transparency. The evaporation rate is much reduced if the lamp is filled with an inert gas such as argon or krypton. Since the glass envelope absorbs some of the light, especially in the IR and UV regions, tungsten lamps from which the UV radiation is required are made with quartz envelopes.

For scientific purposes, tungsten lamps (e.g. Osram Wi15) are available in which the filament is a thin tungsten band and for which the manufacturers provide a specific black-body temperature, T_b, (→ thermal radiator), obtainable under exactly specified operating conditions, e.g. voltage and current. For calculation of the → spectral energy distribution, the true temperature, T_t, must be known and the emissive power of tungsten allowed for. The latter has been determined as a function of wavelength by J.C. de Vos (*Physica*, **1954**, *20*, 690) for the temperature range 1600–2800 K. (See also → thermal radiator).

The tungsten band lamp is used for the calibration of monochromators and radiation detectors. However, it can only be used for the UV spectrum above 300 nm since the spectral energy is insufficient for shorter wavelengths. But it is indispensible for the calibration of fluorescence spectra. The → halogen lamp is a new development.

Ref.: H.-H. Perkampus, K. Kortüm, H. Bruns, *Appl. Spectroscopy*, **1969**, *23*, 105.

Tungsten point lamp, <*Wolframpunktlichtlampe*>, a lamp in which an arc is struck between two small tungsten spheres in a noble gas or nitrogen atmosphere. In a conventional → tungsten lamp the metal filament is heated directly by an electric current. In the tungsten point lamp it is not the emission of the arc which is used but rather the radiation from the tungsten electrodes which are raised to high temperatures by the impact of the electrons. The lamp is characterized by a high and uniform light intensity. In UV technology, the principle of the point lamp is applied in combination lamps where, in addition to the thermal radiation of the hot electrodes, the radiation of the arc which then goes over to a mercury-vapor discharge is used. In these lamps the Hg line spectrum is superimposed on the continuum of the thermal radiator.

Tuning wedge, <*Keil, durchstimmbarer Keil, Tuning-Wedge*>, a device for variation of wavelength used in laser spectroscopy to tune laser wavelengths. An annealed reflecting layer, S_1, is deposited on a glass substrate, followed by a wedge-shaped, optically transparent → dielectric layer and finally a second annealed mirror, S_2. For a particular distance, d, between the two mirrors, the wavelength of maximum transmission in the kth order, λ_k, is, analogously to an → etalon:

$$\lambda_k = \frac{2d}{k}\sqrt{n^2 - \sin^2\alpha} = \frac{2nd}{k}\cos\beta$$

where α = the angle of incidence, n = the refractive index of the wedge-shaped layer and β = the angle of refraction.

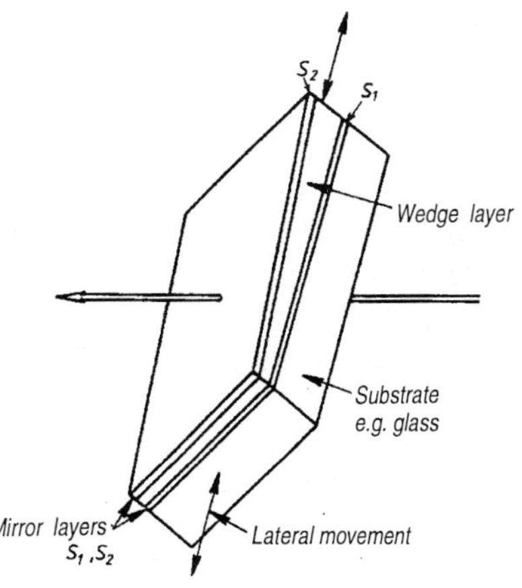

Tuning wedge. Fig. 1. Construction of a tuning wedge

In the case of an etalon, the thickness, d, is constant and the wavelength is tuned by varying the angle of incidence, α. In the tuning wedge, the variable thickness is used to tune the wavelength at a constant angle of incidence, i.e. the unit is moved in a direction perpendicular to the incident beam. Its effect and construction (see Figure 1) correspond to those of a → graded filter. Such a device has a large tuning range from ca. 460 to 900 nm with a line width of 240 GHz, which corresponds to approximately 8 cm^{-1}.

Tunnel effect, <*Tunneleffekt*>, a quantum mechanical effect which explains how electrons or other particles can penetrate or *tunnel* through potential barriers which cannot be overcome classically. In quantum mechanics, the position of a particle cannot be given exactly; for any position, only the probability that the particle will be found there can be stated. This probability decreases exponentially within a potential barrier which is higher than the energy of the particle; but it is not zero. The tunnel effect explains, for example, phenomena such as α decay and field emission.

Turbidity, <*Trübung*>, an optical phenomenon due to → light scattering by particles in a carrier liquid which differ from the liquid in, e.g., their refractive index. It is defined as a measure of the transparency to light and is therefore measured optically. In analogy to the → Bouguer-Lambert-Beer law, the apparent absorbance (→ absorbance, apparent), A_a, caused by the turbidity can be expressed in the form, $A_a = \ln(I_o/I) = S \cdot d$, where d is the path length and S is the scattering coefficient in cm^{-1}, also called the turbidity. I_o is the intensity of the incident light and I that of the transmitted light. The laws which relate the scattering coefficient to the wavelength of the scattered light and the angle at which it is observed are provided by the theories of → Rayleigh scattering and → Mie scattering. See also → turbidity meter.

Turbidity meter, <*Trübungsmesser*>, an instrument for measuring the → turbidity of a solution. Turbidity meters measure either the apparent absorbance, i.e. the scattering coefficient, S, in the direction of the transmitted light or the intensity of the scattered light at some angle of observation, δ_s, to the incident light beam. In most cases the scattering is → Mie scattering, i.e. the scattering particles are of the same size as, or greater than, the wavelength of the scattered light, λ. In these cases, the forward scattering ($\delta_s = 0°$) exceeds the backward scattering ($\delta_s = 180°$) or the scattering to the side ($45° \leq \delta_s \leq 90°$) by orders of magnitude, as the Mie theory predicts. The aim of a turbidity measurement is to establish a relationship between the turbidity and the concentration of the particles.

The forerunner of all modern turbidity meters is Jackson's *candle turbidity meter*. This device consists of a glass tube, closed at its bottom end and held vertically above a candle flame. The solution whose turbidity is to be measured is poured slowly into the tube until the candle flame is no longer clearly discernable. The level of the liquid when this condition is reached depends upon the turbidity. The liquid level thus determined is converted into Jackson candle units (JCU) or Jackson turbidity units (JTU) by means of standard tables. The disappearance of the candle flame depends upon the phenomenon of the forward scattering of light within a very narrow angle. It is very sensitive to the concentration and, at the same time, independent of the color of the light. This technique is the basis of the definition of JTU values which are widely used in industry as a measure of turbidity.

Modern turbidity meters, which are also known as → light-scatter photometers, combine measurement of the transparency with measurement of the forward scattering. Figure 1 shows the optics of the turbidity meter model 210 or 251 from Monitek. The undispersed light from a tungsten-halogen lamp (1) passes through the projection optics (2) and, as a narrow beam, through the entrance window (4) and into the medium to be measured (5). A stop (3) immediately in front of this window limits the narrow beam. The detector (7) which measures the unscattered light is located behind the exit window (6). The light scattered by the medium in the forward direction at an angle of $\delta_s = 12°$ is focused with special detector optics (8) onto the detector (9) where it generates a corresponding signal. The boundary surfaces between the liquid and the windows are critical points for all turbidity measurements. Material deposits, adhering bubbles of gas, streaks, scratches, etc. produce scattering which has nothing to do with the true scattering and therefore causes errors. For this reason, the detector optics (8), also known as a spatial filtering system, are designed such that scattered light which originates in the neighborhood of the entrance window (4) is directed by these optics into the light trap (10). Scattering effects in the region of the exit window are focused at infinity by the optics. Thus, the instrument provides two signals; the transmission signal and the scattering signal. Both signals depend upon the lamp intensity, the quality of the optics, the transparency of the windows and the color, color changes and optical properties of the carrier liquid. In addition, the scattered light (forwards with $\delta_s = 12°$) signal depends upon the light-scattering particles in the sample. If, following suitable amplification, the ratio of the two signals is formed, then all the above factors cancel and a signal is available which is only depen-

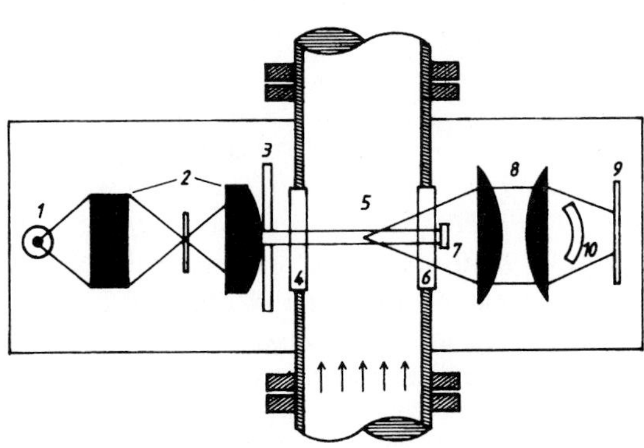

Turbidity meter. Fig. 1. The optics of the Monitek turbidity meters models 210 and 251.

dent upon the number, or better the volume, of the particles in the measured medium.

Figure 1 shows the instrument in its form as an on-line (process) turbidity meter; model 210. If the flow tube is replaced by an appropriate cuvette, then the laboratory turbidity meter, model 251, is obtained. The instrument is calibrated in ppm-kieselguhr by the manufacturers.

The → nephelometer and the Tyndallometer are other instruments for the measurement of turbidity.

Turner filter, <*Filter der verhinderten Totalreflexion, Turner-Filter*>, a filter, first described by Turner and Leurgans in 1947, which uses totally reflecting lamellae in place of the partially transparent metal films which are normally used in → interference filters; especially in → metal-dielectric interference filters. These lamellae are beam splitters with very low absorption. They are formed from a thin layer of a material of low refractive index with an optical thickness which is equal to, or less than, the wavelength of the radiation used. This layer is sandwiched between two materials of higher refractive index so that a part of the totally reflected wave, which travels along the boundary in the optically rarer medium as a damped transverse wave, enters a third medium of higher refractive index. The Turner filter consists of two such totally reflecting lamellae which are effectively absorption-free beam splitters, see Figure 1. The high refractive index layer between the two lamellae also serves as a distance piece. The unit operates in the range of angles near that of total reflection. Theoretically, arbitrarily small half-widths can be obtained at 100% transmission. In practice, limits are imposed by the substances technically suitable for forming the layers. Fundamentally, the filter has two individual maxima with mutually perpendicular polarization. This splittting can be removed, as Billings has shown, if the differences in the phase changes on reflection are compensated by an appropriate choice of birefringent materials for the high refractive index middle layer.

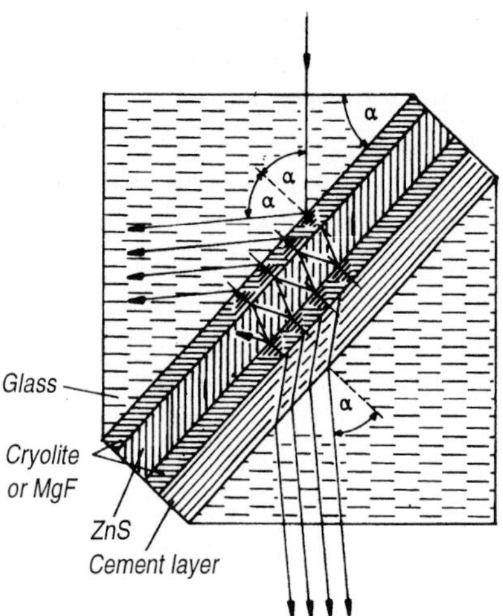

Turner filter. Fig. 1. Enlarged cross section of a Turner filter illustrating its construction

Two-photon spectroscopy, <*Zweiphotonenspektroskopie*> → multiphoton spectroscopy.

Tyndall effect, <*Tyndall-Effekt*> → light scattering by a turbid solution, first observed and described by J. Tyndall in 1869 and named after him. Tyndall found that submicroscopic particles floating in a liquid could lead to a scattering of light. If such a colloidal solution is illuminated with a beam of white light it appears blue from the side. An explanantion of this phenomenon was given by Lord Rayleigh in 1881; → Rayleigh scattering. Tyndall scattering also includes → Mie scattering. Instruments developed to measure the intensity of scattered radiation are known as Tyndallometers or → nephelometers.

Tyndallometer, <*Tyndallometer*> → nephelometer.

U

Ultramat → IR gas analyzer, Ultramat.

Ultraviolet filter, UV filter, <*UV-Filter, Ultraviolettfilter*>, a → filter used to isolate a particular spectral line in the UV region or to select as narrow a region as possible from the emission of a continuum source. → UV reflection filters and → interference filters (line, double-line, band and double-band) are very important for these purposes.

Ultraviolet monitor, UV monitor, <*UV-Monitor, Ultraviolettmonitor*>, a detection system for high-performance liquid chromatography (HPLC) which makes use of the UV absorption of the eluate. The simplest devices of this type are → photometers which operate at a fixed wavelength, λ, in the UV. The light source is a mercury low-pressure lamp (→ mercury-vapor discharge lamp) which emits a line at $\lambda = 253.7$ nm. The light from the lamp is focused through a → UV filter onto the very small aperture of the cuvette through which the eluate flows. Changes in the → transmittance are measured by a → photomultiplier behind the cuvette and recorded against time. The curve produced is the chromatogram. The applicability of this simple device is limited by the fixed wavelength since not all substances can be optimally detected at 253 nm. Thus, UV monitors with variable wavelength have been developed. These instruments are very similar to single-beam spectrometers for the UV-VIS region in their construction, but their optics are specially adapted for microcuvettes. Here too no single wavelength is equally suitable for all substances, although for any particular substance an optimum wavelength can be chosen.
An extraordinary improvement was achieved with the introduction of the → diode array detector and → polychromator in place of a → monochromator. With these → UV-VIS diode array spectrometers an absorption spectrum over a wide spectral range can be simultaneously measured, i.e. many transmittance values as a function of wavelength are available for one and the same moment in time. The temporal sequence of the spectral measurements can be adapted to the temporal sequence of the chromatography. The time scheduling and the electronic reading of the diodes in the array require computer control and this, with the help of appropriate software, brings further advantages to the HPLC technique. Apparatus of this type is now available commercially from a number of suppliers, e.g. Perkin-Elmer, Philips, Hewlett-Packard, Merck-Hitachi, etc.

Ultraviolet photoelectron spectroscopy, UVPS, <*UV-PS, Ultraviolett-photoelektronenspektroskopie*> → photoelectron spectroscopy.

Ultraviolet radiation, UV radiation, <*UV-Strahlen, ultraviolette Strahlen*>, radiation which was discovered in 1801 by J.W. Ritters. It forms a part of the electromagnetic spectrum and runs from the short-wavelength end of the visible into the region of ionizing radi-

ation. Since the short-wavelength limit cannot be precisely defined, it has become customary to regard the region from 100 to 400 nm as the ultraviolet. Koller (1965) has suggested the following further divisions within that region:
Far UV (FUV) 100–200 nm ($100{,}000$–$50{,}000\ cm^{-1}$);
Mid UV (MUV) 200–300 nm ($50{,}000$–$33{,}000\ cm^{-1}$);
Near UV (NUV) 300–400 nm ($33{,}000$–$25{,}000\ cm^{-1}$).
Since UV radiation is absorbed by atmospheric oxygen below 190 nm, measurements must be made in a vacuum. For this reason, the region between 100 and 190 nm is also known as the → vacuum UV. The region between 120 and 190 nm is sometimes called the → Schumann UV. Another division of the UV region has become established, primarily through medical studies of the skin damage caused by UV radiation. It is:
UV A: 315–400 nm
UV B: 280–315 nm
UV C: 200–280 nm.
The region from 200 to 400 nm is readily accessible for spectroscopic purposes and is routinely used in UV spectroscopy (→ spectral regions). Measurements can be extended below 200 nm as far as 160 nm ($66{,}000\ cm^{-1}$) if the instrument is purged with dry nitrogen. The region 200–400 nm is also very important for photochemical investigations.

Ultraviolet radiator, UV radiator, *<UV-Strahler, Ultraviolettstrahler>*, a light source for the ultraviolet spectral region. → Mercury-vapor discharge lamps emit very intense lines in the UV region (→ mercury spectrum) and are therefore used, for example, for the excitation of fluorescence and for photochemical investigations. The mercury low-pressure lamp emits the mercury resonance line at 253.7 nm very strongly and is therefore preferred for excitation in this region. The → hydrogen lamp and the → xenon lamp are continuum UV sources; they are suitable for measuring absorption and fluorescence → excitation spectra.

Ultraviolet reflection filter, *<UV-Reflexionsfilter>*. Mirrors with a high and selective reflection in the UV region can be made using multiple all-dielectric interference layers. This is the ADI-S1 construction of an → interference filter. When evaporated onto black glasses and suitably arranged they provide an effective and highly transparent → UV filter without transmission in the neighboring longer wavelength spectral region. A useful arrangement is one in which two pairs of perpendicularly orientated plates are placed such that the radiation leaves the filter after

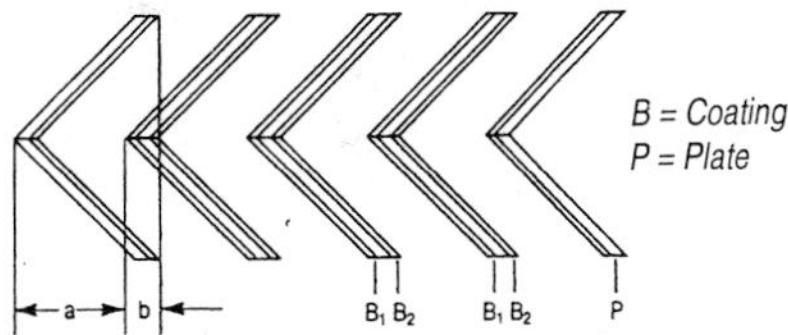

UV reflection filter. Fig. 1a. The arrangement of UV reflection filters

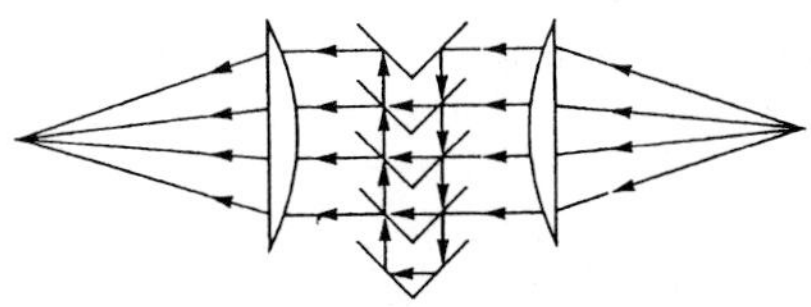

UV reflection filter. Fig. 1b. Light path through a packet of UV reflection filters

fourfold reflection and without a lateral shift, see Figures 1a and 1b. The spectral transmission curves for some UV reflection filters are shown in Figure 2. Schott will supply complete filter systems or single coated plates for laboratory construction. The aperture for the incoming beam should not exceed ±20°. It is worthwhile to protect the filter system from dust and damage due to moisture by enclosing it between quartz plates, quartz lenses or glass filters.

To illuminate larger areas, systems consisting of more than two pairs of plates are suitable. Two plates are brought together at 90° along their long edges. The two outer pairs of plates are coated on one side only, the other pairs of plates on both sides, see Figure 1a. The light path for a system with four channels is illustrated in Fig-

UV reflection filter. Table.

Kind of filter		UV Reflection filter
Available for spectral range [nm]	220–340	
Construction	ADI-4S1	
Filtertype		
UV-R-220	λ_m [nm]	220±10
	HW [nm]	20–40
	τ_{max}	≥0.5
UV-R-250	λ_m [nm]	250±10
	HW [nm]	30–50
	τ_{max}	≥0.7
UV-R-280	λ_m [nm]	280±10
	HW [nm]	40–60
	τ_{max}	≥0.7
UV-R-310	λ_m [nm]	310±10
	HW [nm]	40–60
	τ_{max}	≥0.7
For all types		
$\frac{\text{Tenth-width}}{\text{Half-width}}$		ca. 1.4
Blocked region		above transparent region
Maximum degree of transmission in blocked region		10^{-4}
Covers/blocking filters for short-wavelength region		
For filter UV-R-220 to UV-R-280		Fused quartz, thickness 1mm
For filter UV-R-310		WG 280, thickness 0.7 mm
For filter UV-R-340		WG 295, thickness 1mm

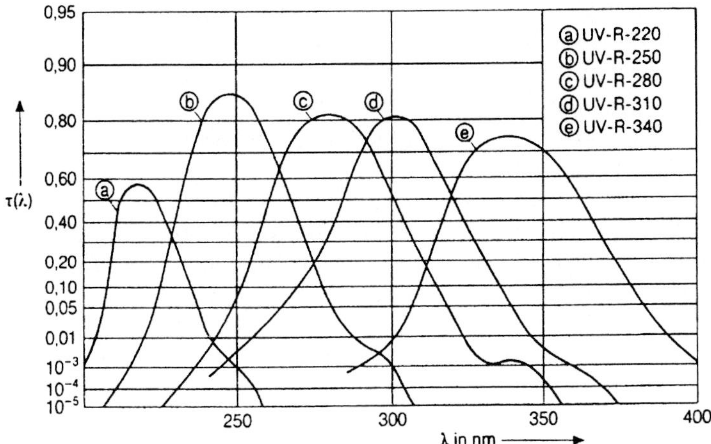

UV reflection filter. Fig. 2. The spectral transmission, T_λ, of UV reflection filters

ure 1b. The table gives the characteristics of some UV reflection filter types provided by their manufacturer Schott. The notation, AI-4S1 indicates a unit containing two plate pairs = 4 mirrors.
The filter UV-R-250 is especially suitable for isolating the line $\lambda = 253.7$ nm of the mercury low-pressure discharge lamp; → mercury-vapor discharge lamp.

Ultraviolet standard lamp, UV standard lamp, <*UV-Normal, Ultaviolettnormal*>, an apparatus introduced into optical measurement and calibration technology by H. Kraft, F. Rössler and A. Rüttenau in 1937. It is a particularly carefully made, 250 W mercury high-pressure lamp with an internal diameter of 1.8 cm. The length of the arc is 10.5 cm. Because the vaporization of the metal reduces the transparency of the quartz envelope in the neighborhood of the electrodes, the tube is masked so that only the central 8 cm of the radiation emerges. The voltage on the arc is some 130 V and the freely radiating 8 cm section of the arc draws a power of 175 W. In 1939 Rössler measured the absolute intensities of the Hg lines and continuous spectrum of this UV standard. The absolute intensities, in Watt, are tabulated for the Hg lines from 577.6 nm to 264 nm. The data for the Hg lines in the visible link the UV standard to the tungsten band lamp → light intensity standard. This has made it possible, for example, to calibrate a luminescence spectrometer down to 264 nm.
Ref.: H.-H. Perkampus, K. Kortüm, H. Bruns, *Appl. Spectrosc.*, **1969**, *23*, 105; F. Rössler, *Ann. Phys. Leipzig*, **1939**, *1*, 34; H. Krefft, F. Rössler, A. Rüttenauer, *Z. Techn. Phys.*, **1937**, *18*, 20.

Uncertainty principle, <*Unschärferelation*> → Heisenberg uncertainty principle.

URAS → IR gas analyzer, URAS.

Useful light, <*Nutzlicht*>, in spectroscopic measurements, that light which lies in the region of the set wavelength, λ_o. If the slit width corresponds to an effective band width of $\Delta\lambda$, then the useful light lies in the range between $\lambda_o + \Delta\lambda$ and $\lambda_o - \Delta\lambda$. A → monochromator set at λ_o should only transmit light in this useful light region, whereby the transmission should decrease linearly from λ_o on both sides. However, because of the presence of → stray light, light of other wavelengths is transmitted by the monochromator and though the proportion of this light is small (approximately 10^{-5}) it can have a large effect because the detector sums up all the stray light throughout the range to which it is sensitive. See → stray light for further details.

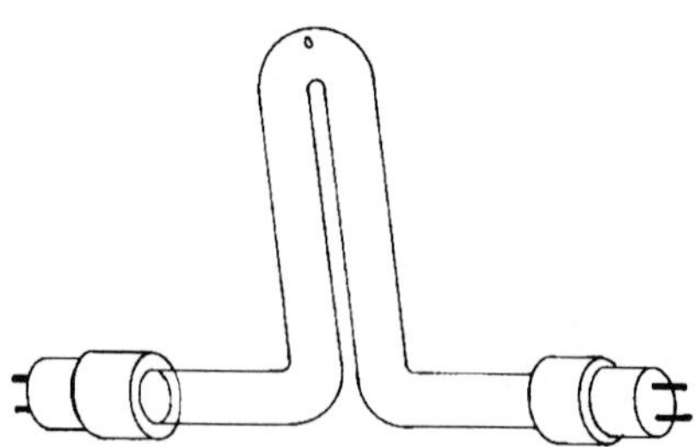

UVAS 247. Fig. 1. The form of the low-pressure lamp in the Perkin-Elmer UVAS 247

UVAS 247, <*UVAS 247*> a nondispersive on-line photometer made by Perkin-Elmer. It is a double-beam instrument and is primarily intended for process and workplace atmosphere concentration monitoring in the ultraviolet spectral region. The light source is a mercury low-pressure lamp (→ mercury-vapor discharge lamp) of a special design with a long discharge, as can be seen from Figure 1. The double-beam technique is illustrated schematically in Figure 2. The light from the source is modulated by the mains frequency and traverses both sample and reference cells simultaneously. The light intensities are measured by two detectors whereby the intensity at the sample detector, I_{sam}, depends upon the concentration of the sample in the sample cell. The gas stream to be measured does not flow through the reference cell and the reference signal, I_{ref}, is independent of the concentration. The measurement:

$$U_{mes} = (I_{ref} - I_{sam})/I_{ref}$$

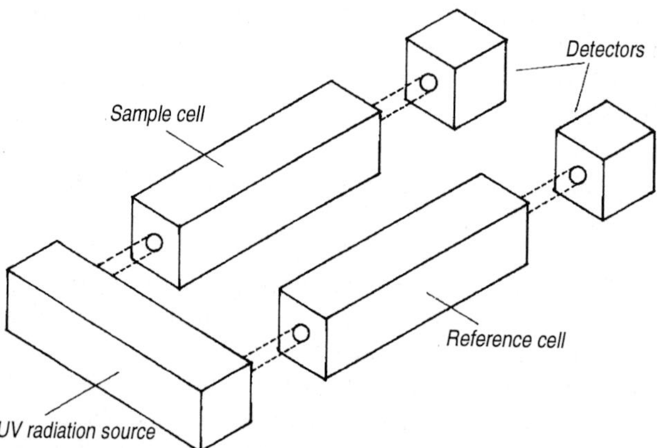

UVAS 247. Fig. 2. Schematic illustration of the double-beam operation in the Perkin-Elmer UVAS 247

is independent of the light intensity and is linearly related to the concentration of the substance to be measured in most cases. If this relationship is not linear, the instrument offers the further option of a linearization of the range of measurement. Typical applications of this equipment are the measurement of traces of benzene, benzene derivatives and mercury in industry.

Uvasole, <*Uvasole*> extremely pure solvents with guaranteed transmission values in both the UV and IR spectral regions, produced by E. Merck for spectroscopic purposes. Similar solvents are produced by other manufacturers, → solvent.

UV-VIS diode array spectrometer, <*UV-VIS- Diodenarrayspektrometer*>, a simultaneously measuring UV-VIS spectrometer of which the most important components are a → polychromator and a → photodiode array as detector. In contrast to a sequential UV-VIS spectrometer, in a diode array spectrometer the cuvette is placed before the polychromator. Therefore, since the full intensity of the radiation falls upon the sample, only photochemically stable compounds can be measured with such an instrument. The principles of sequential and simultaneous measurement are compared in a schematic way in Figure 1. In the sequential measurement (a), the polychromatic light is dispersed by the dispersing element (prism) and passed across the exit slit by rotating the prism. The cuvettes and detector are placed behind the slit. For the simultaneous measurement (b), the cuvette is located in front of the dispersing element and the complete width of the chosen spectrum, as dispersed by the prism, falls on the diode array detector. The advantage of this procedure is the simultaneous measurement at all wavelengths and the correspondingly short time taken to aquire a complete spectrum. UV-VIS diode array spectrometers are always equipped with a grating polychromator which gives linearity with wavelength. The spectral information, briefly stored in the diode array, is read-out electronically and passed to a computer which takes over the remaining processing of the data and the display of the spectrum. This type of spectrometer can only be operated with a computer. The fact that the data is present in digital form makes further processing with appropriate software relatively easy.

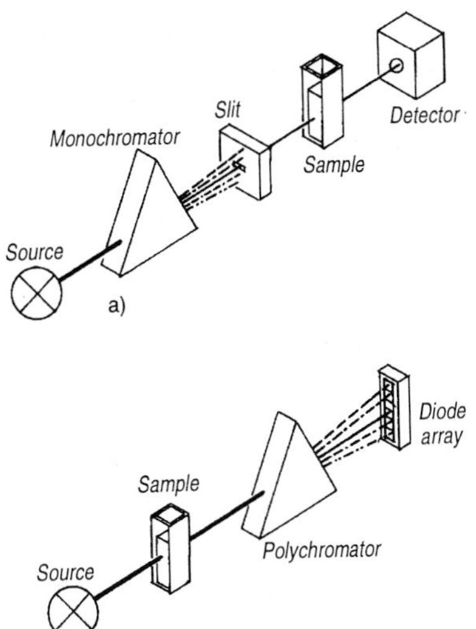

UV-VIS diode array spectrometer. Fig. 1. A schematic comparison of a) sequential and b) simultaneous measurement

Figure 2 shows the Perkin-Elmer Lambda Array 3840 as an example of a diode array spectrometer. The light

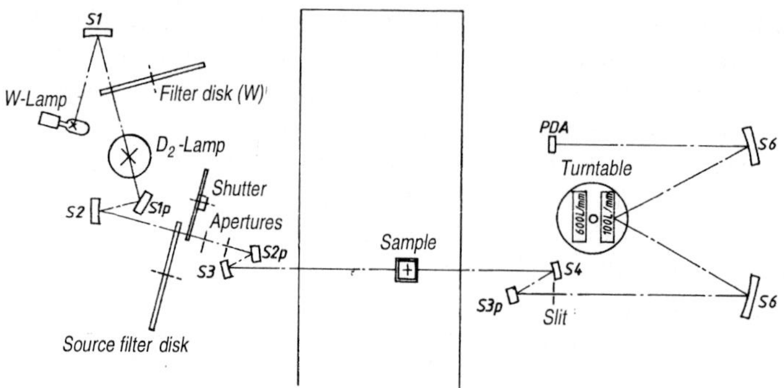

UV-VIS diode array spectrometer. Fig. 2. The optical system of the Perkin-Elmer Lambda-Array 3840

sources are a tungsten lamp and a special deuterium lamp which allows light from the tungsten lamp to pass, via the concave mirror *S*1, through its envelope. The light beam is focused, via the plane mirrors *S*1*p* and *S*2*p* and the concave mirrors *S*2 and *S*3, onto the sample cuvette. From there it passes, via *S*4 and *S*3*p*, through the slit and into the polychromator. The first mirror, *S*6, collimates the light for the grating and the second focuses the dispersed light onto the diode array detector. The polychromator possesses two gratings mounted on a rotatable table. The grating with 100 lines per mm is used to obtain a survey spectrum in the range 190–900 nm. The second grating with 600 lines per mm makes it possible to obtain high-resolution spectra. For this purpose regions of 100 nm are measured in seven consecutive steps. The best resolution is 0.25 nm and some 2 s are required for the measurement of each 100 nm region, so that the acquisition of the entire spectrum (190–900 nm) takes about 14 s. Filter disk *W* is used to block the tungsten lamp and filter disk *S* to select definite spectral ranges from the two sources. The positioning of these two filter disks and the shutter

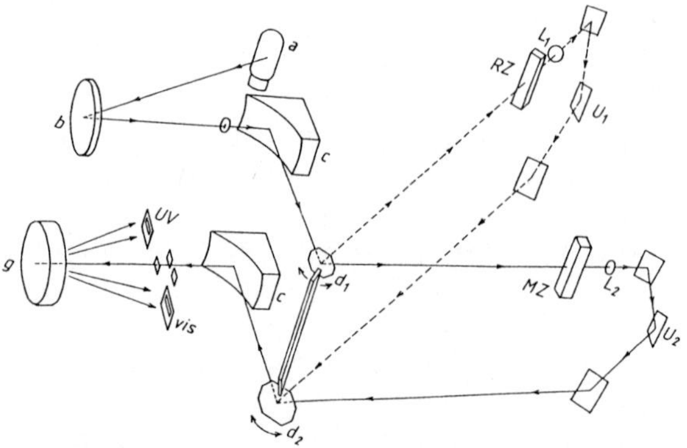

UV-VIS diode array spectrometer. Fig. 3. The optical system of the Hewlett-Packard HP 8450 diode array spectrometer

in setting the conditions of the measurement, particularly the range of wavelengths to fall on the detector, is controlled by the computer. UV-VIS diode array spectrometers are made by many companies. Figure 3 shows the optical system of the Hewlett-Packard 8450. The moving mirrors, *d*1 and *d*2 direct the light to the reference (*RZ*) and sample (*MZ*) cuvettes alternately. However, most instruments of this type are designed for application as diode array detectors for high-performance liquid chromatography (HPLC) and cover the UV range to ca. 400 nm. Because of the small size of the sample compartment, which is determined by the flow-through HPLC cell, these spectrometers are generally not suitable for ordinary UV-VIS spectroscopy.

UV-VIS-NIR spectrometer, <*UV-VIS-NIR-Spektralphotometer, UV-VIS-NIR-Spektrometer*>, a → spectrophotometer for measurement of electronic spectra (→ absorption spectra) from the UV through the visible and into the NIR. When the instrument is purged with dry nitrogen, this is a range from 185 nm to 3200 nm or 54,000 cm^{-1} to 3,125 cm^{-1}. Two gratings are used, one for the UV-VIS and another for the NIR. A → photomultiplier is the detector in the UV-VIS and a PbS cell in the NIR since the spectral sensitivity of a photomultiplier drops to almost zero above 900 nm. Figure 1 illustrates the optics of such an instrument with the example of the Perkin-Elmer Lambda 9. It is a totally reflecting (i.e. no mirrors), double-beam, double-monochromator instrument.

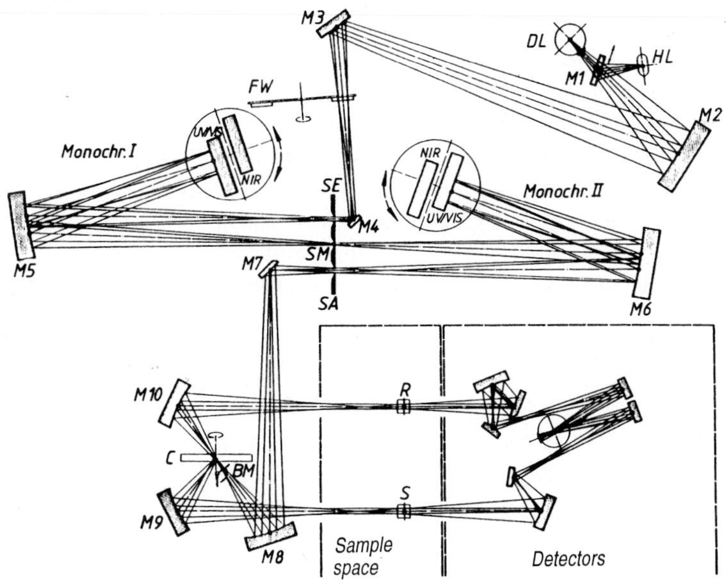

UV-VIS-NIR spectrometer. Fig. 1. Optical system of the Perkin-Elmer Lambda 9, see text

The gratings for the UV-VIS region are holographically blazed and the optical components are coated with SiO_2 to increase their durability.
Two radiation sources, a deuterium lamp (DL) and a tungsten-halogen lamp (HL) cover the required spectral range. When working in the NIR or VIS range, the light from the halogen lamp is directed onto the mirror *M2* by mirror *M1*. *M1* also blocks the radiation from the deuterium lamp. For work in the UV region, *M1* is moved so that the light from the deuterium lamp falls upon *M2*. The change of sources is made automatically as the monochromator scans. The beam path runs via mirrors *M2* and *M3*, through the optical filters of the filter wheel *FW*, to mirror *M4*. The filter wheel is moved synchronously with the monochromator by a stepper motor, filtering the light in a wavelength-dependent manner before it enters the first monochromator. The radiation is directed through the entrance slit, *SE*, of the first monochromator by the mirror, *M4*. The light is collected by the mirror, *M5*, and falls onto the grating of the first monochromator. The appropriate grating for the wavelength region, UV-VIS or NIR, is chosen automatically. Depending upon the setting of monochromator 1, a portion of the spectrum produced falls on the middle slit, *SM*, which is both exit slit for monochromator 1 and the entrance slit for monochromator 2. The light which passes through this slit is almost monochromatic. It passes via *M6* to the appropriate grating of monochromator 2 and via *M6* again to the exit slit, *SA*. The two gratings move synchronously. The light arriving at mirror, *M7*, is of high spectral purity with an extremely small contribution from → stray light. In the UV-VIS region, a choice of fixed slit or an automatic slit program is available. The latter changes the slit width in such a way that the energy falling on the detector is constant. In the NIR region the slit width is controlled automatically. The radiation arrives, via the plane mirror, *M7*, and the concave mirror, *M8*, at the → chopper, *C*. This has four segments, one reflecting, one transparent and two blocking. When the light beam strikes the reflecting segment the sample beam, *S*, is generated via mirror *M9*. The reference beam, *R*, is generated, via mirror *M10*, when the beam falls onto the transparent segment. The two blocking segments generate a dark phase for which the detector gives a dark signal.
Both beams are focused in the plane of the cuvettes, such that the focus extends 10 mm vertically. The width of the beams depends upon the slit width. For an optical slit width of 5 mm the width of both beams is 4.5 mm.
The sample beam and reference beam are directed alternately to the detector (photomultiplier for UV-VIS, PbS for NIR). The detectors are changed automatically. The instrument stops scanning and recording during the changes of filters, light sources and detectors.
A reduction of the height of the beams is recommended when microcuvettes are used. For this purpose a rotatable aperture, *BM*, is installed between the chopper and the mirror, *M9*. This reduces the height of the two beams in the plane of the cuvettes to 5 mm.
The whole instrument is computer-controlled. Apart from the control of the optical system, as outlined above,

the computer has a facility for storing six complete experimental procedures which is particularly valuable for routine analytical work. These methods are protected from loss when the power supply is switched off (or fails) since the methods memory has a battery backup. The battery backup lasts for some 24 hours in the absence of mains power. For further details see the Lambda 9 handbook, Perkin-Elmer.

UV-VIS spectroscopy, electronic spectroscopy, *<UV-VIS-Spektroskopie>*, the spectroscopic method in which → absorption spectra which are primarily due to light absorption resulting from the excitation of electrons in atoms or molecules are measured. These measurements are usually made on solutions, although measurements of gases and solids are possible. For most applications, the spectral range can be said to be ca. 200–800 nm, i.e. 50,000–12,500 cm^{-1}. A spectral range of 180–900 nm can be measured with modern UV-VIS spectrophotometers (UV-VIS-NIR spectrophotometers), but in the wavelength region below 200 nm the instruments must be purged with nitrogen (→ Schumann UV region).

The basis of quantitative absorption measurements is provided by the → Bouguer-Lambert-Beer law which is valid for gases and solutions. UV-VIS and atomic spectroscopy are the oldest spectroscopic methods. The importance of UV-VIS spectroscopy lies in the multiplicity of applications which the technique offers as a branch of molecular spectroscopy; e.g. the provision of information related to quantum chemistry and molecular structure, structure determination, identification, quantitative analysis (photometry), study of chemical equilibria and kinetics, analysis of photochemical reactions, investigation of intermolecular interactions. Special techniques such as derivative spectroscopy, dual-wavelength spectroscopy, reflectance spectroscopy, luminescence excitation spectroscopy and measurements at low temperatures, greatly extend the range of applications.

Ref.: H.-H. Perkampus, *UV-VIS Spectroscopy and Its Applications*, Springer Verlag, Berlin, Heidelberg, New York, **1992**.

UV-VIS spectrum, *<UV-VIS-Spektrum>*. An → absorption spectrum in the UV-VIS spectral region is the result of the excitation of electrons from the ground state, S_o, into higher electronic states of the molecule, S_p ($p \geq 1$). In the case of molecules, it must also be remembered that vibrational and rotational states are superimposed upon the electronic states, so that it is to be expected that electronic excitation will be accompanied by vibrational and rotational excitation. For small molecules in the gas phase, this superimposed structure can be measured and analyzed, with a spectrometer of good resolving power. The UV-VIS absorption spectra of larger molecules, which frequently cannot be obtained in the gas phase without decomposition, are usually measured in solution. Because the rotational quanta are small and the rotation is hindered by the solvent molecules, no rotational structure is seen, even with high-resolution instruments. Thus, any remaining structure on the absorption bands may be attributed to vibrations, as is illustrated for naphthalene in n-

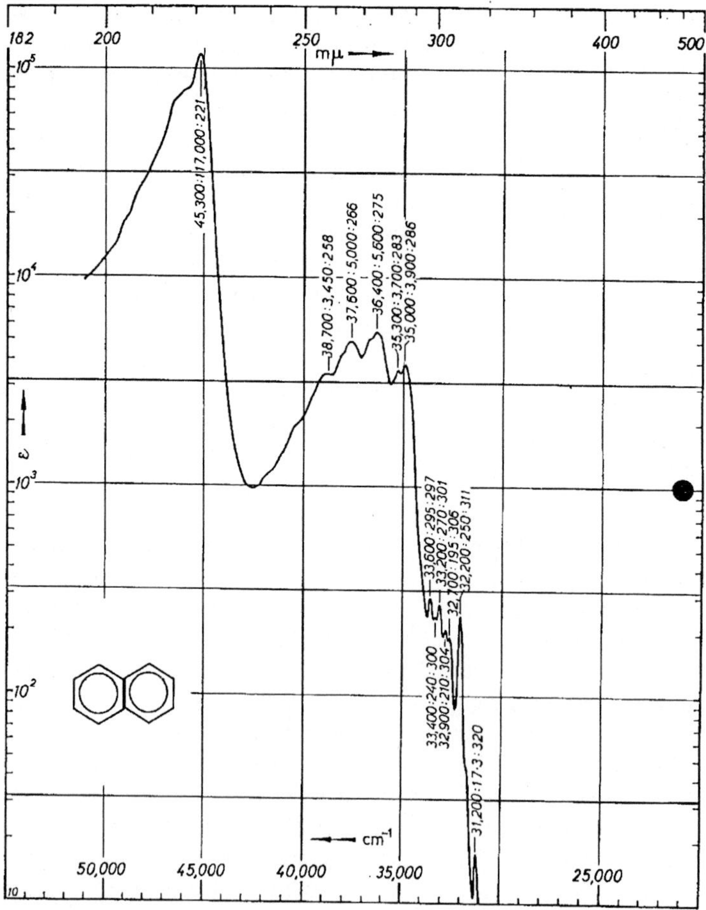

UV-VIS Spectrum. Fig. 1. The UV spectrum of naphthalene in n-heptane

heptane in Figure 1. A UV-VIS absorption spectrum is a plot of the molar decadic → extinction coefficient, $\varepsilon_{\tilde{\nu}}$, as a function of wavenumber, $\tilde{\nu}$. The abscissa in this type of presentation is proportional to excitation energy, but the plot against wavelength, which is still very common, gives a distorted view of the → absorption spectrum. Since the extinction coefficient, $\varepsilon_{\tilde{\nu}}$, can differ in value by several orders of magnitude within one spectrum ($10 < \varepsilon_{\tilde{\nu}} < 10^5$), it has been found useful to plot the logarithm of $\varepsilon_{\tilde{\nu}}$ as a function of the wavenumber. The UV-VIS spectrum thus obtained is characteristic of the molecule which is the reason for the importance of → UV-VIS spectroscopy in structure determination. In Figure 2, a singlet energy-level sheme is assigned to a measured spectrum to make clear that the absorption maxima correspond to quite specific energy states, i.e. excitation energies; $\Delta E = E_p - E_o$ ($p \geq 1$). The diagram also shows that,

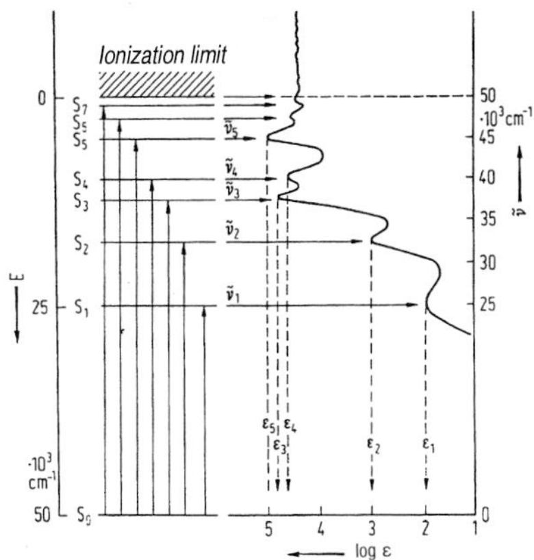

UV-VIS Spectrum. Fig. 2. The connection between energy-level scheme and spectrum

apart from the positions of the absorption maxima, the extinction coefficient is also important in the interpretation of a UV-VIS spectrum since it is related theoretically, through the Einstein *B* coefficient (→ Einstein coefficient), to the → oscillator strength and the → transition dipole moment. If the vibrational structure which overlies the electronic band is also considered, then a UV-VIS spectrum provides the following information:

1. The absorption maxima, $\tilde{\nu}_{max}$, correspond to discrete molecular states which depend upon the structure, geometry and symmetry of the molecule.
2. The extinction coefficient, ε_{max}, or the → integral absorption over an absorption band gives the magnitude of the → transition dipole moment and the oscillator strength.
3. The structure of an absorption band provides information about the normal vibrations which are coupled with the electronic excitation; these are the vibrations in the electronic excited state.

Ref.: H.-H. Perkampus, *UV-VIS Spectroscopy and Its Applications*, Springer Verlag, Berlin, Heidelberg, New York, **1992**.

UV-VIS standard, *<UV-VIS-Standard>*, a standard used to check the accuracy of the wavelength scale and absorbance measurement of a → spectrophotometer.
For wavelength calibration, the lines of a → metal-vapor discharge lamp which are known exactly, e.g. a → mercury-vapor discharge lamp, may be used. The following lines are particularly suitable: 579.0, 576.9, 546.1, 435.8, 404.5, 364.9 and 253.7 nm (→ mercury spectrum). The lamp should be placed as close as possible to the entrance slit of the monochromator. The measurement is then made in single-beam mode (singlebeam instruments) or in the energy mode (double-beam instruments), with as narrow a slit as possible. Routine calibration can be readily made with a holmium (III) solution. This solution shows many sharp bands in the UV-VIS region, which have been measured very precisely by several authors using different instruments. The results agree to within less than 0.5 nm and they are therefore suitable for calibration purposes. The Ho(III) solutions are made by dissolving holmium oxide (Ho_2O_3) in perchloric acid (e.g. 10 g Ho_2O_3 in 100 ml aqueous perchloric acid containing 17.5% $HClO_4$) or hydrochloric acid (e.g. 0.35 g Ho_2O_3 in 10 ml conc. hydrochloric acid containing 32% HCl). The spectrum of a Ho(III) chloride solution is shown in Figure 1 with the wavelengths marked. Samarium(III) in perchloric acid is also a suitable wavelength standard. Absorbance standards have two pur-

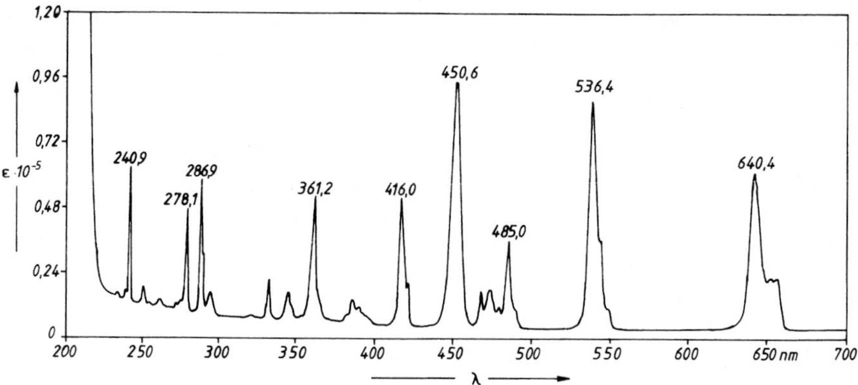

UV-VIS standard. Fig. 1. The spectrum of a holmium(III) chloride solution, see text

poses; they should check both the photometric linearity and the accuracy of the absorbance scale of a spectrophotometer.

Solutions of dyes or colored compounds which obey the → Bouguer-Lambert-Beer law are used as linearity standards. In accord with the equation, $A = \varepsilon \cdot d \cdot c$, the linear dependence of absorbance upon concentration at constant path length is used. Many, exactly defined, solutions for this purpose have been proposed in the literature. At constant concentration, the absorbance depends linearly upon the path length, so that the linearity of an instrument can be tested over one power of ten using a set of cuvettes having pathlengths of 0.5, 1.0, 2.0, 4.0 and 5.0 cm.

A solution of potassium dichromate in 0.005 M sulfuric acid has been found valuable for routine calibration of the absorbance scale. It is useful to prepare two solutions:

Solution A:

50 mg ±0.5 mg in 1 l 0.005 M H_2SO_4 for the range $0.2 \leq A \leq 0.7$.

UV-VIS-Standard. Table 1.

	Absorbance $A = (\varepsilon \cdot d)\, c$	
λ [nm]	Solution A	Solution B
235 (min.)	0.626±0.009	1.251±0.019
257 (max.)	0.727±0.007	1.454±0.015
313 (min.)	0.244±0.004	0.488±0.007
350 (max.)	0.536±0.005	1.071±0.011

Solution B:

100 mg ±1 mg in 1 l 0.005 M H_2SO_4 for the range $0.4 \leq A \leq 1.4$.

The measurements are made in a cuvette of 1 cm path length at a temperature in the range 15–25°C. The reference cell is filled with 0.005 M H_2SO_4. In the table the absorbance values for these two solutions at their two maxima and minima are compiled from data in the literature.

Standard solutions for the measurement of stray light have also been suggested in the literature. They are effectively → long-wave pass filters, i.e. they cut off the short-wavelength part of the absorption spectrum (see

UV-VIS-Standard. Table 2.

Spectral region (nm)	Solvent or soltn.	*d* [cm]
165–173.5	water	0.1
170–183.5	water	1.0
175–200	12 g KCl*	1.0
195–223	10 g NaBr*	1.0
210–259	10 g NaI*	1.0
250–320	acetone	1.0
300–385	50 g $NaNO_2$*	1.0

*All solids to be dissolved in 1 l water.

→ stray light). The table gives a summary.

Ref.: C. Burgess in *Standards in Absorption Spectrometry* (Ed.: A. Knowles), Ultraviolet Spectrometry Group, Chapman and Hall, London, New York, **1981**.

Vacuum spectrograph, <*Vakuumspektrograph*>, a spectrograph for the ultraviolet region below a wavelength of 200 nm. The absorption of the atmosphere from this wavelength downwards makes spectroscopic experiments in this region increasingly difficult. Oxygen and nitrogen begin to absorb at 200 and 150 nm respectively, so that between 200 to 150 nm the spectrometer must be purged with purified, dry nitrogen. Below 150 nm the spectrometer must be evacuated (→ vacuum UV, → Schumann UV region). Evacuated spectrometers, or spectrometers purged with gases which do not absorb, e.g. nitrogen, are also used in the IR spectral region in order to avoid the troublesome absorption of carbon dioxide and water vapor. The first vacuum spectrographs were → prism spectrographs. The optics were made of fluorite which is transparent to about 120 nm. Pure lithium fluoride is also useful down to ca. 110 nm. Concave gratings are used almost exclusively in the vacuum UV region today. These are usable down to the shortest wavelengths since there are no lenses through which the light must pass. The boundaries between the light source and the spectrograph or, in the case of absorption experiments, between the light source, absorbing gas sample and spectrograph, can be closed with fluorite or lithium fluoride windows. For wavelengths, where these are no longer sufficiently transparent, noble gas windows can be used. A stream of circulating noble gas is blown against the vessel in which the absorbing or emitting gas is contained. The vacuum is then maintained by vacuum pumps of very high pumping capacity.

Vacuum ultraviolet spectrum, vacuum UV spectrum, <*Vakuumultraviolett-Spektrum, Vakuum-UV-Spektrum*>, a short-wavelength spectrum beyond visible light. The absorption of the air, due at first to oxygen, begins below 200 nm. As far as about 160 nm, the absorption of oxygen can be eliminated by purging with nitrogen. For lower wavelengths, into the region of X-rays, spectroscopic apparatus must be evacuated. This is the reason for the term *vacuum UV.*

Vacuum wavelength, <*Vakuumwellenlänge*> → wavelengths, λ_v, which apply to a vacuum. In spectroscopy, wavelengths, λ, are usually measured at 288 K and 1 bar in air. Therefore, they require to be corrected with the formula, $\lambda_v = n\lambda$, where n is the refractive index of air at the appropriate temperature, humidity and pressure.
In some monochromators, e.g. the Zeiss M4Q2 and M4Q3, the vacuum wavelengths are given on the wavelength scale.

Valence electron, <*Valenzelektron*>, an electron in the outermost, incompletely filled shell of the atom. The valence electrons are responsible for both the ionic and the covalent chemical bond, and also for the spectra of atoms (→ optical electron).

Valence vibration, <*Valenzschwingung*>, a → vibrational mode in which the atoms involved move either

Valence vibration. Fig. 1. Symmetric and asymmetric valence vibrations of the $-CH_3$ group

towards or away from each other along the line of the chemical bond (valence) between them. They are also called stretching vibrations and are generally indicated by the Greek letter ν, e.g. ν(C-H), ν(C-C), ν(C=C), ν(C≡C), ν(O-H), ν(C-O), ν(C=O), ν(C≡N), ν(C=N), ν(N=N), ν(C-X) where X = F, Cl, Br or I. If the symmetry of the atom grouping involved does not change with a valence vibration, then it is called a symmetric stretching vibration, ν_s. If there is a change of symmetry it is an asymmetric stretching vibration, ν_{as}. This is illustrated in Figure 1 for the C-H stretching vibrations of the methyl group. The situation is analogous for any similar group of atoms, e.g. $-CX_3$ or $-YX_3$.

Vavilov's law, <*Vavilow-Gesetz*>, the fact that the → fluorescence quantum yield of aromatic molecules, q_{FM}, is independent of the excitation wavelength. The law assumes that the → radiationless transition from S_p to S_1 (→ internal conversion) has a quantum yield of 1.

Verdet constant, <*Verdetsche Konstante*> → magnetooptical rotation.

Vibrational mode, <*Schwingungsformen*>, the illustration of the → normal vibrations of a molecule as the motions of the individual atoms with respect to each other. For small molecules, the vibrational modes can be drawn relatively easily. A linear triatomic molecule of the type *Y-X-Y* can execute four normal vibrations which are illustrated in Figure 1. Mode ν_1 is the symmetric → valence vibration (symmetric stretching) which is IR inactive (→ IR active). ν_2 is the asymmetric stretching which is IR active. ν_3 and ν_4 are → bending vibrations (angle bending) which are IR active and degenerate. ν_4 is converted into ν_3 if the molecule is rotated through 90° about the internuclear axis. This simple example shows clearly how the individual normal vibrations may be distinguished and classified in terms of the vibrational modes.

For larger molecules with more complex structures, the illustration of the vibrational modes associated with all the normal vibrations requires an extensive theoretical → normal coordinate analysis. In this process

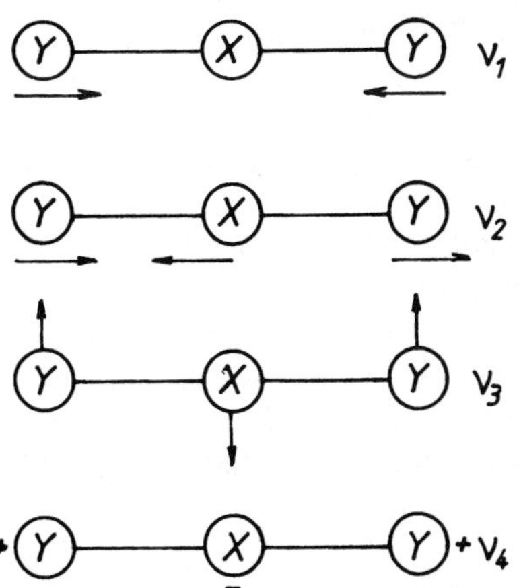

Vibrational modes. Fig. 1. Vibrational modes of a linear triatomic molecule, e.g. CO_2

other vibrational modes, apart from the stretching and bending vibrations described above, may be identified. These modes have been classified in the IR literature and distinguished with special symbols. In principle, all vibrations, apart from stretching vibrations, are → bending vibrations which are classified with respect to the particular form of their angle deformations. This classification is useful because the different bending modes have different excitation energies and different → group frequencies and, consequently, different → expectation regions.

Vibrational spectroscopy, <*Schwingungsspektroskopie*> → IR spectroscopy, → Raman spectroscopy.

Vibrational spectrum, <*Schwingungsspektrum*>. A diatomic molecule can execute only one normal vibration. If this is to be excited by the absorption of electromagnetic radiation, then the molecule must have a permanent dipole moment, i.e. the molecule must be a heteronuclear diatomic molecule. If a harmonic vibration is assumed then, with the help of the → dumbbell model and using the → reduced mass, μ, the vibration can be described in terms of the vibration of the mass, μ, about the center of gravity.
This is the harmonic oscillator model. The vibration is maintained by the effect of a restoring force, F, which is proportional to the displacement, x. The constant of proportionality is the force constant k.

$$F = -kx = \mu \frac{d^2x}{dt^2}.$$

The solution of this differential equation is:

$$x = x_o \sin (2\pi \nu_s t).$$

x_o is the amplitude of the vibration and t is the time. ν_s is the classical vibrational frequency of the harmonic oscillator and is given by:

$$\nu_s = \frac{1}{2\pi}\sqrt{\frac{k}{\mu}};$$

From $F = -kx$, the potential energy, E_{pot}, may be calculated:

$$E_{\text{pot}} = \frac{1}{2}kx^2 = 2\pi^2\mu\nu_s^2x^2.$$

It is a function which is quadratic in the displacement, x, from the equlibrium internuclear distance, r_o, and forms a parabola with its apex at r_o; see Figure 2.
The quantum-mechanical treatment of the harmonic oscillator gives the following equation for the energy states:

$$E(\upsilon) = \frac{h}{2\pi}\sqrt{\frac{k}{\mu}}(\upsilon + \frac{1}{2})$$
$$= h\nu_s(\upsilon + \frac{1}{2})$$

with the vibrational quantum number $\upsilon = 0, 1, 2 \ldots$ and ν_s as the classical vibrational frequency. For $\upsilon = 0$, the → zero-point energy is obtained.

$$E(\upsilon = 0) = \frac{1}{2}h\nu_s$$

The terms (energies in cm^{-1}) of the harmonic oscillator, $G(\upsilon)$, are given by:

$$G(\upsilon) = \frac{E(\upsilon)}{hc} = \frac{\nu_s}{c}(\upsilon + \frac{1}{2})$$
$$= \omega(\upsilon + \frac{1}{2})$$

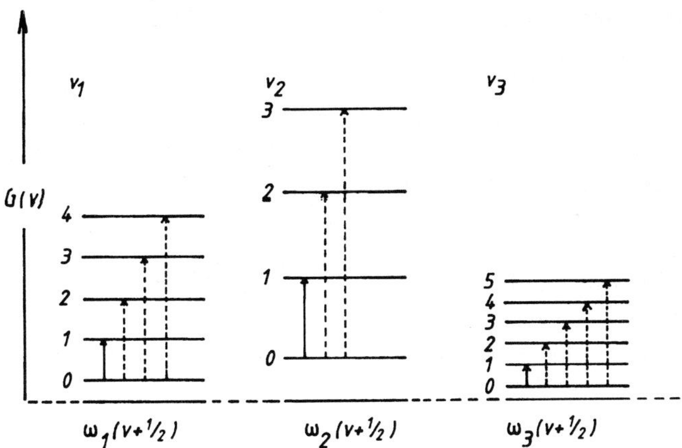

Vibrational spectrum. Fig. 1. Energy-level scheme for three normal vibrations

in which ω is the wavenumber of the classical vibration. This is a sequence of equally spaced energy states. The selection rule is $\Delta v = \pm 1$, which gives for any transition, independent of the two states involved:

$$G(v+1) - G(v) = \omega.$$

All allowed transitions therefore fall at the same position and the same frequency is either emitted or absorbed; see Figure 2. However, the condition for an excitation is that there must be a change of dipole moment associated with the vibration.
The model just described is a simplified one and cannot explain the following observations.

1. The presence of overtones.
2. The disappearance of the attractive force at large amplitudes which leads eventually to the dissociation of the molecule.
3. The increase of the repulsive forces at small amplitudes; see potential curves of the harmonic (→ vibrational spectrum) and → anharmonic oscillators.

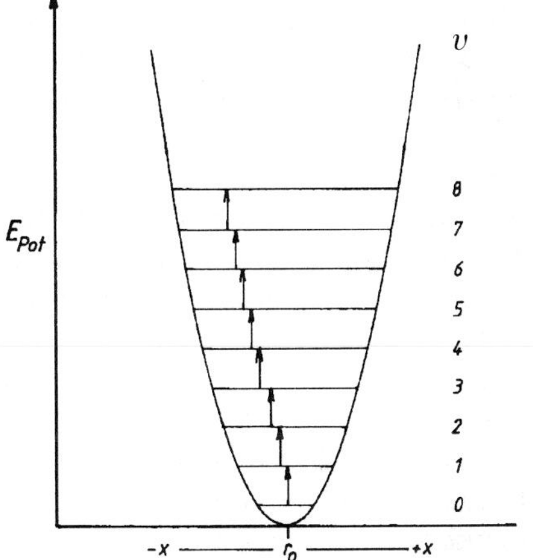

Vibrational spectrum. Fig. 2. Potential energy curve for a harmonic oscillator with equidistant vibrational energy levels

The → anharmonic oscillator provides a better description of the reality. The vibrational spectrum of a polyatomic molecule is the spectrum of a molecule which results from the excitation of quantized vibrational states. It is obtained experimentally as an → IR or → Raman spectrum. Since the excitation energy of a vibration is always greater than that of a rotation, the measured spectrum is basically the → rotation-vibration spectrum because the rotations are always excited with the vibrations. Thus, the individual bands of the IR and Raman spectra of molecules in solution are very broad because the accompanying rotational structure is frequently not resolved. The rotational structure of small molecules in the gas phase can be resolved if the → rotational constants in cm^{-1} are larger than the → resolution of the instrument, $\Delta\tilde{\nu}$, in cm^{-1}. In crystalline solids, the rotational motion is frozen and a vibrational spectrum with sharp lines, i.e. bands of very small → half-widths, are obtained.

A detailed quantum-mechanical calculation of the possible vibrations of a molecule with N atoms of masses m_i ($i = 1, 2, 3 \ldots N$) gives, when the → normal coordinates Q are introduced, a system of $3N$ independent differential equations which correspond to the → harmonic oscillator model. The vibrational energies are found to be:

$$E(\upsilon_1, \upsilon_2, \upsilon_3 \ldots) = h\nu_1(\upsilon_1 + \frac{1}{2}) + h\nu_2(\upsilon_2 + \frac{1}{2}) + h\nu_3(\upsilon_3 + \frac{1}{2}) + \ldots$$

where $\upsilon_i = 0, 1, 2, 3 \ldots$ are the vibrational quantum numbers. If the vibrational frequencies, ν_i, are replaced by the corresponding wavenumbers, $\omega_i = \nu_i/c$, then the vibrational energies can be written as terms:

$$G(\upsilon_1, \upsilon_2, \upsilon_3 \ldots) = \omega_1(\upsilon_1 + \frac{1}{2}) + \omega_2(\upsilon_2 + \frac{1}{2}) + \omega_3(\upsilon_3 + \frac{1}{2}) + \ldots$$

For $\upsilon_1 = \upsilon_2 = \upsilon_3 \ldots = 0$ the → zero-point energy is found to be:

$$E(0, 0, 0 \ldots) = \frac{1}{2}h\nu_1 + \frac{1}{2}h\nu_2 + \frac{1}{2}h\nu_3 + \ldots$$

and the corresponding term is:

$$G(0, 0, 0 \ldots) = \frac{1}{2}\omega_1 + \frac{1}{2}\omega_2 + \frac{1}{2}\omega_3 + \ldots.$$

If other terms are referred to this term as zero then:

$$G(\upsilon_1, \upsilon_2, \upsilon_3 \ldots) - G(0, 0, 0 \ldots) = G_o(\upsilon_1, \upsilon_2, \upsilon_3 \ldots) = \omega_1\upsilon_1 + \omega_2\upsilon_2 + \omega_3\upsilon_3 + \ldots$$

This analysis gives six (for linear molecules five) non-vibrations which represent the rotational and translational → kinetic degrees of freedom of the molecule. In order to take account of degenerate vibrations of any degree of → degeneracy, the terms are written in the form:

$$G(\upsilon_1, \upsilon_2, \upsilon_3 \ldots) = \sum \omega_i(\upsilon_i + \frac{d_i}{2});$$

where d_i gives the degree of degeneracy directly, i.e. $d_i = 1$ for nondegenerate vibrations, $d_i = 2$ for doubly degenerate vibrations, and so on.

The vibrational spectrum may then be derived from the term representation $G_o(\upsilon_1, \upsilon_2, \upsilon_3 \ldots)$ and the selection rule, $\Delta\upsilon_i = \pm 1$. For $\upsilon_1 = 1$ and all other $\upsilon_i = 0$, and correspondingly for $\upsilon_2 = 1$, $\upsilon_3 = 1 \ldots \upsilon_j = 1$, the excitation of the → fundamental vibrations is obtained. This is illustrated in Figure 1 for $N = 3$. When the anharmonicity is included the selection rule includes $\Delta\upsilon_i = \pm 2, \pm 3 \ldots$ The excitations $0 \rightarrow 2$, $0 \rightarrow 3$ etc. then become possible. These are the → overtones which are shown in Figure 1 as dashed lines. In the harmonic oscillator, the energy levels (terms) are equally spaced, but

in the → anharmonic oscillator they move closer together as energy increases so that the overtones are no longer integer multiples of the fundamental. Three series are obtained with the separations ω_1, ω_2 and ω_3. Combinations are also possible, i.e. two or more vibrational modes are simultaneously excited. This is illustrated in the table for $N = 3$ (bent or linear with ω_3 and ω_4 degenerate).
The combinations between the terms $G_o(v_1)$, $G_o(v_2)$ and $G_o(v_3)$ give rise to → combination vibrations. Accordingly, a vibrational spectrum consists of fundamental, overtone and combination vibrations and in polyatomic molecules complex IR spectra can easily arise. In the region 4000–400 cm^{-1}, the only overtones which can arise are those for which the fundamentals lie below 2000 cm^{-1}. And the intensity of the overtones decreases rapidly on account of the sharply decreasing transition probability (→ oscillator strength). The combinations too are mostly of low intensity. Allowance must also be made for the fact that not all the 3*N*-6 (3*N*-5) vibrations can be excited, i.e. are → IR active. For excitation in the IR region there must be a change of dipole moment, or a change in a component of that dipole moment, associated with the vibration. Only then can there be an interaction with the electromagnetic radiation. For the → Raman spectrum, however, there must be a change in the → polarizability during the vibration, if it is to be → Raman active. In the case of simple molecules, it is quite easy to decide, with the help of the → vibrational modes, which vibrations are IR active and which are not. This is much more difficult for larger molecules. For such problems, the use of group theory and symmetry arguments can be of great assistance. The → character tables for the various → symmetry point groups can be used to predict how many of the 3*N*-6 vibrations are IR active and how many are Raman active. Where the symmetry point group possesses a center of symmetry there is a → mutual exclusion rule which says that vibrations allowed in the Raman are forbidden in the IR, and vice versa. The symmetry of a

Vibrational spectrum. Table.

1	2	3		Term/Energy-level
≥ 1	0	0	$G_o(v_1)$	$= \omega_1 v_1$
0	≥ 1	0	$G_o(v_2)$	$= \omega_2 v_2$
0	0	≥ 1	$G_o(v_3)$	$= \omega_3 v_3$
≥ 1	1	0	$G_o(v_2 v_1)$	$= \omega_2 + \omega_1 v_1$
≥ 1	0	1	$G_o(v_3 v_3)$	$= \omega_3 + \omega_1 v_1$
1	≥ 1	0	$G_o(v_1 v_2)$	$= \omega_1 + \omega_2 v_2$
1	0	≥ 1	$G_o(v_1 v_3)$	$= \omega_1 + \omega_3 v_3$
0	≥ 1	1	$G_o(v_3 v_2)$	$= \omega_3 + \omega_2 v_2$
0	1	≥ 1	$G_o(v_2 v_3)$	$= \omega_2 + \omega_3 v_3$
≥ 1	1	1	$G_o(v_2 v_3 v_1)$	$= \omega_2 + \omega_3 + \omega_1 v_1$
1	≥ 1	1	$G_o(v_1 v_3 v_2)$	$= \omega_1 + \omega_2 + \omega_3 v_2$
1	1	≥ 1	$G_o(v_1 v_2 v_3)$	$= \omega_1 + \omega_2 + \omega_3 v_3$

molecule may also dictate that certain bands are forbidden in both the IR and Raman spectra, in which case even the combination of the two spectra will not give all the 3*N*-6 normal frequencies. A more exact treatment of the vibrational spectra of polyatomic molecules can be made if anharmonicity is included. Consideration of overtones and combinations brings the theory closer to reality.

Ref.: G. Herzberg, *Molecular Spectra and Molecular Structure, II. Infrared and Raman Spectra of Polyatomic Molecules*. D. Van Nostrand Co. Inc., Princeton, New Jersey, New York, London, **1945**.

Vibrational structure, <*Schwingungsstruktur*>, the structure seen on the absorption bands of the → UV-VIS spectra of solutions which is due to coexcited vibrational states in the excited electronic state. The excitation always occurs from the vibrational state $v'' = 0$ of the electronic ground state into the vibrational states $v' = 0$, 1, 2, 3 ... of the excited electronic state. They are called $0 \rightarrow 0$, $0 \rightarrow 1$, $0 \rightarrow 2$ ... transitions. The luminescence occurs from $v' = 0$, the lowest vibrational state of the electronic excited state, to the vibrational states of the electronic ground state, $v'' =$ 0, 1, 2, 3 ... the $0 \rightarrow 0$, $0 \rightarrow 1$, $0 \rightarrow 2$ ... transitions, which also explains the mirror-image vibrational structure of the absorption and → fluorescence spectra; Figure 1. The case of → phosphorescence is very similar as far as the vibrational structure is concerned. It differs only in that the triplet electronic state from which the phosphorescence originates lies lower in energy than the lowest singlet and the emission is therefore red-shifted; Figure 1. With gases at low pressure, and spectrometers of high resolving power, more structure due to the unhindered excitation of molecular rotations can be observed. This is the → fine structure of the spectrum.

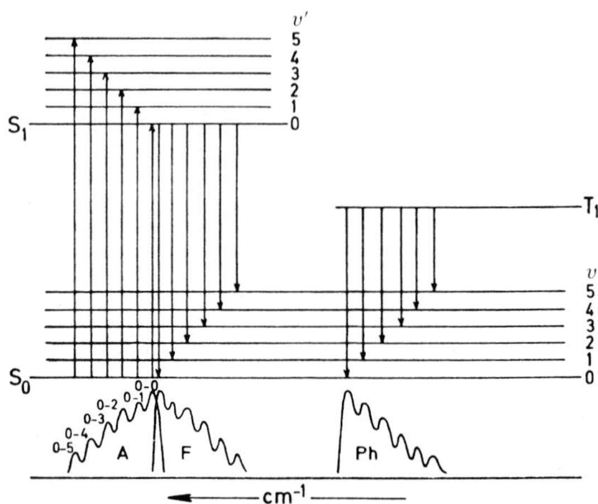

Vibrational structure. Fig. 1. An explanation of the vibrational structure of UV-VIS absorption and luminescence spectra.

Vibrational symmetry species, <*Schwingungsrassen*> → symmetry species.

Vibronic, <*Vibronic*>. In the → electronic band spectra of molecules, the electronic excitation is always accompanied by the excitation of vibrational states of the electronic excited state. The results of such coexcitation are commonly described by the word *vibronic*, a combination of *vibrational* and *electronic*; as in vibronic band and vibronic state. Using the → Born-Oppenheimer approximation, the wave function of a vibronic state may be written as a product of an electronic state wave function and a vibrational state wave function. The vibrational structure of a vibronic transition can then be analyzed by applying the → Franck-Condon principle.

Voigt band shape, <*Voigt-Profil*> → line shape.

Wadsworth mounting, <*Fuchs-Wadsworth-Aufstellung*>, the mounting of a 60° prism which is the one preferred for IR spectrometers with prism monochromators. A plane mirror is placed behind the prism parallel to its base. A beam of radiation which enters the prism at the angle of minimum deviation is only displaced parallel to itself. Since the prism and mirror are rigidly joined and are rotated together about a common axis, the condition of minimum deviation in the prism and a parallel shift of the beam can be achieved for every wavelength → Littrow mounting.

Wave function, <*Wellenfunktion*>, the spatial function, $\Psi(x, y, z)$ in a Cartesian coordinate system, which satisfies the time-independent → Schrödinger equation. As solutions of the time-independent Schrödinger equation, the wavefunctions describe stationary states. The wave function is expressed in a coordinate system suited to the problem; e.g. for the hydrogen atom in polar coordinates, r, ϑ, and ϕ: $\Psi(r,\vartheta,\phi)$.

Wavelength, <*Wellenlänge*>. A periodic plane wave at a particular point may be described by the oscillating function, $s = s_o \sin(2\pi\nu t)$. At a distance, x, from this point the wave may be written:

$$s = s_o \sin 2\pi\nu(t - \frac{x}{c})$$

where ν is the frequency, t the time and c the propagation velocity of the wave, i.e. the wave velocity.
Two points, x_1 and x_2, are then at the same stage of the oscillation if their phase difference (→ phase of oscillation) is an integer multiple of 2π, i.e. $2n\pi$, where n is a positive or negative integer. Therefore:

$$2\pi\nu(t - \frac{x_1}{c}) - 2\pi\nu(t - \frac{x_2}{c}) = 2n\pi$$

or

$$x_2 - x_1 = nc/\nu.$$

Two successive points at the same stage of the oscillation, e.g. two adjacent maxima, $+s_o$, or minima, $-s_o$, are separated by a distance $x_2 - x_1 = c/\nu$. This distance is called the wavelength and is usually given the symbol $\lambda = c/\nu$. In the case of electromagnetic waves, c is the velocity of light $= 3 \cdot 10^8$ m s^{-1}. The dimensions of wavelength are length.
The wavelength range encompassed by spectroscopy stretches from 10^{-2} nm to 10 m; → spectral regions.

Wavelength scale, <*Wellenlängenskala*>, the scale of a → spectroscopic instrument which gives the → wavelengths of the dispersed radiation. The wavelength scale of fixed dispersing systems is also copied onto the spectrum. In instruments with rotating dispersing systems, e.g. → monochromators, the scale is normally engraved on the wavelength drum, or on a read-out device which is coupled to it. In many cases the wavelength scale is preprinted onto the chart paper.
The computer-controlled instruments developed in recent years, display the wavelength scale for the chosen spectral region and the ordinate, in absor-

bance or transmittance, on a screen. The measured spectrum is then shown in this framework and the whole plotted out if required.

Wavelength standard, <*Wellenlängennormal*>, the → wavelength of a particular spectral line which, because of its small line width, is suitable as a basis for length measurement and for which the excitation conditions have been agreed internationally. A wavelength standard must be constant in time, well defined and readily reproducible. A suitable emission line is one with a simple, symmetrical shape (isotope of even atomic number) and small line width (high atomic weight, low emission temperature). The excitation conditions must be chosen such that the line is not shifted, made unsymmetrical, or the → line width increased, by collisions, the → Stark effect or other influences (this suggests a low gas pressure and excitation with weak electrical fields or in an atomic beam). The red cadmium line was used at first. The mean values for this line in normal air are $\lambda_{Cd} = (6.438469 \pm 0.000003) \cdot 10^{-7}$ m, and in vacuum $(\lambda_{Cd})_{vac} = (6.440249) \pm 0.000004) \cdot 10^{-7}$ m.
More recently, the orange line of ^{86}Kr has been frequently used as a wavelength standard. Its vacuum wavelength is $(\lambda_{Kr})_{vac} = (5.6511289 \pm 0.000004) \cdot 10^{-7}$ m. It was agreed at the Tenth General Conference on Measurement and Weight in 1960 that this line should be the wavelength standard. Mercury lines are also useful standards.

Wavenumber, <*Wellenzahl*>, a unit defined as the inverse of the → wavelength ($\tilde{\nu} = 1/\lambda$), which gives the number of wavelengths per meter or per centimeter. (The correct SI unit is m^{-1}, but cm^{-1} is still customary in spectroscopy). The reason for introducing the wavenumber lies in the → Planck-Einstein relation, $E = h\nu$, according to which the energy of electromagnetic radiation is directly proportional to the frequency, ν. Writing $\nu = c/\lambda$, it follows that $E = hc/\lambda$ or $E = hc\tilde{\nu}$. Thus the wavenumber is also directly proportional to the energy, while the wavelength is inversely proportional to it. The wavenumber, $\tilde{\nu}$, is a more practical unit than the frequency, ν, (→ spectral regions).

Wave packet, <*Wellengruppe, Wellenpaket*>, a group of superimposed waves, traveling in the same direction, which differ somewhat in their frequencies and therefore also in their → wavelengths. If the waves differ among themselves in their velocities (→ phase velocity), the amplitude maximum which results from the superimposition of the individual waves propagates with the → group velocity, v_g. Since it is proportional to the square of the amplitude, the intensity maximum propagates with the same group velocity.

Wave plate, lambda plate, <*Halbwellenlängenplättchen, Lambda-Plättchen, Viertelwellenlängenplättchen*> → lambda plate, → half-wave plate, → quater-wave plate.

Wave vector, <*Wellenvektor, Wellenzahlvektor*>. The wave vector, $\vec{k}$, is defined by $\vec{k} = 2\pi/\lambda$ or $\vec{k} = 2\pi\tilde{\nu}$. It points in the direction of propagation of the electromagnetic wave.

Wave velocity, <*Wellengeschwindigkeit*> → phase velocity.

Wedge error, <*Keilfehler*>, an error arising in an optical device because two plates are not plane parallel. It occurs most importantly in → cuvettes (cells) for the → UV-VIS, NIR and → IR → spectral regions.

Wedge-shaped shutter, <*Keilblende*>, a → slit diaphragm by means of which the height of a slit, e.g. in a → monochromator, can be varied.

Wernicke prism, <*Wernicke-Prisma*> → Zenker prism.

White standard, <*Weißstandard*>, a body which ideally has a wavelength-independent reflective power, $R_{st} = \varrho = 1$ in the UV and VIS regions. Because the diffuse reflectance of a sample, R_{sa}, cannot be measured absolutely, it is always referred to that of a white standard. Thus, the quantity measured in → reflection spectroscopy:

$$R'_\infty = R_{sample}/R_{standard}$$

is a relative quantity. If the absolute reflectance, ϱ, of the white standard were equal to one, i.e. if $R_{st} = \varrho = 1$, then the absolute and relative reflectances of the sample would be equal. But there is no known white standard which has this property over the whole spectral region of interest; UV-VIS-NIR. One is therefore obliged to use as standards those materials which approach the ideal value $\varrho = 1$ most closely. To date, MgO has proved to be the most useful white standard in practice, because it is easy to prepare under defined conditions. And for this reason many measurements of the reflective power, ϱ, of MgO as a function of wavelength have been made. Its ϱ values in the visible region are: 0.983 at $\lambda = 420$ nm, 0.986 at $\lambda = 680$ nm, and a maximum of 0.988 at $\lambda = 620$ nm. Kortüm and his co-workers determined the absolute reflectance for frequently used white standards which are of particular interest in physical chemical applications. (G. Kortüm, W. Braun, G. Herzog, *Angew. Chem.*, **1963**, *75*, 653). In addition to MgO, the following substances were measured in the UV-VIS and NIR regions: Li_2CO_3, NaF, NaCl, $MgSO_4$, $BaSO_4$, Aerosil, Al_2O_3, SiO_2 and glucose.
The results showed that the absolute reflectance decreases steeply towards the UV. This is also true of the NIR. Aerosil is an exception in that it shows ϱ values between 0.90 and 0.99, even above 30,000 cm^{-1}.
Ref.: G. Kortüm, *Reflectance Spectroscopy*, Springer Verlag, Berlin, Heidelberg, New York, **1969**.

Wien's displacement law, <*Wiensches Verschiebungsgesetz*> → Wien's radiation law, → spectral energy distribution.

Wien's radiation law, <*Wiensches Strahlungsgesetz*>, a law derived from thermodynamic considerations by W. Wien in 1893. The spectral radiation density, S_λ, within a solid angle, Ω_o, is defined to be:

$$S_\lambda = \frac{2c_1}{\pi\Omega_o} \cdot \lambda^{-5}\, e^{-c_2/\lambda T}.$$

In 1900, the values of the constants, c_1 and c_2, were found through → Planck's radiation law to be:

$$c_1 = c^2h \text{ and } c_2 = hc/k$$

where h is Planck's constant, c the velocity of light and k the Boltzmann constant.

When it is compared with → Planck's radiation law,

$$S_\lambda(T) = \frac{2c_1}{\pi\Omega_o}\lambda^{-5} \cdot \frac{1}{e^{c_2/\lambda T} - 1}$$

Wien's law is seen to be an approximation which reproduces the spectral distribution of the radiation density at lower wavelengths ($\lambda < 800$ nm) very well. If $\exp[c_2/\lambda T] \gg 1$, the 1 in the denominator can be neglected and the Planck formula goes over into the Wien formula. For this reason, Wien's law can be used to calculate the conversion values of → conversion filters.

Wien's radiation law is linked to Wien's displacement law which states that $\lambda_{max} \cdot T$ = constant = 2.8978 10^{-3} m K which is denoted by ω. According to this result, if the temperature of a → black-body radiator is raised, the maximum of the spectral radiation density, λ_{max}, shifts to shorter wavelengths, i.e. in the direction of the visible region. Lowering the temperature shifts the maximum towards the infrared (→ spectral energy distribution).

Wire-grid polarizer, <*Drahtgitterpolarisator*>, a polarizer made of a wire grid which only transmits radiation when the electric field can induce a current in the wires, i.e. when the electric vector is perpendicular to the wires. H. Hertz studied the properties of radio waves using wire-grid polarizers as early as 1888. A grid made from copper wires 1 mm thick and 3 cm apart was used with 66 cm waves which were reflected or transmitted according to whether the wires were orientated parallel or perpendicular to the electric vector of the radiation, respectively.

If the diameter of the wires and the distance between them are made smaller, then such grids also show polarizing effects in the far IR. The separation of the wires must be small in comparison to the wavelength of the light to be polarized, Therefore, the use of wire-grid polarizers in wavelength regions of interest to molecular spectroscopists, (2.5–50 μm) failed because of the difficulty of making grids with such fine wires so close together. But Bird and Parrish fabricated a suitable grid by first pressing an → echelette grating (2160 lines/mm) onto films of polyethylene and polychlorotrifluoroethylene. They then evaporated gold or aluminium onto the impression of the grating, at an angle and from the side. Since the metal deposits only on the high areas of the impression, they obtained a grid of fine parallel metal wires with the same geometry as the grating.

The metal used must be a good electrical conductor and chemically resistant. Gold satisfies both of these requirements. Other substrates including polymethylmethacrylate, polyethyleneterephthalate, irtran-2 and irtran-4, were used. In the case of the last two materials, the lines were scribed rather than pressed onto the substrate.

The great disadvantage of the these polarizers is that they can only be used in part of the infrared region, because they have more or less strong absorption bands or are no longer transparent in the long-wavelength region. These limitations do not apply to the wire-grid polarizers developed by Perkin-Elmer. For these devices, silver

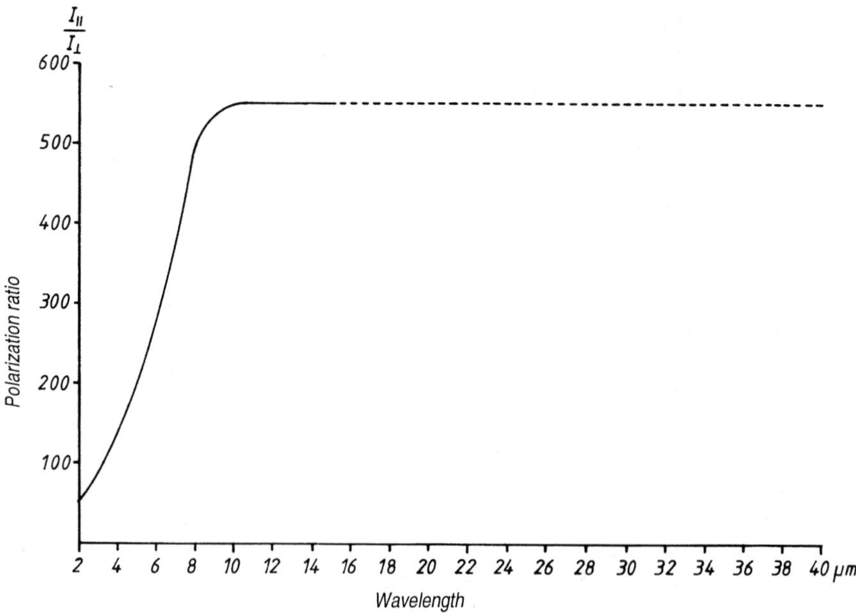

Wire-grid polarizer. Fig. 1. Polarization ratio for a wire-grid polarizer (substrate AgBr) in the region 2.5 to 40 μm

bromide is used as the substrate for the region 2.5 to 40 μm (2880 lines/mm) and polyethylene for the long-wavelength region above 15 μm (640 lines/mm). These polarizers are ca. 0.5 mm thick with a diameter of 16–22 mm and fixed in a plastic ring which can be fitted into the various mountings. Figure 1 shows the wavelength dependence of the degree of polarization, $I_{||}/I_{\perp}$, of an AgBr polarizer, measured with two wire-grid polarizers. $I_{||}$ is the transmission when the two polarizers are orientated parallel to each other and $I_{\perp}$ when they are perpendicular. If the ratio is, say 200, then only 0.5% of the light is transmitted in the crossed orientation. The flat portion of the curve above 8 μm does not mean that the degree of polarization is constant at 500 throughout this region, but that it is not possible to measure accurately the very small value of $I_{\perp}$. Thus, the curve shows only the lower limit of the degree of polarization. The value for the polyethylene substrate grating in the long-wavelength region above 20 μm is greater than 100. In addition to their high efficiency, wire-grid polarizers have a number of advantages over Brewster polarizers.

1. Because of their small size, they can easily be placed in front of the entrance slit of a monochromator, the position most desirable for the measurement. In this position, the polarization properties of the spectroscopic instrument itself are largely eliminated and the sample space is kept free for other accessories.
2. The beam of radiation is not displaced to the side.
3. The convergence of the beam has

no influence upon the degree of polarization which remains constant for changes in the angle of incidence of more than 50°.

4. The polarizers are very easy to operate and, with built-in polarizers, this is possible without opening the housing of the instrument, i.e. without interrupting the purge.

Wollaston prism, <*Wollaston-Prisma*>, a → prism, developed by W.H. Wollaston in 1820, which consists of two 90° calcite prisms cemented together at their bases, as shown in Figure 1. The optic axis of prism I lies in the plane of the drawing and parallel to the entrance surface, *AC*. In prism II it is perpendicular to the plane of the drawing. In prism I, light entering normal to *AC* splits into an ordinary (*o*) and an extraordinary (*e*) ray. These two rays travel to the cemented interface, *BC*, in their original direction, but with different velocities. The *o* ray which is vibrating perpendicularly to the plane of the drawing is refracted towards *AB* on entering prism II because it is now the *e* ray and travels with greater velocity. In contrast, the *e* ray in prism I becomes the *o* ray in prism II where its velocity is decreased and it is therefore refracted away from *AB*. Thus the prism induces a separation of the two mutually perpendicularly polarized rays, and their deviating paths are symmetrically disposed with respect to the direction of the incident light. The angle between them lies between 5° and 20° for → calcite. Wollaston prisms can also be made from quartz, but the symmetrical splitting is smaller; $0.5° < \alpha < 1°$.

The Wollaston prism is preferred for applications where the two rays are to be recombined again later.

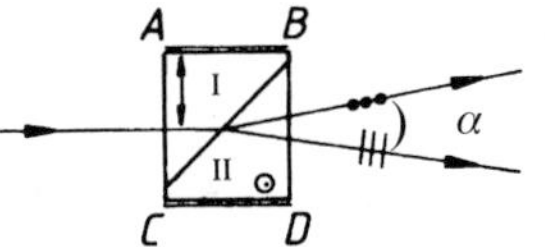

Wollaston prism. Fig. 1. A section through a Wollaston prism showing the ray path

XBO lamp, <*XBO-Lampe*>, the name of a xenon, short-arc, high-pressure lamp (→ xenon lamp) specific to the Osram company. Two or more following figures give the power of the lamp, e.g. XBO 75 W, XBO 450 W ... etc. More details are given in the manufacturer's literature.

Xe lamp, <*Xe-Lampe*> → xenon-pulse discharge lamp, → xenon lamp.

Xenon-ion laser, <*Xenonionenlaser*> → noble-gas ion laser.

Xenon lamp, <*Xenonlampe*>, a → noble-gas discharge lamp operated at high or very high pressure. Because the excitation energies of noble gases are so high, high gas temperatures are required in the discharge to obtain radiative emission. Furthermore, since the excited energy levels lie closer to the ionization limit than they do, for example, with mercury, the high gas temperatures produce a high degree of ionization and a large proportion of continuous radiation, produced by recombination, in the total emission (→ recombination continuum). Xenon is preferred for use in discharge lamps because, as the heaviest of the noble gases, it has the lowest ionization energy. The light yield of the lighter noble gases is also less, and for this reason the xenon high-pressure lamp has proved to be the most valuable for use in the UV-VIS spectral region. Figure 1 shows the → spectral energy distribution of a xenon lamp. The UV-VIS region shows a broad continuum upon which a few pronounced maxima are superimposed, especially in the NIR. With the help of a relatively simple combination of filters, the xenon spectrum can be made to reproduce the energy distribution of the → global radiation quite well and equipment can be built which simulates the radiation of the sun. In practical terms, in a xenon

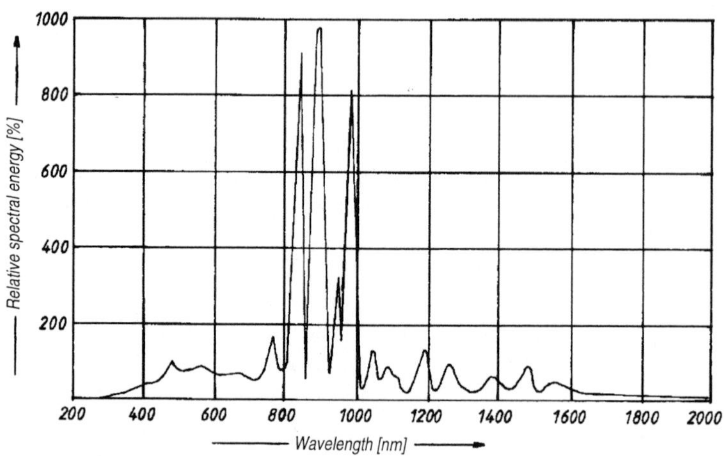

Xenon lamp. Fig. 1. The spectral energy distribution of a xenon lamp

lamp a discharge is taking place in a gas at a temperature of between 7000°C and 7500°C. In addition to its high intensity, the xenon lamp has the advantage that it reaches its full intensity immediately after ignition. This is useful in the construction of → xenon-pulse discharge lamps. Xenon lamps cooled by convection, e.g. type XBO from Osram, are particularly suitable for optical spectroscopy. Like the mercury high-pressure lamps (→ mercury-vapor discharge lamp), they are short-arc lamps, (see Figure 2), and versions with powers between 75 W and several kW are made. Powers up to 30 kW can be achieved with water-cooled electrodes. High-tension igniters are required to start these lamps, but 220 V is sufficient for the continuous supply. The lamps are very unstable when run with alternating current and DC is preferable. For DC operation, the massively built anode should be placed at the top with the more lightly constructed cathode below it.

Xenon lamps are preferred for applications in UV-VIS spectroscopy, especially luminescence spectroscopy, → photoacoustic spectroscopy, optical rotatory dispersion and color measurement; in fact, wherever a high intensity combined with a continuum is required.

Ref.: W.G. Driscoll, W. Vaughan, Eds.:, *Handbook of Optics*, McGraw-Hill Book Co. Inc., New York, San Francisco, Toronto, London, **1978**.

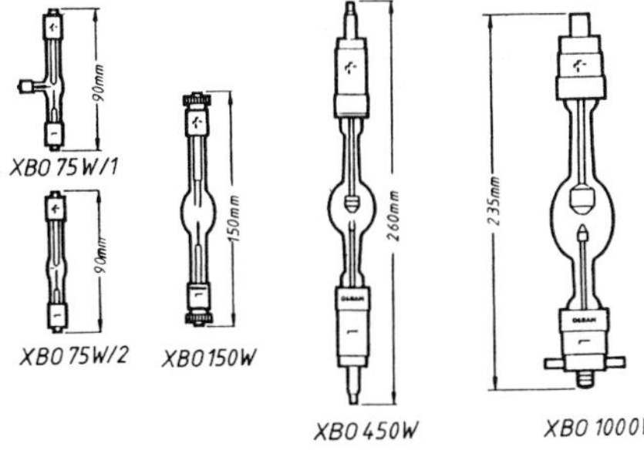

Xenon lamp. Fig. 2. Examples of xenon short-arc lamps

Xenon-pulse discharge lamp, <*Xenon-impulsentladungslampe*>, makes use of the fact that → xenon lamps need no running-in time, i.e. a xenon-pulse discharge lamp reaches its full power immediately after ignition. Therefore, the lamp can be readily modulated and can be used, for example, as a → flashlamp. Its use in the study of photochemical reactions was first described by Norrish and Porter (*Nature*, **1949**, *164*, 658). The xenon-pulse discharge lamps available commercially are models with strengthened walls such as the Xe lamps (Philips) or the XIE lamps (Osram). The XIE lamps have U-shaped discharge tubes and are particularly intended for stroboscopy. The long tube lamps from Noblelight Ltd have variable arc lengths which can be adapted to a particular application. In the short form, they are suitable for exciting luminescence in luminescence spectrometers, (e.g. the Perkin-Elmer LS series). The models with very long arcs find application as pumping sources for lasers or in flash spectroscopy. For higher pulse powers, these lamps can be operated with water cooling. In addition to xenon (lamp types XA, XF, XAD, XAM, XFM, NL), krypton is also used (lamp types KAP, KFP, KFC). The pulse half-widths of these lamps lie between 10 μs and a few hundred μs. They can be adapted to any particular problem by variation of their operating parameters. As examples of the form of these pulse lamps, Figure

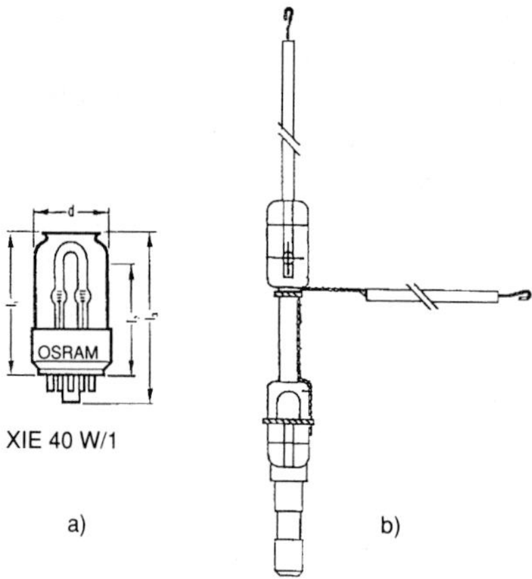

Xenon-pulse discharge lamp. Fig. 1. Two versions of the pulse discharge lamp, a) XIE (Osram) and b) NL (Noblelight)

1 shows an Osram XIE lamp and an NL lamp from Noblelight suitable for application in a → fluorescence spectrometer. Further information is available from the above companies.

XPS, <*XPS*> → X-ray photoelectron spectroscopy.

X-radiation, x-rays, <*Röntgenstrahlung*> → x-ray radiation.

X-ray absorption, <*Röntgenabsorption*>, as with electromagnetic radiation in the optical spectral region, the loss of intensity when → x-ray radiation traverses a material. The loss of intensity may be written:

$$I_\lambda = I_{o,\lambda}\exp[-m_\lambda x]$$

in which m_λ is the linear attenuation coefficient in cm^{-1} and x is the path length in cm. The law corresponds formally to the → Bouguer-Lambert-Beer law and can be derived in an analogous manner. If the linear attenuation coefficient is divided by the density of the material, the mass attenuation coefficient, $\mu = m_\lambda/\varrho_i$, is obtained which is a specific property of the material and also depends upon the wavelength. Its units are $cm^2\ g^{-1}$. Thus, the above law can also be written in the form:

$$I_\lambda = I_{o,\lambda}\exp[-\mu_{i,\lambda}\varrho_i x]$$

If $\mu_{i,\lambda}$ is plotted as a function of wavelength, λ, the x-ray absorption spectrum is obtained. Figure 1 shows the example of lead. In contrast to the optical spectra of atoms where absorption and emission spectra correspond, this is not the case in x-ray spectroscopy. The difference between optical absorption spectra and x-ray absorption spectra can be explained by noting that in optical spectroscopy it is an electron in the outermost shell of the atom which is excited, whereas in x-

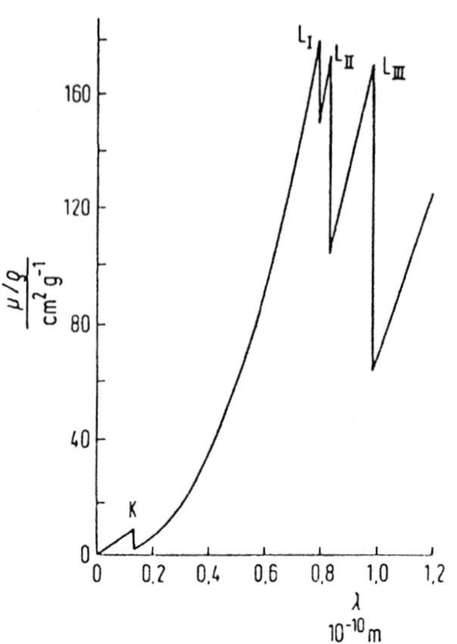

X-ray absorption. Fig. 1. The x-ray absorption spectrum of lead

ray absorption it is an electron in an inner shell. Both rise to a higher energy state in the process and in optical spectroscopy these states are not normally occupied by other electrons. In the case of x-ray excitation, however, the states next in energy are already filled with the electrons of the shells which lie slightly further from the nucleus. Thus, an electron from the inner shells can only find an unoccupied state outside the filled electron shells of the atom. But these free optical levels lie close to the continuum (~ 0.01 eV) so that excitation to such a level is almost the same as removing the electron from the atom entirely. Therefore, the conspicuous edges in the absorption spectrum shown in Figure 1 correspond to the excitation energy required to remove an electron from the → *K* shell or → *L* shell (or from higher shells) of the atom; i.e. they represent the binding energy of the electron in the various states. For → x-ray emission spectra, the situation is different. In this case, an energy state in an inner shell is made vacant by electron impact or other primary radiation and an electron from a shell further out in the atom then jumps into the hole, emitting → x-ray fluorescence. Thus, there is no difference, in principle, between the structure of an emission spectrum in the optical and in the x-ray region.

X-ray diffraction, *<Röntgenbeugung>*. In 1912, M. v. Laue pointed out that crystal lattices might provide a means of observing diffraction phenomena with very short electromagnetic waves. He noted that this would not only answer the question as to the nature of → x-ray radiation but might also allow the wavelength of the radiation to be determined. Following Laue's suggestion, W. Friedrich and P. Knipping, using x-ray radiation and a zinc sulfide crystal, observed a characteristic diffraction pattern which has been known since that time as a Laue pattern.

To derive his diffraction equation, Laue began by considering the crystal lattice points (atoms) arranged periodically with a fixed distance between them, the lattice constant, *a*. Diffraction of the x-ray radiation can only take place when the path difference between two wave trains is a whole number, *h*, of wavelengths. The one-dimensional case is illustrated in Figure 1:

$$\begin{aligned} AB - CD &= a\cos\alpha - a\cos\alpha_o \\ &= a(\cos\alpha - \cos\alpha_o) \\ &= h\lambda \end{aligned}$$

If this relationship is extended to three dimensions the Laue equations are obtained:

$$\begin{aligned} a(\cos\alpha - \cos\alpha_o) &= h\lambda \\ b(\cos\beta - \cos\beta_o) &= k\lambda \\ c(\cos\gamma - \cos\gamma_o) &= l\lambda \end{aligned}$$

a, *b* and *c* are the lattice constants (orthorhombic system), α_o, β_o and γ_o

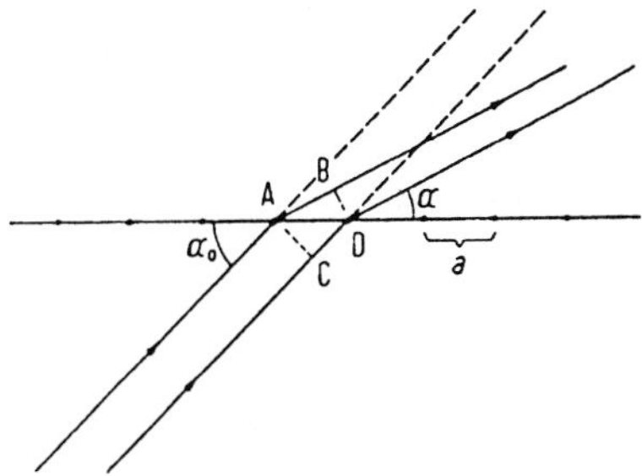

X-ray diffraction. Fig. 1. X-ray diffraction at lattice points according to Laue

the angles of incidence of the incoming radiation and α, β and γ the angles of the diffracted radiation relative to the coordinate axes of the crystal. Diffraction maxima are to be expected only where h, k and l are whole numbers. In 1913, W.H. Bragg and his son W.L. Bragg showed that the diffraction of x-rays could also be interpreted in terms of reflections at the lattice planes of the crystal. This is illustrated for two lattice planes, N_1 and N_2, in Figure 2. Ray 1 is reflected at A_1 on N_1 and ray 2 at point A_2 on N_2. The angles of incidence and reflection, ϑ, should be equal. The two rays have a path difference of $B_1A_2 + A_2B_2 = 2d\sin\vartheta$ and, if the two reflected rays are to show significant interference, this must be a whole number multiple of λ, i.e. for interfering reflection at a group of parallel lattice planes the Bragg condition must hold:

$$2d\sin\vartheta = k\lambda$$

or

$$\sin\vartheta = k \cdot \lambda/2d$$

$k = 1, 2, 3 \ldots$

This means that, for a given separation of the lattice planes, d, only certain wavelengths will be reflected at angle ϑ. Alternatively, for a particular wavelength, λ, and lattice plane separation, d, only certain angles of reflection, ϑ, are possible. The Bragg condition corresponds formally to the equation which describes the diffraction of electromagnetic radiation by a plane → reflection grating. In both cases, the dispersing element, here the crystal, must be rotated about an axis parallel to the entrance slit if different wavelengths are to be detected. This was the principle of the first x-ray spectrograph which was built by Bragg specifically for the analysis of the characteristic x-ray lines.

Both the Laue equations and the Bragg condition offer the possibility of determining crystal structures by x-ray diffraction. Thus, there are two applications of x-ray radiation; x-ray spectral analysis and x-ray structure analysis.

X-ray emission, <*Röntgenemission*>, the → x-ray radiation which is emitted when an electrically conducting, solid sample is bombarded with electrons of very high energy (> 10^4 eV). The emission consists of white x-ray radiation (bremsstrahlung) and the charac-

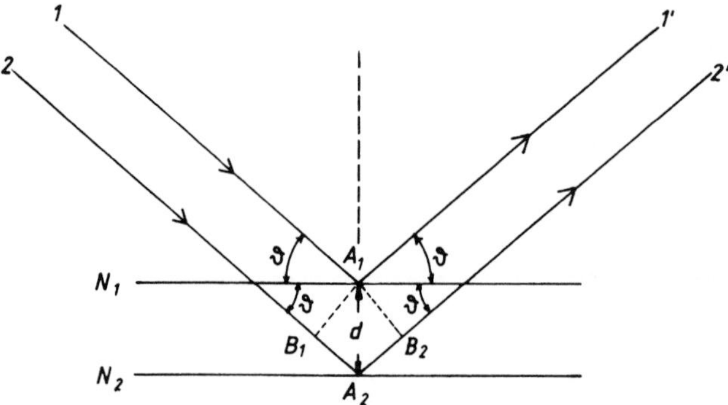

X-ray diffraction. Fig. 2. X-ray diffraction at two lattice planes according to Bragg

teristic → x-ray emission spectrum of the target.

X-ray emission spectrum, <*Röntgenspektrum, Röntgenemissionsspektrum*>, the emission spectrum which is obtained when an anode of a particular metal is bombarded with electrons of very high energy (> 10^4 eV). The spectrum contains white x-ray radiation (bremsstrahlung) and many lines which are characteristic of the anode material. If this radiation is analyzed it is found that, as with → atomic spectra, the lines can be arranged in well-defined series. The spectra can be explained, following Kossel, by arranging the → *K*, → *L*, → *M*, → *N* ... shells of the atom in an → energy-level scheme, as in Figure 1. For example, if an electron is excited out of the *K* shell into the continuum, then a hole is created in the *K* shell. An electron from the occupied higher shells can jump into this hole whereby, according to the fundamental equa-

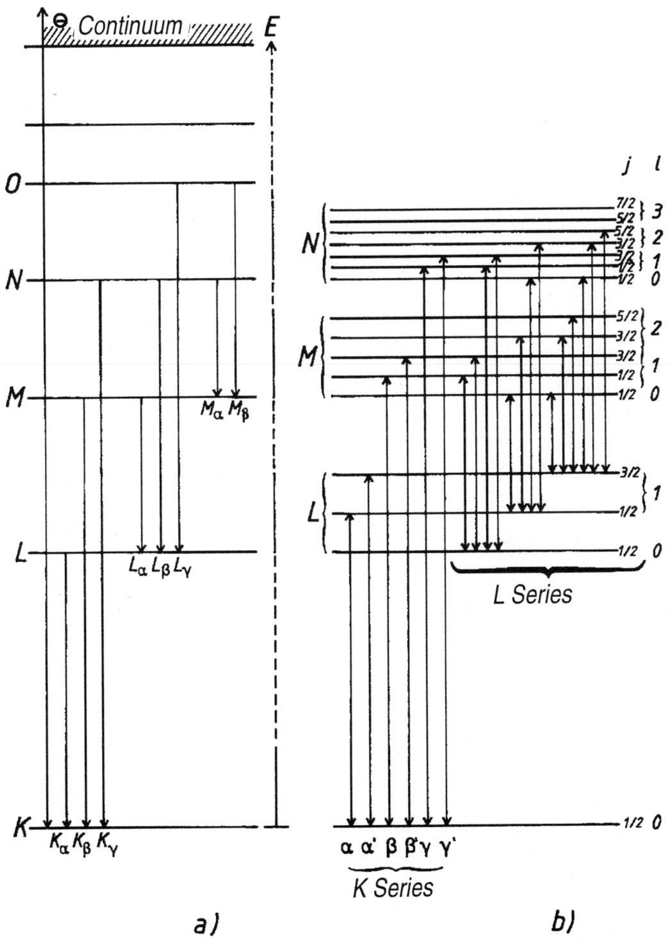

X-ray emission spectrum. Fig. 1. A Kossel term (energy-level) scheme for x-ray emission

tion of spectroscopy, $\Delta E = h\nu$, electromagnetic radiation of energy equivalent to the energy difference, ΔE, must be emitted. The K_α, K_β and K_γ lines in the spectrum then correspond to electrons from the L, M and N shells. Clearly, if the hole is filled by an electron from the L shell, then this creates a hole in the L shell which can be filled by an electron from a higher shell with the emission of electromagnetic radiation. This gives rise to the L series of lines, see Figure 1a. The way of naming the lines is clear.

A more exact analysis shows that the x-ray terms (energy levels) can be regarded as one-electron terms and that their term schemes should be analogous to those of the alkali-metal atoms. The initial state for the emission of an x-ray line is that of a hole in a closed shell of electrons. This state is equivalent to one in which there is just one electron in an otherwise empty shell. Thus we have, as with the → alkali spectra, the terms of a doublet system (→ multiplet structure of atomic spectra) with $1^2S_{1/2}$, (K shell), $2^2P_{1/2,3/2}$ (L shell), $3^2S_{1/2}$, $3^2P_{1/2,3/2}$, $3^2D_{3/2,5/2}$ (M shell) etc. This x-ray term scheme is shown in Figure 1b; the placement of the terms on the energy scale is arbitrary and the term differences are not comparable. Nevertheless, the possible transitions can be used to explain the finer points of x-ray spectra. The selection rules should be noted; as with atomic spectra they are $\Delta l = \pm 1$, $\Delta j = 0, \pm 1$. Each of the three transitions, K_α, K_β and K_γ in the Kossel scheme, Figure 1a, gives rise to two lines, as can be seen from the complete scheme in Figure 1b. Kossel's L_α and L_β lines are each composed of 7 components. The systematic investigations of Moseley (1913/14) showed that the wavenumbers of the K lines of the elements are proportional to $(Z - 1)^2$. In elements with higher nuclear charge, Z, these lines correspond to the first line in the Lyman series (→ Balmer formula) of the hydrogen atom. Moseley found that the wavenumbers of the L lines, which correspond to the first line in the Balmer series, were proportional to $(Z - 7.4)^2$. Thus, the series formulas for the x-ray lines according to Moseley can be written in analogy with the Balmer formula as:

$$\tilde{\nu}_{K_\alpha} = R(Z-1)^2\left(\frac{1}{n_1^2} - \frac{1}{n_2^2}\right);$$

$$n_1 = 1;\ n_2 = 2, 3, 4 \ldots$$

$$\tilde{\nu}_{L_\alpha} = R(Z-7.4)^2\left(\frac{1}{n_1^2} - \frac{1}{n_2^2}\right);$$

$$n_1 = 2;\ n_2 = 3, 4, 5 \ldots$$

Setting $n_1 = 1$, $n_2 = 2$ for the K_α line and $n_1 = 2$, $n_2 = 3$ for the L_α line we obtain:

$$\tilde{\nu}_{K_\alpha} = \frac{3}{4}R(Z-1)^2;$$

$$\tilde{\nu}_{L_\alpha} = \frac{5}{36}R(Z-7.4)^2.$$

The numbers 1 and 7.4 are interpreted as shielding constants. They account for the fact that the inner electrons partially shield the nuclear charge, Z. The shielding effect, and therefore the shielding constant, must increase as the shell corresponding to the particular x-ray term (K, L, M …) gets further from the nucleus, i.e. as the quantum number, n, of the constant term in the x-ray series increases. For the K series, the shielding constant of 1 corresponds to the charge of 1e of the single electron remaining in the K shell. The shielding constant of 7.4 corresponds to the seven electrons which remain in the L shell. The expression for the wavelengths of the K_α lines can be rearranged to give the equation of the → Moseley line:

$$\sqrt{\frac{\tilde{\nu}_{K_\alpha}}{(3/4)R}} = Z - 1.$$

A graph of the square root against Z gives a straight line. Similarly:

$$\sqrt{\frac{\tilde{\nu}_{L_\alpha}}{(5/36)R}} = Z - 7,4.$$

Both equations express the law found empirically by Moseley in 1913 and named after him. The significance of the law lay in the fact that, for the first time, the chemical elements could be arranged in a sequence of increasing atomic number, which made a correct ordering of the periodic table possible. Figure 2 shows the Moseley plots for the K_α, L_α and M_α lines.

X-ray fluorescence, <*Röntgenfluoreszenz*>, the fluorescence which results from the excitation of a sample with the radiation from an x-ray tube (photons). A Coolidge tube (see Figure 1) is normally used as the excitation source. It consists of a closed tube with a heated cathode and a cooled anode (anticathode). The thermionic electrons are accelerated by a voltage of ca. 80 kV and strike the anode. The x-rays which result leave the tube through a side window, d. Au, W, Ag, Rh, Mo and Cr are used as anode materials and the window is made of a thin Be foil. Apart from the side-window design, Coolidge tubes with end windows are also available. A windowless tube has been found advantageous for the excitation of lighter elements with radiation of longer wavelengths. → Gamma radiation

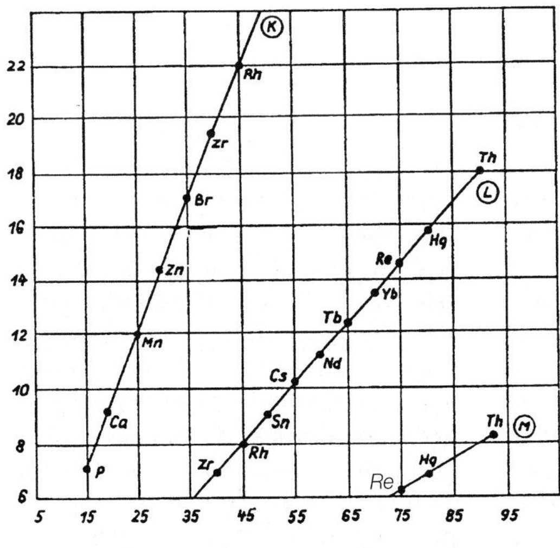

X-ray emission spectrum. Fig. 2. Moseley plots for the K_α, L_α and M_α lines

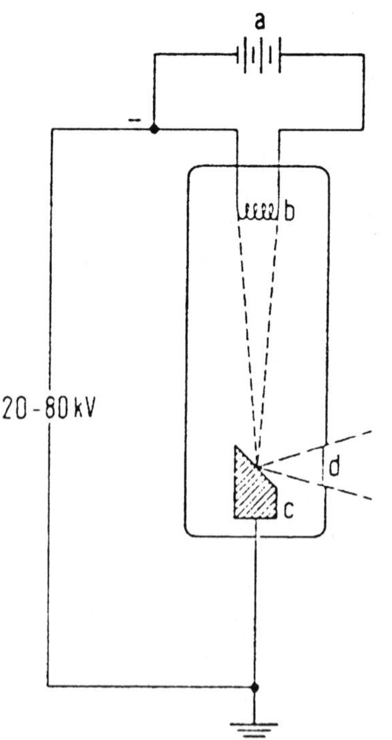

X-ray fluorescence. Fig. 1. A Coolidge tube for the excitation of x-ray fluorescence

from radioisotopes, e.g. ^{59}Fe or ^{109}Cd, can also be used to excite fluorescence. X-ray fluorescence is used in x-ray spectral analysis.

Lit.: R. Tertian, F. Claisse, *Principles of Quantitative X-Ray Fluorescence Analysis*, Heyden and Sons Ltd., London, Philadelphia, Rheine, **1982**.

X-ray photoelectron spectroscopy (XPS), electron spectroscopy for chemical analysis (ESCA), <*Röntgenphotoelektronenspektroskopie, XPS, ESCA-Methode*>, a variant of → photoelectron spectroscopy in which x-ray radiation is used to eject photoelectrons. The fundamental energy equation is the same as in UV photoelectron spectroscopy (UV-PS); $h\nu = E_b + E_{\rm kin}$, where E_b is the binding energy (ionization energy) and $E_{\rm kin}$ the kinetic energy of the ejected electron. Whereas in UV-PS it is the valence electrons which are ejected from the molecule, in XPS it is the deeper lying, more strongly bound electrons in the atomic cores which are ionized. To eject them from the atom, x-ray sources which give x-ray photons of high energy are required. The energy must be so large that the photoelectron is ejected from the atom and given an excess of kinetic energy, $E_{\rm kin}$. If this excess, kinetic energy is measured and the x-ray photon energy, $h\nu$, is known then the binding energy, E_b, can be calculated from the above equation. This is the ionization energy of a core electron and is a significant quantity for the ESCA technique. An important difference between UV-PS and XPS is that for the former the sample must be used in the gas phase whereas for the latter it must be a solid. But gases and liquids can be studied with ESCA if they are cooled to a sufficiently low temperature on a suitable sample holder. The construction of an ESCA spectrometer is shown schematically in Figure 1. X-rays from an aluminium or magnesium cathode strike the sample and eject electrons from it. The electrons are focused with a lens system and then, to increase resolution, delayed before they enter the energy analyzer and eventually reach the detector. The analyzer generally consists of hemispherical condensers with which electrons having large angles of incidence can be focused in two dimensions onto the exit slit. By varying the potential on the condenser, electrons with a particular mean kinetic energy can be

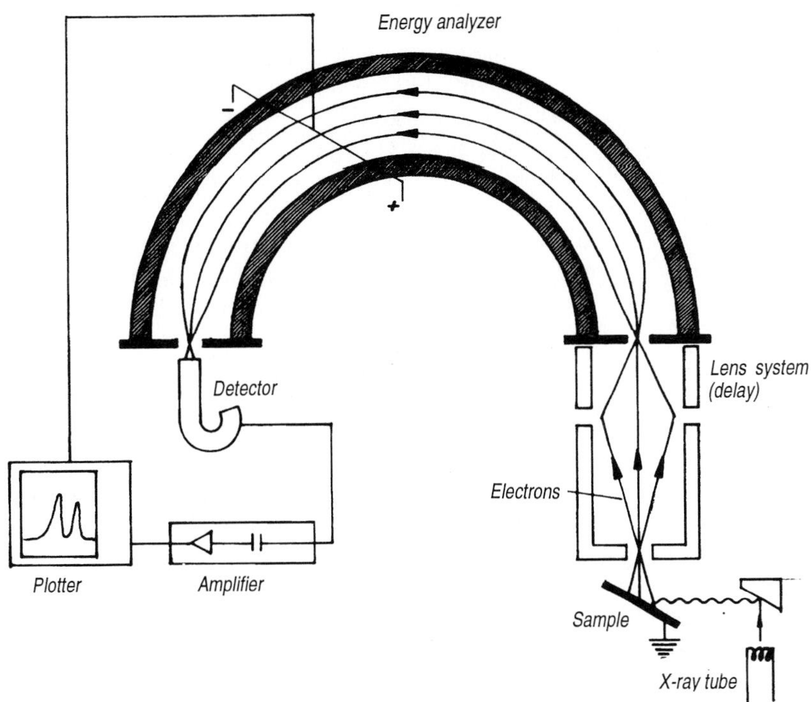

X-ray photoelectron spectroscopy. Fig. 1. The construction of an ESCA spectrometer

brought to a focus at the exit slit. The number of impulses received by the detector in a given time interval, the count rate, is plotted as ordinate against the kinetic energy in eV to give the ESCA spectrum. The energy of the x-ray radiation produced by the cathodes mentioned above is 1458 eV and corresponds to the $K_{\alpha 1}$ and $K_{\alpha 2}$ lines which are separated by only 0.2 to 0.5 eV. The resolution is therefore of the order of 1–2 eV, but it can be improved to ca. 0.5 eV by making the exciting x-ray beam more monochromatic.

The → Auger process and → x-ray fluorescence should be considered as competing processes. But Auger electrons are only important for light elements and for heavy ones the untroublesome fluorescence is dominant. The most important areas for the application of the ESCA technique are:

1) In inorganic and organic chemistry for elemental analysis, structure determination and the investigation of the charge distribution in molecules. The same atom in different chemical structural environments has a measurable difference in ionization energy so that the ESCA peaks occur at different energies and can therefore be used in structure determination. This is often called a chemical shift, though the term is not used consistently in this context.
2) In physical chemistry, industry and technology, ESCA is especially suited to the study of surfaces

where it gives valuable information about e.g. adsorption phenomena, corrosion and the nature of the surface of catalysts. The penetration depth of the radiation is 0.4 to 2 nm in the case of metallic samples and 2 to 10 nm for organic materials. Thus, a surface layer a few atomic diameters thick is probed without damage.

Ref.: C.S. Fadley in *Electron Spectroscopy: Theory, Techniques and Applications*, vol 2, (Eds.: C.R. Brundle and A.D. Baker), Academic Press, London, New York, San Francisco, **1978**.

X-ray radiation, *<Röntgenstrahlung>*, the electromagnetic radiation which has the highest energy. The order of magnitude of the x-ray photon energy is several tens of thousands of eV which corresponds to wavenumbers > 10^8 cm^{-1}. The wavelengths of this short-wavelength radiation are < 1 Å or 0.1 nm. In comparison with optical spectra, very high energies are required to produce x-rays and → x-ray emission spectra. For this purpose an x-ray tube (→ x-ray fluorescence) is used in which the anode (often called anticathode) is bombarded with very energetic electrons, i.e. electrons which have been accelerated by means of voltages of 10^4 to 10^5 V. The emitted radiation is of very short wavelength and consists of two parts; white x-ray radiation, which is also called bremsstrahlung, and the characteristic x-ray lines of the anode. The white x-ray radiation arises because, on entering the anode material, the very energetic electrons are slowed down by the electric fields of the atoms. Their resulting loss of kinetic energy is emitted as radiation, which explains the name *braking radiation*; *bremse* = brake and *strahlung* = radiation in German. The white x-ray emission must have a limit on the short-wavelength side when the electron has lost all its kinetic energy in the braking process, see Figure 1. The characteristic x-ray lines result from a second process. The energetic electrons can eject a bound electron from an inner shell of an anode atom. The resulting hole is then filled by an electron from a shell further out in the atom and, as with optical spectroscopy, radiation must be emitted. The frequency of this radiation corresponds to the energy difference between the two states involved. This process gives the → x-ray spectra or lines. If the x-ray emission is produced by irradiation with x-rays, it is called → x-ray fluorescence. If x-rays are emitted in the course of a nuclear reaction it is called → gamma radiation.

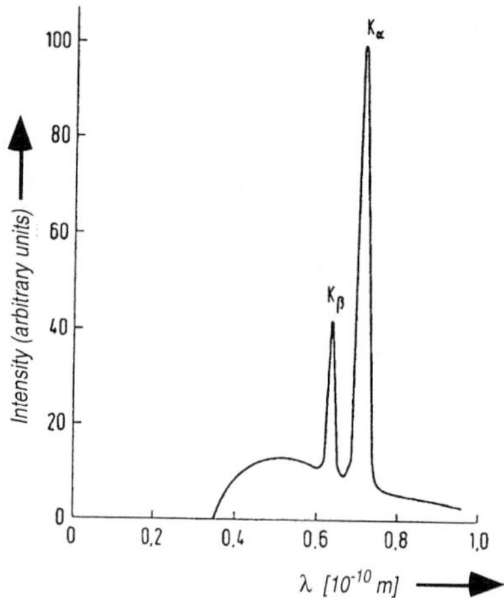

X-ray radiation. Fig. 1. The x-ray emission from a molybdenum cathode

X-ray spectral analysis, *<Röntgenspektralanalyse>*, a method with which the chemical elements can be analyzed qualitatively and quantitatively. The underlying principle of the method is the → x-ray emission spectrum which is characteristic for every element. This characteristic remains, even in the case of chemical compounds, because x-ray spectra are produced by the excitation of tightly bound inner electrons and not by the outer valence electrons of the element. The x-ray spectra used in analysis are produced either by the bombardment of the sample with highly energetic electrons ($E > 10^4$ eV) or by irradiation with high-energy photons, i.e. → x-ray fluorescence. These energetic photons are produced with a primary x-ray source. There is a third possibility, → x-ray absorption, in which the loss of intensity of x-rays passing through the sample is measured.

It is important that x-ray spectral analysis is, in general, nondestructive and that it measures a layer of the sample near the surface. In recent years, the electron beam microprobe and scanning electron microscopy have been developed in combination with → x-ray spectrometers, especially for surface analysis.

Ref.: R. Jenkins, *An Introduction to X-Ray Spectrometry*, Heyden and Son Ltd., London, New York, Rheine, **1974**.

X-ray spectrometer, *<Röntgenspektrometer, Röntgenspektralphotometer>*, an instrument which comprises three components: an excitation source, the actual spectrometer and the detector and measuring electronics. Depending upon the way in which the excitation is brought about, a distinction is made between → x-ray emission, which is usually excited with high-energy electrons ($E > 10^4$ eV), and x-ray fluorescence in which photons are used. The radiation from the sample is dispersed in the spectrometer and two modes of dispersion are possible; the classical dispersion by wavelength using analyzing crystals and the energy dispersive method using semiconductor detectors.

The electronic measuring unit consists essentially of a timer, a scaler (pulse counter) and a rate meter.

In a wavelength dispersive crystal spectrometer, → gas-flow counters or → scintillation counters are used as detectors. The construction of a wavelength dispersive instrument is shown in Figure 1. This is an x-ray fluorescence spectrometer because the sample is excited by photons from an x-ray source, *a*. This form of construction goes back to the classic work of Bragg. A part of the emission from the sample, *b*, passes through the entrance collimator, *c*, a group of parallel plates (Soller slit) which direct the rays, and then falls upon the analyzing crystal, *d*, at an angle of incidence, *θ*. Only those monochromatic wavelengths which fulfill the Bragg condition (→ x-

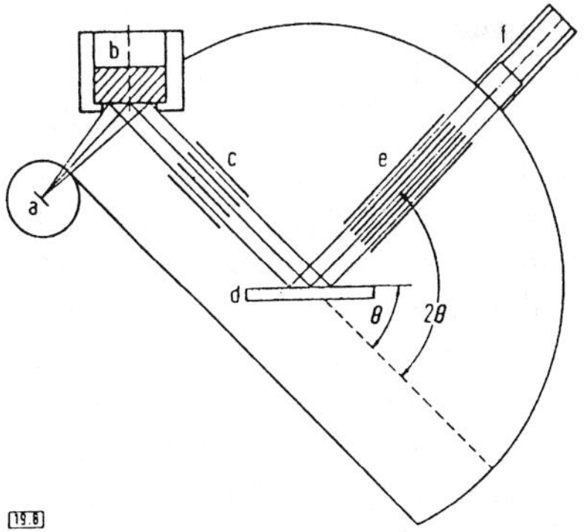

X-ray spectrometer. Fig. 1. Schematic drawing of a wavelength dispersive x-ray spectrometer

ray diffraction) are reflected at the lattice planes of this crystal, and it is these components which, deflected by an angle of 2θ, enter the exit collimator, *e*, and are registered by the detector, *f*. If the analyzer crystal is rotated and the exit collimator and detector, are simultaneously rotated at twice that angular velocity, then the spectrum of the radiation emitted by the sample is recorded at the detector. The analyzer crystal and the collimator plus detector are fixed to a goniometer which performs the required rotation with an accuracy of 0.001°. However, at least three different analyzer crystals are required to cover a wavelength range of $0.05 \leq \lambda \leq 10$ nm. This is a result of the fact that, in practice, θ is only varied between 8° and 75°. According to the Bragg condition, if LiF with a lattice plane spacing of $d = 0.2$ nm is used as analyzer, wavelengths between 0.05 and 0.38 nm can be covered. A d value of 4.95 nm (lead stearate) gives the range $1.5 \leq \lambda \leq 9.5$ nm. An overview of frequently used analyzer crystals is given in the table.

In addition to the use of planar crystals in nonfocusing spectrometers, curved crystals are used in focusing spectrometers. Entrance slit, crystal axis and exit slit are then arranged on a → Rowland circle. This corresponds to the → Paschen-Runge mounting of a → concave grating in a → polychromator. The mounting of a concave analyzer crystal, analogous to the → Eagle mounting of a concave grating, has been realized in the fully-focusing linear spectrometer which is used for sequential measurements. The development of energy dispersive x-ray spectrometers began around 1966 with the development of semiconductor detectors. In these instruments, the x-rays emitted by the sample enter the detector directly, without undergoing reflection from an analyzer crystal (see Figure 2). The special semiconductor detectors not only detect the x-ray photon, they also determine its energy and thus fulfil the function of a

X-ray spectrometer. Table. Analyzer crystals for wavelength-dispersive x-ray spectrometers

Crystal (abbr.)	Full description	Formula	d [nm]	Analytical range K-Radiation	L-Radiation
LiF	Lithium-fluoride	LiF	0.201	$_{20}Ca-_{35}Br$	$_{50}Sn-_{92}U$
PET	Pentaerythritol	$C(CH_2OH)_4$	0.437	$_{14}Si-_{24}Cr$	$_{37}Rb-_{62}Sm$
RHP	Rubidium-hydrogenphthalate	$RbOOC(C_6H_4)COOH$	1.31	$_{9}F-_{14}Si$	$_{24}Cr-_{38}Sr$
RAP	Rubidium-acidphthalate				
KHP	Potassium-hydrogenphthalate	$KOOC(C_6H_4)COOH$	1.33	$_{8}O-_{14}Si$	$_{24}Cr-_{38}Sr$
KAP	Potassium-acid-phthalate				
STE	Lead stearate	$Pb\,[CH_3(CH_2)_{16}COO]_2$	4.95	$_{5}B-_{7}N$	$_{20}Ca-_{23}V$
MYR	Myristate	$CH_3(CH_2)_{12}COOX$	4.03	$_{5}B-_{8}O$	$_{20}Ca-_{24}Cr$

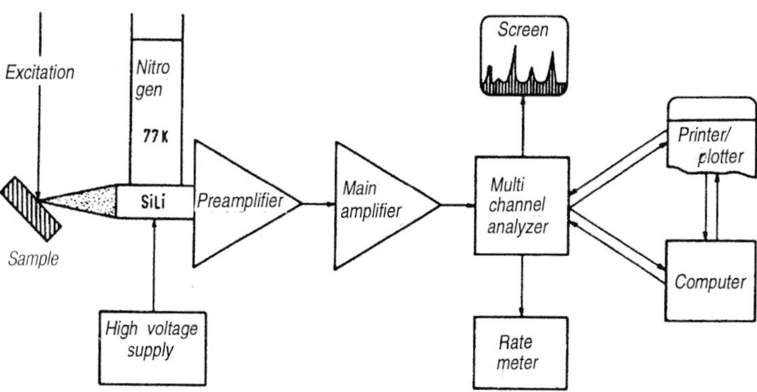

X-ray spectrometer. Fig. 2. Block diagram of an energy dispersive x-ray spectrometer

→ dispersing element. For photon energies above 30 keV, Si(Li) and Ge(Li) detectors are used almost exclusively. An important component of such a detector is a PIN diode in which the sensitive zone consists of a Si crystal in which vacant lattice sites are occupied by Li atoms.

When an x-ray photon strikes the crystal it produces a trail of electron-hole pairs until all its energy, E, has been used up. If ε is the energy required to produce an electron-hole pair, then the total number, υ, of pairs produced is $\upsilon = E/\varepsilon$. ε is 3.8 eV and 2.9 eV for the Si(Li) and the Ge(Li) crystals respectively. If a high voltage is placed on the PIN diode a burst of charge ensues which is converted via a directly connected field effect transistor (FET), preamplifier and main amplifier into a voltage pulse. The amplitude of this pulse is proportional to the number of electron-hole pairs, and therefore to the energy of the incident x-ray photon.

X-ray spectroscopy, <*Röntgenspektroskopie*>, a general expression for the methods in which the very short-wavelength electromagnetic radiation in the region $0.1 > \lambda > 0.01$ nm is used spectroscopically. (→ x-ray emission spectrum, → x-ray spectrometer, → x-ray fluorescence, → x-ray spectral analysis).

Ref.: B.K. Agarwal, *X-Ray Spectroscopy*, 2nd edtn., Springer Series in Optical Sciences, vol. 15, Springer Verlag, Berlin, Heidelberg, New York, **1991**.

X-unit, <*X-Einheit*>, a very small unit of length used in → x-ray spectroscopy; symbol XE. 1 XE = $(1.00202 \pm 0.00003) \cdot 10^{-12}$ m. This number derives from a very accurate measurement of the lattice spacing of calcite at 291 K made by Siegbahn in 1918. d_{oo}^{291} $(CaCO_3)$ = 3029.45 XE.

Yttrium aluminium garnet, YAG, <*Yttrium-Aluminium-Granat, YAG*>, the active medium of the → neodymium laser. Triply positively charged neodymium ions are present in the medium.

Z

Zeeman background correction, *<Zeeman-Untergrundkorrektur>*, background correction in → atomic absorption spectroscopy with the help of the → Zeeman effect. In this application, the magnetic field is perpendicular to the direction of propagation of the light and the effect is called the transverse Zeeman effect. There are, in principle, three ways of carrying out the experiment:

1. With the light source in the magnetic field, using the direct Zeeman effect;
2. with the → atomization apparatus in the magnetic field, using the inverse Zeeman effect with a permanent or DC electromagnet;
3. with the atomization apparatus in a pulsed magnetic field and an AC electromagnet.

Most AAS Zeeman systems utilize the inverse effect and use the graphite furnace (→ graphite furnace technique) as the method of atomization. The principle of the method is as follows. In the most simple case (the normal Zeeman effect), the atomic absorption line profile is split into three components in a magnetic field. The two outer components, known as the σ components, are polarized perpendicular to the magnetic field, while the central π component is polarized parallel to the magnetic field, but is not shifted by it, Figure 1. If the atomization device is situated in a constant magnetic field, then a rotating polarizer in the light beam can be used to distinguish between the two different planes of polarization, Figure 1. During the phase in which the light polarized parallel to the magnetic field is transmitted, both the specific atomic absorption by the parallel polarized π component of the absorption line and also the unspecific background at that wavelength are measured. In the other phase, during which only light polarized perpendicular to the magnetic field is transmitted, no absorption due to the π component can be measured, since this is removed by the polarizer. Therefore, only the unspecific background absorption is recorded and the difference of the two signals provides a background-corrected signal at the resonance wavelength. If the two σ components of the absorption profile are shifted so far from the resonance wavelength that their wings do not overlap the central line, then their contribution to the specific atomic

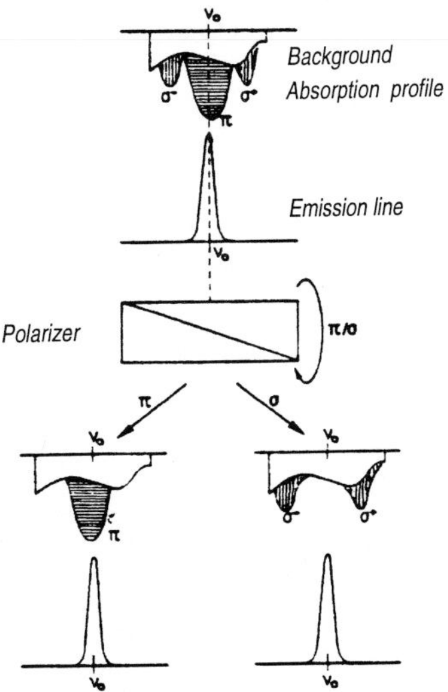

Zeeman background correction. Fig. 1. Background correction using the inverse Zeeman effect and a constant magnetic field

absorption will not be measured. This would be the case in a very strong magnetic field and the use of a pulsed field was suggested by De Loos-Vollebregt and De Galan in 1978. An AC magnet and graphite furnace are used. In contrast to the system described above, the polarizer is fixed so that it transmits only light polarized perpendicular to the magnetic field. The principle of the method is illustrated in Figure 2; it has the important advantage over the technique described earlier that the specific absorption is measured under conditions of zero magnetic field as in normal atomic absorption. If an unspecific, as well as a specific, absorption occurs then, as in normal AAS, the sum of the two is measured. This is the first phase of measurement. In the following phase, the absorption profile is split by the magnetic field and the π component of the line, which remains at the resonance wavelength, is not measured because it is removed by the polarizer and only the unspecific absorption, exactly at the resonance line position, is registered. The difference in the measurements, *field off* minus *field on*, then gives the background-corrected measurement. In the normal Zeeman effect there is only one central π component which is unshifted by the magnetic field, but the anomalous Zeeman effect produces two or more π components. As the magnetic field is increased the σ components are shifted out of the absorption region which, at first, increases the sensitivity of the measurement. But as the field is increased further, the π components are also moved apart and slowly shifted out of the absorption region. The result is that, with increasing field strength, first the absorption of the σ components decreases and then that of the π components, leading to a decreasing sensitivity.

However, if a pulsed magnetic field and graphite furnace are used, as in Figure 2, this problem does not arise.

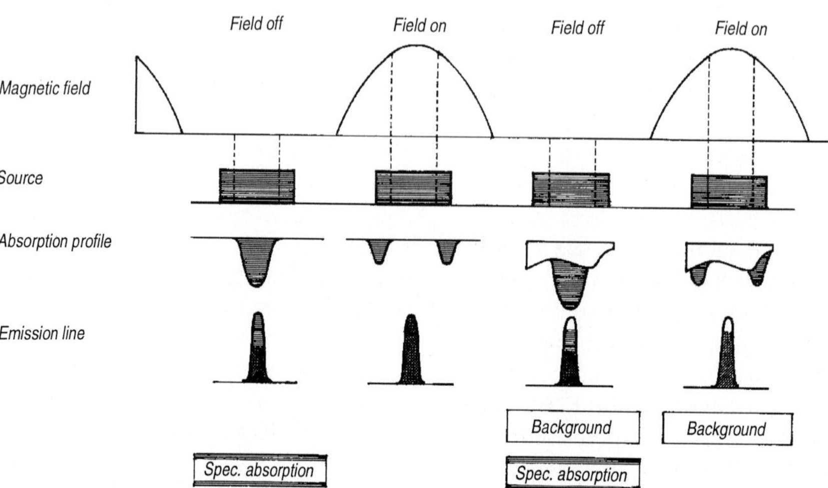

Zeeman background correction. Fig. 2. The principle of background correction with a pulsed magnetic field

In the phase *field off* a normal atomic absorption spectrum is measured, while in the phase *field on* the π components are completely removed since they are not required for the measurement. The task of the magnetic field in this case is simply to shift the σ components sufficiently far from the absorption region that they do not overlap the resonance line. In practice this means that, with a pulsed magnetic field and graphite furnace, the normal AAS sensitivity can be achieved for all elements, provided that the magnetic field is sufficiently large.

Ref.: S.S.M. Hassan, *Organic Analysis Using Atomic Absorption Spectrometry*, Ellis Horwood Ltd./J. Wiley and Sons Ltd., Chichester, New York, Brisbane, Toronto, **1984**.

Zeeman effect, <*Zeeman-Effekt*>, an effect discovered by P. Zeeman in 1896. Each monochromatic atomic line of a light source is split into three components by a magnetic field, and the outer two components are polarized perpendicular to the central component. Also, the splitting of the lines is linearly proportional to the strength of the magnetic field. Faraday had made a similar experiment 50 years earlier, but he was unsuccessful on account of his inadequate apparatus. H.A. Lorentz predicted the Zeeman effect and was able to give the splitting and the degree of polarization of the Zeeman lines a classical theoretical interpretation. The quantum-mechanical interpretation is based upon the → space quantization of electronic angular momentum in a magnetic field. The origin of this is the fact that every atomic state which has a nonzero angular momentum also has a magnetic moment, μ. When such an atom is brought into a magnetic field there is an orientation of these magnetic dipoles with respect to the field. Classically, all orientations are possible, but in quantum mechanics only those orientations are allowed for which the z component of the angular momentum vector, either parallel or antiparallel to the magnetic field, is an integer or half-integer multiple of $h/2\pi = \hbar$. If the total angular momentum of the electrons is characterized by the quantum number, J, i.e. in the case of → Russell-Saunders coupling, then there are $(2J + 1)$ possibilities for the space quantization which are given by the magnetic quantum number, M_J viz: $-J, -J + 1, -J + 2 \ldots J - 1, J$. The potential energy of each of these $(2J + 1)$ states is given by:

$$E = E_o + g\mu_B\mu_o H \cdot M_J.$$

E_o is the energy of the state in the absence of the magnetic field, μ_B is the → Bohr magneton, H the magnetic field strength and g the Landé → g factor which is given by:

$$g = 1 + \frac{J(J+1) + S(S+1) - L(L+1)}{2J(J+1)}$$

(S = total spin = $\sum s_i$; L = total orbital angular momentum = $\sum \ell_i$).

Thus, the consequences of space quantization may be described as follows. A state characterized by the angular momentum quantum number x (x stands for any one of a number of angular momentum quantum numbers, e.g. J, L, S, I) is $(2x + 1)$-fold degenerate in the absence of a magnetic field. In the presence of a magnetic field the degeneracy is removed (lifted) and the state is split into $(2x + 1)$ states of different energies. This is

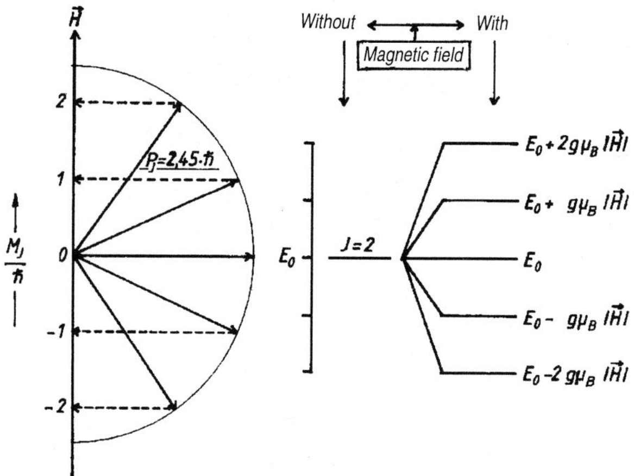

Zeeman effect. Fig. 1. Space quantization in a magnetic field for a total angular momentum $J = 2$

illustrated for $J = 2$ in Figure 1. Naturally, all terms of an atomic term system split in an analogous way, with the exception of ^1S_o for which $J = L = S = 0$. Thus, when account is taken of the selection rule for ΔL (→ atomic spectrum) and the additional rule $\Delta M_J = 0, \pm 1$, a splitting of the spectral lines results. The selection rule also shows that the transitions with $\Delta M_J = \pm 1$ are the σ components (circular or perpendicular polarization), while the transitions for which $\Delta M_J = 0$ are the π components (parallel polarization). These points are illustrated and summarized in Figure 2 for the transition $^1D_2 \leftrightarrow {}^1F_3$.

A distinction is made between the normal and the anomalous Zeeman effect. The normal Zeeman effect occurs with singlets for which the total spin, $S = 0$ and, therefore, $J = L$ always. Also, $g = 1$. Figure 2 illustrates the normal Zeeman effect for the transition $^1D_2 \leftrightarrow {}^1F_3$. Since the selection rule is $\Delta M_J = 0, \pm 1$ and the splitting in each of the two states is the

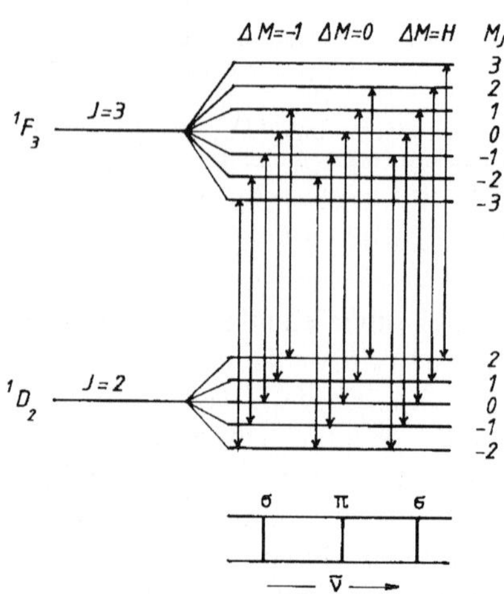

Zeeman effect. Fig. 2. The normal Zeeman effect for the transition, $^1D_2 \leftrightarrow {}^1F_3$

same, there are three lines. For a value of H of 1 T (10^4 G), the splitting, $\Delta\tilde{\nu}_{\text{norm}} = \mu_B \cdot H = 0.4668\ \text{cm}^{-1}$. In general, the splitting is given by the equation $\Delta\tilde{\nu}_{\text{norm}} = 4.668 \cdot 10^{-5}\ H$ with H in Gauss or $\Delta\tilde{\nu}_{\text{norm}} = 4.668 \cdot 10^{-1}\ H$ for H in Tesla. The normal Zeeman effect is found in the singlet terms of Ca and

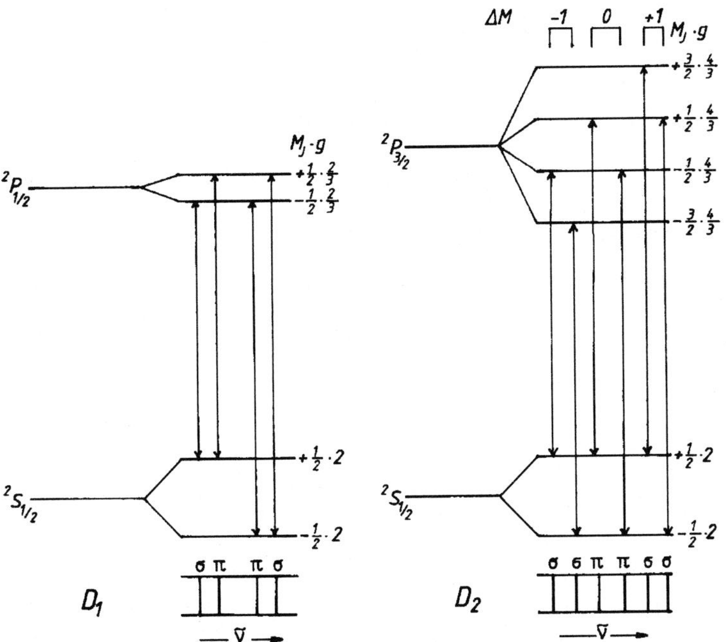

Zeeman effect. Fig. 3. The anomalous Zeeman effect illustrated with the example of the sodium D lines: $^2S_{1/2} \leftrightarrow {}^2P_{1/2}$ and $^2S_{1/2} \leftrightarrow {}^2P_{3/2}$

Zn and other elements of the same configuration.

The anomalous Zeeman effect is found in all other cases. As an example the transitions $^2S_{1/2} \rightarrow {}^2P_{1/2}$ and $^2S_{1/2} \rightarrow {}^2P_{3/2}$, the D lines of sodium, are illustrated in Figure 3. The Landé g factors are calculated to be: $^2S_{1/2} = 2.0$, $^2P_{1/2} = 0.666$ and $^2P_{3/2} = 1.333$ and, as a result, the D line doublet splits into a total of $4 + 6 = 10$ lines.

In addition to its importance in the analysis of the → multiplet structure of atoms, the Zeeman effect is also the basis of → NMR and → ESR spectroscopy. In recent years, it has been successfully introduced into → atomic absorption spectroscopy as a method of background correction; see → Zeeman background correction.

Zenker prism, <*Zenker-Prisma*>, a → direct vision prism formed from two equal prisms of different dispersions. The two prisms, which are made of materials which have the same refractive index for a specific wavelength, are joined together to form a parallelepiped, Figure 1a. Since the incident light falls perpendicularly onto the surface of the first prism and an undeviated beam therefore also leaves the prism in the same way, the reflection losses in the Zenker prism are significantly less than they are with the → Amici prism. The requirement that the refractive index of both prism materials must be the same for some intermediate wavelength, usually means that a liquid medium is used for the second prism. Two Zenker prisms,

a)

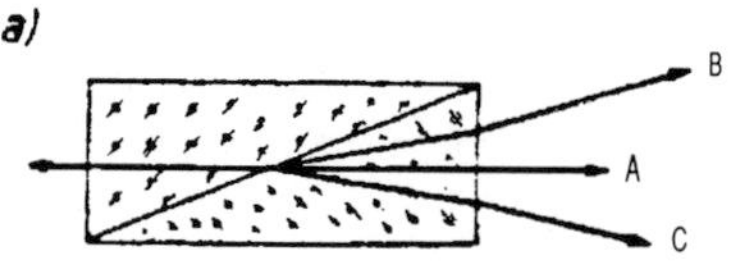

b)

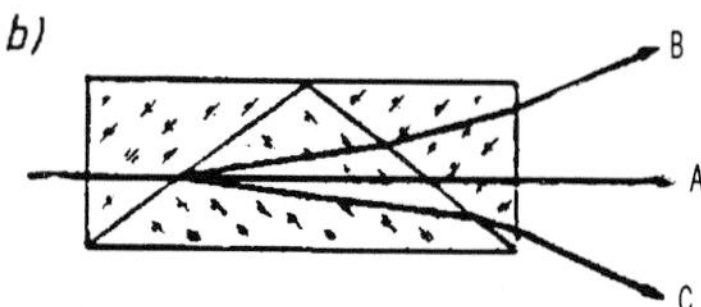

Zenker prism. Fig. 1. a) A Zenker prism, b) a Wernicke prism

one behind the other, constitute a Wernicke prism, Figure 1b, which can have a large angular dispersion because a large prism angle of the order of 100° can be used for the middle prism.

Zerodur, <*Zerodur*>, (a registered trademark of the Schott Glaswerke, Mainz, Germany). An inorganic, pore-free material of the glass-ceramic type; it contains both glassy and crystalline phases. The starting point for the preparation of Zerodur is a glass melt which is purified, homogenized and finally formed while hot. After cooling and removal of strain, the glassy blanks undergo a temperature treatment, following an exact temperature-time program, during which submicroscopically fine crystals are formed. A precondition for this is the addition of substances with high melting points (e.g. TiO_2 and ZrO_2). By their separation in the temperature region of maximum seed formation frequency (T_{sf}), which lies below the temperature region of maximum speed of crystal growth (T_{gr}), these materials act as seed formers. The glass cannot crystallize as the melt is cooled if there are no seeds present. Only when these have been formed in sufficient numbers at T_{sf} can the desired microcrystals be formed in large numbers (up to $10^{17}/cm^3$) which is achieved by warming again to T_{gr}. The crystalline proportion of the material can vary, according to the desired physical properties, between 50 and 90 volume%. Zerodur contains 70–78 weight% of crystalline phase with the β-quartz structure. The mean crystal size of the β-quartz mixed crystals is usually in the region of 50–55 nm. The crystalline phase has a negative temperature coefficient of expansion and the glass phase a positive coefficient. As a result, the linear thermal expansion is effectively compensated and Zerodur is a material which shows almost zero thermal expansion. It is therefore an ideal material for the construction of mirror mountings for astronomical and x-ray telescopes for which changes in the mirrors as a result of variations of temperature can have a significant effect on the quality of the observations. The material is used for the construction of the temperature-independent base for ring-laser gyroscopes. The extraordinary length stability of Zerodur is exploited elsewhere in laser technology. Zerodur rods are used as spacers to ensure the maintenance of exact resonator length throughout all the temperature changes during operation, thereby increasing the productivity of lasers.

Zerodur has a good transparency because the crystals are so small and the difference in refractive index between the glassy and crystalline phases is also very small. The transmission in the VIS and NIR regions is

ca. 90% for a path length of 5 mm, and this is important for spectroscopic applications. For applications in optics it is also important that mirrors of the usual metals can be readily formed on Zerodur surfaces. Because the material has a good chemical resistance, aluminium mirrors can be removed, the polished surface cleaned and a new mirror deposited on it several times. (Schott, Product Information No. 3079). Zerodur can be bonded to itself or to other materials with the help of silicone rubber cements (Schott, Product Information No. 3137d).
Ref.: Schott booklet, *ZERODUR – Precision from Glass Ceramics*. H.G. Pfaender, H. Schröder, *Schott Guide to Glass*, Van Nostrand Reinhold Co. Inc., Princeton, New York, London, Toronto, **1983**.

Zero gap, <*Nullücke*>. For a diatomic molecule with closed electron shells the selection rules for the rotation-vibration transitions are $\Delta v = +1$ and $\Delta J = \pm 1$. Transitions for which there is no change in the rotational quantum number, J, i.e. $\Delta J = 0$, are forbidden. Therefore, the pure vibrational transition, $v'' = 0 \rightarrow v' = 1$ and the Q branch ($\Delta J = 0$), do not occur. Consequently, a gap, the zero gap, is observed between the R and P branches. A zero gap also occurs in the → parallel bands of linear polyatomic molecules.

Zero-point energy, <*Nullpunktsenergie*>, the finite energy of a molecule in the lowest vibrational state. In the case of the → harmonic oscillator, the vibrational energy is given by:

$$E(v) = hc\omega(v + \frac{1}{2}) = h\nu(v + \frac{1}{2}).$$

where, h is → Planck's constant, c the velocity of light, ω the vibrational wavenumber (cm^{-1}) and ν the vibrational frequency (s^{-1}). v is the vibrational quantum number which can take the values 0, 1, 2, 3 ... For $v = 0$ the energy is:

$$E(v = 0) = hc\omega\frac{1}{2} = \frac{1}{2}h\nu.$$

For an N-atomic molecule, the first approximation (harmonic case) to the total vibrational energy is:

$$E(v_1, v_2, ...v_{3N-6}) =$$
$$h\nu_1(v_1 + \frac{1}{2}) + h\nu_2(v_2 + \frac{1}{2}) + ...$$
$$h\nu_{3N-6}(v_{3N-6} + \frac{1}{2}).$$

For $v_1 = v_2 = \ldots v_{3N\text{-}6} = 0$ we have the total zero-point energy of the molecule which is:

$$E(0, 0...0) = \frac{1}{2}h\nu_1 + \frac{1}{2}h\nu_2 + ...$$
$$+ \frac{1}{2}h\nu_{3N-6}$$
$$= \frac{1}{2}\sum_{i=1}^{3N-6} h\nu_i.$$

The zero-point vibrations are the vibrations which the molecule executes which correspond to this energy. The zero-point energy can be understood in terms of the → Heisenberg uncertainty principle.

Zero-zero transition, <*Null-Null-Übergang*>, the transition in an → electronic band spectrum or → UV-VIS spectrum in which the molecule goes from the vibrational state, $v'' = 0$, in the electronic ground state to the vibrational state, $v' = 0$, in the excited electronic state. It is denoted by $\tilde{\nu}_{o,o}$.

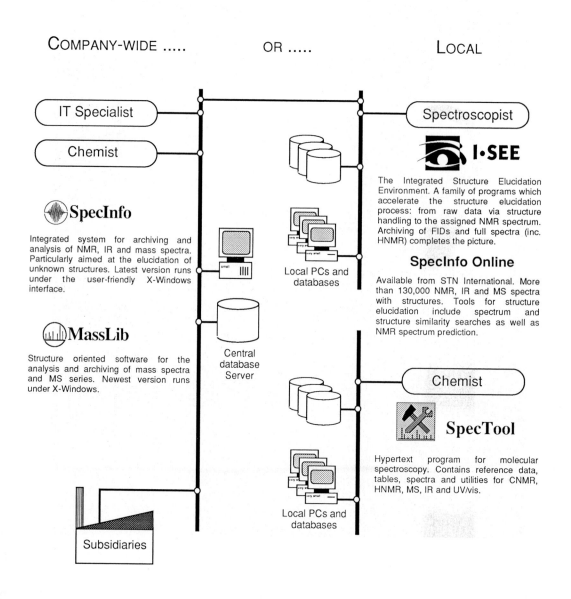